SPACE RESEARCH XI

Vol. 1

COSPAR

SPACE RESEARCH XI

Proceedings of Open Meetings of Working Groups of the Thirteenth Plenary Meeting of COSPAR

LENINGRAD, USSR, 20–29 May 1970

and of

The Symposium on Remote Sounding of the Atmosphere
(jointly sponsored by COSPAR, WMO and IAMAP/IUGG)

LENINGRAD, USSR, 22, 25 and 26 May 1970

Organized by

THE COMMITTEE ON SPACE RESEARCH—COSPAR

and

THE USSR ACADEMY OF SCIENCES

Edited by

K. YA. KONDRATYEV

M. J. RYCROFT

C. SAGAN

Volume 1

AKADEMIE-VERLAG · BERLIN
1971

Executive Editor:
Dr. A. C. Stickland

Library of Congress Catalog Card Number 60-50 878

202 · 100/574/71
Karten: Nr. 45/71, 111/71
Einband und Schutzumschlag: Rolf Kunze
Gesamtherstellung: VEB Druckerei „Thomas Müntzer“, 582 Bad Langensalza
Bestellnummer: 3059/XI · ES 18D1, 18E 2
EDV 761 513 3

FOREWORD

In introduction to this XIth Volume of the Space Research Series, produced by the publishing house of the Academy of Sciences of Berlin, I have the very agreeable duty to express COSPAR's warm gratitude to the Authors of the published contributions and to our very distinguished colleagues, Professor K. Ya. Kondratyev, Dr. M. J. Rycroft and Professor C. Sagan, Co-Editors of the present issue. In this respect, COSPAR's gratitude is equally addressed to Dr. A. C. Stickland, who has helped them in their editorial task in her capacity as Executive Editor.

The present volume contains not only, as usual, the Proceedings of the Open Sessions of the Working Groups concerned with Physical Sciences, but also the Proceedings of the Symposium on 'Remote Sounding of the Atmosphere' successfully co-sponsored by the International Association of Meteorology and Atmospheric Physics of the International Union of Geodesy and Geophysics and the World Meteorological Organization.

For the excellent standard of the Leningrad XIIIth Assembly of COSPAR, the thanks of our Committee are renewed here not only to the host Organization, the Academy of Sciences of the Soviet Union, but also to the effective organizers and especially to Academician A. A. Blagonravov, the eminent President of the Organizing Committee, to Dr. I. A. Zhulin, Secretary of this Committee, and to Mr. G. S. Balayan, Scientific Secretary of the Commission of Outer Space of the Academy of Sciences of the Soviet Union, and finally to Mr. Z. Niemirowicz, COSPAR's Executive Secretary, who contributed to the excellent outcome of all these efforts.

Maurice ROY
President of COSPAR

AVANT PROPOS

En introduction au XIème Volume de la collection Space Research, publié par la maison d'édition de l'Académie des Sciences de Berlin, j'ai le très agréable devoir d'exprimer la vive gratitude de COSPAR aux Auteurs des communications publiées, aux très distingués Collègues, le Prof. K. Ya. Kondratief, le Dr. M. J. Rycroft, le Prof. C. Sagan, Co-Editeurs du présent ouvrage. A cet égard, la gratitude de COSPAR ne s'adresse pas moins au Dr. A. C. Stickland qui les a assistés dans leur tâche éditoriale en qualité d'Editeur Exécutif.

Le présent ouvrage comprend non seulement, et selon l'usage, les comptes rendus des sessions publiques de Groupes de Travail consacrées aux sciences physiques, mais encore les comptes rendus du Symposium du „Sondage à Distance de l'Atmosphère" très heureusement parrainé à la fois par l'Association Internationale de Météorologie et de Physique de l'Atmosphère de l'Union Géophysique et Géodésique Internationale et par l'Organisation Météorologique Mondiale.

Pour l'excellente tenue à Léningrad de la XIIIème Assemblée de COSPAR, les remerciements de ce Comité sont renouvelés ici non seulement à l'Organisation-hôte, l'Académie des Sciences de l'Union Soviétique, mais aussi aux organisateurs effectifs et, en particulier, à l'Académicien A. A. Blagonravov, l'éminent Président du Comité d'Organisation, au Dr. I. A. Jouline, Secrétaire de ce Comité, à Mr. G. S. Balayan, Secrétaire Scientifique de la Commission Spatiale de l'Académie des Sciences concernée et enfin à Mr. Z. Niemirowicz, Secrétaire Exécutif de COSPAR, qui a si bien contribué à un excellent aboutissement de tous les efforts.

Maurice ROY
Président de COSPAR

PREFACE

"One small step for (a) man; a giant leap for mankind"

It is, perhaps, appropriate to start this volume in the COSPAR "Space Research" series with this now famous maxim of Neil Armstrong. This is so since the Thirteenth Plenary Meeting of COSPAR, held from 20–29 May 1970, provided the first opportunity for COSPAR participants to discuss the achievements of the U.S. Apollo 11 mission to the Moon in July 1969. It is also fitting that the COSPAR Meeting took place, for the first time, in the USSR, the country that initiated the era of Man in Space with the single orbit of the Earth made by Yuri Gagarin in April 1961. The continuation of this era is well represented by the long flight of Soyuz 9, the results of which were discussed by scientists and cosmonauts during the COSPAR meeting. The meetings, held in the historic Tavrichesky Palace in Leningrad, were attended by 502 scientists from the USSR and 438 from 30 other countries.

The following Working Groups of COSPAR met to discuss latest significant results:

1, 2, 3 (panels A, B and C), 4 (panels A, B and C), 5, 6 and 7. In addition, a symposium on Remote Sounding of the Atmosphere was held, and four lectures in the Annual Reviews of Space Research series were delivered. Of the papers presented, many dealing with Physical Sciences and Space Research are published in this volume, which is comprised of nine main sections:

The Moon
Planets
Cosmic dust
Solid earth physics
Remote sounding of the atmosphere
The neutral atmosphere
The ionosphere
Solar terrestrial relationships
Astronomical measurements.

Of the remaining papers, those to be published elsewhere are listed at the end of this volume, with appropriate references where available. Papers concerned with Life Sciences and Space Research are published in a companion volume.

In organising the material for publication we, as Scientific Editors, are especially grateful to those authors who complied with the instructions on the preparation of papers, and who delivered them punctually both in manuscript and galley proof form. We acknowledge with gratitude the considerable help given by Dr. J. Kovalevsky, by Mr. Z. Niemirowicz, Executive Secretary of the COSPAR Secretariat, and by other members of the Secretariat. Finally we should like to thank Dr. A. C. Stickland, Executive Editor, for her sterling work and the publishers for the efficient manner in which they have prepared and executed this volume.

K. Ya. Kondratyev
M. J. Rycroft
C. Sagan

Volume 1

The Moon

Lunar Constitution

Lunar Laser and Tracking Experiments

Planets

General

Venus

Cosmic Dust

Solid Earth Physics

Remote Sounding of the Atmosphere

Volume 2

The Neutral Atmosphere

Stratosphere

Mesosphere

Minor Constituents

Theory

Temperature Determinations

Upper Atmospheric Density

Density Variations

Magnetopsheric Physics

Very Low Frequency Radiophysics

Astronomical Measurements

Ultraviolet Radiation

X-Rays

Gamma Rays

Cosmic Rays

The Moon

Space Research XI — Akademie-Verlag, Berlin 1971

APOLLO LUNAR EXPLORATION

G. M. Low and L. R. Scherer

National Aeronautics and Space Administration
Washington, D.C., USA

In successfully carrying out a manned lunar landing and return, with both operational and technological objectives, the Apollo program made possible a variety of significant scientific experiments. This important milestone in the continuing quest for knowledge took the eyes, hands, and mind of man, as well as his instruments, to a new world. The activities of highest priority carried out by the astronauts, once the landing had been successfully completed, were to collect lunar material and data, emplace sophisticated experiments, and record man's impressions and observations.

An examination of both the possibilities and the limitations of the Apollo system as a means for scientific investigation of the moon will be useful to experimenters in the near and distant future.

In the missions ahead, scientific exploration of the moon will be the principal goal. Unique features and sites on the moon will be visited. New experiments, both on the lunar surface and in lunar orbit, will be carried out, as we probe the moon's past and attempt to unravel the early history of the earth. In so doing, we will also be establishing and defining the possibilities and limitations of man as a space explorer as we extend his domain further in space.

The purpose of this paper is to provide a broad overview of the Apollo missions, discussing briefly the Apollo 11 and 12 lunar landings, the problems experienced with Apollo 13, an indication of the major scientific results of these missions, and some of our future plans.

Last summer (1969) sufficient testing on Apollo hardware had been accomplished to provide us with the confidence to attempt man's first landing on the moon. On 16 July 1969, Apollo 11 was launched from Cape Kennedy.

After checkout in earth orbit, Apollo 11 started on its translunar trajectory. The mission proceeded as planned and as Apollo 11 swung into orbit about the moon, the crew made the photograph shown in Fig. 1 of the crater Tsiolkovsky on the lunar farside. This crater, first noted on photographs transmitted from the Russian spacecraft, Luna 3, was named for the pioneer Russian rocket scientist. It is interesting to note the prophetic words on the tombstone of Tsiolkovsky, "Mankind will not stay on the Earth forever."

In lunar orbit, the Lunar Module separated from the Command Module and descended to the lunar surface in the Sea of Tranquility. Astronauts Armstrong and Aldrin remained on the lunar surface for about 18 hours. They were outside of the spacecraft for two hours. Fig. 2 shows Aldrin as he removed the lunar experiments from the spacecraft for deployment on the surface of the moon.

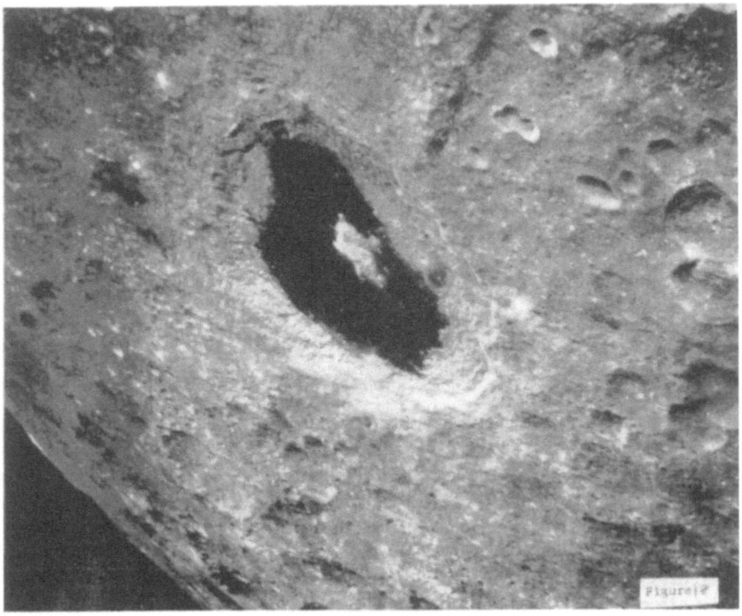

Fig. 1. Crater Tsiolkovsky on Lunar farside (diameter 200 km; 128° E long., 20° S lat.).

Fig. 2. Apollo 11 Lunar Module on lunar surface.

After takeoff, the Lunar Module ascended again to lunar orbit for a rendezvous with the Command Module in which Astronaut Collins had remained. The astronauts returned safely to earth and began their quarantine period.

On 14 November 1969, Apollo 12 was launched on the second lunar landing mission. Again, success was obtained even though the spacecraft was struck by lightning shortly after takeoff. On this mission, it was important to learn how to make a very accurate landing, so that later flights could explore much more difficult sites from an operational point of view — sites that are also more

Fig. 3. Surveyor 3 on lunar surface; Lunar Module in background.

interesting from a scientific point of view. For these, pinpoint accuracy will be required for a successful landing. Scientifically, it was desirable for Apollo 12 to land at a different mare than Apollo 11 in order that we could determine the degree of similarity. With these two points in mind, a site in the Ocean of Storms at which Surveyor 3 had landed in April 1967 was selected. A number of changes were made in a number of procedures and computer programs to improve landing point accuracy. The results are graphically portrayed by the photograph of Surveyor 3 made by the Apollo 12 crew (Fig. 3). The crew inspected and photographed the Surveyor spacecraft in detail and removed and returned the camera, scoop, and other parts. These are currently undergoing detailed analysis to determine the scientific and engineering changes that have occurred in the lunar environment during this precisely known period of time. Such data may be extremely important for future lunar work.

In April of this year, we were ready for Apollo 13. For this mission we selected a site in an upland region known as the Fra Mauro formation. This is thought to be material deposited when Mare Imbrium was formed, perhaps by impact of a smaller moon. In any event, it is distinctly different in appearance from the mare regions in which the previous landings have been made.

About 55 hours into the mission, a muffled explosion was heard by the crew. The events that followed were almost catastrophic. It was immediately obvious that the lunar landing could not be attempted and all remaining efforts were devoted to returning the crew to earth. It was only through outstanding work on the part of the crew, mission operations personnel, and a great many support people that the difficulties were overcome and the crew returned safely.

Fig. 4. Damaged Command Module photographed from Apollo 13 Lunar Module.

An extensive investigation to determine precisely what occurred has been completed* and the results have been reported. As the Lunar Module was separated from the Service Module approaching re-entry to earth, the crew made the photograph in Fig. 4. As can be seen, an entire panel is missing, and there is obvious damage to much of the equipment. The cause was the rupture of an oxygen tank whose design is sketched in Fig. 5. A short circuit occurred within the tank causing combustion of the teflon wire insulation. The pressure and temperature within the tank built up rapidly and the tank finally ruptured. This resulted in an explosive separation of the Service Module panel.

To turn now to a brief look at the science results of these missions. Astronauts Armstrong and Aldrin returned 20 kg of lunar material. In addition, they deployed several experiments on the surface. Shown in Fig. 6 is a seismometer powered by solar cells. For 21 days this instrument provided information on

* The section on Apollo 13 was updated in June 1970, to make available the most recent information.

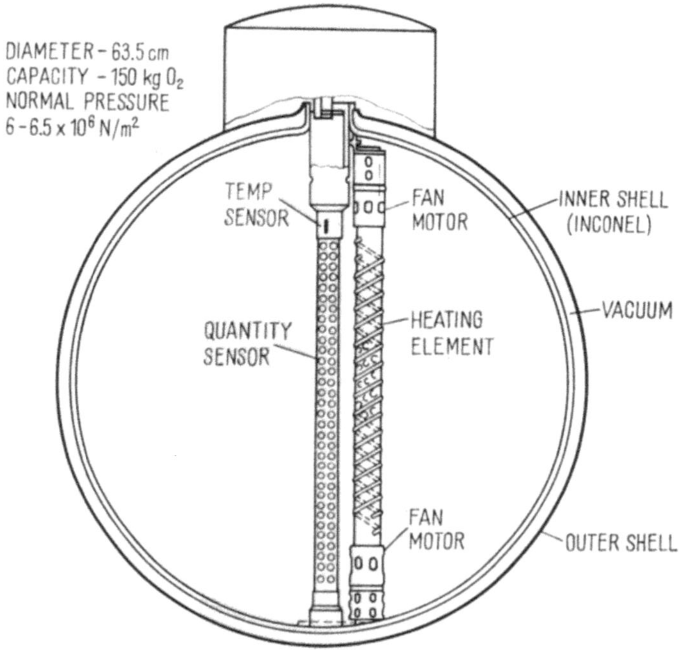

Fig. 5. Oxygen tank.

Fig. 6. Apollo 11 science station.

man-made and natural seismic events. In the background is the laser reflector which is an array of very precise corner reflectors from which laser beams from the earth can be reflected to measure earth–moon distances with extreme precision. Accuracy to 30 cm has been obtained. This improved knowledge of the changes in the lunar distance opens the possibility of increasing our knowledge of the earth, the moon, and the solar system in some very fundamental ways. It is important to note that this reflector (and we plan several others on later missions) is available for all scientists of the world to use. A third experiment

Fig. 7. Apollo 12 science station.

on Apollo 11 is the solar wind composition collector. This is aluminum foil exposed to the sun while the astronauts were on the surface, then returned and analyzed for trapped lighter elements of the solar wind. This is an experiment of Dr. Geiss of Switzerland.

For Apollo 12, a more complicated science station was carried (Fig. 7). Instruments included a seismometer, a magnetometer, a solar wind spectrometer, an atmosphere detector, and an ion detector. The science station is powered by a nuclear generator. At this time, it is still functioning well. In addition, another solar wind composition collector was deployed.

One of the intriguing scientific results is the seismic data. The better to understand this information, as well as for operational reasons, we deliberately impacted the Lunar Module of Apollo 12 after the crew left it. On the Apollo 13 mission, the spent third stage of the Saturn 5 rocket was deliberately impacted on the lunar surface. The geometry of the impacts is indicated in Fig. 8.

The energy of the Lunar Module striking the moon was equivalent to approximately one ton of TNT. Fig. 9 compares the resultant seismic signals on the earth and on the moon from such an explosion. As can be seen, signals were

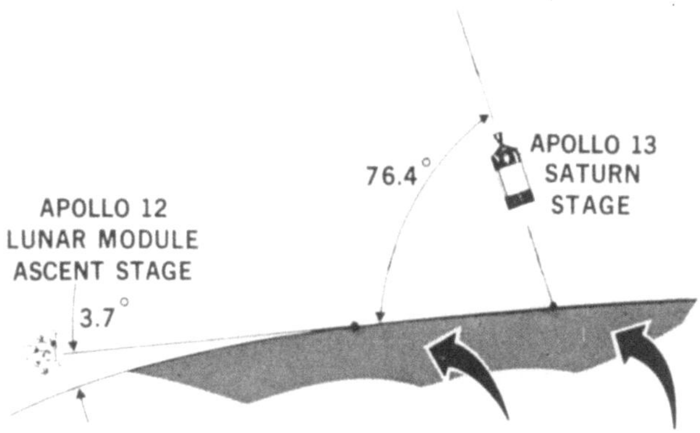

Fig. 8. Man-made impact characteristics.

Weight (kg)	2383	13925
Impact angle	3.7°	76.4°
Distance to seismometer (km)	76	136
Velocity (km s^{-1})	1.68	2.58
Impact energy, TNT equivalent (metric tons)	0.77	11.0

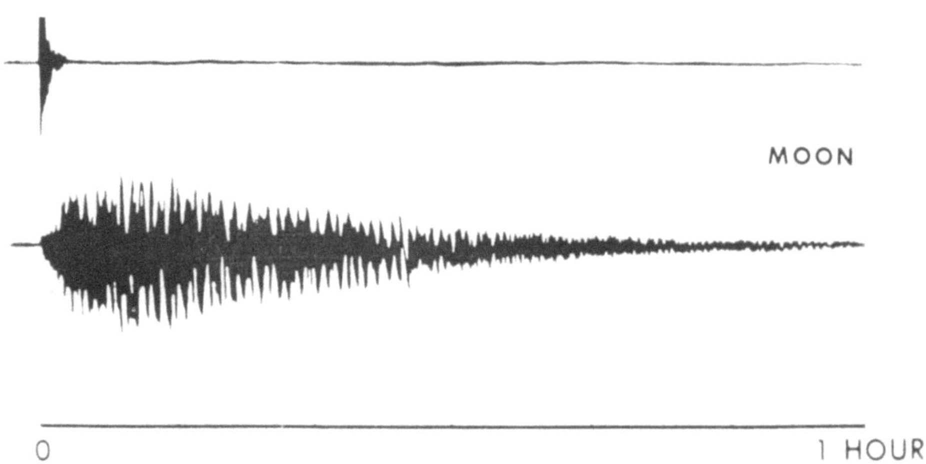

Fig. 9. Comparative seismograms of Lunar Module impact on earth and moon.

received at the seismometer for one hour. The Saturn stage impact was roughly ten times greater in energy than that of the Lunar Module, and the seismic signals were received for more than four hours. It is difficult to hypothesize a lunar interior that would result in such signals.

We have known for hundreds of years that the moon is covered with craters of all sizes at a resolution that could be seen. We conjectured that when we

Fig. 10. Tycho, Lunar Orbiter 5.
8.5×10^4 m diameter.

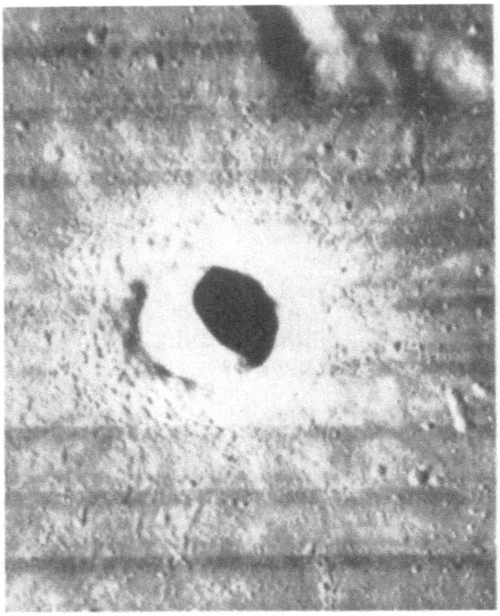

Fig. 11. Mösting C, Lunar Orbiter 3.
3.8×10^3 m diameter.

Fig. 12. Unnamed crater, Lunar Orbiter 3.
2.0×10^2 m diameter.

Fig. 13. Crater at Apollo 12 landing site.
1 m diameter.

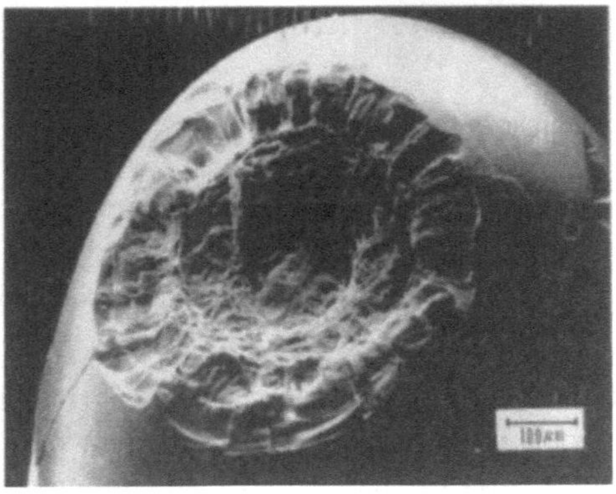

Fig. 14. Crater, Apollo 11 lunar sample.
5.0×10^{-4} m diameter.

finally reached the moon, we would find more and more craters of smaller and smaller size. Figs. 10—16 show a family of craters to illustrate this: Fig. 10 is a massive crater Tycho, almost 10^5 m in diameter; Fig. 11 is Möstung C, several theousand meters in diameter; Fig. 12 is a crater 200 m across, which is below the size that we had been able to see prior to the existence of lunar spacecraft. The next of the series (Fig. 13) is a one-meter crater in the Apollo 12 landing site, and Figs. 14, 15 and 16 made here on earth show craters in lunar samples, 5×10^{-4} m and 3×10^{-6} m in size. This total family runs through a spectrum of eleven orders of magnitude.

There is a very extensive scientific program for analyzing the returned lunar samples. One of the concerns that has required special care is that of possible

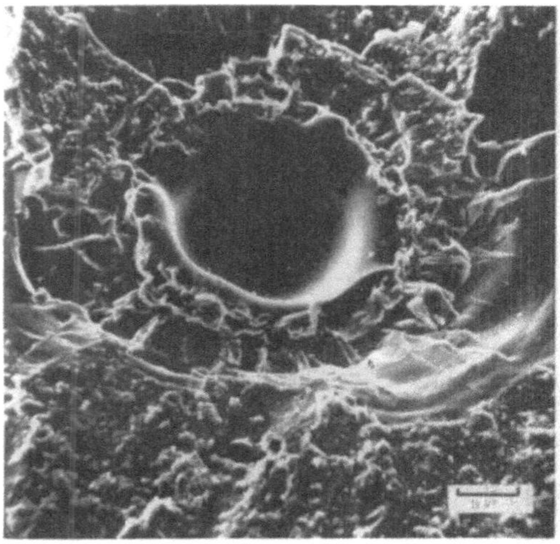

Fig. 15. Crater, Apollo 11 Lunar Sample. 4.0×10^5 m diameter.

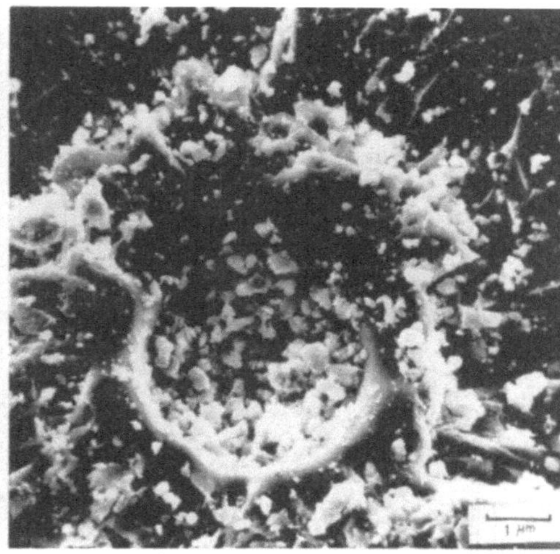

Fig. 16. Crater, lunar sample. 3×10^{-6} m diameter.

contamination of the earth's biosphere. A prudent quarantine program was established, and the astronauts were kept in quarantine for 21 days after leaving the lunar surface. No adverse effects were noted. A large number of biological specimens were placed in contact with the lunar soil and studied carefully to insure that there were no adverse results prior to release of the samples to investigators for detailed analysis. The Mobile Quarantine Facility into which the crew was placed immediately upon recovery was moved by ship and air directly to the Lunar Receiving Laboratory in Houston. The lunar samples were treated equally carefully. They were moved in their sealed containers, and were not opened until they were behind the double quarantine barrier in the Lunar Receiving Laboratory, and in the vacuum chamber.

Fig. 17 is a flow diagrám of the lunar sample operations in the Laboratory. After the containers were opened in a vacuum or in a nitrogen atmosphere, they were subjected to a preliminary scientific examination. Simultaneously, biological testing was undertaken. The samples were retained in vacuum or nitrogen, depending upon the type of later analysis to be conducted. The many investigators were then sent samples in a form most suitable for their investigation. Some were provided with whole rocks, others with thin sections, still others with mineral separates.

A large number of cultures, animal and plant specimens were exposed. No deleterious effects were noted on any of these. An interesting and unexpected

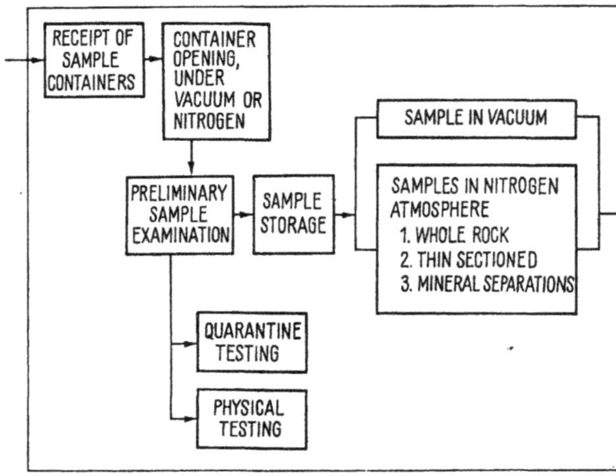

Fig. 17. Lunar sample operations in Lunar Receiving Laboratory.

Lunar soil added

No lunar soil

Fig. 18. Soil effect on plant life.

result was obtained during the botanical testing. It was found that some of the simple plants thrived when a small quantity of lunar soil was added. This effect on the liverwort plant is graphically illustrated in Fig. 18. This result is not fully understood, and considerably more investigations are under way.

We welcome proposals from scientists throughout the world for analysis of the lunar samples and proposals for analysis of material from later missions. At the present time, samples are being supplied to 193 investigation teams in 17 countries of the world. The current breakdown is listed in Table 1.

Table 1.
Apollo 11–12: Lunar sample analysis program – principal investigators

Country	Number
Australia	5
Belgium	2
Canada	8
Czechoslovakia	1
Finland	1
France	3
Federal Republic of Germany	6
India	1
Italy	2
Japan	3
Korea	1
Norway	1
South Africa	1
Spain	1
Switzerland	1
United Kingdom	17
United States of America	139
	193

We can summarize the more important scientific results obtained from our first two landing missions, although, of course, much work is still going on: we now know that parts of the moon are at least four and one-half billion years old; we know that if the moon was once part of the earth it must have split off very early in its history; we have found considerable variation in the composition of the lunar material between the sites of Apollo 11 and 12, even though they both appeared to be very similar terrain; our analyses have uncovered several minerals never before seen; evidence shows that rocks crystallized at a very high temperature; the lunar surface has been continuously modified but at a very slow rate; thus far, we have not found any indisputable evidence of water, organic compounds, or lunar life.

On Apollo 12 an unexpected lunar magnetic field was found. We do not know how widespread it may be. From remnant magnetization in the rocks, the ancient magnetic field must have been stronger than at present. This may infer that the moon was once much closer to the earth than it is now. Natural seismic events on the moon are few and weak compared to the earth, but the seismic transmissions through the moon are considerably different from those found on the earth. Our analyses show much evidence of sun's activity recorded in the lunar material. Our laser reflector has measured the earth–moon distance with an accuracy to 30 centimeters. Improvements in this accuracy are expected soon. Finally, we can state positively that man can function very effectively as a scientific explorer on another planet. We are very pleased with the relative ease in which the crews have been able to adapt to the lunar environment and to perform the many tasks which they were assigned.

Although a great deal was learned on Apollo 11 and 12, the remaining Apollo missions will be increasingly effective. Fig. 19 is a plot of several parameters of Apollo 11 and 12 and what we anticipate for Apollo 14. There is a significant increase in the capabilities of each mission. Most of this is due to learning and increase in confidence.

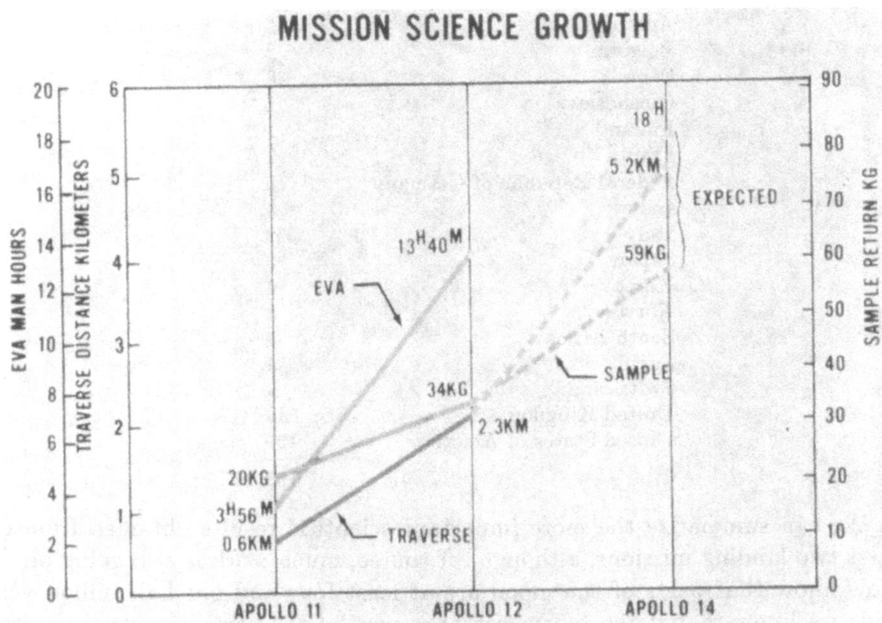

Fig. 19. Mission science growth.

With Apollo 16, there will be some hardware changes that will result in very significant improvements and capabilities. These are summarized in Table 2.

Table 2
Increased capabilities: Apollo 16 and subsequent and spacecraft

Landed scientific payload increased from 200 kg to 450 kg
Surface excursions increased from two to three
Total exploration time increased from 18 to 40 man-hours
Increased range and efficiency of surface operations
Orbital staytime increased from 4 days to 7 days

The landed payload will be increased by over 100%. We anticipate at least three excursions to the lunar surface, with the total man-hours exploring the surface increased from 18 to 40. The range and efficiency of surface operations will be greatly improved by modified space suits and life support systems, and by use of a small roving vehicle. Finally, the spacecraft will be in orbit around the moon for a longer period of time, and a number of orbital experiments for remote sensing of the lunar surface will be added to the Service Module.

For the remaining Apollo missions, the landing sites are under continual review to insure that the limited number of missions are placed in the most important scientific areas. Fig. 20 shows the various candidate sites that are under consideration for the remaining missions. These include areas in the highlands, in apparently fresh volcanic regions, near the mysterious rilles, and

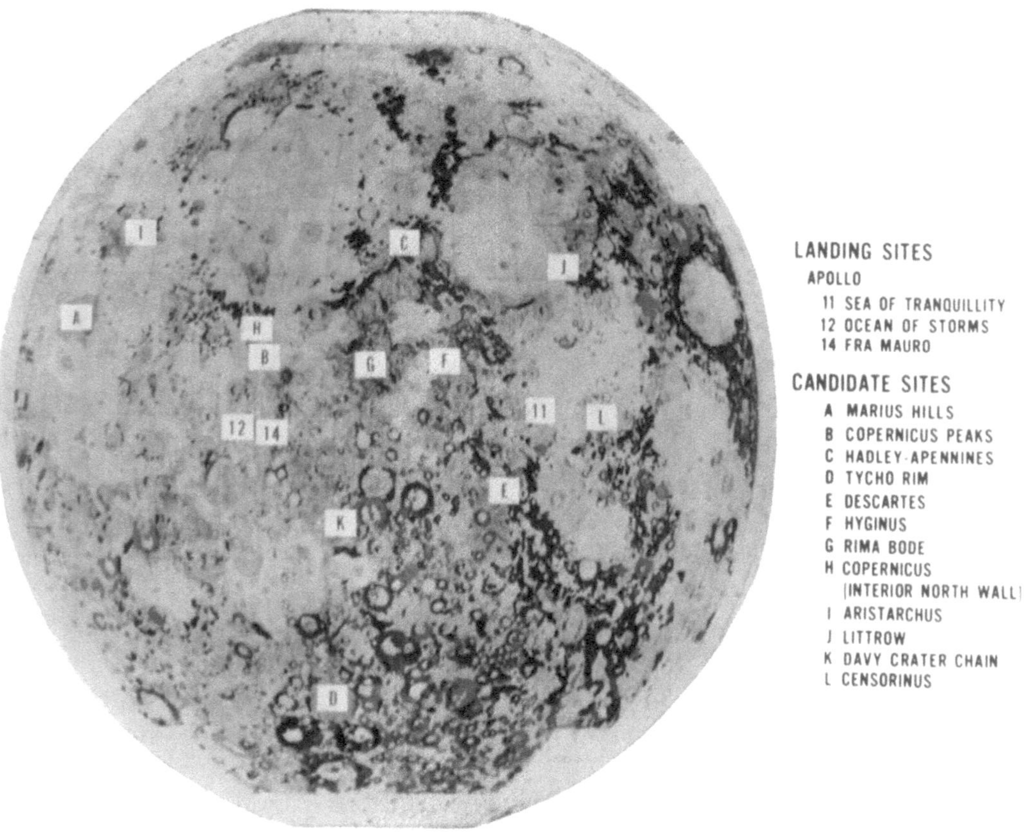

Fig. 20. Geologic map of moon with candidate future landing sites.

in the bottom of deep craters. There are more prime candidate sites than missions to send, so that great care must be taken in the selection. In addition to scientific considerations, there are various operational constraints. For example, it seems doubtful that we will ever consider the risk operationally acceptable for landing near the crater Tycho. A key scientific objective at a landing site near the central peaks of the crater Copernicus would be to bring back samples from the central peaks which would represent material rebounded from deep beneath the lunar surface when this impact crater was formed. A landing site near the Hadley Rille, which extends some 100 kilometers, would provide data on the rille itself and the contact between the mare and the Apennine mountains.

This has been a very broad overview of the Apollo missions and our progress and plans for lunar exploration. We feel we are making substantial progress in unravelling some of the mysteries of man's nearest neighbor.

Space Research XI — Akademie-Verlag, Berlin 1971

LUNAR SURFACE EXPLORATION

NEIL ARMSTRONG
NASA Astronaut

Manned Spacecraft Center, National Aeronautics and Space Administration, Houston, Texas, USA

This paper reviews the experiences of living on the moon, in accordance with observations made by the four men who have walked the lunar surface. The paper begins by describing initial impressions on viewing lunar terrain, optical effects and color changes according to sun position, and adaptation to lunar gravity.

A discussion of the equipment worn during extravehicular activity includes a description of the basic garment assemblies and the backpack and its related equipment. The mobility of a human on the lunar surface is described, as limited by the equipment he wears. Three methods of forward movement are evaluated: walking, hopping, and loping.

A number of tasks are discussed. These include: loading and unloading equipment, carrying and emplacing experiment packages, collecting rock samples, driving core tube samples into the surface, and photography. The paper concludes with a description of workload monitoring and predictions.

1. Introduction

The purpose of this paper is to review the experiences of living on the surface of the moon as well as possible from the observations of the four men who have walked the lunar surface. The performance of such common acts as standing, walking, lifting, and carrying will be discussed. Observations of lighting, color, and lunar surface features will be related. The ability to do tasks and the implication on future work will be analyzed.

2. Initial Impressions

Immediately after touchdown we shutdown our rocket engine, completed our post-landing check list, and enjoyed our first opportunity to take a good look around. The dust kicked up by the rocket exhaust had immediately disappeared after engine shutdown, leaving no residual dust clouds and affording us an extremely clear view. We had landed on a rolling plain (Fig. 1). The area we could see was covered with randomly spaced craters, the largest of which was perhaps 15 meters in diameter. Each square meter of surface included a number of small craters of various sizes down to a diameter of two centimeters. The horizon was irregular, giving the impression of rolling hills that were probably actually portions of larger crater rims. The area appeared to be generally sandy with a wide variety of rocks of every size up to 2 meters in view. They were of angular and semi-angular shapes and were lying on top of the surface or buried to various depths.

The surface was brightly illuminated, giving the appearance of a desert on a summer day. Since the sky is black, the impression is somewhat like being on a sandy athletic field at night that is very well illuminated with flood lights. No stars or planets could be seen with the exception of the earth.

Fig. 1. The Apollo 11 landing site.

While performing the initial post-landing activities, we had the opportunity to notice our reaction to the lunar gravity. Inasmuch as the Lunar Module is operated while standing, our initial observations were confined to that position. There was no difficulty noted in stability or balance. Recovery from incipient falling was easily performed. The tendency to rise up on the toes, noted during lunar gravity simulations on earth, was not so evident as had been expected. In general, the gravity condition was very pleasant and preferable to both weightlessness and earth gravity.

3. Observations

3.1. Lighting and Color

During the time Apollo 11 was on the surface, the sun rose from 10.5° to 22° above the horizon. For Apollo 12 the range was 5.2° to 21.1°. All lighting and color observations should be restricted to these conditions. In general (Fig. 2), the entire illumination level was very bright, somewhat like a cloudless summer day here on earth. Areas of shadow were very dark, but not black, sufficient

Fig. 2. An impression of the range of illuminations visible in and out of shadow of the lunar surface.

back lighting from illuminated hillsides down sun of the shadow existed to permit considerable visibility after a short eye-adaptation time.

The peculiar photometric properties of the moon have been well known for many years. The high reflectance parallel to incident light had led to the prediction that the eye would be unable to discern anything in the area around the "zero-phase" point. Our descent trajectory was carefully structured, in fact, to prevent the requirement for looking along the sun's rays at the intended landing point. As can be seen, a camera can, in fact, see into the region and the eye does somewhat better. Although contrast is certainly degraded in this "wash-out" region, a significant amount of detail can certainly be differentiated.

Little or no color can be seen in this bright "zero-phase" region. No color can be detected anywhere, in fact, at very low sun angles such as that at the time of the Apollo 12 touchdown (5°). As the sun rises to 10° above the horizon, some brown or tan colors begin to appear in the area 30–60° away from down

Fig. 3. The Lunar module sitting on the generally smooth terrain of the Apollo 11 and 12 landing sites.

sun. Cross sun is generally a darker area of high contrast with little or no color detectable. The general area has the brightness of the desert and is in the very light color range characterized by dry cement or beach sand. It is a great surprise, when seeing the rocks and fines at close range, to find that their color is actually a dark or charcoal gray.

3.2. General Topography

The photographs indicate the landing areas of Apollo 11 and 12 to be generally smooth (Fig. 3). The camera, however, de-emphasizes the local roughness. Our appraisal was, in fact, that it was a generally rough area for surface work. I am quite certain that most, if not all, the surface irregularities were parts of crater rims, although that certainly was not obvious at the time, particularly out near the horizon. The horizon was quite irregular, and when combined with the low gravitational pull, made the estimation of the local vertical considerably more difficult than here on earth, probably not being accurate to better than 5°.

The pronounced lunar curvature was evidenced by the closeness of objects on the horizon. Lunar curvature, plus local roughness, gave an impression simi-

Fig. 4. The earth looming above the Lunar Module.

lar to the difficulty of seeing while swimming in a rough ocean. We could not see the rim of West Crater about 400 meters east of us, although its rim appeared to be about 25 meters in height as we flew over it in our final descent In general, there is a tendency to underestimate distances. Both the Apollo 11 and the Apollo 12 crew felt that their initial distance estimates were too small.

3.3. The Earth

No observational report would be complete without mentioning the view of earth (Fig. 4). During the Apollo 11 observations the earth was approximately

32° west of the zenith. It was slightly gibbous and very colorful. The blues of the oceans and the whites of the clouds predominate, but the gray-browns of the continents are easily identified.

The diameter, and hence apparent size as seen from the moon, is four times that of the moon as seen from the earth. Although it still appears quite small, it is a very spectacular sight.

4. Protective Equipment

Before discussing the surface activities, it would be useful to review briefly the characteristics of the protective garment worn, and its related equipment.

The basic garment assemblies provide pressure, fire, thermal, and micrometeorite protection. The backpack and its related equipment incorporate systems for pressurization, ventilation, communications and temperature control. The system provides an environment of 100% oxygen at an operating pressure of 3.75 lb in^{-2} atm. Temperature protection to 250 °F is provided.

The pressure garment assembly consists of a single-piece garment enclosing the torso, arms, legs and feet, a pressure helmet assembly, gloves, protective cover layers and various connectors and controls. The custom-tailored torso-limb assembly is entered through a vertical opening along the spine, filled with a pressure sealing, zipper-type closure and a second restraint closure. Convoluted joints of near constant volume are located at the shoulders, elbows, wrists, hips, knees and ankles. Restraint cables are provided to sustain axial limb loads and prevent ballooning of the convoluted joints. The arms include a circular bearing above the elbow to improve rotational movements. It should be noted that no joints exist in the torso or neck. A duct system provides ventilation paths throughout the assembly.

The pressure helmet is a detachable, transparent bubble which is enclosed in a multi-layered visor assembly. The protective shell holds two adjustable visors for the filtering of ultraviolet, infrared, and visible light. Three eyeshades, one center and one on each side are incorporated to reduce glare.

The pressure glove is basically a bladder molded from a handcast of the individual's hand. Wrist mobility is provided by a convoluted section in conjunction with a wrist bearing, permitting 360° of glove rotation. Fourteen additional layers of various materials provide restraint, thermal, and abrasion protection. Twelve outer layers of assorted fabrics are used over the remainder of the pressure garment to provide thermal, abrasion, and micrometeorite protection. The outer boots provide thermal and abrasive protection with a thick, heavily ribbed sole of silicon rubber and multi-layered tops.

A liquid-cooling garment is worn adjacent to the skin. This garment provides the normal underwear functions, plus thermal transfer between the crewman's body and the cooling media. The garment provides a continuous flow of temperature-controlled water through a network of tubing stitched to the inside surface of the open-mesh fabric.

The backpack, or portable life support system (PLSS), includes expendables, communication equipment, and operating controls and displays. An oxygen circuit takes exhaust gas from the pressure garment, removes heat with a heat exchanger, carbon dioxide by lithium hydroxide, water by a water separator, odors by activated charcoal, and foreign particles by a filter. The clean, cool oxygen is then returned to the inlet of the pressure garment. Compensation for the oxygen loss due to metabolic usage and suit leakage is provided from an oxygen tank holding 0.92 pounds of oxygen charged to 950 lb in^{-2} atm. The oxygen tank may be recharged from the Lunar Module between excursions.

The cooling water in the tubes of the liquid-cooled garment is circulated by a motor-driven pump in the backpack with a flow rate of 4.0 lb min^{-1}. The water is cooled in a sublimator with temperature regulation provided by a manually-controlled, three-position diverter valve. In the minimum position, the valve diverts the flow, allowing the majority of flow to bypass the sublimator. In the maximum position, the entire flow passes through the sublimator for maximum cooling before returning to the liquid cooled garment. The sublimator is of the porous-plate type, and uses water as the evaporative material. The feedwater tank contains 8 lb of water and may be recharged from the Lunar Module. The pumps and fans are powered by a battery which also provides the electrical source for the instrumentation and communications.

The communications system in the backpack provides voice and telemetry in a duplex mode (296.8 and 259.7 MHz) between the two crewmen and the Lunar Module. A relay in the spacecraft converts the signal to S-band and retransmits it to earth through a high-gain antenna.

The temperature-controlling diverter valve, the oxygen valve, and the water valve are located on the lower right side of the backpack. Electrical and communication controls, oxygen pressure gauge, and system status indicators are located on the remote control unit mounted on the chest. This unit also provides a mounting bracket for the 70 mm Hasselblad camera.

Emergency protection against a backpack failure is provided by simple oxygen purge system located on the top of the backpack. Four pounds of oxygen is contained in two spherical bottles at a pressure of approximately 6000 lb in^{-2}. When needed, it operates in a blow-down mode with the exhaust gas regulated to 3.7 lb in^{-2} and temperature controlled by a 116 W heater operated by a separate battery.

5. Mobility

The mobility of a human on the lunar surface must, of necessity, be described within the limitations of the equipment he wears. Due to the high and rearward center of gravity location produced by the backpack configuration, the upright stance is slightly forward of the normal earth posture. Turning from a standing position is slightly slower than on earth due to the lower frictional force available at the sole of the foot, but adequate turning rates can be readily accomplished. Acceleration from a standing position to forward motion is noticeably slower than on earth as a result of the low frictional force available at the sole of the boot. Two methods of acceleration are available: (i) a series of three or four steps adding a small increment of velocity with each step, (ii) leaning very far forward and pushing off vigorously, gaining the desired velocity in the first step. Both of these methods are satisfactory, but the former is usually used unless there is a desire to move quickly.

Three methods of forward movement have been evaluated: walking, hopping, and loping. Walking was used for normal activities while working in the vicinity

Fig. 5.

of the lunar module, while emplacing experiments, and during observations. Speed is limited to about 2 ft s^{-1}. At higher speeds, the force at the foot necessary to maintain the speed will provide sufficient upward force to lift the person off the ground before the other foot comes down. This mode of locomotion, where both feet are separated from the ground at the same time, is called loping, and is the preferred method of traveling over substantial distances. Steady state velocities of 3–5 ft s^{-1} are common with 7 ft s^{-1} being attainable for short distances. Speed is frequently limited to a value suitable for surveying a negotiable path over the rough surface.

The loping pace is like running, in that both feet are off the ground at the same time, but unlike running in that the legs are moved rather showly. It gives one, in fact, the feeling of running in slow motion. Running, as we know it here on earth, cannot be performed on the moon. Hopping, with both feet leaving and hitting the surface simultaneously, was found to be a satisfactory method of movement but was not a preferred method by any of the crewmen.

Fig. 6. A footprint on the soft-textured lunar surface. The fidelity with which the boot markings are reproduced indicates the presence of very fine material.

Stopping from a walk can be achieved in one or two steps, but stopping from the loping mode requires three or four. Side-stepping could be performed, but was somewhat difficult due to poor visibility and poor lateral articulation of the suit leg. In general, movement on the lunar surface requires somewhat more planning and attention than movement here on earth.

Of course, the low gravitational force invites astounding jumping antics. Free jumps could be controlled up to about a meter. Jumps above that height provided an unstabilized period of time that made the guarantee of a stable landing difficult. The maximum height achieved (Fig. 5) was nearly 2 meters in a jump up to the third step of the ladder. In this case, balance was maintained by a loose grip on the handrails of the ladder.

Although several falls have been experienced, they are generally of little consequence. The falling rate is so slow that no damage is expected. Usually, the fall can be prevented once an unbalance is noticed, simply by turning in the direction of the fall and taking a step in that direction. Having fallen, a recovery to a standing position can be performed unassisted fairly easily from a face

Fig. 7.

down position and with somewhat more difficulty from a face-up position. Of course, the recovery with the assistance of another individual is trivial. In general, contact between the suit and the surface is minimized to prevent damage from any sharp rocks or objects that might endanger the suit's pressure integrity.

Of course, walking rates, stability and mobility are somewhat dependent on the nature of the lunar surface itself (Fig. 6). Although the texture seems soft, the footprints are usually less than a centimeter deep. The surface particles were small and cohesive as evidenced by extremely well-defined bootprints with no detectable grains. The footprints were significantly deeper (5 cm) in the rims and floors of small (5 m) craters. The preferred path between points, as a result, is slightly winding so as to avoid local surface irregularities. Continual observation and planning is necessary when walking into new territory.

Fig. 8. Apollo 12 astronaut Aldrin carrying 74.3 kilograms of equipment easily in the low lunar gravity field.

One crewman reported a slippery sensation when walking on a dusty flat rock, but in general, footing is very solid. The 12–20° slopes of the crater in which the Surveyor spacecraft was located caused no difficulty to the Apollo 12 crewman. The steep slopes (40°) of the inner walls of the larger craters have not yet been attempted.

6. Working

A number of tasks have been performed that involve physical work. These include loading and unloading equipment, carrying and emplacing experiment packages, collecting rock samples, and driving core tube samples into the surface.

Carrying of objects (Fig. 8) may be accomplished in a number of ways. Here, Buzz Aldrin is carrying two experiments weighing a total of 74.3 kg earth weight (12.4 kg lunar weight). Lifting the weight is generally less difficult than contending with the bulk. The added weight is, in some cases, an advantage, in that it increased traction and a more solid footing seems desirable when moving large objects. The use of the backpack as a location for attaching and carrying equipment has proved to be very useful, although it tends to lift the center of gravity even higher. This location cannot be reached by the man who wears the backpack and, therefore, is useful only when a second person is available for loading and unloading.

The increasing amounts of equipment that need to be transported around the lunar surface in our forthcoming flights tend to the obvious suggestion of the use of a cart to be pushed or pulled. Pushing and pulling are particularly different operations from here on earth. Relatively small forces can be applied by pulling or pushing unless the body can be inclined at a very large angle from the vertical. Even with a very solid footing, this invites the possibility of a fall. We do however, intend to use a rickshaw-labor cart on our next lunar surface exploration: we have been able to use pulling forces well enough to use this method of transferring equipment up into the lunar module cockpit.

One task of general interest is the driving of rods into the surface. This common task was used for holding of the contingency sample collector handle, the solar wind experiment rod, the flag staff, and the core tube collection. In each of these examples at the Sea of Tranquility (Apollo 11) the rods or tubes could only be pushed or driven down about 20 cm. There did not seem to be a discontinuity, it seemed merely to become increasingly hard with depth. At the Ocean of Storms (Apollo 12), however, after an initial resistance, the core tubes could be driven in to their full length.

The collection of soil samples with a scoop demonstrated an interesting characteristic. The frictional force holding the sample to the scoop was proportional to the lunar weight, which was, of course, quite small. The sample will, therefore, detach itself from the scoop under very low values of acceleration. An open scoop must be handled much more slowly and carefully than earth experience would dictate if one is to hope to arrive at the sample container with any sample remaining. Although we had learned this characteristic in our training, I did lose a lot of material during the bulk sample collection.

A definite shortcoming to effectiveness during surface work is the difficulty of reaching the ground from a standing position. This difficulty is due to our inability to bend the suit's torso and neck. It is rather like working with your back and neck in a cast. The problem is compounded by poor downward visibility. Should an object be dropped, it is a difficult and time-consuming task to recover it. Advanced suit proposals alleviate these shortcomings, but they will continue to be a disadvantage for the explorations of the next several years.

7. Photography

Still photography was performed with a modified 70 mm commercially available camera equiped with motor-driven automatic film advance (Fig. 9). A thin-based film provided 160 exposures in a single film pack. Exposures are manually adjusted, and no viewfinder or lightmeter is used. The camera, attached to the chest-mounted backpack remote control unit, is directed by pointing the chest

Fig. 9. 70 mm motor-driven camera used for surface photography. A hand-held close-up stereo camera, not shown here, was also employed.

toward the object to be photographed. Inasmuch as the exposure varies widely (3 or 4 f-stops from cross-sun to down-sun) an exposure guide for reference is attached to the top of the camera.

8. Workload Monitoring and Prediction

An understanding of the workload or energy expenditure of an individual performing tasks on the lunar surface is clearly an important factor in both the planning and the conduct of lunar exploration. Three factors have been monitored to provide a real-time understanding of workload. The three factors correlated are: heart rate, oxygen usage, and heat production.

In the Mission Control Center, cardiotachs were configured to display 30-second heart rate averages for each crewman. In addition, a strip chart recorder was used to obtain a permanent record of the telemetered output. Conversion from heart rate to metabolic rate (BTU/hr) was accomplished with a calibration curve prepared from preflight exercise response tests using a bicycle ergometer.

The oxygen consumption method is similar to standard laboratory techniques with the exception that carbon dioxide production is not monitored. The calculations of metabolic rate and total energy were based on the telemetered backpack oxygen bottle pressure decay. The respiratory quotient was assumed equal to 0.9. Corrections for the suit leak rate and the oxygen compressibility factor must be included.

The heat production method in estimating metabolic rate consists of solving a single complex linear equation. The equation takes into consideration all heat gains and losses within the suit. The primary measurements are the telemetered values of the change in water temperature across the liquid-cooled undergarment. Additionally, correction estimates must be included for air temperature changes across the suit, radiant gains or losses, and heat removed by perspiration. An important quantity in obtaining the best estimate of the total metabolic expenditure was the determination of the water used by the sublimator. The remaining water in the backpack was drained into a bag and weighed with a spring scale; subtracting this from the initial weight yielded the amount of water vaporized through sublimation. These data, along with more detailed post-flight analysis yielded the best estimate of heat production.

The liquid-cooled garment method gives very good agreement with the integrated estimate; the oxygen method also agrees fairly well. The heart rate method deviates sharply, indicating a significantly higher metabolic consumption. The heart rate method on the second crew member also differed significantly from the best estimate, except that in his case the heart rate indicated a lower metabolic consumption. This disparity may be due to calibration errors, psychogenic effect, heat storage, and deconditioning. Although these factors are being studied in depth, the heart rate method does not seem promising for absolute indicator but can be used to indicate relative rate changes.

The Apollo 11 results indicated that work rates varied between 900 and 1400 BTU/hr with 1100–1200 as a reasonable average. Based on these results, the Apollo 12 excursions were extended to four hours.

9. Conclusions

Two lunar landings and their associated initial surface explorations have provided a good understanding of working in the lunar environment. The hostile vacuum and extreme temperatures are tolerable with the protective equipment that has been developed. The lunar gravity is found to be a particularly favorable condition for the human to operate within. No physical or psychological difficulties have been encountered which might discourage further exploratory planning.

Further improvement in mobility of the pressure garments will improve the effectiveness of persons conducting observations and experiments on the moon. Substantially more ambitious undertakings can be planned during this next decade which will greatly increase our knowledge and understanding of our natural satellite.

Space Research XI — Akademie-Verlag, Berlin 1971

RESULTS OF RECENT MANNED AND UNMANNED LUNAR EXPLORATION

L. D. JAFFE

Jet Propulsion Laboratory, California Institute of Technology, Pasadena, California, USA

Important data about the moon, obtained from spacecraft during the past year, include the age-dating and chemical analyses of samples returned by Apollos 11 and 12. The material of the surface fragmental layer was differentiated 4.6×10^9 y ago. The surfaces of Mare Tranquillitatis and Oceanus Procellarum solidified 3.7×10^9 and about 3×10^9 y ago, from lava of very low viscosity which had undergone extensive chemical fractionation. The lunar interior is of different composition from these surfaces. The moon is tectonically rather quiet. The rate impact of moderate-sized objects was much greater before the mare formation than since. Micrometeorite impact has produced rock erosion, pitting, shock, melting, vitrification, and induration in the surface layer. Pictures, taken by Apollo astronauts, of the lunar surface disturbed by Surveyor 3 show that little change has occurred in 31 months.

1. Introduction

Earlier observations of the moon by unmanned spacecraft and by the Apollo 8 orbiter have been reviewed in [1, 2]. Because of the great quantity of additional data now available, it seems appropriate to compare earlier findings with new results and to note areas in which information is available for the first time.

During the period May 1969 to May 1970 Zond 7 and Apollo 13 flew by the moon and returned to earth; Luna 15 and Apollo 10 orbited the moon; and Apollos 11 and 12 landed, in Mare Tranquillitatis and Oceanus Procellarum, respectively, and returned to earth.

2. Lunar Motion and Mass Distribution

Previous spacecraft provided data for improving the lunar ephemeris and for determining the distribution of mass within the moon. Especially significant was the discovery from Lunar Orbiter tracking of the mass concentrations ("mascons") associated with some of the circular maria [3, 4].

The Apollo 11 laser ranging retroreflector landed in Mare Tranquillitatis and the two-way radio range and Doppler tracking of Apollo orbiters and landers should allow improved determinations of lunar motions and mass distribution. Because of the time required for data acquisition and reduction, however, results reported during the past year are based primarily on analysis of Lunar Orbiter data. Among them is a revised determination of local gravitational accelerations on the near side, including mascon locations and magnitudes [5].

3. Seismic Information

The only lunar seismic information available before manned landings was derived from elastic vibrations of Surveyor spacecraft on the surface [6]. Revised calibration of these results gave the compressional wave velocity for the top half-meter of lunar material as 50 to 90 m s^{-1} and shear wave velocity 20 to 35 m s^{-1} [7].

Apollo 11 and 12 passive seismic experiments have shown very low lunar seismic activity, and very low attenuation of seismic waves. Man-made seismic signals were generated by astronaut activity, firing and impact of the Apollo 12 ascent stage, and impact of the last booster stage of Apollo 13. These indicated a compressional velocity of 108 m s^{-1} for the upper 10 m of surface, within the range indicated by the Surveyor data. For a signal transmitted at depths up to 7 to 8 km, the apparent compressional wave velocity indicated by the travel time was 3.2 km s^{-1}; at somewhat greater depths, 4.6 km s^{-1}. The character of the signals is very different from that on earth. For man-made impacts and the observed natural events, the signal amplitude increases gradually over a time of minutes and then decreases over times of $^1/_2$ to 4 hr [8–11]. Laboratory measurements on samples returned by Apollo 11 give compressional velocities, at 1 atm, of 1.8–3.5 km s^{-1} for crystalline rocks, and 1.2 – 1.5 km s^{-1} for breccia, with lower values for fines. These velocities increase rapidly with pressure, presumably because of the closing of cracks and pores. Velocities for crystalline rock at pressures corresponding to depths of 8 km apparently fit the observed lunar travel time. The material was, however, highly attenuating at low pressure [12–14].

4. Major Surface Features

Earlier pictures from unmanned spacecraft (especially Luna 3, Zond 3 and the Lunar Orbiters) revealed the general character of the far side of the moon: the circular, concentric nature of Mare Orientale and its extensive ejecta, the complex structure of major craters, including slump features, and smaller rilles within such features as the sinuous rille Schroeter's Valley and the relatively straight Alpine Valley.

Direct observations and photographs of major lunar surface features during the past year are essentially a continuation of the earlier photography from unmanned and Apollo 8 orbiters. Astronaut observations and pictures taken at high sun from orbit show a high population of craters with bright interior walls or bright halo and rays, at diameters of 100 to 1000 m [15–18]. The photographs, especially the stereoscopic pairs, should be valuable for mapping work and for studying smaller features (Fig. 1). Multispectral photographs from orbit should permit detection of subtle color variations over smaller surface areas than can be resolved from the earth.

5. Smaller Features

The general nature of the lunar surface layer was clearly determined by earlier unmanned spacecraft such as the Luna and Surveyor landers. The surface layer, or regolith, consists primarily of fine, slightly cohesive particles, with most of the mass in particles smaller than 60 μm. Larger rocks are mixed with the fine matrix and lie on top of it. Craters down to a few centimeters were observed in the rego-

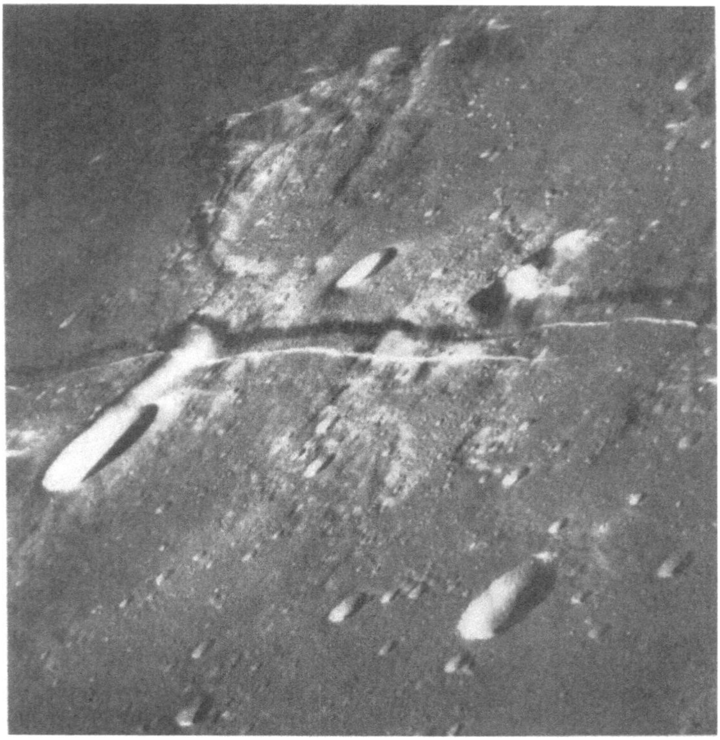

Fig. 1. Ariadaeus Rille. Note apparent elevation changes (Apollo 10 photograph AS 10-31-4645).

lith. Lunar Orbiter photographs that show the shapes and distribution of craters smaller than a few hundred meters provided considerable information about lunar surface processes and about the regolith depth (several meters in the maria, probably deeper in the highlands [19–21]). Explorations made in 1969 have not changed this description.

6. Rocks

Earlier observations showed rocks up to tens of meters across, with different size–frequency distributions in different lunar areas; Apollo 8 photographs revealed individual blocks larger than 100 m in diameter. These rocks are locally more common within one crater radius of the rims of some sharper craters with diameters a few meters to a few kilometers. Exposed rocks are gray, with generally higher optical albedo and polarization than the fine matrix. The rocks show a variety of shapes: some angular, others rounded; angular rocks tend to be associated with sharp craters. Some rocks were found to be rounded on their top exposed surface and more angular at their bottom surface. Some rock surfaces appear pitted, others vesiculated; some rocks are aggregates of rock chips and fine particles (breccia). The density of one rock, near Tycho, was measured

Fig. 2. Fine-grained basalt rock, with vesicles, from Mare Tranquillitatis (LRL specimen 10022; photograph S-49-45209).

as 2.8 ± 0.4 g cm^{-3}; calculations from measured elemental compositions indicated solid densities of 3.0 g cm^{-3} for the Tycho site and 3.2 g cm^{-3} for the maria [17, 22–24].

The Apollo 11 astronauts noted many small surface objects that had a metallic luster and resembled spatter-blobs which occurred at the bottom of craters about 1 m in diameter [25]. These were again observed by the Apollo 12 astronauts and were identified as glass. Thin patches of glass also were found spattered on rock surfaces [26–29].

Laboratory examination of lunar rocks returned by Apollo 11 and 12 astronauts has provided a great deal of information. Many of the rocks are crystalline igneous; the rest are breccias. The crystalline rocks contain gas cavities, spheroidal (vesicles) or irregular (vugs) (Fig. 2). The principal minerals in the samples from Mare Tranquillitatis are clinopyroxene (40–55% by volume), plagioclase (20–35%), ilmenite (10–20%), and olivine (up to 5%). The Oceanus

Procellarum rocks contained these same minerals, but with a considerably greater range of mineralogical composition. Crystobalite, glass, troilite, and metallic iron are also present in rocks from the two sites. A few minerals not known to occur on earth were found; no hydrated minerals were found.

These rocks are classified as basalts and gabbros. A small percentage of anorthosites, composed predominantly of plagioclase, was also found among Apollo 11 samples; their elemental compositions are similar to those analyzed near Tycho by Surveyor 7 and their density is about 2.9 g cm^{-3}, compared to 3.2 to 3.4 for the basalts. The breccias contain rock fragments, mostly angular and smaller than 2 cm, together with fines. Their microstructures show abundant evidence of shock, including shock melting and vitrification at pressures of a few hundred kilobars. The rocks are pitted; pits range from 10 μm or less up to 1 cm in diameter, with a depth/diameter ratio typically 1:5. Pit densities are up to 30 cm^{-2}. They have glassy linings and, usually, raised glassy rims surrounded, on crystalline rocks, with haloes which sometimes whiten the whole surface of the rocks. These haloes apparently are a result of microfracturing [26, 27, 30–42].

7. Fine Particles

Observations from Surveyor spacecraft indicated that most of the fine particles are from 2 to 60 μm in diameter [2, 43, 44]. The bulk density is about 1.5 to 1.6 g cm^{-3} at 5-cm depth [45] and probably lower in the top few millimeters, about 1 g cm^{-3} [44]. The static bearing capacity is about 0.1 N cm^{-2} at 1-mm depth and 5 N cm^{-2} at 5-cm depth; the increase is nearly linear [43–45]. Cohesion is about 0.05 N cm^{-2} at a depth of a few centimeters; the internal friction angle is about 35–40° at this depth [44]. On a level surface, the fine matrix generally contacts larger surface rocks along an almost horizontal plane, with a small fillet at the contact line. On slopes, however, the fillet is larger on the uphill side of the rock [47].

Astronauts noticed appreciable local variations in regolith bearing capacity. Their boots sank less than 1 cm into the fines in some areas, several centimeters in others, especially on the rims of small, fresh-looking craters (Fig. 3); this may be due to local slopes [48]. From the Apollo 11 mission they reported that the top 5 to 15 cm appears stratified with layers of different appearance and mechanical properties; the top layers are 3 to 10 mm thick [25]; core tubes, however, did not show this stratification [48]. Apollo 12 astronauts noted a discontinuous deposit of lighter fines, sometimes on the surface, in other areas under centimeters or more of darker material; the lighter fines were thought perhaps to be part of a Copernican ray. Apollo 12 core tubes, containing material from depths of at least 40 cm, showed considerable stratification, with variations in albedo, texture, and particle size; one layer was quite coarse, with a median grain size near 5 mm; these coarser particles may, however, be aggregates. A core tube was driven to a depth of 70 cm [26].

Under the microscope, the fines show a variety of shapes, angular, irregular, rounded, and spheroidal. Especially interesting is the large fraction of glassy

Fig. 3. Astronaut bootprints in the lunar surface. Note greater depth of sinkage on rim of small crater (Apollo 11 photograph AS 11-40-5942).

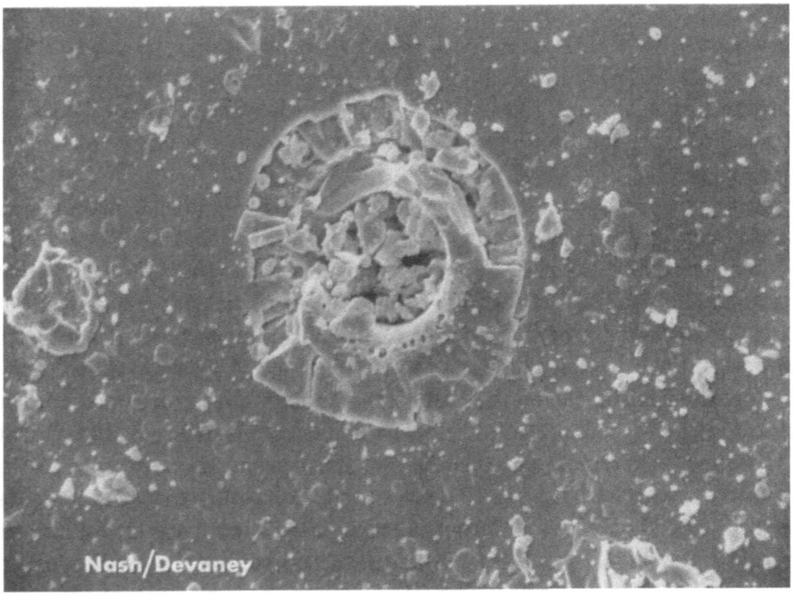

Fig. 4. 9-μm-diameter impact pit and associated fragments on a 1-mm glass sphere from Mare Tranquillitatis (scanning electron photomicrograph, courtesy of D. Nash and J. Devaney).

spheres; most of them are smaller than 20 μm, but some range to several millimeters in diameter. Glassy shards are also present and other rocks and mineral pieces of the same general nature as the larger rocks. One sample of fines consisted mostly of feldspar, but with glass fragments showing flow structure, suggestive of volcanic ash. The microstructures of other samples show extensive evidence of shock, including shock melting. The fine particles themselves show glassy pits like those on larger rocks (Fig. 4) [26, 27].

8. Elemental Composition

Luna 10 gamma-ray spectroscopy from orbit showed some time ago that the U, Th, and K content of the surface material is inconsistent with acidic silicates, and suggested that the content of these elements is lower in the highlands than in the maria. Continued analysis of Surveyor data indicated a deposit of radioactive daughter products of radon on the surface of Mare Tranquillitatis. Complete reduction of the Surveyor data gave elemental analyses (see Table 1) which show a basic rock composition. Particularly interesting is the high titanium content in Mare Tranquillitatis, and the higher iron and lower aluminum and calcium in the maria than at the highland Tycho site [49–54].

Return of the Apollo samples has provided extensive chemical data (see Table 1). The results from Apollo 11 are in accord with those from Surveyor 5, 25 km away in Mare Tranquillitatis. The Mare Tranquillitatis samples are enriched in Ti, Zr, Hf, Sc, Y, Cr, and depleted in Na, K, Rb, Pb, Bi, relative to terrestrial rocks of similar overall basaltic composition. Eu is depleted relative to the other rare earth elements. The soil and breccias differ noticeably from the crystalline rocks, and contain somewhat more Ni, Cd, Zn, Ag, Au, Tl. There is about $^1/_2$ weight per cent metallic iron in the fines, much less in the crystalline rocks. Most of the iron is in the ferrous state; the content of ferric iron is very low. The Oceanus Procellarum rocks have less Ti, Zr, K, Rb and more Fe, Mg, Ni than those from Mare Tranquillitatis. Breccia and fines from this area are more sharply differentiated from the crystalline rocks, containing more Ni, Ti, Zr, K, Rb, U, Th. The Oceanus Procellarum fines, compared to those from Mare Tranquillitatis, are lower in Ti and higher in Mg [26, 27, 55–63].

The Mare Tranquillitatis fines and breccias contain 20 to 40 ppm He, with correspondingly high percentages of other noble gases. Presumably, these gases were implanted primarily by the solar wind, and the elemental and isotopic analyses give good data on solar wind composition. Oceanus Procellarum fines and rocks show less of these elements, and crystalline rocks from both sites show much less [56, 26].

Melting tests on a composition matching a typical crystalline rock from Mare Tranquillitatis indicated that the lava from which these rocks solidified had much lower viscosity than terrestrial basaltic lavas [63].

Table 1

Typical elemental compositions of lunar surface material (by weight)

Element	Mare Tranquillitatis			Sinus Medii	Oceanus Procellarum		Tycho rim	
	Surveyor 5[a)] Fines	Apollo 11[b)] Fines and breccia	Apollo 11[b)] Crystalline rocks	Surveyor 6[c)] Fines	Apollo 12[d)] Fines and breccia	Apollo 12[d)] Crystalline rocks	Surveyor 7[e)] Fines	Surveyor 7[e)] Rock
Si	21 (%)	20 (%)	20 (%)	23 (%)	20 (%)	19 (%)	21 (%)	21 (%)
Fe	9	12	14	10	13	17	4	3
Ca	10	8	7	9	7	8	13	13
Al	8	6	6	8	7	6	11	14
Mg	3	5	5	4	7	7	4	<2
Ti	4	5	6	2	2	2	<0.4	<0.7
Na	0.5	0.4	0.4	0.6	0.3	0.3	0.5	0.3
Cr		0.3	0.4		0.3	0.4		
Mn		0.2	0.3		0.2	0.2		
K		0.1	0.2		0.3	0.05		
Zr		500 ppm	900 ppm		700 ppm	200 ppm		
Ni		200	1 to 50		200	50		
Y		200	200		100	50		
Sr		100	100		200	100		
Ba		90	100		400	70		
Sc		60	100		40	50		
V		30	30		60	90		
Li		10	20		10	6		
Co		10	10		40	40		
Cu		8	5					

Rb	2	2	3	6
Ce	50	60		
Nd	50	60		
Gd	20	20		
Eu	2	2		
F	100	100		
C	100	<50		
H	<100	<50		
He	30	0.08	5	0.02
Ne	2	0.002	0.5	0.0005
Ar	1	0.06	0.1	0.02
Kr	0.002	0.00001	0.0002	0.000001
X	0.002	0.000002	0.0001	0.0000008
Th	2	3	8	0.9
U	0.4	0.7	2	0.2

[a] A. L. Turkevich, E. J. Franzgrote, and J. H. Patterson, Science **165**, 277 (1969); E. J. Franzgrote et al., Science **167**, 376 (1970).

[b] Lunar Sample Preliminary Examination Team, Science **165**, 1211 (1969). The following articles in Science **167** (1970): S. Epstein and H. P. Taylor, Jr., p. 533; I. Friedman, et al., p. 538; H. Hintenberger, et al., p. 543; E. L. Fireman, J. D'Amico, and J. C. DeFelice, p. 566; C. B. Moore, et al., p. 495; J. A. Maxwell, S. Abbey, and W. H. Champ, p. 530; I. R. Kaplan and J. W. Smith, p. 541; A. L. Burlingame, et al., p. 751; J. Oró, et al., p. 765; A. E. J. Engel and C. G. Engel, p. 527; H. B. Wiik and P. Ojanpera, p. 531; L. C. Peck and V. C. Smith, p. 532; T. Kirsten, F. Steinbrunn, and J. Zähringer, p. 571; S. L. Manatt, et al., p. 709; P. W. Gast and N. J. Hubbard, p. 485; L. A. Haskin, P. A. Helmke, and R. O. Allen, p. 487; J. A. Philpotts and C. C. Schnetzler, p. 493; G. H. Morrison, et al., p. 505; A. A. Smales, et al., p. 509; R. A. Schmitt, H. Wakita, and P. Rey, p. 512; H. Wänke, et al., p. 523; and G. W. Reed, Jr., p. 501.

[c] E. J. Franzgrote et al., Science **167**, 376 (1970).

[d] Lunar Sample Preliminary Examination Team, Science **167**, 1325 (1970).

[e] J. H. Patterson et al., Science **168**, 825 (1970).

9. Ages

Isotopic analyses of Apollo returned samples have provided the first absolute age data for the moon. The time since the last chemical fractionation, and presumably solidification, of most of the Mare Tranquillitatis crystalline rocks, is very close to 3.7×10^9 y for both $^{40}K/^{40}Ar$ and $^{87}Rb/^{87}Sr$ distribution; occasional individual rock analyses show times of $2.3–4.4 \times 10^9$ y. The time since fractionation of U/Th/Pb isotopes may be about 4.2×10^9 y. The fines and breccias show an approximate time of 4.6 to 4.7×10^9 y since U/Th/Pb fractionation and of about 4.5×10^9 y since Rb/Sr fractionation. Oceanus Procellarum crystalline rocks Rb/Sr times of about 3.3×10^9 y [26, 56, 64–76, 128].

Measurements of isotopes arising as spallation products of cosmic rays and of nuclear tracks in the minerals show that some of the rocks from Mare Tranquillitatis have been within 1 to 2 m of the surface for 5×10^8 y, and within a few centimeters of the surface for 10^7 y; one was on the surface about 5×10^7 y. These surface rocks have turned over many times. The rate of erosion on their surfaces, when exposed, has been not more than 10^{-7} cm y^{-1}. Fine particles from Mare Tranquillitatis have been within 1 m of the surface for 4 to 10×10^8 y, and within 10 cm of the surface for 1.5×10^7 y. They are thoroughly stirred to a depth of at least 10 cm every 10^4 to 10^5 y. Crystalline rocks and breccia from Oceanus Procellarum have been within 1 m of the surface for 1×10^6 to 2×10^8 y [26, 27, 30, 52, 74, 75, 77–82].

10. Organics and Biology

An upper limit of 0.2 ppm for organic material is set by analyses of samples returned by Apollo 12, for which the contamination level was better controlled than for Apollo 11. No viable lunar organisms were found; there was no micropaleontological evidence of any past life [26, 30].

11. Optical and Thermal Properties

Luna and Surveyor pictures showed that lunar rocks generally appeared brighter than the fines and showed more polarization of reflected light. Fine material, disturbed or ejected by Surveyor spacecraft, appeared darker than undisturbed material; compressed material appeared lighter and more Lambertian. Subsurface material, even a few millimeters down, looked darker than the undisturbed surface [83–88].

Apollo 11 astronauts confirmed the darker appearance of disturbed fines relative to undisturbed, and the brighter appearance of rocks at all phase angles. They also noted that freshly broken rock fragments are darker than other rock surfaces [25]. Sectioning of returned rocks confirmed that their surfaces are often brighter than their interiors, apparently because of microfracturing, related to the light haloes around the glassy pits. Material produced by pulverizing lunar crystalline rocks in the laboratory is much lighter than lunar fines of similar composition. The difference has been attributed to dark glass in the fines, to possible metallic coating on them, or to irradiation darkening. The microfracture of the exposed rock surfaces presumably increases the optical contrast between them and the fines. The overall albedo, photometric function and polarization of large areas of the surface, as seen from the earth, is similar to that of the fines; evidently, the rocks cover too small a fraction of the lunar surface to have much effect [27, 55, 89–93].

Reflection spectra of crystalline rocks returned from Mare Tranquillitatis show absorption features at 0.94 to 1.00 μm and 2.0 μm, with fainter absorption features at 0.5 and 1.3 μm. The stronger features are attributed to ferrous ions in pyroxenes, and the weaker features to Ti. All features are fainter in breccias and fines. The general area of the landing site, as seen from the earth, has spectral reflectivity characteristics similar to that of the returned fines [90, 91, 94, 95].

Laboratory tests of samples from Mare Tranquillitatis showed no detectable luminescence induced by ultraviolet irradiation and very weak luminescence induced by energetic protons. Weak thermoluminescence also was found. These effects are too small to account for lunar transient phenomena that have been reported on the basis of telescopic observations from the earth [91, 93, 96–98].

Thermal conductivity of Mare Tranquillitatis fines in vacuum, at a packing density of 1.26 g cm^{-3}, is 2.1 W m^{-1} $°K^{-1}$ [99], which is very close to that found for terrestrial basalt particles of similar particle size and packing density [100]. Infrared reflection and transmission spectra of lines and rocks, thermal conductivity of rocks, and specific heats also have been measured [99, 101–103].

12. Electrical and Magnetic Properties

Explorer 35 measurements from orbit earlier indicated an electrical conductivity less than 10^{-6} mho m^{-1} for the outer 100 km of the moon [104]. Previous radar data from spacecraft indicated dielectric constants, for about the top 25 cm, of about 2.2 in the maria and about 3.3 near Tycho; for depths of a few meters, the average was about 3 [105–109]. These were in general agreement with earth-based microwave data. Laboratory measurements on returned fines from Mare Tranquillitatis, at a 450-MHz frequency, and packing densities of 1.2 to 1.6 g cm^{-3}, gave a dielectric constant of 2 to 2.5 and an absorption length about ten times the wavelength [92].

Earlier orbiting spacecraft found no lunar magnetic field or magnetosphere [110–114]. Some ferromagnetic or ferrimagnetic material was found in the surface fines with Surveyor spacecraft [115–117]. Some of the fragments returned by Apollo 11 showed remanent magnetism; the highest value was about 10^{-3} emu g^{-1}. This suggests that the material cooled in a magnetic field of at least 1000 gamma (arising from sun, earth, or moon) [27, 30].

The magnetometer placed in Oceanus Procellarum by Apollo 12 found time-variable magnetic field components arising from the interplanetary plasma and the earth's magnetosphere, and, in addition, a time-invariant field of about 30 gamma believed to arise from a localized source near the Apollo site [118].

13. Micrometeorites and Dust Deposition

Pressurized cell micrometeorite detectors on the Lunar Orbiter had given data consistent with zodiacal light measurements of interplanetary dust, with no indication of a secondary contribution from the lunar surface [119, 120]. Preliminary results from the dust detectors left on the moon by Apollo 11 and 12 astronauts showed no definite evidence of dust deposition [121, 122].

14. Material from Surveyor 3

The Apollo 12 astronauts walked to Surveyor 3 and brought back photographs of Surveyor 3 and the nearby lunar surface, together with the Surveyor 3 television camera itself, surface-sampler scoop, painted and unpainted tubing, elec-

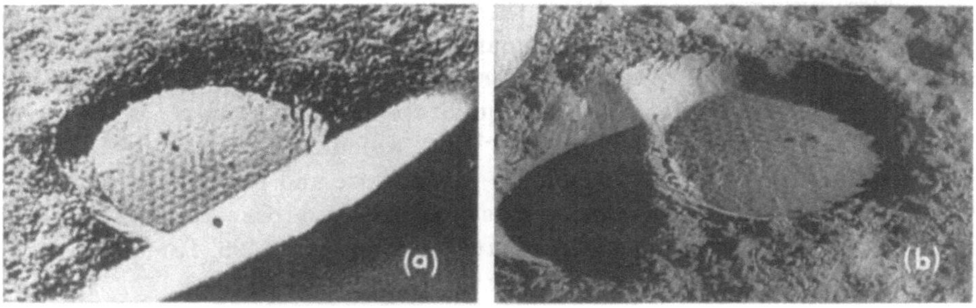

Fig. 5. Imprint in lunar soil made by Surveyor 3 footpad. (a) Part of Surveyor 3 television picture taken April 21, 1967, at 08:24:20 UT (view from north of west; picture is digitally computer-processed). (b) Part of photograph from Apollo 12 hand camera taken November 20, 1969, about 05:22 UT (view from south; from photograph AS 12-48-7110). Waffle pattern in imprint is a replica of that on bottom of Surveyor footpad. Two particles are clearly visible on floor of imprint. In Apollo photograph, another particle is visible to right and down from the right-hand particle. There also appears to be a small shallow pit just above, and to right, of right-hand tip of recticule cross in imprint.

tric cabling, and soil from the scoop. Some of the painted tubing and cabling was returned in vacuum. In a preliminary comparison of the Apollo pictures of the surface with the Surveyor pictures returned months earlier, using about 20 Apollo and 60 Surveyor pictures, only one definite change, other than those obviously produced by the astronauts was found. On the bottom of an imprint made in the surface by a Surveyor footpad, all of the pertinent Apollo pictures show a small particle (about 2 mm in diameter) that does not appear in any of the Surveyor pictures (Fig. 5). If the particle had been present when the Surveyor pictures were taken, its shadow should have easily been detected. This particle may have fallen from the rim of the footpad imprint or may, perhaps, have been kicked in by an astronaut as he approached.

On the floor of this imprint, 20 cm in diameter, any crater as large as 1 mm in diameter should have been visible in the Apollo pictures. Only one pit was noted: it is about 1 mm in diameter, very shallow, and immediately adjacent to a small particle. Most likely, it was produced when the particle fell in as Surveyor landed. Surface areas darkened by ejected regolith fines during the Surveyor landing still appeared dark compared to the undisturbed surface [127].

From the pictures, I found that the sides of several steep walls made by Surveyor footpads and surface sampler were still in place. These include the vertical

Fig. 6. Apollo 12 photograph, taken November 20, 1969, shows trenches dug by Surveyor 3 surface sampler. Far corner of nearer trench preserves vertical wall, about 6 cm deep, dug April 22, 1967 (photograph AS 12-48-7108).

wall of a trench 6 cm deep (Fig. 6). To hold this wall against lunar gravity for an extended time requires a regolith cohesion of at least 0.01 N cm^{-2}; this is consistent with other cohesion data.

Apollo 12 astronauts noted that the white paint and polished aluminum surfaces of Surveyor 3 had turned to tan, and were covered with a dust that adhered tenaciously. Examination of returned parts showed the tan appearance was general. Dark gray lunar fines, up to 150 μm in diameter, adhered lightly in a few small locations. Scanning electron micrographs (Fig. 7) of unpainted tan surfaces revealed particles smaller than 4 μm in diameter [123], with shapes resembling the finest fraction from the lunar regolith. It was known that some material was deposited on Surveyor 3 during its landing. Examination of the returned television camera brought out other directional features in its surface coating; some seem to be associated with the Apollo landing. For example, Fig. 8 shows a component on the side of the Surveyor television camera and the "false shadow" it appears to cast on a painted surface nearby. Tracing back the vector from shadow to component, I find it goes directly through the point where the Apollo 12 descent stage rests, 155 m away. The vertical angle is consistent

Fig. 7. Surface of unpainted aluminum washer from Surveyor 3 television camera, with fine adhering particles (scanning electron photomicrograph; courtesy of W. Carroll and J. Devaney).

with production by particles from the surface below Apollo 12 moving in ballistic trajectories at 10 m s^{-1} or more, and presumably eroded from the surface by the descent engine. The effect has been to lighten the exposed camera paint; micro-Scopy [124] has shown many small chips in the exposed surface.

Scientific areas in which the returned Surveyor material will be useful include lunar surface processes, lunar surface mechanical properties, solar wind composition, energetic particle flux and radiation effects (including induced radioactivity and nuclear tracks), micrometeorite flux, and effect of lunar environment on terrestrial micro-organisms.

15. Interpretation of Findings

Concepts of the moon derived from previous spacecraft and from earth-based measurements, and outlined a year ago [1], still seem valid after return of all the additional data of the past year.

Ratios of the various chemical elements and isotopes are difficult to account for if the moon originally had a composition like that of the earth or its mantle. This weakens support for the theory that the moon originated by fission of the

Fig. 8. Housing of Surveyor 3 television camera. Note "false shadow" cast by rectangular cover of the elevation drive train (left) onto cylindrical motor housing (at center). "Shadow" covers approximately the upper one-third of the cylindrical portion. "Shadow" of the lower-left corner of the cover falls on the electrical wiring.

earth. The depletion of volatile elements and concentration of Ti and Zr suggests that the lunar surface has been at a high temperature and equilibrated with, or underwent, selective loss to a low pressure gas [30].

Chemical and mineralogical compositions indicate that an extensive and large scale process of lunar magma fractionation has occurred. The mare surfaces solidified under extremely low partial pressures of oxygen, water, and sulfur. The lunar interior cannot be of the same composition as the mare surfaces; its density, at the appropriate pressures, would be too high to fit the overall density of the moon. The anorthosite fragments found in Mare Tranquillitatis may be ejecta from highlands with a composition similar to that near Tycho, and a lower density than the mare material [55, p. 140, 27, 30–32, 42].

The material of the Mare Tranquillitatis fines apparently was chemically differentiated 4.6×10^9 y ago; this event may represent the formation of the lunar crust. Since then, fines probably have been differentiated and at least mechanically remixed. Subsequent, more localized, events led to formation of the mare basin and filling; the upper part of Mare Tranquillitatis solidified about 3.7×10^9 y ago, and the area of Oceanus Procellarum in which Apollo 12 landed, about 3.3×10^9 y ago.

The very low viscosity indicated for the mare lavas helps to explain the flatness of mare floors and the sharp contact between them and higher features. It also supports the hypothesis that sinuous rilles originated as lava tubes. The seismic records show that the moon at present is tectonically rather quiet, but there are still major puzzles in interpreting them [8]. The bright interiors of craters a few hundred meters in diameter apparently are due to recent movement of material downslope, exposing fresh surfaces [16].

Differences in impact crater densities between highland and mare areas are too great to be accounted for by the ratio of a possible highland age of 4.6×10^9 y to the indicated age of the mare surfaces, if the impact rate has been constant. Thus, the impact rate of sizable objects must have been much greater during the period between solidification of the highlands and solidification of the maria than it has been since [125].

Some of the regolith fines originally may have been volcanic ash, but detailed examination of the rocks and fines confirms the predominant role of impact in forming this layer. Rocks were knocked from the underlying bedrock by impact, then broken, eroded, pitted, and fractured at the surfaces by continued high-velocity impacts of small particles. This process produced local melting and formation of glass, as well as microfractures that increase the albedo of exposed rock surfaces. The fine particles resulting from erosion were also repeatedly impacted and pitted; impact shock modified their structure and lithified some of them, producing the breccias. 1 to 2% of the regolith may be of direct meteoritic origin, and the solar wind has contributed appreciably to the regolith chemistry [30, 126].

To summarize the results of the past year, I can do no better than quote the conclusion with which I ended my presentation to COSPAR one year ago: "The lunar surface is not a museum, preserving an unaltered record of the early solar system at the time the planets were formed; it has been thoroughly and repeatedly reworked and reshaped over a long period of time to form a complex and varied geological assemblage, whose decipherment may be comparable in complexity to that of the earth" [1, 2].

Acknowledgment

The work reviewed here was performed by very many people.

The author thanks James Burke, Jet Propulsion Laboratory, for making preparation of this paper possible.

The work was sponsored by the National Aeronautics and Space Administration of the U.S.A. under contract NAS 7–100.

17. References

[1] L. D. JAFFE, in: Space Research X, North-Holland Publ. Co., Amsterdam 1970 (p. 989).
[2] L. D. JAFFE, Space Sci. Rev. **9**, 491 (1969).
[3] C. N. CARY and S. L. SJOGREN, Science **160**, 875 (1968).
[4] P. M. MULLER and W. L. SJOGREN, Science **161**, 680 (1968).
[5] L. WONG et al., Trans. Am. Geophys. Un. **51**, 267 (1970).
[6] R. CHOATE et al., Surveyor: Program Results, NASA SP-184 (1969) p. 129.
[7] G. H. SUTTON and F. DUENNEBIER, to be published.
[8] G. V. LATHAM et al., Science **167**, 455 (1970).
[9] G. V. LATHAM et al., Apollo 12 Preliminary Science Report, NASA SP-235, p. 39.
[10] G. LATHAM, F. PRESS, and M. EWING, quoted in: Aviation Week and Space Technology **92**, 20 (1970).
[11] G. LATHAM, F. DUENNEBIER and R. MEISSNER, quoted by: E. BURGESS, Christian Science Monitor (10 April 1970).
[12] H. KANAMORI et al., Science **167**, 726 (1970).
[13] D. R. STEPHENS and E. M. LILLEY, Science **167**, 731 (1970).
[14] E. SCHREIBER et al., Science **167**, 732 (1970).
[15] W. A. ANDERS, J. A. LOVELL, and F. BORMAN, Analysis of Apollo 8: Photography and Visual Observations, NASA SP-201 (1969) p. 1.
[16] R. G. STROM, Analysis of Apollo 8: Photography and Visual Observations, NASA SP-201 (1969) p. 12.
[17] F. EL-BAZ, Analysis of Apollo 8: Photography and Visual Observations, NASA SP-201 (1969) p. 21.
[18] N. J. TRASK and G. A. SWANN, Analysis of Apollo 8: Photography and Visual Observations, NASA SP-201 (1969) p. 29.
[19] Surveyor Scientific Evaluation Advisory Team, Science **152**, 1737 (1966).
[20] L. D. JAFFE et al., J. Geophys. Res. **73**, 3983 (1968).
[21] V. R. OBERBECK and W. L. QUAIDE, Icarus **9**, 446 (1968).
[22] E. M. SHOEMAKER et al., Surveyor: Program Results, NASA SP-184 (1969) p. 19.
[23] R. F. SCOTT and F. I. ROBERSON, J. Geophys. Res. **74**, 6175 (1969).
[24] R. A. PHINNEY et al., J. Geophys. Res. **74**, 6053 (1969).
[25] E. E. ALDRIN, Jr., N. A. ARMSTRONG, and M. COLLINS, Apollo 11 Preliminary Science Report, NASA SP-214 (1969) p. 35.
[26] Lunar Sample Preliminary Examination Team, Science **167**, 1325 (1970).
[27] Lunar Sample Preliminary Examination Team, Science **165**, 1211 (1969).
[28] T. GOLD et al., Apollo 11 Preliminary Science Report, NASA SP-214 (1969) p. 187.
[29] T. GOLD, Science **165**, 1345 (1969).
[30] Lunar Sample Preliminary Examination Team, Science **167**, 449 (1970).
[31] J. A. WOOD et al., Science **167**, 602 (1970).
[32] N. M. SHORT, Science **167**, 673 (1970).
[33] E. C. T. CHAO et al., Science **167**, 644 (1970).
[34] D. S. MCKAY, W. R. GREENWOOD, and D. A. MORRISON, Science **167**, 654 (1970).
[35] J. L. CARTER and I. D. MACGREGOR, Science **167**, 661 (1970).
[36] K. FREDRIKSSON et al., Science **167**, 664 (1970).
[37] N. L. CARTER, I. S. LEUNG, and H. G. AVE'LALLEMANT, Science **167**, 666 (1970).
[38] W. VON ENGELHARDT et al., Science **167**, 669 (1970).
[39] W. QUAIDE, T. BUNCH, and W. WRIGLEY, Science **167**, 671 (1970).
[40] S. B. SCLAR, Science **167**, 675 (1970).
[41] R. F. SIPPEL and A. B. SPENCER, Science **167**, 679 (1970).

[42] A. T. ANDERSON et al., Science **167**, 587 (1970).
[43] R. CHOATE et al., Surveyor: Program Results, NASA SP-184 (1969) p. 129.
[44] L. D. JAFFE, Science **164**, 1514 (1969).
[45] R. F. SCOTT and F. I. ROBERSON, Surveyor: Program Results, NASA SP-184 (1969) p. 171.
[46] L. D. JAFFE, Science **164**, 774 (1969).
[47] D. E. GAULT et al., J. Geophys. Res. **73**, 4115 (1968).
[48] N. C. COSTES et al., Apollo 11 Preliminary Science Report, NASA SP-214 (1969) p. 85.
[49] A. VINOGRADOV et al., Kosm. Issled. **5**, 874 (1967).
[50] A. VINOGRADOV et al., COSPAR, Moon and Planets II (1968) p. 77.
[51] A. L. TURKEVICH, E. J. FRANZGROTE, and J. H. PATTERSON, Science **165**, 277 (1969).
[52] A. L. TURKEVICH et al., Science **167**, 1722 (1970).
[53] E. J. FRANZGROTE et al., Science **167**, 376 (1970).
[54] J. H. PATTERSON et al., Science **168**, 825 (1970).
[55] Lunar Sample Preliminary Examination Team, Apollo 11 Preliminary Science Report, NASA SP-214 (1969) p. 123.
[56] W. N. HESS and A. J. CALIO, Apollo 11 Preliminary Science Report, NASA SP-214 (1969) p. 1.
[57] C. L. HERZENBERG and D. L. RILEY, Science **167**, 683 (1970).
[58] A. H. MUIR Jr., et al., Science **167**, 688 (1970).
[59] S. K. RUNCORN et al., Science **167**, 699 (1970).
[60] A. LAROCHELLE and E. J. SCHWARZ, Science **167**, 700 (1970).
[61] T. NAGATA et al., Science **167**, 703 (1970).
[62] R. A. WEEKS et al., Science **167**, 704 (1970).
[63] T. MURASE and A. R. MCBURNEY, Science **167**, 1491 (1970).
[64] M. TATSUMOTO and J. N. ROSHOLT, Science **167**, 461 (1970).
[65] A. L. ALBEE et al., Science **167**, 463 (1970).
[66] G. TURNER, Science **167**, 466 (1970).
[67] L. T. SILVER, Science **167**, 468 (1970).
[68] K. GOPALAN et al., Science **167**, 471 (1970).
[69] W. COMPSTON et al., Science **167**, 474 (1970).
[70] V. RAMA MURTHY, R. A. SCHMITT, and P. REY, Science **167**, 476 (1970).
[71] R. K. WANLESS, W. D. LOVERIDGE, and R. D. STEVENS, Science **167**, 479 (1970).
[72] T. P. KOHMAN et al., Science **167**, 481 (1970).
[73] D. HEYMANN et al., Science **167**, 555 (1970).
[74] J. G. FUNKHOUSER et al., Science **167**, 561 (1970).
[75] T. KIRSTEN, F. STEINBRUNN, and J. ZÄHRINGER, Science **167**, 571 (1970).
[76] P. W. GAST and N. J. HUBBARD, Science **176**, 485 (1970).
[77] J. H. REYNOLDS et al., Science **167**, 545 (1970).
[78] R. W. STOENNER, W. J. LYMAN, and R. DAVIS, Jr., Science **167**, 553 (1970).
[79] P. EBERHARDT et al., Science **167**, 558 (1970).
[80] G. CROZAZ et al., Science **167**, 563 (1970).
[81] E. L. FIREMAN, J. D'AMICO, and J. C. DEFELICE, Science **167**, 566 (1970).
[82] R. L. FLEISCHER et al., Science **167**, 568 (1970).
[83] YU. I. EFREMOV, First Panoramas of the Lunar Surface (1966) Science Press, Moscow.
[84] E. M. SHOEMAKER et al., Science **158**, 642 (1967).
[85] J. J. RENNILSON, H. E. HOLT, and E. C. MORRIS, J. Opt. Soc. Am. **58**, 747 (1968).
[86] E. M. SHOEMAKER et al., J. Geophys. Res. **73**, 3989 (1968).
[87] R. F. SCOTT and F. I. ROBERSON, J. Geophys. Res. **73**, 4045 (1968).
[88] E. M. CHRISTENSEN et al., J. Geophys. Res. **73**, 4081 (1968).
[89] E. M. SHOEMAKER et al., Apollo 11 Preliminary Science Report, NASA SP-214 (1969) p. 41.
[90] B. W. HAPKE et al., Science **167**, 545 (1970).
[91] D. B. NASH, J. E. CONEL, and R. T. GREER, Science **167**, 721 (1970).
[92] T. GOLD, M. J. CAMPBELL, and B. T. O'LEARY, Science **167**, 707 (1970).

[93] J. E. GEAKE et al., Science **167, 717** (1970).
[94] J. B. ADAMS and R. L. JONES, Science **167**, 737 (1970).
[95] B. O'LEARY and F. BRIGGS, Trans. Am. Geophys. Un. **51**, 338 (1970).
[96] G. B. DALRYMPLE and R. R. DOELL, Science **167**, 713 (1970).
[97] J. A. EDGINGTON and I. M. BLAIR, Science **167**, 715 (1970).
[98] N. N. GREENMAN and H. G. GROSS, Science **167**, 720 (1970).
[99] R. C. BIRKEBAK, C. J. CREMERS, and J. P. DAWSON, Science **167**, 724 (1970).
[100] E. C. BERNETT et al., AIAA J. **1**, 1402 (1963).
[101] J. A. BASTIN, P. E. CLEGG, and G. FIELDER, Science **167**, 728 (1970).
[102] K.-I. HORAI et al., Science **167, 730** (1970).
[103] R. A. ROBIE, B. S. HEMINGWAY, and W. H. WILSON, Science **167, 749** (1970).
[104] N. F. NESS, Space Research X, North-Holland Publ. C., Amsterdam 1970 (p. 969).
[105] D. O. MUHLEMAN et al., Surveyor: Program Results, NASA SP-184 (1969) p. 203.
[106] O. I. YAKOLEV and A. I. YEFIMOV, Dokl. Akad. Nauk SSSR **174**, 583 (1967).
[107] G. L. TYLER et al., Science **157, 193** (1967).
[108] G. L. TYLER, Nature **219, 1243** (1968).
[109] G. L. TYLER, J. Geophys. Res. **73, 7609** (1968).
[110] SH. SH. DOLGINOV et al., Dokl. Akad. Nauk SSSR **170**, 574 (1966).
[111] L. N. ZHUZGOV, SH. SH. DOLGINOV, and E.C. EROSHENKO, Cosmic Res. **4**, 758 (1966).
[112] C. P. SONETT, D. S. COLBURN, and R. G. CURRIE, J. Geophys. Res. **72**, 5503 (197).
[113] N. F. NESS et al., J. Geophys. Res. **72**, 5769 (1967).
[114] K. W. BEHANNON, J. Geophys. Res. **73**, 7257 (1968).
[115] J. N. DE WYS, Science **158, 632** (1967).
[116] J. N. DE WYS, J. Geophys. Res. **73, 6915** (1968).
[117] J. N. DE WYS, Surveyor: Program Results, NASA SP-184 (1969) p. 223.
[118] Apollo 12 Mission Report, MSC-01855, p. 3.
[119] C. A. GURTLER and C. W. GREW, Science **161**, 462 (1968).
[120] S. J. PEALE, J. Geophys. Res. **73, 3025** (1968).
[121] J. R. BATES, S. C. FREDEN, and B. J. O'BRIEN, Apollo 11 Preliminary Science Report, NASA SP-214 (1969) p. 199.
[122] Apollo 12 Mission Report, MSC-01855, p. 3 – 7.
[123] W. F. CARROLL, quoted in: Apollo 12 Preliminary Science Report NASA SP-235 (1970) p. 217–2.
[124] B. COUR-PALAIS, Apollo 12 Preliminary Science Report, NASA SP-235 (1970). p. 217.
[125] D. E. GAULT, Radio Science **5, 273** (1970).
[126] R. R. KEAYS et al., Science **167, 490** (1970).
[127] L. D. JAFFE, Science **170, 1092** (1970).
[128] D. A. PAPANASTASSIOV and G. J. WASSERBURG, Earth Planet. Sci Lett. **8**, 269 (1970),

Space Research XI — Akademie-Verlag, Berlin 1971

THE NATURE OF THE SURFACE OF THE MOON

T. Gold

Center for Radiophysics and Space Research, Cornell University, Ithaca, New York, USA

The new observational material concerning the moon shows it to be very different from the earth in major respects. The bulk electrical conductivity appears to be very low; the seismicity is very low; the seismic transmission properties are different from any observed on earth. A "cold" moon with very little internal activity is suggested by these data. The radioactive age determinations have shown the dust component to be older than any of the rocks investigated, and this, together with the chemical information, can be understood in terms of the moon having formed from material that had suffered chemical differentiation already in earlier phases of the solar system. The younger rocks may have been manufactured on the moon in major impact events.

1. Introduction

Information concerning the moon's surface is now very extensive, chiefly as the result of the US and USSR lunar research programs. The information is very clear about some points, and poses several other novel problems. It is clear chiefly in pointing out the great differences between the moon and the earth. Every one of the significant measurements and observations gives a result very different from that which would have been obtained on the earth in similar circumstances, even after making an allowance for the obvious differences arising from the effects of the atmosphere and the biosphere of the earth.

Furthermore, the observations of the moon raise new problems with regard to the surface processes that have taken place and for which also we know of no counterpart on the earth.

2. The Material on the Lunar Surface

It is now quite certain that most of the moon is covered with a soil of finely divided rock particles. The optical appearance is dominated by grains that are smaller than 10 μm, and most of the mass is in particles smaller than 50 μm [1]. The material is thus for the most part a fine rock dust. The depth of this deposit is not known from any direct observation, although some inferences have been attempted. The particle size is small enough to allow the possibility that transportation over the surface and deposition on low ground has taken place by means of forces that are connected with electrical or plasma physical phenomena. There are clear indications that some downhill transportation has taken place and that level deposition of transported material is common.

3. Electrical Conductivity

The electrical conductivity of the major part of the moon's body is exceedingly low, much lower than that of the earth [2, 3]. The measurements of the currents induced by the solar wind in the body of the moon lead to almost no detectable result, while the value of the conductivity of the earth would have led to clearly recognizable and characteristic patterns of the induced magnetic field. Since the electrical conductivity of rocks increases steeply with temperature, a remarkably cold lunar interior seems to be implied, with most of the volume below 1000 °K.

4. Seismic Observations

The seismic information [4] obtained by the Apollo 11 and 12 missions is most remarkable. Firstly, it is clear that the seismicity of the moon is less than that of the earth by a very large factor. At a level of sensitivity at which a seismograph on the earth would have given a continuous output due to underground disturbances, the lunar result shows just a few events per day, and those may all be due to meteorite impacts.

The low level of seismicity is in accord with the knowledge that the horizontal deformations of the surface features on the moon are exceedingly small by comparison with such deformations on the earth. It appears that in geologic time intervals during which on the earth the growth and positioning of the continents has occurred, accompanied by folding and shear displacements of hundreds or thousands of kilometers, on the moon displacements of more than three kilometers are rare, and they would readily be recognized as dislocations of older crater rims.

The other important result of the seismic observations concerns the structure of the signals that are received. The class of signals thought to be of natural origin and at some distance from the instrument are all similar to the record of the impact of the lunar module. They are all totally different from seismic signals that would be observed from impact events on the earth. There is no clear indication of the different types of waves that can be propagated on a solid near the surface. The first signal arrived from the Lunar Module impact 76 kilometers away at a point-to-point speed of 3 km s^{-1}. But instead of a signal amplitude quickly reaching a maximum and then decaying, there was a period of 7 minutes during which the signal increased, followed by a very gradual decay taking another 50 minutes before the signal became unrecognizably small. No terrestrial seismic signals are known resembling this. What is clearly implied is that the seismic energy, for some reason, spreads out only very slowly from the source of the impact, in a medium which is, however, a good conductor of sonic waves. The observed signal lacks significant indications of transverse waves, which would have been seen on the earth in similar circumstances. There is no recognizable frequency dispersion, and there is no correlation between orthogonal seismometers, which is normally produced by the polarization of a transverse oscillation. It seems that almost only compressional waves were observed on the moon, while in similar circumstances on the earth transverse waves would have been prominent.

On the earth there is in almost every locality a subsurface of solid rock at comparatively shallow depth. A seismic signal transmitted over a distance of 76 km would always have a major component travel in that rock, in part as a transverse wave. The signature of the arrival of such a wave at the receiving site would be clearly identifiable and was not seen in the Apollo 12 observation. The explanation is not that it was obscured by other signals that are present there, for the major rise of the other signals occurs only at a later time. The conclusion is that on the moon, in the region of the Apollo 12 site, there is not a solid rock at a shallow depth making a seismic connection between the impact and receiving sites. (The seismic information is discussed elsewhere [16].)

5. Appearance of the Surface Features

In the general appearance of the surface features, as seen on many excellent photographs obtained by the Orbiter and Apollo missions, some structural information emerges, which also suggests a very different subsurface for the moon from that of the earth. Firstly, there is the extreme rarity of horizontal shear displacements, which, if we could only see old features on the earth, would be very prominent here. Secondly, there is the extreme scarcity of indications of stratification. The steep sides of large young craters show mostly a uniform texture with no indications of layers that would result if such layers possessed differences of chemical or physical properties leading to differences of color, albedo or terracing in the subsequent erosion. Almost anywhere on the earth a deep cut of several kilometers would show strata. This would be particularly true in places where large outpourings of lava have successively flooded the area. The interior slopes of the crater Copernicus, for example, as seen in the Orbiter picture (Fig. 1), represents a difference of level of more than 3 km. Yet unlike slopes of comparable magnitude and steepness on the earth (Fig. 2), no widespread layers can be found in the Copernicus picture. Terracing is seen, but only on a very large scale, that suggests slumping of the bulk medium under excessive stress rather than any material stratification. The same general absence of strata is seen in many young craters that exhibit the subsurface to a depth of several kilometers. The occasional ledges that are seen are of very limited extent. There are different regional properties for the depth of craters at which thrown out rocks appear, and this implies a stratification of some kind; but it may be no more than a variability of the degree of compaction of the soil.

Many fresh looking mare craters are seen in Orbiter and Apollo pictures (Fig. 3), which appear to be of remarkably regular shape. They are almost exactly circular, and the interior is a very smooth rounded bowl. By comparison the Arizona meteor crater (Fig. 4) is extremely irregular (even after making an allowance for the subsequent effects of erosion). Rock does not generally have isotropic properties for its mechanical strength, and the anisotropies show up

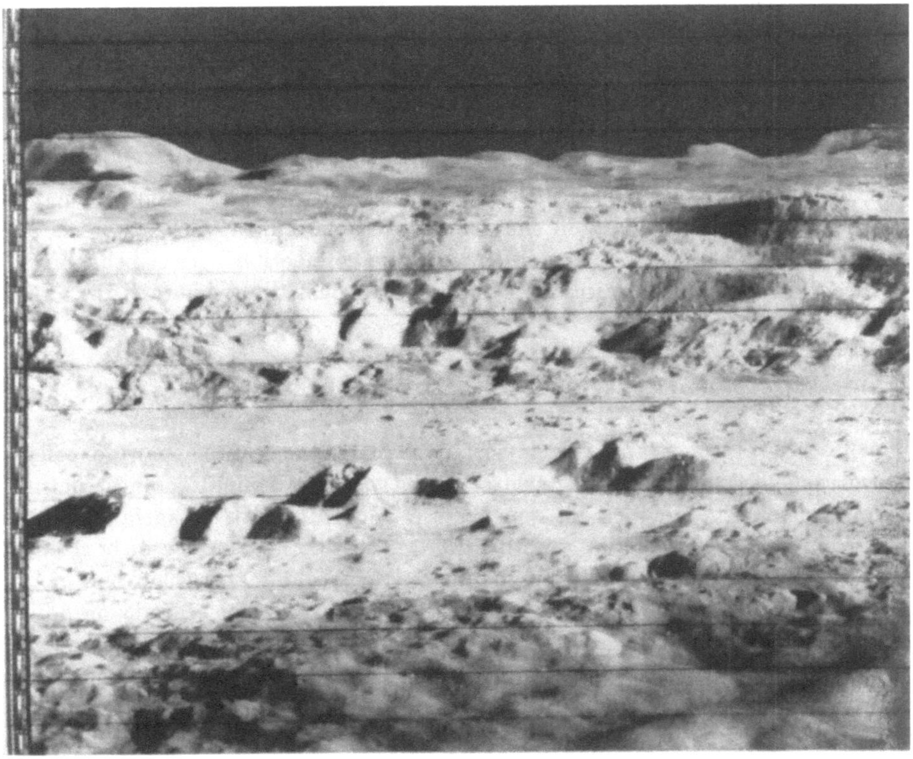

Fig. 1. The crater Copernicus, photograph taken by Lunar Orbiter 2's telephoto lens. The crater is 113 km in diameter and 3.2. km deep, mountains rising from the flat floor of the crater are about 305 m high.

in the shape of the region excavated by an explosion. The perfect shapes that are frequently seen on the moon imply that the mare material down to a depth of kilometers is often very homogeneous, isotropic in its mechanical properties in the horizontal directions, and with no abrupt change in the mechanical properties in the vertical direction. If there were in general an abrupt change from the powdered soil on the surface to layers of bedrock, this would have been clearly recognizable; it thus seems that we are dealing in most regions with the same fine material at depth, modified by compaction and perhaps quite firm in some cases, but without any sharp discontinuity.

The distribution of rock types found in both the Apollo 11 and 12 mission also suggests the absence of a continuous sheet of rock at a shallow depth. The spray patterns of rocks surrounding craters are frequently clearly seen, and it is obvious that in any locality the distribution of surface rocks is dominated by the excavations of a few nearby craters. If there were a continuous layer of rock at a shallow depth, then one would expect to see in any one locality a great dominance of that particular type. The fact that a variety of rocks with different formation histories are seen (though with striking similarities in the

Fig. 2. A part of the Dry Falls precipice, Grand Coulee, looking upstream. The horizontal stratification of the basalt can be seen on all steep slopes.

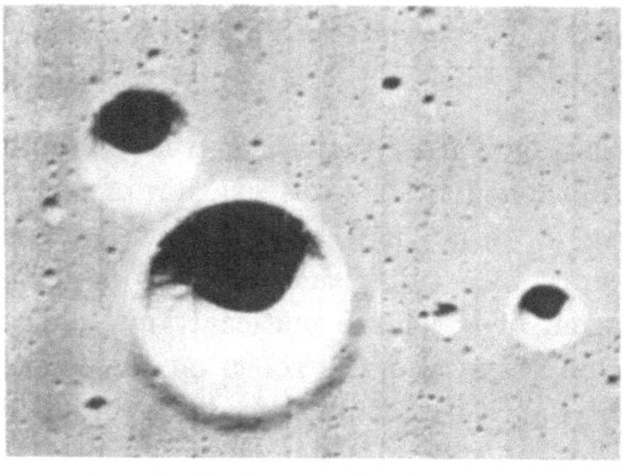

Fig. 3. Craters in the flat region of the Oceanus Procellarum, photographed by Lunar Orbiter 2. The diameter of the largest crater on the picture is about 3 km.

Fig. 4. Aerial photograph of the Arizona meteor crater. The diameter of the crater is about 1 km.

chemical composition) suggests that the thrown out material already had a variety of rocks imbedded in it. In each cratering event the fine material is thrown out and widely dispersed while the rocks fall in a recognizable pattern. In addition, the larger cratering events will be associated with pressures and temperatures at which new rocks can be formed. It is possible that the present surface represents the equilibrium between rock forming and erosion processes, and that the surface at an earlier epoch looked similar in general.

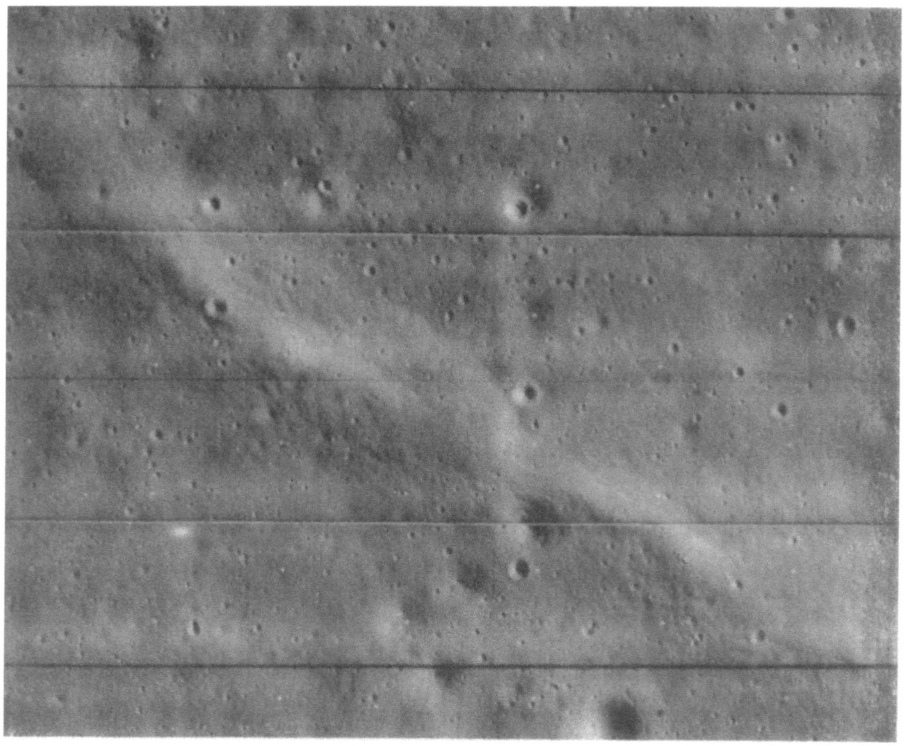

Fig. 5. Lunar Orbiter photograph of a ridge with concentrations of rocks on top. Note especially the lower right part of the ridge.

The picture of a deep deposit of fine powder with a distribution throughout its volume of some rocks is greatly strengthened by the observation of ridges with a concentration of rocks on top (Fig. 5). Since it is hard to envision any mechanism for placing rocks along the crest of a ridge, one can only suppose that the ridge was raised by a compression of the mare material and that it subsequently suffered erosion. The fine material and the rocks on the slopes were transported away, while the rocks that became uncovered on the crest remained in place and therefore became concentrated in the course of time.

The evidence does not only favor the suggestion that the maria are deep deposits of fine material [5], but the same appears to be true for the highlands. Also in the highlands one sees very little evidence of stratification or of anisotropic mechanical strength. The highland surface material, it is understood from optical [6], infrared [7] and radio [8] observations, is very similar to the lowland material and is undoubtedly also a fine powder. If, however, there were a layer of rock at a shallower depth in the highlands than in the maria, then the cratering processes would have exhibited and distributed more rocks on the surface. Thermal maps of the lowest temperatures reached in the lunar night or during an eclipse [9] are a very sensitive indication of the proportion of the surface area covered by exposed rocks, for these cool much more slowly.

Fig. 6. Rock covered area on lunar surface near the Apollo 11 landing site. Photograph taken by the Apollo 11 astronauts.

While individual young craters and various other regions show such slower rates of cooling, there is no systematic difference between highlands and lowlands, and therefore probably no widespread subsurface rock exists in either location.

On a very small scale one has clearly seen that the fine material is subject to some surface transportation process. Rocks strewn over the surface do not have in their vicinity the furrows and surface markings they must have produced in the soft soil as they fell. These markings must have been eradicated, and now most rocks are partially imbedded and have a sharply defined junction line with the ground (Fig. 6). Whatever surface transportation process brought about this change did not cause the tops of most rocks to be dust covered. Rock surfaces extending as little as 10 cm above the surrounding surface are seen to be generally completely free from any dust deposit [10]. We are seeing here the evidence of some surface transportation process which may have been effective over long periods of time and be responsible for the level surfaces which so characteristically occur in almost all low areas on the moon.

6. Mascons

The discovery that the large circular maria have mostly positive gravity anomalies associated with them can be understood if these maria are basins created by major impacts and subsequently filled in by a sediment. The impact event would apply pressures amply enough to squeeze out all porosity of the material and to dig a basin far deeper than could subsequently be maintained by the mechanical strength of the material. Consequently after the event the ground in the bottom of the basin must rise, but only to a level below that of the surrounding ground, which possesses more porosity. Porosity on the moon is not squeezed out by static pressure until one comes to depths of more than 50 km or so, and in the absence of a general percolation by water a large porosity would in general be maintained in infallen material where it was not subjected to internal melting or to the pressure and heat of a major impact.

A basin left by a major impact, after the hydrostatic adjustment of the ground, will then be a low region with denser material beneath it but with no significant gravity anomaly. An initial porosity of 20% down to a depth of 20 km would imply that an impact basin would hydrostatically adjust to a level 4 km below the surrounding ground. If any sedimentation process then fills in such low regions a positive gravity anomaly will result. The observed mascon anomalies would be matched by a filling of between 1 and 2 km (leaving the surface, in our example, still 2 to 3 km below the surrounds) [11].

The impacting body may of course in any one case have brought a denser material with it, but the effect of this would be minor compared with the extra density created as the pressure pulse removes the initial porosity.

7. Chemical Composition

The chemical composition shows the lunar material to have significant differences from most materials on the surface of the earth. The terrestrial material that is most similar in general is the widespread deep ocean basaltic layer. It is material that appears to have undergone some chemical differentiation associated with crystallization from a melt. It would, however, be quite wrong to conclude that all differentiated material that is found in the inner solar system must have been differentiated where it is now found. The chemical composition of the moon is indeed somewhat similar to that of a class of meteorites, the Eucrites. It has been argued that the Eucrites perhaps originated on the moon and were blasted off into space by impacts. Present data on isotopic abundances show this to be improbable. On the other hand, it is possible that the surface of the moon accumulated mainly from material that had also, like the Eucrites, been differentiated elsewhere. We see in the meteoritic material much evidence of differentiation having taken place, going even so far as to make an iron core, and of such material having then been scattered again. There is no evidence to suggest that the moon must have formed from undifferentiated "first generation" material rather than from any of the material made available by the earlier formations of the solar system, traces of which are still found as meteorites. One of the chief discoveries to be made in the lunar investigation is indeed the type of material from which the moon formed. We must not at this stage assume that we already know that answer.

The comparison of the chemical information from three Surveyor [12] and two Apollo [13] sites is most interesting. Firstly, there is a remarkable degree of uniformity. On the earth it would have been very hard to find five widely spaced sites that showed so little difference of chemical composition. If differentiation had taken place on the moon's surface with not much subsequent mixing, as appears to have been the case over much of the earth, then much larger regional differences would have been expected.

The small difference in some of the constituents that has been noted between the Apollo 11 and Apollo 12 sites is, however, most interesting. These differences show up with some similarity in both the rocks and the fine material. It therefore has to be argued that there are slight regional differences of composition, and that in each region either the dust derives from the rock, or the rock from the dust. What differences there are between dust and rock must have been introduced in the process that made the one from the other. The regional chemical differences on the moon may be representative of infall patterns of different materials on a variety of orbits, or they may represent differences of the depth to which cratering processes had mixed the material, together with some chemical difference of the materials contributed at different epochs. It will be important to gain some information regarding the regional scale of the patterns.

8. Ages of Rocks and Dust

Dating methods from radioactive decay processes have been used with apparently very good accuracy on both the dust and the rocks in both Apollo sites. The uranium–thorium–lead [14] and the rubidium–strontium [15] methods used by different investigators all give quite concordant results. The ages obtained date the final solidification of the material. The answers are approximately as follows; the dust appears to be the oldest component, giving an age for the most part around 4.6 billion years in both the sites. The rocks in the Apollo 11 and 12 sites are slightly but significantly younger, giving an age of around 3.8 billion (3.8×10^9) years.

Has the dust derived from the rocks or the rocks from the dust? A straightforward reading of the evidence of the ages would suggest that we have to search for a mechanism that would make a variety of rocks of similar age from the dust and that in general the dust is the primary constituent. It is not very difficult to see how this could have come about, for in each region a few major impacts would have dominated in the production of subsurface liquid rock leading to slow cooling and crystallization (including some further chemical differentiation). Subsequent smaller impacts would then regionally distribute such rock material, leaving the local chemistry of dust and rock associated but giving the rocks a younger age since solidification. It could well be that the very large impacts that made the maria are themselves responsible for most of the rocks and that the dating is therefore a means of dating essentially the formation of the mare basin. Equally, it could be that the pattern is on a smaller scale and dominated by a size range of impacts whose evidence has not been preserved to the present.

9. Conclusions

The evidence altogether seems to favor the view that the moon has been essentially cold, with very little local differentiation, and displaying mainly a surface that is the final one produced by the accumulation process. Earlier surfaces may have been quite similar in general character. The indications are for the absence of any general "bedrock" but rather for a deep deposit, gradually compacting with depth, of a powder within which are distributed rocks of different types. These appear to be regionally made, no doubt at a variety of depths, and their age may represent the major impact events that caused melting in the originally finely divided material. The origin of this finely divided material has then to be sought in the early processes that took place in the solar system, where, as we know from other evidence, all kinds of chemical sorting processes seem to have occurred.

The major questions concerning the derivation of the lunar surface have to be solved first by the lunar exploration program, before detailed interpretations of the lunar features are possible. The fact that we do not have at present a clear understanding of the surface phenomena on the moon does not argue for the absence of widespread erosion and transportation processes. This may have occurred continuously at a slight rate or linked to the rate of infall and therefore concentrated in the past; or it may be related to other events on the moon or in the solar system of which we are not aware at present.

Future Apollo missions may well be able to confirm the point of view presented here. On this basis it would be expected that the seismic behavior is similar in most flat regions, that the age dating continues to show the dust generally as the oldest component and that major chemical differences occur only on a large scale pattern or in relation to deep craters. The ages of rocks would be expected to show a regional distribution perhaps associated with major basins. One would also on this basis expect to see no evidence of widespread volcanism, although of course the escape of gases and volatiles from the interior of the moon at a comparatively low temperature is to be expected.

Acknowledgments

Work on lunar studies at Cornell University is supported under a NASA Grant Number NGL-33-010-005.

References

[1] T. Gold, M. J. Campbell and B. T. O'Leary, Science **167**, 707 (1970).

[2] N. F. Ness, K. W. Behannon, C. S. Scearce and S. C. Cantarano, J. Geophys. Res. **72**, 5769 (1967) see also N. F. Ness, in: Space Research IX, North-Holland Publ. Co., Amsterdam 1969 (p. 678).

[3] T. Gold, in: The Solar Wind, Pergamon Press, New York 1966 (p. 381).

[4] G. V. Latham et al., Science **167**, 455 (1970).

[5] T. Gold, Mon. Not. Roy. Astron. Soc. **115**, 585 (1955).

[6] B. W. Hapke, in: The Nature of the Lunar Surface, Johns Hopkins Univ. Press, Baltimore 1966 (chap. 7).

[7] J. M. Saari and R. W. Shorthill, NASA CR-855, 1967.

[8] B. Gary, J. Stacey and F. D. Drake, Astrophys. J. Supplement Series, **12**, 239 (1965).

[9] J. M. Saari and R. W. Shorthill, Nature **205**, 964 (1965).

[10] T. Gold, Icarus **12**, 360 (1970).

[11] J. E. Gouel and G. B. Holstrom, Science **162**, 1403 (1968).

[12] A. L. Turkevich et al., Surveyor Program Results, NASA SP-184, (1969) p. 271.

[13] The Lunar Sample Preliminary Examination Team, Science **165**, 1211 (1969) and **167**, 1325 (1970).

[14] L. T. Silver, Science **167**, 468 (1970).

[15] A. L. Albee et al., Science **167**, 463 (1970).

[16] T. Gold and S. Soter Science **169**, 1071 (1970).

Space Research XI — Akademie-Verlag, Berlin 1971

LUNAR SURFACE LAYER DENSITY FROM SPACECRAFT RADAR MEASUREMENTS

N. N. Kroupenio

Institute for Space Research, Academy of Science USSR, Moscow, USSR

The characteristics of radio wave scattering by lunar regions at centimetric, decimetric and metric ranges were obtained using radar experiments on automatic lunar stations. The data processing of some of these experiments allow a determination of the effective dielectric permeability as well as of the surface layer density. Such calculations were performed for the regions of Luna and Surveyor landings and for some lunar regions using the data of experiments on Explorer 35, Luna 11, 12 and 14.

The values of density and porosity on the lunar upper surface layers for a depth of not more than 30–50 cm for the regions of the Luna and Surveyor landings varies from 0.8 to 1.8 g cm^{-3} and from 45 to 75% respectively.

These results are compared with the data of the radiative and direct mechanical measurements performed by automatic stations. The difference from radar measurements is a factor of 1.5 to 2, with a maximum error in the radar determination of density of ± 60 to 70%.

Radar data from lunar artificial satellites indicate that the density of the lunar upper layers of thickness up to 5—10 metres can differ 1.5 to 2 times from its average value in some lunar regions.

Radar experiments carried out on most automatic stations (ASC) landed on the lunar surface and on a series of lunar artificial satellites (LAS) made it possible to determine the characteristics of radio wave scattering by the lunar surface in the centimetre and metre wavelength ranges. These characteristics were measured by the biostatic radar method from LAS at 13.5 cm, 1.7 and 2.2 m wavelengths using non-directional on-board antennae. Because of this the dimensions of the reradiating regions of the lunar surface were determined by the satellite position relative to the surface, by wavelength and by radiated surface roughnesses and ranged from 10 to 50 km. On the automatic stations Luna and Surveyor radar experiments have been carried out by the autonomous radar method at 2 and 3 cm wavelengths using narrow-beam on-board antennae with coverage determined by the dimensions of the resolved regions of the surface; these ranged from some tens of metres to some tens of kilometres.

Radar experiments on lunar satellites were carried out using several somewhat different techniques. Thus, in the Luna 11 and Luna 12 experiments only the intensity of the signal scattered by the moon was recorded by the ground station equipment [1, 2]; in the Luna 14, Lunar Orbiter 1 and Explorer 35 experiments the spectrum of the scattered signal was recorded by the ground station equipment. From the spectral measurements of the signal scattered by the moon it was possible to determine the position and dimensions of reradiating regions with the aid of the LAS external trajectory measurement system.

In the Luna 11 and Luna 12 experiments the dependence of the ratio of the power scattered by the lunar surface to the radiated power η on the angle of incidence θ was obtained, and this made it possible to determine the effective dielectric permeability of the surface layer ε and the root-mean-square angle of inclination σ_α of the surface by correlation of these experimental curves with the theoretical ones. Similar data were obtained in the Luna 14 and Explorer 35 experiments, but in the latter the variation of the effective dielectric permeability in different regions of the moon was estimated by spectral analysis and more delicate study of the experimental dependence $\eta(\theta)$.

In the Lunar Orbiter 1 experiment, at $\lambda = 13.5$ cm, a radar map of the lunar region near the crater Kestner was obtained. It was noted that the intensity of the scattered signal for different radio bright spots varied within the range of 2 dB [3], which, assuming the usual law of radio wave scattering, may correspond to a 1.5–2 fold variation in the density of the surface layer of thickness up to 1.5 m.

Assuming low conductivity of the lunar surface layer, a density ϱ_n can be estimated from the dielectric permeability measurements using Krotikov's equation [4]

$$\varrho_n = \frac{\varepsilon^{1/2} - 1}{a} . \quad (1)$$

In calculation of the density from formula (1) the frequency dependence parameter a should be taken into account. For lunar-type rocks parameter a changes from 0.5 to 0.6 with a wavelength increase from 1 cm to 20 m [5]. The results of the determination of the parameter ε from LAS experimental data and the corresponding calculated density values are given in Table 1. The

Table 1

Artificial satellite	Wave-length (cm)	Dielectric permea-bility (ε)	Density (g cm^{-3}) ϱ_n	$\bar{\varrho}_n$	$\varrho_n / \bar{\varrho}_n$	Region of measurements
Luna 11, Luna 12	170	2.7–3.0	1.14–1.28	1.15	0.98–1.11	Maria and mainlands
Luna 14	170	2.8–3.0	1.18–1.28	1.15	1.02–1.11	Maria and mainlands
		2.0–6.0	0.72–2.55	1.15	0.62–2.20	Max. variations in individual regions
Explorer 35	220	2.8	1.18	1.16	0.975	Mainlands
		3.2	1.39	1.16	1.180	
		5.5–6.1	2.38–2.58	1.16	2.02–2.18	Crater, Flamsted's region

ϱ_n, from LAS measurements; $\bar{\varrho}_n$, from earth-based radar measurements

values of ε determined in the same wavelength range from earth-based lunar radar data are also given [6]. These data indicate that the density of the lunar upper layer of thickness up to 5–10 m can differ by a factor of 1.5 to 2 from its average value in some lunar regions.

Radar experiments which have been carried out on "Luna" and "Surveyor" automatic stations [7–9] made it possible to determine a specific normal cross section (for normal incidence) σ_{0N} at the landing sites of these stations. Because of considerable oscillation around the centre of gravity during descent the angular law of back-scattering $B_0(\theta)$ was measured at the landing sites of Surveyor 3, 5, 6 and 7 for angles of incidence up to 60°; $B_0(\theta)$ can be expressed approximately by Beckmann's law [10] as:

$$B_0(\theta) = (\cos^4 \theta + C \sin^2 \theta)^{-3/2} . \tag{2}$$

The coefficients C are found to be: $C = 10$ for Surveyor 3, 5 and 6 landing sites and $C = 5$ for Surveyor 7 landing site. According to Beckmann and Klemperer's [10] calculations the lunar experimental angular law of back-scattering obtained from the earth-based radar data can be approximated by the expression (2) with $C = 10$ for $\lambda = 2.25$ cm and $C = 5$ for $\lambda = 1.3$ cm. Thus the law of scattering in the Surveyor landing sites located in the lunar mare regions coincides with the law of radio wave scattering at the same wavelength in the case of earth-based radar observations of the moon.

From σ_{0N} and $B_0(\theta)$ data in the local lunar regions the coefficient of reflection of the upper layer at normal incidence ϱ_{3N} can be determined [11]:

$$\varrho_{3N} = \frac{\sigma_{0N}}{2} \int_0^{\Pi/2} B_{30}(\theta) \sin \theta \cos \theta \, d\theta , \tag{3}$$

where $B_{30}(\theta)$ is a quasi-specular component of the back-scattering diagram which can be approximated by the law:

$$B_{30}(\theta) = (\cos^4 \theta + C_{30} \sin^2 \theta)^{-3/2} . \tag{4}$$

Calculations carried out assuming that the diffusion component of a scattered signal satisfies Lambert's law gave the values $C_{30} = 2.3$ for the Surveyor 3, 5 and 6 landing sites and $C_{30} = 1.5$ for the Surveyor 7 landing site.

Table 2 gives measurements of σ_{0N} and calculated values of ϱ_{3N} for the Luna and Surveyor landing sites. It was assumed in the calculations that the angular law of back-scattering for the local mare regions near the Luna 9, Luna 13 and Surveyor 1 landing sites coincides with the angular law of back-scattering for the moon determined from earth-based radar data at corresponding wavelengths for Surveyor 3, 5 and 6 landing sites. Table 2 also gives the values of $\bar{\sigma}_{0N}$ determined by processing the earth-based radar data [11].

The average values of dielectric permeability of the lunar upper layer at the Luna and Surveyor landing sites are calculated from the average values of

$$\varepsilon = \left(\frac{1 + \varrho_{3N}^{1/2}}{1 - \varrho_{3N}^{1/2}}\right)^2 . \tag{5}$$

The calculated values of ε and the corresponding values of ϱ_n (Eq. (1)) are given in Table 2.

Assuming that the chemical composition of the lunar upper layer does not vary with the depth, i.e. that the density changes only with the variation of porosity, one can estimate the porosity of lunar soil Π. The density of a non-processed lunar layer is considered to be equal to $\varrho_{nM} = 3.3$ g cm^{-3}, which

Table 2

Automatic station	σ_{ON}(dB) average	σ_{ON}(dB) exact	$\left(\frac{\sigma_{ON}}{\bar{\sigma}_{ON}}\right)_{av}$	C_{30}	ϱ_{3N}	ε	ε_{av} average	ϱ_n (g cm^{-3})	$(\varrho_n)_{av}$ (g cm^{-3})	Π_{av}(%)
Luna 9	0.5	±3	0.72	3.5	0.023 – 0.094	1.85 – 3.6	2.4	0.75 – 1.8	1.1	66.6
Luna 13	1.0	±3	1.43	3.5	0.048 – 0.19	2.4 – 6.5	3.6	1.1 – 3.1	1.8	45.3
Surveyor 1	0.25	±3	0.63	2.3	0.02 – 0.08	1.8 – 3.25	2.25	0.7 – 1.6	1.0	67
Surveyor 3	0.2	±3	0.5	2.3	0.015 – 0.062	1.7 – 2.75	2.05	0.64 – 1.34	0.88	73.3
Surveyor 5	0.16	±3	0.4	2.3	0.012 – 0.05	1.85 – 2.5	1.9	0.6 – 1.2	0.8	75.7
Surveyor 6	0.2	±3	0.5	2.3	0.015 – 0.062	1.7 – 2.75	2.05	0.64 – 1.34	0.88	73.3
Surveyor 7	0.4	±3	1	1.5	0.04 – 0.16	2.25 – 5.3	3.25	1.0 – 2.62	1.6	51.5

corresponds to the maximum density of lunar stones determined from Surveyor 7 and Apollo 11 data [9, 12, 13]. Then:

$$\Pi = 1 - \frac{\varrho_\Pi}{\varrho_{\Pi M}}. \tag{6}$$

The values of density and porosity of the lunar upper layer given in Table 2 pertain to the upper layer of thickness not more than 30–50 cm.

The density of the lunar upper layer was determined at the Luna 13 and Surveyor landing sites by radiative analysis and by measurement of soil bearing capability [9, 13, 14]. The average density in a layer of thickness 5–10 cm was found to be 0.8 g cm^{-3} [14] at the Luna 13 landing site; at the Surveyor landing sites the soil density near the surface was 1–1.2 g cm^{-3} and at depths of 2.5 cm and 5 cm 1.3–1.5 and 1.5–1.7 g cm^{-3} respectively [13]. Examination of lunar soil samples gathered in the Apollo 11 landing region carried out under earth laboratory conditions gave a value for density of 1.2 g cm^{-3} [12].

Comparing these data with those of radar measurements we can see that the latter differ from the radiative and mechanical data by a factor of up to 1.5–2, and that at the Luna 13 landing site the radar measurements give a higher density value and at the Surveyor landing sites a lower density value than the other methods of measurement. Taking account of errors in reflection coefficient determination, the error in the determination of density can reach $\pm$60–70%.

References

[1] O. I. Yakovlev and A. I. Yefimov, Dokl. Akad. Nauk USSR **174**, 583 (1967).
[2] O. I. Yakovlev, A. I. Yefimov and S. S. Matyugov, Kosm. Issled. **6**, 432 (1968).
[3] G. L. Tyler, V. R. Eshleman, G. Fjeldbo and H. T. Howard, Science **157**, 193 (1967).
[4] W. D. Krotikov, Radiofizica **5**, 1057 (1962).
[5] R. Belov et al., report at Symposium on Lunar and Planetary Physics, Kiev USSR (1968).
[6] N. N. Kroupenio, Inst. Space Res., Acad. Science USSR, Preprint No. 4 (1969).
[7] N. N. Kroupenio, report on XII COSPAR Meeting, Prague (1969).
[8] G. A. Zonnenshtrall and Yu. I. Feldman, Kosm. Issled. **8**, 588 (1970).
[9] JPL Technical Report, 32-1265 (1968).
[10] P. Beckmann and W. K. Klemperer, J. Res. Natl. Bur. Std. **69D**, 1669 (1965).
[11] N. N. Kroupenio, Radar Investigation of the Moon, Nauka, Moscow 1971.
[12] T. Gold, report on Meeting in Goddard Space Research Institute New York (1969).
[13] L. D. Jaffe, Science **164**, 1514 (1969).
[14] I. I. Cherkasov et al., Kosm. Issled. **5**, 746 (1967).
[15] O. I. Yacovlev, S. S. Matjugov, N. N. Kroupenic and K. M. Shvathkin, Report Symposium on Radiowaves propagation, Kharkow, USSR (1969).

Space Research XI — Akademie-Verlag, Berlin 1971

THE BIMODALITY OF THE MICROSTRUCTURE OF LUNAR GROUND

T. E. SHVIDKOVSKAYA, G. A. LEIKIN and V. A. KRASNOPOLSKY

Academy of Sciences, Moscow, USSR

The analyses of infrared and ultraviolet spectra of the continental part of the surface of the moon received from Zond-3 spacecraft shows that the surface microstructure is characterized by bimodal distribution of inhomogeneities much more and much less than several micrometres in size. The comparison of ultraviolet spectra with those calculated according to the Mie theory opens up a possibility of evaluating the dimensions of small inhomogeneities and of the portion of the surface covered by them.

It has been shown [1] that the angular function of the continental part of the lunar surface may be sufficiently accurately described by scattering on inhomogeneities large compared with the wavelength in the infrared spectral range. However, such representation appears to be impossible in the ultraviolet range. It is natural to suppose that in the ultraviolet range the angular function can be explained by aggregate scattering on inhomogeneities large compared with the wavelength and by diffraction processes on small "particles" which are not noticeable in the infrared range.

Two approaches to the mathematical study of the diffraction mechanism are possible. One of them is based on works by Tversky [2] who showed that the scattering from a surface covered with small spherical or cylindrical inhomogeneities which are far from each other could be described by the method known as the image method according to Rayleigh. In this case the vector of the electric field strength is the sum of the vectors of electric field strengths scattered on spherical or cylindrical inhomogeneities from the real light source and from the source reflected in the plane under consideration. This theory has been developed for the ideal conductor. The second approach is reduced to the consideration of the inhomogeneity "heap". In this case one should use the theory of radiative transfer including scattering of higher orders. In the case studied by us the reflection coefficient is very small (not more than 2%). Thus, within an accuracy corresponding to the accuracy of measurement it is possible, in terms of Tversky's theory, to neglect the reflected component in considering the surface as a ideal conductor. On the other hand, the smallness of the reflection coefficient makes the inclusion of higher order reflection practically senseless. Hence, in both cases the techniques of comparison of theory with experiment appear to be the same.

Analysis of the observational material is reduced to the determination of the refraction coefficient and of the dimensions of micro-inhomogeneities together with the density of their distribution. It was assumed that the inhomogeneities were spheres. The angular function of such spheres was calculated on the basis of the Mie theory for various values of the refraction coefficient and of the dimensions of micro-inhomogeneities. Then the calculation results were compared with observation.

It was assumed in this case that the measured radiation flux, as related to the solar flux in a given wavelength range, was the sum of two components determined by scattering on large micro-inhomogeneities with characteristic scale of about 40 μm and by diffraction on small inhomogeneities with a scale comparable with the wavelength:

$$I_0 = (1 - \sigma)\, p_\lambda\, y_{\rm H} + \sigma D\,,$$

where σ is the portion of the surface covered with micro-inhomogeneities, p_λ is the effective normal albedo in a given wavelength range, $y_{\rm H}$ is the angular function of scattering on large inhomogeneities according to Hapke [3], and D is the angular function of radiation scattering on spherical micro-inhomogeneities calculated according to Mie and related to a unit of area; D is a function of the refraction coefficient, particle radius and wavelength.

It was assumed that n was real and the particle radius did not exceed 1 μm since otherwise the angular function of infrared radiation could not be described on the basis of pure geometrical consideration [1].

For each wavelength it was possible to construct a system of equations for the real values of the angles observed. By solving this system of equations one could estimate the range of the values of n and particle radius a satisfying the observations for the given value of λ. The same procedure was repeated for other wavelengths. It was assumed that n was independent of λ in the range under consideration, which is admissible because of the approximate character of the estimates. It must be mentioned that for wavelengths 2200–2400 Å the assumption of the constancy of the refractive index is not true: the refractive index for wavelengths 2400–2700 Å is 1.45, but for 2200–2400 Å it decreases apparently to 1.4–1.3. It appeared that the values $n = 1.45$ and $a = 0.6$ μm were most probable. The portion of the area covered with such inhomogeneities is about 5%. On the basis of these equations it was also possible to estimate the effective normal albedo p_λ. If one assumes that reflection coefficient is connected with the refraction coefficient by the Fresnel formulae it is possible to estimate the mean inclination of the surface to the scale of "large" inhomogeneities. It appeared to be close to 40°.

Thus, the angular function of the continental part of the lunar surface may be explained as a result of the sum of the effects of scattering on comparatively large inhomogeneities with characteristic dimensions $a \approx 40$ μm and mean density $\sim 10^5$ cm^{-2} and diffraction on small inhomogeneities with characteristic dimensions 0.6 μm and density $(1–2) \times 10^7$ cm^{-2}. It is not impossible that the large micro-inhomogeneities are microcraters and the small ones fragments thrown away when these craters were formed. It should be noted, of course, that this representation is very generalized.

References

[1] G. A. Leikin, T. E. Shvidkovskaya and V. A. Krasnopolsky, in: Space Research X, North-Holland Publ. Co., Amsterdam 1970 (p. 1011).
[2] P. Beckmann and A. Spizzichino, The Scattering of Electromagnetic Waves from Rough Surfaces, Pergamon Press, Oxford 1963 (p. 99).
[3] B. Hapke, J. Geophys. Res. 68, 4571 (1963).

Space Research XI — Akademie-Verlag, Berlin 1971

STRUCTURAL–MECHANICAL PROPERTIES OF LUNAR SOILS AND THEIR TERRESTRIAL ANALOGUES

I. I. CHERKASOV, V. V. SHVAREV and G. S. STEINBERG

Institute of Space Research, Moscow, USSR

Luna 10, 12, Surveyor 5, 6, 7, and Apollo 11 experiments show that lunar surface soils have a volcanic origin and in composition are close to the earth's basalts. The lunar regolith consists of fine and silty sands with some admixture of coarse grains.

Some of the contemporary friable volcanic deposits on the earth have analogous composition. Lunar panoramas obtained with the aid of Luna 9, 13, Surveyor 1–7, and Apollo 11 resemble the volcanic landscapes of Kamchatka in their relief, microrelief and the exterior of big rock fragments. The astronaut's footprints on the moon's surface are difficult to distinguish from men's footprints on the earth's volcanic sand.

Despite differences in chemical composition of lunar and earth's soils the friable volcanic deposits can be regarded as close analogues of lunar regolith in their granulometry, density, shear and penetration resistance. The closest resemblances are fine and silty basaltic volcanic sands.

The characteristic features of the latter are the friable structure, high compressibility, high internal friction, local deformations under an exterior load, weak cohesion in capillary-wet condition.

The results of the analysis and tests are presented in the paper.

Radiation-chemical determinations of the composition of lunar rocks by Luna 10 and 12 and Surveyor 5, 6 and 7 unmanned spacecraft have shown that the presence of basalt rocks is the most probable on the territory of lunar maria. The analysis of lunar samples obtained by Apollo 11 has shown that by their mineralogical and chemical composition lunar soils are in all probability of volcanic origin.

Hence volcanogenic rocks may be the closest analogues of lunar rocks. Of special interest are modern depositions originating during recent eruptions and not subject to considerable changes under the effect of weathering, erosion and soil formation. Among these are depositions of directed explosions, pyroclastic flows, eruptive clouds, pumice and slag ejections, as well as washed-off and redeposited volcanic sands which are characterized by finer granulometric composition and homogeneity.

The present paper describes studies of Kamchatka volcanoes during 1967 to 1969. The base for the expedition was the Volcanologic Station of the Volcanology Institute of the USSR Academy of Sciences in the settlement Klyuchi in the central part of Kamchatka, and investigations were also made with USSR civil aviation helicopters.

In choosing the sections for study the age of deposits was first taken into account. This age varied from a few years to a few centuries. Also, sections were chosen so their landscapes resembled moonscapes most closely. It has been found that terrestrial pyroclastic flows, slag fields and eruptive cloud depositions correspond more closely to lunar maria while guided explosions correspond to lunar highlands.

Volcanic explosions take place due to the rise in the pressure of gases in the volcanic pipe, leading to the destruction and ejection of the lava plugs and part of the old structure of the volcano itself. During the explosion of the Bezymyanny Volcano iu 1956 about 1.2 cubic kilometres of rocks forming the top of the volcanic structure were transported over a distance of about 12 km. The thickness of the depositions of the guided explosion of the Shiveluch Volcano in 1964 was not less than 20 metres, the lava debris and the largest heaps reaching 8 to 15 metres in height. Similar mass ejections of crushed rocks might occur during the infall of large meteorites on the lunar surface. During a volcanic explosion and impacts of debris incident upon the earth many rocks and stones are formed outwardly resembling similar objects detected on the lunar surface [1–3]. Large fragments lie on the loose volcanic sand layer. The deposits studied are grey with pink and greenish tints. Large fragments are black, grey, reddish and brown (Figs. 1 and 2).

Deposits in a group of craters of "Yubileiny Proryv" of the Klyuchevskoy Volcano are more homogeneous. They were formed during the eruption of 1945 and are represented by volcanic slags and sand of basaltic composition 10 m thick. On the surface there are also ashes of the nearby Bezymyanny Volcano ejected during repeated eruptions over the period 1956–1966.

Individual lava fragments 30–50 cm and more in size lie above, as well as slag and lapilli. Volcanic sand containing lava and slag fragments lies beneath them. On the surface the deposits are grey.

Materials forming the pyroclastic flow of the Shiveluch Volcano of 1964 are characterized by relatively acid composition and a high content of volatile components. The flow appeared when a mass of sand, slag and lava fragments saturated by volcanic gases was ejected to the surface with a definite slope. The mass was very movable and moved downhill at a high speed. With the loss of gases the movement slowed down and at last the flow stopped, forming thick sand and gravel and also stony deposits of a wide granulometric spectrum. One may trace longitudinal walls and rilles on the surface.

The deposits of the pyroclastic flow of the Shiveluch Volcano represent a mixture of various fractions of pumiceous material of andesite composition beginning from rounded boulders and ending with dust. The thickness of the flow is 10–15 m. A thick layer of other volcanogenic deposits lies beneath. The surface is bright grey.

The eruptive cloud of the Shiveluch Volcano of 1964 was formed out of crushed volcanic material: sand, lapilli and pumice ejected upward from the volcano. The cloud was carried away by wind and on its way solid particles precipitated to the earth. In the vicinity of the volcano the eruptive cloud depositions have the greatest thickness and the coarsest granulometric composition. With distance their thickness decreases and the composition becomes finer and finer. Volcanic bombs of different size lie at individual points.

The topography of the deposits where it is unaffected by erosion is ideally smooth. The thickness of the layer is from 0.2 to 2.0 m. With increase of distance from the crater the size of grains and the thickness of the layer decrease.

Fig. 1. Left, deposits of the guided explosion of the Bezymyanny Volcano (photo by I. I. Cherkasov) 1956. Right, lunar surface in the vicinity of the crater Tycho (Surveyor 7.)

Fig. 2. Left, large fragments on the surface of the deposits of the guided explosion of the Bezymyanny Volcano of 1956 (photo by I. I. Cherkasov). Right, stones on the lunar surface in the vicinity of the crater Tycho (Surveyor 7).

The deposits are represented by pumices and sand of andesite composition. The granulometric composition is more uniform than that of the pyroclastic flow of the same volcano. The particles have a shape with sharper facets and very loose packing. In some places moss and shrubs are preserved in the mass of the depositions. The branches of the shrubs support a thin pumice layer. The deposits are bright grey. There are scarcely any large blocks on the surface (Fig. 3).

A slag field formed by lapilli was investigated at the Plosky Tolbachik Volcano. The section is situated on a smooth hillside with a slope of 0.01 to 0.12, in some places nearly horizontal. The microrelief is very smooth; there are no plateaux, depressions, trails of washout and large stones. The deposits were formed during past centuries and were not subjected to noticeable erosion. The average thickness is about 4 to 5 metres and in some places may be greater. The deposits are basaltic in composition and are dark brown.

In the caldera of the Krasheninnikov Volcano volcanic sands were studied. These sands contain slag lapilli and larger fragments of slag and basaltic lavas. The surface of the deposits is horizontal and smooth and only individual fragments of lavas and slags are encountered on it. The deposits are reddish and dark brown, their composition being andesitic–basaltic.

At the Helicopter Platform of the Volcanologic Station of the Volcanology Institute of the USSR Academy of Sciences at Klyuchi studies were made of volcanic sand 4.5 metres thick formed during numerous eruptions of the Klyuchevskoy Volcano. This thick sand is of stratified structure in which one can trace the succession of individual eruptions. Because of the fairly large distance from the crater (35 km) there are no large fragments here, the sand is of more homogeneous composition and slaggy chips occur only at a depth of about 3 metres. The sand is brown and grey, its composition is basaltic.

Similar volcanic sand lies at the flood-lands of the Kamchatka river, nearby Klyuchi. Apparently it also originates from the eruptions of the Klyuchevskoy

Table 1

Results of complex studies of Kamchatka friable volcanic depositions

1	2	3	4	5	6	7	8	9	10	11
Klyuchevskoy Volcano	Coarse friable volcanic sand	Plagioclaso-olivine slags impregnated with pyroxene, rarer amphibole	53.1	1.1	68	4.4	44	45	2.71	1.29
Shiveluch Volcano, pyroplastic flow depositions	Pumiceous friable land-waste	Pumiceous andesite impregnated with amphibole and plagioclase	53.8	0.9	68	4.4	42	44	2.75	1.07
Shiveluch Volcano, eruptive cloud depositions	Gravel volcanic sand on buried shrubs	Pumiceous andesite impregnated with amphibole and plagioclase	60.8	0.9	8	0.5	42	45	2.75	0.89
Shiveluch Volcano, eruptive cloud depositions	Pumiceous friable landwaste on buried moss	Amphibolic andesite with rare plagioclase and ore mineral crystals	62.3	—	2.7	0.2	38	54	2.58	1.00

1	2	3	4	5	6	7	8	9	10	11
Plosky Tolbachik Volcano	Slaggy friable landwaste	Plagioclase basaltic slag impregnated with olivine and pyroxene	50.3	1.0	6.8	4.4	44	45	2.78	0.82
Krasheninnikov Volcano	Volcanic medium-size friable sand	Andesitic-basaltic slag impregnated with plagioclase, olivine and pyroxene	52.3	0.9	62	4.0	38	40	2.70	1.25
Shiveluch Volcano, depositions of guided explosion	Dusty friable volcanic sand	—	—	—	2.7	0.2	34	—	2.72	0.95
Shiveluch Volcano, depositions of guided explosion	Friable pumiceous chips	—	—	—	25	1.7	33	48	2.84	0.91
Bezymyanny Volcano, depositions of guided explosion	Gravel volcanic medium-density sand with stones	Andesite-basalt impregnated with plagioclase, amphibole and ore mineral crystals	58.4	—	11.6	7.7	36	59	2.73	1.38
Klyuchi Helicopter Station	Dusty friable volcanic sand	—	—	—	51	3.3	36	—	2.54	1.03
Kamchatka river	Fine friable volcanic sand	Basalt with rhombic pyroxene and plagioclase crystals and single amphibole and magnetite crystals	53.7	— —	51 26 18	3.3 1.7 1.2	28	36	2.78	1.01

The bulk of the soils investigated consists of volcanic glass. Chemical composition is given according to the data of the Volcanology Institute of the USSR Academy of Sciences and Fundamentprojekt. Mineralogical composition and all other data are given by the results of determinations of the Department of the Mechanics of Soils of the Moscow Kuibyshev Civil Engineering Institute.

1, test site; 2, construction classification of soils of the USSR; 3, mineralogical composition; 4, SiO_2 content (%); 5, TiO_2 content (%); 6, 7 mechanical parameters of the soil at 1 kg cm^{-2} in natural bedding; 6, modulus of total deformation E_0 (kg cm^{-2} at $\mu_0 = 0.1$); 7, coefficient of bed C_z (kg cm^{-3}); 8, angle of inner friction of air–dry soil Φ (°); 9, angle of repose of air–dry soil Φ_r (°); 10, specific weight γ (g cm^{-3}); 11, bulk density of dry soil in natural bedding γ_0 (g cm^{-3}).

Volcano, but has been subjected to washout, sorting and repeated redeposition by water flows. Ground water was detected at the spot of investigations at a depth of 60 cm.

Table 1 summarizes the results of chemical and mineralogical analysis, as well as studies of ground conditions. All materials studied by us contain groups of minerals typical of volcanic rocks. Mostly they belong to middle rocks with SiO_2 content of 50.3–62.3%. Rocks of andesite–basalt composition prevail. The specific weight of mineral substance is 2.54–2.84 g cm^{-3}. Natural humidity varies from 7 to 16%; only in one case does it reach 28% by weight.

The analysis of the first samples of lunar rocks brought to earth in 1969 [2] has shown that their SiO_2 content varies from 36 to 46%, i.e. the rocks are basic or ultrabasic. The mineralogical composition of fine fractions of lunar soil is characterized by the presence of volcanic glass, plagioclase, clinopyroxene, ilmenite, olivine and, very rarely, small iron–nickel balls. Lunar samples differ from terrestrial basalts by an unusually high titanium content, a higher iron content and a small alkali metal content. They are also enriched by zirconium and chromium. The specific weight of mineral substance is 3.1 to 3.5 g cm^{-3}. Lunar soils do not contain free water, no trails of moisture have been detected in them and the presence of organic substances is measured by millionths.

It should be noted that radiation-chemical analysis of lunar rocks made earlier by Soviet and American unmanned mooncraft gave somewhat different results; in particular, SiO_2 content was determined as 50% and the composition of lunar rocks was found to be close to that of terrestrial basalts [3, 4]. Petrographic studies of the samples of lunar rocks show that in mineralogical composition and structure they are close to basalt. They differ from terrestrial basalts by the enhanced content of ore mineral represented mainly by titanium ilmenite.

According to the description given in [5], lunar soils consist of particles of three types: fine dust, grains of a beach-sand type, and porous and dense stone fragments. The same may be said of the bulk of loose terrestrial volcanic rocks studied by us.

On the landscape photos volcanic deposits often look like pebble beaches (see Fig. 3). This is only due to the fact that sandy–dusty fractions are washed out by precipitation from the surface and penetrate voids between larger debris which protrude outside. Such a process is ruled out on the moon.

Granulometric analysis has made it possible to name all soils investigated according to the construction classification adopted in the USSR (SNIP II-B.1.62) listed in Table 1 with the addition of words "volcanic", "pumiceous", etc. indicating their origin.

By their size the soils studied belong mainly to the fine gravel, "landwaste" (over 50% of grains larger than 2 mm), coarse sandy (over 50% larger than 0.5 mm) and sandy–dusty (less than 75% of grains larger than 0.1 mm). The samples of fine-grained lunar soils brought to earth by Apollo 11 belong to dusty sands. Fine and dusty–sandy volcanic deposits of the guided explosion of the Shiveluch Volcano, the Helicopter Platform of the Klyuchi station and the flood-lands of the Kamchatka river are close to lunar soils by granulometry.

Fig. 4 gives curves of granulometric composition of two such sands plotted against the background of grain composition of lunar samples. The content of clay fractions with a size less than 0.005 mm in all the soils investigated, as

Fig. 3. Depositions of the eruptive cloud of the Shiveluch Volcano of 1964. Friable volcanic sand lies under the layer of large pebbles. In the foreground the trails of the helicopter wheels are seen (photo by I. I. Cherkasov).

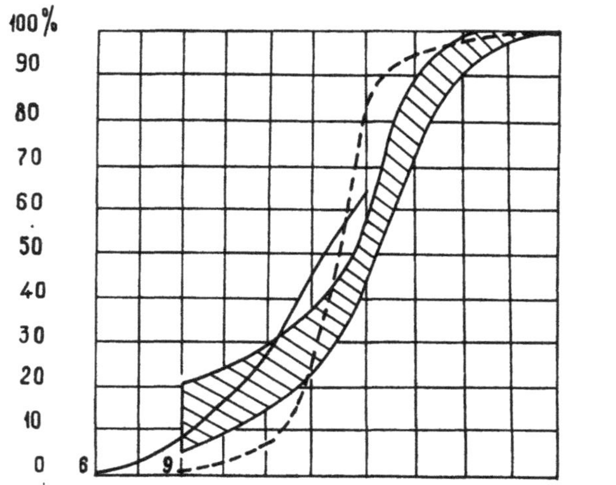

Fig. 4. Granulometric composition of friable volcanic sands and lunar soils. The size of grains in mm is given along abscissa, the content of grains larger than the prescribed value in percentage by weight is given along the ordinate. The area of granulometric composition of the first samples of lunar soils recovered to earth by Apollo 11 is shaded.
6 — depositions of the guided explosion of the Shiveluch Volcano, dusty sand.
9 — depositions in the flood-lands of the Kamchatka river, fine sand.

well as lunar soils, is very low. For dusty volcanic sand from the flood-lands of the Kamchatka river it is 8.9% from the overall weight of grains, in other soils it varies from 0 to 1.5% and in lunar samples it is lower than 1%.

Table 1 gives the bulk density of dry skeleton of soil investigated in natural bedding which varies in terrestrial volcanic deposits within the limits 0.82 to 1.38 g cm^{-3}. This is far lower than ordinary values of bulk density of the skeleton of terrestrial alluvial sands which varies within 1.5 to 1.75. The bulk density of the samples of lunar soils in a friable filled state is 1.36 g cm^{-3} and in sampling pipes 1.54–1.66 g cm^{-3}. (It should be noted that earlier Luna 13 and Surveyors obtained the values of bulk density within 0.8–1.5 g cm^{-3} [1, 6].)

The coefficient of porosity of the soils studied varies within 0.95 to 2.3, i.e. is much higher than 0.70–0.80, when, according to Soviet norms, sandy soils are already classified as friable. The coefficient of porosity of lunar soils brought to earth by Apollo 11 is 0.87–1.01 in sample collectors' pipes and 1.65 in a filled state. Hence, fine-grained lunar soils lie in a friable state.

All the volcanic soils studied contained an amount of moisture. However, the latter does not fill all pores: on the one hand it plays a role of lubrication between the grains and on the other hand it imparts low capillary adhesion to sandy soils. Weak adhesion of lunar soils is due to molecular interaction of pure surfaces of grains in high vacuum.

The mechanical properties of volcanic deposits in natural bedding were determined by the method of pressing-in circular plates 20.0 cm in diameter specially designed for operation from the helicopter MI 4. The plates were loaded by a mechanical or hydraulic jack with a spring dynamometer standing at the jacking points of the chassis of the helicopter. Fig. 5 gives the curves of vertical movement of the plate as a function of mean pressure at the foot during loading and unloading.

All volcanic deposits described above have a local character of deformation: the surface settles only beneath the footpad of the plate or under the helicopter wheels without bending of adjacent sections and without protrusion of the soil sideways from under the plate or the wheel even for a pressing-in depth of many centimetres. With deformations of this nature the ratio of the stress beneath the plate and the depth of pressing in is characterized by a coefficient of the bed (modulus of subgrade reaction) C_z. By the pressing-in of the plate the coefficient of the bed is calculated by the formula

$$C_z = p/S \quad \text{kg cm}^{-3} .$$

From the same data it is possible to calculate also a modulus of overall deformation E_0 from the formula

$$E_0 = 0.79\,(1 - \mu_0^2)\, p(D/S) \quad \text{kg cm}^{-2} ;$$

where p (= 1 kg cm^{-2}) is the mean pressure at the plate footpad in kg cm^{-2}, S is the plate settling (cm), D (= 20 cm) is the diameter of the plate (cm) and μ_0 (= 0.1) is the Poisson coefficient for friable soils.

The values obtained for C_z and E_0 are very low for all the soils, except gravel sand from the deposits of the guided explosion of the Bezymyanny Volcano.

But there were many large stones in this sand which reduced its compressibility.

Table 2 shows results of determining the compressibility of lunar soils by Luna 13 and Surveyor spacecraft [3, 6].

The deposits of the eruptive cloud of the Shiveluch Volcano are characterized by the lowest value of the coefficient of the bed. However, the influence of the buried shrubs and moss might be felt here. Friable volcanic sand in the deposits of the guided explosion of the Shiveluch Volcano has $C_z = 0.2$ kg cm^{-3} which is already comparable with the mean value of $C_z = 0.13$ kg cm^{-3} for lunar soils. If one takes into account that with lunar gravitation the depth of the indentor's

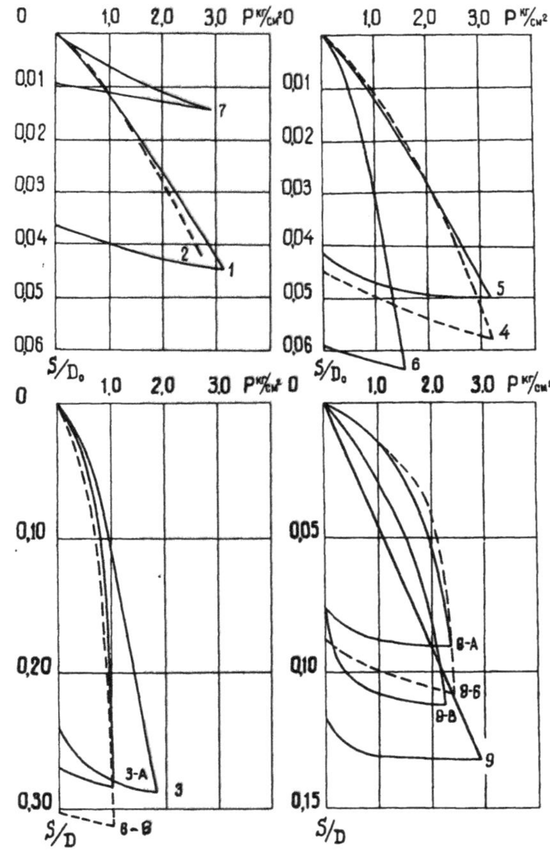

Fig. 5. Mean pressure in the plate footpad (kg cm^{-3}) as a function of relative settling of the plate in fractions of its diameter: *1*, Klyuchevskoy Volcano. Coarse-grained friable sand. *2*, Shiveluch Volcano, pyroclastic flow. Pumiceous friable landwaste (fine gravel). *3*, Shiveluch Volcano, depositions of the eruptive cloud. Gravel friable sand on buried shrubs. *3*A, Shiveluch Volcano. Pumiceous friable landwaste on buried moss. *4*, Plosky Tolbachik Volcano. Slaggy friable landwaste. *5*, Krasheninnikov Volcano. Medium-grained friable sand. *6*, Shiveluch Volcano. Pumiceous friable chips. *7*, Bezymyanny Volcano. Depositions of the guided explosion. Gravel sand with large stones. *8*, Klyuchi Helicopter Platform. Dusty friable sand. *9*, Flood-lands of the Kamchatka river. Fine friable sand.

Table 2

1	2	3	4	5
Luna 13 (indentor)	2.9	0.68	3.5	0.18
Surveyor 1 (footpads)	4.0	0.42–0.55	25	0.12
Surveyors 6, 7 (bumper)	1–2	0.18	18	0.12
Surveyor 5 (footpad)	6	0.39–0.59	25	0.08
			Average	0.13

1, source; 2 mean depth of pressing-in S (cm); 3, mean pressure p (kg cm^{-2}); 4; diameter of a footprint D (cm); 5, coefficient of bed C_z (kg cm^{-3}).

pressing in sandy-landwaste friable soils increases at the same pressure approximately by a factor of 4 [7], then fine volcanic sand from the flood-lands of the Kamchatka river with gravity reduced six times assumes the value $C_z = 0.3$ while for dusty volcanic sand from the Helicopter Platform $C_z = 0.4$–0.8 kg cm^{-3}, which approaches the figure found for lunar soils. The great similarity of man's footsteps on friable volcanic sand and on the lunar surface (Fig. 6) may serve as indirect proof of this.

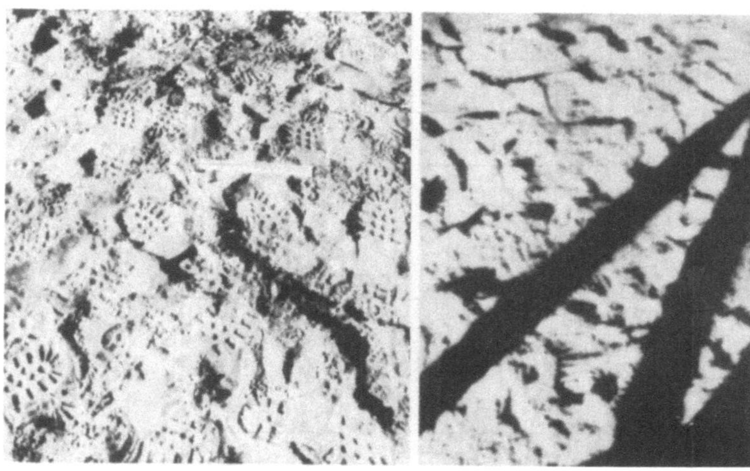

Fig. 6. Left, the footsteps of the explorer on the surface of volcanic sand in the flood-lands of the Kamchatka river (photo by I. I. Cherkasov). Right, the footsteps of the astronaut on the surface of the lunar crater near to Apollo 11 landing site.

According to the data obtained from the experiments with the aid of unmanned spacecraft sent to the moon [1, 3], the angle of internal friction of lunar soils is 32–37° and cohesion is $c = 0.0035$–0.007 kg cm^{-2}. From experiments with lunar samples [5] cohesion is 0.035–0.14 kg cm^{-2}. Volcanic soils in a dry form on the average have an angle of internal friction of 38° while true cementation is absent. With watering of up to 10% the angle of friction is reduced on the average by 8°, while capillary cohesion is of the order of 0.01–0.02 kg cm^{-2}. In their mechanical properties moistened friable volcanic sands are closer to fine-grained lunar soils, although the nature of cohesion is absolutely different in the two cases.

It is seen from the data given above that the modern friable volcanic deposits of Kamchatka, despite considerable differences of chemical compositions and conditions of origin and natural bedding, may, by their structural–mechanical properties, serve as terrestrial analogues of lunar soils. The analogy is traced in volcanic origin, landscapes and topography, as well as heterogeneity of granulometric composition, dense and porous grain content, very low clayish fractional content, friable packing and high compressibility. The weak molecular cohesion typical of lunar soils is ensured in volcanic deposits on the earth by their natural humidity. The presence of a large number of similar features gives ground to deepen and expand studies into volcanic deposits as analogues of moon conditions and to use them in various experiments associated with the estimate of probable properties of lunar soils under external mechanical influences.

References

[1] I. I. Cherkasov et al., Kosm. Issled. **5**, No. 5 (1967).
[2] Preliminary Examination of Lunar Samples from Apollo 11, Science **164** (1969).
[3] Surveyor Project Final Report, Part II. Science Result, NASA Technical Report 32-1265 (1968).
[4] Yu. A. Surkov, New Data on Lunar Matter, Zemlya i Vselennaya, No. 6 (1967).
[5] K. W. Wylie, Moonflight for Soils and Rocks, The Testing World, No. 23, winter 1969–1970.
[6] L. D. Jaffe, Science **164**, 774 (1969).
[7] I. I. Cherkasov et al., Osn. F-ty i Mekh. Gruntov, No. 1 (1970).

Space Research XI — Akademie-Verlag, Berlin 1971

ELECTRICAL CONDUCTIVITY AND THE AGE OF THE MOON

T. NAGATA, T. RIKITAKE and M. KONO

University of Tokyo, Tokyo, Japan

The electrical conductivity σ of an Apollo 11 lunar crystalline igneous rock is represented by

$$\sigma = 7.9 \times 10^{-2} \exp(-0.51 \text{ eV}/k\,T) + 3.1 \times 10^{4} \exp(-1.25 \text{ eV}/k\,T) \ (\Omega \text{ cm})^{-1} .$$

This observed value of σ is much larger than any assumed or reduced value of the lunar conductivity proposed hitherto.

1. Introduction

The distribution of electrical conductivity within the moon has been discussed by several workers [1, 2], mostly based on the observed dependence of the electrical conductivity of terrestrial olivines upon temperature and hydrostatic pressure. Since various petrological and physical properties of Apollo 11 lunar materials are significantly different from those of terrestrial olivines and chondritic meteorites [3], characteristics of the electrical conductivity of the lunar rocks also may be considered to be different from that of terrestrial rocks.

On the other hand, possible effects of the electrical conductivity of the moon on the magnetic field of the solar wind have been theoretically studied [4–9]. One of general conclusions of these theoretical studies on the interaction between the moon and the solar wind is that a strong disturbance (or a shock wave) in the solar wind must take place on the sunlit side of the moon, if the electrical conductivity of the moon is sufficiently large. However, no shock wave has been observed by a sequence of Lunar Explorer 35 experiments [10]. This observed result has implied that the bulk electrical conductivity of the moon should be less than 10^{-7} $(\Omega \text{cm})^{-1}$. Ness [11] has recently observed a beautiful example of the effect of the moon on the propagation of magnetic field reversal transient in the solar wind. Attributing this effect to electromagnetic induction effects in the lunar interior, Ness has estimated the lunar Cowling time to be about 20 s, which corresponds to a bulk lunar conductivity of about 10^{-6} $(\Omega \text{cm})^{-1}$. Since the electric conductivity of metallic silicates depends mainly on temperature, the estimated value of the bulk lunar conductivity can indicate an allowable limit of the thermal history of the moon. Thus, based on the lunar thermal models suggested by England et al. [2], Ness has tentatively concluded that the moon must be a relatively young body, its age being about 1×10^{9} years, if it possesses a homogeneous heat source composition similar to that of chondritic meteorites. However, the result of actual age determination of Apollo 11 lunar materials has given the age of the moon to be 4.66×10^9 years [12]. It seems thus that examinations of the electrical conductivity of the lunar materials themselves and of the abundance of heat source elements in them become highly necessary.

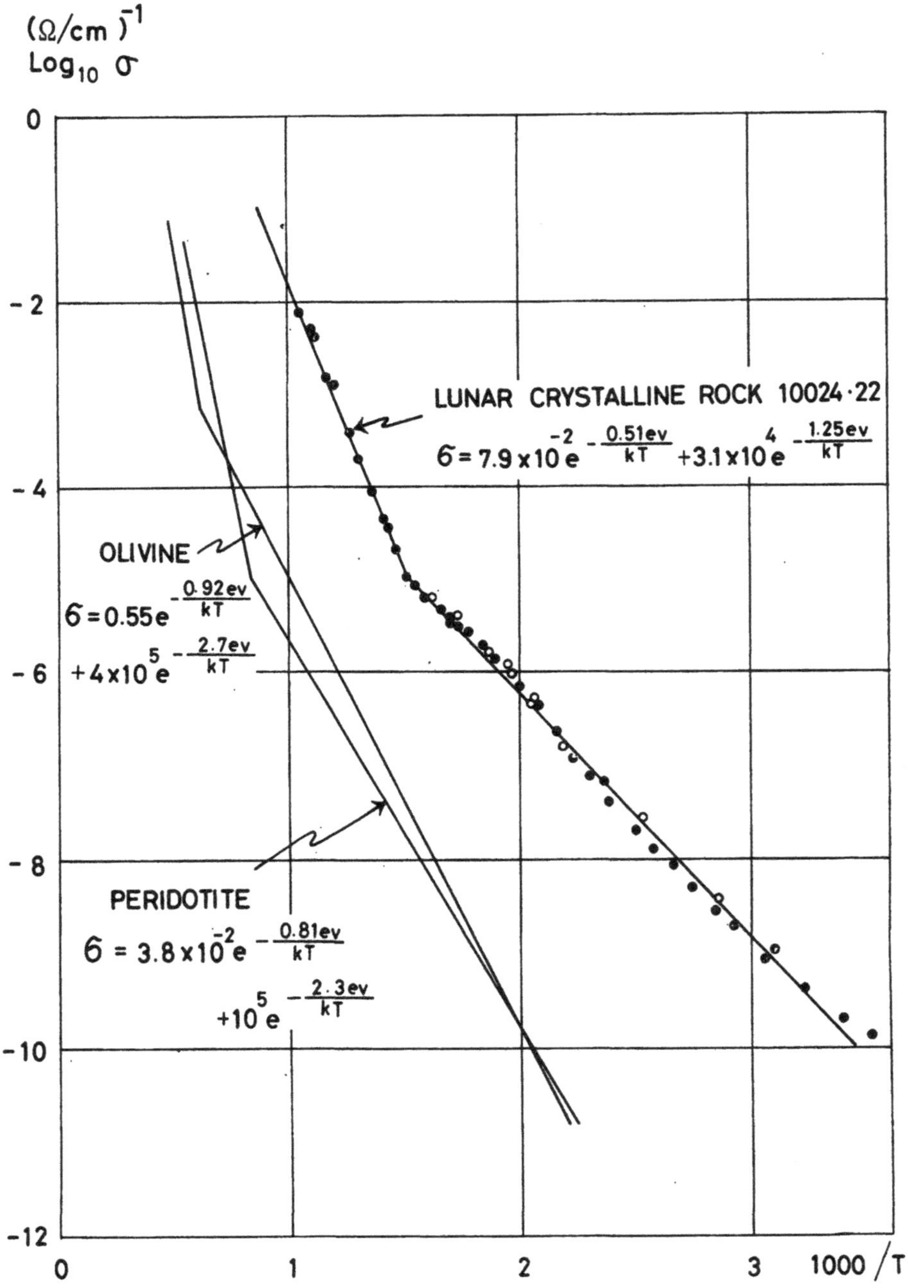

Fig. 1

The present paper describes the result of an measurement of the electrical conductivity of an Apollo 11 lunar igneous rock dependent on temperature.

2. Sample and Experimental Procedure

The sample examined is a piece of size 4.705 mm $\times$ 5.720 mm $\times$ 1.069 mm cut from an Apollo 11 lunar crystalline igneous rock, No. 10024.22 [13]. The magnetic properties of this sample have been well studied [14]; a remarkable feature of this lunar rock sample may be that it contains stoichiometric ilmenites ($FeTiO_3$) of about 20% in weight and metallic irons of about 0.1% in weight.

For measuring the electrical conductivity, electrodes of silver paste are attached on two smoothed parallel surfaces (4.705 mm $\times$ 5.720 mm). The measurements were carried out in a dry argon gas chamber using the ordinary dc method in a temperature range from 11 °C to 680 °C (284–953 °K). Unfortunately, the electrodes were broken at temperatures above 700 °C.

3. Results of Measurements

The observed values of the electrical conductivity of the lunar rock at various temperatures are plotted in Fig. 1. For a temperature range from 280 °K to 1000 °K, the electrical conductivity σ of this sample can be represented by

$$\sigma = 7.9 \times 10^{-2} \exp(-0.51\ \mathrm{eV}/k\,T) + 3.1 \times 10^{4} \exp(-1.25\ \mathrm{eV}/k\,T)\ (\Omega\ \mathrm{cm})^{-1},$$

where k and T denote respectively the Boltzmann constant and the absolute temperature.

For reference, typical examples of the dependence of electrical conductivity of olivine and peridotite of terrestrial origin upon temperature [2] are reproduced in Fig. 1. It may be clearly seen that the electrical conductivity of the lunar rock is much greater than that of olivine or peridotite for the whole range of temperature. Other experimental data on the electrical conductivity of olivines [15, 16] are reasonably close to the olivine curve and definitely smaller than the lunar rock curve shown in Fig. 1.

It may thus be tentatively concluded that the electrical conductivity of the lunar rock is extremely large (by 2 or 3 orders of magnitude) compared with that of olivine, which has been assumed to represent the lunar conductivity. The reason for this large conductivity has not yet been clarified. It seems likely, however, that anomalously large abundances of Fe (about 20% in weight) and Ti (about 8% in weight) may play the essential role.

References

[1] D. C. Tozer and J. Wilson, Proc. Roy. Soc. A **296**, 320 (1967).
[2] A. W. England, G. Simmons and D. Strangway, J. Geophys. Res. **73**, 3219 (1968).
[3] Science **167**, No. 3918 (1970).

[4] T. GOLD, in: The Solar Wind, Pergamon Press, New York 1966.
[5] C. P. SONETT and D. S. COLBURN, Nature **216**, 340 (1967).
[6] J. V. HOLLWEG, J. Geophys. Res. **73**, 7269 (1968).
[7] J. L. BLANK and W. R. SILL, J. Geophys. Res. **74**, 5175 (1969).
[8] K. SCHWARTZ and G. SCHUBERT, J. Geophys. Res. **74**, 4777 (1969).
[9] W. R. SILL and J. L. BLANK, J. Geophys. Res. **75**, 201 (1970).
[10] N. F. NESS, K. W. BEHANNON, C. S. SCEARCE and S. C. CANTARA, J. Geophys. Res. **72**, 5769 (1967).
[11] N. F. NESS, NASA-GSFC Preprint X-616-69-191 (1969).
[12] M. TATSUMOTO and J. N. ROSHOLT, Science **167**, 461 (1970).
[13] NASA, Apollo 11 Preliminary Science Report, NASA SP-214 (1969).
[14] T. NAGATA et al., Science **167**, 703 (1970).
[15] K. NORITOMI, J. Mining Col. Akita Univ., **A. 1**, 27 (1961).
[16] R. M. HAMILTON, J. Geophys. Res. **70**, 5679 (1965).

Space Research XI — Akademie-Verlag, Berlin 1971

THE DISTRIBUTION OF RADIOACTIVE ELEMENTS IN THE MOON'S INTERIOR

V. S. TROITSKY
Radiophysical Research Institute, Gorky, USSR

Radioastronomical investigations allow the determination of the total content of radioactive elements in the whole of the lunar material. The Apollo 11 mission also gave the opportunity to estimate the content of radioactive elements on the lunar surface. From the comparison of these data it is concluded that for an exponential decrease of the content with the depth, the characteristic thickness of the radioactive layer is 60–70 km. This appears to contradict the calculations of the heat history of the moon, indicating that at some given thickness, there must take place at present the process of heating of the planet. The possible causes of the apparent contradictions are discussed.

1. Introduction

A chemical analysis of the moon rocks from Apollo 11 makes it possible to determine the concentration of radioactive uranium, thorium and potassium 40 in the moon's surface layer [1]. Also, radioastronomical observations establish the magnitude of the total heat flux from the moon's interior which is believed to result in the radioactive decay of the above-mentioned elements [2, 3]. The data obtained enable one to measure the effective thickness of the radioactive layer which contains all the moon's radioactive elements responsible for the heat release. The present paper is devoted to this subject.

The determination of the effective thickness of the radioactive layer enables one to deduce the present thermal condition of the moon's interior. Indeed, at the initial stage of the moon's formation the radioactive elements are uniformly distributed. The calculations of the moon's heat history have shown [4, 5] that in the relatively short period of time of about a milliard of years, melting of the material takes place due to heat release with radioactive decay and the temperature continues to increase indefinitely. To stop the growth of the temperature the radioactive elements must be ejected to the surface. This process occurs and is closely associated with a differentiation in the interior which is accompanied by the more light-weight rocks coming to the surface. Convection of the radioactive elements to the surface is due to their chemical bonds with light silicate rocks. The convection of the radioactive elements to the surface leads further to the cessation of interior heating since all the heat release occurs now near the surface and heat disperses into space. The result of this interior differentiation is an equilibrium thermal state or slow cooling of the planet.

The thickness of the radioactive layer (equilibrium depth) at which the heating of the planet terminates depends on the concentration of the radioactive elements. Fig. 1 shows the calculated curve of the dependence of this layer thickness on the value obtained for the total heat flux from the lunar radioactive elements [5].

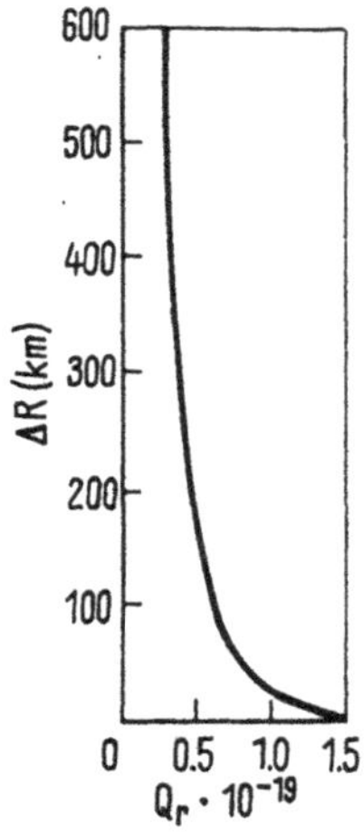

Fig. 1

2. The Estimation of the Effective Thickness of the Radioactive Layer of the Moon

According to the radioastronomical observations the mean density of the heat flux from the lunar interior over the visible hemisphere of the moon is

$$q = (0.8 \pm 0.1)\ 10^{-6}\ \text{cal cm}^{-2}\ \text{s}^{-1}\ .$$

The same flux is naturally assumed to occur on the back side of the moon. At present the total heat flux for a year is 10^{19} cal y^{-1}.

If we assume that the total heat flux is associated with the radioactive decay, it turns out that the mean concentration of the radioactive elements in the whole mass of the moon must be four times that in the earth. However, the observed heat flux is not completely associated with the radioactive decay, i.e. there may be other energy sources, for example, gravitation, recrystallization and so on. Yet, we shall not be beyond the scope of the assumption that the source of the lunar heat energy now and earlier has only been the decay of the radioactive elements. Nevertheless, this does not mean that the total heat flux from the moon's surface at the present time is equal to the flux generated by the radioactive elements of the moon, for this only occurs when a stationary thermal state is achieved. The heat flux density in this state is determined only by the present rate of the radioactive heating. Calculations show [5] that heat is continuing to escape through the lunar surface which was earlier accumulated during warming-up, i.e. there exists a residual part of the flux which accounts for 10–15% of the total flux. Thus only a part of the flux, equal to $q_r = 0.7 \times \times 10^{-6}$ cal cm^{-2} s^{-1}, is associated with the present radioactive decay. So, the heat release by the radioactive elements will be $q_r = 0.88 \cdot 10^{19}$ cal/year.

The measurements of the radioactive content of the lunar rocks from Apollo 11 show that the uranium content accounts for about $0.6 \cdot 10^{-6}$ g/g and thorium about $2.5 \cdot 10^{-6}$ g/g. The content of the radioactive isotope of potassium 40 (K 40) may be estimated if the ratio of the radioactive isotope content to the non-radioactive is accepted to be $1.2 \cdot 10^{-4}$ both for the moon and the earth. Then, assuming the value of potassium content estimated for the moon to be equal to 0.002 g/g, we obtain the value $2.4 \cdot 10^{-7}$ g/g for the content of potassium 40. With this content of the radioactive elements the heat release per gramme of the lunar rock for a year will be

$$q_0 = 0.6 \cdot 10^{-6} \cdot 0.8 + 2.5 \cdot 10^{-6} \cdot 0.2 + 2.4 \cdot 10^{-7} \cdot 0.22 = 10^{-6} \text{ cal g}^{-1} \text{ y}^{-1} .$$

Here the coefficients 0.8, 0.2 and 0.22 are the heat release in calories of a gramme of uranium, thorium and potassium 40 for a year, respectively.

Note, that the obtained heat release for the lunar rocks is similar to the heat release of earth basalt which is known to be about $1.6 \cdot 10^{-6}$ cal g^{-1} y^{-1}. It is interesting, that according to the chemical analyses of lunar rock [1], the material of the upper layer of the moon is also identified with earth basalt. But the latter contains almost 5 times as much potassium as the lunar basalt taken from the surface. This is partially responsible for the lowered heat release of the "lunit" (the lunar material on the surface). It should be taken into account that potassium is likely to evaporate from the upper lunar layer as outpouring magma in vacuum. The decrease of the potassium and sodium content was detected in the experiments while cooling down the fusion material in vacuum [6]. In view of the above, it is likely that the potassium content in the depth of the layer will be similar to that of the earth. Then the heat release of such lunar basalt is $q_0' = 1.25 \cdot 10^{-6}$ cal g^{-1} y^{-1}.

From the available data of the total heat flux and the radiogenic heat release in the upper layer we can estimate the effective thickness ΔR of the radioactive basalt layer of the moon. It is evident that the total radioactive heat release of the moon is

$$Q_r = 4 \pi \int_0^R q(r) \varrho(r) r^2 \, dr ,$$

where r is the distance from the lunar centre, R is the lunar radius, $q(r)$ and $\varrho(r)$ are the heat release and density distributions respectively. The same magnitude expressed in terms of the effective thickness ΔR is equal to

$$Q_r = 4 \pi R^2 \Delta R \varrho_0 q_0 ,$$

where ϱ_0 and q_0 are corresponding values on the surface and

$$\Delta R = \frac{1}{R^2 \varrho_0 q_0} \int_0^R q(r) \varrho(r) r^2 \, dr .$$

On the other hand, the magnitude

$$Q_r = q_r 4 \pi R^2$$

is known from experiment. Comparing the two expressions for Q_r and assuming $\varrho_0 = 3.1$ g cm^{-3} with $q_0 = 10^{-6}$ cal g^{-1} y^{-1} and $q_r = 0.7 \cdot 10^{-6}$ cal cm^{-2} s^{-1}, we obtain the following value of the effective thickness of the radioactive layer:

$$\Delta R = \frac{q_r}{\varrho_0 q_0} = 70 \text{ km} .$$

It is not difficult to show that if there is the exponential decrease of heat release, according to the law

$$q(r) = q_0 \exp\left(-\frac{R-r}{\Delta R'}\right).$$

So the effective thickness will be $\Delta R'$. Assuming the homogeneous distribution of the radioactive elements in the layer thickness, one may state that the basalt layer of the moon is at no more than about 70 km.

3. Discussion of the Results

For the value obtained for total heat flux generated due to radioactive decay, $Q_r = 0.88 \cdot 10^{19}$ cal y^{-1}, according to Fig. 1, the equilibrium thermal state will take place at a radioactive layer thickness of about 40 km. The heating of the moon's interior at 70 km must have lasted until it began melting nearer to the surface and the radioactive layer became shorter. It is most unlikely that this process would occur at the present time. As the calculations of the moon's heat history show, this process must have been completed 2–3 milliards of years ago. Thus, the values of ΔR obtained somehow contradict the heat history of the moon.

This disagreement is attributed to the errors of measurements of Q_r and q_0 as well as to the fact that the radioactivity of the lunar material, determined only for the single place of the spacecraft landing, is not characteristic of the whole moon. The contradiction disappears if the content of potassium in the deep layer of the lunar basalt is assumed to be the same as in the earth, the evaporation of the alkaline oxides at depth being absent. In this case $q_0 = 1.25 \cdot 10^{-6}$ cal g^{-1} y^{-1} and the effective thickness of the radioactive layer is $\Delta R = 50$ km. This value is in good agreement with the heat history of the moon. The treatment of the subject of interest shows that it is necessary to specify the mean content of the radioactive elements in the upper layer of the moon, as well as to carry out more strict and detailed analyses of the theory of the moon's thermal history.

References

[1] Science **165**, 1211 (1969).
[2] V. D. Krotikov and V. S. Troitsky, Astron. Zh. **40**, 1076 (1963).
[3] V. S. Troitsky and T. V. Tikhonov, Astron. Zh. **46**, 159 (1969).
[4] C. V. Maeva, Dokl. Akad. Nauk USSR **159**, 294 (1964).
[5] O. I. Ornatskaya and Ya. I. Alber, Astron. Zh. **44**, 158 (1967).
[6] Ts. M. Roytburg, Problemy Geologii Luny, Trudy Geolog. Inst. Akad. Nauk USSR **204**, 165 (1969).

Space Research XI — Akademie-Verlag, Berlin 1971

GLOBAL MAPPING OF THE MOON

Y. N. Lipsky, Y. P. Pskovsky, J. F. Rodionova, V. V. Schevchenko, V. I. Chikmachev and L. I. Volchkova

Sternberg Astronomical Institute, USSR

A more precise version of the entire map of the moon, scale 1:5000000, and the entire globe of the moon, scale 1:10000000, were prepared and edited in the USSR. New publications represent 99.5% of lunar surface. The photographs obtained by Luna 3 and Zond 3 space vehicles and partly by Lunar Orbiter were used in the process of construction of the coordinate grid of the map and the globe of the far side of the moon and for the drawing of relief.

Representation of the visible side of the moon was made from original photographs and photographic material from atlases based on telescopic observations from the earth. The precision of the coordinate grid values of features on the visible side is to 0.5° on both coordinates. The corresponding precision on the far side is to 1–2°.

Cartographic grid referencing of photographs of the far side of the moon was carried out by the method of rectification on a spherical screen and was controlled by an astronomical method of representation of the earth over the horizon of the moon.

During the drawing of the relief, great attention was given to details of systems of rays on both hemispheres and to the tinted distinction of maria areas etc.

The map of the moon is accompanied by a complete list of the names of features of the moon. On the globe the names of the largest features are plotted.

The first complete map of the moon at a scale of 1:5000000 and the entire globe of the moon, scale 1:10000000, covering 95% of the lunar surface were published in 1967 and presented at the XIII General Assembly of the International Astronomical Union in Prague. Lately some new material has been received and the method of producing a cartographic grid for the photographs of the moon's far side has been improved. All these data made it possible to start drawing a new more precise version of the complete map of the moon and the globe of the moon. Both the complete map of the moon and the globe were edited in 1969. The new version represents 99.5% of the moon's surface. The whole number of features represented on the new edition of the complete map is as much as 50000. Photographs from the automatic interplanetary stations (AIS) Luna 3 and Zond 3 and some from Lunar Orbiter were used for representation of the moon's far side. Mapping of the moon's front side was made from earth-based photography obtained through the various observatories around the world.

Lunar surface pictures photographed by AIS are usually perspective. To prepare these materials for mapping it is necessary to eliminate both an inclination of the optic axis of the camera relative to the vertical and distortions due to the surface curvature. These distortions are usually eliminated by the methods of phototransformations or projection of an image on a spherical screen.

The catalogue of 500 points of Golosseyevo served as a basis and for transmission of the coordinate grid on the far side. In the Sternberg Astronomical Institute a method has recently been developed for combining photographs of the moon into a unified selenographic coordinate system by means of projection equipment. In addition to yielding corrected images, in which perspective distortions are practically eliminated, this method simultaneously provides a cartographic combination in one plot of a selenographic coordinate grid with corrected images (compilation of a photomap). The mean square error in the position of the objects on Zond 3 photographs does not exceed $\pm 0.5°$. A cartographic combination of part of the Lunar Orbiter photographs was carried out with an error of $\pm 2°$, because the overlapping was not sufficient.

One of the methods of producing an independent cartographic grid is to determine the selenographic coordinates of a survey point and visual features according to the location of the earth's disk above the moon's horizon. The information concerning the image itself is used as initial data. The Lunar Orbiter photograph of the edge region of the west section of the far side and the AIS Zond 6 frame of the east region were "gridded" by this method. In the first case the error in the coordinates of the moon's features is as much as $\pm 0.2°$, in the second, $\pm 0.4°$. When compared with the method of projection on a spherical screen, the astronomic gridding method appeared to have a random deviation in the coordinates of features up to $\pm 0.6°$.

Artistic rendering of the relief features was accomplished by shading. Some latitude in the choice of illumination was accepted to increase the range of information given on the map. The mean value of the angle of incidence was taken to be 60° for drawing large features of relief. Ray systems, gentle slopes in maria etc. were shown with more detail than would have been possible otherwise, i.e. such details were drawn under optimal conditions. The contrast between maria and highlands was specially emphasized and it was for this purpose that preliminary models showing maria boundaries were made.

The colour of the map and the globe was chosen so as to take into account not only the natural lunar surface colour but also to achieve the best expressiveness of a polygraphic production.

The complete map of the moon is displayed on nine sheets. Six of them are developed on a modified cylindrical projection for the central zone, extending to latitudes 60° N and 60° S, the scale along latitudes $\pm 30°$ being unchanged. Three sheets cover the polar areas, the entire list of nomenclature, and brief information about the moon. The polar regions are developed on an azimuthal projection on the same scale as that of the central zone. In the list of nomenclature the names of features are given in Russian and Latin transcriptions and indexes are provided which facilitate finding the named features on the map.

Having in mind the smaller scale of the globe, topographic images and lunar nomenclature here were subjected to some kind of generalization in the process of production.

The complete (global) map of the moon drawn on a navigation projection at a scale 1:5000000 is still the only edition of its kind in the world. This map has an advantage over maps of lunar hemispheres which represent the edge areas with large perspective distortions and the libration zones practically impossible to observe.

In order to increase the amount of information which can be put in a map, a method of physical mapping was worked out. The essential feature of such a method is a gradual division into districts. This allows different complexes of physical properties of the lunar surface and the regions in which they exist to be distinguished. Depending on the types of physical information which are to be charted, topological and individual divisions into districts are used. The topological division means choosing types of surfaces characterized by different parameters studied in the astronomical observations. An individual division means determining relationships between astrophysical properties of the lunar surface on the one hand, and its structural, mechanical and chemical characteristics on the other hand.

Studying the distribution of physical properties on the moon's surface on a global scale will be a significant step towards the understanding of the moon's nature. Physical charting of a selected lunar area was undertaken using a sheet of the complete map of the moon as a topographic basis. The optical properties of lunar surface in the visible spectrum were chosen as a physical parameter. A region subjected to physical charting covers the portion of the eastern sector of the moon's far side. AIS Zond 3 photographs were used as initial material. From corresponding photometric measurements, isophotes representing the photometric properties of the area were plotted.

An analysis of these properties led us to a conclusion that "improved Hapke's formulae" can be used as the most suitable analytic expression of the photometric function, including parameters of micro-relief (porosity) and a shaded function of a relief. The region under examination was divided into elementary sections and the parameters mentioned above were determined by a solution of the corresponding system of uniform linear equations for each of these elementary sections. The discovery of typical combinations of photometric parameters and regions, with primary expansion of these parameters, led to topological charting of the area.

Individual division into districts accompanied by interpretation of data enabled us to classify the regions according to mean statistical values of the porosity of lunar soil and different kinds of irregularity. Six types of photometric relief were distinguished in the part of the highlands charted. This result is of interest because it shows the inhomogeneity of the properties of such a characteristic type of the moon's surface as highlands.

An estimate of the photometric accuracy shows that the topographic basis used — sheets of the complete map of the moon — fully satisfies the requirements of physical charting on the present scale.

References

[1] Yu. N. Lipsky, Vest. Akad. Nauk SSSR, No. 4 (1968).
[2] Yu. N. Lipsky and V. I. Chikmachev, Astron. Zh., No. 3 (1969).
[3] V. V. Shevchenko, Kosm. Issled. 4, No. 6 (1968).
[4] Yu. N. Lipsky and V. V. Shevchenko, Astron. Zh. 47, No. 3 (1970).
[5] V. V. Shevchenko, Astron. Zh. 47, No. 3 (1970).

Space Research XI — Akademie-Verlag, Berlin 1971

PRELIMINARY RESULTS OF LASER RANGING TO A REFLECTOR ON THE LUNAR SURFACE

J. D. MULHOLLAND[a], C. O. ALLEY[b], P. L. BENDER[c], D. G. CURRIE[b], R. H. DICKE[d], J. E. FALLER[e], W. M. KAULA[f], G. J. F. MACDONALD[g], H. H. PLOTKIN[h] and D. T. WILKINSON[d]

[a]Jet Propulsion Laboratory, California Institute of Technology, Pasadena, California, USA
[b]University of Maryland, College Park, Maryland, USA
[c]Joint Institute for Laboratory Astrophysics, Boulder, Colorado, USA
[d]Princeton University, Princeton, New Jersey, USA
[e]Wesleyan University, Middletown, Connecticut, USA
[f]University of California, Los Angeles, California, USA
[g]University of California, Santa Barbara, California, USA
[h]Goddard Space Flight Center, Greenbelt, Maryland, USA

The first Lunar Ranging Experiment (LURE) retroreflector array was placed on the lunar surface during the Apollo 11 mission. Prior to this event, a special high-precision lunar ephemeris (designated LE 16) was developed by means of a composite numeric/analytic process, so as to provide more accurate predictions for use at the telescope. First returns were observed at the Lick Observatory on 1 August 1969 and at the McDonald Observatory shortly thereafter. The observing program is to continue for several years. Preliminary use of these data consists of their comparison with the LE 16 ephemeris, preparatory to a differential correction of the lunar elements. Present indications are that the ephemeris must undergo order-of-magnitude improvements before the full power of the laser data can be utilized. Other parameters in the predictive process are also capable of being corrected; an improvement in the coordinates of the Lick Observatory 120-inch telescope is already indicated.

1. Introduction and Background

On 20/21 July 1969 an estimated 800 million people watched as the first men to set foot on the moon performed a number of tasks of scientific importance. Among these tasks was the deployment of a rack structure carrying an array of 100 fused silica retroreflectors designed to return some of the light of a pulsed laser beam to the same telescope that transmitted it. The array was erected so that its face was normal to the mean direction to earth from Tranquillity Base, to ensure that the retroreflectors be usable at all lunar libration angles.

The purpose of the Apollo Lunar Laser Ranging Experiment (LURE) is, briefly, to obtain extremely high accuracy measures of the range from known locations on the surface of the earth to known locations on the lunar surface, and to continue a steady observing program over several years. The immediate uses to which these data are to be put will be the improvement of geocentric telescope coordinates, of the selenocentric coordinates of the reflectors, and of our representation of the motion of the moon about the earth. Ultimately, the extreme resolution of these data will permit improvements in our knowledge of the rotation of the earth and the moon about their centers of mass, measurements of solid body tides, examination of tidal acceleration in the lunar motion and of continental drift, and a discrimination between models of relativistic dynamics. Some of these goals require the placement of more reflectors, while others require that ranging operations be conducted from several sites distributed around the world. A continuing laser ranging program has also been started by the US Air Force Cambridge Research Laboratory at a site near Tucson, Arizona. In addition, several groups in other countries have conducted or are preparing to conduct such observing programs.

2. Prediction and Observation

The data-taking process requires that fairly precise range predictions be available at the telescope, to "range gate" the signal detection electronics. This system activates the detector at a preselected interval before the predicted arrival time of the return pulse. The detector is then active for only a few microseconds; if the prediction is not correct to within the half-width of the range gate, the observation will be missed. The choice of prediction ephemeris can be nearly as critical as the design of the ranging apparatus, simply because there are some poor choices available.

The basic ephemeris currently printed in the national almanacs, the *Improved Lunar Ephemeris* [1], or $j = 0$ ephemeris, will cause prediction errors as large as 10 km (70 μs) if used with the currently accepted constants. The almanac volumes for 1969–1971 contain corrections to be applied to the $j = 0$ ephemeris to account for the changes in the System of Astronomical Constants, but even this may be expected to have errors as large as 2 km, arising from inadequacies that yet remain in the astronomical constants, the transformations determined by Eckert, Walker and Eckert [2], the planetary theories implicit in the lunar theory [3], and series truncation in the planetary perturbations [4, 5].

A special high-precision lunar ephemeris, developed at the Jet Propulsion Laboratory [6] and designated LE 16, has been adopted as the initial prediction ephemeris. This is the result of a composite numeric/analytic process in which the aspects of the lunar theory that are at variance with the equations of motion on which it is based (e.g. tidal dissipation) are treated by algebraic processes, while the equations of motion are used in a high-order numerical integration process, differentially corrected to fit synthetic observations computed from an analytic theory. At the same time, the corrections to the elements of the lunar orbit determined by Van Flandern [7] were applied, so as to place the final ephemeris on the FK4 equinox. Center-to-center range errors are believed to be no larger than a few hundred meters. Results from analysis of Surveyor spacecraft tracking [8], as well as order-of-magnitude analogies with the apparent range precision, indicate that the rate of drift between LE 16 and the real lunar range is not likely to exceed from 10 to 20 meters per day.

The topocentric range is dependent on the coordinates of both the observatory and the reflector array. The first task in the application of the laser observations is the improvement of the ephemeris, the telescope coordinates, and those of the reflector; the ultimate goals of the experiment require that the residuals be reduced two orders of magnitude from their present levels. At this time, we have acquired enough data to permit some preliminary remarks on the locations of the Lick Observatory and the McDonald Observatory.

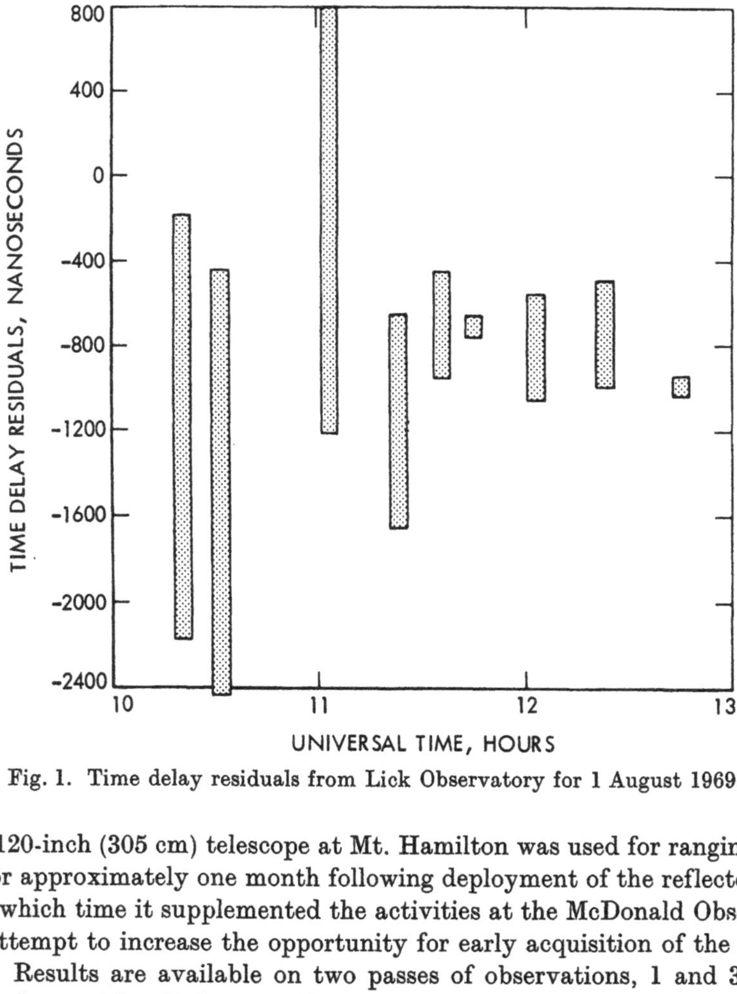

Fig. 1. Time delay residuals from Lick Observatory for 1 August 1969.

The 120-inch (305 cm) telescope at Mt. Hamilton was used for ranging operations for approximately one month following deployment of the reflector array, during which time it supplemented the activities at the McDonald Observatory in an attempt to increase the opportunity for early acquisition of the reflector [9–11]. Results are available on two passes of observations, 1 and 3 August 1969, and one night, 21 August. The 1 August returns are particularly interesting, both because they were the first observations of the reflector array and because they still represent the strongest return detected.

The electronic detection system used on the night of 1 August was such that the observation consisted of a preselected time interval during which the return signal was detected. There is no information as to *when* during the interval the event occurred, but the interval resolution could be altered from one observation run to the next. The residuals from the observations of 1 August are shown in Fig. 1, with the computed range based on a recently completed reduction to the North American Datum of the telescope coordinates, as provided to us by D. W. Dunham (ACIC) and confirmed by J. Brenkle (JPL). Note that the resolution intervals vary between 2000 and 100 nanoseconds. We shall return to these data shortly.

The primary observation station for the LURE team is the McDonald Observatory of the University of Texas. First acquisition at that site occurred [12] on

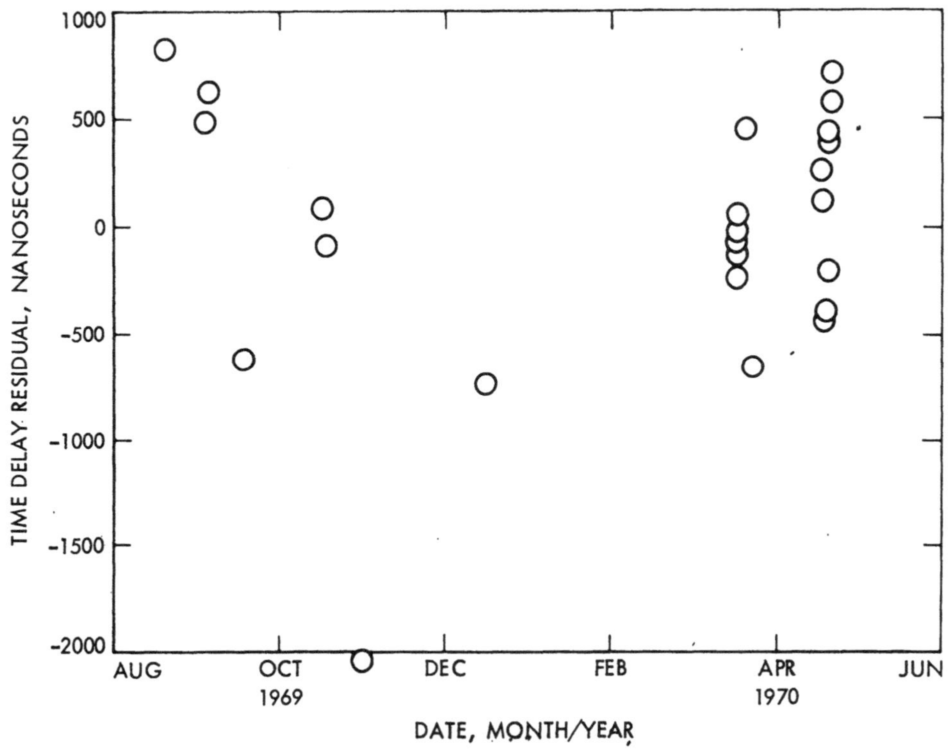

Fig. 2. Time delay residuals from McDonald Observatory for August 1969–April 1970.

20 August 1969. During the first few months operations were severely hampered by bad weather and technical problems, but confirmed observations have been made with some regularity and, in Spring 1970, the introduction of a new observing schedule and the solution of some of the operating problems has led to a satisfyingly high observation density. Fig. 2 shows the residuals for the definite observations up to 20 April 1970, although some of the more recent ones are still subject to minor changes due to equipment calibration analysis. The predictions to which these residuals are referred apply to a new reduction of the geocentric coordinates of the 107-inch (272 cm) telescope, based on the best available geodetic data. These observations are still in various stages of reduction to eliminate instrumental effects; their current uncertainties range from 2 to 50 ns, or 30 to 750 cm. The eventual uncertainties will be established by the duration of

the laser pulse and the number of returns per run; for most of these observations, the pulse duration is 4 ns.

One should note, in passing, that even though we speak of ranging and sometimes use linear distance units, the real observable is two-way time delay. The question of the uncertainty in the speed of light is quite irrelevant to this, as well as to any similar, work: the IAU/IUGG value of 299792.5 km s^{-1} has been adopted.

3. Preliminary Results

The data taken at the Lick Observatory on 1 August are not adequate to supply definitive information, but they do permit some speculative discussion of the absolute coordinates of the 120-inch telescope. During the observing pass the lunar declination was nearly zero so that the partial derivatives of range with respect to geocentric radius and latitude of the telescope reduce approximately to the cosine of the local hour angle, apart from a constant factor. Since over the same interval, the derivative with respect to longitude is approximately a constant times the sine of the hour angle, the residual signature from a telescope coordinate error can be expressed

$$\Delta\tau = k \sin (H + \sigma) ,$$

where σ is a phase angle depending on the ratios of the errors in the thre coordinates.

From the 1 August observations at 1145 and 1245, it is evident that thee observed range is drifting very significantly with respect to the predictions, although this drift is entirely subject to the condition that the apparent resolution of 100 ns is justified. It is not thought that the LE 16 ephemeris can contribute more than about 15 ns of drift over the entire observing pass, which is an order of magnitude too small to account for the observations. These data are evidence of erroneous coordinates for the telescope and/or reflector. For the present discussion, it has been assumed that the drift is entirely due to the telescope coordinates. Since meridian transit occurred at 1108, the residuals are consistent with a negative k and a phase angle σ near zero, i.e. the principal error appears to be in the telescope longitude. This is also the condition that minimizes the amplitude of the harmonic curve required to fit the data, and thus also minimizes the absolute magnitude of the error vector. One can draw limit lines through the observation bars, assuming no error in radius and latitude, and obtain the result $5.1 \pm 1.0''$ as the indicated sum of the corrections to the right ascension of the reflector and to the assumed east longitude of the 120-inch telescope. The corresponding error at transit in the predicted geocentric distance to the reflector is 92 ± 5 meters.

Comparison with meridian circle observations indicates that the LE 16 ephemeris is reliable in right ascension to within $1''$, while a $1''$ error in the differential right ascension from the center of mass to Tranquillity Base implies a selenocentric longitude error of about 4 arc minutes, which gives a range error whose constant term is of the order of 800 meters. It does not appear likely that these effects can fully account for the size of the required correction, unless the run at 1145 is affected by a very peculiar noise distribution, which must be considered a possibility. It seems likely that the absolute longitude of the Lick Observatory is shifted significantly westward from the location based on the North American Datum.

The detection equipment at the McDonald Observatory is such that the observatory quantity is the two-way time delay from the beginning of the transmitted pulse to the detection of the returning signal. The timing circuits are designed to operate with a subnanosecond precision.

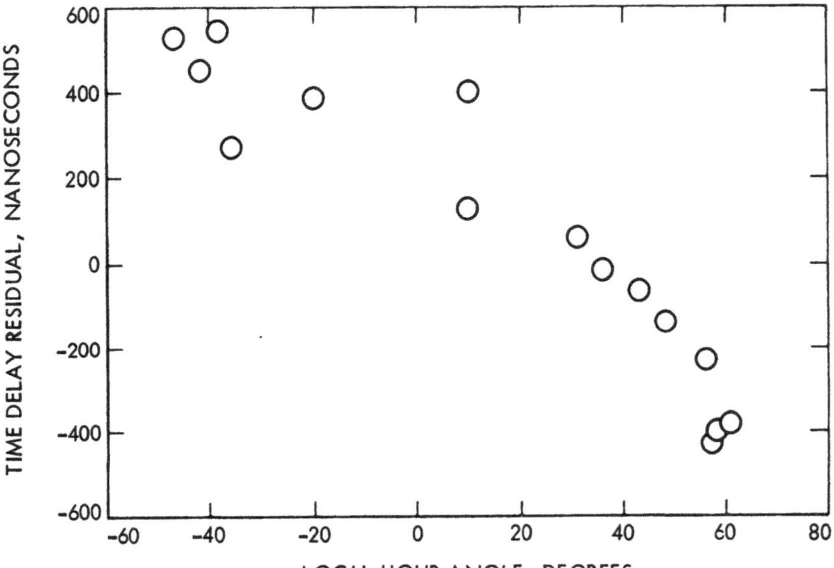

Fig. 3. Time delay residuals from McDonald Observatory as a function of local hour angle. (All observations at declinations of 19° and higher over the interval September 1969–April 1970.)

One of the results that seems to be emerging from the operations at McDonald is that West Texas has cloudy skies whenever the moon is near zero declination! One cannot yet apply the type of simplified procedure that has been described for the Lick observations. The collection of data at high declination is now large enough to examine for evidence of telescope coordinate errors; range points for declination of 19° to 28° are shown in Fig. 3 plotted against local hour angle. While it certainly appears that there are range variations due to both the ephemeris and to the declination effect, there can be little question but that the geocentric coordinates of the telescope require some significant corrections. We do not yet have a set of preliminary corrections, but judging from the residuals, the length of the correction vector would be of the order of 100 meters if no other errors were present. The phase angle σ appears to be about 50° to 70°, which implies that the dominant error is in the radius and/or latitude. The required correction will bring the telescope closer to the spin axis.

Since it is not yet possible to ensure that the adopted coordinates of the reflector are not a major factor in the observed residuals, it seems appropriate to comment upon these values. We have adopted the Tranquillity Base coordinates that resulted from trajectory determination solutions* from the radio tracking of the Lunar Excursion Module as it descended from orbit. The longitude and latitude are slightly inconsistent with the selenographic coordinates obtained from photographic feature identification, but it should not be surprising if the optical selenographic system were distorted in some degree. On the other hand, the trajectory calculations were performed with a different lunar ephemeris, and we do not know what effect this may have had on the location of the landing site. It seems unlikely that either effect can account for the residuals thus far obtained. The telescope coordinate errors are real, even if the definitive corrections are yet to be found.

With the data now available to us, we are unable to determine if the required corrections arise from distortions of the geodetic net, translations of the North American Datum, or a rotation of the entire geodetic system with respect to the "absolute" frame based on the planetary system.

Rigorous data analysis for determination of improvement of the ephemeris and the determination of definitive coordinates for the McDonald Observatory and the retroreflector array will soon be in progress.

Acknowledgments

We are indebted to the many people at our individual institutions, and also at the University of California and the University of Texas, who have contributed materially to the success of this experiment. The work has been supported by the National Aeronautics and Space Administration, through grants and contracts to our institutions.

Note added in proof:

It is not clear from the above discussion, but the residuals presented there are referred to coordinates based on the North American Datum for both telescopes. (The phrase "best available . . ." is an unfortunate choice.) Reduction of these coordinates to the 1966 Smithsonian Standard Earth requires significant corrections to two coordinates in each case. For the Lick Observatory, the radial distance is decreased by 113 m, while the longitude is shifted 104 m westward. When these corrections are applied, the residual envelope indicates limits

$$\Delta H = -1.7'' \pm 1.4$$

on the error in hour angle, with a topocentric range error of -185 ± 7 meters.

The reduction for the McDonald Observatory requires that the longitude be

* Private communication from J. MacPherson, NASA Manned Spacecraft Center, 22 July 1969.

shifted 77 m westward and the latitude 219 m north. Application of this adjustment to the data of Fig. 3 reduces the amplitude of the variation with hour angle by more than an order of magnitude.

It follows that the observations are now consistent with the predictions to within a few tenths of an arc second in topocentric direction, the error being composed of contributions from several sources. It can be stated, then, that our results appear to accord with the corrections to the North American Datum obtained by the Smithsonian Astrophysical Observatory and other groups.

References

[1] W. J. ECKERT, R. JONES and H. K. CLARK, Improved Lunar Ephemeris 1952–1959, US Government Printing Office, Washington (1954).
[2] W. J. ECKERT, M. J. WALKER and D. ECKERT, Astron. J. **71**, 314 (1966).
[3] J. D. MULHOLLAND, Nature **223**, 247 (1969).
[4] J. D. MULHOLLAND and C. J. DEVINE, Science **160**, 874 (1968).
[5] C. N. CARY and W. L. SJOGREN, Science **160**, 875 (1968).
[6] K. GARTHWAITE, D. B. HOLDRIDGE and J. D. MULHOLLAND, Astron. J. **75**, 1133 (1970).
[7] T. C. VAN FLANDERN, Celest. Mech. **1**, 163 (1969).
[8] F. B. WINN, JPL Space Programs Summary 37–56 II, 74 (1969).
[9] J. FALLER et al., Science **166**, 99 (1969).
[10] J. E. FALLER and E. J. WAMPLER, Scient. Amer., March 1970, p. 38.
[11] C. O. ALLEY et al., Science **167**, 458 (1970).
[12] C. O. ALLEY et al., Science **167**, 368 (1970).

Space Research XI — Akademie-Verlag, Berlin 1971

DETERMINATION OF ASTRODYNAMIC CONSTANTS AND A TEST OF THE GENERAL RELATIVISTIC TIME DELAY WITH S-BAND RANGE AND DOPPLER DATA FROM MARINERS 6 AND 7

J. D. Anderson[a], P. B. Esposito[a], W. Martin[a] and D. O. Muhleman[b]

[a]Jet Propulsion Laboratory, California Institute of Technology, Pasadena, California, USA

[b]California Institute of Technology, Pasadena, California, USA

Range and Doppler tracking data from Mariners 6 and 7 have been used to obtain information on the ratio μ^{-1} of the mass of the earth to the mass of the moon and on the mass and ephemeris of Mars. Based on data from five Mariner and two Pioneer spacecraft, the best value for μ^{-1} is 81.3007 ± 0.0005. The best value of the ratio of the mass of the sun to the mass of Mars was obtained by Null from analyses of the Mariner 4 data. His value (3098714 ± 5) has been substantiated by the Mariner 6 data which give 3098697 ± 72. By combining the Mariner 6 range data with optical and radar data of the planets, it is possible to deduce that the mean equatorial radius of Mars is 3393.0 ± 1.7 km.

Range data around the time of solar superior conjunction have yielded a test of the general relativistic time delay to the level of $\pm 10\%$. Additional post-conjunction data will significantly improve the accuracy of this test.

1. Introduction

Range and Doppler tracking data from Mariners 6 and 7 can be used to obtain new information on the ratio of the mass of the earth to that of the moon, the gravity field of Mars and the ephemerides of Mars and the earth. In addition, it is possible to perform a relativity propagation and solar coronal plasma experiment with the range data during solar superior conjunction. New developments in the ground-based ranging equipment have produced measurements within one arc degree of the sun at a geocentric distance of 4×10^8 km. A transmitted power of 200 kW has been used to obtain these measurements. Also, a new Doppler rate-aiding technique, which was conceived by Dr. R. Goldstein of JPL, has been implemented in the ranging equipment. The accurate tracking of the Mariners is scheduled to continue until the end of 1970. This extended coverage will permit a very accurate determination of the orbit of the Mariners so that the relativistic time delay can be measured during the time of solar conjunction (29 April 1970 for Mariner 6 and 10 May 1970 for Mariner 7).

The total arc of tracking data will extend over a period of almost two years (i.e. from launch on 24 February 1969 to the end of 1970). Therefore, it makes sense to split the experiment into three orbital arcs, each with its own objectives. In addition there is a real criterion for the separation of arcs. During the encounters of Mariners 6 and 7 with Mars, an infrared spectrometer (IRS) was operated in a cryogenic environment. This environment was produced by expelling nitrogen and hydrogen gas through the IRS into space. As a result, a total velocity change of the order of 10 cm s^{-1} was added to the encounter orbit by the gas jetting. Consequently, it is practically impossible to fit one orbital arc all the way from launch through encounter and beyond. It is primarily for this reason that we have split the analysis of the tracking data into three orbital arcs.

The first arc extends from injection into the earth–Mars transfer orbit to a somewhat arbitrary point which defines the start of the encounter trajectory. This point is selected at a time five days before the closest approach of the spacecraft to Mars. In a similar fashion to the beginning of the encounter orbit, we arbitrarily define it to end at a point 13 days after the closest approach to the planet. The third orbital arc, the post-encounter orbit, extends from 5 days after encounter with Mars to the end of the scheduled tracking of Mariners 6 and 7 on 1 January 1971. The overlap between the end of the encounter orbit and the beginning of the post-encounter orbit does not present any problems. Because the three arcs of data are chosen to satisfy basically different objectives of the experiment, the same data are never used twice to determine the same parameters.

The parameter that has been obtained so far from the first orbital arc, the pre-encounter orbit, is the ratio μ^{-1} of the mass of the earth to that of the moon. We have not yet used Mariner data to obtain a definitive correction to the ephemeris of the earth. This must await later solutions, when we combine the Mariner data with data from other spacecraft and with radar data and optical observations of the planets.

From the second orbital arc, the encounter orbit, we have been able to extract the mass of Mars from Mariner 6. Because of the IRS gas jetting, the uncertainty on the mass of Mars is about an order of magnitude greater than that obtained from Mariner 4 by Null [1]. The encounter orbit for Mariner 7 is practically worthless for celestial mechanics purposes because of non-gravitational forces from the on-board systems.

This discouraging situation with respect to the encounter orbits of Mariner 6 and 7 is, fortunately, not applicable to the post-encounter orbits. A very significant test of the general relativistic propagation delay in range is possible from both of them. When the final analysis of the data is completed, probably late in 1971, the test will be accurate to at least the 4% level in the total relativistic delay. In addition, by tracking both Mariners in a ranging mode for a period of more than a year, it will be possible to obtain very significant information on the ephemeris of the earth. At this time, we have begun to obtain solutions for the relativistic propagation parameter in combination with parameters characteristic of a static solar corona, and parameters of the spacecraft orbit. The Einstein relativistic delay has been verified to $\pm$ 10%. However, this level of accuracy can also be achieved by radar tracking of the planets; the Mariner experiment is unique only at the $\pm$ 4% level or less.

In order to separate the general relativistic propagation delay from parameters of the spacecraft orbit and the earth's orbital elements, it will be necessary to track the two spacecraft for long intervals of time on either side of solar conjunction; accurate orbits must be established for both of them. In this way, the actual coordinate positions of both spacecraft will be known during solar conjunction, and it will be possible to compare these actual positions with the apparent positions and the apparent positions will be descriptive of the general relativistic propagation delay.

2. The Ratio of the Mass of the Earth to that of the Moon

By means of a very accurate determination of the lunar inequality, it has been possible to obtain the ratio μ^{-1} of the mass of the earth to that of the moon from a number of Mariner and Pioneer spacecraft.

Table 1
Values for the ratio of the mass of the earth to that of the moon, μ^{-1}

Spacecraft	Value of μ^{-1} with with formal uncertainty	Deviation from the mean value ($\mu^{-1} - \bar{\mu}^{-1}$)
Mariner 2 (Venus)	81.3001 ± 0.0013	-0.0006
Mariner 4 (Mars)	81.3015 ± 0.0017	0.0008
Mariner 5 (Venus)	81.3013 ± 0.0002	0.0006
Mariner 6 (Mars)	81.3005 ± 0.0002	-0.0002
Mariner 7 (Mars)	81.3005 ± 0.0002	-0.0002
Pioneer 8	81.3004 ± 0.0001	-0.0003
Pioneer 9	81.3008 ± 0.0001	0.0001
Radar Data	81.302 ± 0.002	

Table 1 shows the results for μ^{-1} as determined by Null [2] from Doppler and range pre-encounter data of Mariners 6 and 7. Also his recent determinations from range and Doppler data of Mariner 5 are given along with his determinations from Doppler data of Pioneers 8 and 9. Earlier determinations from Doppler data of Mariner 2 [3] and Mariner 4 [4] are also given. In addition, a determination from radar tracking of the planets is given in Table 1 [5]. The mean of the values of μ^{-1} determined by spacecraft is 81.3007. The deviations from the arithmetic mean of the Mariner and Pioneer values are tabulated for each spacecraft in the last column of Table 1. The rms deviation of 0.0005 is a good indication of the accuracy in this mean value of μ^{-1}.

3. The Mass and the Ephemeris of Mars

Clearly, the non-gravitational forces acting on the Mariners 6 and 7 spacecraft during encounter complicate the determination of the mass and ephemeris of Mars from those data. To date, no attempts have been made to use the Mariner 7 encounter orbit for celestial mechanics purposes. Previous solutions for the mass of Mars are reported in [6]. These solutions include data from the start of the encounter orbit to the time of the initiation of the IRS gas jetting about 35 min before encounter. The previous value for the mass expressed in terms of GM was $42828.48 \pm 1.38\ km^3\ s^{-2}$ (G is the Newtonian gravitational constant). Corrections to the ephemeris of Mars were not reported previously.

In order to obtain corrections to the ephemeris of Mars and to extract as much information as possible about its mass, data have been fitted from 26 July to 13 August 1969. The effects of the IRS gas jetting have been accounted for by means of a six degrees of freedom discontinuity in the trajectory. This discontinuity is introduced in such a way that it can be included in the differential correction procedure.

The ephemeris that has been used as a basis for the Mariner 6 corrections is Development Ephemeris 71 (DE 71) produced by D. O'Handley of JPL for the Mariner Mars 1969 project. This ephemeris includes the radar tracking of Mars from Goldstone in the summer of 1969, and it represents the best estimate of the geocentric distance to Mars during the Mariner 6 encounter.

The results of the fit to encounter data from 26 July to 14 August 1969 are that GM for Mars is equal to 42828.67 ± 1.0. The correction to the distance given by the ephemeris (DE 71) is 1.04 ± 1.71 km at the time of the Mariner 6 encounter with Mars. The conclusion of this result is that DE 71 can be used without correction for purposes of analyzing the 1969 radar data for topographic information. Also, the mean equatorial radius of Mars given by DE 71 (3393.0 km) is accurate enough to satisfy the Mariner range data at the $\pm$ 1.7 km level.

Residuals for the range data used in the Mariner 6 fit are shown in Table 2. The rms Doppler residuals from this same fit are shown in Table 3 for the seven stations used to obtain Doppler. All Doppler residuals are referred to a ten minute integration time.

Table 2
Mean planetary range residuals
(1 ns $\approx$ 15 cm in distance)

Data interval (UTC)		Mean residual (ns)
27 July	0112–0636	−0.21
8 August	0132–0609	−138.5
9 August	0021–0554	387.6
12 August	0600–0600	149.3
13 August	0021–0604	133.3

Table 3
Statistical properties of encounter Doppler residuals
(1 Hz $\approx$ 67 mm s^{-1} in range rate)

Tracking station	No. of points	Data interval (UTC)	Mean residual (Hz)	Root-mean-square residual (Hz)
Australia (DSS 42)	12	11 Aug. 0650 – 11 Aug. 1434	−0.000772	0.0029
Australia (DSS 41)	95	26 July 0649 – 12 Aug. 1352	0.000306	0.0030
South Africa (DDS 51)	14	27 July 1628 – 6 Aug. 2257	0.000345	0.0019
Spain (DSS 62)	78	26 July 1739 – 14 Aug. 1914	0.000358	0.0028
California (DSS 12)	58	26 July 0240 – 11 Aug. 0547	0.000027	0.0021
California (DSS 14)	45	27 July 0112 – 13 Aug. 0538	0.000567	0.0026

Expressed in terms of the ratio of the mass of the sun to the mass of Mars, the Mariner 6 determination of the mass is equal to 3098697 ± 72. This should be compared with the only other accurate determination of the mass of Mars, that from Mariner 4 [1]. The Mariner 4 value for the sun/Mars mass ratio is 3098714 ± 5. The two determinations from Mariners 4 and 6 are in excellent agreement when compared with their respective standard errors.

4. A Test of the Theory of General Relativity

According to the Newtonian theory of gravitation, the propagation of electromagnetic waves near massive bodies is unaffected by the gravitational field. However, in the theory of general relativity, the waves appear to be propagated at speeds less than the velocity of light. The exact nature of the effect of the solar gravitational field on such signals has been computed by Muhleman and Reichly [7] and by Shapiro [8] for the case of the Schwarzschild solution of the Einstein field equations. The possibility of a measurement of the general relativistic delay in the propagation of radio signals was first suggested several years ago by Shapiro [9]. The observational reality of the delay was subsequently analyzed very carefully by Ross and Schiff [10].

The magnitude of the propagation delay in range can be expressed in several different coordinate systems. The simplest equation has been obtained in isotropic coordinates by Tausner [11] and Holdridge [12]. In order to parameterize the magnitude of the relativity effect, we have used a general isotropic, static, spherically symmetric metric with two free parameters, β and γ as suggested by Eddington [13] and Robertson [14]. Orbital calculations, time transformations, and propagation effects have been computed within the framework of this generalized metric. It is given by

$$\mathrm{d}s^2 = \left[1 - 2\frac{m}{r} + 2\beta\left(\frac{m}{r}\right)^2 + \cdots\right] c^2\,\mathrm{d}t^2 - \left[1 + 2\gamma\frac{m}{r} + \cdots\right](\mathrm{d}x^2 + \mathrm{d}y^2 + \mathrm{d}z^2)$$

In terms of this metric, the total one-way light time is written in the form

$$c\,\Delta t = \varrho + (1 + \gamma)\, m \ln \frac{r_0 + r_1 + \varrho}{r_0 + r_1 - \varrho}$$

where ϱ, r_0 and r_1 are coordinate distances representing respectively the observer–spacecraft distance, the sun–observer distance, and the sun–spacecraft distance. All distances are computed from orbits (geodesics) defined by the generalized metric. As such, they depend implicitly on both β and γ.

The parameter m is equal to one-half the gravitational radius of the sun and is equal to $m = \mathrm{GM}_{\odot}/c^2 = 1.4766$ km.

The magnitude of the relativistic delay is plotted in Fig. 1 for the actual Mariner 6 trajectory. The maximum effect occurs when the ray path makes its closest approach to the sun on 29 April, at a distance from the center of the sun of about 0.95° or a distance from the limb of a little less than 3 solar radii. At the maximum, the relativistic delay is 200 microseconds in round trip range.

The range measurements available at the time of writing (11 May 1970) are indicated on the curve of the relativistic delay. It is evident that at least to this point in time a very good coverage of the relativistic curve has been obtained.

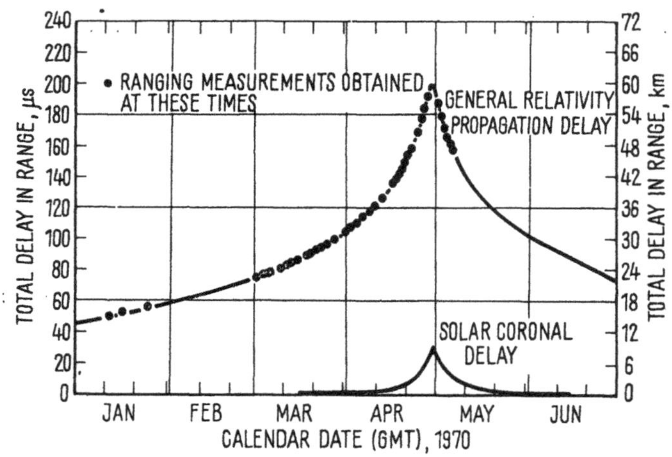

Fig. 1. General relativistic and coronal delay in range for Mariner 6.

4.1. Post-Encounter Trajectories of Mariners 6 and 7

Trajectories have been computed for Mariners 6 and 7 which show the relative geometry of the earth and the spacecraft with respect to the sun. The post-encounter trajectory of Mariner 6 projected on to the plane of the ecliptic is shown in Fig. 2. The trajectory for Mariner 7 is similar. Mariner 6 passed over the north pole of the sun within an angle of about 0.95° on 29 April 1970 at a distance of about 4.02×10^8 km from the earth. Mariner 7 passed within about 1.59° of the sun on 10 May, at a distance of about 3.87×10^8 km from the earth.

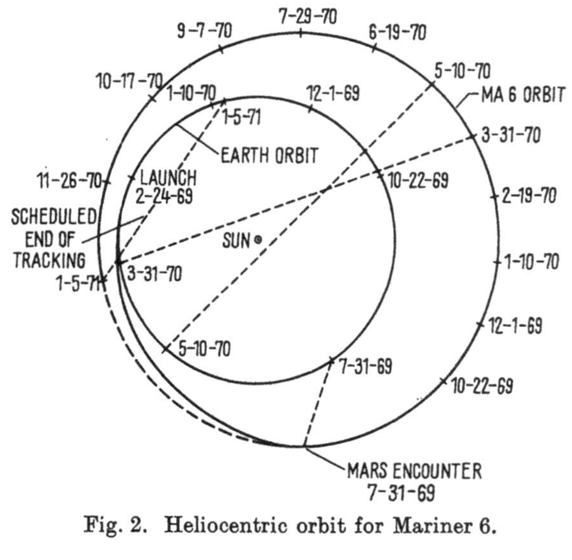

Fig. 2. Heliocentric orbit for Mariner 6.

4.2. Interplanetary and Coronal Plasma Effects

Range and Doppler observations are affected significantly by the electron density in the interplanetary medium and the outer solar corona. However, the associated range delays are small compared with the relativistic delay (see Fig. 1).

Considerable information concerning the electron density profiles is available from solar eclipse measurements made during the last few decades. The intensity of the scattered light seen around the sun's limb during a total eclipse has been used to determine electron density profiles out to distances of about 40 solar radii from the sun [15]. Recent work [16] yields the profile

$$N_e(r) = \frac{A}{r^6} + \frac{B}{r^{2+\varepsilon}}; \qquad 2.5 \leqq r$$

where A and B are electron densities, $\varepsilon \approx 0.3$, and r is expressed in solar radii ($r = 215$ at earth's distance). The value of ε in this equation is not well known but is apparently greater than zero based on spacecraft *in situ* measurements and solar wind theory. The coefficient A apparently varies by a factor of 2 or 3 depending on the solar latitude and the solar activity, i.e. it is a function of the sunspot cycle. The coefficient B is much less variable and is apparently only a weak function of solar activity. A compilation of values of A and B taken from the literature is shown in Table 4. All of the values are from eclipse measurements except line 6. The numerical values of this line were obtained from measurements of ray bending near the sun with radio interferometry [17].

Table 4
Coefficients for Solar Corona Model

A (cm)$^{-3}$	B (cm)$^{-3}$	ε	N_e at 1 AU	State	Ref.
1. 0.96×10^8	—	—	—	Solar minimum, equatorial	[15]
2. 1.17×10^8	—	—	—	Solar maximum, equatorial	[15]
3. 1.18×10^8	1.46×10^6	0.3	6.4	Solar minimum, equatorial (July 1963)	[18]
4. 1.01×10^8	2.01×10^6	0.33	7.4	Average solar minimum, equatorial	[16]
5. 2.62×10^8	2.07×10^6	0.33	7.6	Strong solar maximum equatorial and polar	
6. 0.60×10^8	0.8×10^6	0.33	2.9	Measured 2–14 Oct. 1969, 13° solar latitude	[17]
Expected for Mariner 6 and 7					
1.3×10^8 $\pm 0.9 \times 10^8$	1.15×10^6 $\pm 0.7 \times 10^6$	0.3 ± 0.3			

Also shown in Table 4 are the values of the parameters and their *a priori* uncertainties that we expect to encounter during the Mariner 6 and 7 experiments. The plan of our experiment is to estimate A, B and ε simultaneously with the relativity parameters from the range and Doppler observations. It may be necessary to make a sequential estimate of a series of these parameter sets over the total tracking period of the experiment. However, their values within a month of the solar conjunction are of primary significance to the relativity experiment.

We have performed a theoretical covariance analysis in order to estimate our ability to estimate simultaneously the relativity parameter γ and A and B (ε fixed at 0.3). We conclude from these studies that the plasma effects will not be the major source of error in the relativity determination; they contribute less than 1% to the total expected error. The calculations also indicate that highly significant results will be obtained on the electron density profile of the interplanetary medium for the region from 5 solar radii to the orbit of the earth. We expect that the limiting source of error in the determination of γ will be random variations in the non-gravitational forces acting on the spacecraft.

Acknowledgments

The authors wish to express their appreciation to the Mariner Mars 1969 project team at JPL for the many hours of time devoted to the extended operations of the Mariner 6 and 7 spacecraft. Special thanks are due the project manager, John R. Casani, and also Victor C. Clarke, Jr., who has shown a very real interest in the general relativistic propagation experiment ever since its inception in early 1966. Also, the support given to the experiment, financial and otherwise, by the Office of Space Science and Applications of the National Aeronautics and Space Administration is sincerely appreciated.

References

[1] G. W. NULL, paper presented at the 130th Meeting of the American Astronomical Society, Albany, New York, 1969.
[2] G. W. NULL, paper presented at the meeting of the Division of Planetary Astronomy, American Astronomical Society, Austin, Texas, 1970.
[3] J. D. ANDERSON, Thesis, University of California at Los Angeles (1967); Jet Propulsion Laboratory Technical Report No. 32-816 (1 July 1967).
[4] G. W. NULL, H. J. GORDON and D. A. TITO, Jet Propulsion Laboratory Technical Report No. 32-1108 (1967).
[5] M. E. ASH, I. I. SHAPIRO and W. G. SMITH, personal communication.
[6] J. D. ANDERSON, L. EFRON and S. K. WONG, Science **167**, 277 (1970).
[7] D. O. MUHLEMAN and P. REICHLEY, Jet Propulsion Laboratory Space Program Summary No. 37-29, **4** (1964).
[8] I. I. SHAPIRO, Phys. Rev. Lett. **13**, 789 (1964).
[9] I. I. SHAPIRO, Bull. Astron. **25** (1965).
[10] D. K. ROSS and L. I. SCHIFF, Phys. Rev. **141**, 1215 (1966).
[11] M. J. TAUSNER, MIT Technical Report No. 425 (7 October 1966).
[12] D. B. HOLDRIDGE, Jet Propulsion Laboratory Space Program Summary No. 37-48, **3**, 2 (1967).
[13] A. S. EDDINGTON, The Mathematical Theory of Relativity, Cambridge University Press, Cambridge, England 1960, p. 105).
[14] H. P. ROBERTSON, in: Space Age Astronomy, Academic Press, New York 1962 (p. 228).
[15] H. C. VAN DE HULST, Bull. Astron. Inst. Netherlands **11**, 135 (1950).
[16] D. E. BLACKWELL, D. W. DEWHIRST and M. F. INGHAM, Advances in Astronomy and Astrophysics, **5**, 25 (Academic Press, New York 1967).
[17] D. O. MUHLEMAN, R. D. EKERS and E. B. FOMALONT, Phys. Rev. Lett. **24**, 1377 (1970).
[18] D. E. BLACKWELL and A. D. PETFORD, Mon, Not. R. Astron. Soc. **131**, 383 (1966).

Space Research XI — Akademie-Verlag, Berlin 1971

UTILIZATION OF ZVENIGOROD'S LARGE SATELLITE CAMERA FOR TRACKING DEEP SPACE PROBES

A. Losinsky

Astronomical Council, Academy of Sciences, Moscow, USSR

Zvenigorod's satellite tracking camera has a mirror-lens objective designed by D. Maksutov and M. Sosnina of the Pulkovo Observatory. The main features of the camera are: aperture 500 mm, mirror diameter 1100 mm, F = 700 mm, d/F = 1:1.4. The focal surface is spherical. A field of $5° \times 30°$ can be photographed on a frame 60 mm $\times$ 360 mm. The camera has triaxial equatorial mounting which permits it to be pointed to any part of the sky and to track a deep space probe in any direction with speeds from 0 to 60 arcsec s^{-1}. It is equipped with a rotating sector shutter which operates only for ordinary satellites and is inoperative during tracking of deep space probes, and with a clamshell shutter whose operation is recorded by a chronograph. The accuracy of coordinate determination for slow-moving deep space probes depends mainly on the astrometric properties of the optical elements and distortion of the film. To estimate accuracy five regions of diameter 4° were selected along the long axis of a frame taken in the zenith with an exposure time of 20 seconds. Measurements were made of the positions of 18–20 reference stars in each of these regions and in addition of five control stars in the central part of each region that had not been included among the reference stars. Tha data thus obtained were treated by Terner's six-constant method without corrections excepting for the star's proper motion.

Accuracy determinations were made twice: (i) by using residuals for the reference stars and (ii) from the deviations of the positions of the five central control stars from those in the SAO Catalogue.

The mean square error by the reference stars residuals method is $\pm$ 1.1″, and by deviation of central star positions is $\pm$ 0.9″. If the size of the reference star region is reduced from 4° to 2°, the error diminishes to $\pm$ 0.6″. For deep space probes a sufficient number of reference stars can usually be found still closer to the probe and the effect of film distortion can be reduced still more. An exposure of 20 seconds will yield images of stars of 12–13 magnitude.

Space Research XI — Akademie-Verlag, Berlin 1971

OBSERVATIONS PHOTOGRAPHIQUES D'APOLLO 12

P. MULLER et R. FUTAULLY

Observatoire de Meudon, France

A l'occasion du lancement d'Apollo 12, les stations de l'OHP et de Nice ont obtenu respectivement 12 et 10 clichés dont plusieurs avec deux poses successives. Ceux du 14 novembre montrent des traces ou des images de l'étage S IV B, de petits objets associés et la formation du nuage d'oxygène. Deux clichés obtenus le 15 à l'OHP portent l'image de la capsule Apollo à 200000 km de la Terre.

Des mesures ont pu être faites sur un certain nombre de points par rapport aux étoiles du champ; on donne les positions conclues.

Le début de la trajectoire translunaire d'Apollo 12 était visible en France dans de bonnes conditions. Désireux de répondre à l'appel adressé en 1969 par Ch. E. Lundquist, nous avons demandé à la Bellcomm Inc. et obtenu des prévisions topocentriques très complètes pour nos stations de Meudon, de Nice, de

Fig. 1. Apollo 12, Observatoire de Haute-Provence, 14 novembre 1969, Télescope de Schmidt de 30 cm.
Deux poses successives sur les étoiles (en tout de 2125 à 2129). On constante l'extension progressive du nuage autour de l'objet.

Haute-Provence et de Lyon. Les trois premières pouvaient prendre des photographies, Lyon est engagé dans un programme de photométrie. Le temps n'a pas été favorable à Meudon ni à Lyon; Nice (camera Antares, 300/900 mm) a obtenu 10 clichés divers le 14 novembre entre 2008 et 2139, l'OHP (Schmidt de 300/600 mm) en a réussi 12 dont 10 le même jour entre 2009 et 2200, et deux le 15 à 1908 et à 2009. La série de l'OHP paraît la plus intéressante dans l'ensemble. Certains clichés montrent des objets annexes: ainsi le N° 3169 (de 2119 30 à 2122 30) et surtout le N° 3170 (de 2125 à 2129) (Fig. 1) pendant lequel l'obser-

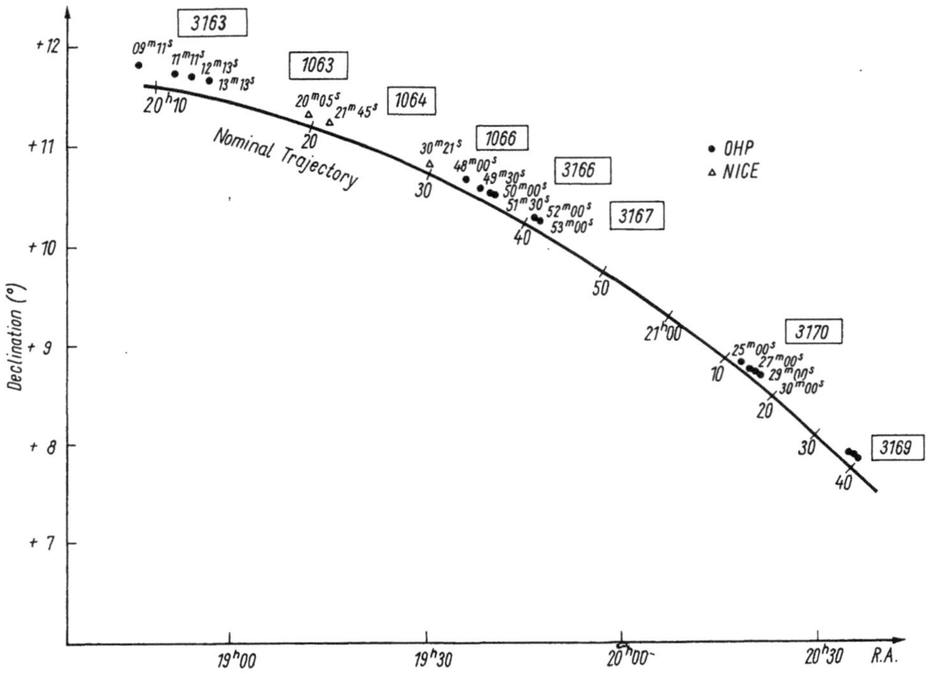

Fig. 2. Apollo 12, 14 novembre 1969 TU.

vation visuelle simultanée montrait jusqu'à six objets brillants accompagnant le nuage d'oxygène déjà bien formé, on suit aussi sur ce même cliché la progression en diamètre de ce dernier.

Nous nous attacherons ici surtout aux positions qui ont pu être déduites de certains de ces clichés, généralement pris dans ce but précis et à cet effet guidés sur les étoiles. Cela a été possible finalement pour dix clichés dont trois de Nice. Portant deux traces ou poses successives, quelques clichés ont fourni plusieurs points. En règle générale, quand une trace est longue on a mesuré les deux extrémités et le centre, et adopté une moyenne générale; pour les poses très courtes (10 s par exemple) on les a traités comme des images ponctuelles. Nous avons appliqué nos méthodes habituelles avec une vingtaine d'étoiles de référence, prises dans le catalogue SAO, la measure étant faite avec notre comparateur Zeiss et la réduction sur l'ordinateur de Meudon.

Voici les positions conclues:

N° du cliché	Date	Station	Temps TU	α1950	δ1950
3163	14. 11. 69	O.H.P.	2010 11	18 48 54.9	+11°45′36″
3163	14	O.H.P.	2012 43	18 55 30.2	11 40 07
1063	14	NICE	2020 05	19 12 14.3	11 19 09
1064	14	NICE	2021 45	19 15 23.9	11 14 43
1066	14	NICE	2030 21	19 30 39.1	10 49 32
3166	14	O.H.P.	2048 45	19 37 32.1	10 37 39
3166	14	O.H.P.	2051 15	19 40 20.6	10 31 46
3167	14	O.H.P.	2052 30	19 47 13.2	10 16 25
3170	14	O.H.P.	2125 00	20 18 21.8	8 50 13
3170	14	O.H.P.	2127 00	20 19 51.9	8 45 22
3170	14	O.H.P.	2129 00	20 20 43.1	8 42 41
3170	14	O.H.P.	2130 00	20 21 07.0	8 41 26
3169	14	O.H.P.	–	20 34 44.7	7 54 01
3169	14	O.H.P.	–	20 35 42.8	7 50 32
3177	15	O.H.P.	1908 00	22 34 11.8	– 1 34 25
3178	15	O.H.P.	2000 00	22 34 06.5	– 1 44 09

Si l'on reporte ces positions sur un même diagramme avec la courbe de l'éphéméride (ici celle de l'OHP, les coordonnées étant très voisines pour Nice), on constate un très bon accord géométrique, mais une discontinuité dans les temps après 2030 qui montre que les positions se réfèrent à deux objets différents, l'un avant cette heure, l'autre après qui doit être l'étage S IV B.

Les positions observées le 15 concernent évidemment Apollo, déjà à 200000 km de la Terre et de magnitude apparente 12 à 13. Sur le cliché 3177 on voit une petite traînée continue, sur le 3178 au contraire quatre petites condensations alignées; dans les deux cas on a pris des positions moyennes.

Planets

Space Research XI — Akademie-Verlag, Berlin 1971

SOME PROBLEMS OF ATMOSPHERIC PHYSICS FOR TERRESTRIAL PLANETS

A. M. Oboukhov and G. S. Golitsyn

Institute of Atmospheric Physics of the Academy of Sciences of the USSR, Moscow, USSR

The methods of investigating planetary atmospheres are discussed for the nearest planets Mars and Venus. Principal attention is paid to the general circulations of their atmospheres. Problems of energetics of the atmospheres are considered, including estimates of mean winds and characteristic temperature differences which drive the circulations. These considerations are carried out by similarity and dimensional arguments as well as by using the results of some numerical modelling experiments. The main features of the atmospheres of Mars and Venus are noted. Some problems are formulated for further theoretical and experimental studies.

Practical studies of the earth's nearest planets, Mars and Venus, require at first hand knowledge of the pressure, density and temperature in their atmospheres. Estimates of wind speeds are important for soft landing on a planet. Also one needs the chemical composition and a knowledge of illumination conditions if photographs are to be taken.

Besides these practical purposes, the studies of the planetary atmospheres have great general scientific value. Comparison of their properties with those of the earth will allow better understanding of the evolution of our own atmosphere and the whole solar system.

We shall consider the atmospheres of Mars and Venus mainly from the point of view of the general circulation and some mean climate characteristics. The cause of an atmospheric circulation is always the same — the differential heating by solar radiation which, through the action of gravity on the inhomogeneous density field, produces buoyancy forces. The intensities of atmospheric motions and temperature distribution are determined by many factors: energy input into an atmosphere, its mass and heat capacity, size of a planet and the angular velocity of rotation, the angle between the axis of rotation and the ecliptic plane, absorbent properties of atmospheric gases, gravity acceleration and so on. All these factors differ sometimes very strongly for various planets. Therefore the study of specific actions of general laws governing the wind and temperature regimes on various planets will provide a better and deeper understanding of those laws for purely earth purposes.

Let us consider shortly first of all the state of the general circulation problem of our own atmosphere. Though there has been tremendous progress in the growth of our knowledge of the general circulation, we are nevertheless still far from a complete understanding of the causes and physical laws governing it and making its definite patterns, as was most clearly demonstrated by Lorenz [1]. Moreover, we still know rather little of some of the most important numerical characteristics of the circulation such as the total kinetic energy of atmospheric motions and the mean rate of generation or dissipation of kinetic energy. We know them only with an accuracy of a factor of the order of 2 or even 3 [1]. What are the methods and the ways of studying of the earth's atmosphere?

The first and basic one is by observations of temperature, wind, pressure, humidity, cloudiness and radiation performed every few hours at several thousands of meteorological stations. At present these observations are quite insufficient since they do not cover even a fifth part of the globe and now we have in full swing preparations for the global atmospheric research programme GARP. During the implementation of this programme wide use of a multi-satellite system is planned with new methods of remote sensing of the atmosphere which will allow the whole atmosphere and the underlying surface to be observed. After the conclusion of this programme meteorologists may hope that they will understand the atmosphere somewhat better.

The second method of study is the numerical simulation of the behaviour of a planetary atmosphere by long-time integration of finite-difference equations describing motions and energy transformations in an atmosphere. Starting from the first numerical experiment made by Phillips [2] this method is becoming more and more a powerful tool of study. Contemporary numerical experiments reproduce well all the main features of the general circulation of the earth's atmosphere: there are two numerical experiments for the simulation of the circulation of the Martian atmosphere [3, 4] and there are in preparation numerical experiments for Venus in the USSR [5] and in the USA [6]. This will apparently be a main method for study of the detailed picture of temperature and velocity fields and for the determination of quantitative characteristics of circulation in planetary atmospheres. However, this method has one very great disadvantage: change of a single atmospheric parameter necessitates the repetition of all the lengthy, time- and money-consuming calculations. And in the study of other planets practically all the basic atmospheric parameters are still very far from being known exactly. Numerical simulation also requires precise specification of the properties of the optically active gases and their concentrations. These properties are often insufficiently known, especially at the high temperatures and pressures existing on Venus. Therefore, it is clear that one has to perform in the case of Venus numerical experiments for several altitude distributions of the direct radiation absorption.

Another method is to attempt to find some simple approximate quantitative regularities governing the motion of planetary atmospheres. These attempts may be based on energetic and thermodynamic considerations and on the use of dimensional analysis and similarity theory. In this way one could obtain only estimates of some average characteristics of an atmospheric circulation but not the patterns of velocity or temperature fields. These we know sometimes from observations or can guess them from experience in geophysical fluid dynamics.

Quite simple though very approximate estimates were suggested in [7]. They are based on the following formula for the rate of generation (or dissipation) of kinetic energy on unit mass

$$\varepsilon = \mathrm{k}\,\frac{\delta T}{T_1}\,\frac{q_A}{4\,M} \tag{1}$$

Here $q_A = q(1 - A)$, q is the solar constant for a planet, A is its albedo, M is the mass of an atmospheric unit column, T_1 is a characteristic temperature of the most heated parts of an atmosphere, δT is a characteristic temperature difference causing atmospheric circulation, k is a numerical factor of the order of 0.1 for the earth's atmosphere [7]. The quantity $\eta = k\,\delta T/T_1$ plays the role of the efficiency of an atmosphere as a heat engine transforming the solar energy into kinetic energy of atmospheric motion. As a characteristic scale L of synoptic motions the Oboukhov scale $L_0 = (g H)^{1/2}/l \approx c/l$ was used in [7] for the earth and Mars and the radius r for Venus. Here g is the gravity acceleration, H is the atmospheric scale height, l is the Coriolis parameter, c is the velocity of sound. Expressions were assumed for: velocities $U \approx (\varepsilon L)^{1/3}$, time $\tau_U \approx L/U$ and large-scale eddy mixing coefficients $K \approx 0.1\, \varepsilon^{1/3} L^{4/3}$ (Richardson–Oboukhov law). The value $k = 0.1$ was used for all the planets (the estimated values are proportional only to $k^{1/3}$!), δT and T_1 were taken from observations [8]. The estimates obtained agree with observations for the earth, and for Mars with the results of numerical experiment [3].

But in any closed theory of the general circulation the values of δT and T_1 must be determined by the theory itself. They could be estimated by the following (for more details see [9, 10]):

$$T_1 = T_e/\alpha \approx (q_A/4\,\sigma)^{1/4} \tag{2}$$

where σ is Stefan's constant, $\alpha \lesssim 1$. For the, not very thick, atmospheres of the earth and Mars $\alpha \approx 1$, for Venus $\alpha \approx 1/3$.

The heat balance equation in an atmosphere averaged over height may be written in the following simplest way:

$$\mathrm{M} c_p\, U_i \frac{\partial T}{\partial X_i} \approx \sigma\, T_e^4 \,. \tag{3}$$

Such an equation is justifiable for sufficiently thick atmospheres where there is no local radiative equilibrium; the temperature regime is determined mainly by the dynamical heat advection and the temperature distribution on the planetary disk differs only slightly from the equilibrium temperature T_e. The conditions for such a regime are considered in detail in [11], which we shall discuss later. Approximating $U_i\, \delta T/\delta X_i$ by $(\varepsilon r)^{1/3}\, \delta T/r$ we may get for a slowly rotating planet, taking into account (1) and (2),

$$\delta T \approx \frac{(q_A/4)^{9/16}}{k^{1/4}\,\sigma^{1/8}\,c_p^{3/4}} \left(\frac{r}{M}\right)^{1/2} . \tag{4}$$

Substituting this into (2) and (1) we may get estimates for ε, U, τ_U and so on through, so to speak, the "external parameters". We write here the expression for the total atmospheric kinetic energy

$$E = \frac{1}{2} M \cdot 4\,\pi\, r^2\, U^2 \approx 2\,\pi\, r^2\, M(\varepsilon\, r)^{2/3} \approx 2\,\pi\, K^{1/2} \frac{\sigma^{1/8}}{c_p^{1/2}} \left(\frac{q_A}{4}\right)^{7/8} r^3 \tag{5}$$

The simple structure of this formula, particularly its independence of mass M, suggests the use of dimensional and similarity considerations for the description of the general circulation.

Physical assumptions which form the base of such an approach could be formulated as follows: The intensity of the circulation and its driving temperature difference are interconnected and self-consistent and are determined by the energy input to the atmosphere, by its size, mass and heat capacity. The heat balance is established by long-wave radiation to space.

Let us consider first a slowly rotating planet. Then the determining parameters essential for the description of the circulation are the surface-averaged energy input $q_A/4$ [g s^{-3}], the heat capacity per unit mass of the atmosphere C_p [cm^2 s^{-2} deg^{-1}], the mass of a unit column of the atmosphere M [g cm^{-2}], the gravity acceleration g [cm s^{-2}], the radius of a planet r [cm] and Stefan's constant $\sigma = 5.67 \times 10^{-5}$ g s^{-3} deg^{-4}. Among these six parameters only four have independent dimensions therefore two non-dimensional quantities can be formed

$$\mathfrak{H} = \frac{c_p (q_A/4\,\sigma)^{1/4}}{g\, r} \approx \frac{H}{r}$$
$$\mathfrak{M} = \frac{\sigma^{3/8}}{C_p{}^{3/2}} \left(\frac{q_A}{4}\right)^{5/8} \frac{r}{M} \tag{6}$$

The first has a clear meaning. It is the ratio of the atmospheric scale height to the planetary radius. Note that always $\mathfrak{H} \ll 1$ if g is sufficiently large, which is usual. Therefore the exact value of g is not essential at the first approximation and it could be excluded from the number of determining parameters. This is a kind of self-similarity for circulation regimes with respect to g. Its meaning is as follows: if g is changed by some factor the circulation regime will remain the same to the first approximation if we have still $\mathfrak{H} \ll 1$.

The meaning of the second quantity $\mathfrak{M}$ will be discussed later; here we calculate its value for the terrestrial planets. For the earth $\mathfrak{M} \approx 10^{-3}$, for Venus $\mathfrak{M} \approx 10^{-5}$, for Mars $\mathfrak{M} \approx 3 \times 10^{-2}$. We see that this similarity parameter is also small, i.e. there should not be to the first approximation a dependence on its exact value, at least, for some characteristics of the general circulation. The smallness of $\mathfrak{M}$ is equivalent to the largeness of M, i.e. if $\mathfrak{M} \ll 1$, the exact value of M is not essential and it can be excluded from the set of determining parameters. These two self-similarity hypotheses may be verified by comparison of the conclusions reached by using them with observations or results of numerical experiment.

From the remaining four parameters T_e, velocity, time and energy we can form combinations with the dimensions of temperature

$$w^{1/2} = C_p{}^{1/2} (q_A/4\,\sigma)^{1/8} = (C_p\, T_e)^{1/2} = (\varkappa - 1)^{1/2}\, C_e, \tag{7}$$

$$\tau_e = r/C_e\,, \tag{8}$$

$$E = B\, \sigma^{1/8}\, C_p{}^{-1/2}\, (q_A/4)^{7/8}\, r^3 = \frac{B}{4\,\pi} (\varkappa - 1)^{1/2} - Q_A\, \tau_e\,, \tag{9}$$

where w is the enthalpy or sensible heat per unit mass, B is a numerical coefficient, Q_A is the total energy input into an atmosphere. Within the adopted assumptions the coefficient B should be a universal one. However, it is difficult to expect everywhere their exact fulfilment, therefore one may hope only for the order of the value estimated. Also, B can depend on such dimensionless parameters as $\varkappa = C_p C_v$, the angle of declination of the polar axis to the ecliptic plane, i.e. on the illumination conditions or season and so on. We see that (9) coincides with (5) if

$$B \approx 2\pi k^{1/2} \tag{10}$$

i.e. (9) expresses the total mean kinetic energy of a planetary atmosphere. In paper [9] and especially [10] is given a detailed discussion of this formula and many of its consequences in application to the earth, Mars and Venus. In [10] there are also possible modifications of the theory for the cases of very rarefied atmospheres and rapidly rotating planets. Note that if we add ω [s^{-1}] then we have a new non-dimensional parameter

$$\lambda = \omega\, r / C_p^{12} (q_A / 4\sigma/)^{18} \approx \omega\, r / C_e \approx r / L_o \tag{11}$$

which is of the order of unity for the Earth and Mars (and 10^{-2} for Venus). The constant B should be a function of λ.

It is shown in [9, 10] that the parameter $\mathfrak{M}$ is of the order of the squared Mach number, therefore there is no wonder that it is small. A parameter, equivalent to $\mathfrak{M}$ is also appearing in [11] though in a somewhat different role. It is shown there that the small $\mathfrak{M}$ means that the temperature regime in the atmosphere is completely determined by large-scale dynamics and there is no local radiative equilibrium. The case of $\mathfrak{M} \gg 1$ corresponds to establishing local radiative equilibrium which determines the temperature distribution on the planet's surface. Large-scale motions do not play there any appreciable role and are determined by temperature distribution. Let us make now a short survey of various planets.

Mars

The surface pressure is about 5–8 mb, the atmosphere consists mainly of CO_2; no other gases are found in appreciable quantities. Water vapour was found but its content shows diurnal and seasonal variations from 0 to 30 μm of condensed water per cm^2 [12]. The planetary albedo $A \approx 0.2$, therefore the quantity $q_A/4$ is about half that for the earth. Two numerical experiments performed for $p_S = 5$ mb [3] and 7.5 mb [4] give a fairly full picture of the general circulation and its seasonal variations. We shall list the main conclusions.

The first basic feature of the Martian meteorology caused by the small mass of the atmosphere and insufficient "greenhouse" properties of CO_2 is that the surface of the planet and the lower part of its atmosphere are quite close to the local radiative equilibrium state. The large scale advection cannot transfer enough heat to change appreciably the temperature distribution determined by radiation. The main part of the incident solar radiation is radiated back to space at the same place where it is received. As a result the efficiency of the Martian atmosphere is an order of magnitude less than that of the earth's atmosphere.

The inevitable consequence of all of this is the second basic feature, that in the polar regions the carbon dioxide atmosphere has to condense. At $p_S = 5$–8 mb the CO_2 condensation temperature is 144–147 °K. The latent heat released does not allow the atmosphere to cool further. If there were no condensation the temperature in the winter polar region would drop down to something like 50 °K as can be estimated from the formula analogous to (4) (see [10]), if according to [4] one determines that $K \approx 10^{-2}$, i.e. an order of magnitude less than on earth. According to Leighton and Murray [13] who were the first to propose the idea of CO_2 condensation on polar caps, up to 25% of the whole mass of the atmosphere may go into the polar cap during the winter. At the equinox periods the condensation rate is an order of magnitude less than during solstice.

Infrared radiometric measurements by Mariner 7 give a surface temperature of about 150 °K in the polar cap region [14] which is very close to the CO_2 condensation temperature. One may hope that further analysis and new data will specify this value.

Due to condensation there is a large meridional mass flux to the winter pole. It is accompanied by the near eastward Coriolis torque which produces strong zonal mean west winds with velocities 40–70 m s^{-1}. In the summer hemisphere and near the equator there are moderate and weak east winds, and on the equator winds up to 20 m s^{-1}. There are wave cyclones in the winter hemisphere. During the equinox period the circulation is more like that in the earth's atmosphere. In both hemispheres there are west winds (10–20 m s^{-1}) in the latitudes with wave cyclones in the middle and higher latitudes, and east winds (5–10 m s^{-1}) in the tropics [4]. There are large diurnal variations of wind and temperature.

It is interesting to note a strong seasonal change of the circulation of kinetic energy [4]; for the solstice the mean kinetic energy per unit mass $\frac{1}{2} U^2 \approx 500$–$560$ m^2 s^{-2}, while for the equinox it is 160–180 m^2 s^{-2}. This can be related with the noticeable decrease of temperature meridional gradients. Note that in the experiment [3] with $p_S = 5$ mb there was for the solstice $\frac{1}{2} U^2 = 650$–800 m^2 s^{-2}, i.e. a hypothesis on the independence of mass of the total kinetic energy [9, 10] is verified by computation [3, 4].

Of the theoretical problems of Martian meteorology there is unclear the question of the role of the melting of the spring polar cap which requires much heat, in the character of temperature and dynamical regime of the atmosphere. This should introduce a certain asymmetry in the regimes of the spring and autumn hemispheres. The problem can be clarified only by the numerical simulation of the whole year's climate cycle.

The complexity of the Martian surface topography revealed by radar methods and optical observations of CO_2 content above certain regions of the planet may have an essential influence on the wind and temperature patterns in various Martian regions.

An appreciable role in the radiative and temperature regimes of the atmosphere may be played by aerosols. Its layers are clearly seen on many photographs of the planetary limb made by Mariners 6 and 7.

Among other important problems of the physics of the Martian atmosphere one can note the exact determination of the atmospheric composition which could be made by a mass spectrometer on a soft landed automatic station.

Venus

The measurements of the Soviet automatic stations Venera 4, 5 and 6 have established that the atmosphere of the planet consists almost entirely of CO_2, the surface pressure might be of the order of a hundred atmospheres and the surface temperature is about 750 °K. Due to the great thickness of the atmosphere it cannot cool noticeably even during the long two-month night; this is confirmed by the almost complete absence of the phase effect in the temperature of radio-emission by the planet. Completed numerical experiments on the general circulation of the atmosphere of Venus are still lacking and ideas on the winds and temperature distribution may be obtained only from estimates based on the similarity theory [9, 10]. These estimates indicate that the mean atmospheric winds should be of the order of 1 m s^{-1} and $\delta T \approx 2$ °K. The last estimate agrees with the results of radio interferometric measurements [15], which showed no temperature difference within an error limit of 15° or even less.

The parameter $\mathfrak{M} \approx 10^{-5}$ for Venus, i.e. the temperature regime is fully controlled by large-scale dynamics [11] which perform an effective mixing of the atmosphere in the horizontal as well as in the vertical direction. The last should be by followed a vertical adiabatic temperature gradient which explains the existence of high surface temperature on the planet.

The theory of similarity alone cannot say what kind of the circulation exists on Venus [10]. The numerical simulation of the three-dimensional flow may answer this question. In this respect it could be very useful to make photographs of the cloudiness from sufficiently close to the planet, with various resolutions and filters.

The most difficult problem for Venus is now, without doubt, the nature and composition of its clouds. From the measurements of water vapour content made by the Venera stations it would seem that the clouds should be of liquid water at the lower levels and of ice particles in the upper part [16]. However, spectroscopic determinations of the H_2O content give concentrations two or even four orders of magnitude less. Polarization measurements [17] suggest spherical particles of the order of 1 μm in diameter with refractive index $n = 1.45$–1.50 which excludes water or ice. There are purely chemical difficulties with the water [18]; if there were much water then we could not observe HCl in the quantities existing there. Other proposed compounds like $FeCl_2 \cdot 2\,H_2O$ or mercurous halides also introduce various chemical or observational difficulties. One should not exclude the possibility that there are several cloud layers on Venus of various natures. The problem could be resolved by direct mass-spectrometric or other measurements within the atmosphere of the planet during the descent of an automatic station.

There is another problem related to this one which is important for understanding the circulation pattern of the Venus atmosphere and its illumination regime. Where is the incident solar radiation absorbed — on the solid surface of the planet, in the volume of its atmosphere or somewhere in the very uppermost layers of it? Laboratory modelling of optical paths several kilometres long at high pressure and temperature is impracticable, the precision of the quantum mechanics calculations and their relation to reality will be always questionable. Therefore, an answer can be obtained only by direct measurements of illumination during the descent of an automatic station on the day side of the planet.

Of the, so to speak, micrometeorological problems a study of turbulence, i.e. the velocity and temperature fluctuations, is of large practical and scientific importance. Estimates of possible orders of magnitude of these fluctuations are given in [10]. The temperature fluctuations cause fluctuations of the refractive index of electromagnetic waves, which appear in the fluctuations of amplitude (and phase) of the radio signals passing through the atmosphere of the planet. Such amplitude fluctuations were recorded for the signals of Venera 4 [19] and Mariner 5 [20]. Direct measurements of velocity and temperature fluctuations within the atmosphere would be highly desirable.

References

[1] E. N. LORENZ, The Nature and Theory of the General Circulation of the Atmosphere, WMO Publ. (1967).
[2] N. A. PHILLIPS, Quart. J. Roy. Met. Soc. **82**, 153 (1956).
[3] C. LEOVY and Y. MINTZ, RAND Corp. RM-5110-NASA (1966).
[4] C. LEOVY and Y. MINTZ, J. Atmos. Sci. **26**, 1167 (1969).
[5] S. S. ZILITINKEVICH, Izv. Akad. Nauk SSSR, Ser. Fiz., Fiz. Atmos. Okeana **5**, No. 11 (1969).
[6] J. CHARNEY, Contribution to 4th Arizona Conf. on Planetary Atmospheres, Tucson 1970.
[7] G. S. GOLITSYN, Izv. Akad. Nauk SSSR, Ser. Fiz., Atmos. Okeana **4**, 1132 (1968).
[8] V. I. MOROZ, Physics of the Planets, Nauka, Moscow 1967.
[9] G. S. GOLITSYN, Dokl. Akad. Nauk SSSR **190**, 323 (1970).
[10] G. S. GOLITSYN, Paper presented at 4th Arizona Conf. on Planetary Atmospheres, Tucson 1970 see also Icarus **13**, 1 (1970).
[11] P. GIERASCH, R. GOODY and P. STONE, J. Geophys. Fluid Dynamics **1**, No. 1 (1970).
[12] R. A. SCHORN, Paper presented at 40th IAU Symp. on Planetary Atmospheres, Marfa, Texas 1969.
[13] R. B. LEIGHTON and B. C. MURRAY, Science **153**, 136 (1966).
[14] G. C. PIMENTEL, Paper presented at 40th IAU Symp. on Planetary Atmospheres, Marfa, Texas 1969.
[15] D. MORRISON and J. POLLACK, Paper presented at 40th IAU Symp. on Planetary Atmospheres, Marfa, Texas 1969.
[16] A. M. OBOUKHOV and G. S. GOLITSYN, Kosm. Issled. **6**, 759 (1968).
[17] J. HANSEN, Contribution to 4th Arizona Conf. on Planetary Atmospheres, Tucson 1970.
[18] J. S. LEWIS, Paper presented at 4th Arizona Conf. on Planetary Atmospheres, Tucson 1970.
[19] M. A. KOLOSOV, O. I. YAKOVLEV and A. I. YEFIMOV, Dokl. Akad. Nauk SSSR **182**, 93 (1968).
[20] A. S. GURVICH, Izv. Akad. Nauk SSSR, Ser. Fiz., Fiz. Atmos. Okeana **5**, 1172 (1969).

Space Research XI — Akademie-Verlag, Berlin 1971

CHEMICAL COMPOSITION OF THE VENUS ATMOSPHERE

A. P. Vinogradov, Yu. A. Surkov, B. M. Andreichikov,
O. M. Kalinkina and I. M. Grechishcheva

V. I. Vernadsky Institute of Geochemistry and Analytical Chemistry, Moscow, USSR

Results of experiments on the determination of the chemical composition of the Venus atmosphere carried out on the automatic stations Venera 4, 5 and 6 are considered.

For the determination of the chemical composition gas analysers were installed on each of these automatic stations. Measurements were made in the range of pressure from 0.6 to 10 atm and in the range of temperatures from 25 to 220 °C.

As a result of experiment the CO_2, H_2O, O_2 and N_2 content (together with that of the inert gases) has been determined, as well as the high-altitude distribution of water vapour density.

The data obtained on the atmosphere of Venus are presented, with a discussion on the composition and intermixing of the atmosphere and the structure of the cloud layer, and consideration of the origin and evolution of the atmosphere.

1. Introduction

In earlier communications [1, 2] we briefly gave data on the composition of the Venus atmosphere according to measurements from the interplanetary station Venera 4. In the present paper we give data which are partly already known [3], having been obtained from the interplanetary stations Venera 5 and 6, and consider the composition of the atmosphere on the basis of all measurements.

Before the launching of the automatic station Venera 4 there was extremely contradictory information about the chemical composition of the Venus atmosphere. As is well known, this information was obtained by investigators either from spectral observations from the earth, or by the various calculations, comparisons and model constructions of the Venus atmosphere [4].

Data on the content of CO_2, H_2O, O_2, N_2 and other gases gave a very wide range of values. Therefore the first experiment on station Venera 4 was limited to threshold determinations of these components. The results of measurements were considered at the symposium on the atmosphere of planets and have been cited in a number of papers [5].

Using data obtained on Venera 4, it was possible during further investigations on the stations Venera 5 and 6 to carry out the determination of the content of these gases more precisely, in narrower ranges of the CO_2, H_2O, N_2, O_2 concentrations found earlier. This could be achieved thanks to the fact that the optimum operating regimes of the devices were selected in accordance with the assumed conditions of their operation in the planet's atmosphere which to a considerable extent was realized in the experiments.

2. Organization of Experiments

In June 1967 the automatic station Venera 4 was launched, which four months later for the first time accomplished a soft descent into the Venus atmosphere and carried out an investigation of the physico-chemical and structural parameters of the atmosphere. The temperature, pressure and composition of the atmosphere were measured. During the descent of the station the temperature and pressure changed from 0.7 atm and 25 °C to 18 atm and 280 °C. As a result of the experiment carried out there were obtained for the first time data of direct measurements in the atmosphere of the planet which made it possible to narrow considerably the range of various assumptions and to create a definite idea about the Venus atmosphere, its origin and evolution [1, 2].

At the same time, the uncertainty in the initial data on the parameters being measured, which, as has already been mentioned, existed before the flight of the station Venera 4, led to the fact that on Venera 4 there was installed wide-range apparatus which relatively roughly, but reliably, determined the chief parameters of the atmosphere.

In January 1969 two more stations were sent to Venus: Venera 5 and 6. These stations were launched with the aim to carry out further and more detailed investigations of the planet's atmosphere. As is known, these stations penetrated deeper into the Venus atmosphere. Measurements were also carried out in the area of the parachute descents, the pressure and temperature having changed from 0.6 atm and 25 °C to 27 atm and 320 °C. With the aid of these stations, besides the investigation of temperature, pressure and composition, the depth of the atmosphere was also determined.

As has been proved, the course of temperature change with height approaches the adiabatic over the whole range of measurements. If we assumed that the temperature up to the surface changes adiabatically, then at the surface level determined by the altimeter of Venera 6 the temperature and pressure must be $\sim$ 440 °C and $\sim$ 60 atm and at the surface level determined by the altimeter of Venera 5 $\sim$ 550 °C and $\sim$ 140 atm [6].

These differences in the temperature and pressure in the area of descent of the stations Venera 5 and 6 follow from the differences (about 14 km) of the surface levels measured by the altimeters of the stations and may, apparently, be explained by the great unevenness of the planet's surface relief.

Since we have results of measurements made at two points we may speak of a certain average surface level to which must correspond a pressure of about 100 atm and a temperature of 500 °C (this level does not contradict the data of radioastronomic measurements [7]). At the same time it will doubtless become more precise as data on other regions are obtained.

For the investigation of the Venus atmosphere on all three stations special analysers were installed which determined the content of carbon dioxide, oxygen, water and nitrogen, together with the inert gases. The general appearance of the gas analyser is shown in Fig. 1. On each station two devices were working at definite heights (pressures). The command for device operation was given by the programme provisional control. The weight of each device is about 1 kg and the electroenergy consumption about 30 W during 10 seconds.

During the work of all stations in the Venus atmosphere the content of carbon dioxide, oxygen, water, nitrogen (together with the inert gases) was determined several times; information from each device was transmitted many times.

Fig. 1. General view of the gas analyser.

3. Methods of Measurements and Results

The determination of the content of the chief components of the atmosphere was by the manometric method which is distinguished by its simplicity and reliability. According to their construction pressure sensors of the membrane type measure the absolute pressure or the difference of pressures between two volumes in the case of absorption of a definite component. For each component of the atmosphere a corresponding absorber was selected. By this method the content of carbon dioxide, nitrogen, oxygen and water vapour was determined. In addition, an electrolytic method was used in the devices for the determination of small amounts of moisture and a thermochemical method for determination of oxygen content. Measurements of some atmospheric components were duplicated to increase the reliability of results.

The selection of chemical absorbers was limited by the following conditions: Absorbers with sufficient completeness and selectivity must secure complete gas adsorption for a definitely established time interval. At the same time the rate of adsorption reaction must not be too high in order to avoid appreciable distortions in the gas composition during gas sampling before closing the device.

The reagents must be stable over a sufficient temperature range determined by technical conditions of storage and operation of the device and must meet the indicated requirements of reaction rate over this whole range. The reagents must not interact with one another and must not change their properties when stored for a long time under vacuum. During the reaction process no gaseous products must arise which interfere with the determination of the given component in this volume. The reagents must stand substantial overloads (which makes the use of liquid and pasty materials difficult).

3.1. Determination of Carbon Dioxide

During the selection of a carbon dioxide absorber the absorbing properties of askarite, natroncalk, caustic soda, caustic potash and calcium hydroxide

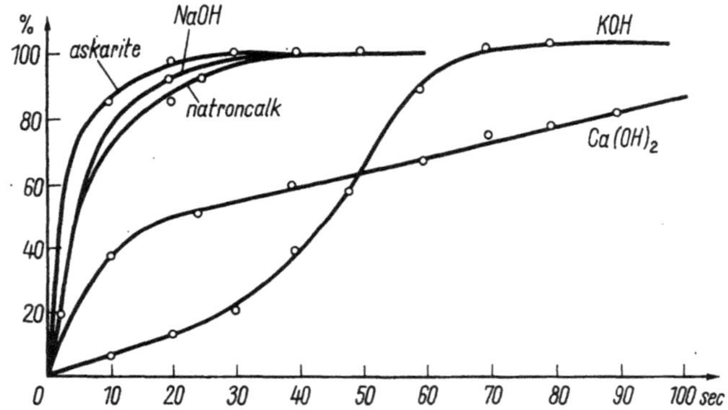

Fig. 2. CO_2 absorption by various absorbers.

were compared. As is known, when using these absorbers, acid smokes (HF, HCl) will also be absorbed. However, the HF and HCl detected in the Venus atmosphere make up respectively not more than 5×10^{-9} and 6×10^{-7} of the CO_2 content, i.e. they are by 6–8 orders lower, and, consequently, may be neglected.

In Fig. 2 the dependence on time of carbon dioxide absorption by various absorbers is shown. It is seen from the figure that calcium hydroxide is not applicable because of the low reaction rate. Caustic potash has a distinctly pronounced period of reaction delay depending on the dampness of the reagent. The adsorptive properties of askarite, caustic soda, natroncalk are almost equivalent. For their use a minimum sampling time is required. The most suitable absorber proved under our conditions to be caustic potash. A lithium absorber of carbon dioxide (about 3%) was added to caustic potash in order to eliminate the dependence of reaction rate on the degree of drying. The degree of drying did not influence the course of carbon dioxide absorption and complete absorption was reached within 35–40 seconds.

The content of carbon dioxide in the Venus atmosphere was determined from the difference of pressures between the cell compartments into one of which was placed the carbon dioxide absorber. In order to avoid any error due to caustic soda absorption by water vapour, calcium chloride was placed in the other compartment to absorb the moisture. The pressure difference was recorded by a sensor of the membrane type as indicated above. Results of carbon dioxide determination by this method are listed in Table 1.

Table 1
Results of CO_2 determination in the Venus atmosphere

No.	Pressure ($kg\ cm^{-2}$)	Temperature (°C)	CO_2 (vol. %)	Station
1.	0.6	~25	97 ± 4	Venera 5
2.	0.7	~25	90 ± 10	Venera 4
3.	2.0	85	>56	Venera 6
4.	5.0	150	>60	Venera 5
5.	10.0	220	>30	Venera 6

As is seen from the table, up to 97 ± 4% of carbon dioxide is contained in the planet's atmosphere. The large error in the determination of carbon dioxide by Venera 4 (90 ± 10%) occurs because the general pressure was not measured at the moment of sampling and has been determined by extrapolation.

In order to reduce the error of determination in the devices on Venera 5 and 6 measurement of general pressure directly at the point of sampling for analysis was provided for. This reduced the error of determination by up to 4%.

Some devices operated at a pressure substantially greater than the ranges assumed for the operation of pressure sensors; therefore threshold values of carbon dioxide content have been obtained which do not contradict the quantitative determinations. Reduction of threshold determination of carbon dioxide with height is associated with a pressure increase, the sensitivity being the same.

Thus, according to direct quantitative determinations the average carbon dioxide content in the Venus atmosphere may be accepted to be 97 ± 4%.

3.2. Determination of Elementary Nitrogen

Nitrogen was determined by two methods: on Venera 4 by the difference of pressures between two compartments, in one of which were absorbed all active gases together with nitrogen and, in the other, carbon dioxide, oxygen and water vapour; on Venera 5 and 6 from the residual pressure after absorption of the chief atmospheric components CO_2, H_2O and O_2.

For nitrogen absorption its ability to form nitrides is usually utilized. When metals easily forming stable nitrides were considered, zirconium proved to be the most suitable. It is sufficiently stable under normal conditions and, when heated, interacts well with various gases. The optimum temperature of nitrogen absorption by zirconium is 800 °C.

As a heater of zirconium a quartz capillary was used, inside which a tungsten spiral is placed for heating.

In order to obtain a sufficiently firm absorbing layer on the quartz capillary, zirconium powder (grain size 1–5 μm) was cemented by a 20% solution of sodium silicate. For acceleration and intensification of the reaction a basic mixture was used. Best results were obtained when using aluminium and barium peroxide in a 2:1 ratio. The weight of the spark mixture was 1/15 of the zirconium absorber weight.

Verification has shown that the presence of small amounts of carbon dioxide, oxygen and water vapour does not influence the work of the zirconium absorber. As oxygen, carbon dioxide and water vapour are absorbed simultaneously with nitrogen, absorbers were placed into the comparison chamber of the gas-analytical cell: for carbon dioxide and water caustic potash was used and for oxygen, phosphorus.

By this method the nitrogen content was determined only on the first station Venera 4 as initially a high nitrogen content in the planet's atmosphere was assumed: at that time there were no more precise data on nitrogen. Measurements on Venera 4 have shown that there is little nitrogen [1, 2]. Therefore in the following experiments (on Venera 5 and 6) the nitrogen content (together with possibly existing inert gases) was determined from the residual pressure after absorption of carbon dioxide, oxygen and water vapour. This method was selected because zirconium absorbers in mixtures with high carbon dioxide content do not have sufficient stability.

Results of all nitrogen determinations in the Venus atmosphere are listed in Table 2. As is seen from the table, nitrogen has been determined eight times

Table 2
Results of N_2 determination in the Venus atmosphere
(together with possibly existing inert gases)

No.	Pressure (kg cm^{-2})	Temperature (°C)	N_2 (vol. %)	Station
1.	0.6	~25	<3.5	Venera 5
2.	0.7	~25	<7*	Venera 4
3.	2	85	<2.5*	Venera 4
4.	2	85	<9.5	Venera 6
5.	5	150	<4	Venera 5
6.	5	150	<4	Venera 5
7.	10	220	<2.5	Venera 6
8.	10	220	<2	Venera 6

* In these measurements nitrogen was determined from zirconium absorption; in the others from residual pressure.

and all measurements show its low content. As the error of determination is possible only at the expense of incomplete adsorption, the data may be considered as reliable upper limits of the nitrogen content.

Thus, the nitrogen content in the Venus atmosphere may be accepted as not higher than 2%. At the same time it has to be noted that the amount of inert gases, chiefly argon, is apparently very little. In the atmosphere of the earth the content of argon, the most abundant of the inert gases, makes up about 1% of the nitrogen content.

3.3. Determination of Oxygen

As mentioned above, two methods were used for oxygen determination: the manometric and the thermochemical.

The simplest and best studied oxygen absorber is white phosphorus. The curves of oxygen absorption by white phosphorus on pumice are shown on Fig. 3. As is seen from the figure, the reaction of adsorption by phosphorus depends strongly on temperature and at low temperatures does not proceed sufficiently rapidly. This circumstance forced us to find a means of heating it.

As absorber red phosphorus was used which during sublimation passes into the white modification, intensely absorbing oxygen. This ensures inertness of the reagent during its storage and independence of the reaction rate on the

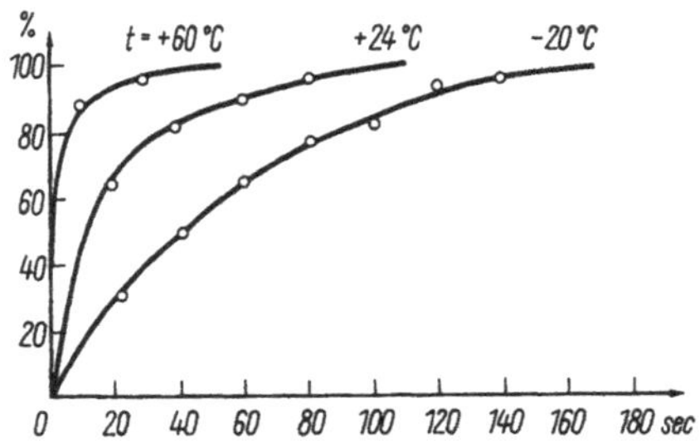

Fig. 3. O_2 absorption by white phosphorus on pumice, at various temperatures.

temperature of the environment. In order to secure mechanical strength of the absorber a bond was used (solution of potassium bichromate and sodium silicate).

This method was used only during the first experiment on Venera 4, when the possibility of a high oxygen content in the planet's atmosphere was not excluded. However, Venera 4 has already shown that the oxygen content in the Venus atmosphere is low. Therefore on the following stations only the thermochemical method was used which is intended for threshold determination of small amounts of oxygen.

In view of the great chemical activity of oxygen a whole number of solid substances may be selected which react with it during heating. Of greatest interest are the refractory metals for which there exists a large gap between the temperatures of oxidation and of fusion. For oxygen determination they may be used in the form of a thin wire (filament or coil). The resistance of the glowing wire increases during interaction with oxygen owing to temperature increase as the result of a substantial thermal effect or the reaction and to the decrease of the diameter of the conductor during oxidation.

Furthermore, a refractory wire can be used to achieve heating. Tungsten proved the most suitable metal: its interaction with oxygen in the 700–1000 °C range shows a sufficiently selective character. As is known, at these temperatures in pure nitrogen and carbon dioxide no burning out of the tungsten coil occurs. Addition of a small amount of oxygen (partial pressure of some mmHg) leads to sensor operation for the required time interval. For oxygen determination we used a tungsten wire with a diameter of 12 μm.

Results of measurements of oxygen content in the Venus atmosphere are listed in Table 3. As is seen from the table, oxygen was determined seven times. All measurements except one show that only very small oxygen amounts are possible in the Venus atmosphere. One measurement (more than 0.4%), obtained on Venera 4 is possibly connected with less precise determinations compared with the following ones (Venera 5 and 6).

Table 3
Results of O_2 determination in the Venus atomsphere

No.	Pressure (kg cm^{-2})	Temperature (°C)	O_2 (vol. %)	Station
1.	0.7	~25	≧0.4	Venera 4
2.	2.0	85	≦1.5*	Venera 4
3.	2.0	85	≦0.3	Venera 6
4.	5.0	150	≧0.1	Venera 5
5.	5.0	150	≦0.1	Venera 5
6.	10	220	≦0.1	Venera 6
7.	10	220	≦0.2	Venera 6

* This measurement has been obtained by the manometric method, the others by the thermochemical method.

Thus, the oxygen content in the Venus atmosphere does not exceed 0.1%. At the same time this content is at the lower limit of detection of our devices. Therefore there was no possibility of carrying out determinations of still smaller amounts of O_2.

3.4. Determination of the Moisture

For moisture determination two methods were used: the manometric for large amounts, and the electrolytic for small amounts.

In the case of the manometric method the absorber was selected after a comparison of their drying efficiency: P_2O_5, anhydrous KOH, $CaCl_2$ etc. In the final construction of the device $CaCl_2$ was used as a drier; this has the advantage of its neutrality with respect to the other gaseous components and to their absorbers. The pure calcium chloride was first melted, finely crushed and screened without access of moisture. When the screened portion, of 1–3 mm grain size, was used complete moisture absorption was practically attained in the device within 30–60 seconds (Fig. 4).

In the electrolytic method a sensor was used based on the change in electrical resistance of the conducting layer of the moisture-absorbing substance. The moisture sensor is made in the form of a glass bobbin with external duplex winding of platinum wires on the rifling. As a moisture-absorbing material hydrated phosphorous pentoxide, in the form of a 20% P_2O_5 solution in water, was used; this was put on the working surface of the sensor and dried by electrolysis, the metaphosphoric acid being practically completely electrolysed, decomposing according to the reaction:

$$2\,HPO_3 \rightarrow H_2 + \frac{1}{2}\,O_2 + P_2O_5\,.$$

During this the water concentration decreases and the P_2O_5 concentration increases giving rise to an increase in the resistance of the absorbing layer and to a decrease in the current power. In case of a sufficiently high P_2O_5 concentration the layer becomes practically an insulator. During moisture absorption the formation of metaphosphoric acid occurs again, accompanied by a lowering of resistance determined by the humidity of the environment.

The residual moisture above the phosphorous pentoxide is very small (at 20 °C it is 2.5×10^{-5} mg l^{-1}). It is extremely complicated to maintain such a low moisture in the device for a long time in the presence of other chemical absorbers. Therefore to preserve the sensor in a state of definite moisture, it was placed above the alkali. When a state of equilibrium was reached, the resistance of the coil was preserved in predicted ranges. The behaviour of the moisture sensor under various conditions is shown in Fig. 5.

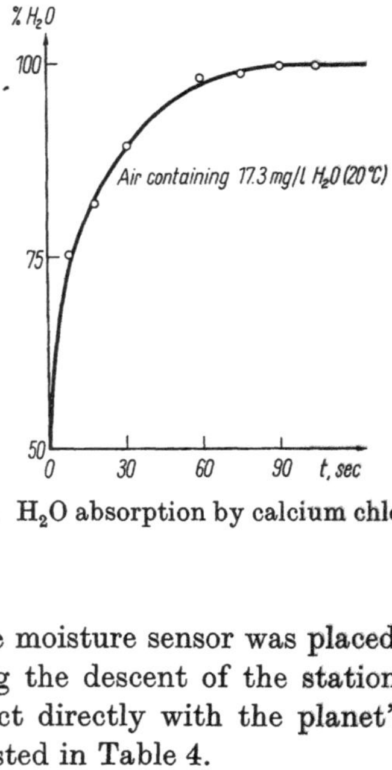

Fig. 4. H_2O absorption by calcium chloride.

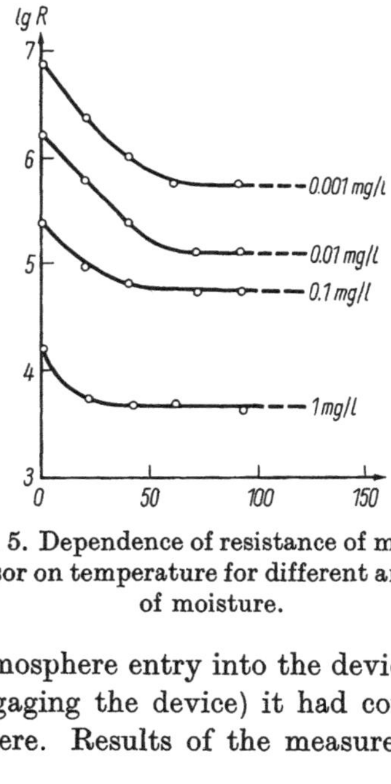

Fig. 5. Dependence of resistance of moisture sensor on temperature for different amounts of moisture.

The moisture sensor was placed at the atmosphere entry into the device and during the descent of the station (after engaging the device) it had constant contact directly with the planet's atmosphere. Results of the measurements are listed in Table 4.

As is seen from Table 4, all data show the presence of water in the Venus atmosphere, some tendency towards a decrease of its content with pressure

Table 4
Results of H_2O determination in Venus atmosphere

No.	Pressure (kg cm^{-2})	Temperature (°C)	H_2O (mg l^{-1})	Absorber	Station
1.	0.6	~25	$\geqq 4$	P_2O_5	Venera 5
2.	0.6	~25	~11*	$CaCl_2$	Venera 5
3.	0.7	~25	>0.7	P_2O_5	Venera 4
4.	2	85	$\leqq 8$*	$CaCl_2$	Venera 4
5.	2	85	~6*	$CaCl_2$	Venera 6
6.	5	150	>0.7	P_2O_5	Venera 5
7.	10	220	>0.7	P_2O_5	Venera 6

* Data obtained by the manometric method, the others by the electrolytic method.

increase being observed. This tendency we have also noticed from the moisture change of the sensors as the station was descending in the Venus atmosphere. The character of the moisture change in the sensors in the upper layers of the atmosphere gives ground to assume that the water vapour content at condensation level approaches 11 mg l^{-1}, and the temperature of the lower part of the cloud layer must be about 286 °K (dew point). Then the atmospheric pressure at the condensation level may be estimated by extrapolation from the measured value to higher atmospheric layers. For the above mentioned chemical composition of the atmosphere and a quasi-linear temperature gradient from the level of the first moisture determination ($P_1 = 0.6$ atm, $T_1 = 25$ °C) to the condensation level ($T_2 = 13$ °C), the pressure at the condensation level will be about 0.5 atmosphere.

Using the known barometric dependence, the height of the lower boundary of the cloud layer above the level of the first measurement and the temperature gradient in this section may be also estimated; they are ~1.3 km and ~7.5 deg km^{-1} respectively.

When analysing the atmosphere within the cloud layer, it may be assumed that the ratio between the water vapour pressure and the pressure inside the cloud layer remains constant. Then the release of latent heat, occurring during water vapour condensation, will change the temperature gradient according to the expression:

$$\frac{dT}{dH} = \left(\frac{dT}{dH}\right)_{ag} \left(a + M_{H_2O} \frac{L\,E}{P\,R\,T}\right)\left(1 + \frac{M_{H_2O}{}^2}{M} \frac{L}{P\,C_p} \frac{dE}{dT}\right)^{-1},$$

where a is the ratio between the temperature gradient under the cloud and the dry adiabatic gradient, M_{H_2O} is the molecular weight of water vapour, L is the latent heat of phase transition, E is the elasticity of saturated vapour, M is the molecular weight of the atmosphere, C_p is the heat capacity of the atmosphere at constant pressure.

On the basis of this change of temperature and pressure in the upper layers of the atmosphere the amount of vapour condensed to water or ice may be determined from the formula

$$Q = \frac{M_{H_2O}}{R\,T}(e - E)\,,$$

where e is the partial pressure of water vapour in the atmosphere.

The change of the cloud layer water content with height is shown on Fig. 6. It is seen from the figure that the thickness of this layer is about 15 km. Up to a height of 7 km from the level of condensation the clouds most probably consist of water droplets, above that they are probably of ice crystals. The maximum water content of the clouds is about 2.4 g m^{-3} at atmospheric temperature from −5 to −25 °C and is dispersed 5–10 km above the level of condensation. The upper boundary of clouds must be vague..

Apparently a region of intense moisture circulation exists in the subcloud layer. In this region there occurs precipitation of the water condensed in the cloud layer which, descending, reaches higher temperatures, evaporates and again ascends. During this its amount in the atmosphere must decrease with depth.

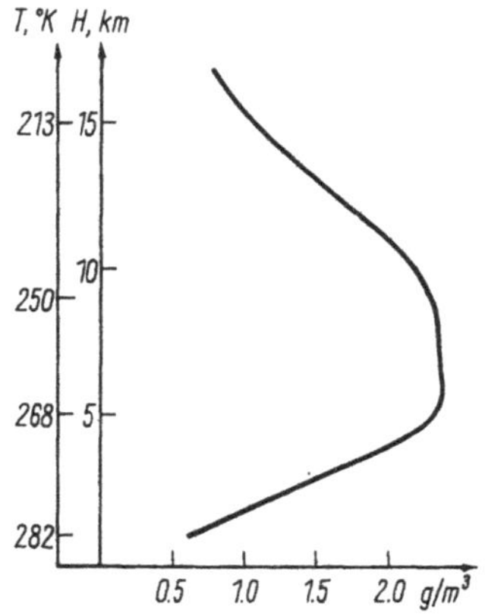

Fig. 6. Height distribution of condensed H_2O in the cloud layer.

4. Composition and Origin of the Planet's Atmosphere

As is known, we have on Venera 4 determined the following composition of the Venus atmosphere: CO_2, 90 ± 10%; O_2, 0.4–1.5%; N_2, not more than 7% (probably not more than 2.5%); H_2O, 1–8 mg l^{-1}.

As a result of more precise and repeated determinations on Venera 5 and 6, mentioned above, this composition has been more precisely investigated and may be adopted as follows: CO_2, 97 ± 4%; O_2, not more than 0.1%; N_2, not more than 2%; H_2O, 6–11 mg l^{-1} (at a pressure of 2–0.6 atm).

Thus the Venus atmosphere mostly consists of carbon dioxide. The nitrogen content (together with inert gases which can make up only a small part of the nitrogen amount) does not exceed 2%. The oxygen content proved to be noticeably less than was found on Venera 4 and does not exceed 0.1%. This amount does not differ from the estimates made from radioastronomic measurements of concentration of the molecular and other forms of oxygen in the Venus ionosphere. The character of water distribution in the Venus atmosphere has already been discussed in considering the moisture content.

In the course of all the experiments carried out on automatic interplanetary stations, we did not notice the presence in the atmosphere of any other gases in any appreciable amounts. Naturally this does not exclude the possibility of finding in the Venus atmosphere CO, HF, HCl, Ar, and, possibly, other gases, in amounts of very small parts of one per cent. As is known, their presence is indicated by data on ionospheric investigations.

Returning to the problem of the origin of the Venus atmosphere [8], it may be said that new data on its composition, pressure and temperature indicate that it is connected with the greenhouse effect. We already have considered the formation of large amounts of CO_2 in the Venus atmosphere of high equilibrium temperature as a consequence of the planet's proximity to the sun. It seems to us that under these conditions an intensive erosion of rocks is occurring and the composition of the atmosphere is chiefly determined by wollastonite equilibrium of, for instance, the form:

$$CaSiO_3 + CO_2 \rightleftarrows CaCO_3 + SiO_2 \, .$$

Many other reactions of wollastonite formation are known, for which experimental data on equilibrium at different temperatures and pressures are available. This reaction at a pressure about 50 atm leads to wollastonite and CO_2 formation at temperatures above 350 °C.

The analogy of the possible CO_2 content on the earth and on Venus has already been mentioned. If we assume that the general CO_2 reserves on the surface of the two planets are equal, the amount of CO_2 on Venus must from these data be about 50 atm. This, in general, does not contradict the experimental determinations of the amount of CO_2, and the pressure and temperature observed at the Venus surface.

Oxygen may be formed in the Venus atmosphere at the expense of carbon dioxide and water. However, the low oxygen content indicates that it is intensively recombined and is bounded by rock matter. Hydrogen, arising during photodissociation, dissociates. It is known that Venus has, if any, only a small hydrogen corona, i.e. a process of water, or more precisely of hydrogen, loss is occurring.

As we have seen, the Venus atmosphere has a small water content. According to our data, its maximum content is in the cloud layer. It may be assumed that some of the water content in the cloud layer is present in the solid phase. However, this requires direct determination.

The nitrogen source, ammonia, had apparently, as on the earth, degassed in consequence of volcanic activity which without question exists on Venus and further experiments will be able to confirm it. We already have considered the advantage in the rate of nitrogen formation from ammonia under the conditions on Venus, compared with those on the earth.

As calculations show, the possible share of heat supplied from the inner parts of the planet to the surface is very small in comparison with the solar heat absorbed by the atmosphere from carbon dioxide and water. Therefore the greenhouse effect is the chief mechanism which had led to self-heating of the surface of Venus and to the formation of the heavy carbonic acid atmosphere.

References

[1] A. P. Vinogradov, Yu. A. Surkov, K. P. Florensky and B. M. Andreichikov, Dokl. Akad. Nauk SSSR **179**, 37 (1968).
[2] A. P. Vinogradov, Yu. A. Surkov and K. P. Florensky, J. Atmos. Sci. **25**, 535 (1968).
[3] A. P. Vinogradov, Yu. A. Surkov et al., Dokl. Akad. Nauk SSSR (in press).
[4] The Solar System in: Physics of the Planets, Moscow 1967.
[5] J. Atmos. Sci., **25**, No. 4 (1968).
[6] V. S. Avduevsky et al., Dokl. Akad. Nauk SSSR (in press).
[7] A. D. Kuzmin, in: Itogi Nauk (Reviews of Science), ser. Fizika, VINITI, Moscow 1967.
[8] A. P. Vinogradov, Vestnik MGU, No. 4, Geologia, 1969.

Space Research XI — Akademie-Verlag, Berlin 1971

LOWER ATMOSPHERE OF VENUS FROM RADIO ASTRONOMICAL AND SPACE MEASUREMENTS

A. D. Kuzmin, A. P. Naumov, T. V. Smirnova and Yu. N. Vetukhnovskaya

Academy of Sciences, Moscow, USSR

Three independent evaluations of pressure and temperature of the Venus atmosphere at the level of average surface level based on space and ground-based radio astronomical and radar measurements lead to two internally consistent models: (i) an adiabatic model with relative abundance of water vapour 0.5%. The atmosphere pressure and temperature at the average surface are 65 ± 20 atm and 700 ± 50 °K; (ii) a model with an isothermal near-surface region $T = T_s = 650$ °K and relative abundance of water vapour zero. The atmospheric pressure at the average surface is equal to 100 ± 20 atm. These two models may be extreme and some intermediate models may be acceptable.

Radio astronomical measurements seen to favour a lower relative abundance of water vapour in the Venus atmosphere.

The successful flight of Venera 4, 5 and 6 Soviet interplanetary stations started a new phase in the exploration of the planet Venus. The first direct measurements of chemical composition [1, 2], temperature, pressure and density [3, 4] along the 35 km descent paths were made. Mariner 5 measurements provided the temperature, pressure and density at higher altitudes. As a result the altitude distribution of these parameters in a range of altitude of about 50 km and a range of temperature of 240–600 °K and pressure of 0.03–27 atm were obtained. However, the measurements of Venera 5 and 6 space probes ended at the altitude about 20 km above the planet's surface [4], principally because Mariner 5 could not measure the level lower than the pressure of 5 atm due to critical refraction. The surface and near-surface atmosphere were measured only by radio astronomy and radar.

Moreover, the space measurements investigate a relatively small local region of the planets. The parameters of a local region may be different from the average parameters of the planets. On the other hand, the ground-based radio astronomical and radar measurements provided average parameters which are more typical for the planet as a whole. Therefore, the space and ground-based measurements are complementary. Taking from space measurements the precise data about chemical composition and local temperature and pressure at some atmospheric level one can estimate from radio astronomy and radar data the temperature and pressure at the average planet's surface.

The evaluation of an average surface pressure equal to 50–70 atm based on radio astronomy, radar and Venera 4 data made by us [5] was confirmed by the work published in [6, 7, 8]. The present paper contains a more complete analysis of the radio astronomical and radar measurements based on new data of Venera 5 and 6 [2, 4]. The analysis is based on the following foundations and assumptions: (i) the surface is isothermal and homogeneous; the dielectric constant $\varepsilon = 4.7 \pm 0.5$ [9]; (ii) the atmosphere is spherically symmetrical; (iii) the microwave absorption coefficient of carbon dioxide was that adopted by [10]; (iv) the microwave absorption coefficient of water vapour was calculated by a quantum mechanics formula [11]; this suggested that H_2O–CO_2 collisions are 1.6 times more effective than H_2O–N_2 collisions; (v) the effect of molecular oxygen on absorption is negligible; (vi) the temperature of the near-surface atmosphere is equal to the temperature of the surface; (vii) the relative abundance of carbon dioxide in the atmosphere is equal to 95% and is independent of the altitude; (viii) the altitude distributions of temperature and pressure in the above-mentioned 50 km range were taken from the measurements of Venera 4, 5, 6 and Mariner 5; (ix) the level of 1 atm pressure was adopted as zero altitude level.

The aim of our analysis was to find the temperature and pressure at an average surface which best fits the radio astronomical and radar measurement data.

Three independent parameters of these measurements were under investigation: the brightness temperature as a function of wavelength $\overline{T}_{B♀}(\lambda)$, the polarization of radio emission and the radar crosssection as a function of wavelength $\sigma(\lambda)$. On this ground we have utilized three independent methods of finding the temperature and pressure at an average surface. The first method is based on the dependence of the calculated $\overline{T}_{B♀}(\lambda)$ on the surface level. Therefore we find a surface level which fits better the measured radio astronomical data. The second method is based on the dependence of the polarization of Venus radio emission on the absorption in the planet's atmosphere. Therefore we find the pressure and the absorption which fit the radio astronomical polarization measurements [12]. The third method is based on the dependence of the radar cross section on atmospheric absorption. We find again the pressure and absorption which fit the radar measurements of $\sigma(\lambda)$.

Several models of the lower atmosphere which differ one from another by altitude distribution of temperature and pressure and by relative abundance of water vapour were examined. Two models were selected for further consideration as internally consistent. The first modelhas an adiabatic lower atmosphere down to the surface and constant relative abundance of water vapour equal to 0.5%. The second model has an isothermal near-surface region. In order to obtain internal consistency we adopt in this model a relative abundance of water vapour in the lower atmosphere equal to zero. The results of our analysis are given in Table 1 which contains the atmospheric temperature and pressure at an average surface and the altitude of zero level ($p_0 = 1$ atm) above the average surface. There are good internal consistencies of three independent evaluations of temperature and pressure at average surface for both models under consideration.

Figs. 1 and 2 show the results of measurements of $\overline{T}_{B♀}(\lambda)$ and $\sigma(\lambda)$ and calculations based on parameters given in Table 1. The good consistency of the measurements and calculations supports the earlier [5] conclusion that carbon dioxide is the principal absorbent for microwave radio emission. This consistency

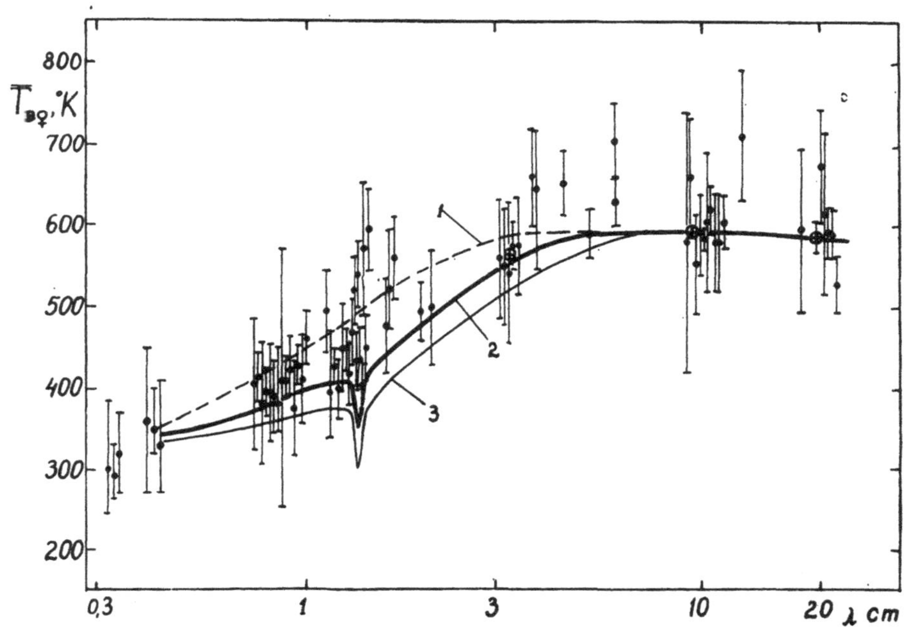

Fig. 1. Measured and calculated dependence of $\overline{T}_{B♀}(\lambda)$ for the adiabatic lower atmosphere, $T_s = 705\,°K$ and $f_{CO_2} = 0.95$; 1, $f_{H_2O} = 0$; 2, $f_{H_2O} = 0.005$; 3, $f_{H_2O} = 0.01$.

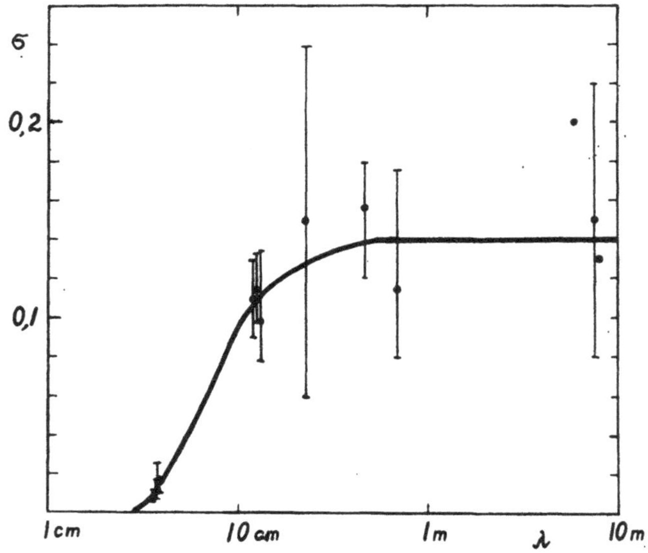

Fig. 2. Measured and calculated dependence of $\sigma(\lambda)$ for $f_{H_2O} = 0.005$.

Table 1

Model	Method of evaluation	T_s (°K)	P_s (atm)	h_0 (km)
1	$T_B(\lambda)$	705 ± 30	$66 {+20 \atop -15}$	46.5 ± 4
	polarization	680 ± 20	55 ± 10	43 ± 3
	$\sigma(\lambda)$	685 ± 20	58 ± 6	44 ± 2
	averaged	700 ± 50	65 ± 20	45 ± 5
2	$T_B(\lambda)$	650	100 ± 20	51.5 ± 3
	polarization	650	110 ± 15	53 ± 2
	$\sigma(\lambda)$	650	100 ± 20	51.5 ± 2
	averaged		100 ± 20	52 ± 3

excludes the necessity of suggesting some additional absorbing constituents. Moreover it is possible to evaluate the upper limit of these constituents. Between the known and suggested constituents of the Venus atmosphere there are water vapour and liquid water which absorb microwaves. Fig. 1 shows that 1% of water vapour (through the lower atmosphere) does not agree with the radio astronomical measurements. This consistency is good for relative water vapour abundance less than 0.5%, that is, the radio astronomical data appear to favour the lower (0.1%) rather than the upper (1%) limit of Venera space probe water vapour measurements.

The integral amount of liquid water in Venus clouds is

$$\int M_0(h) \exp [0.04 (273 - T)] \, dh \leqq 0.3 \text{ g cm}^{-2} .$$

It is of interest to point out that the isothermal model shows a significant decrease of $T_{B♀}$ at decimetre wavelengths. This decrease agrees qualitatively with the radio astronomical measurements [13], but quantitatively the calculated brightness temperatures are much higher than those at wavelengths of 30–70 cm.

In both models under examination it was suggested that the relative water vapour abundance was constant at any altitude. But the real distribution of water vapour in the Venus atmosphere must be more complicated. It follows from the fact that liquid water cannot exist in the Venus atmosphere below 430 °K and 5.5 atm. Therefore the water vapour in the Venus atmosphere does not generate on the planet's surface as it does on the earth, but at a fairly high altitude above the surface. As a result of such a situation it must be expected that the relative abundance of water vapour will decrease below the level of 430 °K. Such a model, being intermediate between the two models, needs additional data on the altitude distribution of water vapour.

References

[1] A. P. VINOGRADOV, YU. A. SURKOV, K. P. FLORENSKY and B. M. ANDREICHIKOV, Dokl. Akad. Nauk SSSR **179**, 37 (1968).
[2] A. P. VINOGRADOV, YU. A. SURKOV and B. M. ANDREICHIKOV, Dokl. Akad. Nauk SSSR **190**, 552 (1970).

[3] V. S. Avduevsky et al., Dokl. Akad. Nauk SSSR **179**, 310 (1968).
[4] V. S. Avduevsky, M. Ya. Marov and M. K. Rozhdestvensky, Report at the Symposium of Atmospheres and Surfaces of the Planets, Woods Hole, USA 1969.
[5] A. D. Kuzmin and Yu. N. Vetukhnovskaya, J. Atmos. Sci. **25**, 546 (1968).
[6] N. N. Krupenio and A. P. Naumov, Report at the Symposium on the Moon and Planets, Kiev, USSR 1968.
[7] O. N. Rhziga, Report at the Symposium on the Moon and Planets, Kiev, USSR 1968.
[8] A. T. Wood, Jr., R. B. Wattson and J. B. Pollack, Science **162**, 114 (1968).
[9] Yu. N. Vetukhnovskaya and A. D. Kuzmin, Astron. Vest. **4**, No. 1, 1970.
[10] W. Ho, I. A. Kaufman and P. Thaddeus, J. Geophys. Res. **71**, 5091 (1966).
[11] S. A. Hzevakin and A. P. Naumov, Izv. Vuz'ov, Radiophysika, **6**, 674 (1963).
[12] A. D. Kuzmin and B. G. Clark, Astrophys. J. **142**, 23 (1965).
[13] A. D. Kuzmin, The Radiophysical Investigations of Venus, VINITI, Moscow (1967).

Space Research XI — Akademie-Verlag, Berlin 1971

RADIO WAVE FLUCTUATIONS AND REFRACTION COEFFICIENT VARIATIONS IN THE ATMOSPHERE OF VENUS

M. A. Kolosov, O. I. Yakovlev, G. D. Yakovleva and A. I. Yefimov

Institute of Radioengineering and Electronics, Academy of Sciences of the USSR, Moscow, USSR

Data are given on the decimetre-wave fluctuations from the parachuted Venus 4, 5 and 6 spacecrafts. The fadings in the radio test of the atmosphere of Venus by Mariner 5 are analysed. Statistics of the refraction coefficient in the planet's atmosphere have been determined.

1. Introduction

The basic parameters of the atmosphere of Venus have so far been determined by direct measurements of the temperature, pressure and gaseous composition performed by the soft-landed Soviet spacecraft [1–4]. Valuable data on the upper atmosphere have been obtained in the USA by the radio-wave occultation method performed by Mariner 5 [5, 6]. Study of the radio-wave fluctuations can give more data on the inhomogeneities and traffic of the atmosphere of Venus. The variations of the refraction coefficient may be regular, for instance due to the inversion of the coefficient at the lower edge of the clouds. Radioscopy of the inversion part of the atmosphere should detect variations in field intensity. The statistical inhomogeneities in the refraction coefficients due to large-scale meteorological formations or to turbulent processes in the atmosphere will also lead to fluctuations both in the case of parachuted spacecraft and in radioscopy of the planet's atmosphere. The preliminary analysis of the radio signals received from Venus 5 and Mariner 5 gave estimates of the inhomogeneities in the refraction coefficient of the atmosphere of Venus [7–9]. Venus 5 and 6 gave new data on radio-wave fluctuations in the planet's atmosphere. This paper will analyse the fluctuations of the decimetre waves from Venus 4, Mariner 5, Venus 5 and Venus 6 and determine the parameters of the inhomogeneities of the atmosphere of Venus.

2. Radio-wave Fluctuations in the Atmosphere of Venus

The interplanetary stations Venus 4, 5 and 6 have recorded the fluctuations of the radio-wave field intensity when the spacecraft descended from the level where the pressure is 0.5 atm to the altitude where the pressure is 27 atm. The total period of the radio-field intensity recordings amounted to 3h 14 min. The records were processed statistically in order to determine the values of rapid fadings. The statistical analysis of noise recorded simultaneously with the signals from the descending spacecraft permitted the elimination of the noise component. Small fluctuations emerging when the waves propagated through the earth's atmosphere and the interplanetary plasma were measured during the pre-landing telemetry report thirty minutes before the signals from the descending spacecraft were received.

Fig. 1a shows the experimentally obtained depth of rapid fluctuations in the field intensity for waves of the order of $\lambda = 32$ cm at different altitudes in the atmosphere of Venus. The vertical axis represents the quantity

$$\eta_1 = [\langle E_i - E_0 \rangle^2]^{1/2}/E_0 ,$$

where E_i are the instantaneous values of the field intensity every 0.05 s, E_0 are the values of the level averaged over the intervals of 5 s. Each experimental

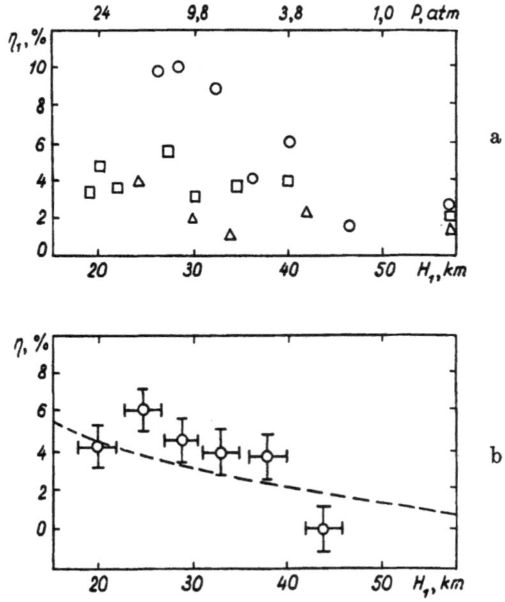

Fig. 1. Radio-wave field intensity fluctuations versus the altitude of the spacecraft parachuted to the surface of Venus. ○, Venus 4; △, Venus 5; □, Venus 6.

point in Fig. 1a was obtained by averaging the quantity η_1 over the intervals of 4 to 5 minutes that corresponded to a certain altitude of the craft. The data from Venus 4, 5 and 6 are denoted by circles, triangles and squares respectively. The abscissae at the bottom give the average altitudes of the craft relative to an arbitrary reference level where the pressure is 92 atm, those at the top the pressures corresponding to the given altitudes. The values of η_2 obtained in the pre-landing telemetry report are shown conditionally for the altitude $H = 60$ km. The values of the depth of radio-wave fluctuations η with allowance for the interplanetary medium and the earth's atmosphere are given by $\eta = (\eta_1^2 - \eta_2^2)^{1/2}$ averaged over altitudes in the vicinity of the craft. The relation of η versus H_1 thus obtained describes the variations in the degree of rapid radio-wave scintillations when the source of radio waves is immersed in the atmosphere. The plot is shown in Fig. 1b. With the craft at altitude $H_1 = 20$ km the probable value of the depth of rapid fluctuations is 6% while the possible minimal value is about 3%. At $H_1 = 40$ km this quantity is $\eta \approx 2\%$.

The fluctuation depth depends on the time interval T for which E_i and E_0 are found. Fig. 2 shows the variation of fluctuation depth with time interval T. The fluctuations are apparently best pronounced at $T \approx 12$ s. This means that the conditional fluctuation period is about 1–2 s.

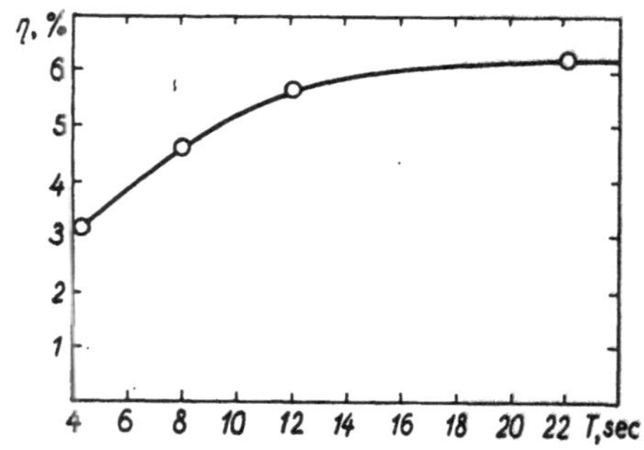

Fig. 2. Fluctuation depth at different periods T.

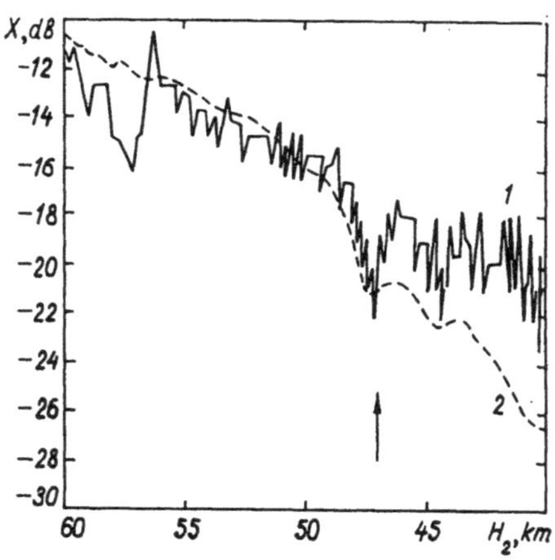

Fig. 3. Variations of the field intensity obtained by radioscopy of the atmosphere of Venus.

The radio-wave fluctuations were observed when the atmosphere of Venus was radioscoped by Mariner 5. The results of the field intensity measurements obtained by the same technique in the upper atmosphere are given in Fig. 3 [5, 6]. The solid line was obtained in the region of $\lambda = 71$ cm, while the broken line represents the smoothed results for $\lambda = 13$ cm.

It is seen from Fig. 3 that at beam altitude $H_2 = 57$ to 49 km above the planet's surface there are shallow fadings of the field intensity; at $H_2 = 49$ to 46 km the field level decreases uniformly. This phenomenon will be analysed in § 4. With further decrease in the beam altitude H_2 strong fading appears that may be caused either by variations in the refraction coefficient or by instrumental factors. Thus to reveal rapid fluctuations observed by radioscopy a small part of the record at $H_2 = 57$–49 km obtained for a 30 s observation interval may be used. It follows from Fig. 3 that the fading depth for $\lambda = 71$ cm in this interval amounts to ± 1 dB, i.e. $\eta = 0.12$; at $\lambda = 13$ cm $x \geqq 0.5$ dB, or $\eta \geqq 0.06$. Larger fadings were observed on the day-time side of Venus.

3. Statistics of the Inhomogeneities in the Radio-wave Refraction Coefficient of the Atmosphere of Venus

The radio-wave fluctuations described in the previous section are related to the effect of the inhomogeneities in the refraction coefficient of the Venus atmosphere. These inhomogeneities can be caused by large formations such as clouds, variations in humidity and meteorological factors, and the turbulence of the planet's atmosphere. Therefore the estimates of the inhomogeneities are made with two assumptions: in the first place we will assume that the variations can be described by a Gaussian auto-correlation function with characteristic scale l and the r.m.s. value ΔN of the refraction coefficient fluctuations. This approach is an approximate one, but it does not require the introduction of assumptions on unknown causes for the variations in the refraction coefficient. In the second place it will be assumed that the variations are the effect of the turbulence in the atmosphere and are thus caused by temperature fluctuations. The variations in refraction coefficient should theoretically decrease with altitude. Note that the magnitude of decrease in the earth's atmosphere can vary from day to day but on the average is very weakly pronounced in the range from 1 to 8 km. Therefore it will be tentatively assumed that in the Venus atmosphere the refraction coefficient variations are not related to the altitude within $H = 30$–70 km. It will be assumed that at $H_{max} \geqq 70$ km there are no inhomogeneities.

To determine the magnitude of ΔN a certain value of l must be assumed. The conditional inhomogeneity scale l in the earth's atmosphere is about 50 m; this value can also be presumed for the Venus troposphere. In estimating ΔN we should bear in mind that the radioscopy of the Venus atmosphere by decimetre waves from Mariner 5 did not reveal rapid fading at $H_2 > 70$ km. Since at $H = 70$ km the pressure $P = 0.03$ atm, the upper atmosphere does not contribute much to the fading of decimetre waves. The magnitude H_{max} may differ slightly from the one assumed because at $H = 80$ km, $P = 0.01$ atm. Therefore the effect of these altitudes on the fluctuations can be neglected.

Table 1

Radio-wave fluctuations and statistics of the refraction coefficient in the Venus atmosphere

Spacecraft	Wave length (cm)	Fluctuations (%)	Craft or beam alt. (km)	ΔN_{av}	ΔN (40)
Venus 4, 5, 6	32	5 ± 2	20	$(6.6-3.8) \cdot 10^{-6}$	$6.3 \cdot 10^{-6}$
Venus 4, 5, 6	32	2 ± 1	40	$(3.9-2.4) \cdot 10^{-6}$	$5.2 \cdot 10^{-6}$
Mariner 5	71	≈ 12	$57-49$	$2.1 \cdot 10^{-6}$	$3.1 \cdot 10^{-6}$
Mariner 5	13	$\geqq 6$	$57-49$	$\geqq 2 \cdot 10^{-7}$	$\geqq 2.8 \cdot 10^{-7}$

Table 1 represents the experimental values of the depth of decimetre wave fluctuations and the average values of ΔN_{av} found by various workers [9–11]. It follows that for the altitudes 60–50 km $\Delta N_{av} \approx (1-2) \times 10^{-6}$ which corresponds to the values of ΔN in the earth's atmosphere at comparable pressures.

The radio-wave fluctuations observed by the parachuted spacecraft at an altitude of 20 km led to the values $\Delta N_{av} = (4$–$6) \times 10^{-6}$. This implies that the fluctuations of the refraction coefficient increase with decrease of altitude. This value of ΔN_{av} is valid for the altitude of 40–50 km.

Now, let us consider a linear model of decrease of ΔN: $\Delta N \sim H^{-1}$. This relation will be assumed valid for altitudes of 70–20 km. Table 1 gives the values of $\Delta N(40)$ for $H = 40$ km found in this way. These values determined in different situations are close, which implies that the approximate relation $\Delta N \sim H^{-1}$ is realistic. In Fig. 1b the broken line was calculated for $\Delta N(40) = 5.5 \times 10^{-6}$ and is a satisfactory representation of the experimental evidence.

4. Fading of Radio Waves and the Inversion of the Atmospheric Refraction Coefficient

During the radioscopy of the Venus atmosphere by Mariner 5 regular fadings at both frequencies on both the day-time and the night-time side of the planet at beam altitude $H_2 = 47$–49 km were recorded as well as random fluctuations [5, 6]. These regular fadings, denoted by an arrow in Fig. 3, can hardly be explained by the absorption since they are independent of frequency. We believe that this can be most naturally explained by the refraction effects on the inversion of the radio-wave refraction coefficient. Such an inversion in the presence of solid cloud cover is always to be found in the earth's atmosphere and can be caused by a specific distribution of temperature or humidity with altitude. To determine the contribution made by the inversion we will consider the fading found in the radioscopy of the Venus atmosphere.

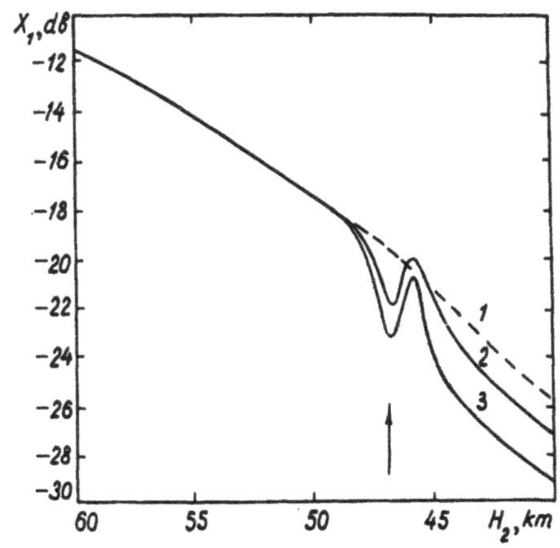

Fig. 4. Refraction attenuation in the presence of an inversion stratum.

The refraction attenuation in the Venus atmosphere can be calculated on the basis of the beam concept by analysing the deformation of beam cross sections. The variation of refraction coefficient with distance $n(H)$ can be determined by using the results obtained by direct measurements of pressure, temperature and gas composition reported in [1–4] and by using the magnitude of $n(H)$ found by the Doppler method during radioscopy [6]. The numerical analysis of these data led to the following approximation for the relation between refraction coefficient and altitude

$$n(H) = 1 + \exp(-a\,H^2 - b\,H - c)\,, \qquad (1)$$

$$a = 8.33 \times 10^{-4}\,\mathrm{km}^{-2}\,, \quad b = 3.48 \times 10^{-2}\,\mathrm{km}^{-1}\,, \quad c = 4.11\,.$$

The altitude H here is taken, as above, from a certain arbitrary reference level where the pressure is $P_0 = 92$ atm. It is important to indicate that in determining the relation $n(H)$ in the form of Eq. (1) we did not use the experimental data on the refraction fading in the Venus atmosphere. In Fig. 4 curve 1 represents the results for $n(H)$ in the form of Eq. (1). The radius of Venus was assumed to be $A = 6052$ km and the radius of the spacecraft orbit $R = 10{,}220$ km. The latter value was true for the time when Mariner 5 flew beyond Venus. A comparison of the calculated curve 1 in Fig. 4 with the experimental values of the refraction fadings represented in Fig. 3 has clearly demonstrated that the average profile of Eq. (1) is a fair representation of the refraction coefficient at $H > 50$ km. However, the characteristic fading denoted by an arrow in Fig. 3 is not described by a smooth profile (1). Now, let us introduce the inversion of the refraction coefficient and use instead of Eq. (1) an expression of the form

$$n(H) = 1 + \exp(-a\,H^2 - b\,H - c) + M \exp[-m\,(H - H_\mathrm{i})^2]\,. \qquad (2)$$

Eq. (2) differs from Eq. (1) in an additional term which describes the symmetrical inversion of the radio-wave refraction coefficient. The parameter M characterizes the deflection of the refraction coefficient in the inversion stratum, while the "depth" of the stratum is represented by the value $\Delta H \approx 2\,m^{-1/2}$. In Fig. 4 curves 2 and 3 represent the variation of refraction fading with the minimal altitude H_2 of the beam in the presence of an inversion stratum for $M = 10^{-5}$ and 2×10^{-5} respectively at $H_\mathrm{i} = 46$ km and $m = 1\,\mathrm{km}^{-2}$. A comparison of the experimental and calculated values indicates that the fading at $H_2 = 47$ km is caused by the presence of a regular inversion with deflection of the refraction coefficient from the smooth profile $M = (1\text{–}2) \times 10^{-5}$ and the semiwidth $\Delta H = 2$ km. To assume that this feature of the coefficient is due to the temperature inversion would be quite natural. The deflection in the inversion centre from the smooth value is described by the relation

$$M = -\,0.13\,\Delta T\,\frac{P}{T^2}\,. \qquad (3)$$

At altitude 47 km the pressure $P = 1.6$ atm and the temperature $T = 350$ °K; therefore, for an inversion with value $M = (1\text{–}2) \times 10^{-5}$ to appear, a negative temperature inversion with $T = 6\text{–}12°$ would suffice.

5. Conclusions

The fading indicates the presence of statistical variations of the refraction coefficient in the Venus atmosphere. At altitudes of 50–60 km these amount to $\Delta N = (1\text{–}2) \times 10^{-6}$ which is close to observations in the earth's atmosphere at comparable pressures of 1–0.2 atm. With decrease in altitude the variations increase and at 40–50 km where the pressure is 4–1 atm, the magnitude of ΔN is about 5×10^{-6}. The statistical variations of the refraction coefficient indicate the turbulence of the Venus atmosphere. The radiowave fluctuations on the day-time side are greater than on the night-time side where the turbulence is less developed.

The analytical representation of the (refraction coefficient, altitude) profile agrees with the direct measurements of the pressure, temperature and gas content and is a fair description of the smoothed experimental curve of refraction attenuation. This curve has a regularity at altitude 47 km over the day-time side that is easily explained by an additional refraction fading which is caused by the inversion of the refraction coefficient. The thickness of the inversion stratum is about 1 km, the deflection of the refraction coefficient from the smoothed value in the centre of the inversion stratum is about $(1\text{–}2) \times 10^{-5}$. This deflection can appear if the value of the temperature negative inversion is $\Delta T = 6\text{–}12°$. This inversion is associated with the lower edge of the cloud stratum. On the day-time side this feature is less pronounced because the inversion is less stable due to the greater turbulence of the atmosphere.

References

[1] V. S. Avduevsky, M. Ya. Marov and M. K. Rozhdestvensky, Kosm. Issled. **7**, 233 (1969).
[2] V. V. Mikhnevich and V. A. Sokolov, Kosm. Issled. **7**, 220 (1969).
[3] A. P. Vinogradov, Yu. A. Surkov and B. M. Andreichikov, Dokl. Akad. Nauk **190**, 552 (1970).
[4] Pravda, 4 June 1969.
[5] G. Fieldbo and V. R. Eshleman, Radio Sci. **4**, 879 (1969).
[6] A. Kliore, D. Gain, G. Levy, G. Fieldbo and S. Rasool, Space Research IX, North-Holland Publ., Co., Amsterdam 1969 (p. 712).
[7] M. A. Kolosov, O. I. Yakovlev and A. I. Yefimov, Dokl. Akad. Nauk SSSR **182**, 93 (1968).
[8] A. S. Gurvich, Izv. Akad. Nauk SSSR, Ser. Fiz., Fiz. Atmos. Okeana, **5**, 1172 (1969).
[9] D. A. Wolf, private communication.
[10] L. A. Chernov, Wave Propagation in a Medium with Random Inhomogeneities, Izd. Akad. Nauk SSSR, Moscow 1958 (in Russian).
[11] V. V. Pisarev, Astron. Zh. **35**, 112 (1958).

Space Research XI — Akademie-Verlag, Berlin 1971

MARINER 6 AND 7 TELEVISION RESULTS

B. A. SMITH

Department of Astronomy, New Mexico State University, Las Cruces, New Mexico, USA

Mariners 6 and 7 encountered Mars on 31 July and 5 August 1969, taking a total of 200 pictures throught their approach and encounter. During the near-encounter phase a wide-angle camera recorded pictures in red, green and blue light with surface resolutions of a few kilometers, while a narrow-angle camera took minus-blue pictures with resolutions of a few hundred meters. Three distinct types of martian terrain are recorded in the photographed region that extends from the equator to the south pole. Cratered terrains are found to be widespread; "chaotic" terrains appear in certain areas near the equator, and Hellas is observed to be "featureless". Two distinct populations of primary craters are present. An episodic surface history is indicated. Identification of several "canals" is probable. Optically thin scattering layers are seen in several approach and near-encounter photographs, but the classical optically thick "blue haze" is found to be non-existent. The size and reflectivity of Phobos have been determined.

Mariners 6 and 7 encountered Mars on 31 July and 5 August 1969, carrying the second and third successful optical imaging experiments to that planet. Television pictures transmitted back from Mariners 4, 6, and 7 have clearly shown that Mars is neither like the earth nor the moon, but is apparently unique with its own peculiar martian characteristics. Each of the three spacecraft has discovered a particular type of terrain: Mariner 4 in July 1965 sent back pictures of heavily cratered terrain, much like the moon, but showing the effects of significant erosional processes. Mariner 6 revealed what has been described as "chaotic" terrain, and Mariner 7 detected the enigmatic "featureless" terrain of Hellas.

Unlike Mariner 4, Mariners 6 and 7 each carried two television cameras: a narrow-angle camera with a field of view of 1.1° by 1.4° and a fixed minus-blue filter, and a wide-angle camera with a field of view of 11° by 14° and a repeating filter sequence of four filters: blue-green-red-green. The television vidicons and their associated electronics were identical for both narrow and wide-angle cameras. Each frame contained 704 television lines with 945 samples along each line. Thus, each individual frame contained 665,280 picture elements or "pixels". The scene brightness of each pixel was stored on an analog tape recorder after passing through an automatic gain control (AGC) and contrast pre-emphasis circuit. To preserve photometric accuracy, every seventh sample of each line was also given eight-bit encoding without passing through either the AGC or contrast pre-emphasis. The six least significant bits were then stored on a separate digital tape recorder. Both digital and analog signals were later transmitted to the 64 meter Goldstone antenna as six-bit digitally encoded signals.

The two most significant bits were later added and the AGC and pre-emphasis effects corrected by computer imaging processing.

The Mariner 69 encounters were divided into two distinct phases: (i) The approach or "far-encounter" sequence began at E-48 hours for Mariner 6 and E-67 hours for Mariner 7, at which times the spacecraft were 1,241,000 kilometers and 1,716,000 kilometers from Mars respectively. Television picture frames were taken at intervals of approximately 40 minutes, totaling 50 and 93 far-encounter frames for Mariners 6 and 7. During far encounter only the narrow-angle camera was used. The phase angle remained nearly constant at 20–25° with the morning terminator visible throughout this sequence for both spacecraft. (ii) The "near-encounter" sequence began as the limb of Mars crossed the field of view of the wide-angle camera, at which time a picture was taken through the blue filter by each spacecraft. From the time of limb crossing the camera operation was automatic with alternating wide and narrow-angle frames, and the filters stepping in sequence in the wide-angle camera as described above. The timing of the individual pictures was such that wide-angle frames slightly overlapped and the narrow-angle frames were nested in the overlapping areas. A total of 26 near-encounter pictures were taken by Mariner 6, and 33 by Mariner 7. At closest approach (approximately 3400 kilometers) the resolution of the surface was about 3 kilometers and 0.3 kilometers for the wide-angle and narrow-angle cameras, respectively.

Long before launch the encounter dates had been so chosen as to photograph particular regions of the planet, regions which would include the widest variety of classical martian features. It was decided that Mariner 6 should sweep equatorially along the border between the dark area Aurorae Sinus and the bright region Chryse, then across the "oasis" Oxia Palus, then Meridiani Sinus, and finally along the border between the dark Sabaeus Sinus and the variably lighter Deucalionis Regio. Meridiani Sinus, a feature of particular interest, would be seen under a relatively low solar elevation angle. Mariner 7 was to sweep in a southerly direction across Meridiani Sinus, up to and across part of the south polar cap, then step northward and make a west to east pass across Hellas. Meridiani Sinus would be rephotographed by Mariner 7 with the sun nearly directly overhead.

Many of the following preliminary results have been published elsewhere [1–5].

The far-encounter series has given us the highest resolution ever seen of the full planetary disk of Mars. Most "canals" were conspicuously absent, and the dark areas or "maria" were mottled with soft or diffuse boundaries, very unlike the sharp, angular edges usually depicted by visual observers. In the latter or closest part of the far-encounter sequence, at least some of the dark-area mottling was found to be caused by dark-floored craters with lighter rims. Curious white streaks covered the Tharsis-Candor region, and parts of the famous "W-cloud" were seen to brighten appreciably as they approached the late afternoon limb. Nix Olympica, a feature long known for its conspicuous diurnal brightening, was found surprisingly to be a bright rimmed crater some 600 kilometers in diameter, with a nearly concentric inner bright ring of 200 kilometers diameter. Several brilliant finger-like projections were seen at the edge of the north polar "hood" in the regions of Tempe and Arcadia. Of particular interest was the appearance of the south polar cap. Several dark spots, identified as craters, were seen daily along the edge of the cap. The persistence of their visibility suggested strongly that the south cap was a surface and not an atmospheric phenomenon. Also of considerable interest was the pronounced darkening of the south polar cap near the limb and terminator, apparently caused by either an atmospheric haze or highly unusual surface scattering properties. Further evidence of haze was found in stratified aerosol layers above the limb in the middle southern latitudes over Mare Cimmerium, Mare Sirenum, Eridania, and Electris, and at higher northern latitudes over Arcadia and Tempe. The upper surface of these layers ranged from 20–40 kilometers above the evening limb.

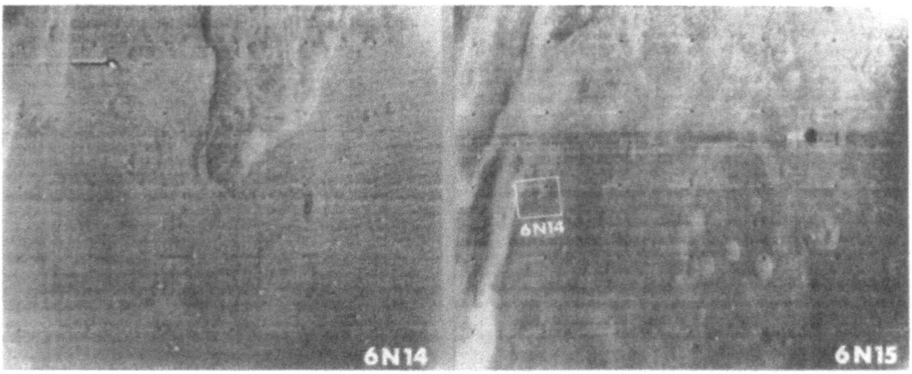

Fig. 1. Chaotic terrain along the northern edge of Aurorae Sinus. Note the ridge network within the depressed area in the narrow-angle photograph (6N14). As seen in the wide-angle frame (6N15), the chaotic terrain appears to be lighter than that surrounding it.

The early near-encounter pictures taken by Mariner 6 recorded what we now call "chaotic" terrain, extending over 10^6 square kilometers within a strip 1000 kilometers wide and 2000 kilometres long, from Aurorae Sinus to Margaritifer Sinus. The chaotic terrain is composed of a complicated network of ridges and depressions, characteristically 1–3 kilometres wide and 2–10 kilometers long. This highly irregular structure seems to be depressed with respect to its surroundings, and has an appearance that suggests localized withdrawal of permafrost or magma (Fig. 1).

The Meridiani Sinus area consists of irregular dark areas and dark-floored, bright-rimmed craters, many polygonal in shape. Some of the dark-floored craters had bright intrusions along the northern extremities of their floors, perhaps indicative of a preferentially wind-blown dust. At certain points along the irregular border of Meridiani Sinus the edge is particularly sharp, with the transition from dark to light taking place within a kilometer or so.

The series of pictures along Deucalionis Regio (Fig. 2) shows an unbroken cratered terrain with the craters exhibiting a bimodal distribution in both the diameter-number distribution and morphology. Craters larger than 7 kilo-

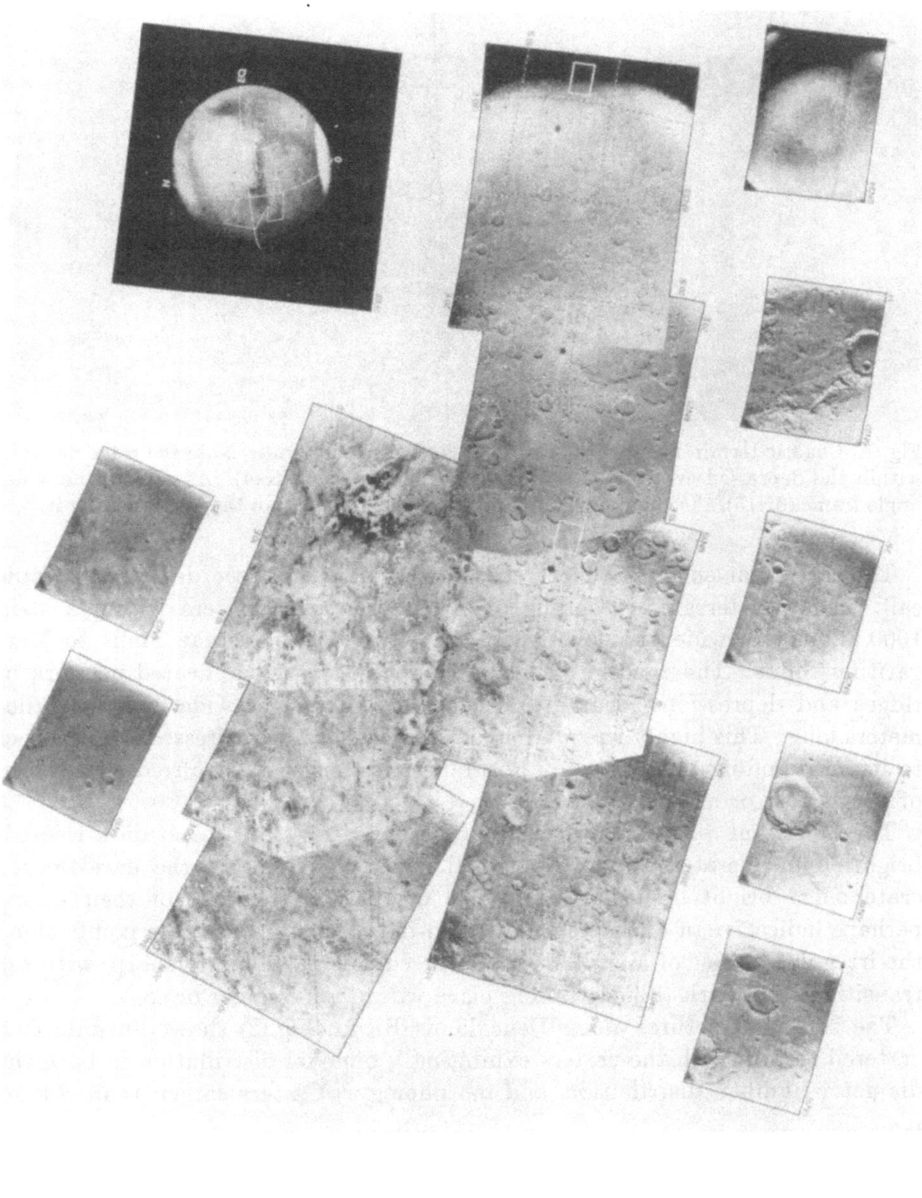

Fig. 2. The region of Meridiani Sinus and Deucalionis Regio (cratered terrain). Large-scale brightness variations were removed by the television AGC (see text), and have not been restored in this montage of wide-angle photographs. The locations of the nested narrow-angle frames are indicated.

metres are generally flat-bottomed and rimless with diameter-to-depth ratios of about 100:1. They exhibit a wide range of morphology and state of preservation, although no fresh craters in this class have been found. The diameter-number distribution is similar to that of the primordial lunar surface around Tsiolkovsky. Craters larger than 0.7 kilometer (but smaller than those described above) tend to be bowl shaped and show very little variation in ageing. The diameter–number distribution is similar to that of primary craters on the lunar maria. The absence of strongly eroded bowl shaped craters is indicative of episodic bombardment or erosion. Unlike the moon, crater visibility shows no appreciable change over a range of solar elevation angle from a few degrees to over 40 degrees. The mean slopes of crater walls on Mars are very small, further evidence of significant erosion and dust transport.

The stratified haze layers were again seen at higher resolution and in three colors over the Chryse-Eden region during the near-encounter limb crossing of Mariner 7. The layers were about 10 kilometers thick, with their upper surfaces ranging from 25–40 kilometers above the surface. The optical thickness has been estimated as 0.01–0.03 with an apparent λ^{-2} wavelength dependence.

Sinus Meridiani and Deucalionis Regio were again observed, this time under high solar lighting. The path of the cameras then moved down and across the south polar cap, recorded the south areographic pole, and moved across the terminator while still on the cap (Fig. 3). The polar coverage was actually more than had originally been planned, because the interesting polar phenomena seen in the far-encounter pictures demanded a change in the near-encounter strategy for Mariner 7. The transition zone along the edge of the cap was only about 2 degrees of latitude in width, further evidence for very small topographic slopes on Mars. There is no visible evidence of the dark collar or "melt band" that has been alleged to surround the receding south cap. Many of the larger craters well within the polar cap show darkened or partially darkened floors and this suggests a tendency for the CO_2 snow to be preferentially removed from low areas and deposited on high areas. This might be accomplished either by wind or by sublimation–condensation processes. Close to the areographic pole are a series of nearly parallel, quasi-linear bright streaks of unknown physical explanation. Also found within the cap are several oddly shaped depressions of a form seen nowhere else on the planet.

After swinging the cameras back onto the illuminated part of Mars, a series of pictures was taken across Noachis, Hellespontis, and Hellas (Fig. 4). Noachis presents a cratered terrain similar to Deucalionis Regio, but Hellespontis was found to be a transition zone with decreasing crater densities and a series of scarps dropping down to the Hellas basin. Within Hellas itself, no readily identifiable physiographic features are seen down to the resolution limit of about 300 meters. This great circular feature Hellas, 1800 kilometers in diameter, is composed of "featureless" terrain. It is difficult to imagine the highly efficient yet selective erosional processes that have worked in the past and must be continuing even today.

Two phenomena of unusual controversy have been associated with Mars for many decades — the "blue haze" and "canals". The question of the sometimes

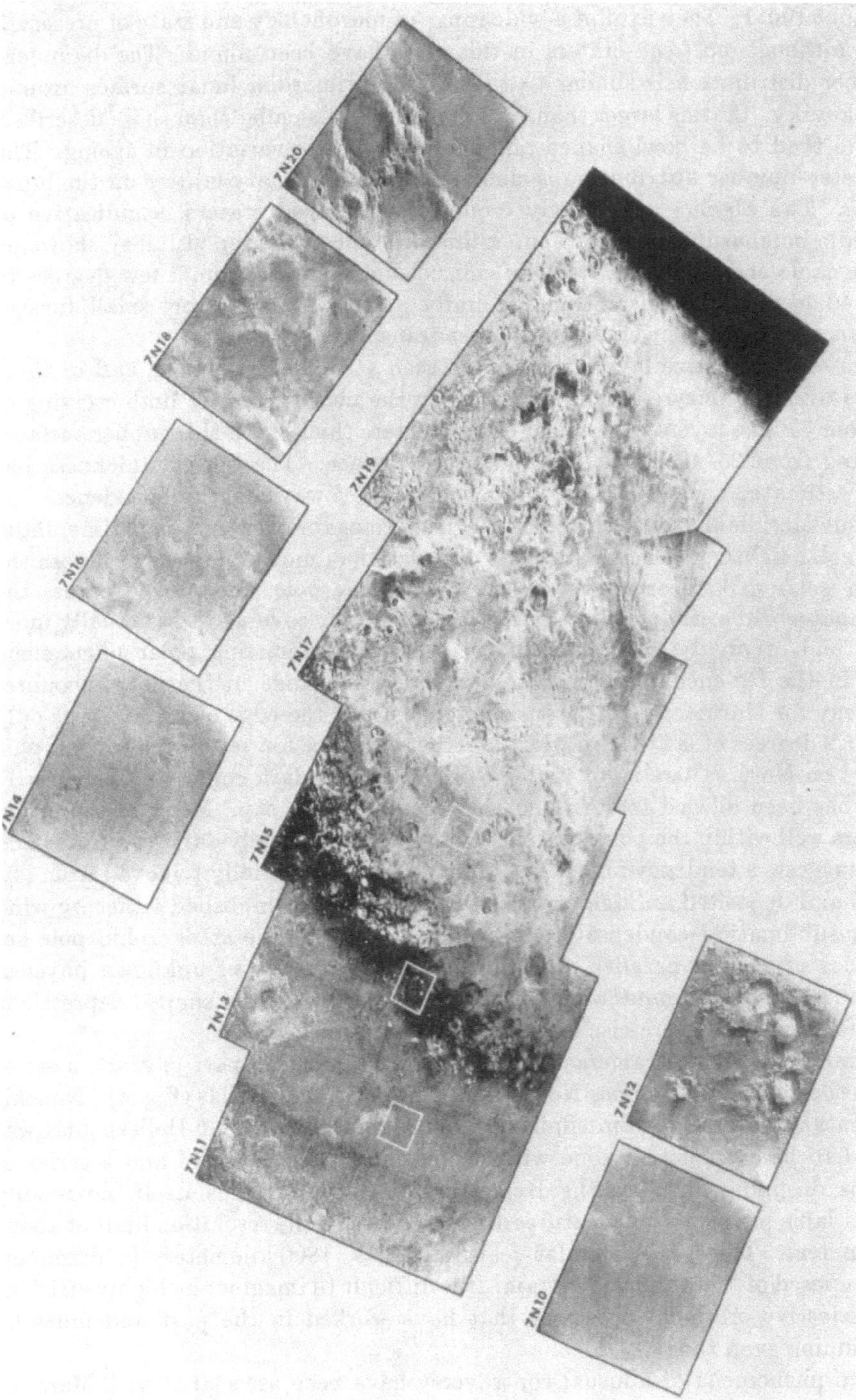
7N20
7N18
7N16
7N14
7N19
7N17
7N15
7N13
7N11
7N12
7N10

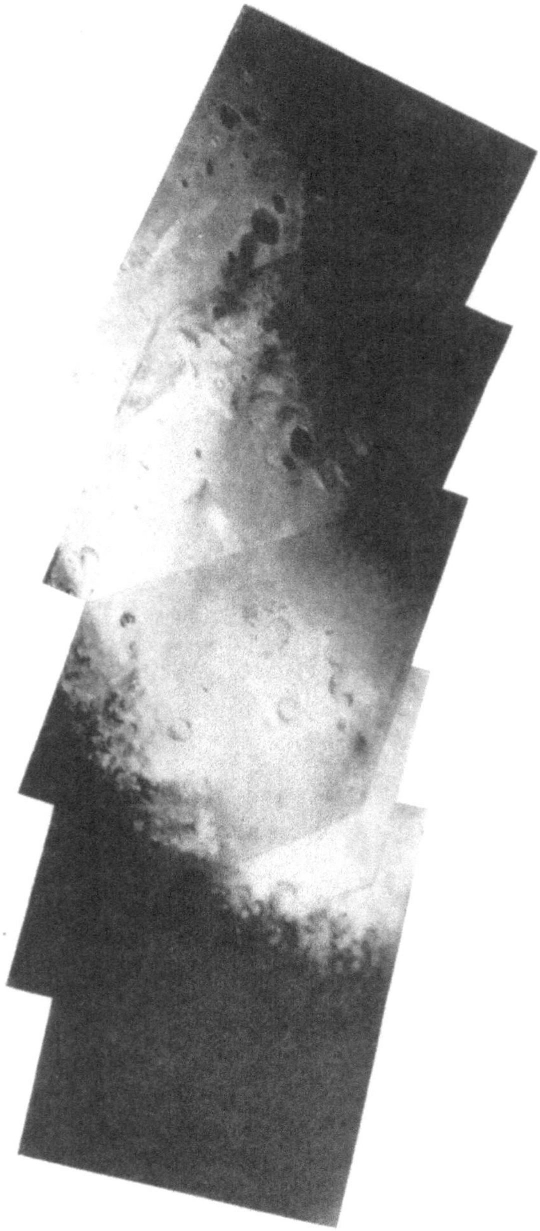

3 b

Fig. 3. The south polar cap of Mars uncorrected a) and corrected b) for the television AGC. Picture 7N19 was taken in blue light and clearly shows surface features.

Fig. 4. The region of Noachis, Hellespontis and Hellas. The scarps along the edge of the Hellas basin are clearly seen in 7N26. Featureless terrain covers much of 7N27 and all of 7N29.

thick, sometimes thin blue haze can be answered directly – it does not exist. All pictures taken through the blue filter on both spacecraft show geological features as clearly as those taken through the green and red filters. Albedo features, including dark crater floors, however, do not show up well in the blue pictures. There is simply no difference in the reflectivity between "light" and "dark" areas at the shorter wavelengths, and this as well explains the ground-based observations. Several "canals", including Gehon and Cantabras, have been identified as a quasi-linear arrangement of dark floored craters. Other canals

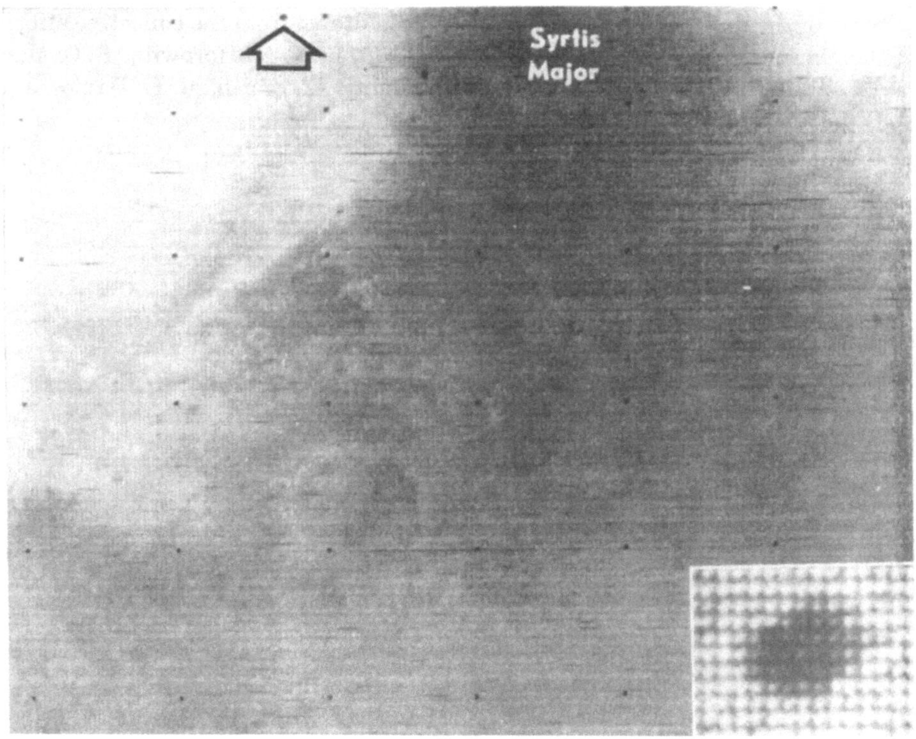

Fig. 5. Phobos as recorded on picture 7F91, with a 15× enlargement (insert). A part of Syrtis Major is seen near the top. The phase angle is 22° and the range is 130,900 kilometers to Phobos and 137,000 kilometers to the martian surface.

seem to be associated with alignments of diffuse dark patches that are found throughout the light areas on Mars. Most of the "Lowellian" canals, however, appear to be nothing more than the products of overactive imaginations of many visual observers, both contemporary and past.

On one of the last frames of the far-encounter series of Mariner 7, we were fortunate to catch Phobos silhouetted against Mars (Fig. 5). An analysis of this image has shown Phobos to be elongated along the plane of its orbit with major and minor diameters of $22\frac{1}{2}$ and $17\frac{1}{2}$ kilometers. The average visual geometric albedo of Phobos is only 0.065, the smallest known for any object in the solar system. Both the shape and reflectivity suggest that Phobos did not form as a product of accretion within the planetesimal cloud around primordial Mars, but was probably captured at some later time.

The results of Mariners 6 and 7 have answered some old questions, but have posed many new ones. It is hoped that some of these new questions will be answered by the orbiting spacecraft to be sent to Mars next year.

Acknowledgments

Much of the material presented here has been drawn from the collective efforts of the Mariner 69 Television Team: R. B. Leighton, N. H. Horowitz, B. C. Murray, R. P. Sharp, A. G. Herriman, A. T. Young, B. A. Smith, M. E. Davies, and C. B. Leovy.

References

[1] R. B. LEIGHTON et al., Science **165**, 684 (1969).
[2] R. B. LEIGHTON et al., Science **165**, 787 (1969).
[3] R. B. LEIGHTON et al., Science **166**, 49 (1969).
[4] B. A. SMITH, A. T. YOUNG and C. B. LEOVY, Science **167**, 908 (1970).
[5] B. A. SMITH, Science **168**, 828 (1970).

Space Research XI — Akademie-Verlag, Berlin 1971

SUMMARY OF MARINER 6 AND 7 RADIO OCCULTATION RESULTS ON THE ATMOSPHERE OF MARS

A. J. Kliore, G. Fjeldbo and B. L. Seidel

Jet Propulsion Laboratory, California Institute of Technology,
Pasadena, California, USA

During the close fly-by of Mars by Mariners 6 and 7 in the summer of 1969, their S-band radio beams were used to probe the atmosphere of Mars at four locations. These measurements indicate surface pressures ranging from a high of 7.3 mb in the Amazonis/Arcadia area to a low of 4.2 mb near the southern end of Hellespontus, indicating a range of local elevation differences of about 6 km. The surface temperatures range from about 250 °K measured near the equator in the afternoon, to about 173 °K in the north polar region at night. The temperature profiles suggest that condensation of CO_2 is probable at an altitude of about 15 km near the north pole, and at altitudes ranging from 27 to 38 km in equatorial and temperate latitudes. The daytime measurements also show the existence of an ionosphere with a primary layer of about 1.7×10^5 el cm^{-3} at an altitude of about 135 km. The topside scale height indicates a plasma temperature of about 400–500 °K if CO_2^+ is assumed to be the dominant ion.

1. Introduction

The United States Mariner 6 and 7 space probes were aimed to fly by Mars in such a manner that their S-band radio beams to the earth would be interrupted by the atmosphere and surface of Mars. After making its closest approach to Mars at about 0519 06 UT on 31 July 1969, Mariner 6 went into occultation about 15 minutes 25 seconds later and emerged 35 minutes 15 seconds later. Its trajectory was aimed to provide occultations at different latitudes, and hence entry into occultation occurred at 3.7° North in the area of Meridiani Sinus, and exit took place at a latitude of 79.3° North in the area north of Boreosyrtis. The other parameters pertinent to the occultation are given in Table 1. Mariner 7, which made its closest approach at 0500 49 UT on 5 August 1969, was aimed in order to obtain television images of the south polar cap, but the occultation geometry was also quite favorable. About 13 minutes and 18 seconds after closest approach, its radio beam was interrupted by the surface of Mars near the southern end of Hellespontus, and it emerged again about 43 minutes later in the area between Amazonis and Arcadia, thus affording a sampling of conditions at moderate latitudes. The longitudes, local times, and solar zenith angles for all four occultations are shown in Table 1.

The locations of the occultation points on a map of Mars are shown in Fig. 1. The Mariner 6 entry point at 3.7° North appears in near-encounter television pictures taken by both Mariner 6 and Mariner 7, and an analysis of these pic-

Table 1
Occultation parameters

	Time past closest approach	Area	Latitude	Longitude	Local time	Solar zenith angle
Mariner 6						
Entry	15 min 25.2 s	Meridiani Sinus	3.7° N	355.7° E	1534	57
Exit	35 min 14.7 s	North of Boreosyrtis	79.3° N	87.1° E	2210	107
Mariner 7						
Entry	13 min 18.8 s	Southern end of Hellespontus	58.2° S	30.3° E	1430	56
Exit	42 min 48.1 s	Arcadia/Amazonis	38.1° N	211.7° E	0310	130

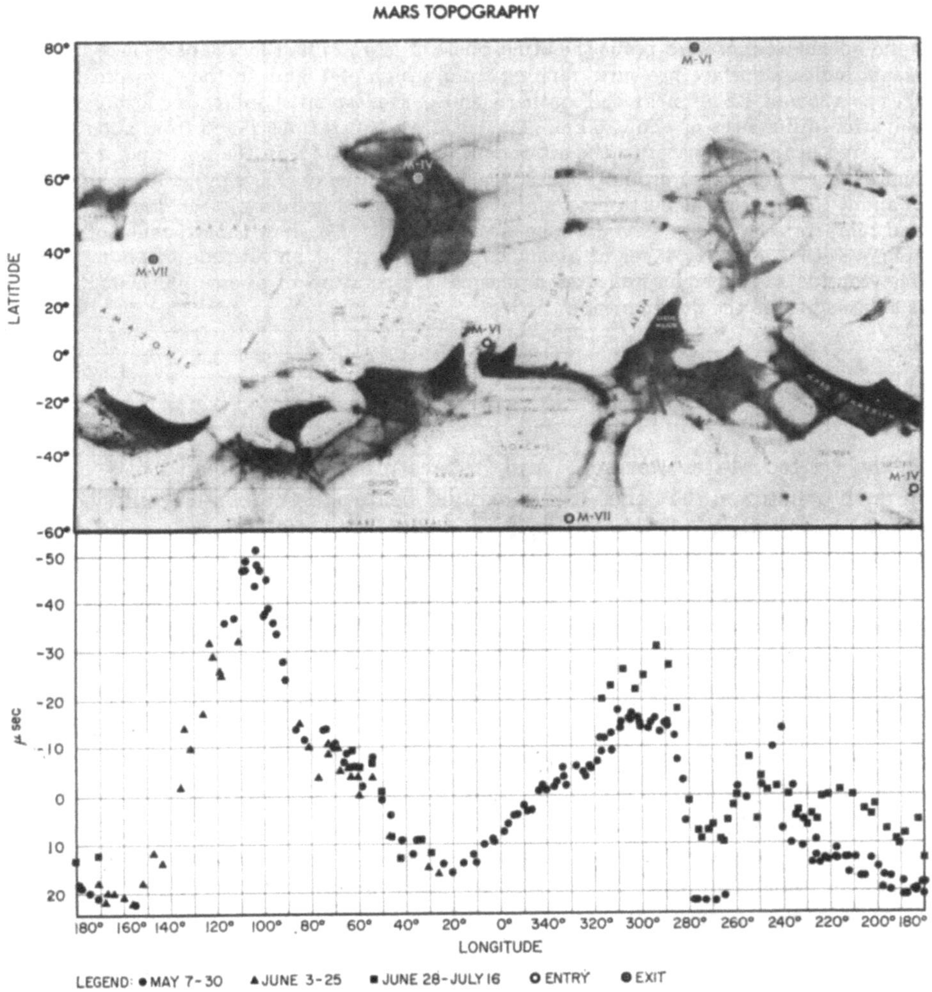

Fig. 1. Areographical locations of Mariner occultation points. The Mariner 6 entry point appears in near-encounter television pictures and is in the path of earth-based planetary ranging sub-earth traces for 1969. The lower portion shows range residuals (μs of delay) from planetary radar ranging measurements taken at the JPL/Goldstone facility in 1969 [1].

tures should provide a better insight into the nature of the local terrain over which the occultation measurements were made. The lower portion of Fig. 1 shows planetary radar ranging residuals as a function of longitude from measurements taken at the JPL Goldstone tracking station from May to August 1969. Zero delay corresponds to a planetary radius of 3393 km [1]. The latitudes examined by the planetary radar varied from 3° North to 12° North. In the area of about 5° West longitude, the radar measurements were taken at a latitude closely corresponding to the 3.7° North of the Mariner 6 entry occultation point. A comparison of the measured pressure in the Meridiani Sinus area with the associated range residual implies that the pressure in the north equatorial regions of Mars should vary from a high of about 6.7 mb in the Amazonis region (which is in fairly good agreement with the measurement of Mariner 7 during exit) to a low of about 2 mb in the Tharsis region.

2. Instrumentation and Data Analysis

The raw data for the occultation experiment consist of the measurement of the perturbations on the phase, frequency, and amplitude of the S-band spacecraft signal produced by propagation through the planetary atmosphere and ionosphere. These effects are discussed in detail in publications describing previous radio occultation experiments [2]. The basic methods to obtain the data have also been described [3], and the following is an abbreviated description of the instrumentation used for the Mariner 1969 experiment.

For both Mariners 6 and 7, the entry into occultation occurred in a two-way mode of operation, in which a frequency referenced to a rubidium frequency standard at the Deep Space Network (DSN) station is transmitted to the spacecraft, where it is coherently retransmitted. The exit data, however, were obtained in the one-way mode of operation in which the spacecraft's transmitter is referenced to its on-board crystal oscillator. The received signal was passed through the standard phase-locked-loop receiver, and the record of non-destructively counted Doppler signals was recorded. This constituted the closed loop data. Simultaneously, the signal was passed through an open-loop receiver having an audio passband of approximately 5 kHz. It produced a frequency-translated version of the received signal, which was recorded on an analog tape recorder, as well as being digitized in real time and recorded. These data are referred to as the open-loop data.

For the Mariner 6 and 7 occultation experiments, Deep Space Station (DSS) 14 (Mars) and DSS 12 (Echo) at the Goldstone, California, complex were instrumented with open-loop receivers, and the open-loop data from DSS 14 were digitized in real time and recorded. Closed-loop Doppler data were also taken at DSS 14 and DSS 12, as well as DSS 41 in Woomera, Australia.

Two major differences exist between the system that was used for the 1969 mission and that which had been used for the previous occultation experiments. First, the open-loop receivers were equipped with local oscillators that were completely independent of those used by the phase-locked-loop receivers. This simplified operations and improved the reliability of the total receiving subsystem. The second difference was that for the first time the open-loop occultation data were digitally recorded in real time. This improved the quality of the received data by not introducing time-base instabilities.

The closed-loop data, on which all of the results described in this paper are based, were then compared with predictions generated by an orbit determination program to obtain Doppler residuals. These residuals ideally should represent the effects of the ionosphere and atmosphere of Mars only, with all

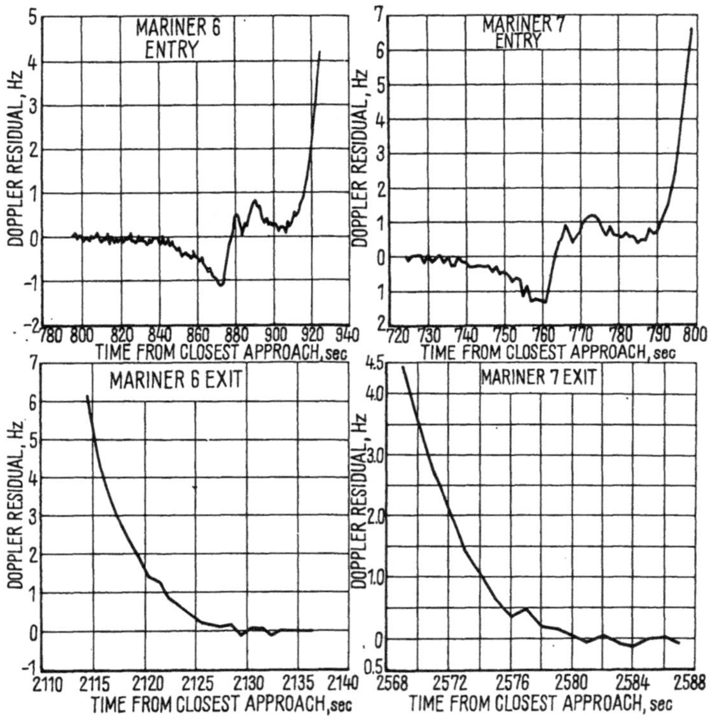

Fig. 2. Corrected Doppler residuals computed from closed-loop data. These residuals form the basic observable for the experiment. High-frequency fluctuations are representative of random noise in the data.

other sources of apparent change in Doppler effect removed. In reality, however, because of uncertainties in the spacecraft orbit, as well as drifts in reference oscillator frequency, bias and drift terms are usually present in these residuals. These unwanted components are removed by fitting a straight line to a specified portion of the Doppler residual data prior to or, in the case of exit, after, the regions of data in which the effects of the atmosphere of Mars appear, and then using that line as the zero reference for the residuals in the atmosphere-affected regions. Examples of such corrected residuals for all four occultation measurements of Mariners 6 and 7 are shown in Fig. 2. It should be pointed out that the entry residuals reflected a two-way passage through the atmosphere and ionosphere of Mars, whereas the exit residuals are the result of only a one-way passage. The effect of the main peak of the ionosphere, as well as a minor peak, is readily apparent from the entry residuals; no such effect is seen in the exit data. These residual data are not smooth and hence the appearance of the graphs in Fig. 2 gives a fairly realistic indication of the random uncertainty of the measurements.

3. Summary of Results

The preliminary results of Mariner 6 and 7 measurements of the martian atmosphere have been described in previous publications [4–6]. The results described here represent the state of data analysis as at March 1970.

After the corrected residuals described in § 2 are integrated, the result is the total phase delay, or advance in the case of the ionosphere, caused by the refrac-

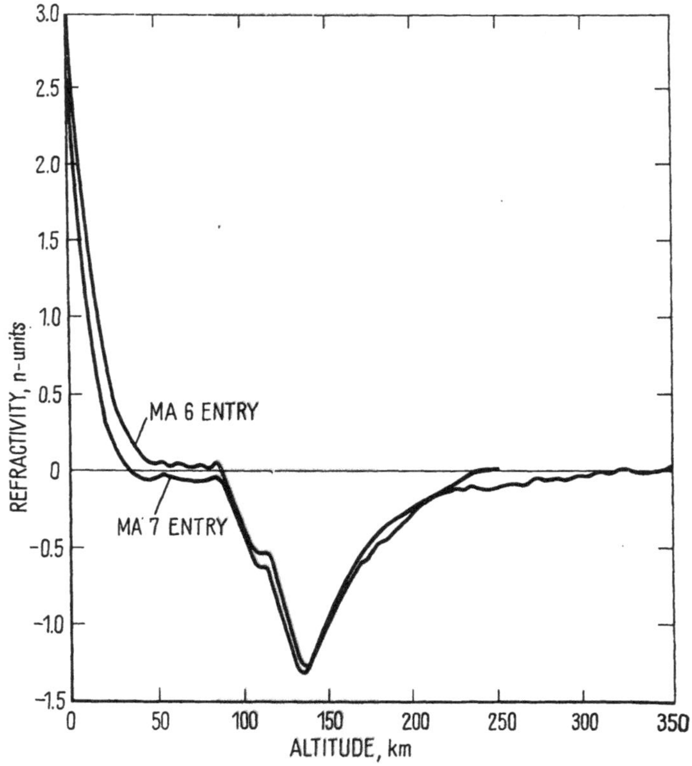

Fig. 3. Refractivity as a function of altitude from entry data of Mariners 6 and 7. Note the similarity of the ionospheric effects in the two profiles.

tive effects. A special ephemeris of the spacecraft is then used to compute and remove that portion of the phase change due to the deflection of the path of the signal by refractive bending. The resulting phase propagation delay quantity is then inverted by using the Abel integral transform [7] to obtain profiles of the refractivity as a function of the distance of closest approach of a ray from the center of the planet. The refractivity profiles for the two entry measurements are shown in Fig. 3. The distance axis on this figure, as well as on all other figures, is labelled with altitude, which is defined here as the difference in radial distance from the center of the planet between the point in question and the last point of measurement. The actual radial distances are not used because it is felt that the uncertainty in the knowledge of the orbits of both Mariners 6 and 7 is still large enough to make the relevance of that quantity questionable. The negative refractivities represent the effects of the charged particles in the

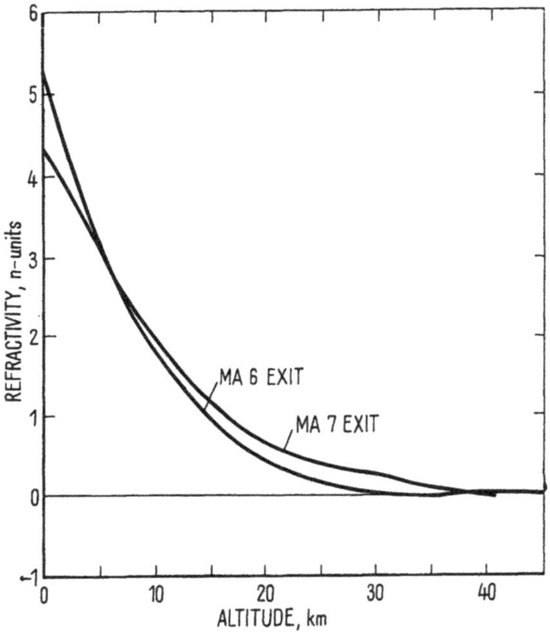

Fig. 4. Refractivity as a function of altitude from exit data. Note the different slopes, indicative of widely differing temperatures.

ionosphere, while positive refractivity is due to neutral molecules in the lower atmosphere. The region between the two does not contain a sufficient amount of either charged or neutral particles to affect the S-band radio signal, and hence the refractivity there should be zero. However, a slight bias due to processing of the data is evident in Fig. 3. This bias is removed before further processing. Refractivity profiles obtained from the exit measurements are displayed in Fig. 4. Here there is no evidence of an ionosphere, and the differing slopes of the two profiles indicate widely different temperatures in the regions that these profiles represent.

To obtain other quantities, such as temperature and pressure, from the refractivity results, the composition of the atmosphere must be known. In the absence of evidence for other significant constituents in the results of the Mariner 1969 ultraviolet spectrometer experiment [8], the assumption is here made that the atmosphere of Mars consists mostly of carbon dioxide. All temperatures and pressures shown in this paper are computed under the assumption of a 100% CO_2 atmosphere. Once the composition is assumed, one directly obtains the mass density from the refractivity. Then a direct application of the hydrostatic equation and the perfect gas law yields temperature and pressure. However, if the temperature obtained at the pressure computed by the hydrostatic equation falls below the saturation temperature of CO_2 at that pressure, it is assumed that the saturation law must hold and the temperature can not lie below the saturation temperature for any given pressure. In that case, an iteration is

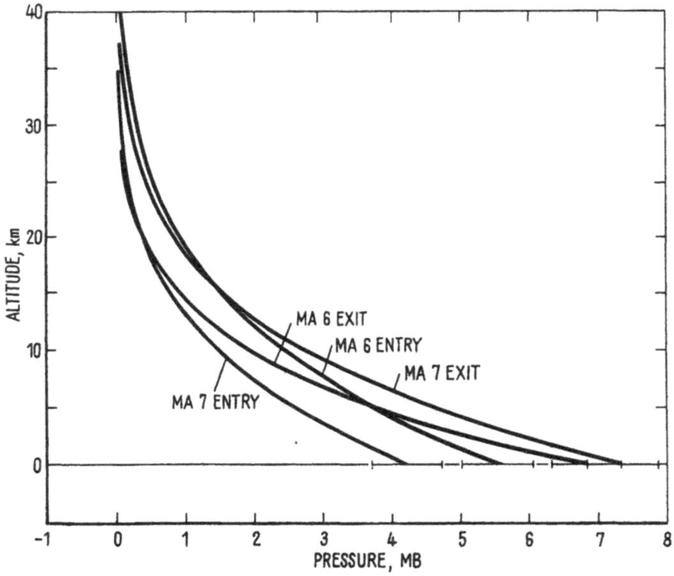

Fig. 5. Pressure as a function of altitude from Mariner 6 and 7 radio occultation measurements.

performed until a pressure–temperature point lying on the saturation boundary is obtained which is also consistent with the observed refractivity.

The profiles of pressure as a function of altitude for all four measurements are shown in Fig. 5. The uncertainty boundaries shown are based on an assumed random uncertainty of 0.05 *n*-units in the refractivity data, as well as possible systematic errors introduced by uncertainties in the orbit of the spacecraft. The variations of the surface pressure are probably due to variations in local elevation, but no attempt was made to plot these pressures against equipotential altitude based on a constant-pressure surface, because of the uncertainty of the radial distance from the center of the planet of one profile relative to another. The rather high surface pressure of 7.3 mb observed for the Mariner 7 exit measurement is consistent with the observation from planetary radar ranging that the general area of Amazonis is lower relative to the other equatorial regions. However, the Mariner 6 entry pressure of about 5.5 mb indicates that this area, in the vicinity of Meridiani Sinus, is not particularly low; in fact, it is more consistent with the observations of martian topography from Mariner 6 and 7 infrared spectra [9], which indicate that Meridiani Sinus is a rather high region. The low value of 4.2 mb obtained at the Mariner 7 entry point at the south end of Hellespontus clearly indicates that this region is elevated relative to the mean equipotential surface of Mars.

The profiles of temperature as a function of altitude for the four Mariner 1969 occultation measurements are shown in Fig. 6, labelled with the latitude of the occultation point and the local time. The sharp breaks in the profiles are indicative of noise in the refractivity data, as are the lengths of the uncertainty boundaries, which were again computed on the basis of a 0.05 n-unit random uncertainty in refractivity.

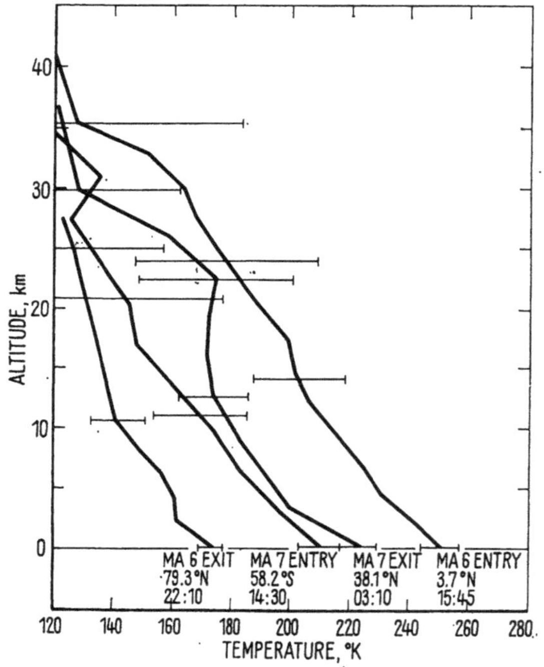

Fig. 6. Temperature as a function of altitude from Mariner 6 and 7 radio occultation measurements. Uncertainty increases with altitude.

The "inversion" in the Mariner 7 exit profile between about 15 and 22 km is apparently real, at least in nature if not in magnitude, as it persists when data from different DSN stations are used. The most significant difference between present results and those reported previously appears in the nature of the Mariner 6 entry temperature profile [6]. In the previous reduction, because of improper removal of bias from the Doppler residuals, an artificial region of high constant temperature was observed from about 10 to 40 km. This was understood to be in conflict with theoretical predictions, and a careful re-analysis of the data has produced the results reported herein, which display no such anomalous behavior. Also, the surface temperature of 250 °K indicated at the point of Mariner 6 entry is in good agreement with that predicted theoretically on the basis of a surface temperature of about 275 °K observed by the infrared radiometer experiment of Mariner 6 [10].

A crossplot of the temperature against pressure is shown in Fig. 7, again labelled with the measurement, latitude and local time. The uncertainty boundaries are not shown for clarity but one should not forget that they are rather large in temperature, as shown in Fig. 6. The smooth line at the left of the diagram represents the saturation temperature curve for carbon dioxide. Examination of Fig. 7 clearly indicates that conditions favorable to the condensation of carbon dioxide are reached at some altitude in the case of all four measurements. For the Mariner 6 exit measurement, occurring in the north polar region at night, the CO_2 condensation altitude is about 15 km. For the Mariner 7 entry occurring

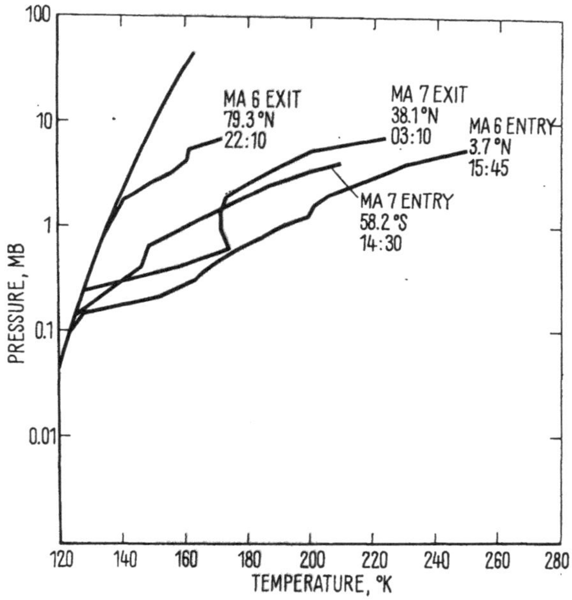

Fig. 7. Pressure as a function of temperature. The smooth solid line on the left is the carbon dioxide saturation curve.

near the south polar cap in early afternoon, and for the Mariner 7 exit, occurring at a moderate northern latitude at night, the altitudes of CO_2 condensation are 27 and 30 km, respectively. Even in the region of Mariner 6 entry, near the equator in the martian afternoon, the temperature profile reaches the saturation temperature at about 38 km. These results are in relatively good agreement with the observation of solid CO_2 reflection features in the infrared spectra taken off the limb of Mars by the Mariner 6 infrared spectrometer experiment [11].

The Mariner 69 radio occultation measurements also provided new data on the structure of the upper atmosphere of Mars [5]. The main ionization layer was observed at 135 km altitude with a peak density of approximately 1.7×10^5 el cm^{-3}; a minor layer was located 25 km below the main peak.

Interpreting the main ionization peak as an F1 layer consisting primarily of CO_+^2 ions yields a topside plasma temperature of 400 to 500 °K and a neutral density of approximately 10^{10} molecules cm^{-3} at an altitude of 135 km. The topside plasma temperature obtained in this manner appears to be in good agreement with the preliminary results deduced from the Mariner 69 ultraviolet spectrometer experiments [12]. Thus, the question concerning the exospheric temperature of Mars may finally have been settled. A number of other intriguing details in the puzzle are still missing, however. For example, it is not yet clear why photodissociation of CO_2 by solar ultraviolet fails to generate enough atomic oxygen to make this the principal constituent at ionospheric heights. The Mariner results have, therefore, brought about a re-evaluation of the chemistry of CO_2 and O. In time, this new work may also lead to greater understanding of the CO_2 chemistry of our own environment.

A summary of all results appears in Table 2.

Table 2
Summary of results

(1) Measurement; (2) latitude; (3) local time; (4) surface pressure (mb)*; (5) surface temperature (°K)*; (6) CO_2 condensation altitude (km); (7) maximum electron density (cm^{-3}); (8) altitude of maximum (km); (9) plasma temperature (°K)**

(1)	(2)	(3)	(4)	(5)	(6)	(7)	(8)	(9)
MA 6 entry	3.7°N	1545	5.5 ± 0.5	250.3 ± 7	38	1.7×10^5	135	400–500
MA 6 exit	79.3°N	2210	6.9 ± 0.5	172.7 ± 3	15	–	–	–
MA 7 entry	58.2°S	1430	4.2 ± 0.5	210.3 ± 7	27	1.7×10^5	135	400–500
MA 7 exit	38.1°N	0310	7.3 ± 0.5	223.8 ± 4	30	–	–	–

* For 100% carbon dioxide
** For CO_2^+ ions

Acknowledgments

The authors thank the NASA Program Office under Program Manager N. William Cunningham and the JPL Mariner 1969 Project Office with Project Manager H. M. Schurmeier, and the entire staff of the project, for assuring the successful performance of the spacecraft and the success of the mission. Special recognition is also due David L. Nixon for his contributions in the acquisition and analysis of the open-loop data, as well as to M. P. Milane and Joan E. Jordan for unselfish help with the computation under adverse circumstances. As a co-investigator, S. I. Rasool, with R. Stewart, J. Hogan and L. Russell provided helpful interpretations of observations.

This paper presents the results of one phase of research carried out at the Jet Propulsion Laboratory, California Institute of Technology, under Contract No. NAS 7-100, sponsored by the National Aeronautics and Space Administration.

References

[1] D. A. O'Handley, W. G. Melbourne, R. M. Goldstein, G. A. Morris and G. S. Downs, Paper No. LP 1 Am. Astronomical Soc. Meeting, New York, N.Y., December 1969.

[2] A. Kliore et al., Science 49, 1243 (1965).

[3] G. S. LEVY, T. OTOSHI and B. L. SEIDEL, IEEE Trans. Instrum. Meas. IM-16 (1967).
[4] A. KLIORE, G. FJELDBO and B. SEIDEL, Radio Science **5**, 373 (1970).
[5] G. FJELDBO, A. KLIORE and B. SEIDEL, Radio Science **5**, 381 (1970).
[6] A. KLIORE, G. FJELDBO, B. L. SEIDEL and S. I. RASOOL, Science **166**, 1393 (1969).
[7] G. FJELDBO and V. R. ESHLEMAN, Planet. Space Sci. **16**, 1035 (1968).
[8] C. A. BARTH, paper presented at the IAU Symposium 40, Marfa, Texas, October 1969.
[9] K. C. HERR, D. HORN, J. MCAFFEE and G. C. PIMENTEL, Astrophys. J., in press.
[10] G. NEUGEBAUER et al., Science **166**, 98 (1969).
[11] K. C. HERR and G. C. PIMENTEL, Science **167**, 47 (1970).
[12] A. I. STEWART, paper presented at 1969 Fall Meeting of the Am. Geophys. Union, San Francisco, California 1970.

Space Research XI — Akademie-Verlag, Berlin 1971

COMPARISON OF PARAMETERS OF THE MARS IONOSPHERE ACCORDING TO MARINERS 4, 6 and 7 MEASUREMENTS AND TO CALCULATIONS OF RADIOWAVE ABSORPTION

A. N. KAZANTSEV

USSR

The integral coefficients of radiowave absorption in the Mars ionosphere are determined by comparison of ionosphere parameters obtained under different conditions. The ionosphere profiles and the law of electron concentration distribution with altitude for each ionized region have been studied. The main layer seems to correspond best to model F. 1 The altitude profiles of the effective frequency of electron–gas molecule collisions are given, taking into account that the effective collision frequency of electrons with CO_2 molecules is much greater than for the air in the earth's atmosphere, other things being equal.

The integral coefficients of absorption in the main ionized region, in the lower adjacent layer and in narrow layers of sporadic type have been calculated. The absorption curves are given for frequencies from 1000 MHz to 10 MHz, which is important for the case when the wavelength is not limited by the earth's ionosphere.

For ionospheric profiles of Mariner 6 and Mariner 7 the total number of electrons in a vertical column with a 1 cm^2 cross section is found to be 2.5 times greater than for the Mariner 4 profile.

The purpose of this paper is to determine the radio-wave absorption characteristics in the Mars ionosphere on the basis of comparison of the ionosphere parameters obtained under different conditions. In the paper we used publications [1–7] by the authors of the experiments.

According to these publications the maximum electron concentration in the Mars ionosphere, as recorded by Mariner 6 and 7, is about 1.7×10^5 el cm^{-3} and about 0.95×10^5 el cm^{-3}, according to Mariner 4 data. For convenience these values are translated to the zenith solar angle 0°, which gives $N_0 \approx 2.3 \times$ $\times 10^5$ el cm^{-3} for Mariner 6 and Mariner 7, and $N_0 = 1.5 \times 10^5$ el cm^{-3} for Mariner 4.

The profile of the main ionized region satisfies the Chapman law quite well. The lower region of the main layer (up to the maximum) may be approximated by a parabola, if the region is separated from the immediately adjacent layer which gives the second maximum. The main layer seems to correspond best to model F 1.

To calculate the integral coefficient of radio-wave absorption in the Mars ionosphere one must know the law of distribution of electron concentration with altitude for each ionized region, and the altitude profile of the effective frequency of electron–gas-molecule collisions.

In this paper we assume the Mars atmosphere to consist of 100% CO_2 (which is the most unfavourable case in terms of electron–gas-molecule collisions and radio-wave absorption).

In plotting graphs of the distribution of number of collisions with altitude we took into account the fact that the effective frequency of collisions of electrons with CO_2 gas molecules is approximately 20 times greater than that for the earth's atmosphere consisting of N_2 and O_2, other things being equal. This is because the effective cross section of elastic scattering is 20 times greater for CO_2 than for the earth's atmosphere within the electron energy interval corresponding to the thermal motion in the ionosphere of Mars [8]. Curves of the altitude distribution of collisions from the data of Mariners 4, 6 and 7 are given in Fig. 1.

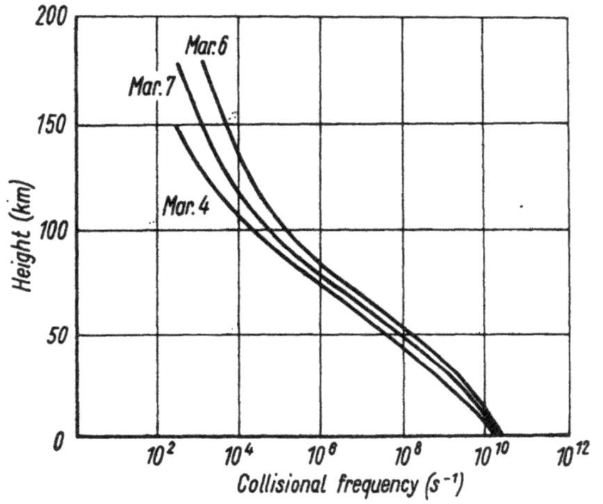

Fig. 1. Variation of collisional frequency with altitude.

Integral coefficients of absorption in separate regions of the Mars ionosphere have been calculated. Absorption in the lower region of the main layer (up to the maximum of electron concentration [5]) is calculated from the formula for parabolic distribution of electron concentration with altitude [9]. The integral coefficient of absorption in the lower layer (immediately adjacent to the main one) which has the maximum electron concentration $N_{max} \approx 7 \times 10^4$ el cm^{-3}, is found to be half that in the lower region of the main layer. This layer might correspond to the earth's E layer. It is interesting that the maximum electron concentration in this additional layer had increased in 1969 (Mariners 6, 7) by 2.5 to 3 times in comparison with 1965 (Mariner 4), though the maximum electron concentration in the main layer had increased by less than 60%.

Integral coefficient of absorption in the upper region of the main layer was found on the assumption that variation in electron concentration and number of collisions are governed by an exponential law.

One can observe in the lower part of ionization profile obtained from the Mariner 7 data [5] a narrow ionized layer (probably somewhat resembling the Earth's D layer). The integral coefficient of absorption in this layer can be readily found from the formula for a linear law of variation in electron concentration with altitude.

Fig. 2 shows the graphs of variation of integral coefficient of absorption with frequency, from 1000 MHz to 10 MHz; this range is significant for the case when the space ship frequencies are not limited by the earth's atmosphere. As can be seen on the curve referring to the equatorial ionosphere (Mariner 6) the absorption coefficient for the decametre range can reach a considerable value.

Absorption coefficients are calculated for a double path through the Mars ionosphere (direct and reverse paths) for directly incident radio waves. The values of the absorption coefficients for the equatorial area of the planet are transformed to the subsolar point (angle $\chi = 0$), while the coefficients for middle latitudes are transformed to local noon.

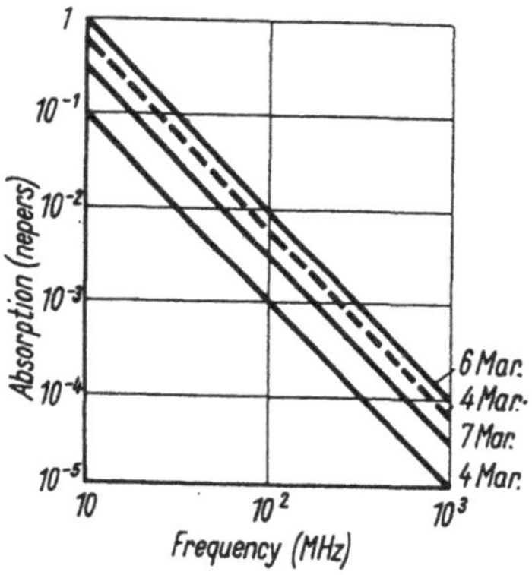

Fig. 2. Variation of absorption coefficient with frequency.

It is interesting to point out that on the ionosphere profile obtained from Mariner 4 [3] data there are several narrow layers of a sporadic type, the lowest being observed at an altitude of only 50 km. Due to the large number of collisions the absorption in this layer (if any) is almost an order of magnitude greater than in the main ionized region (broken curve in Fig. 2).

The total number of electrons in a vertical column of the ionosphere with a 1 cm^2 cross section is determined on the basis of ionosphere profiles. For the profiles derived from the Mariner 6 and 7 data this is equal to 1.25×10^{12} el cm^{-2}, that is, 2.5 times greater than for a profile derived from the results of Mariner 4 (5×10^{11} el cm^{-2}).

References

[1] A. Kliore et al., Science **149**, 1243 (1965).
[2] A. Kliore et al., Report presented at the COSPAR VII Int. Space Science Symp., Vienna 1966.
[3] G. Fjeldbo and V. R. Eshleman, Planet. Space Sci. **16**, 1035 (1968).
[4] A. Kliore et al., Report presented at IAU–URSI Symp. on Planetary Atmospheres and Surface, Pasadena, California 1969.
[5] G. Fjeldbo et al., Report presented at IAU–URSI Symp. on Planetary Atmospheres and Surface, Pasadena, California 1969.
[6] A. Kliore et al., Science, **166**, 1393 (1969).
[7] A. Kliore, Space Research XI, Akademie Verlag, Berlin 1971 (P. 165).
[8] E. W. McDaniell, Collision Phenomena in Ionized Gases, 1964.
[9] A. N. Kazantsev, Trud. IRE Akad. Nauk SSR, No 2, 36 (1956).

Space Research XI — Akademie-Verlag, Berlin 1971

ON GAMMA-RADIATION OF THE ATMOSPHERE AND SURFACE OF MARS

YU. A. SURKOV, L. P. MOSKALEVA, A. N. KHALEMSKY and V. P. KHARYUKOVA

V. I. Vernadsky Institute of Geochemistry and Analytical Chemistry, Academy of Sciences of the USSR, Moscow, USSR

In experiments on high-altitude air-balloons and on accelerators spectra of γ-radiation have been obtained. This radiation originates during the interaction of high-energy protons with different rocks and rock-forming elements. Cross-sections of γ-quanta of different energies have been determined.

Calculations have been made of the intensity and spectra of γ-radiation expected on the orbit of an artificial satellite of Mars assuming that the surface of the planet is formed of silicate and iron-containing rocks or has a moisture-bearing cover.

For some models of chemical composition and density of the martian atmosphere the attenuation of spectra of γ-radiation of the surface has been computed, as well as the contribution of the atmosphere's own γ-radiation to the γ-radiation expected on the orbit of the satellite.

The possibility of determining the chemical composition of the planetary atmosphere and surface has been discussed. This determination has been made by cross sections of emission of γ-quanta from nuclei of elements forming the atmosphere and rocks.

1. Introduction

The study of the composition of planetary atmospheres and surfaces is one of the major problems solved with the aid of automatic stations. The study of radioactivity and in particular γ-radiation makes it possible to determine the type of rocks forming the lunar surface [1, 2]. The present paper gives the calculation of the intensity and spectral composition of γ-radiation due to interaction of cosmic rays with the atmosphere and surface of Mars, as well as γ-radiation originated from decay of natural radioactive elements contained in martian rocks. The possibility of determining the type of martian rocks from artificial satellites of Mars is also discussed in the paper.

The calculation is based on the results of measuring γ-radiation of the lunar surface from a satellite of the moon [2] and on experimental data obtained from high-altitude air-balloons under the effect of cosmic rays and at accelerators with proton energies of 150 and 660 MeV [3, 4].

2. Gamma-Radiation of Rocks forming the Surface of Mars

To obtain experimentally the spectra of γ-radiation similar to those which are expected on Mars the silicate rocks of the type of granite, basalt and dunite which are most widely distributed on the earth were irradiated by cosmic rays. With this aim in view a container with a layer of rock about 50 g cm^{-2} thick was lifted on air-balloons to a height of 33–34 km. With this thickness of the rock layer the γ-radiation on its surface was over 80% of the total intensity for an infinitely thick layer for energies $E \geqq 1$ MeV.

The γ-radiation measured in these experiments is caused by two processes: (i) decay of natural radioactive elements (uranium, thorium and potassium), (ii) nuclear reactions originated during the interaction of cosmic rays with the rock. The latter lead mainly to the formation of excited nuclei-products and to the formation of cosmogenic isotopes with different half-lives. Bremsstrahlung and π^0 meson decay makes a considerable contribution to the γ-radiation recorded. For instance, the yield cross section of bremsstrahlung does not exceed 10^{-31} cm^2 sr^{-1} MeV^{-1} for light nuclei at a proton energy of 150 MeV [5]. The cross sections of production of π^0 mesons on nuclei at $E_p = 660$ MeV amount to a value of less than 10 mbn. The spectrum of γ-quanta from the decay of π^0 mesons falls sharply with the decrease of the energy of γ-quanta. Table 1 gives

Table 1
Cross sections of γ-quanta emission at interaction of cosmic rays with rocks

E (MeV)	σ (mbn)		
	Dunite	Basalt	Granite
1.3	205	160	185
1.7	145	123	116
2.1	80	78	85
2.6	120	59	79

the cross sections of emission of γ-quanta (main lines) produced at the interaction of cosmic radiation with various rocks.

However, air-balloon experiments do not give exhaustive information on induced γ-radiation. For instance, a relatively small irradiation time and low intensities of cosmic rays prevent us from detecting the contribution of long-lived isotopes. Also, as a result of the latitudinal cut-off effect only reactions with cosmic rays of energy over 2 GeV have been studied in these experiments.

To obtain more information various rocks (granite, basalt, dunite) and rock-forming elements (C, O, Al, Mg, Si, etc.) were irradiated by protons with energies 150 and 660 MeV on the synchrocyclotron of the Joint Nuclear Research Institute at Dubna . It was found that γ-quanta emission cross sections were practically constant over a wide range of proton energies. The contribution of γ-radiation from the decay of long-lived isotopes was of the order of 1–3%, so that this process can be ignored in our calculations. Thus, γ-quanta emission cross sections measured on air balloons can be used for plotting γ-spectra expected near the martian surface.

The final correction of data obtained on air balloons consisted in reducing them to martian conditions. The first stage of this calculation consisted in the normalization of γ-spectra measured on air-balloons to the intensity of γ-radiation of lunar rocks measured by Luna 10 [2]. The normalization coefficient was chosen under the assumption that the integral intensity of γ-radiation with energies of over 2.7 MeV is determined only by the interaction of cosmic rays with the rock and is practically independent of its composition. This normalization permits one to take into account some factors specific for the space experiment (the indefinite rock mass as compared with model experiments, the continuous spectrum of primary cosmic radiation, the effect of solar cosmic radiation, etc.). The second stage consisted in the recalculation of γ-spectra normalized in this way to the conditions of measurements from the orbit of a satellite of Mars. The intensity of γ-radiation recorded by the detector is determined by the following parameters:

$$I = B\, n_0\, S\, G\, \sigma \tag{1}$$

where I is the intensity of recorded γ-radiation (s^{-1}), $B = N\, d/A$ is the number of nuclei per cm^2, n_0 is the proton flux in particles $s^{-1}\, cm^{-2}$, S is the surface (cm^2) on which the proton flux is incident and from which γ-radiation is recorded, A is the mean atomic number of the rock, G the geometric factor, N Avogadro's number, d the thickness of the rock layer (g cm^{-2}) from which γ-radiation is recorded, σ is the γ-quanta emission cross section.

Since the proton fluxes incident upon unit of the surface of the moon and Mars can be regarded as equal (the thickness of the martian atmosphere will lead only to some distortion of the primary spectrum of cosmic rays, which should not affect essentially the magnitude of the cross section σ and the thickness d), if $\sigma_1 = \sigma_2$ and $d_1 = d_2$ the intensity of γ-radiation which will be recorded by the sensor from the orbits of artificial satellites of the moon and Mars can be expressed as:

$$\begin{aligned} I_1 &= q\, S_1\, G_1 \\ I_2 &= q\, S_2\, G_2\,, \end{aligned} \tag{2}$$

where $q = B\, n_0\, \sigma$ is the intensity of γ-radiation from a surface unit identical for the moon and Mars. Index 1 refers to the moon and index 2 to Mars.

The value $S\, G$ is determined as follows. When the distance from the detector to the surface of the planet is comparable with the radius of the planet, its surface can be considered a sphere, and the thickness of the rock layer from which γ-radiation is recorded can be ignored while calculating G. Thus, it is only necessary to determine the intensity of γ-radiation from the impenetrable sphere covered by the thin layer of radiating substance. [6] gives the solution of the similar problem for the hollow sphere; however, in the case of the hollow sphere the integral is taken over the whole surface of the sphere, whereas in the present case it is taken only over the surface from which γ-radiation is recorded. Hence, while solving this problem one can write down the magnitude of the intensity of γ-radiation recorded by the detector in the form:

$$I = \int_0^{S_{max}} q\, G_S\, dS$$

where G_S is the geometrical factor taking into account the portion of γ-radiation from each area unit recorded by the detector, and S_{max} is the maximum area from which γ-radiation is recorded.

Integrating, we find :

$$I = q \frac{R}{2(R+H)} \ln\left(1 + \frac{2R}{H}\right) \tag{3}$$

where R is the radius of the planet and H the distance from the detector to the surface of the planet.

From Eqs. (1), (2) and (3) a formula is obtained from which one can transform intensities from the moon to intensities from Mars:

$$I_2 = I_1 \frac{R_2(R_1 + H_1)}{R_1(R_2 + H_2)} \frac{\ln(1 + 2R_2/H_2)}{\ln(1 + 2R_1/H_1)} \tag{4}$$

These calculations permitted one to plot spectra of γ-radiation of various rocks for conditions of recording at different distances from the martian surface, for example, Fig. 1 gives spectra of γ-radiation due to interaction of cosmic rays

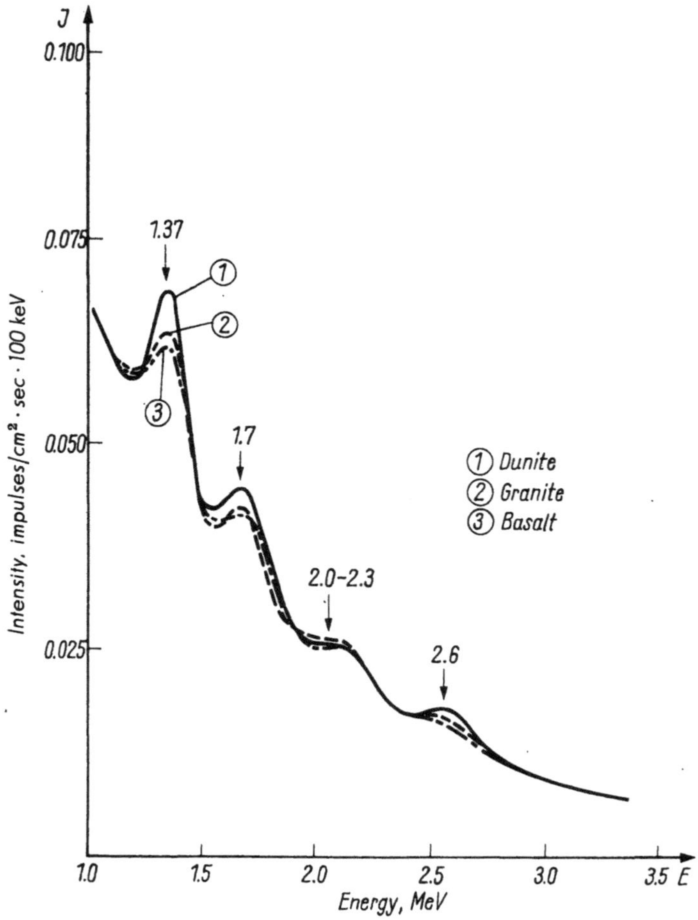

Figs. 1 and 2. Spectra of γ-radiation from cosmic rays computed for the γ-spectrometer located at a height of 1000 km above the surface of Mars.

Fig. 1. Radiation due to interaction with various rocks.

with the surface of the planet at a distance of 1000 km from Mars, assuming that the martian surface is formed of silicate rocks of the type of granite, basalt and dunite. For the same rocks Fig. 2 gives the spectra of γ-radiation both due to interaction of cosmic rays with the rock and due to decay of natural radioisotopes contained in the rock. γ-spectra were calculated for the scintillation γ-spectrometer with a 40 mm $\times$ 40 mm Na I (Tl) crystal and with resolution of about 9% (for ^{137}Cs).

For distances of 40000–50000 km from the martian surface the $S\,G$ factor becomes three orders of magnitude lower than for 1000 km. Hence, measurements at distances of more than 4000–50000 km can be used to estimate the background due to the satellite's own γ-radiation.

At present, there are no convincing data on the chemical composition of martian rocks. According to different sources [7], one can expect on Mars the

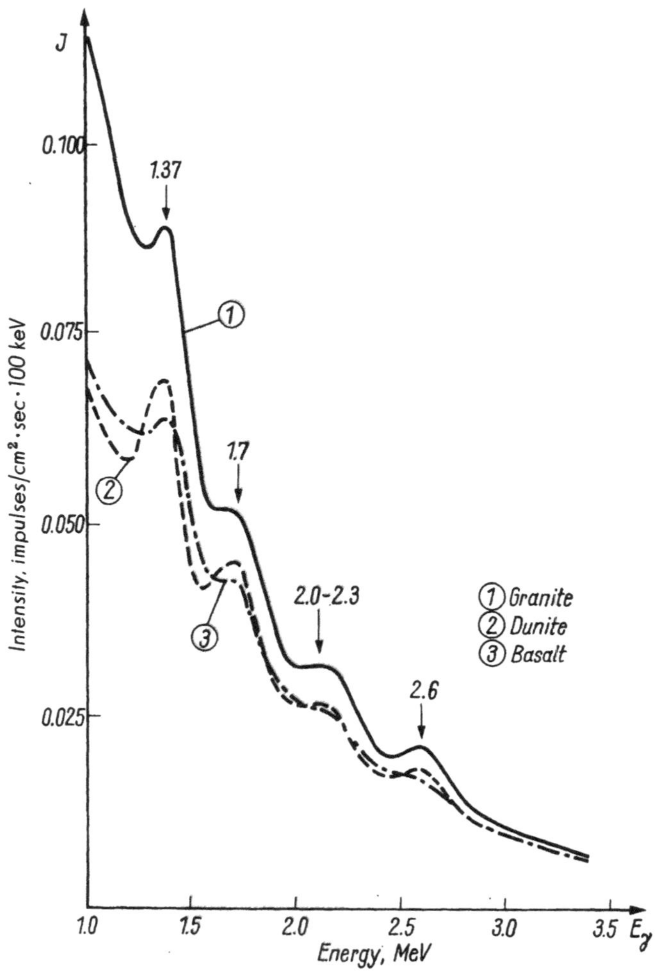

Fig. 2. Radiation due to interaction with rocks and to decay of natural radioisotopes contained in these rocks.

existence of silicate rocks, various iron compounds, in particular, hematite (Fe_2O_3) and limonite ($2\,Fe_2O_3 \cdot 4\,H_2O$). Vast ice and snow-covered expanses are expected. From estimates of the integral intensity of induced γ-radiation for different rocks and compounds, it appears that there is a strongly pronounced correlation with effective mass-number (A_{eff}) of rocks or compounds. This correlation is given in Fig. 3 where the ordinate is the integral intensity of γ-radiation ($E_\gamma = 1$–3 MeV) calculated for the same conditions as the data of Fig. 1.

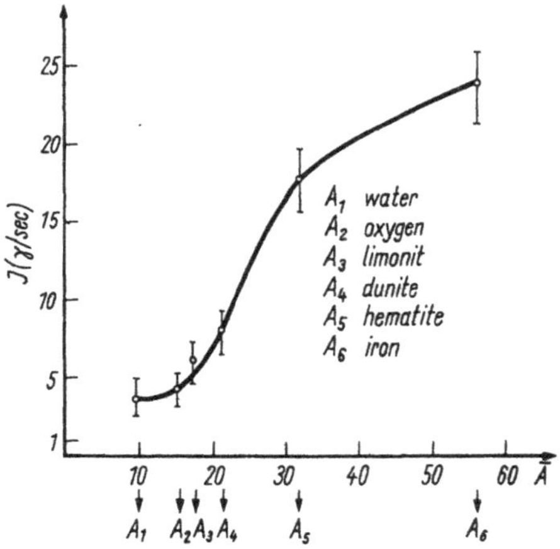

Fig. 3. Integral intensity of induced γ-radiation of different rocks and compounds depending on the effective mass number.

Thus, in principle, one can determine the character of the cover from the integral intensity, although the γ-radiation spectra will give, of course, more exact information. In particular, integral intensities of γ-radiation of granite and hematite are identical, but the ratios I_γ (1–2 MeV)/ I_γ (2–3 MeV) are about 1 and 10, respectively.

3. The Influence of the Martian Atmosphere on the Gamma-Radiation Spectrum from an Artificial Satellite of Mars

In measuring the spectrum of γ-radiation of Mars one should take into account the influence of the planetary atmosphere. In the first place, it absorbs γ-radiation of rocks. Secondly, under the influence of cosmic rays in the atmosphere nuclear reactions take place which also lead to the production of γ-quanta.

The calculation of absorption by the atmosphere was carried out by an ordinary method using the mass coefficient of attenuation assuming different compositions and pressures of the atmosphere [8]. As an example Fig. 4 gives spectra of γ-radiation of basalt type silicate rock taking into account absorption by an atmosphere of possible composition [7] for four extreme cases:

1) CO_2 $5\,g\,cm^{-2}$,
2) Ar $5\,g\,cm^{-2}$,
3) CO_2 $20\,g\,cm^{-2}$,
4) Ar $20\,g\,cm^{-2}$.

Corresponding spectra for granite and dunite are of similar character. One may conclude from the data obtained that the resulting spectrum is determined mainly by the thickness of the atmosphere and depends little on its composition.

The contribution of atmospheric γ-radiation to the recorded γ-spectrum was estimated as follows. As indicated above, emission cross sections of γ-quanta with energies 1–3 MeV are practically constant over a wide range of proton energies for $E_p > 100$ MeV. Therefore, to estimate the relative contribution of

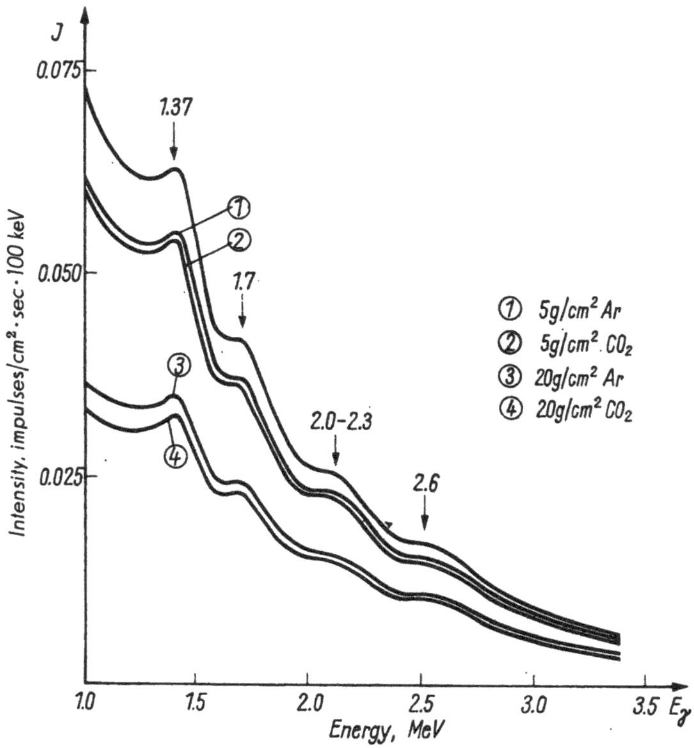

Fig. 4. γ-radiation spectra of basalt computed for the height of 1000 km above the Martian surface, taking into account attenuation of γ-radiation of the atmosphere.

γ-radiation of the atmosphere and rocks of the planet use was made of cross sections for $E_p = 660$ MeV. The composition of martian rock was assumed as basalt one, while the atmosphere was regarded as consisting of CO_2 and argon. Let the integral γ-quanta emission cross section in the interval $E_\gamma = 1$–3 MeV be

for rock $$\sigma_2 = \frac{I_2 A_2}{d_2 n_2 N G_2} \qquad (5)$$

for atmosphere $$\sigma_3 = \frac{I_3 A_3}{d_3 n_3 N G_3} . \qquad (6)$$

From experiments at the accelerator with proton energies of 150 and 660 MeV it follows that $\sigma_2 = 3\,\sigma_3$. It may be assumed that $n_2 = n_3$ since the cosmic radiation intensity is practically the same for rock and atmosphere. Calculation of geometrical factors G for rock and atmosphere shows that if the thickness of the atmosphere is 5–20 g cm^{-2} then, for measurements at heights not less than 1000 km, $G_2 = G_3$. Hence, the sought-for integral intensity of atmospheric γ-radiation is

$$I_3 = \frac{I_2\, A_2\, d_3}{3\, A_3\, d_2} \tag{7}$$

where A_2, A_3 are average mass numbers of rock and atmosphere, d_2, d_3 thicknesses of rock and atmosphere from which γ-radiation is recorded and I_2 is the integral intensity of γ-radiation for the interval 1 MeV $< E_\gamma <$ 3 MeV determined by Eq. (5) and equal to 7.5 γ-quanta per second.

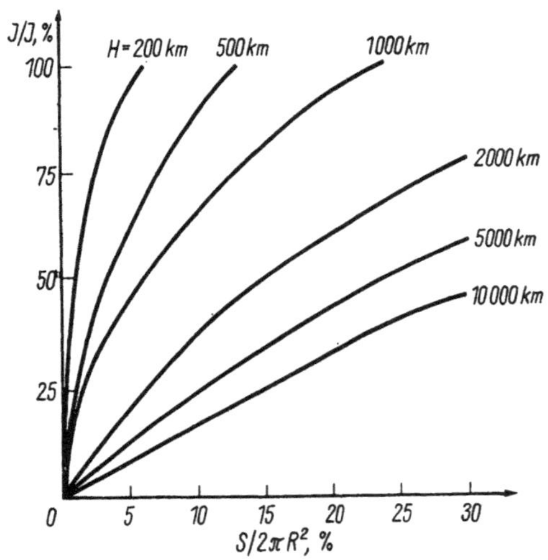

Fig. 5. Dependence of γ-radiation intensity on the dimensions of the area from which it is recorded for different distances of the detector from the surface.

The bulk of γ-radiation comes from distances of the order of the free path of γ-quanta $L = 1/\mu$. For γ-quanta with energies 1–3 MeV the mean value $L \approx d_2$ in rocks is 20 g cm^{-2} [6]. From the ratio (7) for the atmosphere consisting of CO_2 (100%) we obtain values of I_3 of 0.7 and 2.8 γ-quanta per second respectively for $d_3 = 5$ g cm^{-2} and $d_3 = 20$ g cm^{-2}.

If the martian atmosphere consists of 20% argon and 80% CO_2 the integral intensity of γ-radiation of the atmosphere can be estimated only through extrapolation of some experimental data. This is due to the absence of experimental data about the γ-radiation of inelastic interaction of high-energy protons with argon nuclei. The total cross section of γ-radiation of inelastic interaction of high-energy protons σ_γ increases with the growth of the Ar target [3]. Using the extrapolated value for argon and carrying out analysis similar to that given above (Eqs. (5–7)), we obtain for the γ-radiation intensity in the interval E_γ = 1–3 MeV for an atmosphere consisting of 20% Ar and 80% CO_2 the values $I_3 = 1.1$ γ-quanta per second for thickness 5 g cm^{-2} and $I_3 = 4.7$ γ-quanta per second for thickness 20 g cm^{-2}. Thus, to a first approximation the intensity of γ-radiation of the martian atmosphere of thickness 5 g cm^{-2} can be ignored for measurements performed from a height of 1000 km above the planet's surface. With the thickness of the martian atmosphere 20 g cm^{-2} it is necessary to take into account the contribution of its γ-radiation to the resulting γ-spectrum measured on the orbit.

Thus, measuring γ-radiation on an artificial satellite of Mars it is possible to obtain definite integral and spectral characteristics which in general will show the types of rocks forming different areas of the martian surface and will enable one to estimate the level of natural radioactivity of these rocks. (It must be noted that minimal dimensions of the areas for which the rock type may be determined depend essentially on the parameters of the artificial satellite orbit (Fig. 5).)

4. Determination of the Chemical Composition of Planets from their Gamma-Radiation

A different approach to the γ-radiation of the atmosphere and rocks of the planet is possible which enables one to obtain data on their chemical composition. The principle of this method is as follows. If several photopeaks are detected in the recorded γ-spectrum the intensity of each is dependent on the intensity of γ-radiation of nuclei — elements — components of the atmosphere and the rock (or only the rock for planets devoid of an atmosphere).

Since the formation of γ-quanta is mainly due to two processes, the formation of γ-quanta due to the effect of cosmic radiation and the decay of natural radioisotopes, it is possible to write in a general form the expression for the recorded intensity of γ-quanta of the ith energy

$$I_i = \sum_{j=1}^{m} \sigma_i^j \, n \, Q^j + \sum_{k=1}^{p} q_i^k \, Q^k$$

where i is the number of energies of γ-quanta recorded in the spectrum, m is the number of elements determining induced γ-radiation, n the flux of incident cosmic particles, σ_i^j the cross section of the production of γ-quanta of the ith energy from the jth nucleus, Q^j the number of nuclei of the jth element, p the number of elements determining γ-radiation of natural radioisotopes, Q^k the number of nuclei of the kth element and q_i^k the output of γ-quanta of the ith energy per nucleus of the kth element.

Obviously the system of equations is solved only when $i \geqq m + p$. However, while solving the system of equations it will be difficult to determine coefficients σ_i^j. Although by the present time vast experimental material on nuclear reactions has been accumulated, it is undoubtedly necessary to obtain additional data on cross sections of the generation of nuclei in ground and excited states in a wide range of proton energies (10 MeV–70 GeV). Taking into account different values of cross sections for the whole spectrum of cosmic protons, it is possible to solve in a general form the problem of determining the chemical composition (in atomic quantities) of rocks forming the planetary surface. The accuracy of the final result will be determined by the accuracy of measuring the cross sections. Apparently this method will be of some interest for investigating the composition of the rock of far-away planets devoid of atmosphere or having a rarefied atmosphere of the type of Mars.

References

[1] A. P. VINOGRADOV, YU. A. SURKOV and L. P. MOSKALEVA, Proc. 7 th Int. Space Science Symp., Vienna 1967.
[2] A. P. VINOGRADOV et al., Kosm. Issled., **4**, 871 (1966, 1967).
[3] YU. A. SURKOV et al., Presented at the XXIth Conference on nuclear spectroscopy and the structure of the atomic nucleus, Moscow, January 1971.
[4] K. L. FOLEY et al., Nucl. Phys., **31**, 43 (1961); **37**, 23 (1962).
[5] D. COHEN et al., Phys. Rev., **130**, 1505 (1963).
[6] T. V. GORSHKOV, Gamma-Radiation of Radioactive Bodies, Publishing House of Leningrad State University 1956 (p. 59).
[7] V. I. MOROZ, The Physics of Planets, Nauka, Moscow 1967.
[8] New Data on Mars and Venus, Collection of Papers, Mir Publishing House, Moscow 1968.

Space Research XI — Akademie-Verlag, Berlin 1971

THE MAIN PROBLEMS OF STUDY OF THE PLANET JUPITER

V. G. Teifel

Astrophysical Institute, Academy of Sciences Kazakh SSR, Alma-Ata, USSR

The main problems of the study of Jupiter at optical frequencies, including the determination of quantitative chemical composition, the thermal regime and the vertical structure of its atmosphere by observation from the earth and by space probes are discussed.

1. Introduction

The study of Jupiter, the largest planet of the solar system, has considerably increased during the last ten years. We are very far from a full understanding of the physical conditions and the nature of the processes in different layers of the tenuous atmosphere of this planet. The current observational information suggests a very complicated structure for the Jovian atmosphere and non-trivial character for many of the phenomena observed at optical and radio frequencies.

It is more probable that Jupiter has retained the primary chemical composition of protoplanetary matter. The atmosphere of Jupiter is a natural chemical laboratory where photochemical reactions and organic synthesis processes may be occurring at present. The study of Jupiter is very important not only for planetary physics but for planetary cosmogony and biochemistry. There are many problems in the study of Jupiter as a source of non-thermal radio-emission.

Jupiter, as is seen, for example, from the published programs of NASA, will be included in the number of projects to be studied by space techniques. It is very important to obtain the most complete information about Jupiter using ground-based instrumentation. The space programs of Jovian study are in general analogous to programs of study of other planets and based on present ideas about the structure and physical properties of Jupiter. These programs included the study of magnetosphere, Van Allen belts around Jupiter, cosmic rays, solar wind and meteoric matter in the neighbourhood of the planet, temperature measurements on the Jovian disk, photopolarimetry and television observations, ultraviolet and infrared spectrometry, the determination of the high atmosphere parameters by the radio-occultation method, the study of radio-emission and radar study of Jupiter.

We shall consider shortly some problems and results of the study of Jupiter in the optical range (including the thermal infrared).

2. The Chemical Composition of the Atmosphere

Fig. 1 is a summary of our knowledge about the chemical composition of Jovian atmosphere. Only three gases have been identified in the spectrum of Jupiter: hydrogen, methane and ammonia, and indirect data indicate the existence of helium. Ther have been some attempts [1–3] to detect more complex substances but these attempts are unsuccessful and we can speak only about the probable higher limits of these compounds. These limits, as Fig. 1 shows, are very low compared with the abundances of the main components of the Jovian atmosphere.

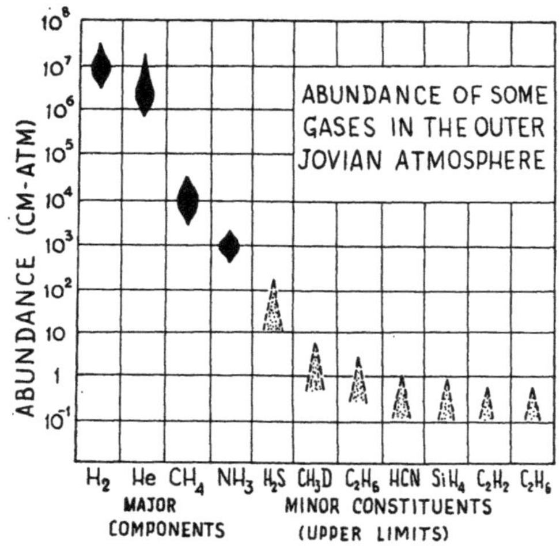

Fig. 1. Diagram of the present estimates of the main components and minor constituent abundances (in a model of simple reflection from the cloud surface of Jupiter).

The ratio H_2/He is a fundamental characteristic of the atmospheric composition for the major planets.

Owen and Mason [4] have showed that the ratio H_2 to He is more than 5. This is a more probable result than the data of Baum and Code [5] derived from observations of σ Arietis occultation by Jupiter. As McElroy noted [6] the noise level on the brightness curve during the star occultation was high, and the observed curve may be fitted by a series of theoretical curves with different values of mean molecular weight.

There are no precise data about the abundance of hydrogen, methane and ammonia. It is clear now that the spectral determination of the gas abundances in terms of simple reflection from the cloud surface of Jupiter can not give definite and reliable results. Two main effects have emphasized the necessity of taking into account the influence of the multiple scattering in Jovian aerosol layers on the formation of absorption lines and bands. The first is the observed change of the band intensity towards the Jovian limb from the centre of the disk. That change does not follow the normal increase of atmospheric mass along the line of sight (Figs. 2 and 3). The second fact is the very great difference

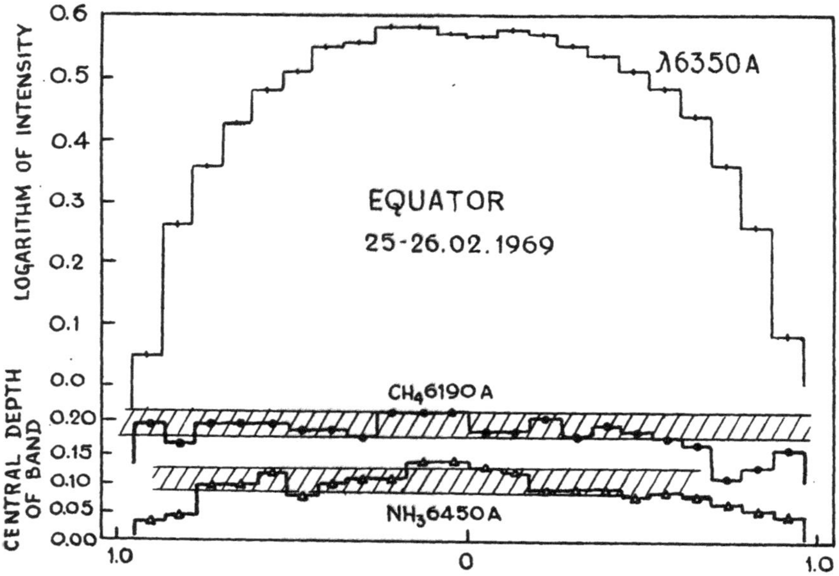

Fig. 2. Distribution of the brightness and molecular absorption of CH_4 and NH_3 along the Jovian equator, 25–26 February 1969.

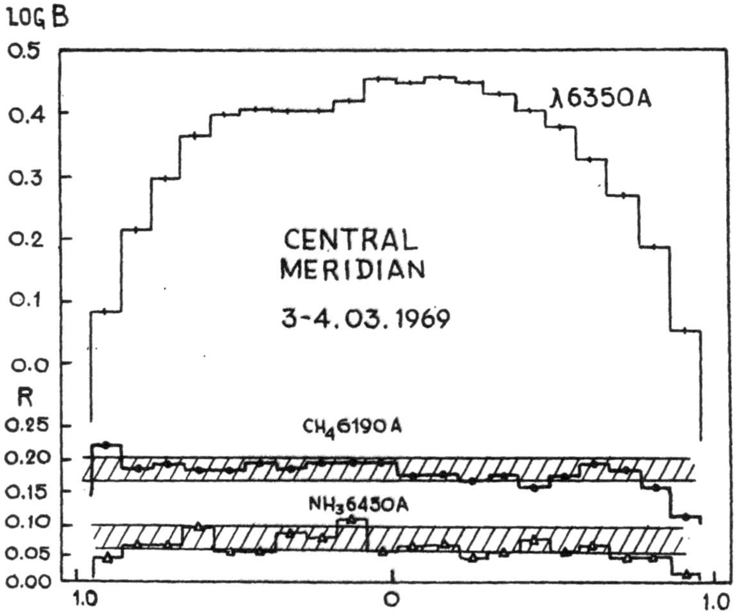

Fig. 3. Distribution of the brightness and molecular absorption of CH_4 and NH_3 along the central meridian of Jupiter, 3–4 March 1969.

in gas abundances as determined by the strong and weak bands. The latter give the abundance a factor of ten more than the strong bands. We can not say to what layer of the atmosphere the observed abundance must refer if the effect of multiple scattering is not taken into consideration.

The theoretical calculations of the equilibrium abundances in the Jovian atmosphere [7] are in good agreement with the observational data and have assumed the existence in the observable atmospheric layers of only two minor constituents (H_2O and H_2S) with very small concentration, about 10^{-12} at temperature 150 °K.

It is scarcely probable that the first flyby probes will be able to improve our knowledge of the chemical composition with the exception of the determination of the helium abundance by the resonance line λ 584 Å. The search for hydrogen and helium emission on the dark side of Jupiter is very important. Some information may be obtained during the observation of the sun's occultation by Jupiter. But these experiments will be available only during the closest approach of the space probe to Jupiter. They need orbital vehicles from small distances so that the optically effective zone of the Jovian atmosphere on the limb will have angular dimensions more than 6′ (the solar diameter as seen from Jupiter).

The basic solution of the chemical composition problem can only be achieved by direct atmospheric probes with gas analysers, mass spectrometers etc. The study of the high atmosphere by probes is very interesting for the comparison of predicted and determined products of the photochemical reactions.

3. The Temperature and Vertical Structure of the Atmosphere

The most impressive data, from our point of view, have been obtained during the past years in the range of infrared thermal emission from Jupiter. The bolometric measurements on a jet plane in the wavelength range 1.5–200 μm [8] have given the effective temperature of Jupiter as 134 ± 4 °K. This has confirmed earlier estimates at 8–14 μm and we can now say more definitely that Jupiter really has an excess of heat radiation. The thermal flux from Jupiter is more than the flux caused only by the equilibrium temperature 105 ± 3 °K (if only solar radiation is absorbed) by a factor 2.7 ± 0.6.

A more detailed study of the infrared spectrum of Jupiter at 2.8–14 μm [9] led to the discovery of some peculiarities, in particular a strong peak of emission at $\lambda = 4.6$–5.2 μm with brightness temperature about 230 °K, and higher emission with brightness temperature about 145 °K at $\lambda = 7$–8 μm, where the strong absorption band of CH_4 is present. This last peculiarity may be caused by the temperature inversion in the outer atmosphere of Jupiter. According to [10] the temperature profile of the atmosphere can be presented as in Fig. 4. The temperature decreases upwards in the adiabatic zone to 115 °K and after that increases to about 145 °K. The heating of the higher atmosphere is probably cause by the absorption of solar radiation within the other methane band near $\lambda = 3.3$ μm [9, 10].

In the region 4–6 μm there is not the methane or ammonia absorption, and emission can radiate from the deeper layers. Westphal has discovered [11] that the thermal emission at $\lambda = 4.8$ μm is formed only in the dark cloud belt near the

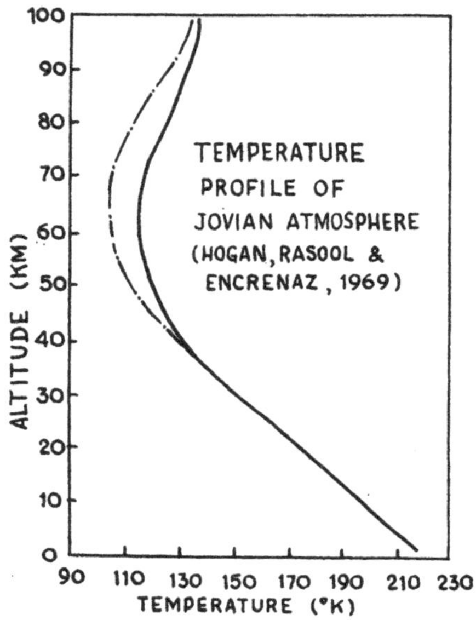

Fig. 4. Temperature profile of the Jovian atmosphere after Hogan et al. [10] in the case of model 3 (hydrogen pressure 4 atm at level of $T = 220$ °K) and model 5 (the pressure is 6 atm on the same level).

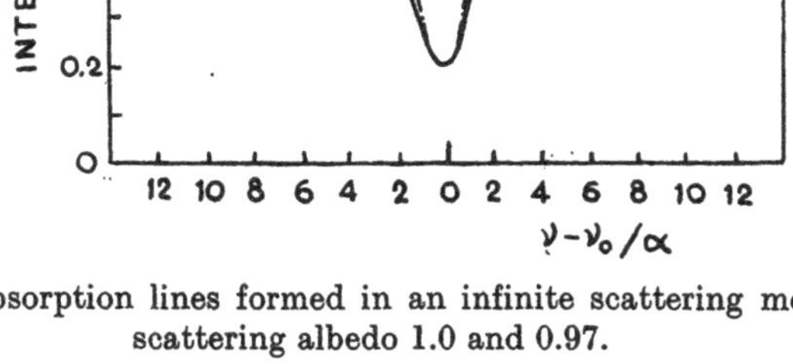

Fig. 5. Profiles of absorption lines formed in an infinite scattering medium with single scattering albedo 1.0 and 0.97.

equator of Jupiter. In that belt the brightness temperature reaches 240 °K and in localized regions may be 300 °K or more.

According to Low [12] the brightness temperature of the Jovian equatorial region is 150 °K and that of the temperate and polar regions about 130 °K. It is very necessary to observe Jupiter in the thermal infrared more regularly.

The thermal measurements from space probes are very interesting. They give greater space resolution and it is possible to detect the local thermal irregularities in the atmosphere of Jupiter. The space probes would probably make possible measurements of the thermal emission and temperature of dark (night) side of Jupiter and a study of the role of radiation heating of the atmosphere and the inner heat of the planet in the heat balance of the atmosphere. It will also be possible to obtain the temperature profile of the Jovian atmosphere.

It is very important to know the vertical structure of the Jovian atmosphere for entry probe braking calculations and other practical purposes. We can only say definitely that the density of the Jovian atmosphere decreases with altitude more slowly than does the density of the earth's atmosphere, thus the braking effect on the orbital probes with low perigee will be stronger than on the earth.

The main difficulty in the model calculations for Jupiter is the absence of definite data on the levels for which observable effective pressures and temperatures were determined. Thus, according to the measurements of some authors the effective pressures obtained by spectral observations are in the range 1.6 to 2.8 atm. The absorption lines used for these purposes can be formed not only in the outer atmosphere above the clouds but also inside the scattering cloud layer. The effective depth of their formation depends on the cloud properties and the line strength. Moreover, as has been shown (e.g. [13, 14]) the profile of the line may be widened if the line is formed in the scattering medium with no absorption in continuum or with very small absorption. Some cases are shown in Fig. 5. For the single scattering albedo $\omega_c = 0.97$ the profile of the line is closely approximated to the Lorentz profile but the profile for $\omega_c = 1$ is appreciably wider.

The process of line formation in the inhomogeneous cloud layers is very complicated, and it is difficult to make an unique working model of the vertical structure of atmosphere and clouds using estimates of chemical abundances, brightness and rotational temperatures and effective pressures simultaneously.

Some information about the atmospheric structure may be obtained from observations of radio occultation of the probe by Jupiter [15], but these observations can not give independent estimates of the mean molecular weight and temperature of the high atmosphere of Jupiter. However it would be very interesting to carry out this experiment and to determine the differences between the theoretical predictions and the observational data.

4. The Nature and Structure of the Aerosol Layers

The atmosphere of Jupiter (in the more or less observable zone) contains three aerosol layers of different physical nature. A schematic picture of the position of these layers is shown in Fig. 6.

The mean layer is observed in the visible region of the spectrum. It consists apparently of solid ammonia crystals. The presence of some chemical compounds in this layer is possible but their chemical nature is unknown. We observe very clear differences and changes of albedo and colour on the visible surface of the clouds.

The lower and most dense cloud layer, as follows from the calculations [16] exists at the depth where the temperature is more than 200 °K; this may be consistent with the several reciprocity self-penetrating layers of different composition (ammonia solution in water $2\,NH_3 \cdot H_2O$ and NH_4SH).

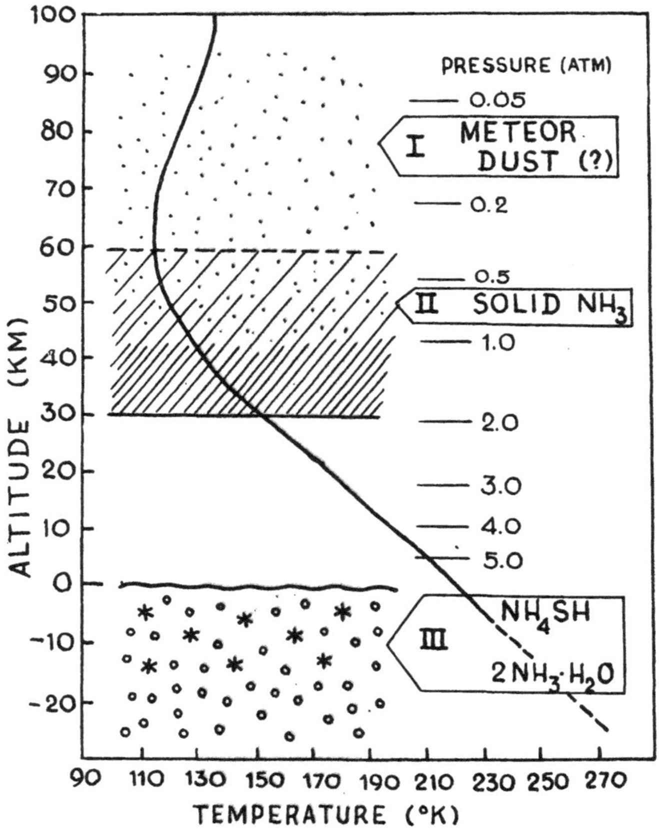

Fig. 6. The probable structure of the aerosol layers in the Jovian atmosphere (schematic picture).

Rocket measurements of the Jovian geometric albedo at 0.18–0.30 μm [17–19] have shown that the reflectivity of Jupiter in the ultraviolet is very low. The quantity of hydrogen as determined by comparison with the calculations for the Rayleigh scattering gas layer is very small (10–15 km atm or less). The ammonia abundance according to rocket ultraviolet measurements is no more than 2×10^{-3} cm atm. These results are in contradiction to the estimates for infrared spectra (H_2 about 65–85 km atm, NH_3 about 5×10^2 cm atm).

Apparently the rocket results in the ultraviolet are confined to the higher levels of the Jovian atmosphere, as lower layers are obscured by the high and very strong absorbing aerosol haze, which is transparent for visible light. This high aerosol layer can not have the nature of ammonia because the concentration of ammonia in this level is very low. It seems more probable that this haze consists of meteor dust originating in the disruption of the meteor bodies penetrating the Jovian atmosphere. The metallic and silicate particles fall with a velocity depending on their size and the atmospheric density. By our calculation particles with radius about 0.05 μm will come within a layer of 1 km thickness in the ultraviolet haze zone during a time more than 500 years. In the high atmosphere there may be concentrated about 10^8 tons of fine dispersed aerosol matter.

It may be possible that the absorption in the near ultraviolet ($\lambda < 4200$ Å) is caused fully or partly by high aerosol haze. As observations [20, 21] have shown, this ultraviolet absorption in the Jovian spectrum has a global character. Meteor dust can also play the role of condensation nucleus in lower layers where the ammonia haze or clouds can exist.

The mean cloud layer must have an influence on the process of formation of the absorption lines. The study of this layer structure is very important for many questions of Jovian atmosphere physics and of radiation transfer in the atmosphere. Also spectroscopic measurements of the absorption lines and band intensities across the Jovian disk could give information about the vertical structure and the horizontal inhomogeneity of the cloud layer. For this it is necessary to carry out theoretical calculations of the radiative transfer in the inhomogeneous scattering layer.

The first attempts to explain the observed character of the molecular absorption distribution over the Jovian disk were based on the model of simple reflection from a dense cloud layer [22, 23]. In terms of the model of the scattering medium Moroz [24] and Teifel [25] have obtained for an infinite layer a mean scattering coefficient about 10^{-6}–10^{-7} cm^{-1}. But it is necessary take into account the probable inhomogeneity of the clouds. The vertical extension of the ammonia cloud layer depends strongly on the ammonia concentration and the atmospheric regime of circulation.

In the convective zone one may suggest full mixing. In this case the concentration of ammonia must be changed upwards proportionally to the total atmospheric pressure just as the saturation pressure of NH_3 decreases rapidly with temperature decrease. Then the quantity of sublimated ammonia changes upwards slowly (see Fig. 7).

At the levels above the tropopause where convective mixing is very weak or absent the vertical transfer of ammonia may be caused by turbulence. Its intensity in the Jovian atmosphere is unknown. Calculations show that if the sizes of the turbulent cells are less than 1 km the density of the ammonia clouds decreases strongly upwards in the range about 10–20 km of cloud layer thickness.

The ammonia haze has apparently not been formed above the level with $T \approx 115$ °K.

The ammonia cloud layer can influence the formation of strong and weak absorption bands of CH_4 and NH_3 differently. It has been found [25–28] that the change in intensity of relatively weak bands of CH_4 (0.62 μm) and NH_3 (0.64 μm) across Jovian disk is very small. The character of the changes of stronger infrared bands [29–31] is more complex. Simultaneous observations of the absorption distribution on the Jovian disk in the strong and weak bands are very necessary. It is not clear what is the optical thickness of the ammonia

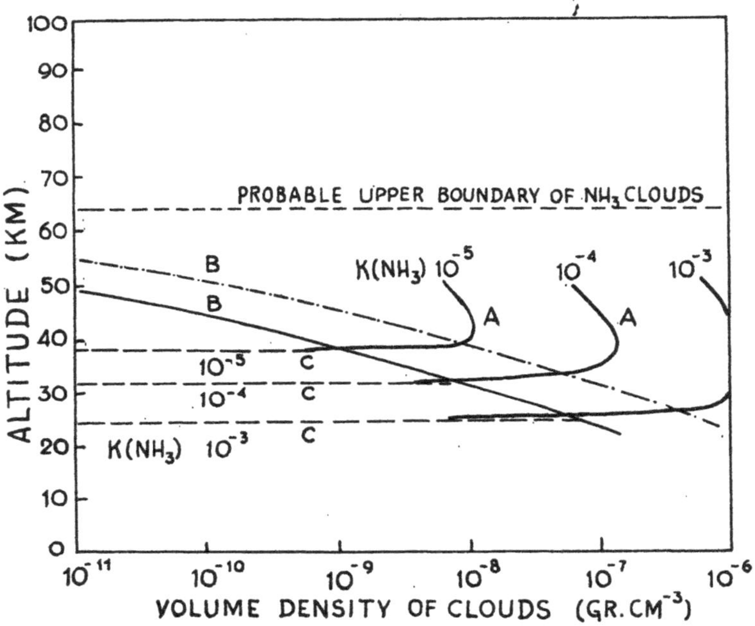

Fig. 7. Distribution of cloud density in convective (A) and turbulent (B) cases of atmospheric mixing for some average NH_3 concentrations.

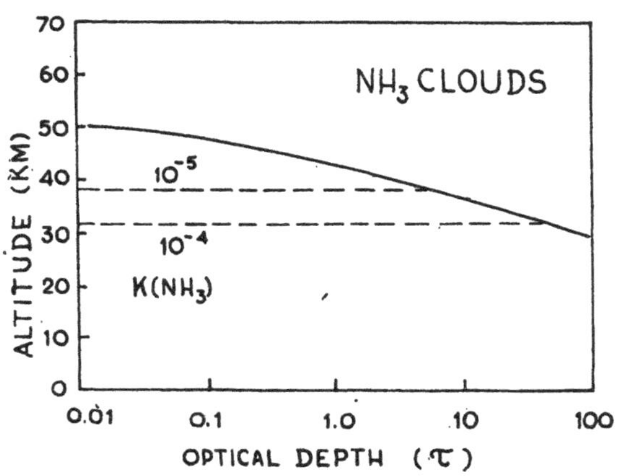

Fig. 8. Vertical changes of the optical depth of the ammonia clouds when the sizes of the turbulent mixing cells are about 0.1 km.

cloud layer. The thickness of clouds depends on the ammonia concentration, atmospheric lapse rate and regime of vertical and horizontal circulation. Fig. 8 shows the optical thickness of the ammonia cloud layer for the case of turbulent mixing with cell sizes about 0.1 km.

The study of the Jovian cloud structure by measurements in the absorption bands may be carried out from nearby space probes. This method may be useful (with television pictures obtained simultaneously) for the determination of sizes of the horizontal cloud structure elements and for detecting the connection of these elements with the circulation processes and vertical thickness of clouds.

It is very difficult to hope that observations from flyby probes could determine the physical and chemical nature of Jovian clouds. Apparently the solution of this problem will be achieved only by using entry probes. Nevertheless photometric and polarimetric measurements from flyby probes could be used to determine some optical parameters of the clouds and outer atmosphere. The possibility of observing Jupiter from space vehicles at different phase angles will be very valuable especially for polarimetry and photometry. The interpretation of the ground-based polarimetric measurements of Jupiter [32] using only the Rayleigh scattering model of the outer atmosphere have lead to a result which is in contradiction to the measurements of methane absorption. The latter decreases near the poles but the polarimetric measurements were interpreted as an indication of the increase in the Rayleigh optical thickness from 0.05 near the equator to 0.5 in the polar regions of Jupiter.

Taking into account the difficulties and continuance of the space flights to Jupiter we must not overestimate the possibilities and results of Jovian study from space probes at least during the next decade. It may be confidently said that the ground-based means and methods of study of this planet have yet many possibilities and will be the main source of information on some problems for a long time. We must expect most progress of our knowledge of Jupiter only with the closest complex of ground-based and space researches. This should open up very wide opportunities for fruitful international cooperation.

References

[1] T. C. OWEN, Comm. Lunar and Planetary Lab. **2**, 65 (1963).
[2] D. P. CRUIKSHANK and A. B. BINDER, Astrophys. Space Sci. **3**, 347 (1969).
[3] T. C. OWEN, Icarus **6**, 138 (1967).
[4] T. C. OWEN and H. P. MASON, J. Atmos. Sci. **26**, 870 (1969).
[5] W. A. BAUM and A. D. CODE, Astron. J. **58**, 108 (1953).
[6] M. B. MCELROY, J. Atmos. Sci. **26**, 798 (1969).
[7] J. S. LEWIS, Icarus **10**, 393 (1969).
[8] H. H. AUMANN, C. M. GILLESPIE and F. J. LOW, Astrophys. J. **157**, 69 (1969).
[9] F. C. GILLETT, F. J. LOW and W. A. STEIN, Astrophys. J. **157**, 925 (1969).
[10] J. S. HOGAN, S. I. RASOOL and T. ENCRENAZ, J. Atmos. Sci. **26**, 898 (1969).
[11] J. A. WESTPHAL, Astrophys. J. **157**, 63 (1969).
[12] F. J. LOW, Astron. J. **71**, 391 (1966).
[13] J. LENOBLE, C. R. Acad. Sci. Paris **261**, 4633 (1965).
[14] V. G. TEIFEL, Astron. Vestn. **3**, 85 (1969) (in Russian).
[15] B. A. SODEK, J. Spacecraft and Rockets **5**, 461 (1968).
[16] J. S. LEWIS, Icarus **10**, 365 (1969).
[17] T. P. STECHER, Astrophys. J. **142**, 1186 (1965).
[18] E. B. JENKINS, Icarus **10**, 379 (1969).

[19] R. C. ANDERSON, J. G. PIPES, A. L. BROADFOOT and L. WALLACE, J. Atmos. Sci. **26**, 874 (1969).
[20] V. G. TEIFEL, Izv. Akad. Nauk Kaz. SSR Astrofizika **16**, 74 (1963).
[21] V. V. AVRAMCHUK, in: Fizika Luny i Planet, Kiev, 1966 (p. 83).
[22] S. L. HESS, Astrophys. J. **118**, 151 (1953).
[23] P. SQUIRES, Astrophys. J. **126**, 185 (1957).
[24] V. I. MOROZ, Fizika Planet (Nauka, Moscow 1967).
[25] V. G. TEIFEL, J. Atmos. Sci. **26**, 854 (1969).
[26] V. V. AVRAMCHUK, in: Astrometria i Astrofizika I, Kiev 1968 (p. 126).
[27] V. G. TEIFEL, Trudy Astrofiz. Inst. Akad. Nauk Kaz. SSR, **9**, 59 (1967).
[28] V. G. TEIFEL, Atmosfera Planety Jupiter (Nauka, Moscow 1969).
[29] T. OWEN, Icarus **10**, 355 (1969).
[30] V. G. TEIFEL, Astron. J. of USSR, **43**, 154 (1966).
[31] V. I. MOROZ and D. P. CRUIKSHANK, J. Atmos. Sci. **26**, 865 (1969).
[32] T. GEHRELS, B. M. HERMAN and T. OWEN, Astron. J. **74**, 190 (1969).

Space Research XI — Akademie-Verlag, Berlin 1971

A REVIEW OF NEW DATA ON SATURN'S SYSTEM

M. S. Bobrov

The Astronomical Council, USSR Academy of Sciences, Moscow, USSR

The international co-operative observations of Saturn's system during the period of very narrow opened rings (1966) are considered, as well as radio brightness temperature measurements.

The international co-operative observations led to the discovery of the tenth Saturn's satellite (Janus), to the detection of the fourth ring of Saturn ("ring D"), to the more detailed investigation of the rarefied envelope of the rings and to the first estimation of the rings' thickness (some 3 km).

New findings give rise to new problems, in particular, to the problem of the movement of three inner Saturn's satellites in direct contact with the matter of ring D and to the problem of the energy replenishment mechanism able to prevent full flattening of the rings during the lifetime of the solar system.

Radio brightness temperature (T_B) measurements, now ranging from 1 mm to 21.3 cm, show a pronounced increase of T_B with λ; in addition, at $\lambda = 10$ cm at least 90% of radio-emission originates from Saturn's visible disk. Very likely it is the thermal emission due to the temperature gradient of the deep atmosphere whose opacity depends on wavelength. Then the radio-emission measurements may determine the law of temperature growth with depth in Saturn's atmosphere well below the layers accessible to optical observations.

The contribution of the rings to the total radio-emission flux from Saturn is small, possibly because of greater transparency of the rings at radio frequencies than at optical frequencies. This may be used for estimation of the upper limit of the typical ring particle size.

In current literature we find projects for unmanned spaceflights not only to Jupiter but also to more distant giant planets. In this connection it seems to be useful to review briefly some new data concerning Saturn's system.

1. Optical Observations of Events in Saturn's System in 1966

In 1966 Saturn's rings were seen nearly or quite edgewise. As seen from Fig. 1 there were three transits of the earth through the ring plane, and one of the sun. Also there were two periods of visibility of the dark side of the rings. The observation of these rare events not only supplied new information on the nature of Saturn's rings, but the almost complete absence of stray light from the rings permitted a study of the disk of the planet and of Saturn's satellites in more detail than is possible in normal conditions.

In accordance with the proposal and program submitted by the Astronomical Council of the USSR Academy of Sciences, the International Astronomical

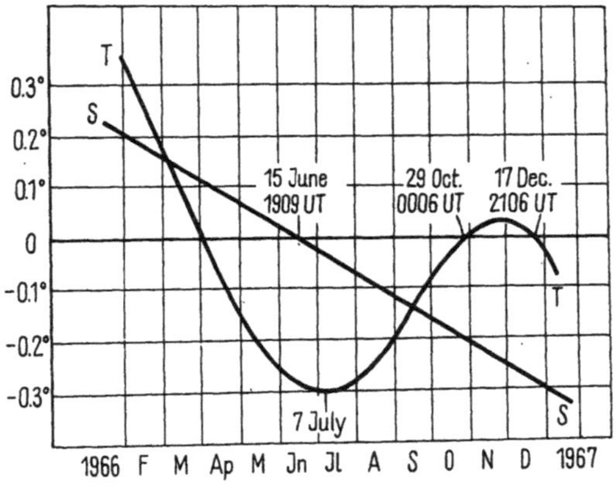

Fig. 1. The time variation of the elevation angles A and A′ of the sun (S) and earth (T) above the plane of Saturn's rings in 1966.

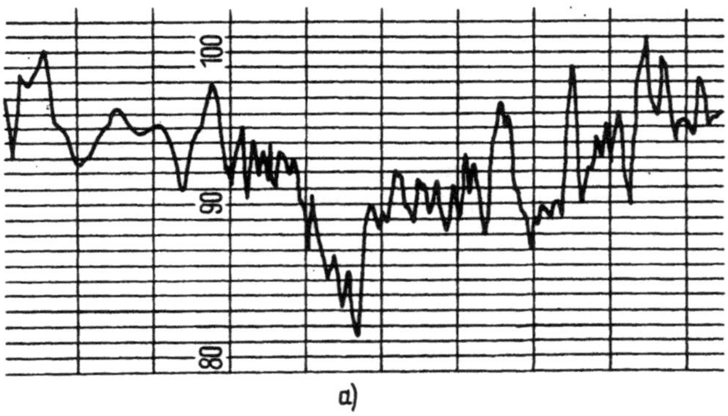

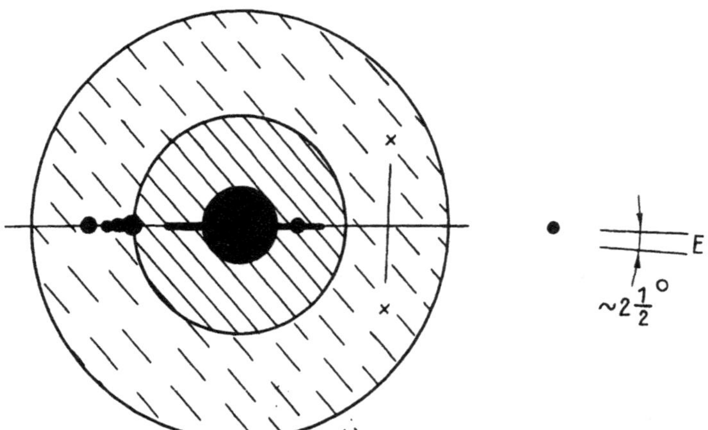

Fig. 2. The detection of Saturn's D ring by Feibelman [1]. Below: aspect of Saturn, 14 November 1966. The inner hatched circle indicates the extent of the greatly overexposed image of Saturn; the outer circle represents the photographic halo. Above: Microdensitometer trace along path X–X. The central dip corresponds to the location of the visible thin line of the D ring.

Union organized international cooperative observations of Saturn's system during 1966. Observatories of four continents, Europe, Asia, America and Australia, took part in the observations.

Fig. 2 demonstrates one of the interesting results of the observations [1]: the first instrumental detection of the so-called ring D, or the continuation of the ring system outside ring A. The plane of ring D coincides with that of the other rings. The ring is extremely rarefied and can be detected only when the rings are turned with their dark side to the earth. Ring D extends to more than twice the diameter of the known ring system.

Fig. 3 is the photograph, taken by Dollfus [2], in which Saturn's tenth satellite was discovered. This satellite, named Janus, is only three earth's radii away from the outer boundary of ring A and has brightness 14 mg. Assuming

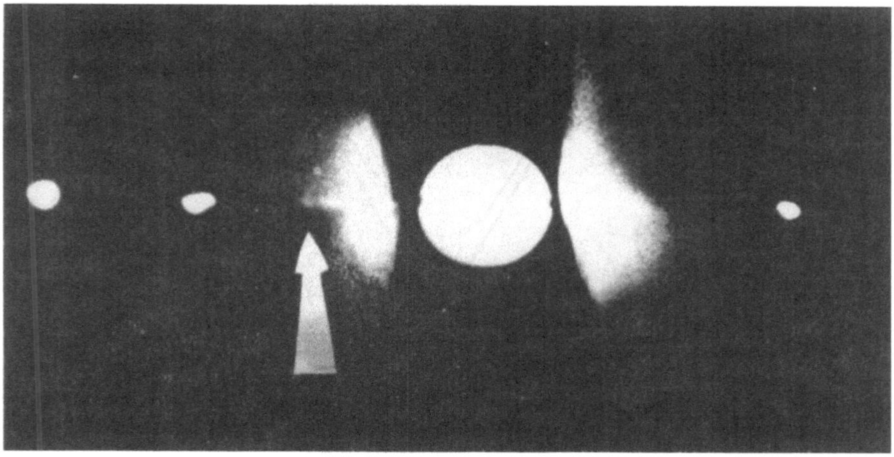

Fig. 3. The discovery of the 10th Saturn's satellite by Dollfus [2]. The photograph taken on 15 December 1966 with 43-inch reflector of the Pic-du-Midi Observatory. The brightness of Saturn's disk is reduced to approximately 10^{-2} of its actual value with the aid of an absorbing stripe. The arrow indicates the position of the new satellite.

the reflecting power of the surface of Janus to be 0.6, its diameter would be about 350 km. Like ring D, Janus is also an object discernible only when the rings have their dark sides turned to us.

Janus as well as the more distant satellite, Enceladus, moves around Saturn immersed in the matter of ring D. The velocities of ring particles in relation to the satellites must be near to zero, for we may expect that the rotation of ring D is also Keplerian. For Mimas, intermediate between Janus and Enceladus, the situation is somewhat different because its orbit is inclined 1.5° to the ring plane. Therefore Mimas is not permanently immersed in ring D but only intersects it periodically.

As well as ring D there is the radial continuation of the ring system; the rarefied "atmosphere" possibly enveloping the rings may be called the continuation of the ring system normal to its middle plane. The existence of such an atmosphere was first suggested by Maggini [3] on the basis of his observations carried out in 1936–1937. In 1966 Prof. Kozyrev [4] attempted to study this envelope spectroscopically. Since the rings were seen nearly edgewise, the path of solar rays inside the envelope would be long enough to reveal some of its spectral features. The slit was oriented parallel to the ring plane, and the spectra of Saturn's disk chords near and far from the rings were obtained. The observer noticed that near the rings the methane band λ 6190 Å is weaker and the ammonia bands are stronger than far from the rings. In other words, the spectrum of the part of Saturn's disk illuminated by the sun through the envelope of the rings appeared to be nearer to the spectrum of Jupiter than that of the part illuminated by direct solar radiation. It was interpreted as the greenhouse effect of the envelope of the rings. Prof. Kozyrev also noticed the strengthening of the lines

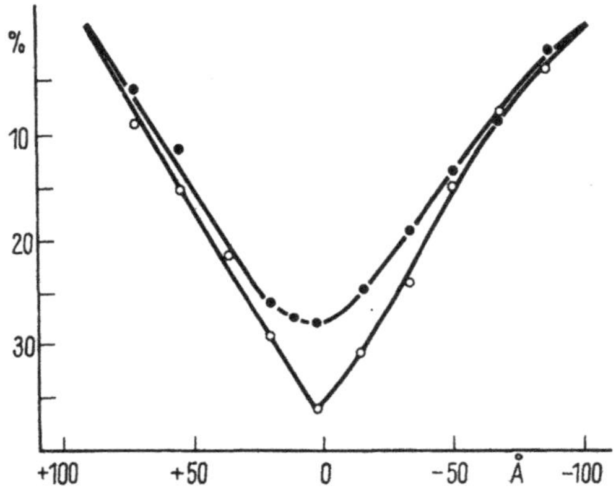

Fig. 4. The comparison of the methane band 6190 Å profiles in the spectra of Saturn's disk taken near the plane of the ring (full circles) and far from it (open circles) [4].

of water vapour and the course of spectral intensity typical of ice crystals. He concluded that the envelope of Saturn's rings consists of fine ice crystals and water vapour, the latter being formed by photosublimation. Ice crystals may, in turn, originate from collisions of ring particles (which, as shown by Kuiper and Moroz, are covered with hoar-frost).

In their movement the molecules of water vapour and ice crystals must intersect the ring plane and therefore most of them would be soon captured again by ring particles. But accidentally some of the molecules and crystals may follow trajectories lying in free space between the inner boundary of the rings and Saturn's globe. In this region they may exist for a long time, and therefore we may expect to find a cloud of gas and fine crystals not above the plane of the rings but nearer to Saturn than ring C. There are several facts supporting this viewpoint. Barabashov and Semejkin [5] showed that the space between ring C and Saturn had noticeable brightness in blue light. Further evidence of the presence of some rarefied matter in this region was given by the photograph taken in October 1969 at Pic-du-Midi by P. Guerin [12]. This time the faint trace of the cloud was detected in yellow light.

The absence of significant atmosphere above rings A and B was recently demonstrated by observations of Franklin and Cook [6]. They obtained spectra of Rhea, Saturn's satellite, when the pencil of light reflected by the satellite to the earth was nearly tangential to the ring plane. The spectra showed no trace of absorption, discrete or continuous, which could be attributed to the atmosphere of the rings. The upper limit of Na concentration is estimated by the authors as 100 cm^{-3}.

One of the main tasks of the 1966 Saturn system observations was the estimation of the geometrical thickness of the rings. From earlier observations it was only known that the value of this important parameter probably lay well below 10 km. It must be emphasized that at the distance of Saturn even 10 km corresponds to a very small angle, 1.4×10^{-3} s of arc, whereas the best telescopic

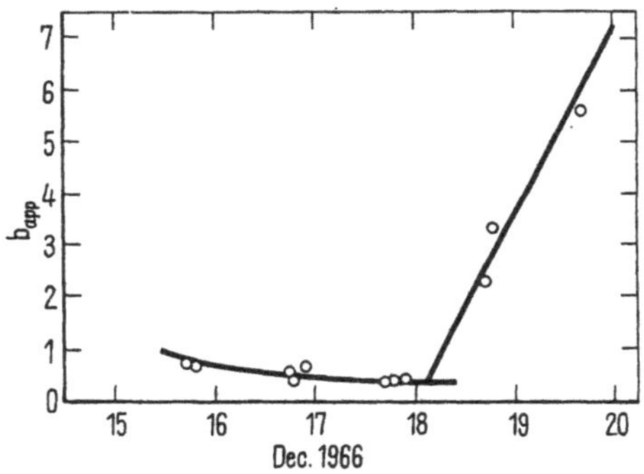

Fig. 5. The estimation of the thickness of Saturn's rings [7]. The variation of the apparent brightness of the rings' image in ansae during the passage of the earth from the dark to the bright side of the rings in December 1966.

resolution hardly reaches 1.0×10^{-1} s of arc, i.e. nearly two orders poorer. Therefore optical telescopes cannot resolve even the upper limit of the thickness of Saturn's rings.

But if the order of the edge brightness of the rings is known from general considerations we may determine their thickness by measuring the apparent brightness of the ring's image at the moment of transit of the earth through the ring plane. Indeed, it is evident that the apparent brightness

$$b_{app} = \frac{z_0}{w} b_e \,, \tag{1}$$

where b_e is the true brightness of exactly edgewise rings, z_0 is their thickness, and w is the width of the image of the ring at the photographic plate.

Such photometric measurements were carried out in December 1966 at the Pic-du-Midi Observatory by Dollfus and Focas [7]. Fig. 5 shows the course of b_{app} before and after transit of the earth through the ring plane from the dark side to the sunlit side. Unfortunately, the very moment of transit could not be observed and the minimum value of b_{app} was obtained by interpolation. With usual assumption that $b_e = b_B$, where b_B is the brightness of the most intense zone of the widely opened ring B, the authors find for the geometrical thickness the value

$$z_0 = 0.44 \times 10^{-3}\,\text{s} \quad \text{of} \quad \text{arc} = 2.8 \pm 1.4\ \text{km}\,. \tag{2}$$

Examination of the data of Dollfus and Focas reveals shows that their photometry is of high accuracy. Moreover, the value $z_0 \sim 3$ km may be independently obtained from the data of Dollfus and Focas in another way [13], the results agreeing with each other very satisfactorily. Hence we may conclude that the geometrical thickness of Saturn's rings, a parameter which for so many years has escaped quantitative estimation, is now determined.

This result is directly connected with the dynamics of the rings. A number of data indicate that the size of a typical ring particle is many orders below 3 km so that, to form rings 3 km thick, the orbits of particles must have significant inclinations to the middle plane of the rings, and every particle will intersect this plane periodically. This, taking into account the opacity of the system, will cause frequent collisions of particles with each other. The impacts cannot be perfectly elastic, and so the energy of oscillations of particles normal to the middle plane will be dissipated. To keep a constant thickness of the rings there must exist some source of continuous energy replenishment.

The analysis of the problem [8] leads to the conclusion that such a source is possibly the energy of differential rotation of the rings. The collisions of particles give rise to the transfer of angular momentum which must be accompanied by partial transformation of the energy of differential rotation into the energy of oscillation of the particles about their mean positions. Since the directions of oscillations are random they must have a component normal to the middle plane of the rings.

It is of interest to estimate the effectiveness of this mechanism assuming the lifetime of the rings to be 5×10^9 years. Computations for the simplest model of equal particles yield a value of the thickness sufficiently smaller than that observed, namely 0.1 km. The real case of unequal particles has not been considered but it will clearly result in stratification of particle size normally to the ring plane and in considerable increase in the value of ring thickness compared with that given above. It may be added that one of the consequences of this process must be secular diminution of the radial dimensions of the rings.

2. Radioastronomical Observations

Saturn's radio-emission brightness temperature T_B was measured in the wavelength range 1.2 mm to 21.3 cm. A full list of references may be found in [9]. The values obtained as a function of wavelength are displayed in Fig. 6. In all cases the measured radio-emission flux S was referred to the solid angle of the full visible Saturn's disk, Ω. Then, according to the Rayleigh–Jeans law,

$$T_B = S\lambda^2/2k\Omega , \qquad (3)$$

k being Boltzmann's constant. Such normalization of the radio-emission flux is equivalent to the hypothesis of cold fully transparent rings and of the absence of radiation belts. The accuracy of radioastronomical measurements is not yet high enough to check the validity of this hypothesis; we may only say that it appears a satisfactory approximation for all the available data.

Indeed, the radio interferometric observations of Berge and Read [10] showed that at wavelengths 10.7 and 9.0 cm at least 90% of Saturn's radio-emission is due to the optically visible disk of the planet, and, within the accuracy of measurements, is uniform across the disk; polarization was not detected. The contribution from the radiation belts and from the rings may amount to less than 10%. All these facts lead to the conclusion that very likely the predominant part of Saturn's radio-emission originates in its atmosphere.

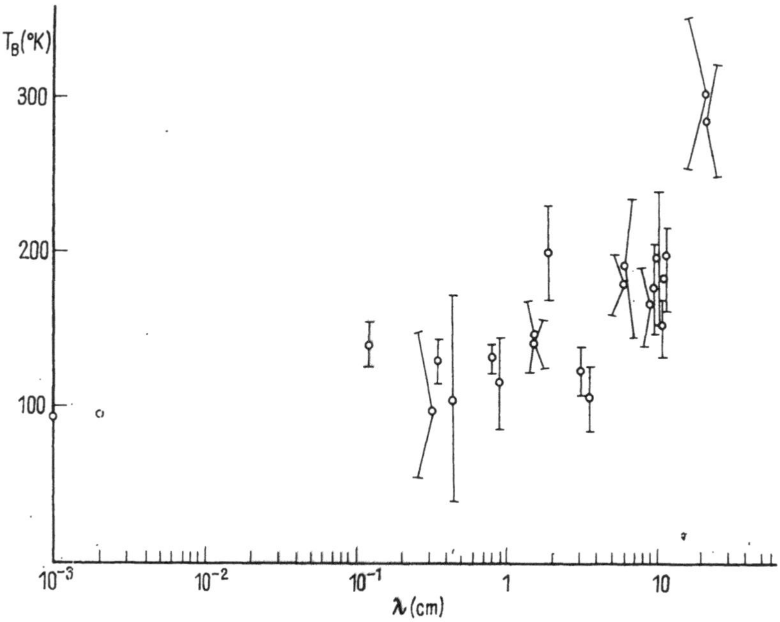

Fig. 6. Saturn's brightness temperature versus wavelength as derived from infrared and radioastronomical observations (see text).

From the plot of measurements (Fig. 6), we see the obvious increase of the temperature with wavelength, though the large scatter of the points prevents a determination of the exact law of increase. We also cannot judge whether the scatter is random or is due to real spectral features.

It is very attractive to suggest that radio astronomers detect from Saturn the thermal radio-emission of a deep atmosphere with temperature gradient depending on wavelength. The greater the wavelength, the deeper is the layer responsible for the radiation detected. If so, radioastronomical observations may be an effective means for sounding Saturn's atmosphere below the layers accessible to optical telescopes. It must be emphasized that for Jupiter such sounding would face great difficulties because of the intense non-thermal emission of the radiation belts.

There is, however, another possible explanation of Saturn's radio spectrum, namely, the emission of the ionosphere. In this connection Gulkis, McDonough and Craft [9] discussed a model of free–free emission from a hot isothermal ionosphere and found the required electron density to be 3×10^7 cm^{-3}, whereas for Jupiter the value is only 10^6. Since the solar ionizing flux is less at Saturn than at Jupiter, the authors conclude that the ionospheric mechanism is unlikely to play an important role in the explanation of Saturn's microwave spectrum. Hence at the present stage the thermal interpretation of Saturn's radio-emission seems to be more reasonable than the non-thermal.

Radioastronomical observations of Saturn may also supply some information on the size of the ring particles. As we have seen, the contribution of the rings to the total flux of radio-emission from Saturn appears to be small. Consequently, optical thickness in radio frequencies may be sufficiently smaller than in optical range. For instance, Berge and Read [10] estimated that at $\lambda = 10.7$ cm the mean brightness temperature of the rings $\overline{T}_{\text{br}} \leqq 40$ °K. Taking into account that

$$\overline{T}_{\text{br}} = T_{\text{part}}\, \varepsilon \left[1 - \exp\left(-\frac{\tau_{\text{or}}}{\sin A'}\right)\right], \tag{5}$$

ε being the emissivity of the particle and T_{part} the equilibrium temperature of a particle, and putting $T_{\text{part}} = 68$ °K, $\varepsilon = 1$, $A' = 5.4°$, Berge and Read obtained $\tau_{\text{or}} \approx 0.08$ against 0.8 in the optical range. Correction for shielding of the particles from the sun by each other and by Saturn's disk diminishes T_{part} to approximately 54° and yields $\tau_{\text{or}} \leqq 0.13$ but even this value is sufficiently lower than that for optical range.

Remembering now that the optical thickness of a cloud of particles becomes sufficiently smaller for the wavelength

$$\lambda = 2\pi\varrho\,, \tag{6}$$

ϱ being the radius of the particle, we may estimate the upper limit of radius of a typical particle of Saturn's rings as

$$\varrho_{\max} = \frac{\lambda}{2\pi} = \frac{10.7}{6.28} = 1.7 \text{ cm}\,. \tag{7}$$

The lower limit of ϱ may be estimated using the well-known Kuiper's result [11] (absorption equivalent water filter 2/3 mm thick). Since for Saturn's rings multiple scattering is small, this figure yields

$$4\varrho_{\min} \approx \frac{2}{3}\text{ mm} \quad \text{or} \quad \varrho_{\min} \approx \frac{1}{6}\text{ mm}\,. \tag{8}$$

Therefore

$$0.017 \text{ cm} \lesssim \bar{\varrho} \lesssim 1.7 \text{ cm}\,, \tag{9}$$

where the upper limit appears to be not very certain and needs verification on the basis of more precise radioastronomical observations. More accurate estimates of $\varrho_{\max}$ must, of course, include the particle size distribution function.

References

[1] W. A. Feibelman, Nature **214**, 793 (1967).
[2] A. Dollfus, Sky and Telescope **34**, No. 3 (1967).
[3] M. Maggini, Ric. Scient., Ser. 2, No. 5–6 (1937).
[4] N. A. Kozyrev, Izv. Pulkovo Obs., No. 184, 99 (1968).
[5] N. P. Barabashov and B. E. Semejkin, Astron. Zh. **10**, 381 (1933).
[6] F. A. Franklin and A. F. Cook, Icarus **10**, 417 (1969).
[7] A. Dollfus and J. Focas, Astron. Astrophys. **2**, 251 (1969).
[8] M. S. Bobrov, Izv. Astrofiz. Inst. Akad. Nauk Kazakh SSR **9**, 83 (1967); see also [13], pp. 438–441.
[9] S. Gulkis, T. R. McDonough and H. Craft, Icarus **10**, 421 (1969).
[10] G. L. Berge and R. B. Read, Astrophys. J. **152**, 755 (1968).
[11] G. P. Kuiper (Ed.), The Atmospheres of the Earth and Planets, Chicago 1948 (p. 340).
[12] P. Guérin, l'Astronomie, **84**, 253–254 (1970).
[13] M. S. Bobrou, in „Surfaces and Interiors of Planets and Satellites", ed. by A. Dollfus, Acad. Press, 1970, Ch. 7, pp. 429–430.

Cosmic Dust

Space Research XI — Akademie-Verlag, Berlin 1971

THE DUST POPULATION IN THE ASTEROID BELT

W. Kokott

Messerschmitt-Bölkow-Blohm GmbH, Space Division, Ottobrunn bei München, FRG

Problems of the dust population in the asteroid belt are of basic importance for understanding the cosmogony of the solar system as well as for adequate planning of space missions beyond the orbit of Mars. The possibility of observing asteroids from the earth is limited to bodies of a minimum diameter of the order of kilometres. At present, data on the distribution of micro-asteroids, meteoroids and dust are necessarily based on extrapolations and theoretical considerations. Various models of the dust population are under discussion, accounting for cosmological, dynamical and observational evidence on the asteroid belt and its interrelations with the solar system as a whole.

Both upper and lower limits of dust flux may be established as a basis for spacecraft reliability and experiment design considerations. The dust distribution does not coincide with the spatial distribution of known asteroids.

1. Introduction

During the next years, the asteroid belt will become a region of major interest to both space scientists and spacecraft designers. While, because of the limits inherent in earth-based observational methods, our knowledge of the asteroids up to now has been limited to bodies larger than 1 km in diameter, it will soon be possible to obtain information about the small fragments of matter which are undoubtedly associated with them. Experimentation and analysis *in situ* will take the place of theoretical and statistical extrapolations which are as yet our only means of dealing with the problem of the dust population in the asteroid belt.

However, these extrapolations are by no means a vain effort. Adequate planning of experiments as well as optimum design of spacecraft for trans-martian missions depend on a careful assessment of the dust environment to be expected.

2. State of Observational Knowledge on the Asteroids

2.1. Orbits

The main bulk of the asteroids or minor planets is located between the orbits of Mars and Jupiter. Out of the 1647 asteroids whose orbits were known some years ago with adequate accuracy, 94% have semi-major axes between 2.22 AU and 3.30 AU; the median value is 2.77 AU, the arithmetic mean slightly larger.

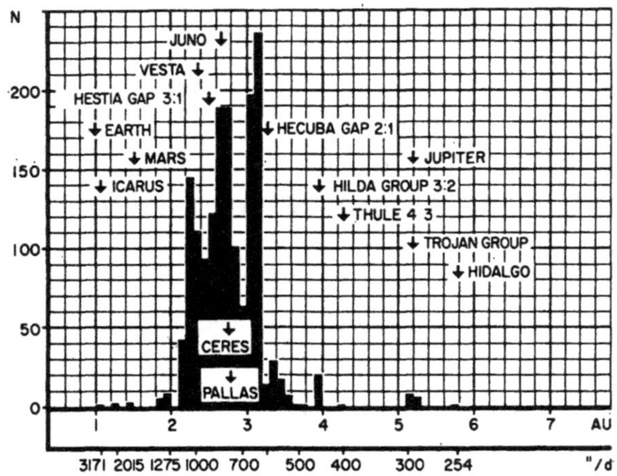

Fig. 1. Distribution of semi-major axes of numbered asteroids.

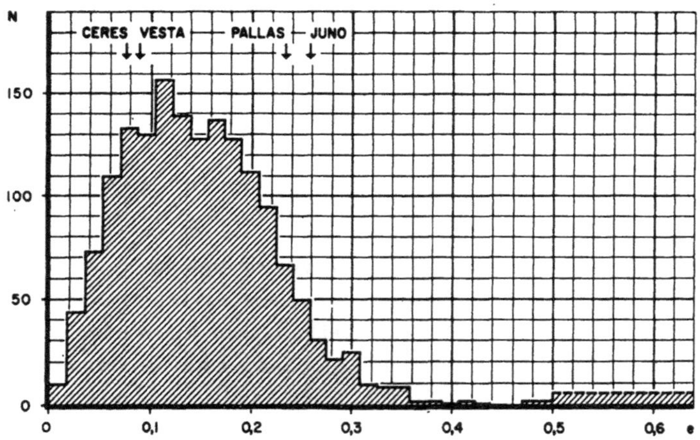

Fig. 2. Distribution of eccentricities of numbered asteroids.

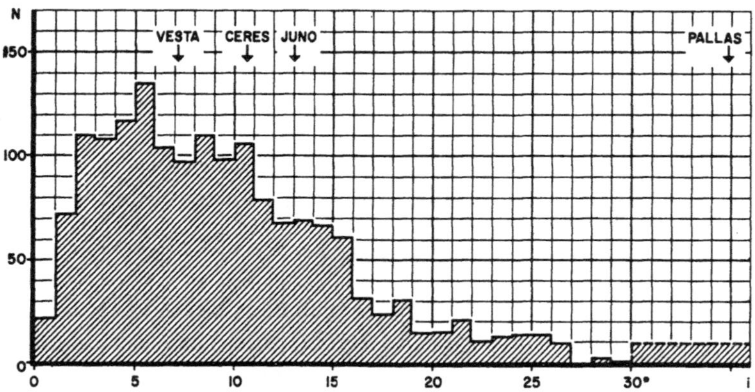

Fig. 3. Distribution of inclinations of numbered asteroids.

Some asteroids (whose cosmological relationship with the belt is frequently questioned) reach as small perihelion distances as 0.19 AU (Icarus) or far-out aphelia (e.g. Hidalgo with 9.59 AU). The median value of orbital eccentricities is 0.14; the maximum value known thus far is 0.827 (Icarus). Ecliptical concentration is moderate: the median inclination is 8.6°, while values up to 52° (Betulia) are known.

Figs. 1–3 show the distribution of 1637 asteroidal orbits according to [1]. Fig. 1 is a histogram of semi-major axes with an abscissa step size of 0.1 AU. The main commensurability gaps and clusterings (due to Jupiter's accumulated perturbations in the case of simple integer ratios of orbital periods with Jupiter's period) are marked; they are not as conspicuous as they would be in case of smaller abscissa intervals. Fig. 2 shows the eccentricities, Fig. 3 the inclinations.

It is not quite clear why only few asteroids are known beyond 3.3 AU. While undoubtedly some observational bias exists (asteroids which may be detected from the earth must be larger with increasing distance), at least any moderate-sized asteroids would have been detected in this region if they did exist. While orbits with semi-major axes in excess of 4.6 AU may be excluded due to Jupiter's excessive perturbations, even semi-major axes greater than 4.04 AU may be unstable on account of secular perturbations [2, 3]. However, the region between 3.3 AU and 4.0 AU seems to be theoretically satisfactory and it is not warranted to exclude it entirely from consideration. Only the near absence of large to medium-sized asteroids in this region is known with certainty.

2.2. Spatial Distribution

Since the heliocentric coordinates of bodies moving in elliptical and inclined orbits do not depend linearly on the orbital elements, the actual distribution of the asteroids in space may not be deduced trivially from Figs. 1–3. Therefore, we consider for three selected dates (1 Jan. 1965, 20 March 1973, 19 April 1973) the distribution of asteroids in radius vector and latitude according to Narin ([4]). The radial distribution (Fig. 4) between 1.9 AU and 3.8 AU is with very good accuracy a Gaussian distribution. The latitude distribution (Fig. 5) is approximately Gaussian.

Due to the eccentricity of their orbits, the asteroids spend more time near their respective aphelia than in the vicinity of the perihelia; therefore, the median value of the radius vector distribution is slightly greater than the median semi-major axis (Fig. 1). During the interval of 30 years (1965–1995) on which the discussion in [4] has been based, the median value of latitudes is oscillating between + 0.3° and − 0.5°. The distributions of 445 asteroids with preliminary orbits behave accordingly; however, the amplitude of the temporal oscillation of the latitude distribution is greater than in Fig. 5. This is probably only a statistical effect caused by the comparatively low number of preliminary orbits considered; however, the possibility that these asteroids (which are generally either smaller or more distant than the main bulk) may belong to families with orbits intersecting outside the ecliptical plane must not be dismissed prematurely.

The longitudinal distribution shows a slight concentration towards Jupiter's perihelion. However, there are irregularities which are far in excess of the expected statistical deviations. This effect seems to be due to the existence of Hirayama's asteroidal families [5–7]. Evidently, the time which has elapsed since the generation of these families has been too short for an equal distribution to be shaped by planetary perturbations.

The parameters of the Gaussian distribution of asteroids in radius vector are approximately $\bar{r} = 2.84$ AU and $\sigma = 0.48$ AU. The distribution of asteroids

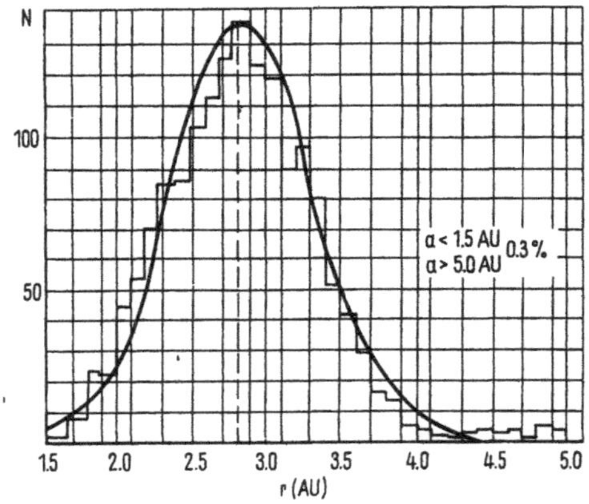

Fig. 4. Radial distribution of numbered asteroids.

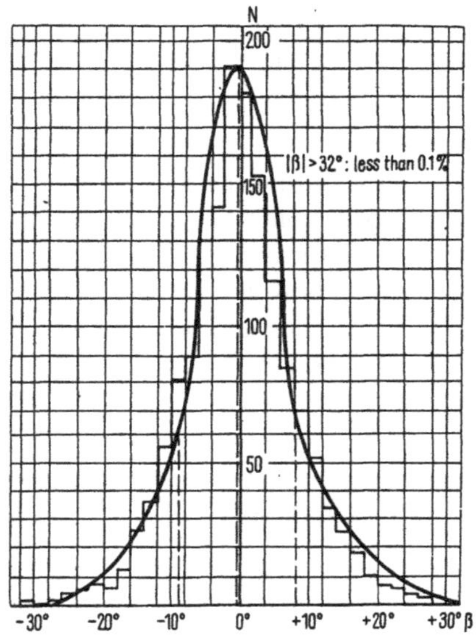

Fig. 5. Latitudinal distribution of numbered asteroids.

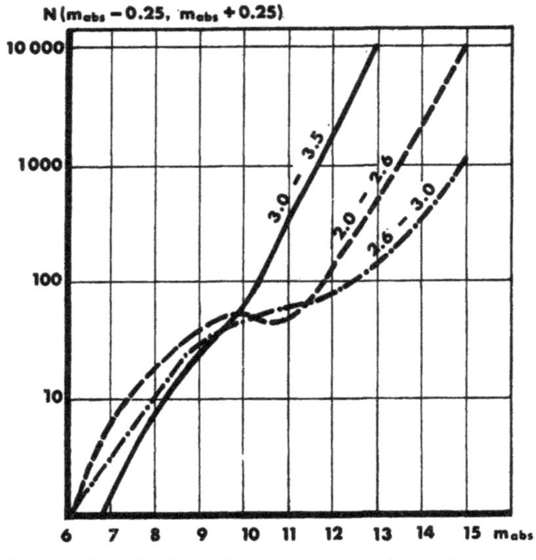

Fig. 6. Brightness distribution of survey asteroids, radius vector as parameter.

with their absolute magnitudes as a parameter has been investigated in the course of a survey initiated by Kuiper [8]. This distribution is shown in Fig. 6. With absolute magnitude as a parameter, the following radial distributions of asteroids are Gaussian:

m_{abs}	4–8	8–9	9–10	10–11
$\bar{r}$ (AU)	2.74	3.02	3.09	3.21
σ (AU)	0.40	0.46	0.47	0.38

The distribution for $m_{abs} = 11–12$ is not Gaussian, because of the steep increase of the number of asteroids between 3.0 AU and 3.5 AU and its constancy between 2.6 and 3.0 AU.

No sufficient data are available thus far for fainter asteroids beyond $r = 3.0$ AU.

If we denote in the usual way the number of asteroids in a set of equidistant magnitude intervals m_{abs} as $N(m_{abs})$ there exists the well-known relation

$$\log N(g) = A + B\, m_{abs} .$$

The inhomogeneous slopes in Fig. 6 are most readily explained by means of the assumption that the distributions for asteroids with absolute magnitudes brighter than $m_{abs} = 9$ represent a primordial Gaussian distribution of magnitudes, while for fainter asteroids the distribution is logarithmic due to frequent collisional grinding.

The original Kuiper survey [8] leads to the following values of B for the logarithmic part of the distributions: 2.0–2.6 AU, $B = 0.60$; 2.6–3.0 AU, $B = 0.35$; 3.0–3.5 AU, $B = 0.73$. By applying some completeness corrections to the data, Anders [9] gets for the middle region 2.6–3.0 AU a B value in the vicinity of 0.60. This would lead to a relatively homogeneous distribution of small asteroids inside 3.0 AU, without the deficiency of objects fainter than 11th magnitude which shows up in the original data. This could point a way to construct a normal distribution in radius vector for the fainter asteroids.

This distribution would call for a denser population of faint asteroids in the region 3.5–4.3 AU than envisaged in the survey. However, an extension of the survey [10] has been interpreted in the sense of an identical value $B = 0.53$ for all radius vector intervals [9]. This would seem to require a homogeneous distribution of asteroids between 2.0 and 3.5 AU with virtually very few bodies elsewhere. *

2.3. Magnitude–Size Relation

Only for a few large asteroids have diameter measurements been obtained so far. Since diameter (in km), albedo and absolute magnitude are related by the equation

$$\log_{10} d = 3.25 - 0.5 \log_{10} A - 0.2\, m_{abs} ,$$

or its converse

$$m_{abs} = 16.25 - 2.5 \log_{10} A - 5 \log_{10} d ,$$

* *Note added in proof:* After publication of the extended survey [10] which occurred during the printing of this volume, the spatial distribution of asteroidal matter will have to be rediscussed in a later paper. The average dust model presented in § 3 seems to remain satisfactory.

these diameter determinations can be used for calibrating the magnitude–size relationship. According to Richter [11], albedo values vary between 0.03 (Ceres) and 0.25 (Vesta). An appropriate average is 0.16. Since it is safe to assume the density of the asteroids to be of the same order of magnitude as that of meteorites recovered on the earth, an adequate assumption will be $3.5\ \mathrm{g\ cm^{-3}}$. Values below $1\ \mathrm{g\ cm^{-3}}$ and above $8\ \mathrm{g\ cm^{-3}}$ are not expected to occur. For a density ϱ, we get from the above equations the mass M (in g) as

$$\log_{10} M = 8.47 - 1.5 \log_{10} A + \log_{10} \varrho - 0.6\, m_{\mathrm{abs}}$$

and

$$m = 15.79 - 2.5 \log_{10} A + \frac{5}{3} \log_{10} \varrho - \frac{5}{3} \log_{10} M\,,$$

whence an asteroid with an absolute magnitude of 10 will have a diameter of 45 km, corresponding to a mass of 1.6×10^{20} g (with one order of magnitude maximum uncertainty).

3. Extrapolation

3.1. General Considerations

Table 1 shows the logarithmic part of the distributions according to Fig. 6 related to magnitude, diameter and mass. Since it has been demonstrated by

Table 1
Asteroid population function (logarithmic branch)

Zone r (AU)	(1)		(2)				(3)			
	A_m	B_m	A_d for $A = 0.02$	0.16	0.252	B_d	A_M for $A = 0.032$, $\varrho = 8$	0.16, 3.5	0.252, 1	B_M
2.0–2.6	−6.25	0.60	4.85	3.79	3.51	3.0	21.35	20.35	19.15	1.0
2.6–3.0	−5.5	0.35	0.97	0.36	0.19	1.75	10.6	9.6	9.3	0.58
3.0–3.5	−5.65	0.73	7.27	6.07	5.73	3.65	25.95	24.95	23.95	1.22
2.0–3.5	−5.0	0.70	7.92	6.72	6.38	3.5	26.6	25.6	24.6	1.17

(1) $\log_{10} N_m (m_{\mathrm{abs}} \pm 0.25) = A_m + B_m\, m_{\mathrm{abs}}$
(2) $\log_{10} N_d (\log_{10} \frac{1}{2}d \pm 0.05) = A_d - B_d \log_{10} \frac{1}{2}d$
(3) $\log_{10} N_M (\log_{10} M \pm 0.15) = A_M - B_M \log_{10} M$
m = magnetude; d = diameter (km); M = mass (g).

many authors that a steady-state distribution with mutually equivalent production and depletion rates necessitates $B_M = 2/3$ (e.g. [9]), it follows that most of the slopes (except in the central region, 2.6–3.0 AU) are too steep. Therefore, extrapolating on the basis of these distributions will lead to an excessively dense population of small asteroids. This would undoubtedly be consistent with the "explosion" hypothesis but not at all with any grinding mechanism. In any case, this kind of extrapolation will provide us with maximum population densities for micro-asteroids and meteoroids, and we may be certain that the resulting fragment densities will not be surpassed in reality.

Anders [9] demonstrates from the mass distribution of asteroid families in the inner belt that the mass distribution of known asteroids with semi-major axes below 2.6 AU may be traced back to an original normal distribution in magnitude, with virtually no asteroids brighter than 6th or fainter than 14th magnitude. According to this investigation, the mass of micro-asteroids in this region will be limited, while fragmentation beyond 3.0 AU will be very predominant due to the more dense population and Jupiter's perturbations in this region. Unfortunately, no similar analyses have been performed thus far for asteroidal families beyond 2.6 AU. However, Dohnanyi [7] has recently formulated a collisional model for the whole asteroid belt. The model is based on the Survey of Asteroids [8], and, while employing a size–magnitude relationship slightly different from the formulae in § 2.3, the deviation is within the margin allowed for by our assumptions on albedo and densities. Dohnanyi's model employs an equal-density distribution of asteroids within a toroidal space of 2.0 AU inner radius, 3.5 AU outer radius, and $\pm$ 10° heliocentric latitude. This is consistent with the Narin distribution of bright asteroids (Figs. 1 and 2) and leads to the required results as long as the overall number and behaviour of particles are investigated. The real geometry of the dust distribution will of course differ markedly from these assumptions, e.g. the zone of most dense population will be located somewhere between 3.0 and 4.3 AU, for the reasons pointed out above.

Dohnanyi's theoretically derived distribution has a B_M of 0.837, which is consistent with the mean gradient of the bright asteroids in the Survey, but not with the slopes of the logarithmic branches according to [8]. Since B_d for the above-mentioned van Houten distribution [9] would be 2.9, the Dohnanyi model seems to be consistent with observation also in this respect.

3.2. Lifetime and Related Phenomena

When discussing any distribution of small particles in the asteroid belt, we must take into account their possible lifetime. While for a long time the Poynting–Robertson effect has been thought to be the limiting factor for the lifetime of asteroidal dust, we know now that for most particle sizes other effects are predominant. Dohnanyi [7] has shown that, for particles below 1 mm radius, erosion by smaller particles will possibly be the crucial effect, limiting lifetime to between 10^2 and 3×10^6 years, depending on size and density distribution. Catastrophic collision, which is of prime importance for large meteoroids, may also be significant for the dust particles. The Poynting–Robertson effect works by some orders of magnitudes more slowly than the depletion mechanisms mentioned above. This effect will have no major importance for the dust population beyond 3.0 AU but is an important mechanism for replenishing the dust in the inner asteroidal belt and in the zodiacal cloud.

Particles with diameters less than 1 μm will be swept out of the solar system by the solar radiation pressure. It has been postulated by some authors that under certain assumptions particles of very small size may survive this effect for a relatively long time. However, even extreme assumptions about the physical constitution of particles and electric and magnetic effects lead only to lifetimes of the order of 10 years [12, 13]. We are thus faced with a lower limit of about 1 μm diameter or 10^{-13} g (varying slightly with particle density and structure) for any distribution of small particles in the belt.

Another effect which may be of importance is the change of orbits by gravitational interaction. Sitte [14] has shown that a considerable fraction of meteoroids in the asteroid belt will not be destroyed by catastrophic collisions but will suffer significant changes of their orbits by near passes in the vicinity of larger meteoroids and micro-asteroids. In most cases, the orbital eccentricity of the smaller meteoroids will increase appreciably. This leads to a secular flattening of the distribution function in radius vector which may result in a denser population of small fragments in the inner belt as well as beyond 4.3 AU than would be expected *a priori* from any extrapolation of the empirically known asteroid population. Any steady-state mass distribution of bodies will therefore also for this reason be valid only for the belt as a whole but not for any particular zones within it.

3.3. Number Density of Dust Particles in the Asteroidal Belt

Fig. 7 shows the extrapolated number density of micrometeoroids for mass ranges of one order of magnitude for the whole asteroid region. The "upper limit" graph is extrapolated from the logarithmic branch of the original Survey for the whole region between 2.0 and 3.5 AU, assuming an albedo of 0.16 and a density of 3.5 g cm^{-3} and near-steady-state conditions only for particles with

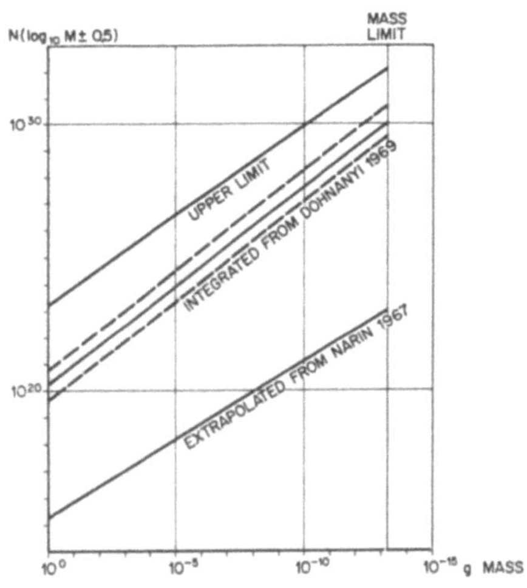

Fig. 7. Mass distribution of asteroidal dust.

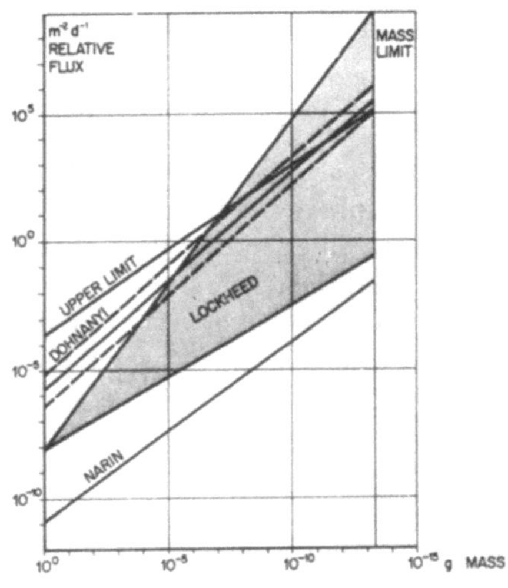

Fig. 8. Flux densities of asteroidal dust.

lifetimes less than some 10^9 years (masses less than 10^6 g), in accordance with some assumptions made in [15]. The distribution function of Dohnanyi [7] with its uncertainty range has been integrated for our abscissa definition, and Narin's assumptions [16] have been extrapolated accordingly. As has been shown above, the spatial distribution of the asteroidal dust is subject to a number of heterogeneous effects. While production of small fragments occurs mainly in the region between 3.0 and 4.0 AU, the inner belt (2.0–3.0 AU) will not only contain the (comparably few) fragments following from an extrapolation of the distribution of Survey asteroids in this region, but also a considerable number of objects originating in the outer belt. On the other hand, beyond 4.0 AU there may be expected a zone which is replenished continuously from the 3–4 AU zone but depleted by Jupiter's perturbations and radiation pressure effects. The latitudinal distribution will also be somewhat broader than the distribution of bright asteroids, i.e., more than 32% of the dust will be located beyond $\pm 9°$ of latitude, especially in the outer belt, due to the peculiarities of the fragmentation and perturbation processes outlined above.

However, there are at present no quantitative data available which would permit construction of a realistic spatial distribution. Therefore, we assume the following average distribution: 68% of the dust population as shown in Fig. 7 is located within a trapezoidal torus limited by 2 and 4 AU in radius vector and $\pm 9°$ in heliocentric latitude. The volume of this torus is approximately 36 AU^3 or 10^{35} m^3. Therefore, the number density of asteroidal fragments within the torus is found by multiplying the values in Fig. 7 by 0.68×10^{-35} m^{-3}.

3.4. Total Mass of Asteroidal Matter

With the above assumptions, the upper limit of the total mass of the asteroid belt is 1.5×10^{25} g (0.0025 earth masses); Dohnanyi [7] finds 5×10^{23} g, while some earlier values are: 1.7×10^{24} g (Kiang), 6×10^{24} g (Watson), 3×10^{25} g (Orlov, including all comets and interplanetary matter), 6×10^{25} g (Schütte). Considering the different methods and assumptions employed, the consistency of the results is remarkably good. Furthermore, they are also consistent with the observed absence of perturbations due to the asteroidal belt in the motion of the planets; such perturbations would reach significant values only for a total mass greater than 10^{26} g.

3.5. Flux Density

The practical application of the present considerations on the dust population in the asteroid belt is the impending exploration of this region of space by means of interplanetary spacecraft. For designing both spacecraft and experiments, the flux density of particles rather than their overall number is of interest. Since any foreseeable spacecraft will move in direct orbits, i.e. according to the general direction of revolution of most bodies in the solar system, relative velocities between spacecraft and asteroidal debris will in general be less than 10 km s^{-1} (or 8.6×10^8 m d^{-1}). When multiplying the values from § 3.3 by this factor (or the values shown in Fig. 7 by 5.85×10^{-27} m^{-2} d^{-1}), we find upper limits for the average flux density within the torus defined in § 3.3. The actual flux density varies in proportion to both the actual spatial density of particles and their relative velocities with respect to the spacecraft. The average flux density is shown in Fig. 8. In addition, a flux density derived by Lockheed Company [17] is plotted for comparison.

3.6. Cometary Matter

Due to the clustering of comets and meteor streams of cometary origin in the neighbourhood of Jupiter additional dust will be produced beyond 4.0 AU. However, the flux density of these particles will be very significantly lower than that of the asteroidal dust population. Any spacecraft experiments exploring the asteroidal belt will also gather data on these particles. They will quite conveniently be discriminated from the asteroidal particles by their different density (less than 0.5 g cm^{-3}) and structure.

4. Conclusions

Spacecraft traversing the asteroid belt should be shielded against micrometeoroid punctures corresponding to the "Dohnanyi" curve in Fig. 8. Experiment designers should regard the "Narin" curve as a conservative minimum of particles to be detected easily.

The dust population is expected to vary considerably in the region of space under consideration, but dust is expected to occur everywhere between 2.0 AU and Jupiter's orbit, as well as in comparatively high heliocentric latitudes.

Due to the current very great research activity in related fields, significant theoretical results are expected to furnish additional and more accurate data on the asteroidal dust population in the very near future. Exact results which are free of the bias of simplifying assumptions can, however, be gathered only by means of spacecraft-borne experiments *in situ*.

References

[1] Landolt-Börnstein, Numerical Data and Functional Relationships in Science and Technology, VI/1, Springer, Berlin 1965.
[2] D. Brouwer, Astron. J. **68**, 152 (1963).
[3] S. A. Dole, J. Am. Rocket Soc. **31**, 214 (1961).
[4] F. Narin, J. Spacecraft Rockets **3**, 1438 (1966).
[5] H. Alfvén, Astrophys. Space Sci. **4**, 84 (1969).
[6] L. Danielsson, Astrophys. Space Sci. **5**, 53 (1969).
[7] J. S. Dohnanyi, J. Geophys. Res. **74**, 2531 (1969).
[8] G. P. Kuiper et al., Astrophys. J., Supp. Ser. **3**, 289 (1958).
[9] E. Anders, Icarus **4**, 398 (1965).
[10] C. J. van Houten, I. van Houten-Groeneveld, P. Herget and T. Gehrels, Astron. Astrophys. Suppl. **2**, 339 (1970).
[11] N. Richter, Sterne **36**, 125 (1960).
[12] Th. Schmidt and H. Elsässer, in: The Zodiacal Light and the Interplanetary Medium (NASA SP-150) p. 287.
[13] H. Elsässer and Th. Schmidt, Z. Naturf. **21 a**, 1116 (1966).
[14] K. Sitte, Z. Naturf. **21 a**, 231 (1966).
[15] W. Kokott, Raumfahrtforschung **12**, 75 (1968).
[16] D. L. Roberts and F. Narin, in: Advances in Space Science and Technology 9, Academic Press, New York 1967 (p. 123).
[17] Lockheed Missiles and Space Corp., NASA CR-64621 (1965).

Space Research XI — Akademie-Verlag, Berlin 1971

MORE THAN TWO YEARS OF MICROMETEORITE DATA FROM TWO PIONEER SATELLITES

O. E. BERG and U. GERLOFF

NASA Goddard Space Flight Center, Greenbelt, Md, USA

Identical cosmic dust experiments on board the Pioneer 8 and 9 heliocentric satellites have each yielded almost continuous data for more than two years. The experiment measures the kinetic energy, momentum, speed, and direction of particles having energy 1.0 erg or greater, and measures the kinetic energy and direction of particles of energy 0.2 erg or greater.

Only nine events by particles having kinetic energy greater than 1.0 erg have been recorded. This gives a particle flux of 2×10^{-4} m^{-2} s^{-1} $(2 \pi$ s r$)^{-1}$. The average flux for particles having KE of 0.2 ergs or greater is 1×10^{-3}m^{-2} s^{-1} $(2 \pi$ sr$)^{-1}$. The measured influx rates seem to confirm the zodiacal light predictions of particle fluxes rather than agree with earlier data from direct measurements. The data have shown no significant indications of meteor shower activity.

Considerations of the pre-launch tests and calibration of the experiment using a 2-million volt electrostatic dust particle accelerator and considerations of the consistency of the data from the two experiments justifies a high degree of confidence in the data. The probability of an "accidental" velocity measurements is shown to be 2×10^{-7}.

The persistent lack of agreement between cosmic dust particle fluxes and mass densities as measured by rocket and satellite experiments is somewhat understandable if one considers the following argument: If an *in situ* cosmic dust experiment is to record a reasonable number of micrometeor impact events its sensors must be designed to respond to kinetic energies and/or momenta very near the upper limits of sensitivity of the fast pulse amplifiers. Further, the flux of micrometeorites is only in the order of a few particles m^{-2} day^{-1}. Accordingly, the signal-to-noise ratio for cosmic dust experiments in satellites is very low. It seems obvious, then, that reliable data having good statistical value can be acquired only from relatively long-term satellite measurements.

Identical experiments in each of the two heliocentric satellites, Pioneers 8 and 9, have transmitted cosmic dust impact data for a combined total period of more than three years. Both the instruments are still performing normally, as indicated by in-flight controls and electronic monitors. All the measurements were obtained at heliocentric distances varying from 0.75 to 1.1 AU. A brief description only of the experiment is given here; for a full description see [1].

Fig. 1 is a photograph of the flight unit for Pioneer 8. A box approximately 15 cm cube contains the entire experiment including sensors, amplifiers, logic etc. The basic sensor is shown in Fig. 2. As a cosmic dust particle enters the experiment sensor system it produces a plasma at the front, very thin film, sensor. The plasma is separated into ions and electrons and collected on the

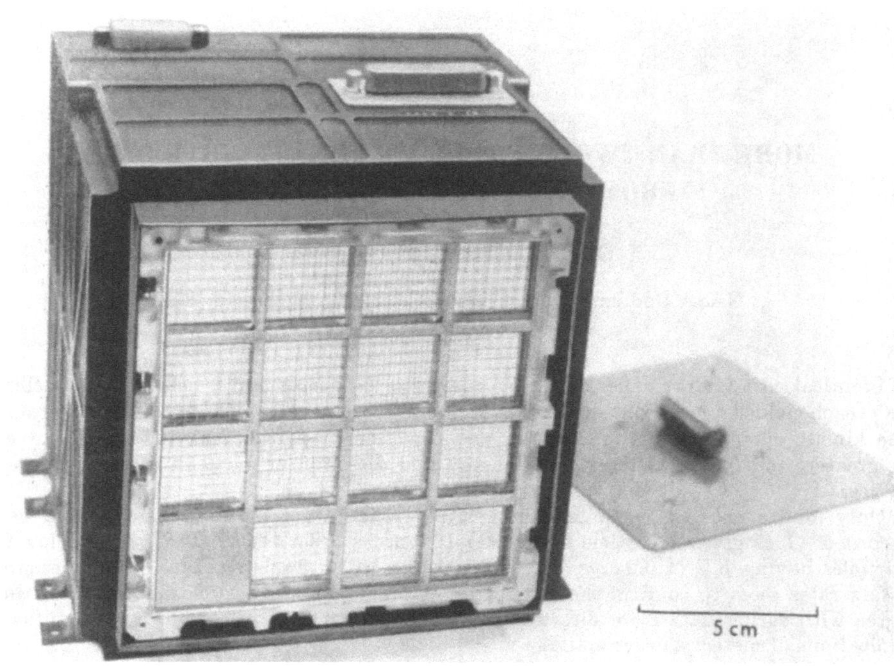

Fig. 1. Flight unit for Pioneer 8.

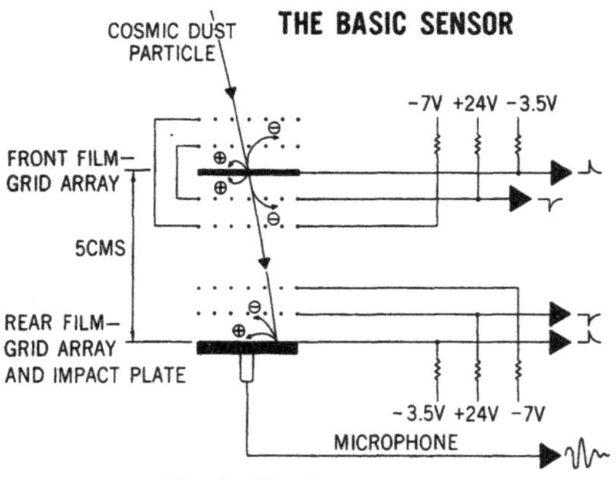

Fig. 2. The basic sensor.

appropriate film and grid, producing positive and negative pulses, as shown. As the particle proceeds to impact on the rear plate sensor it produces a second set of plasma pulses; if it has sufficient momentum its impact is also registered by the acoustical sensor.

There are five separate means of acquiring data from this experiment:

The front film
The front collector grid
The rear film
The rear collector grid
The acoustical sensor

The time of flight (TOF) array serves the principal objective of the experiment: to obtain information on the speed, direction, and kinetic energy of meteoroids in order to determine their orbits. It requires response from both the front film and the rear film in proper sequence and within a limited time window. The TOF array fully utilizes all the unique sensor controls and electronic monitors that are features of the Pioneer experiment. The probability of an accidental TOF event is 6.2×10^{-7}.

The basic sensor shown is one of 256 similar sensors which compose the Pioneer cosmic dust experiment. Fig. 3 shows an exploded view of the overall

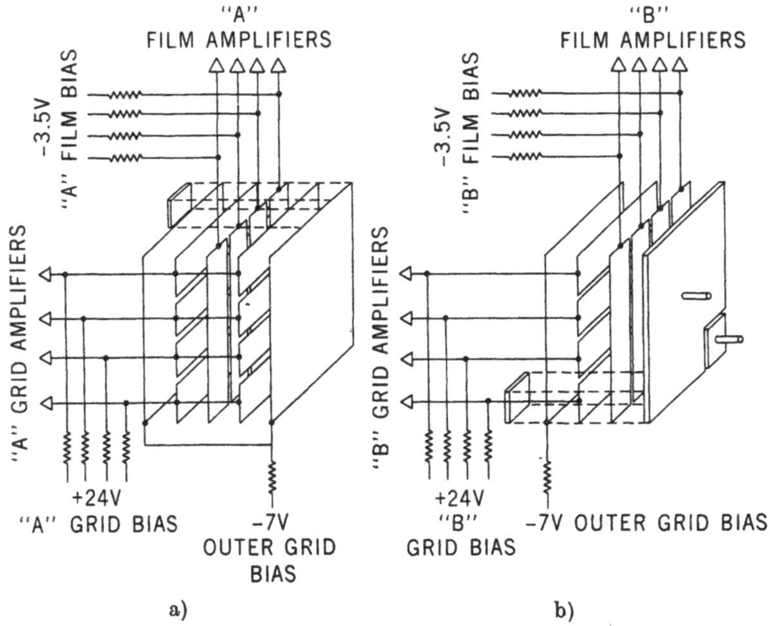

Fig. 3. a) Front sensor array; b) back rear sensor array.

experiment sensor system. The front sensor array and the rear sensor array are each composed of 16 squares, formed by 4 film strips crossed by 4 grid strips. By proper utilization of amplifiers and logic, one can determine which of the 16 squares in each of the arrays was affected by an impact. Thus the path direction of the meteoroid within the experiment can be determined. The position of the sensor axis relative to the sun at the moment of impact is given by a solar aspect device. The front and rear film controls are designed to be unaffected by particle impacts.

Fig. 4 shows the total number of sensor output pulses per day for the front film, the microphone, and the rear sensor for a sample period of 150 days. There are two significant features exhibited in this figure: (i) the extremely low data rate observed in the Pioneer 8 and 9 measurements; (ii) typical examples of the correlation observed between periods of high cosmic ray activity and the microphone and rear film output rates.

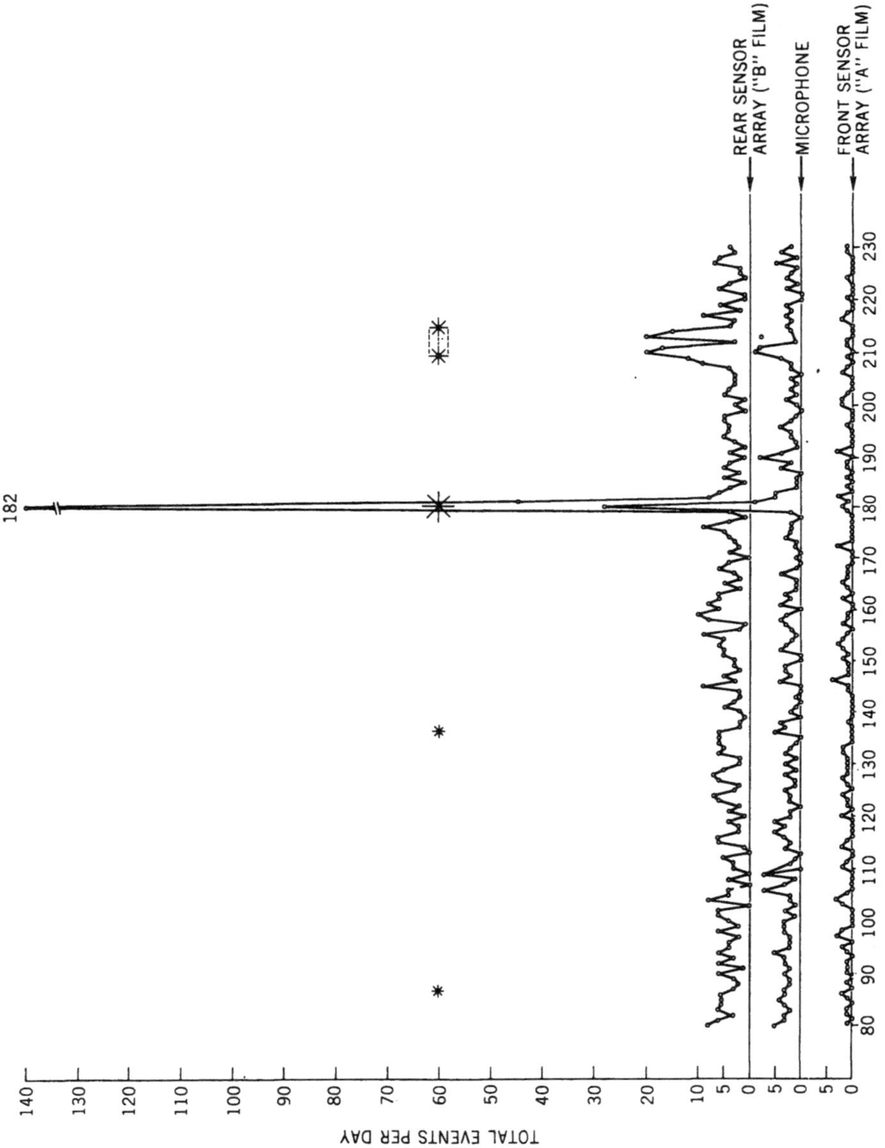

Fig. 4. Cosmic ray effects.

The front film sensor output is extremely quiet, with an average of about 0.8 impact per day. The microphone and rear film sensor output pulse rates are somewhat higher, due to their rate correlation with cosmic rays. The stars denote periods of high cosmic-ray activity as measured by two independent cosmic-ray experiments on board the Pioneer spacecraft. The size of the star is roughly in proportion to the intensity of the cosmic-ray activity. This interaction is similarly evident in the data from both Pioneer instruments although they are millions of kilometers apart. Eleven correlative peaks of this type have been recorded to date. It is noteworthy that the front film sensor has never shown a response to cosmic-ray activity.

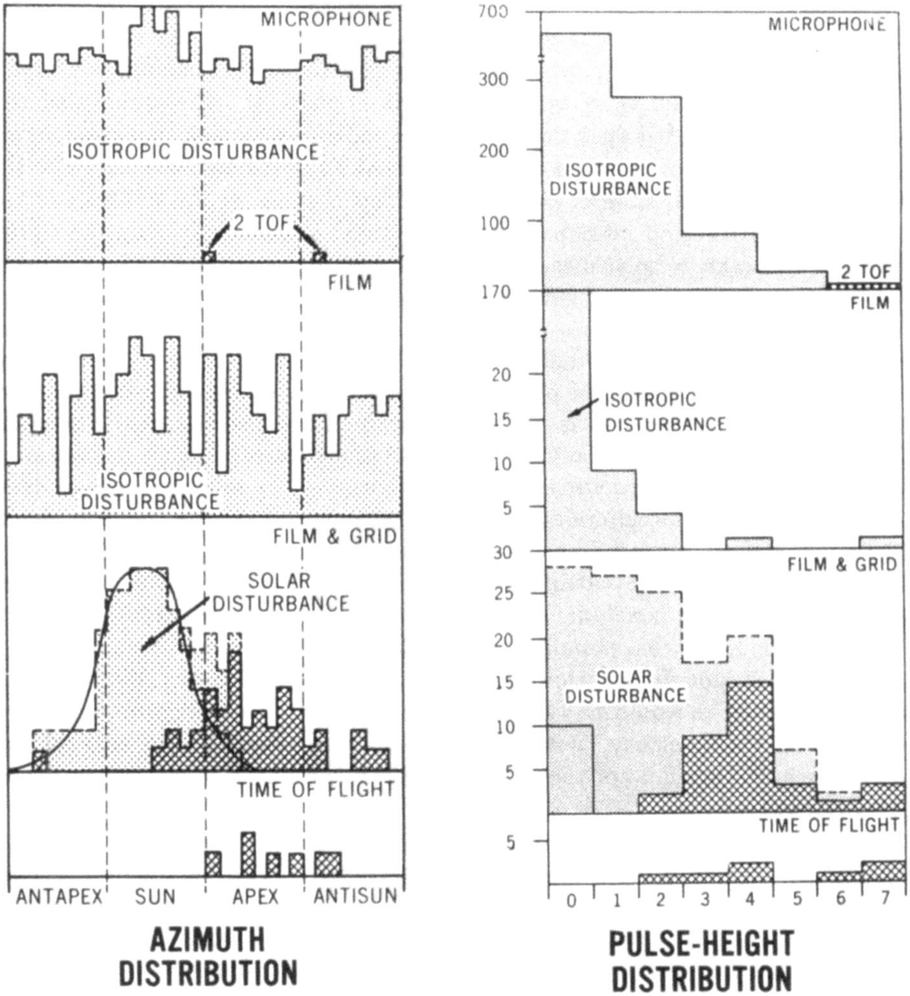

Fig. 5. Meteoroid impacts as measured by Pioneer 8 during the first 20 months in orbit.

Azimuth distribution and sensor pulse height distribution are shown in Fig. 5. As in Fig. 4, the plots include all the meteoroid sensor output pulses for the period shown.

Only nine TOF events have been recorded in more than three years of data acquisition from Pioneers 8 and 9. Seven of them are presented in this paper. The TOF events, whose impact direction and velocity are known very well, show a pronounced accumulation around the apex direction of the satellite. This apex peak, caused by the aberration of the original meteoroid radiants due to the satellite velocity, must occur for any meteoroid satellite experiment where the satellite velocity is comparable in magnitude with the average meteoroid velocity. It thus provides a unique criterion for distinguishing true impact events from noise pulses of any kind. It is, in fact, the first time that this theoretically demanded apex peak could be verified by a satellite meteoroid experiment. In view of this, the apex peak must be postulated for any true signals among the remaining events measured by Pioneer 8. An analysis of the azimuthal distribution of signals from the front sensor array (film and grid) reveals two peaks, a solar disturbance peak and an apex peak. The reality of a solar disturbance peak of such small magnitude (0.2 event per day) was not unexpected since the sensor's principle of detection is to respond to pulses of ions and electrons. The source of the extraneous pulses of electrons and ions is probably a combination of photoemission and the solar wind. The shape of the solar disturbance curve is readily determined because the sensor axis–sun angle is known to an accuracy of $\pm 6°$, and the field of view of the sensor is known. Now, if those specific events which compose the solar disturbance peak are removed from the adjoining film and grid pulse height distribution plot of Fig. 5, the darker shaded area remains. The 40 impact events represented, as shown, substantiate the cut-off exhibited by the TOF data and with better statistics. We must conclude, therefore, that this cut-off represents a general feature of the meteoroid population at 1 AU distance from the sun.

The pulse height distribution of TOF events does not show a large number of small signals, as would be expected for a "falling" size distribution of meteoroids. On the contrary, none of the seven particles has an energy that approaches the threshold of the detector. A particle size cut-off is suggested by this distribution.

Further proof of the cut-off is provided by those events which were registered by the front film only (no grid). From hypervelocity impact simulation studies it is known that these signals can be produced only by impacts below a certain energy that approximately coincides with the cut-off energy. All impacts of higher energy are "seen" also by the front grid. The "film-only" events thus provide an exclusive and sensitive test for the existence of very small particles of mass $< 10^{-12}$ gram.

Although the pulse height distribution of film-only events seems to indicate the existence of a large number of small particles, the azimuth distribution of these events clearly disproves this interpretation. The assumption of more than a few submicron particles, that might be unresolved in the statistical variations of the azimuth distribution, is inconsistent with the evidence of the apex peak discovered. The film-only events thus further confirm the conclusion of a cut-off at about 1 m size. The occurrence of the extraneous front-film only pulses is not surprising if one considers again their extremely low occurrence rate of approximately 0.2 event per day. It is, in fact, a measure of the excellent design of the sensors and electronics.

The next pair of plots in Fig. 5 is concerned with the acoustical sensor output pulses which are already suspect because of the previously shown interaction between cosmic rays and the acoustical sensor. Again, we see an isotropic distribution which, if interpreted as caused by meteoroid impacts, is totally inconsistent for a meteoroid sensor having its axis in the ecliptic plane and mounted in a spinning heliocentric satellite moving at 30 km s^{-1}. This type of response

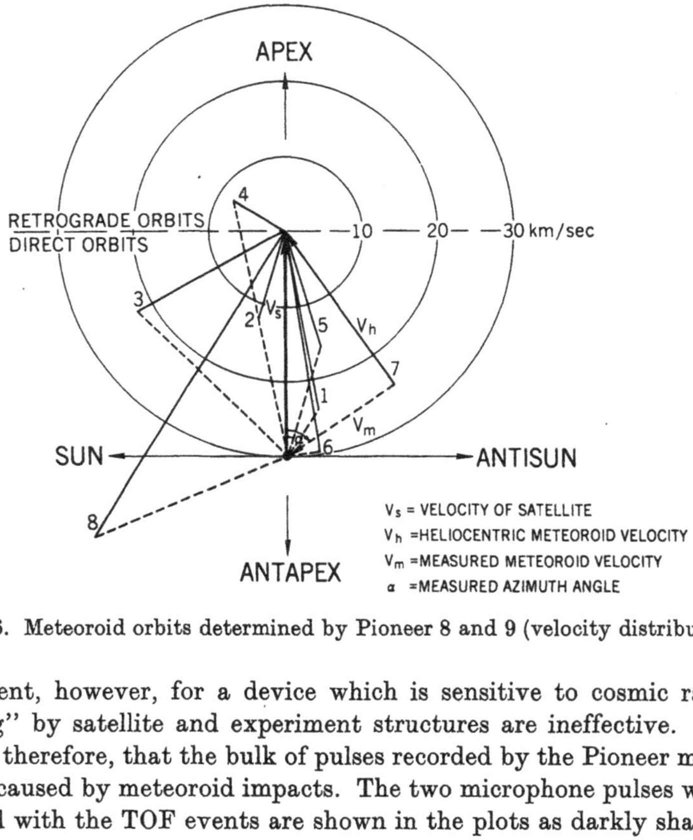

Fig. 6. Meteoroid orbits determined by Pioneer 8 and 9 (velocity distribution).

is consistent, however, for a device which is sensitive to cosmic rays where "shielding" by satellite and experiment structures are ineffective. One must conclude, therefore, that the bulk of pulses recorded by the Pioneer microphone were not caused by meteoroid impacts. The two microphone pulses which were associated with the TOF events are shown in the plots as darkly shaded areas. Although statistically poor, their position in the plots is interesting in that they tend to substantiate both the apex peak and the cut-off described earlier.

A vectorial presentation of the eight TOF events is presented in Fig. 6 and clearly shows a majority of direct orbits. A solid vertical line, V_s, represents the velocity vector of the satellite. Measured velocity vectors and heliocentric velocity vectors for the eight TOF particles are represented by the dotted lines and solid lines respectively. Since the Pioneer spacecraft are moving in direct heliocentric orbits at relatively high velocities, the measurements favor the detection of retrograde orbits. A similar plot showing proper compensations for this orbital bias will be presented in a later publication, but the evidence of a preponderance of direct orbits is clearly shown here.

A comparison between trends of orbital parameters derived from the Pioneer measurements and those of ground based measurements is shown in Fig. 7. This type of presentation has been used for ground-based measurements and has been expanded here to include the Pioneer "*in situ*" data for smaller particles. The abscissa gives the absolute brightness of the meteor and the corresponding

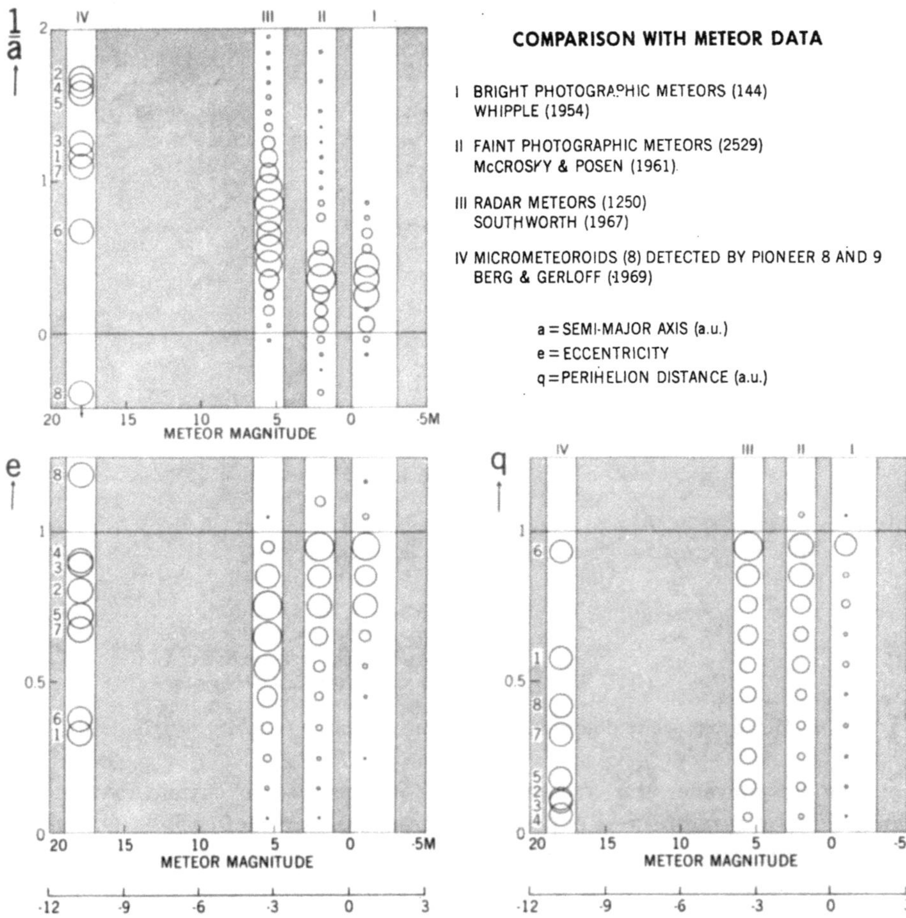

Fig. 7. Meteoroid orbits determined by Pioneer 8 and 9.

mass of the meteoroid. The ordinate gives three orbital parameters: the semi-major axis a, the eccentricity e, and the perihelion distance q. The results of three major meteor surveys are compiled so that the diameter of the circles shown is proportional to the number of meteors with orbital parameters within a certain interval. The meteor surveys are shown in the strips I, II, and III and relate to progressively smaller particles from right to left. The eight Pioneer TOF events are plotted to the extreme left. The trend toward smaller semi-major axes, slightly reduced eccentricities, and smaller perihelion distances indicated by the successive meteor measurements, is clearly confirmed by the

Pioneer results. The orbits of micron-sized meteoroids thus differ markedly from the orbits of meteor-producing objects: they are much smaller, slightly more round, and reach much closer to the sun. They are similar neither to short-period comet orbits, nor to circular orbits, which, for lack of better information and for simplicity, have been used to understand the dynamics of the zodiacal cloud. The measured orbits of micrometeoroids can be interpreted as the result of the Poynting–Robertson effect acting on originally highly eccentric cometary

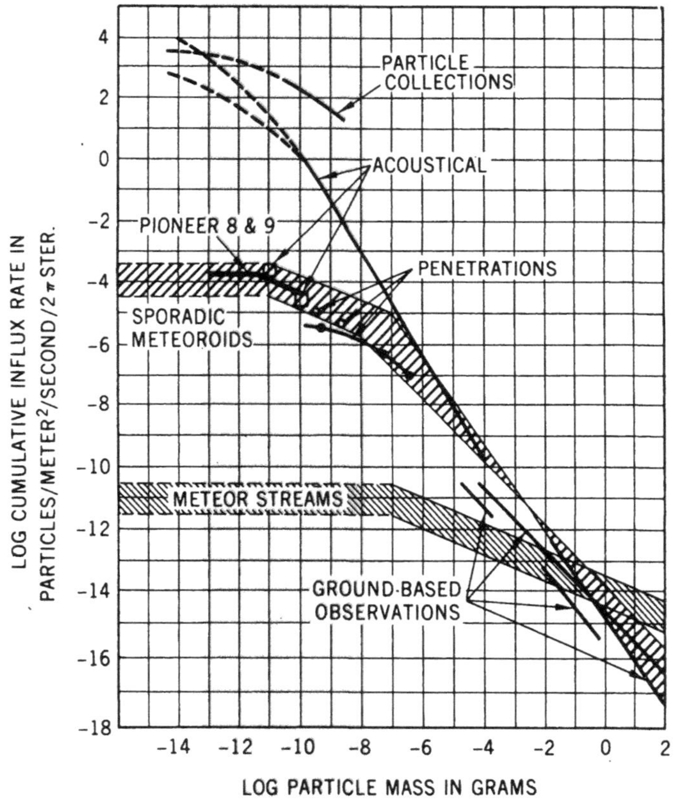

Fig. 8. Flux of meteoroids (measurements and theory).

orbits. They cannot easily be understood on the basis of an asteroidal origin for bulk of meteoroids.

Interesting physical parameters related to the Pioneer TOF particles are given in Table 1. As mentioned earlier in this paper, only two of the eight TOF events produced a response from the microphone. They are shown as event numbers 3 and 7. Their mass, as derived, was considerably greater than those of impacting particles which did not activate the microphone. Particle numbers 3 and 8 fragmented in the front film as evidenced by the response from two of the 16 rear film squares. These two particles have been associated with the known comets Grigg-Skellerup and Encke. Correlation of orbital parameters for meteoroid number 3 and meteoroid number 8 and the comets is shown in

Table 1
Meteoroid orbits determined by Pioneer 8 and 9
(Results of the time-of-flight measurement)

TOF-Event No.	Satellite date of event	Microphone response $I>2$ (cgs · 10^{-5})	Meteoroid momentum 1 (cgs · 10^{-5})	Meteoroid enegy E (erg)	Meteoroid velocity rel. to satellite V (km · s^{-1})	Meteoroid mass m (g · 10^{-11})	Meteoroid diameter d (μ)	Meteoroid radiation parameter $\beta (\varrho = 3)$	Meteoroid fragmentation
1	13 Mar. 1968	NO	1.8	6.8	7.7	2.4	2.4	0.16	NO
2	29 Mar. 1968	NO	1.6	14.9	19.0	0.8	1.7	0.22	NO
3	18 Apr. 1968	YES	10.9	155.9	28.2	3.9	2.9	0.13	YES
4	20 Apr. 1968	NO	4.0	71.3	35.9	1.1	1.9	0.20	NO
5	13 Sept. 1968	NO	1.9	15.0	16.1	1.2	1.9	0.20	NO
6	15 Feb. 1969	NO	1.1	3.1	5.7	2.0	2.3	0.17	NO
7	17 Jun. 1969	YES	17.9	155.9	17.4	10.3	4.0	0.10	NO
8	11 Feb. 1969	NO	4.6	71.3	31.3	1.5	2.1	0.18	YES
Average	—	—	5.5	61.8	20.2	2.9	2.4	0.17	—

Table 2
Meteoroid orbits determined by Pioneer 8 and 9
(Correlation to comets)

	Solar distance of node (AU) $R_M < R_C$	Longitude of node (°) $\Omega_M \approx \Omega_C$	Inclination of orbit (°) $i_M \approx i_C$	Eccentricity of meteoroid orbit	Radiation parameter of meteoroid as function of eccentricity e and mass-density ϱ (g · cm^{-3})	Characteristic feature of meteoroid
Meteoroid No. 3	$R_3 = 1.05$	$\Omega_3 = 208$	$i_3 = 0 - 54$	$e_3 < 1$	$0.13 < \beta_3 < 0.27$ for $3 > \varrho_3 > 1$	Fragmentation
Comet Grigg-Skellerup	$R_{GS} = 0.856$	$\Omega_{GS} = 215$	$i_{GS} = 17.6$	—	$0.15 < \beta < 0.25$ for $e > 1$	—
Meteoroid No. 8	$R_1 = 0.9$	$\Omega_8 = 322$	$i_8 = 0 - 14$	$e_8 > 1$	$0.4 < \beta_8 < 0.9$ for $1 > \varrho_8 > 0.3$	Fragmentation
Comet Encke	$R_E = 0.34$	$\Omega_E = 334$	$i_E = 12.4$	—	$0.6 < \beta$ for $e > 1$	—

Table 2. Both of the comet nodes lie inside the satellite orbits as required for this correlation. The longitude of the ascending node Ω and the inclinations i of the comets and their respective meteoroids are approximately the same. The radiation parameters β_3 and β_8 derived from the measurement of meteoroid mass aboard Pioneers 8 and 9, and the radiation parameters β postulated by radiation pressure theory for particles related to the comets discussed, can be matched for a reasonable mass density between 3 and 0.3 g cm^{-3}.

It is necessary also to consider the position of these measurements of meteoroid impacts on a meteoroid flux curve. The Pioneer results have been plotted on the well known flux curve shown in Fig. 8, although that flux curve was originally proposed as a near-earth flux curve. The Pioneer contribution to the curve reflects the cut-off as represented by the horizontal portion of the curve and the flux as derived from 47 impacts. Specifically, the displayed flux of $(2 \pm 0.5) \times 10^{-4}$ m^{-2} s^{-1} $(2\pi\ sr)^{-1}$ refers to a sensor with its axis in the ecliptic plane having a half-angle field of view of 60° and performing in a spinning heliocentric satellite moving at 30 km s^{-1} at distances ranging from 0.75 to 1.1 AU from the sun.

Conclusions

The Pioneer 8 and 9 Cosmic Dust Experiments have reliably measured the following results:

(i) Microphone meteoroid detectors respond to cosmic rays associated with solar flares. The enormous rate of these pulses is a main reason for the high flux data reported by earlier acoustic measurements. The potential use of this effect for investigating solar flares should be considered.

(ii) The flux of micrometeoroids at a distance between 0.7 and 1.1 AU from the sun is $2 \pm 0.5 \times 10^{-4}$ particles m^{-2} s^{-1} $(2\pi\ sr)^{-1}$ and shows a cut-off at a meteoroid mass of 5×10^{-12} g. This result, which was obtained by considering the entire data measured by Pioneer 8 and 9 until August 1969, now provides a firm basis for future meteoroid experiments.

(iii) Among eight particles, for which a complete orbit could be evaluated, two travel in the orbital planes of two comets (Encke and Grigg-Skellerup). The orbits of future meteoroid satellites should be selected with respect to the orbit of comets.

Acknowledgment

This research was carried out while one of the authors (U. G.) held a National Research Council Postdoctoral Resident Research Associateship supported by NASA.

References

[1] O. E. Berg and F. F. Richardson, Rev. Scient. Instrum. 40, 1333 (1969).

Space Research XI — Akademie-Verlag, Berlin 1971

COSMIC RAY EFFECTS ON INTERPLANETARY DUST AND DUST DETECTORS

K. Sitte

Laboratorio di Cosmo-geofisica del C.N.R., Torino, Italy, and
Department of Physics, University of Freiburg, Freiburg i. Breisgau, FRG

Galactic and solar cosmic radiation form a part of the environment of interplanetary dust which has not been studied extensively so far. Its interactions with microparticles are discussed, based on the known data of cosmic-ray intensity and energy distribution. It is shown that despite the high individual energy of galactic cosmic-ray particles the direct effects on the orbits of dust particles are negligible.

However, cosmic-ray induced sputtering may contribute appreciably to the interplanetary ablation of micrometeoroids. Radioisotope production is insignificant during the life-time of dust particles in orbit, but could be measurable if that was preceded by an extensive "storage" phase. Finally, the possible effects are discussed by which the impact of cosmic-ray particles could cause spurious counts in the principal micrometeoroid sensors, microphones and capacitor detectors. It is shown that certainly capacitor sensors, and possibly microphones as well, will record spurious counts especially at times of high solar activity.

1.

Microparticles orbiting about the sun move in a rather complex medium. There is, dominant with regard to energy density, the sun's thermal radiation. There is also the steady outward flow of solar plasma, carrying with it a magnetic field. There is, finally, cosmic radiation: high-energy corpuscular and photon radiation with its two components, that of galactic origin pressing inward against the solar wind and in part penetrating to the inner regions, and the other of lower energy emitted by the sun. We believe that we have a fair understanding of the interactions which define the balance between solar wind, magnetic fields and cosmic rays; likewise the effects of sunlight, solar plasma and magnetic fields on the motion of interplanetary microparticles have been studied in great detail.

So far, however, little attention has been paid to possible effects of cosmic radiation on these particles. Apart from occasionally inducing nuclear reactions which might produce identifiable radionuclides, cosmic rays have been assumed to exert no influence. In the following we wish to subject this assertion to a somewhat closer check. We must begin with a brief review of our knowledge of the properties of cosmic radiation, and of the interactions its particles will undergo. Its intensity and energy spectrum have been studied for a long time and by many methods, beginning with the physically simplest of using the earth's magnetic field as analyser. This has shown that the flux of cosmic-ray particles with energies in the GeV region and above is of the order 1 particle $cm^{-2}\,sr^{-1}\,s^{-1}$, and its differential energy distribution roughly follows a power law

$$s(E)\,dE = K\,E^{-\gamma}\,dE \tag{1}$$

with an exponent γ close to 2.5 between 1 and 15 GeV, steepening to about 2.7 at higher energies. More recently space probes have made possible an ex-

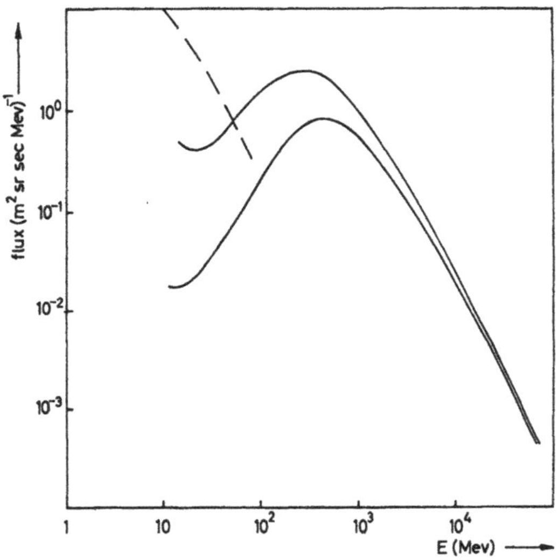

Fig. 1. Differential energy spectrum of the cosmic radiation at minimum solar activity (upper curve), and at solar maximum (lower curve).
Broken line: energy spectrum of an average flare.

tension of the measurements to much lower energies. Fig. 1 reproduces a summary of those data. Here the upper line refers to values obtained at times of low solar activity, and the lower line to those from periods of solar maximum. Attention should be given to three points: Firstly, for all but the highest energies there is a marked variation over the solar cycle, due to the varying outward flow of solar plasma. In correlation one must expect, and has observed, an intensity gradient of cosmic radiation within the solar system, with increasing particle density as one moves outward from the sun. Beyond the boundaries of the solar wind region the flux of low-energy cosmic rays, in the range of 1–10 MeV, may be larger by orders of magnitude than that measured near earth.

Secondly, we note the shape of the spectrum which passes through a maximum at a few 100 MeV, and a minimum at lower energies. The presence of low-energy particles has now been confirmed at periods of low and high solar activity, and it is likely that not all these particles are emitted by the sun but that the galactic component contributes as well, probably not much in our vicinity but significantly in the outer regions of the planetary system. This leads us to the third point, the production of energetic particles by the sun. In spectacular flare events the sun can more than match during a short period the energy influx of galactic cosmic rays for a year. However, the energy spectrum of flare particles, though tremendously variable, is always very steep and usually steepening with time. Few particles are found with energies above 10 MeV. The broken line in Fig. 1 represents a typical energy distribution of an average flare. Estimates of the average emission rate of the sun are hard to make and not yet reliable. But values as high as 10^{30} to 10^{31} particle s^{-1} have been derived; we shall use those later on, though with some misgivings. This would amount to a low-energy flux well in excess of that of the more energetic galactic radiation up to distances large compared with the earth's orbit. We must mention, finally, that both the galactic and the solar cosmic radiation consist of about 93% of protons, with some 7% of helium nuclei and a small fraction of heavier species. But it has to be realized that while protons strongly dominate in numbers, the ratio is less one-sided if the flux of nucleons is considered. This has some implications especially concerning the rate of nuclear interactions induced by the cosmic radiation.

To sum up our review: the cosmic-ray part of the environment of interplanetary dust consists of a flux of the order of 1 particle $cm^{-2}\,sr^{-1}\,s^{-1}$ in very energetic particles (at $E = 3\times10^{9}$ eV the energy of a single cosmic-ray proton roughly equals that of a micrometeoroid of 0.1 μm diameter orbiting near earth!), and a probably much higher flux of lower energy solar particles. Its energy density is around 10^{-12} erg cm^{-3}, considerably in excess of that of dust particles. How does this ambient component effect the micrometeoroids?

2.

We are dealing with a beam of electrically charged, nucleonic particles. Two kinds of interactions are therefore to be expected: those connected with *atomic* collisions of charged particles, and those resulting from *nuclear* collisions. The first, electromagnetic, interactions result in energy loss by ionization, the second give rise to nuclear transmutations. We propose to concentrate on the first.

To a very good approximation the energy loss due to atomic collisions of a particle of charge Ze passing through matter is a function of its velocity $\beta = (v/c)$ only:

$$-\,(\mathrm{d}E/\mathrm{d}x)_{\mathrm{coll}} = C\,Z^2 f(\beta)\,, \tag{2}$$

where C varies only slightly with the properties of the material traversed. This is illustrated in Fig. 2 in which the collision loss $\mathrm{d}E/\mathrm{d}x$, measured in MeV g^{-1} cm^2, is plotted as a function of $p/m = \beta/(1-\beta^2)^{1/2}$ for a heavy absorber (lead), and for a light material (carbon). We note that for $p/m > 0.5$ the energy loss becomes nearly constant, typically of the order 1.5–2 MeVg^{-1} cm^2, while it rises sharply at low energies. In consequence the range of relativistic particles will be very nearly proportional to their energy, but will shorten rapidly for slower particles. Let us apply these data to collisions of cosmic-ray particles

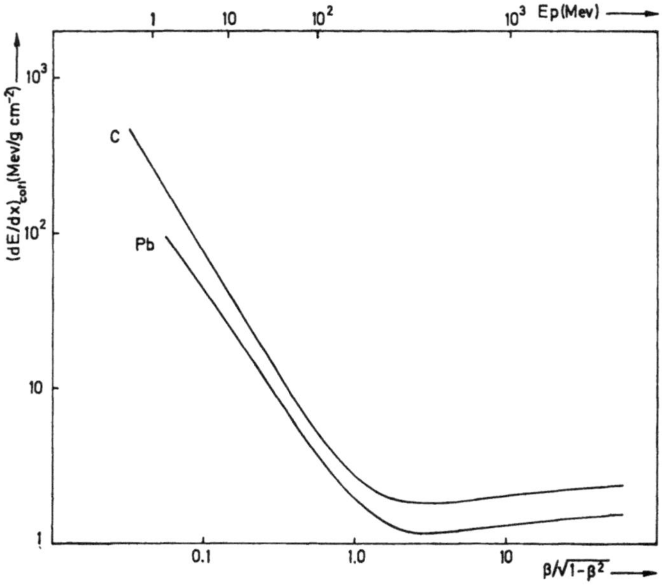

Fig. 2. Energy loss by collision $(\mathrm{d}E/\mathrm{d}x)_{\mathrm{coll}}$, in MeV $\mathrm{g}^{-1}\,\mathrm{cm}^2$, in carbon and in lead, as a function of $\beta/(1-\beta^2)^{1/2}$.

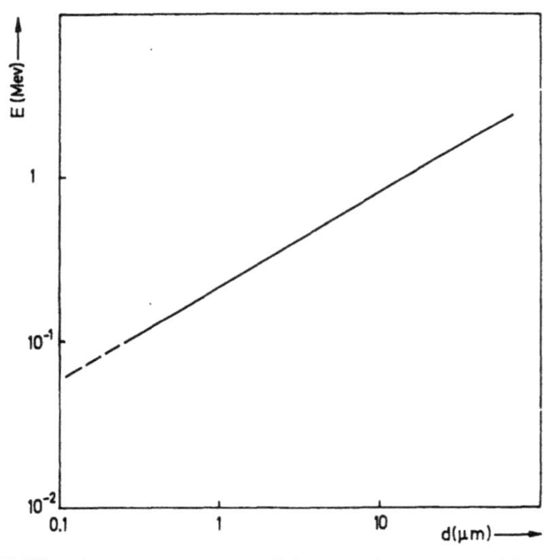

Fig. 3. Energy E (MeV) of protons stopped in a micrometeoroid as a function of the particle diameter d (μm).

with micrometeoroids. We take as typical diameters $d = 0.1$ to $10\,\mu\mathrm{m}$, and for convenience assume a density $\varrho = 2.5\,\mathrm{g\,cm^{-3}}$.

We ask, first of all, at what energy a cosmic-ray particle will be stopped in an average collision with a dust grain, that is, in a collision with a path length $\frac{3}{2}d$. The results are presented in Fig. 3, and it is seen that protons with energies exceedinging 1 MeV will traverse all but the largest of our microparticles. However, that is no longer true for the heavy nuclei of cosmic radiation. It is evident from (2) that at equal velocities, and therefore at equal energies per nucleon, the range $R_{A,Z}$ of a particle of mass A and charge Z is related to that of a proton, R_p, by

$$R_{A,Z}(\beta) = (A/Z^2) R_p(\beta) \tag{3}$$

or since $A \approx 2Z$ for most of the more abundant nuclides, $R_{A,Z} \approx (2/Z)\, R_p$. Such particles will come to rest in a micrometeoroid, transferring to it all of their energy. In general it is this energy transfer which is of interest, not the

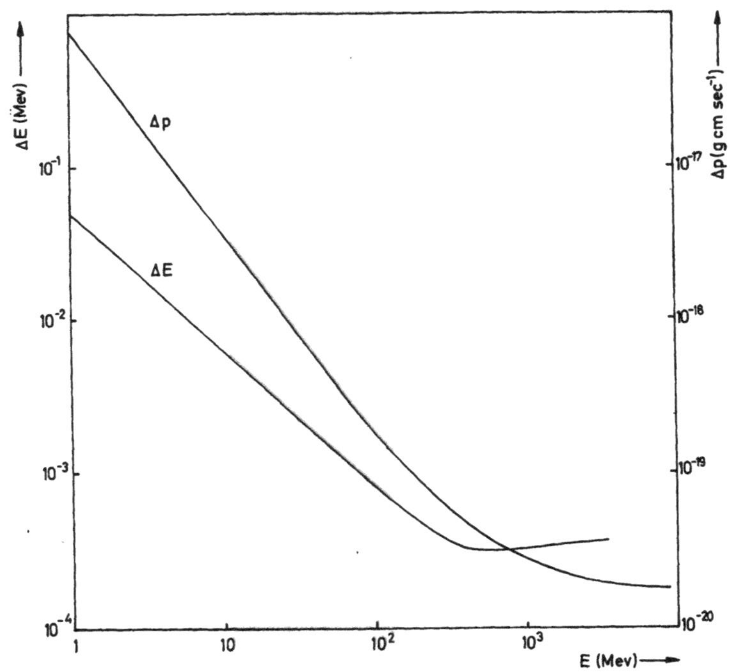

Fig. 4. Energy loss ΔE (MeV), and momentum loss Δp (g cm s^{-1}); in a micrometeoroid of 1 μm diameter, as a function of the particle energy E (MeV).

question whether the particle is stopped. The data of Fig. 2 allow us to calculate the total energy loss ΔE of a cosmic-ray particle in an average collision with a dust grain. We show in the curve marked ΔE of Fig. 4 results for a 1 μm micrometeoroid, noting that except for the lowest energies and the largest particles, the losses $\Delta E(d)$ can be taken as a linear function of the particle size. How does this energy absorption affect the microparticle? Heating, momentum transfer, and radiation damage including sputtering will have to be considered. As to the first, take for the moment the extreme and rather naive view that the entire energy is quickly degraded to uniform heating of the particle. Since the absorbed energy ΔE MeV corresponds to about $3.8 \times 10^{-14}\, \Delta E$ calories, a temperature rise δT given by

$$\delta T = 3.8 \times 10^{-14} \frac{\Delta E}{(4\pi/3)\, r^3\, p\, c} \tag{4}$$

will result (c = specific heat). Using $c = 0.2$ one finds that massive heating, $\delta T \geqq 100°$, which might lead to evaporation or breakup, occurs only for extremely small particles, $d \leqq 0.01$ µm. Or to put it differently: the equilibrium temperature difference ΔT between a dust grain and its surroundings which is defined by balance of radiative losses and collision gains,

$$\Delta T = \int \Delta E\,(\mathrm{cal})\, s(E)\, \mathrm{d}E / 4 T_0^3 \sigma \tag{5}$$

(T_0 = "outside" temperature, σ = Stefan–Boltzmann constant) is extremely small at all particle sizes. We conclude that the effect of uniform heating is negligible at least for the bulk of the dust population. This will hold even for collisions with heavy nuclei, despite the increase by a factor Z^2 of the energy loss. But bear in mind that the conclusion refers to uniform heating only; we shall return to this point later. Next, the momentum transfer per collision can be approximated by

$$\Delta p = p\,(\Delta E/E)\,, \tag{6}$$

where p is the momentum of the incident cosmic ray particle. The resulting values are reproduced in curve Δp of Fig. 4. Again we notice that low-energy particles are much more effective; at high energies Δp approaches a nearly constant value. Besides, owing to the near-isotropy of the galactic component the net effect of its impact will be small even if the individual transfers were larger. They would merely produce a kind or random walk superimposed on the normal orbital motion. But the solar cosmic radiation has an essentially outward directed flow, and by its cumulative effects could influence the micrometeorvid orbits just as sunlight and solar wind. To obtain the momentum transfer Δp per second, we have to multiply the transfer $\Delta p(E)$ per collision with the collision frequency, and to integrate over the energy:

$$\Delta P \approx (d^2/4\,\pi) \int \Delta p(E) \cdot s_s(E)\, \mathrm{d}E \tag{7}$$

where $s_s(E)$ stands for the differential energy distribution of the solar cosmic radiation. As we have pointed out above, this quantity is very poorly known. But fortunately we do not need good figures to arrive at definite conclusions. Recalling the data of Fig. 4 we see that for our 1 µm microparticles only the collective effect of about 10^{10} collisions with protons of energies around 1 MeV will produce a ΔP somewhat comparable with the momentum of the particle in orbit. Now even if we accept for the total emission of solar cosmic rays the value of $10^{30}-10^{31}$ particle s^{-1}, that is, a flux of the order $10^3\ \mathrm{cm}^{-2}\,\mathrm{s}^{-1}$ in our vicinity or $10^5\ \mathrm{cm}^{-2}\,\mathrm{s}^{-1}$ at 0.1 AU from the sun, and let them all have that low energy, our 1 µm dust particle will experience collisions with a frequency of about $10^{-5}\ \mathrm{s}^{-1}$ or $10^{-3}\,\mathrm{s}^{-1}$, respectively, at these distances. During the life-time in orbit of a micrometeoroid the total momentum transfer is, therefore, negligible. (The estimate can, of course, be refined, but only with somewhat arbitrary assumptions on intensity and spectrum. If plausible parameters are chosen, the result is the same as ours.) Notice also that the result is independent of the particle size, since both ΔP and the momentum of the grain are proportional to d^3.

It is true that in spectacular flares the flux rate rises to values several orders of magnitude higher that that used above. But these events last for hours or days only. If the emission extends to well below 1 MeV with strongly rising intensity, they just might produce an effect in the immediate vicinity of the sun, just conceivably observable in zodiacal light experiments.

We can now briefly turn to the third item, the effects of radiation damage and sputtering. Once more our report must remain qualitative and semi-speculative, this time for lack of detailed information on structure and composition of the micrometeoroids. Essentially the cosmic-ray particle leaves behind a trail of ions: a faint one consisting of about ten ion pairs per μm at density 2.5 if the particle energy is high, but nearly hundred times more at 1 MeV and still denser in the ratio Z^2 for a heavy nucleus. We also know that dense ionization can do permanent damage to a lattice: witness, for instance, the 'fossile tracks' for cosmic-ray nuclei found in suitable substances. It is very likely that the impact of such a nucleus will have a catastrophic effect on a small fragile dust grain; but how can we treat it quantitatively if we know nothing about the nature of the grain? In this frustrating situation we shall make a rather conservative guess about the magnitude of the effect, taking into account merely local heating.

Also, and again conservatively, we shall assume that permanent damage is restricted to the passage of particles of energies below 20 MeV/nucleon for medium–heavy elements (C, N, O), or correspondingly higher energies for the heavier nuclei. This once more practically reduces the effect to the impacts of solar cosmic rays: only a minute fraction of the galactic component could contribute, and its collisions are too rare to count. On the other hand, virtually *all* solar particles are effective. At 20 MeV/nucleon or less, the rate of energy loss of a nucleus of the (C, N, O) group exceeds 500 MeV g^{-1} cm^2, and all this energy is deposited along a narrow trail. In a simple-minded model, let us transform it into heating of a cylinder with the particle trail as its axis, and ask up to what distance Δr the grain temperature will rise by some 100°, or sufficiently to affect lattice disruptions. We find $\Delta r \gtrsim 5 \times 10^{-7}$ cm, and conclude from this that, as a minimum effect, the "evaporation" per collision of a volume ΔV is of the order $\gtrsim 10^{-18}$ cm^3. With regard to the flux of heavy solar nuclei we know that, as far as it was measured, it follows the abundance ratio of the elements in the sun's chromosphere. This would give us somewhat less than 1% of the total solar output in "effective" heavy nuclei; say, an integral flux $\Phi \approx 10$ cm^{-2} s^{-1} in our vicinity. Accordingly we expect a life-time T_e:

$$T_e = r/3\,\Phi\,\Delta V \approx r \times 10^9\,y \qquad (r \text{ in cm}) \tag{8}$$

shortened in the ratio of the square of the distance as the microparticles approach the sun. In the words, if the solar emission is indeed of the order 10^{30}–10^{31} particles s^{-1} and contains heavy nuclei in their normal abundance ratio, we must expect a quite appreciable erosion effect even on the basis of the crudest model. This "ablation" will be largest near the sun, but may increase again at far distances – if microparticles spend some time there – because of the rising flux of galactic cosmic rays.

Of course the estimate could be refined, and should include also other sputtering effects which occur already at the passage of relativistic particles. We have pointed out that they leave a trail of ions with energies around 30–40 eV every 10^{-5} cm or so. When impinging on a surface such ions are known to induce considerable sputtering on metal foils, depending essentially on the vaporization energy per atom of the target material. This may be as low as 0.2–2 eV in an ice mantle, and not larger by orders of magnitude in other likely materials. We can picture, therefore, the surfaces of our dust grains as bombarded from the inside by ions of their own substance, and thus sputtering on many impacts. Again, a safe quantitative valuation is impossible because of our lack of knowledge of the composition of micrometeoroids, but it appears likely that the result will be an erosion rate comparable with that estimated for heavy ions. This has an evident consequence for the size spectrum of dust particles. Suppose their history can be described by constant injection from a source characterized by a power law $S_0\, r^{-n}$, and mass-independent losses perhaps due to capture by major bodies characterized by a constant life time T_0. Then the equilibrium distribution of the micrometeorid density $N(r)$ is given by

$$S_0\, r^{-n} - N(r)/T_0 - (\Phi\, \Delta V) \cdot (\mathrm{d}N/\mathrm{d}r) = 0\,. \tag{9}$$

Of the two loss terms, the first dominates for large r and the second for small particles. Therefore the approximate solutions

$$N(r) \approx S_0\, T_0\, r^{-n} \tag{10}$$

and

$$N(r) \approx \frac{S_0}{(n-1)\,\Phi\, \Delta V} \cdot r^{-(n-1)} \tag{11}$$

are valid in the ranges $r \gg \Phi\, \Delta V\, T_0/n$ and $r \ll \Phi\, T_0/n$ respectively. Thus a "break" in the size spectrum of dust particles could give us information about the life time T_0.

A similar change in the size distribution could also result if other size-dependent losses occur, such as speeding up of the spiralling towards the sun for smaller particles. If that causes a "spiralling life" T_s proportional to $1/r$, one finds again a spectrum of the shape (ii) and (iii). But there is one essential difference: since this a *solar* flux, the "erosion" mechanism will become increasingly effective as the particles approach the sun. There may not be time to establish complete equilibrium, but the "knee" must be expected to shift towards larger r at distances closer to the sun. This is a prediction which perhaps could be tested on zodiacal light data.

We have omitted a consideration of nuclear interactions induced by cosmic-ray particles in dust grains. This has two good reasons: shortage of time to do it justice, and the fact that it is inevitably connected with the question of the origin of micrometeoroids which cannot be a subject of our discussion. For it is evident from the figures we have quoted that nuclear interactions will be rare during the short life time of a micrometeoroid in orbit and spiralling towards the sun. This leaves us little chance to identify radionuclides in collected or recorded microparticles, unless they spend much more time "stored" somewhere in the outer regions of the planetary system but already more or less in their final shape. Indeed, the search for radioactivity induced in dust grains appears as an attempt to investigate their "prehistory".

Two brief remarks may be added here. Firstly, it is easily seen that on the average, radionuclides will be found only in comparatively large grains. This follows from the condition that the mean time T_n within which a nuclear collision occurs must be shorter than the erosion life time T_e for which we have derived an upper limit above. Now in our notation T_n is

$$T_n = \frac{\lambda}{\varrho r} \frac{1}{4 \pi r^2 \Phi} \tag{12}$$

(λ = interaction mean free path of cosmic ray particles). Therefore our condition reads

$$\frac{\lambda}{\varrho r} \frac{1}{4 \pi r^2 \Phi} < \frac{r}{3 \Phi \Delta V} \tag{13}$$

or

$$r^4 > 3 \lambda \Delta V / 4 \pi \varrho . \tag{14}$$

The interaction length λ is of the order 100 g cm^{-2} if all collisions are counted, but larger in the ratio of the total inelastic cross section σ_{in} to the partial cross section σ_{rn} for formation of a radionuclide if observations are confined to the results of those events. Thus only particles with radii well in excess of 10^{-4} cm will, on the average, survive for sufficient time to suffer a nuclear interaction.

Secondly, and in consequence, a collection experiment, designed to accumulate an amount of micrometeoroid material large enough to permit a radioactivity analysis, will have to gather up substantially more than a simple calculation neglecting erosion effects would indicate. The small-sized grains which constitute a considerable fraction in weight contribute very little to the activity. We conclude the discussion of nulear interactions with these remarks, stressing again that our failure to deal with it in more detail does not imply that it is judged unimportant.

3.

We can deal briefly now with the effects which the impacts of cosmic-ray particles may have on dust detectors. Fortunately in recent years the workers in the field of micrometeoroids have simplified our task by confining themselves in most cases to two kinds of sensors, microphones and impact or capacity detectors.

3.1. Microphones

If microphones in all instances recorded only the momentum deposited in the impact plate, as was originally thought, they would be quite safe from cosmic-ray interference. Their sensitivity lies at about 10^{-5} to 10^{-6} g cm s^{-1} many orders of magnitude higher than the momentum transfer a cosmic-ray particle could induce, as can be seen by comparison with the data of Fig. 4. But it is known now that above velocities of a few km s^{-1} the microphone response rises with mv^2 rather than with mv. This leads to an intriguing possibility: suppose for the moment, and without reasoned justification, that it does not matter how that

mv^3 is deposited, whether shallowly on a comparatively large area (as by a micrometeoroid impact), or in a narrow channel (as by a cosmic-ray particle). In this case, and if the cosmic-ray particle were stopped – a point to be checked later – the two will be equally effective if

$$(M/AM_p) = v/v_c)^3 \tag{15}$$

where M is the micrometeoroid mass, v its velocity, and AM_p and v_c the mass and velocity of the impinging nucleus. To "calibrate" the relation we assume the "knee" in the sensitivity curve to occur at $v = 3$ km s^{-1}, and the detectable mass as this point to be $M = 10^{-11}$ g. Then we find that the same transfer is obtained from a cosmic-ray particle of $v_c \approx 5.4 \times 10^9/A^{1/3}$, or at a kinetic energy a little above 2 MeV/nucleon. Such particles will indeed be stopped in the impact plate. It follows that if our assumption is correct, a microphone will count a good fraction of solar medium-heavy nuclei – in fact microphones might be a very useful detector for these species.

Naturally we have no assurance that this is so. It seems that experimental work on the impact of nuclei in the MeV range on microphones has never been done; it might well be worth a check. In the meantime our result should be taken as a warning. Even if cosmic-ray particles are only fractionally effective, the rate of spurious counts induced by them could be quite comparable with the genuine rate micrometeorid impacts, especially on a mission towards the sun or at times of solar flares.

3.2. Capacitor Detectors

We find a very similar situation with respect to capacitor detectors. Take as example a foil of 0.2 μm thickness (it is worse for thicker membranes). This will be traversed even by a medium-heavy nucleus of a few MeV/nucleon energy but in this traversal the particle will lose energy at a rate Z^2 times higher than that presented in Fig. 2. That energy is quickly distributed to ion pairs which take, on the average, some 30–40 eV/ion depending on the material. Thus the 2 MeV/nucleon particle considered above will create some $10^2 Z^2 \varrho_f$ ion pairs in a foil of density ϱ_f. If the sensitivity threshold of the detector is of the order of 10^4 elementary charges, it will just begin to react upon the traversal of a medium-heavy nucleus of that energy. Again we conclude, therefore, that galactic cosmic rays will not contribute spurious counts in a serious measure, but solar particles might completely dominate in the recording, especially on a solar probe, if solar emission is as high as we have tentatively assumed, or during a flare. Certainly a close cooperation between dust experimenters and their colleagues who register energetic particles must be recommended!

4. Conclusions

Summing up our remarks compels us to a rather apologetic note. Of the three subjects which we had to bring into relation, micrometeoroids, galactic cosmic rays, and solar energetic particles, we have safe knowledge only about one, the galactic cosmic radiation. And this we have found to be of negligible or very minor influence, except perhaps for some ablation effects. On the others the values of important, often decisive, parameters are not much better than guess work. Our conclusions indicating a considerable erosion rate and the possibility of grave contributions in spurious counts of capacitor detectors and perhaps microphones as well, must be taken as a warning rather than as a firm result. But their verification by laboratory tests and by collaboration with the particle physicists seems certainly worth while.

Literature

All relevant information not pertaining to interplanetary dust is found in the following books or review articles:

On cosmic rays:

W.R. Webber, The Spectrum and Charge Composition of the Primary Cosmic Radiation, Hdb. d. Phys. 46/11, Springer Verlag, Berlin–Heidelberg 1967.

P. Meyer, Cosmic rays in the galaxy, Ann. Rev. Astron. Astrophys. 7, 1 (1969).

On high-energy interactions:

B. Rossi, High Energy Particles, Prentice-Hall Inc., New York 1965.

On solar energetic particles:

C. E. Fichtel and F. B. McDonald, Ann. Rev. Astron. Astrophys. 5, 351 (1967).

C. E. Fichtel, Energetic Solar Particles, Proc. Xth Int. Conf. Cosmic Rays, Calgary, Part A, 194 (1968).

On sputtering effects:

G. K. Wehner, Advanc. Electronics and Electron Phys., 7, 239 (1955).

B. T. Lynds and C. N. Wickramasinghe, Ann. Rev. Astron. Astrophys. 6, 215 (1968).

Space Research XI — Akademie-Verlag, Berlin 1971

THE ZODIACAL LIGHT LINES IN THE PARTICLE FLUX DIAGRAM

C. LEINERT

Landessternwarte, Heidelberg, FRG

The spatial density of interplanetary dust particles as derived from zodiacal light measurements by different authors is converted to particle fluxes. The corresponding curves in the flux diagram do not give a well-defined zodiacal light line, so that a comparison between particle measurements and zodiacal light data is difficult to make. If there exists a lower limit of particle size at a mass of about 10^{-11} g, reasonable agreement between zodiacal light and micrometeorite data may be expected.

1. Introduction

Micrometeorites in interplanetary space are observed either directly by counting and analysing individual particles or indirectly in the zodiacal light which is made up by the integrated effect of their scattering. The results of both methods are usually compared in a flux diagram, the zodiacal light mesaurements being represented by two lines close to each other. With the recently measured particle fluxes approaching these lines a more detailed presentation of the zodiacal light data seems justified and necessary in order to make the comparison as conclusive as possible. As a first step the rather large differences that exist among published models of zodiacal light might be discussed to get obtain a better idea of the accuracy of the seemingly well-defined zodiacal light lines. Therefore for six typical models particle numbers were converted to fluxes. The assumptions used for this simple calculation and the discussion of the results are given in the following sections.

2. Flux Values from Zodiacal Light Data

Most zodiacal light data were interpreted in terms of the scattering of spherical particles with radii in the interval $a_1 < a < a_2$ and a distribution of radii

$$n(a)\, \mathrm{d}a = C\, a^{-p}\, \mathrm{d}a \qquad \text{particles cm}^{-3}\,. \tag{1}$$

To obtain flux values depending on particle mass assumptions on the density ϱ of the particles and their mean velocity v relative to an observer have to be made. Following [1] and [2] comparatively low values of density and velocity are assumed: $\varrho = 1.0$ g cm^{-3}, $v = 10.0$ km s^{-1}. From the equation $m = (4\pi/3)\, \varrho\, a^3$

masses may be calculated, m_1 being the minimum and m_2 the maximum mass. The number of particles per cm³ with a mass greater than m is given by

$$n_m = \int_{a(m)}^{a_2} n(a)\, \mathrm{d}a = \frac{C}{(p-1)\,[a(m)]^{p-1}}\,[1 - (a\,(m)/a_2)^{p-1}]\,. \tag{2}$$

The flux is $F = n_m v \times 0.25 \times 10^4$ particles $\mathrm{m}^{-2}\,\mathrm{s}^{-1}$ per half-sphere, where the factor 0.25 is valid for an isotropic velocity distribution relative to the observer and 10^4 gives the transformation from cm² to m². Inserting the adopted value for v we have

$$F = 2.5 \times 10^9\, n_m \qquad \text{particles m}^{-2}\,\text{s}^{-1} \text{ per half-sphere.} \tag{3}$$

If $m < m_1$, $F = F(m_1)$; if $m > m_2$, $F = 0$. Fluxes were calculated for the following size distributions most of which have the form of Eq. (1) and all of which refer to earth orbit:

Source	C	p	a_1 (cm)	a_2 (cm)
Blackwell [3]	size distribution proportional to that given in [4]			
Elsässer [5]	1.6×10^{-18}	2.0	3.0×10^{-4}	1.0×10^{-2}
Gillett [6]	6.3×10^{-22}	3.0	2.5×10^{-5}	2.5×10^{-3}
Ingham [7]	2.5×10^{-26}	4.0	4.0×10^{-5}	$\approx 10^{-1}$
Powell [8]	2.0×10^{-40}	≈ 7	$\approx 1 \times 10^{-5}$	$\approx 2 \times 10^{-5}$
Weinberg-Giese [9]	2.3×10^{-26}	4.0	8.0×10^{-6}	2.1×10^{-4}

The resulting six lines are shown in the particle flux diagram (Fig. 1) They have to be shifted to the right for higher values of ϱ and upwards for higher values of v. The envelope of these lines which tends to approximate the particle measurements has no physical meaning. The comparison has to be made between the particle measurements and one of the zodiacal light lines. These, however, differ from each other by orders of magnitude, although the brightness of zosdiacal light is certainly known to better than $\pm$ 30%, thus preventing a satifying comparison. The zodiacal light values are in general lower than the particle measurements but the quantitative results depend strongly on the assumptions made about scattering properties, large-scale distribution in space and size distribution of the particles. It can be seen from the diagram that at least the size distribution should be normalized to some extent before a reasonable comparison can be possible. With the size distribution fixed the general influence of the remaining free parameters on the comparison between zodiacal light data and particle measurements can be shown. This will not solve the problem but it helps to clarify the situation.

3. Influenec of Free Parameters on Predicted Flux

For the purpose of this section we describe the interplanetary dust by a size distribution $n(a)\,\mathrm{d}a \propto a^{-4}\,\mathrm{d}a$ taken from the particle measurements for ease of comparison, a minimum particle radius a_{min}, a spatial distribution $n(r) \propto r^{-\delta}$ and a simplified scattering function $\sigma(\vartheta)$ which is the sum of a diffraction and a reflection component

$$\sigma(\vartheta) = a^2 \cdot \frac{J_1^{\,2}(2\,\pi\,a\,\vartheta/\lambda)}{\vartheta^2} + A\,\frac{a^2}{4}\,\mathrm{cm}^2\,\mathrm{sr}^{-1}\,, \tag{4}$$

J_1 being the first Bessel function, A the albedo of the particle, $\lambda = 500$ nm the wavelength of scattered light and ϑ the scattering angle. For convenience J_1 was approximated in a way given by Allen [10]. The steep size distribution enhances the importance of the small particles in light scattering. If, for example, we consider totally reflecting particles with a very small a_{min}, 90% of the

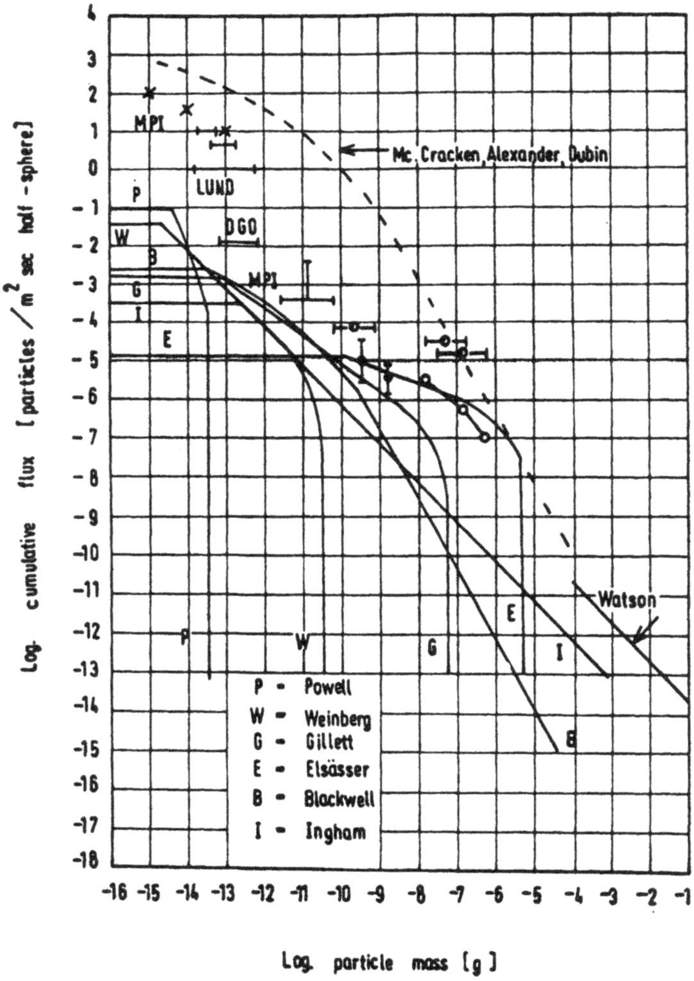

Fig. 1. Particle fluxes predicted by some models of zodiacal light. Particle measurements by Pegasus (○), Explorer (●), Gemini-Agena (⊙), OGO and some results of particle collections are shown for comparison.

scattered light is contributed by particles smaller than 0.5 μm which is the radiation pressure limit [11]; Rayleigh scattering by particles with $a < 0.07$ μm is included with a contribution of 25%. This suggests that a_{min} probably is the most critical of the free parameters and may change the predicted flux values by an order of magnitude. The quantitative results for our model are given in Fig. 2. A minimum particle size $a_{min} = 0.46$ μm, an albedo $A = 0.20$ and a spatial distribution according to $\delta = 1.0$ are used to calculate a cumulative flux for particles with $m \geqq 10^{-10}$ g, and the change of this flux resulting from changes in one of the three parameters is studied. The effects of spatial distribution and especially particle albedo cannot be neglected, but the minimum particle size seems to be the main factor of uncertainty when calculating fluxes from zodiacal light data, larger values of $a_{\min}$ giving appreciably higher fluxes for the larger particles. Although the result was deduced with a simplified scattering function and for only two points of zodiacal light, with the brightness (2200 S10 for an

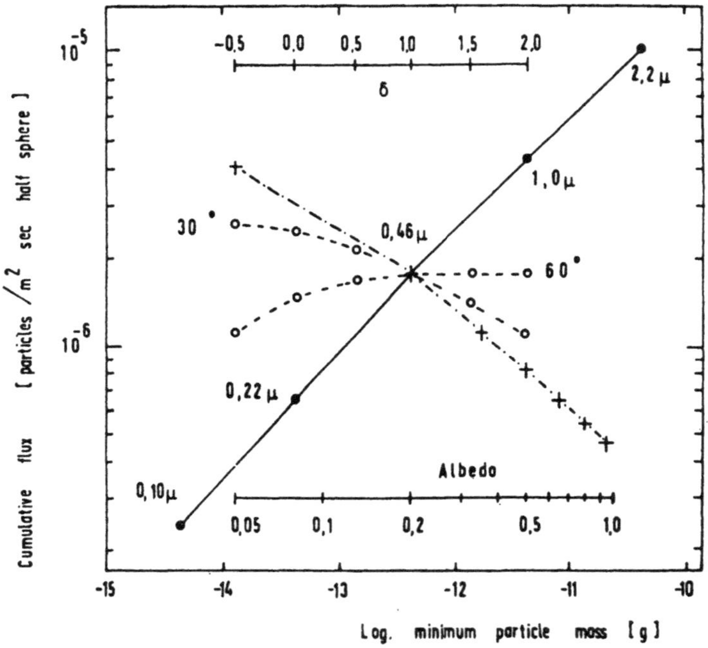

Fig. 2. Dependence of cumulative particle flux for $m \geqq 10^{-10}$ g on albedo A (+ · — · +), spatial distribution $n(r) \propto r^{-\delta}$ (○——○) and minimum particle size $a_{\min}$ (●——●) for our model. The central point corresponds to $A = 0.20$, $\delta = 1.0$, $a_{\min} = 0.46$ μm. Scales for varying albedo, δ or minimum particle mass are given; in addition the values of $a_{\min}$ are written in the diagram. When varying δ the curves for elongation $\varepsilon = 30°$ and $\varepsilon = 60°$ differ and both are shown.

elongation of 30°, 520 S10 for an elongation of 60°) taken from [9], it also should be representative of models describing the whole phenomenon. We have assumed the interplanetary particles to be spherical. This is not certain to be a good approximation to real particles, which introduces an additional uncertainty in the fluxes derived from zodiacal light data. Taking this into account, Elsässer's [12] conclusion that the calculated fluxes are not better defined than to a factor of ten is upheld.

4. Conclusion

A reasonable comparison of particle measurements and zodiacal light data is difficult to achieve until spatial density, scattering properties and size distribution of interplanetary dust are better known. A major improvement of this situation may be obtained by determining a lower limit of particle size. In view of this, accurate measurements of the colour of zodiacal light which is correlated to particle size, or direct determinations of the minimum particle size by space probes are very valuable. Fig. 3 demonstrates how the calculations may be

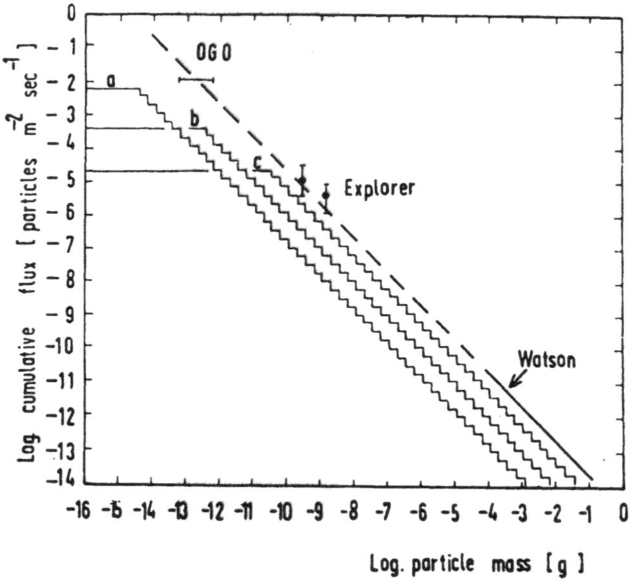

Fig. 3. Effect of minimum particle size on cumulative flux values for our model. Assumed albedo $A = 0.20$. Spatial distribution $n(r) \propto r^{-1.0}$. The curves a, b, c correspond to a cut-off at 0.10 µm, 0.46 µm and 2.2 µm respectively. The staircase appearance is due to numerical steps.

more closely fitted to the observations for larger values of $a_{\min}$. If the cut-off in the particle size distribution near 10^{-11} g indicated by the Pioneer measurements [13] is confirmed, the discrepancy between fluxes derived from zodiacal light observations and direct flux measurements will probably disappear.

References

[1] F. Verniani, Space Sci. Rev. **10, 230** (1969).
[2] J. J. Kornblum, J. Geophys. Res. **74, 1911** (1969).
[3] D. E. Blackwell, M. F. Ingham and A. D. Petford, Mon. Not. R. Astron. Soc. **136**, 325 (1967).
[4] W. M. Alexander et al., NASA Techn. Note No. 1669 (1963).
[5] H. Elsässer, Mitt. Astr. Inst. Tübingen Nr. 35, p. 73 (1958).
[6] F. C. Gillett, Thesis, Univ. of Minnesota, p. 101 (1966).
[7] M. F. Ingham, Mon. Not. R. Astron. Soc. **122**, 171 (1961).
[8] R. S. Powell et al., in: The Zodiacal Light and the Interplanetary Medium, NASA SP-150, Washington D.C. 1967 (p. 233).
[9] J. L. Weinberg, Ann. d'Astrophys. **27**, 718 (1964).
[10] C. W. Allen, Mon. Not. R. Astron. Soc. **106**, 137 (1946).
[11] C. Schalén, Nova Acta Reg. Soc. Sci. Upsaliensis, Ser. IV, Vol. 11, No. 6 (1939).
[12] H. Elsässer, in: Space Research X, North-Holland Publ.Co., Amsterdam 1970 (p. 244).
[13] O. Berg and U. Gerloff, in: Space Research XI, Akademie-Verlag, Berlin 1971 (p.225).

Space Research XI — Akademie-Verlag, Berlin 1971

MODEL COMPUTATIONS CONCERNING ZODIACAL LIGHT MEASUREMENTS BY SPACE MISSIONS

R. H. Giese

Ruhr-Universität Bochum, Bereich Extraterrestrische Physik, Bochum, FRG
and Max-Planck-Institut für extraterrestrische Physik, Garching, FRG

By use of simple two-parametric models of the interplanetary dust distribution predictions are made concerning the appearance of zodiacal light observable from space probes at positions in and out of the ecliptic plane. Variation of the absolute intensity up to a factor of 20 may be expected when the probe approaches the sun in the ecliptic plane from the distance of 1 to 0.3 AU. The relative brightness distribution changes or remains constant depending on the type of the model adopted. For out-of-ecliptic missions an observable asymmetry of the brightness pattern with respect to the ecliptic at the celestial sphere is expected even for moderate (0.36 AU) elevations of the position of the probe above the invariable plane at solar distances near 1 AU.

1. Introduction

Photometry of the zodiacal light by space probes in and outside the ecliptic plane will provide a powerful tool to dissolve ambiguities in the theoretical interpretation of earth-based measurements. To obtain an idea about the significance of space measurements a series of computations has been started which

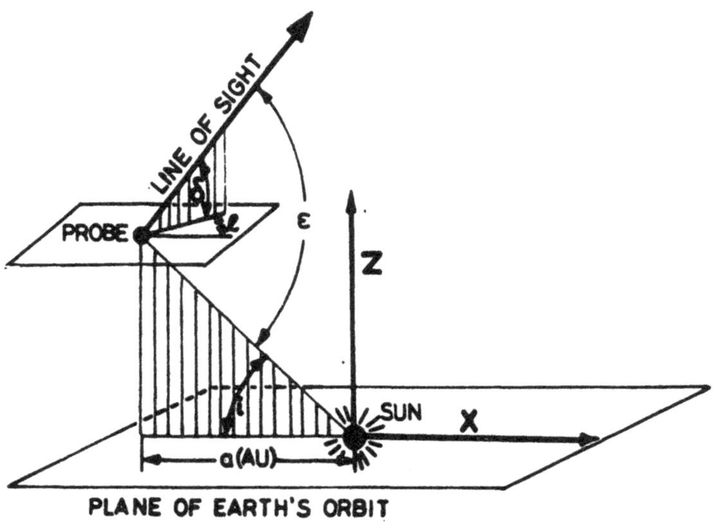

Fig. 1. Basic geometry.

provides, for different model assumptions, the expected pattern of surface brightness observable at the sky from different positions in the solar system.

2. Formalism

The geometry involved is shown in Fig. 1.

For the distribution law of the particle number density $n(\mathrm{r})$ as a function of the position vector $\mathrm{r}(x, y, z)$ a term proportional to $r^{-\beta}$ (parameter β) was used in all cases. It was modified by a multiplying factor dependent on z, involving a second parameter γ in order to take into account the concentration of dust towards the ecliptic (approximately invariable) plane. This factor was defined for the first type of model (I) in such a way that surfaces of equal density became ellipsoids. In the second case (II) a factor decreasing exponentially with increasing z was assumed. In the third type of model (III) a linear dependence on z/r was adopted.

The resulting surface brightness of the zodiacal light I as a function of the average scattering function $\sigma(\theta)$ (cf. [1]), the viewing direction (l, δ), the probe position (a, i), and of the parameters defining the particle density distribution law (β, γ) is

$$I(l, \delta; a, i) = \mathrm{const} \left(\frac{\cos i}{a \sin \varepsilon}\right)^{\beta+1} \int_{\varepsilon}^{\pi} \sigma(\theta) \cdot (\sin \theta)^{\beta} \cdot F(\theta) \, \mathrm{d}\theta \, . \tag{1}$$

The laws of spatial particle number density distribution and $F(\theta)$ are, for models of

Type I

$$n \propto r^{-\beta} \cdot \left[1 + \left(\gamma \frac{z}{r}\right)^2\right]^{-\beta/2} ; F = \left[1 + \gamma^2 \cdot \left(\frac{\sin \theta \sin i + \sin (\theta - \varepsilon) \sin \delta}{\sin \varepsilon}\right)^2\right]^{-\beta/2}$$

Type II:

$$n \propto r^{-\beta} \cdot \exp\left[-\left(\gamma \frac{z}{A}\right)^2\right]; F = \exp\left[-\left(\gamma \frac{a}{\cos i}\right)^2 \cdot \left(\frac{\sin \theta \sin i + \sin (\theta - \varepsilon) \sin \delta}{\sin \theta}\right)^2\right]$$

Type III:

$$n \propto r^{-\beta} \cdot \left[1 - \gamma \left|\frac{z}{r}\right|\right]; F = \left[1 - \gamma \left|\frac{\sin \theta \sin i + \sin (\theta - \varepsilon) \sin \delta}{\sin \varepsilon}\right|\right].$$

Here A stands for one astronomical unit (AU) and θ for the scattering angle.

3. Results

3.1. Computed Examples

The model parameters used in the computations are summarized in Table 1. In each case three considerably different scattering functions $\sigma(\theta)$ were used, one for isotropic scattering, one for very small ('Rayleigh'), submicrometre particles, and one for larger ('Micron'), dielectric particles. Using these assumptions $I(l, \delta)$ was computed in steps of 15° as seen from the earth's orbit ($a = 1$ AU,

Table 1
List of model parameters used in the calculations

Models of type I		Modells of type II		Models of type III	
β	γ	β	γ	β	γ
0.1	300	0.5	0.33	0.1	1.0
0.3	50	0.5	0.66	0.5	0.8
0.5	10	1.0	0.33	0.5	1.0
1.0	6	1.0	0.66	1.0	0.8
1.5	5	1.0	1.5	1.0	1.0
1.5	0	1.5	0.33	1.5	0.8
		1.5	0.66	1.5	1.0

$i = 0$), from the ecliptic plane ($i = 0$) at $a = 0.3$ AU, and from two positions out of the ecliptic ($a = 1$ AU, $i = 20°$, $z = 0.36$ AU and $a = 1$ AU, $i = 40°$, $z = 0.84$ AU, respectively).

3.2. Example

One example is shown in Fig. 2. It involves a rather smooth scattering function, typical for mixtures of small particles (of $s = 0.16$ to 4 μm in diameter) with a steep ($\propto s^{-4}\, ds$) decrease in the size distribution function. The brightness pattern (relative units, $I\,(90°, 0°) = 1$) is displayed in such a way that the table gives a rough impression of the isophotes. There are very significant effects if the observer's position is moved from the earth's orbit (Fig. 2a) to another place. At $a = 0.3$ AU ($i = 0$) the relative brightness distribution (Fig. 2b) is quite different from that in Fig. 2a, especially along a great circle towards the pole between $l = 60°$ and $90°$. From the out-of-ecliptic position $a = 1$ AU, $i = 20°$ there is marked asymmetry in brightness between the northern ($\delta > 0$) and southern ($\delta < 0$) hemisphere (Fig. 2c).

These effects are maintained in principle also for mixtures of larger (2 μm $\leqq s \leqq 20$ μm) dielectric particles with a lower ($\propto s^{2.5}\, ds$) decrease in size distribution although the pattern is more complicated due to scattering effects like the "haze bow" and enhanced backward scattering ("Gegenschein").

From the formulae and case studies the following results are obvious:

a) Position of the space probe in the ecliptic plane:

(i) The relative brightness distribution observable along the ecliptic is independent of a. The absolute brightness, however, increases as $a^{-(\beta+1)}$ when the probe approaches the sun [2, 3]. From $a = 1$ AU to $a = 0.3$ AU this means an increase of I by a factor of 6.1 or 20.3 for $\beta = 0.5$ or 1.5, respectively.

(ii) The relative brightness distribution in other ($\delta \neq 0$) viewing directions remains independent of the position of the probe a only for models of type I and type III. In the case of type II the relative brightness pattern over the sky is changing with a (cf. Fig. 2).

(iii) There is another difference between models of types I and III and type II. If one compares for example the intensity ratio $I(l, 0)/I(0, \delta)$ for $l = \delta = \varepsilon$ by scanning along a circle in an angular distance ε around the sun, one obtains at small elongations lower values for models of type II than for comparable models

$a = 1 \quad i = 0°$

δ \ ε	0	15	30	45	60	75	90	105	120	135	150	165	180
δ = 90	5.14E-01	5.14E-01	5.14E-01	5.14E-01	5.14E-01	5.14E-01	5.14E-01	5.14E-01	5.14E-01	5.14E-01	5.14E-01	5.14E-01	5.14E-01
75	7.52E-01	7.41E-01	7.11E-01	6.68E-01	6.19E-01	5.70E-01	5.25E-01	4.87E-01	4.56E-01	4.32E-01	4.15E-01	4.05E-01	4.02E-01
60	1.33E 00	1.27E 00	1.13E 00	9.56E-01	7.93E-01	6.60E-01	5.59E-01	4.84E-01	4.30E-01	3.92E-01	3.67E-01	3.53E-01	3.49E-01
45	3.52E 00	2.82E 00	2.11E 00	1.50E 00	1.08E 00	7.99E-01	6.22E-01	5.07E-01	4.32E-01	3.83E-01	3.52E-01	3.36E-01	3.31E-01
30	1.11E 01	8.70E 00	5.13E 00	2.51E 00	1.51E 00	9.98E-01	7.22E-01	5.61E-01	4.62E-01	4.01E-01	3.65E-01	3.47E-01	3.41E-01
15	6.66E 01	3.07E 01	1.04E 01	4.07E 00	2.07E 00	1.26E 00	8.65E-01	6.51E-01	5.25E-01	4.50E-01	4.08E-01	3.87E-01	3.82E-01
0		7.31E 01	1.46E 01	5.44E 00	2.45E 00	1.47E 00	1.00E 00	7.49E-01	6.04E-01	5.17E-01	4.68E-01	4.44E-01	
-15	6.66E 01	3.07E 01	1.04E 01	4.07E 00	2.07E 00	1.26E 00	8.65E-01	6.51E-01	5.25E-01	4.50E-01	4.08E-01	3.87E-01	3.82E-01
-30	1.11E 01	8.70E 00	5.13E 00	2.51E 00	1.51E 00	9.98E-01	7.22E-01	5.61E-01	4.62E-01	4.01E-01	3.65E-01	3.47E-01	3.41E-01
-45	3.52E 00	2.82E 00	2.11E 00	1.50E 00	1.08E 00	7.99E-01	6.22E-01	5.07E-01	4.32E-01	3.83E-01	3.52E-01	3.36E-01	3.31E-01
-60	1.33E 00	1.27E 00	1.13E 00	9.56E-01	7.93E-01	6.60E-01	5.59E-01	4.84E-01	4.30E-01	3.92E-01	3.67E-01	3.53E-01	3.49E-01
-75	7.52E-01	7.41E-01	7.11E-01	6.68E-01	6.19E-01	5.70E-01	5.25E-01	4.87E-01	4.56E-01	4.32E-01	4.15E-01	4.05E-01	4.02E-01
-90	5.14E-01	5.14E-01	5.14E-01	5.14E-01	5.14E-01	5.14E-01	5.14E-01	5.14E-01	5.14E-01	5.14E-01	5.14E-01	5.14E-01	5.14E-01

2a

$a = 0.3 \quad i = 0°$

δ \ ε	0	15	30	45	60	75	90	105	120	135	150	165	180
δ = 90	8.39E-01	8.39E-01	8.39E-01	8.39E-01	8.39E-01	8.39E-01	8.39E-01	8.39E-01	8.39E-01	8.39E-01	8.39E-01	8.39E-01	8.39E-01
75	1.25E 00	1.23E 00	1.18E 00	1.10E 00	1.01E 00	9.24E-01	8.45E-01	7.78E-01	7.24E-01	6.82E-01	6.53E-01	6.36E-01	6.31E-01
60	2.16E 00	2.06E 00	1.82E 00	1.53E 00	1.26E 00	1.03E 00	8.65E-01	7.40E-01	6.51E-01	5.88E-01	5.48E-01	5.25E-01	5.18E-01
45	5.05E 00	4.20E 00	3.15E 00	2.23E 00	1.58E 00	1.16E 00	8.95E-01	7.22E-01	6.09E-01	5.36E-01	4.90E-01	4.66E-01	4.58E-01
30	1.42E 01	1.11E 01	6.47E 00	3.27E 00	1.96E 00	1.30E 00	9.34E-01	7.22E-01	5.92E-01	5.12E-01	4.64E-01	4.40E-01	4.32E-01
15	7.24E 01	3.33E 01	1.13E 01	4.49E 00	2.30E 00	1.41E 00	9.75E-01	7.35E-01	5.94E-01	5.09E-01	4.61E-01	4.37E-01	4.32E-01
0		7.31E 01	1.46E 01	5.44E 00	2.45E 00	1.47E 00	1.00E 00	7.49E-01	6.04E-01	5.17E-01	4.68E-01	4.44E-01	
-15	7.24E 01	3.33E 01	1.13E 01	4.49E 00	2.30E 00	1.41E 00	9.75E-01	7.35E-01	5.94E-01	5.09E-01	4.61E-01	4.37E-01	4.32E-01
-30	1.42E 01	1.11E 01	6.47E 00	3.27E 00	1.96E 00	1.30E 00	9.34E-01	7.22E-01	5.92E-01	5.12E-01	4.64E-01	4.40E-01	4.32E-01
-45	5.05E 00	4.20E 00	3.15E 00	2.23E 00	1.58E 00	1.16E 00	8.95E-01	7.22E-01	6.09E-01	5.36E-01	4.90E-01	4.66E-01	4.58E-01
-60	2.16E 00	2.06E 00	1.82E 00	1.53E 00	1.26E 00	1.03E 00	8.65E-01	7.40E-01	6.51E-01	5.88E-01	5.48E-01	5.25E-01	5.18E-01
-75	1.25E 00	1.23E 00	1.18E 00	1.10E 00	1.01E 00	9.24E-01	8.45E-01	7.78E-01	7.24E-01	6.82E-01	6.53E-01	6.36E-01	6.31E-01
-90	8.39E-01	8.39E-01	8.39E-01	8.39E-01	8.39E-01	8.39E-01	8.39E-01	8.39E-01	8.39E-01	8.39E-01	8.39E-01	8.39E-01	8.39E-01

2b

$a = 1$ $i = 20°$

2c

l = 0	15	30	45	60	75	90	105	120	135	150	165	180	
2.40E-01	2.40E-01	2.40E-01	2.40E-01	2.40E-01	2.40E-01	2.40E-01	2.40E-01	2.40E-01	2.40E-01	2.40E-01	2.40E-01	2.40E-01	δ = 90
3.06E-01	3.04E-01	2.97E-01	2.87E-01	2.74E-01	2.61E-01	2.48E-01	2.37E-01	2.27E-01	2.19E-01	2.14E-01	2.10E-01	2.09E-01	75
4.49E-01	4.39E-01	4.13E-01	3.77E-01	3.39E-01	3.04E-01	2.74E-01	2.49E-01	2.30E-01	2.16E-01	2.07E-01	2.02E-01	2.00E-01	60
7.99E-01	7.62E-01	6.69E-01	5.61E-01	4.63E-01	3.84E-01	3.26E-01	2.83E-01	2.52E-01	2.31E-01	2.18E-01	2.12E-01	2.10E-01	45
1.68E 00	1.70E 00	1.33E 00	9.66E-01	7.06E-01	5.34E-01	4.23E-01	3.49E-01	3.00E-01	2.69E-01	2.52E-01	2.44E-01	2.45E-01	30
7.11E 00	5.66E 00	3.58E 00	1.94E 00	1.22E 00	8.29E-01	6.11E-01	4.80E-01	4.00E-01	3.50E-01	3.22E-01	3.09E-01	3.22E-01	15
3.60E 01	2.30E 01	9.74E 00	4.15E 00	2.27E 00	1.43E 00	1.00E 00	7.61E-01	6.19E-01	5.32E-01	4.82E-01	4.57E-01	4.50E-01	0
9.88E 02	8.90E 01	1.92E 01	7.19E 00	3.24E 00	1.91E 00	1.29E 00	9.55E-01	7.61E-01	6.46E-01	5.79E-01	5.44E-01	5.33E-01	-15
2.27E 02	7.06E 01	1.95E 01	7.65E 00	3.48E 00	2.06E 00	1.38E 00	1.02E 00	8.17E-01	6.92E-01	6.18E-01	5.79E-01	5.67E-01	-30
2.54E 01	1.94E 01	1.09E 01	5.89E 00	3.13E 00	2.02E 00	1.43E 00	1.09E 00	8.86E-01	7.60E-01	6.84E-01	6.43E-01	6.30E-01	-45
7.24E 00	6.63E 00	5.12E 00	3.46E 00	2.52E 00	1.88E 00	1.46E 00	1.18E 00	1.00E 00	8.83E-01	8.08E-01	7.66E-01	7.53E-01	-60
2.71E 00	2.64E 00	2.45E 00	2.20E 00	1.93E 00	1.69E 00	1.48E 00	1.32E 00	1.19E 00	1.10E 00	1.04E 00	1.00E 00	9.91E-01	-75
1.49E 00	1.49E 00	1.49E 00	1.49E 00	1.49E 00	1.49E 00	1.49E 00	1.49E 00	1.49E 00	1.49E 00	1.49E 00	1.49E 00	1.49E 00	-90

Fig. 2. Relative brightness distribution at the celestial sphere as a function of l and δ (cf. Fig. 1).

Model: "Rayleigh"

particle size $0.16\ \mu m \leqq s \leqq 4\ \mu m$

size distribution $\propto s^{-4}\,ds$

refractive index $m = 1.27 - 1.37\,i$

number density distribution type II $\beta = 1, \gamma = 1.5$

Position of the probe:

a) at the earth's orbit ($a = 1$ AU, $i = 0$)

b) in the ecliptic plane at $a = 0.3$ AU

c) above the ecliptic plane at $a = 1$ AU; $z = 0.36$ AU; $i = 20°$.

of type I and type III. The reason is, that for sufficiently small elongations the line of sight passes the inner parts of the solar system, where $n(\mathbf{r})$ for models of type II approaches spherical symmetry. Therefore this effect should become more pronounced if the space probe approaches the sun.

b) Position of the space probe out of the ecliptic plane:

(i) The asymmetry of the brightness distribution with respect to $\delta = 0$ can be characterized by the ratio $I\,(0, -90°)/I\,(0, +90°)$ between the intensities at the opposite poles of the ecliptic. Depending on the choice of the model these values were found in the region between 1.3 and 7 for a position of the probe at $a = 1$ AU, $i = 20°$, $z = 0.36$ AU and between 2 and 92 for $a = 1$ AU, $i = 40°$, $z = 0.84$ AU.

(ii) A second characteristic intensity ratio observable from out-of-ecliptic positions is $I\,(180°, 0)/I\,(90°, 0)$. It was found between 1.15 and 1.35 for 'Micron' and between 0.4 and 0.7 for 'Rayleigh' models. No marked dependence on the type of the distribution function (I to III) nor on the position of the probe ($a = 1$ AU , $i = 20°$ or $40°$) was detected.

(iii) The intensity ratio $I\,(0, 0)/I\,(90°, 0)$ is fairly dependent on the model and on the scattering function. For $a = 1$ AU, $i = 20°$ the values were found between 5 and 112. This sensitivity is partially lost, if the space probe is moved to $i = 40°$, probably due to the shift of the lower boundary in the integral (1) away from the forward scattering (i.e. from $\varepsilon = 20°$ to $40°$). In this case $I\,(0, 0)/I\,(90°, 0)$ was obtained in the region between 1.4 and 12.

4. Conclusions

The results referred to above show in principle that deep space missions can contribute in many ways to dissolve the present ambiguities. For out-of-ecliptic missions trajectories with inclinations of about 20° seem tobe sufficient in this respect.

No attempt was made in this study to approximate zodiacal light measurements by the extremely simple models referred to above. Further investigations with more flexible models taking into account observations of brightness, polarization and colour are necessary to arrive at realistic predictions for zodiacal light studies by specific space missions (Helios, Jupiter probe).

References

[1] D. E. Blackwell, D. W. Dewhirst and M. F. Ingham in: Advances in Astronomy and Astrophysics, vol. 5, Academic Press, New York 1967 (p. 1).

[2] R. H. Giese and C. v. Dziembowski, Planet. Space Sci. **17**, 949 (1969).

[3] C. Leinert, Bundesministerium für wissenschaftliche Forschung, Forschungsbericht BMwF-FB 69-18 (1969).

Space Research XI — Akademie-Verlag, Berlin 1971

CONFIGURATION PARTICULIERE DES RAPPORTS D'ABONDANCES DES ELEMENTS DANS LES SPHERULES

T. GRJEBINE[a], Y. YOKOYAMA[b] et P. BRISTEAU[a]

[a]Service d'Electronique Physique, CEA, Saclay, Gif-sur-Yvette, France
[b]Centre des Faibles Radioactivités, CNRS, Gif-sur-Yvette, France

Le regroupement de l'ensemble des analyses effectuées sur les sphérules microscopiques collectées dans les neiges polaires ou dans l'atmosphère, montre que les abondances des éléments ne correspondent pas à celles des roches terrestres ni à celles des roches lunaires, ni même à celles des météorites.

Les abondances se présentent par ailleurs avec moins de variations que pour ces roches, il semble que les sphérules soient constituées d'une matière qui a subi moins de ségrégation que la terre ou la lune.

Les premières poussières à qui on a attribué une origine extra-terrestre étaient des sphérules, collectées pendant une pluie par Mayer et Van Stoop en 1819. Des études extrêmement nombreuses ont été faites depuis sur les sphérules collectées, soit avec la pluie, soit par filtration d'air en avion, ou bien extraites de la neige et des sédiments, ou bien encore recueillies à la limite de l'atmosphère terrestre. Certains auteurs ont suggéré que la plupart des sphérules ne devaient pas être d'origine cosmique parce qu'elles ne présentaient pas les rapports fer-nickel des sidérites. Cependant, la plupart des auteurs qui ont étudié les sphérules leurs attribuent une origine extra-terrestre.

En 1965, Grjebine [1] a attiré l'attention sur les associations particulières d'éléments que l'on rencontrait dans les sphérules, associations qui ne se conformaient pas aux règles de la géochimie terrestre.

Dans ce travail, nous avons regroupé l'ensemble des analyses quantitatives qui ont été faites sur des sphérules dont le mode de collection excluait toute différentiation ou triage particulier. Ainsi, ont été écartées toutes les analyses faites sur des sphérules extraites magnétiquement.

Après sélection, 256 analyses étaient disponibles effectuées par:

77 sphérules collectées dans l'Antarctique [2, 3]
14 sphérules collectées dans l'Antarctique [4]
132 sphérules collectées au Groenland [2, 3]
4 sphérules collectées au Mt. Rainier Glacier [2]
23 sphérules collectées dans la stratosphère [2]

Toutes ces analyses ont été faites avec des microsondes électroniques en étudiant la surface des sphérules.

Les sphérules montrent de très grandes différences de composition chimique, aussi est-il indispensable de considérer la moyenne de toutes les analyses disponibles. Les quantités minima détectables en pratique avec les microsondes électroniques sont de l'ordre de 0,5%. Nous n'avons donc pas pris en considération les éléments détectés à l'état de traces ou en quantités inférieures à cette limite.

Pour calculer la composition moyenne et connaître l'erreur statistique que l'on pouvait introduire, nous avons procédé de la manière suivante:

Le fer par exemple, est détecté dans 193 sphérules et n'apparait pas dans 63 sphérules. Pour ces dernières, on peut donc formuler deux hypothèses extrêmes qui vont permettre de fixer les limites de l'erreur qui peut être commise en effectuant les moyennes arithmétiques:

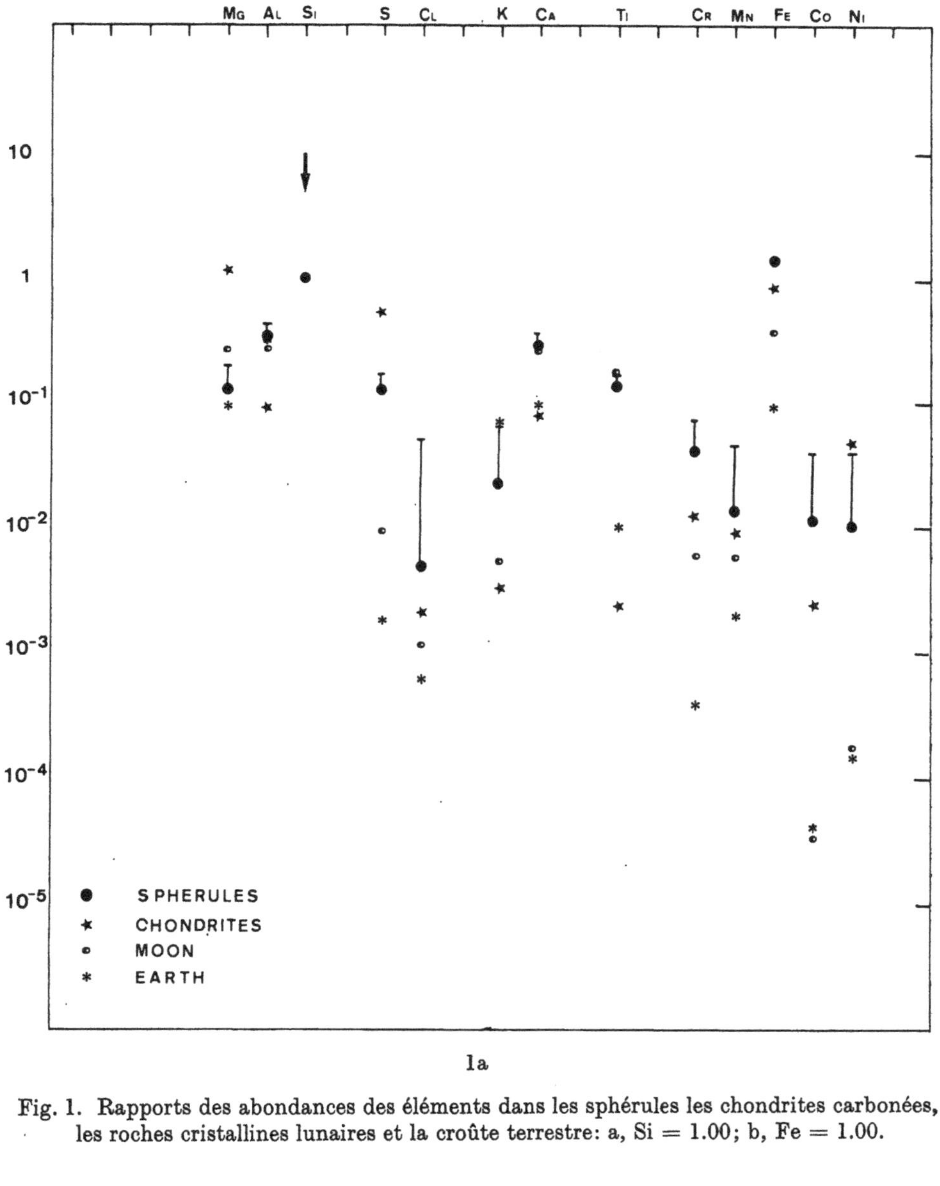

1a

Fig. 1. Rapports des abondances des éléments dans les sphérules les chondrites carbonées, les roches cristallines lunaires et la croûte terrestre: a, Si = 1.00; b, Fe = 1.00.

(i) on peut supposer que dans les 63 sphérules où le fer n'est pas détecté, la teneur en fer était réellement nulle

(ii) on peut supposer au contraire que cette teneur était juste en dessous de la limite de détection, c'est-à-dire $\sim 0{,}5\%$.

Avec la première hypothèse, la moyenne arithmétique générale donne une limite inférieure possible de la teneur de l'élément:

les 193 cas où le fer est présent totalisant 6267%;

la moyenne générale devient donc $\frac{6267 + 0}{256} = 24{,}5\%$.

Avec la deuxième hypothèse qui maximalise les possibilités d'existence de l'élément dans les cas où il n'est pas pondéralement détecté, cette moyenne devient:

$$\frac{6267\% + (63 \times 0{,}5\%)}{256} = 24{,}6\%\,.$$

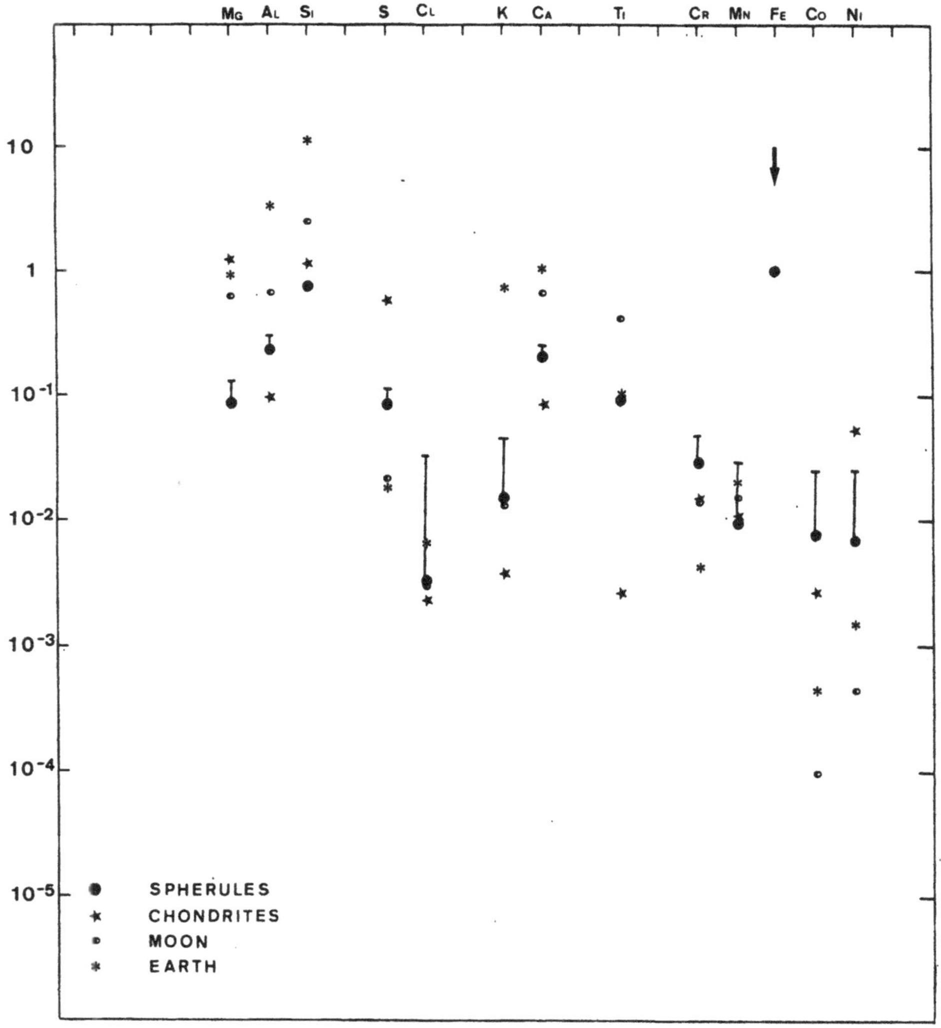

1b

Comme on le voit, le fait que le fer n'a pas été détecté à sa teneur exacte dans 63 cas, ne constitue pas une interdiction pour une évaluation statistique quantitative de son abondance dans les sphérules, l'erreur statistique étant du même ordre de grandeur que les erreurs des meilleures analyses.

En procédant de la même manière, on peut établir des moyennes d'abondances maxima et minima pour tous les éléments détectés par les microsondes, suffisamment de fois et avec des quantités telles que la correction qui doit être introduite reste dans des limites tolérables (voir graphique 1a et b). En fait, on peut donc établir un *modèle statistique de la composition des sphérules.*

Tableau 1
Composition moyenne de 256 sphérules

Elément	Nombre de sphérules détecté	pas détecté	Teneur moyenne en pourcentage pondéral: pourcentage dans les sphérules ou l'élément a été détecté %	dans toutes les sphérules minimum %	maximum %
Mg	15	241	15,0	0,9	1,4
Al	59	197	11,2	2,6	3,0
Si	164	92	13,0	8,3	8,5
S	50	206	5,6	1,1	1,5
Cl	10	246	1,5	0,06	0,5
K	21	235	2,9	0,24	0,7
Ca	102	154	8,4	3,4	3,7
Ti	62	194	7,3	1,8	2,1
Cr	40	216	3,8	0,6	1,0
Mn	27	229	2,2	0,23	0,7
Fe	193	63	32,5	24,5	24,6
Co	18	238	2,7	0,19	0,7
Ni	10	246	4,4	0,17	0,7

Tableau 2
Rapport des abondances des éléments dans les sphérules les chondrites carbonées, les roches cristallines lunaires, la croûte terrestre
(Normalisation par rapport au silicium)

Elément	sphérules minimum	maximum	chondrites carbonées	croûte terrestre	roches cristallines lunaires
Mg	$1{,}2\times10^{-1}$	$1{,}9\times10^{-1}$	1,1	$8{,}7\times10^{-2}$	$2{,}4\times10^{-1}$
Al	$3{,}3\times10^{-1}$	$3{,}8\times10^{-1}$	$8{,}5\times10^{-2}$	$3{,}1\times10^{-1}$	$2{,}6\times10^{-1}$
Si	1	1	1	1	1
P	–	–	$1{,}3\times10^{-2}$	$3{,}9\times10^{-3}$	$3\ \times10^{-3}$
S	$1{,}2\times10^{-1}$	$1{,}6\times10^{-1}$	$5{,}1\times10^{-1}$	$1{,}7\times10^{-3}$	$8\ \times10^{-3}$
Cl	$6\ \times10^{-3}$	$5\ \times10^{-2}$	$2{,}0\times10^{-3}$	$6\ \times10^{-4}$	$1\ \times10^{-3}$
K	$2{,}1\times10^{-2}$	$6\ \times10^{-2}$	$3{,}2\times10^{-3}$	$6{,}7\times10^{-2}$	$5{,}0\times10^{-3}$
Ca	$2{,}8\times10^{-1}$	$3\ \times10^{-1}$	$7{,}4\times10^{-2}$	$9{,}2\times10^{-2}$	$2{,}6\times10^{-1}$
Sc	–	–	$3{,}3\times10^{-5}$	$1\ \times10^{-5}$	$2{,}3\times10^{-4}$
Ti	$1{,}3\times10^{-1}$	$1{,}5\times10^{-1}$	$2{,}3\times10^{-3}$	$9{,}3\times10^{-3}$	$1{,}6\times10^{-1}$
V	–	–	$9\ \times10^{-4}$	$2{,}2\times10^{-4}$	$1{,}5\times10^{-4}$
Cr	$3{,}9\times10^{-2}$	$7\ \times10^{-2}$	$1{,}2\times10^{-2}$	$3{,}9\times10^{-4}$	$5{,}6\times10^{-3}$
Mn	$1{,}4\times10^{-2}$	$4\ \times10^{-2}$	$8{,}8\times10^{-3}$	$1{,}9\times10^{-3}$	$5{,}3\times10^{-3}$
Fe	1,4	1,4	$8{,}9\times10^{-1}$	$9{,}1\times10^{-2}$	$3{,}6\times10^{-1}$
Co	$1{,}1\times10^{-2}$	$4\ \times10^{-2}$	$2{,}3\times10^{-3}$	$4\ \times10^{-5}$	$3\ \times10^{-5}$
Ni	$1{,}0\times10^{-2}$	$4\ \times10^{-2}$	$4{,}6\times10^{-2}$	$1{,}4\times10^{-4}$	$1{,}7\times10^{-4}$

Tableau 3
Rapport des abondances des éléments dans les sphérules les chondrites carbonées, les roches cristallines lunaires et la croûte terrestre
(Normalisation par rapport au fer)

Elé-ment	sphérules minimum	maximum	chondrites carbonées	croûte terrestre	roches cristallines lunaires
Mg	$8,8\times10^{-2}$	$1,4\times10^{-1}$	1,2	$9,5\times10^{-1}$	$6,6\times10^{-1}$
Al	$2,3\times10^{-1}$	$2,7\times10^{-1}$	$9,6\times10^{-2}$	3,4	$7,1\times10^{-1}$
Si	$7,0\times10^{-1}$	$7,2\times10^{-1}$	1,1	$1,1\times10$	2,8
P	—	—	$1,4\times10^{-2}$	$4,3\times10^{-2}$	8×10^{-3}
S	$8,3\times10^{-2}$	$1,1\times10^{-1}$	$5,7\times10^{-1}$	$1,9\times10^{-2}$	2×10^{-2}
Cl	4×10^{-3}	3×10^{-2}	$2,2\times10^{-3}$	7×10^{-3}	3×10^{-3}
K	$1,5\times10^{-2}$	4×10^{-2}	$3,6\times10^{-3}$	$7,4\times10^{-1}$	$1,4\times10^{-2}$
Ca	$2,0\times10^{-1}$	$2,2\times10^{-1}$	$8,3\times10^{-2}$	1,0	$7,2\times10^{-1}$
Sc	—	—	$3,7\times10^{-5}$	1×10^{-4}	$6,3\times10^{-4}$
Ti	$9,1\times10^{-2}$	$1,1\times10^{-1}$	$2,6\times10^{-3}$	$1,0\times10^{-1}$	$4,5\times10^{-1}$
V	—	—	1×10^{-3}	$2,4\times10^{-3}$	$4,0\times10^{-4}$
Cr	3×10^{-2}	5×10^{-2}	$1,4\times10^{-2}$	$4,3\times10^{-3}$	$1,5\times10^{-2}$
Mn	1×10^{-2}	3×10^{-2}	$1,0\times10^{-2}$	$2,1\times10^{-2}$	$1,5\times10^{-2}$
Fe	1	1	1	1	1
Co	8×10^{-3}	3×10^{-2}	$2,6\times10^{-3}$	4×10^{-4}	9×10^{-5}
Ni	7×10^{-3}	3×10^{-2}	$5,1\times10^{-2}$	$1,5\times10^{-3}$	$4,8\times10^{-4}$

Tableau 4
Comparaison des rapports des abondances des éléments entre les sphérules et des mélanges pondérés de différentes manières entre les chondrites carbonées, les roches lunaires et la croûte terrestre

Elément	sphérules	Composition de: chondrites ×0,2, roches lunaires ×0,6, croûte terrestre ×0,2		chondrites ×0,2, roches lunaires ×0,8	
		abondance	rapport	abondance	rapport
	A	B	A/B	C	A/C
Mg	0,12	0,38	0,32	0,41	0,29
Al	0,33	0,24	1,4	0,23	1,4
Si	1	1	1	1	1
S	0,12	0,11	1,1	0,11	1,1
Cl	0,006	0,001	6	0,001	6
K	0,021	0,017	1,2	0,005	4,2
Ca	0,28	0,19	1,5	0,22	1,3
Ti	0,13	0,10	1,3	0,13	1
Cr	0,039	0,003	13	0,003	13
Mn	0,014	0,005	2,8	0,006	2,3
Fe	1,4	0,41	3,4	0,46	3,0
Co	0,011	0,0005	22	0,0005	22
Ni	0,010	0,010	1	0,010	1

Les rapports d'abondances des sphérules apparaissant très souvent entre les rapports des chondrites de la terre et de la lune. Il était intéressant d'examiner avec quelle pondération de ces trois sources il serait possible de reconstituer la composition des sphérules. Le tableau V constitue une tentative de différentes pondérations possibles, comme on peut le voir pour les éléments légers, les modèles pondérés sont possibles mais aucune pondération ne convient pour les éléments lourds.

Les résultats sont regroupés dans le tableau 1 où les abondances des éléments sont portées en pourcentages massiques, et dans le tableau 2 où ces teneurs maxima et minima sont comparées aux roches terrestres, aux roches cristallines lunaires et aux chondrites carbonées, en normalisant l'ensemble des résultats par rapport au silicium. Les valeurs pour les chondrites carbonées sont celles de Cameron [5], pour la croûte terrestre celles de Mason [6] et pour les roches cristallines lunaires celles de Schmitt et al. [7], Morrison et al. [8], Maxwell et al. [9] et Keays et al. [10]. Les abondances normalisées par rapport au fer figurent dans le tableau 3.

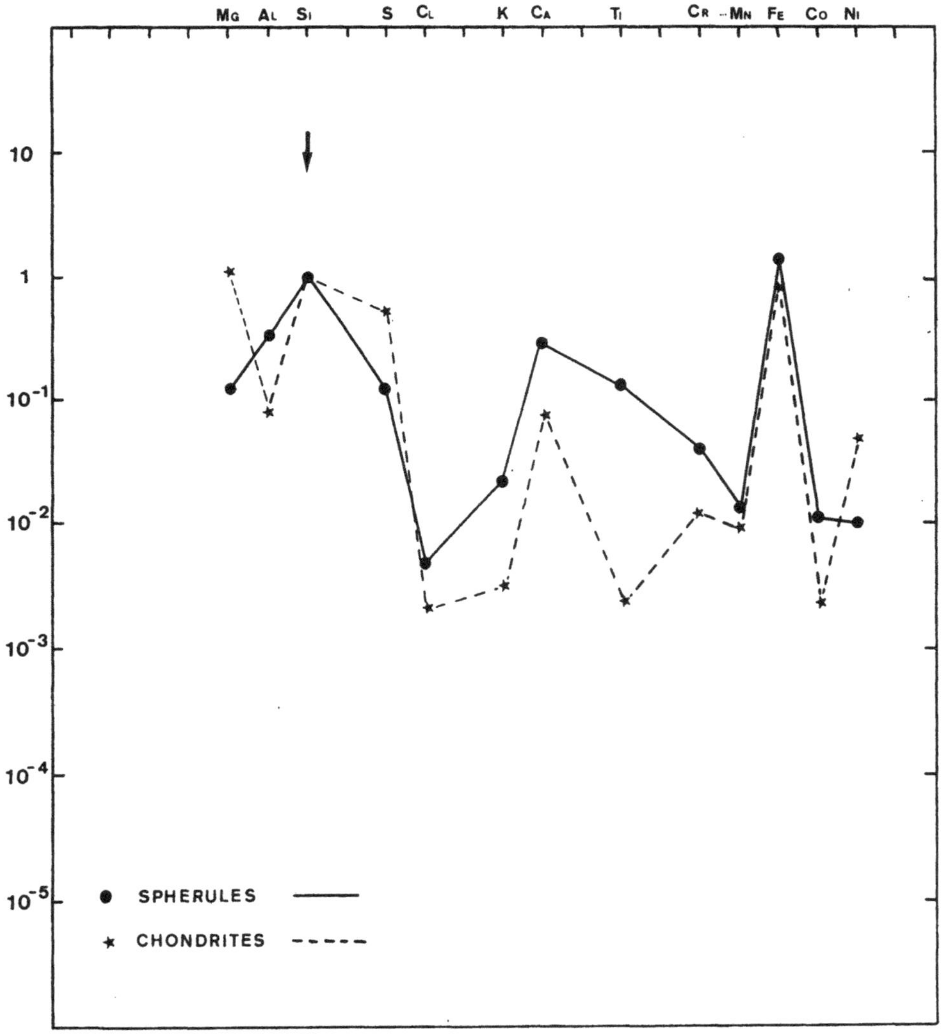

2a

Fig. 2. Comparaison des rapports des abondances des éléments entre les sphérules et les chondrites carbonées: a, Si = 1.00; b, Fe = 1.00.

Dans les figures 1 a et b, les rapports d'abondances sont comparés à ceux des chondrites carbonées, de la croûte terrestre et des roches cristallines lunaires. Ces graphiques suggèrent les remarques suivantes: la valeur des rapports d'éléments pour les sphérules se situent pour la plupart des éléments analysés entre les valeurs relatives aux chondrites et aux roches terrestres; le calcium et les éléments lourds sont plus abondants dans les sphérules. Il faut rappeler que ces différences d'abondances ne peuvent être expliquées par une sélection magnétique, puisque tous les cas où celle-ci avait été employée ont été rejetés.

Les cas du titane et du calcium sont très intéressants parce qu'ils se présentent exactement dans le même rapport que pour les roches cristallines lunaires.

L'abondance du fer, par rapport au silicium fait penser que l'on se trouve en présence d'un matériau qui pourrait être produit par une sélection du type

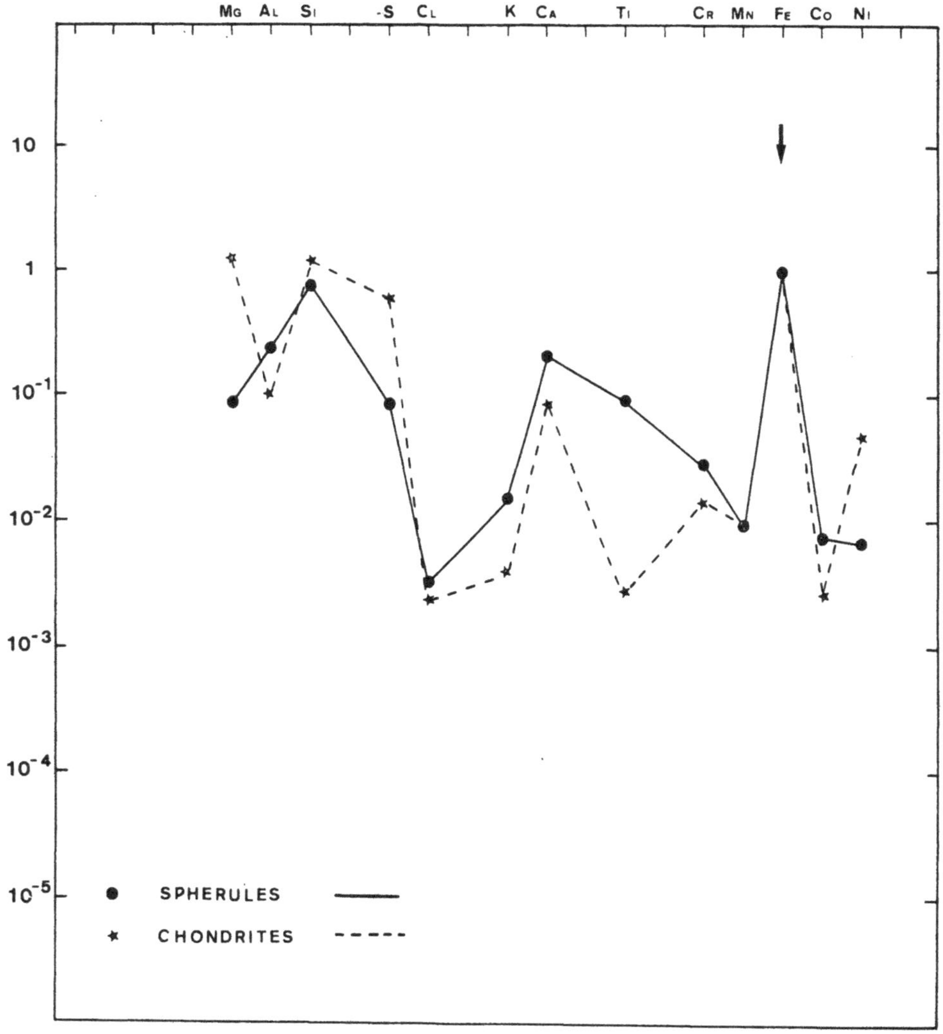

2b

sidérophilique. Les grandes abondances du nickel et du cobalt peuvent confirmer cette observation. Mais nous ne pensons pas que cette observation est déterminante parce que si la grande abondance du fer, du nickel et du cobalt devait être expliquée par un processus de ségrégation sidérophilique, elle n'aurait pas été accompagnée par une haute abondance du titane, du chrome et du calcium qui ne sont pas des éléments sidérophiliques. Dès lors, l'explication pour ces enrichissements doit être recherchée ailleurs, soit dans une autre possibilité de ségrégation ni aqueuse ni fondue, soit dans une constitution plus élémentaire qui n'aurait pas été affectée par des ségrégations.

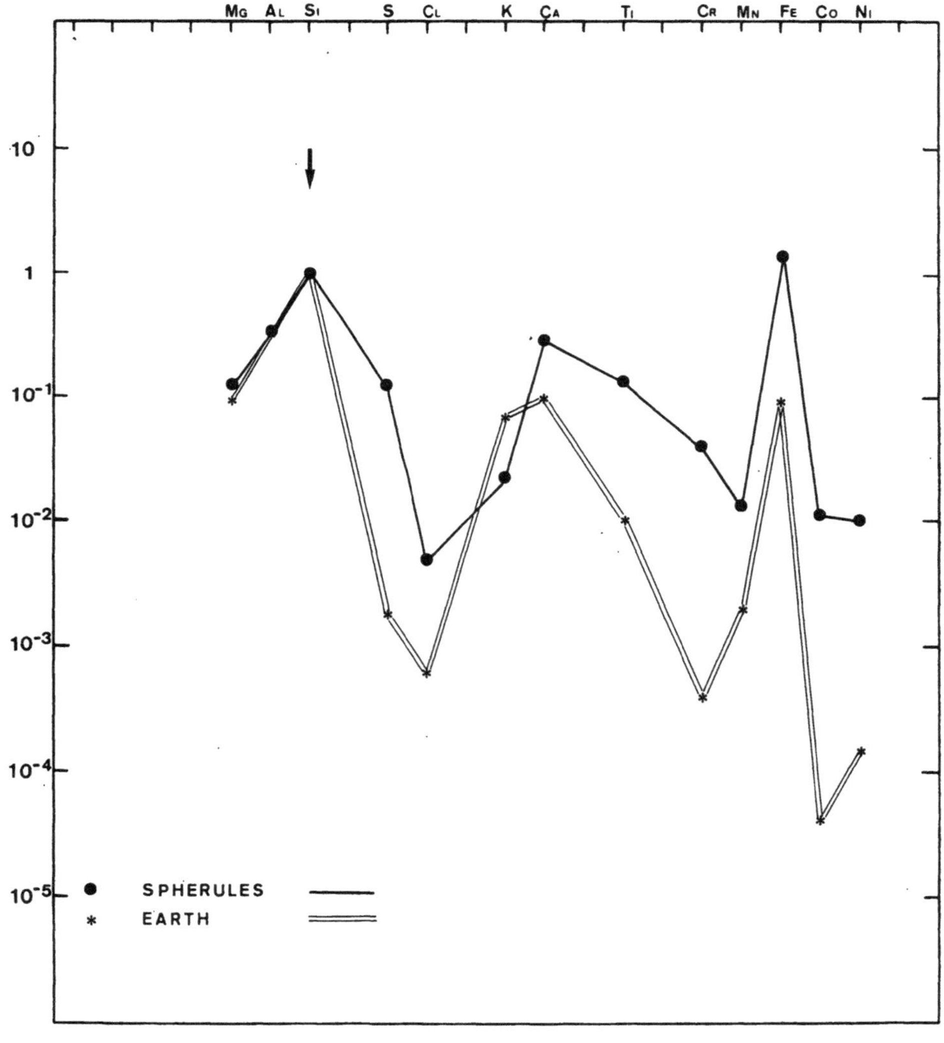

3a

Fig. 3. Comparaison des rapports d'abondance des éléments entre les sphérules et la croûte terrestre: a, Si = 1.00; b, Fe = 1.00.

La comparaison de la composition des sphérules avec celle des chondrites carbonées (fig. 2a, b) montre en général de plus grandes analogies, sauf pour quelques cas. La principale différence se présente pour le magnésium et le titane, avec deux tendances différentes. Cependant, il faut remarquer que, en ce qui concerne les dosages du magnésium, des erreurs expérimentales d'analyses sont possibles: en effet, cet élément était, à l'époque des analyses, à la limite de détection des microsondes électroniques. Ainsi la faible teneur en magnésium a peut-être une explication expérimentale.

La comparaison avec les chondrites carbonées fait penser au prime abord, que la composition des sphérules pourrait fort bien s'expliquer par un mélange de matière lunaire et de matière chondritique (voir tableau 4). Mais un examen

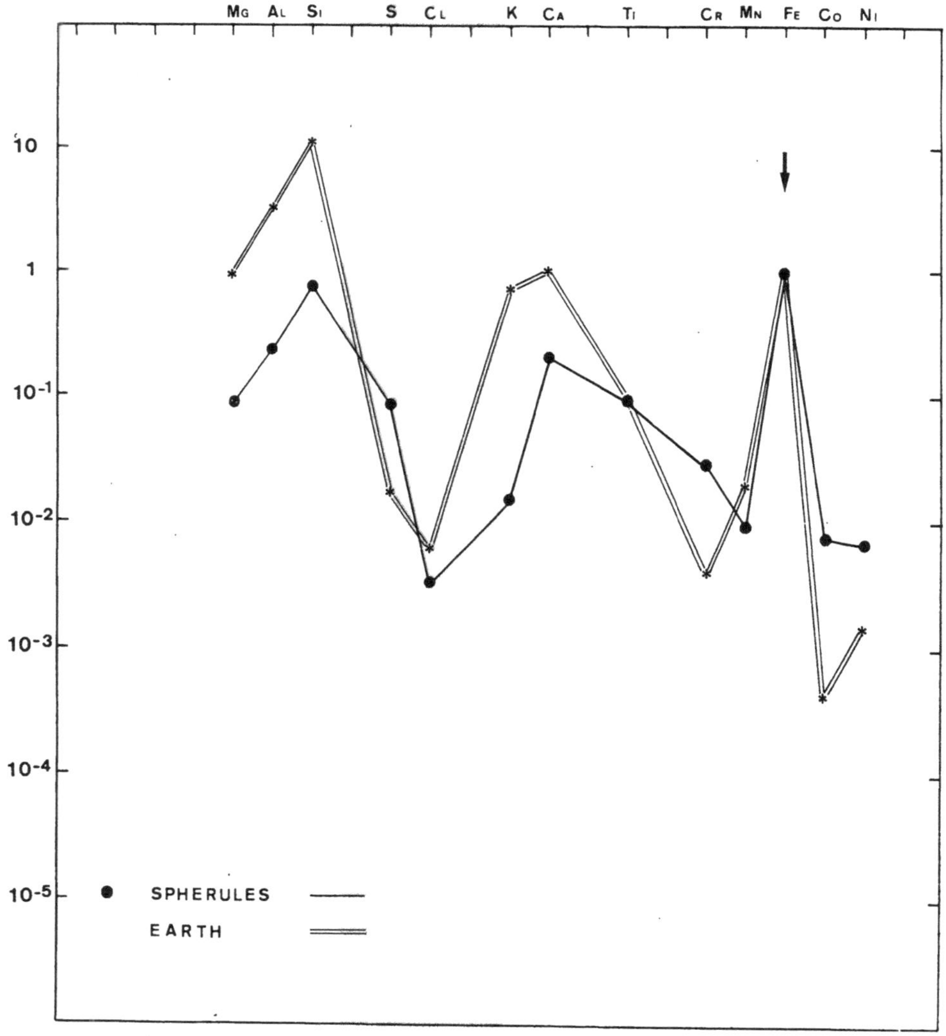

3b

plus approfondi montre que cette double origine n'explique pas tous les écarts.

Sur les figures 3a et b, la comparaison avec les abondances des éléments sur la terre montre que la croûte terrestre est très fortement ségrégée, les rapports des éléments sur la terre variant de plusieurs puissances de dix. La même chose peut être dite des roches cristallines lunaires (fig. 4a, b) qui montrent également de très fortes variations.

Nous avons tracé, par la méthode des moindres carrés pour chacun des graphiques, une droite qui représentait la tendance moyenne de l'évolution de l'abondance des éléments. Nous avons mesuré ensuite les écarts réels des abondances

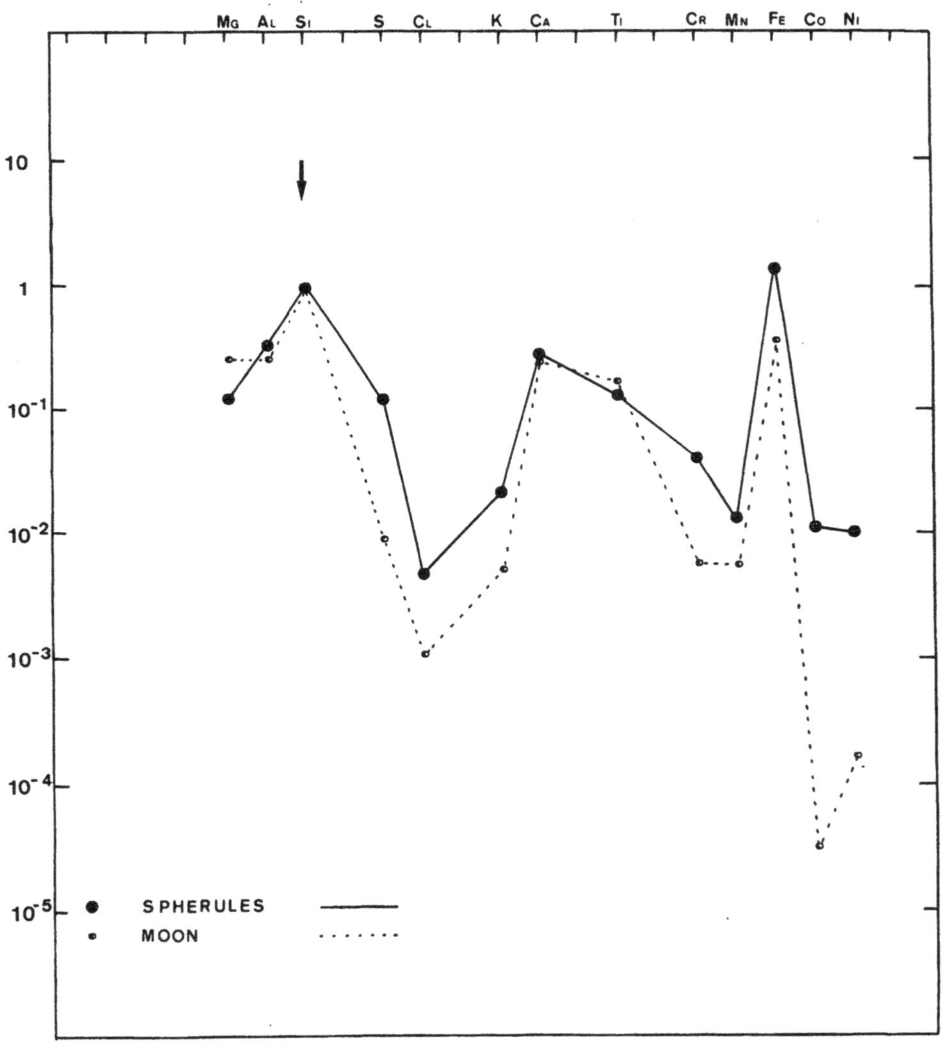

4a

Fig. 4. Comparaison des rapports des abondances des éléments entre les sphérules et les roches cristallines lunaires: a, Si = 1.00; b, Fe = 1.00.

des éléments pour chaque graphique, par rapport à cette tendance moyenne. En sommant l'ensemble des carrés des écarts nous aboutissons à un nombre qui peut être représentatif, entre autres choses, du degré de ségrégation qui a affecté les éléments dans le corps considéré. Ainsi la croûte terrestre et les roches cristallines lunaires apparaissent avec de fortes sommes des carrés des écarts, 1384 pour la terre et 1609 pour la lune. Les chondrites carbonées ne totalisent que 1154 et dénotent ainsi un degré de ségrégation bien moindre. Il est finalement important de noter que cette somme d'écarts n'est que de 626 pour les sphérules. La somme des écarts établie pour ces quatre cas ne subit aucune modification, selon que les normalisations sont faites avec le fer ou avec le silicium. Jusqu'à présent les chondrites carbonées étaient considérées comme le matériau le moins

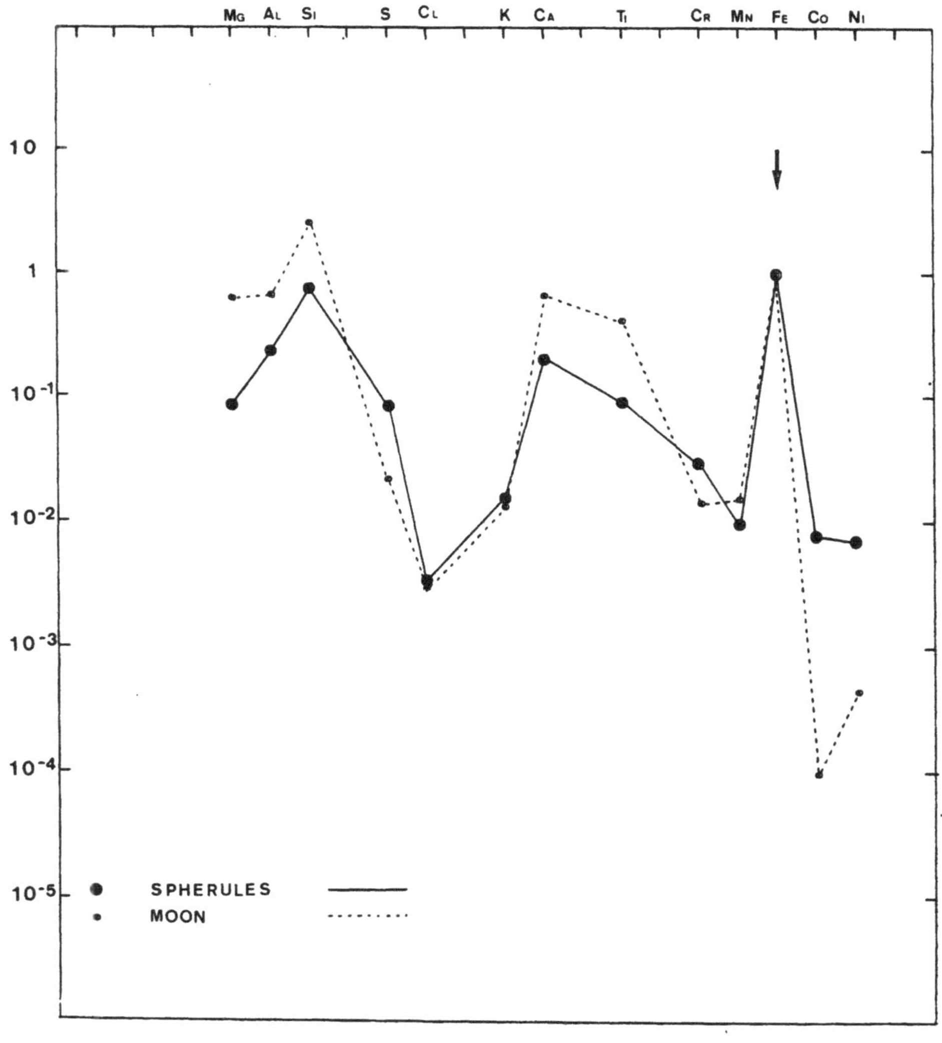

4b

ségrégé et le plus primitif que nous connaissions, et il faut se souvenir qu'il existe de très grandes différences entre les météorites, qu'elles soient carbonées, chondritiques ou achondritiques.

D'autre part, les abondances solaires ont été révisées beaucoup de fois et, là encore, il y a un certain danger à considérer le modèle solaire comme le modèle le plus élémentaire que nous connaissions.

A la suite des remarques que nous venons de faire, en ce qui concerne notamment les écarts quadratiques des quatre catégories de corps étudiés, deux possibilités se présentent: (i) ou bien les chondrites carbonées continuent à demeurer la matière la plus primitive que nous connaissions dans le système solaire et,

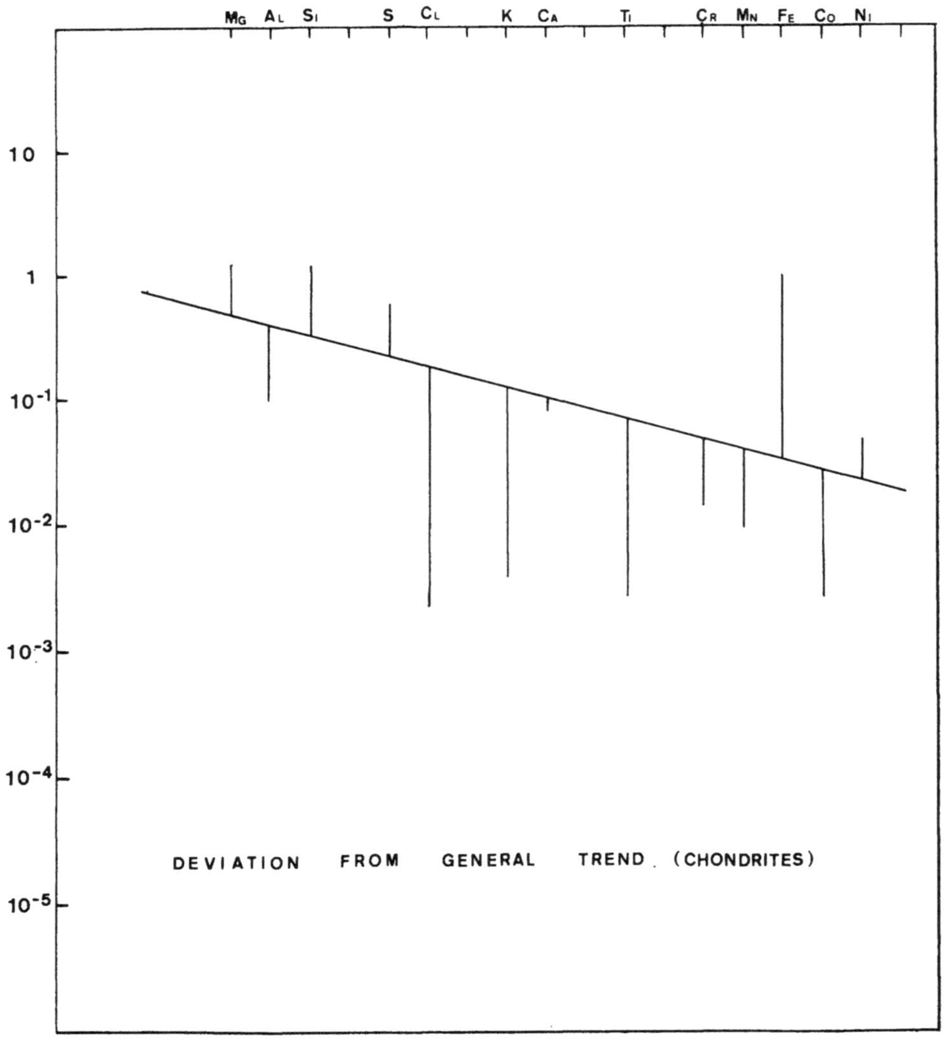

Fig. 5. Schéma montrant comment, après le tracé par la méthode des moindres carrés de la tendance générale, sont faites les sommations des écarts et des écarts quadratiques.

dans ce cas, les sphérules ont subi en moyenne une ségrégation, mais dans un sens opposé à celui de la ségrégation de la croûte terrestre et des roches cristallines lunaires, (ii) ou bien les chondrites ne représentent pas la matière la plus primitive du système solaire, et seraient, elles aussi, ségrégées. Dans ce cas les sphérules, si elles sont d'origine cosmique, bien que ne représentant pas la totalité de la poussière cosmique collectable, sont les représentants possibles de la matière la plus primitive du système solaire en particulier.

Références

[1] T. Grjebine, Communication présentée au Symposium Micrométéorites et poussière cosmique à Heidelberg 1965.
[2] F. W. Wright, P. W. Hodge et C. C. Langway Jr., J. Geophys. Res. **68**, 55 (1963).
[3] P. W. Hodge, F. W. Wright et C. C. Langway, J. Geophys. Res. **69**, 2919 (1964).
[4] R. A. Schmidt, Res. Rept. Ser. 63-3 (1963) p. 65, Geophysical and Polar Research Center, University of Wisconsin.
[5] A. G. W. Cameron, in: Origin and Distribution of the Elements, Pergamon Press, Oxford 1968 (p. 676).
[6] B. Mason, Principle of Geochemistry, John Wiley, New York 1966 (p. 45).
[7] R. A. Schmidt, H. Wakita et P. Rey, Science **167**, 512 (1970).
[8] G. H. Morrison et al., Science **167**, 505 (1970).
[9] J. A. Maxwell, S. Abbey et W. H. Champ, Science **167**, 530 (1970).
[10] R. R. Keays et al., Science **167**, 490 (1970).

Space Research XI — Akademie-Verlag, Berlin 1971

LUNAR DUST POTENTIAL

J. W. Rhee

Department of Physics, Rose Polytechnic Institute, Indiana, USA

The equilibrium potential of a lunar dust particle is determined by the condition that the net current to the surface area vanishes. The potential is found to range from +5 to +10 volts, depending on the flux of solar wind particles and photoefficiency of the dust particles on the lunar surface.

1. Introduction

The lunar surface seems to be completely covered with dust. A thick layer of dust is present everywhere as shown by the recent space experiment (Apollo series), although it may be thin on mountain slopes. It may be a few millimeters thick on hard rock, or it could be thicker. It can be expected that surface dust particles have a certain electric charge due to the solar wind and solar radiation. In such cases, the electromagnetic field carried in the solar wind will exert a Lorentz force on the dust particles [1].

The importance of such study to lunar science is threefold. Firstly, adequate analysis is necessary to understanding the history and properties of the lunar surface and interior; secondly, there are unique physical phenomena associated with the moon in interplanetary space; and thirdly, the moon provides a useful platform from which to measure and study various interplanetary phenomena.

The moon very likely has an electric field for the first 1–100 m altitude due to the charged layer of surface dust particles, but the magnitude of this field is difficult to estimate. Previous studies show that the moon as a solid body is charged [2, 3, 4]. The electric field is clearly related to the interaction of the solar wind with the lunar surface particles [7, 8].

2. Procedure

Lunar surface dust particles are expected to have an electric charge and corresponding equilibrium potential which is determined by the following processes:

(i) photoelectric emission
(ii) capture of solar wind particles
(iii) field and secondary emission.

Only the first two processes are important [5]. Our main objective here is to derive a simple relationship between the equilibrium potential and solar wind density.

Let us consider a spherical dust particle of radius S on the sunlit side of the lunar surface. The frequency ν_t of a photon that can eject an electron from the particle is

$$\nu_t = \frac{1}{h}\left[W + \frac{Q\,e}{S}\right] \cdots, \tag{1}$$

where h is Planck's constant, W the work function, e the electronic charge; Q (esu) is related to the equilibrium potential φ (volt) by

$$\varphi = \frac{Q}{S} \cdots \tag{2}$$

The number of electrons ejected by the incident solar radiation is equal to

$$N_1 = \pi\, S^2 \left(\frac{R_\odot}{r}\right)^2 \int\limits_{\nu_t}^{\infty} \frac{Y(\nu)\, P(\nu)\, \mathrm{d}\nu}{h\,\nu} \cdots, \tag{3}$$

where $R_\odot$ is the radius of the sun, r is 1 AU, $Y(\nu)$ is the photoefficiency, and $P(\nu)$ is Planck's radiation function. A cut-off wavelength of 2000 Å will be used in our computation. Along with this, we must investigate the rate at which the solar wind particles stick to the dust particles. If we assume that there are N_e electrons per cm³ in cislunar space, and these electrons are characterized by the kinetic temperature T_e according to the Maxwellian velocity distribution function $f(V)$, then the electron capture rate N_2 is given by

$$N_2 = \pi\, S^2 \left(\frac{8\,k\,T_e}{\pi\, M_e}\right)^{1/2} N_e \left[1 + \frac{e\,\varphi}{k\,T_e}\right] \cdots \tag{4}$$

Similarly, we can write the proton capture rate from the solar wind:

$$N_3 = \pi\, S^2 \left(\frac{8\,k\,T_p}{\pi\, M_p}\right) N_p \left[1 - \frac{e\,\varphi}{k\,T_p}\right] \cdots, \tag{5}$$

where N_p is the proton density, T_p proton temperature, and M_p is the mass of proton. A stationary solution is achieved if

$$N_1 = N_2 - N_3$$

and

$$N_p = N_e\,, \quad T_e = T_p\,. \tag{6}$$

3. Discussion and Conclusion

Using Eq. (6), we have computed the equilibrium potential as function of solar wind density for an electron temperature of 10^5 °K. The solar wind temperatures deduced from Mariner 2 data are about 6×10^4 °K and 10^6 °K, and the median is close to 10^5 °K [6]. The current best estimate for N_e ranges from 1 to 10. The results are plotted in Fig. 1. Since the photoefficiency η is unknown at the present time, we assumed four different values of η and plotted N_e as a function of η also.

An examination of Fig. 1 shows that the potential is about 4–6 volts at 10^5 °K if an average photoefficiency of 0.01 is assumed. The equilibrium potential seems to be insensitive to great changes in the electron density and the photoefficiency. At high electron densities, the higher temperature results in lower positive potential and for low density the reverse is true. This is because of the fact that the capture cross section decreases rapidly as the particle energy goes up.

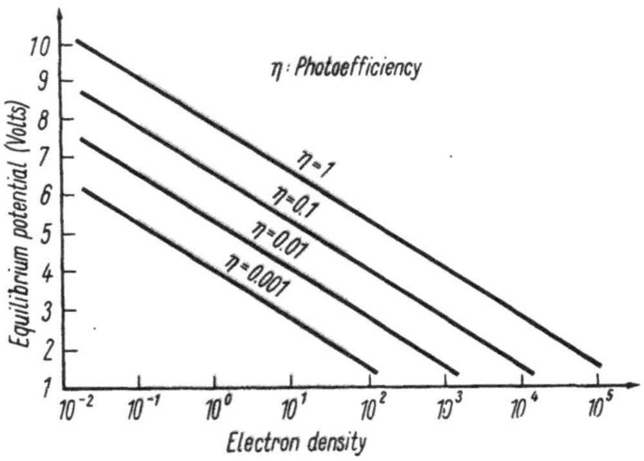

Fig. 1. Equilibrium potential as a function of electron density (N_e/cm^3) at an electron temperature of 10^5 °K.

It must be pointed out that at 10 volts, which is roughly the sum of the work function and electrostatic potential, the blackbody approximation for the solar spectrum does not agree very well with observations. The observed spectrum has a greater photon flux at energies $\geqq$ 10 volts than does the Planck spectrum. Taking this into account would increase our computed values for the potential and raise it to perhaps + 10 volts.

In summary, an attempt has been made to estimate the equilibrium potential of a lunar dust particle even though it is rather difficult to attack the problem satisfactorily. It is shown that the potential is positive for the sunlit surface dust particles and it ranges from 5 to 10 volts at most. The existence of a high value of positive potential is unlikely. The dust particles in the shadow region of the lunar surface must have negative charge due to the absence of the solar radiation.

Acknowledgment

This research was supported by grant NSF-GY-5389 from the National Science Foundation.

References

[1] NASA 1965 Summer Conference on Lunar Exploration and Science, NASA SP-88, Washington, D.C.
[2] E. J. Öpik, Planet, Space Sci. **9**, 221 (1962).
[3] E. J. Öpik and S. F. Singer, Escape of Gases from the Moon, J. Geophys. Res. **65**, 3065 (1960).
[4] J. L. Blank and W. D. Grobman, J. Geophys. Res. **74**, 3943 (1969).
[5] J. W. Rhee, The Zodical Light and Interplanetary Medium, NASA SP-150 (1967).
[6] M. Neugebauer and C. W. Snyder, The Solar Wind, Pergamon Press, Oxford 1965.
[7] S. F. Singer and —. Walker, Icarus **1**, 7 (1962).
[8] —. Walbridge, Icarus **10**, 342 (1969).

Space Research XI — Akademie-Verlag, Berlin 1971

LUNAR EXPLORER 35 AND OGO 3: DUST PARTICLE MEASUREMENTS IN SELENOCENTRIC AND CISLUNAR SPACE FROM 1967 TO 1969

W. M. ALEXANDER[a], C. W. ARTHUR[a] and J. L. BOHN[b]

[a]Baylor University, Waco, Texas, USA
[b]Temple University, Philadelphia, Pa, USA

The OGO 3 and Lunar Explorer 35 dust particle experiments have continued to produce information concerning the flux of picogram particles in cislunar and selenocentric space. A re-evaluation of previous OGO 3 data reduces the mean cumulative rate by a factor of two to 5×10^{-3} pm^{-2} s^{-1} π^{-1} sr^{-1}. Additional Lunar Explorer 35 dust particle experiment measurements in selenocentric space now extend through September 1969. A total of 1146 events have been detected. The mean cumulative flux for this 26-month period is 5.9×10^{-4} pm^{-2} s^{-1} π^{-1} sr^{-1}, with fluctuations of approximately an order of magnitude. The major variations in the flux occur during periods of major meteor showers. A study of the data, year to year, indicates a strong correlation of the time of occurrence of peak impact rates and a fairly strong correlation, year to year, of the magnitude of the rates.

1. Introduction

Initial data for 1967 and parts of 1968 from the dust particle experiment on Lunar Explorer 35 have been reported by Alexander et al. [1]. Additional data have been received from this experiment for the remainder of 1968 and part of 1969. Furthermore, OGO 3 has continued to transmit data from its dust particle experiment which provides additional information to combine with that previously reported [1, 2] for the study of dust particle flux in cislunar space. It is the intention of this paper to present the new data from the dust particle experiments on Lunar Explorer 35 and OGO 3, to discuss the major characteristics of these measurements, to compare the new data with previous data from the two experiments, and to review the current mean cumulative flux-mass distribution model for microparticles in selenocentric, cislunar, and interplanetary (near 1 AU) space with respect to the Lunar Explorer 35 and OGO 3 measurements.

2. OGO 3: 1966–1967

A description of the OGO 3 dust particle experiment instrumentation has been published [1, 2]. A tube containing three sensor elements is the basic detector in the instrument. The first sensor, at the entrance of the tube, is a thin film (1500 Å); the second sensor is a metallic disk to which a lead zirconate acoustic transducer is bonded; the third sensor is a metallic grid in front of the disk. When a particle traverses the tube, it is detected as it passes through the thin film. The ceramic transducer detects the impulse delivered to the plate by the impacting particle, and the grid detects the ionization produced by the hypervelocity impact of the particle on the plate. It is possible to obtain a time-of-flight measurement from this tubular detector, as well as information related to the energy and momentum of the particle. Since the acceptance angle for the entrance sensor is much larger than that of the tube, particles are detected by this sensor alone. The experiment instrumentation recognizes all events from each of the three sensors, as well as the time-of-flight event.

In low data rate experiments, such as dust particle experiments, one has to be extremely concerned about the mechanisms which may cause anomalous signals from the electronic instrumentation. Mechanical noise sources, such as vibration and thermal creaks, have been a concern since the rocket flights of 1957 [3]. Since Explorer 8 was launched in 1960 the satellite experiments have contained more than one sensor in each detector unit in order to use double or triple coincidence detection of a single event to increase the reliability of the data. Considerable doubt was raised concerning possible spurious events from the ceramic acoustical transducer caused by thermal gradients within the sensor [4]. Several laboratory studies were conducted from which conflicting results have been reported [4–7]. Bohn et al. [5] carried out the most exhaustive studies reported and concluded that each dust particle sensor and satellite system was unique in itself. There were no cases of noise problems within any of the sensors except those exhibiting prior mechanical damage (hairline cracks, etc.). Electromagnetic interference and large current transients in the spacecraft system may cause anomalous experiment information, and spacecraft testing prior to launch is designed to identify or eliminate these interference noise sources. However, it is often the case that, after launch, anomalous events occur which can be definitely identified as interference noise.

Four triple-coincidence tubular detectors are used in the OGO 3 dust particle experiment. A high level of confidence is attached to any triple-coincident event. Since the front film sensor on any tube can detect particles which do not traverse the tube, a mechanical block was placed at the entrance of one tube in order to increase the confidence level of the data from these sensors. Prior to launch no interference of any nature was found to be caused by spacecraft functions. The same tests showed that the switching on and off of experiment 17 produced anomalous data in the dust particle experiment. After launch, this expected interference was evident. In addition, a large count rate occurred from all four front film sensors during the malfunctioning of another experiment. These noise bursts were not seen in prelaunch testing and were no longer present when the experiment was turned off.

The OGO 3 spacecraft contained tape recorders which would record data from experiments at a low bit rate and then, upon ground command, dump this information at a high bit rate. Because of the very eccentric orbit of OGO 3, a large amount of real-time data was also obtained. In addition there were some times for which real-time data and playback data from the onboard tape recorders were both available. The redundancy of this operation provided a check on the two systems of obtaining data. Upon comparison of the two types of data, two important aspects of the playback data emerge: (i) while real-time events are sometimes present in playback data, no playback events are found in the real-time data, and (ii) front film events on the control tube occur only in the playback data. Thus, a re-evaluation of the data is underway in which no OGO 3 playback data is considered reliable. Table 1 shows 450 hours of

Table 1
OGO 3 reevaluated data
(Days 168–202, 1966; days 35–76, 1967)

Tube	+Y	+X	−X (blocked)	−Z
Events	6	6	0	4

$\varphi = 5 \times 10^{-3}$ pm^{-2} s^{-1} π^{-1} sr^{-1}

re-evaluated data which contains only real-time data and from which noise burst periods have been eliminated. The − X tube is the blocked control tube. Playback data have been included in OGO 3 papers [1, 2]. A much higher confidence level can be attached to the information in Table 1. The three time-of-flight (TOF) events previously reported [1] were obtained from playback data and are not considered reliable. Two possible TOF events are contained in the revised data, but further analysis is necessary before they can be accepted as real, and they are not quoted in Table 1.

3. Lunar Explorer 35: 1967–1969

A description of the Lunar Explorer 35 dust particle instrument is given in [8]. The experiment contains a large area circular detector and a triple coincidence tubular detector. A brief description of the large area circular detector is given here because the data to be described are those obtained from this detector. The detector comprises two sensors: an acoustical ceramic transducer bonded to the plate and a large area capacitor sensor deposited on the impact side of the plate. The acoustical wave produced by the impact of a dust particle on the plate is detected by the ceramic transducer. A pulse height analysis is performed on the resulting signal, and the event data are stored in the electronics. Hypervelocity calibration studies indicate that the nominal mass threshold of this sensor is 5 picograms. The penetration of the capacitor sensor's dielectric by a dust particle also produces an electrical signal which is analyzed before the data are stored. The nominal mass threshold of this sensor is 100 picograms. A coincidence event in two sensors is detected by the logic circuitry and stored as data.

As stated above, the following data are from the large area detector only. Data have been received for the first twenty-six months of operation, from day 201, 1967 to day 266, 1969. There are several gaps during this period for which no data are available, and there are bursts of noise in the data which are caused by the calibration sequence of another experiment on the spacecraft. When these noise bursts were excluded from the data the remaining time generally constituted 90–95% of the actual day and was in no case less than 85% of any two-day period.

Table 2
Lunar Explorer 35 mean cumulative fluxes

Time period	Flux ($pm^{-2}\,s^{-1}\,\pi^{-1}\,sr^{-1}$)
First twelve months	
(day 201/1967–day 200/1968)	5.7×10^{-4}
Shower (201–79)	7.8×10^{-4}
Non-shower (8–200)	1.12×10^{-4}
Second twelve months	
(day 201/1968–day 200/1969)	5.9×10^{-4}
Shower (201–76)	6.9×10^{-4}
Non-shower (85–200)	4.6×10^{-4}
Fall 1969 (201–266)	6.34×10^{-4}
Total shower time	7.24×10^{-4}
Total non-shower time	2.97×10^{-4}
Total time	5.87×10^{-4}

Table 2 shows the values of cumulative flux for several different divisions of the data. Some aspects of these fluxes merit further discussion. First, the flux during shower periods is much higher than that during non-shower periods in every case. Also, the flux during the non-shower period of 1968 (days 80–200) agrees with that determined by Mariner 4 in interplanetary space near 1 AU [2]. However, the flux during the non-shower period of 1969 (days 85–200) is somewhat higher than the interplanetary rate. The most probable explanation for this is suggested by the data itself: when four-day event totals are plotted, definite peaks emerge which coincide with two permanent daylight showers and with some unnamed southern hemisphere showers detected by Nilsson [4]. These peaks are similar to those which occur during shower periods corresponding to major showers.

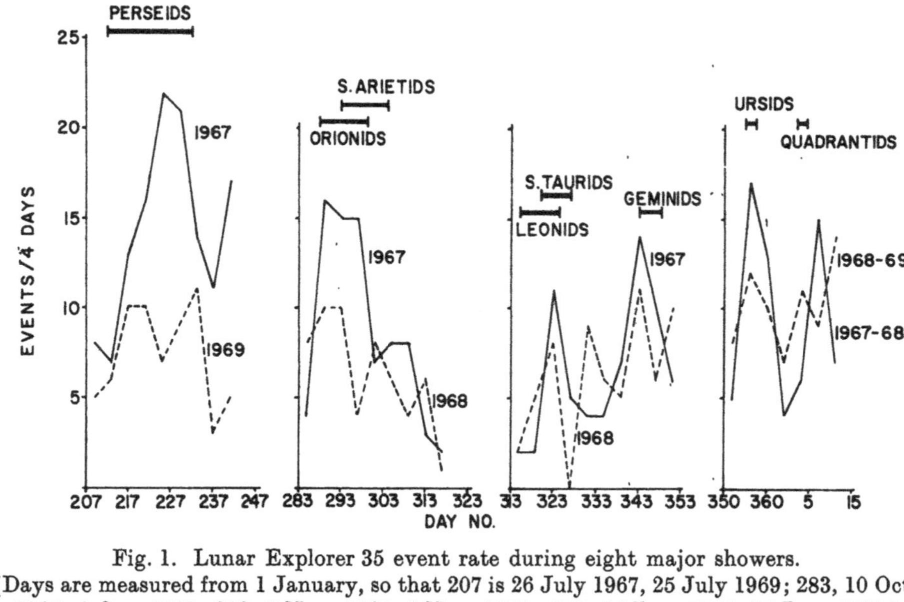

Fig. 1. Lunar Explorer 35 event rate during eight major showers.
(Days are measured from 1 January, so that 207 is 26 July 1967, 25 July 1969; 283, 10 Oct. 1967, 9 Oct. 1968; 313, 9 Nov. 1967, 8 Nov. 1968; 350, 16 Dec. 1967, 15 Dec. 1968.)

A statistical analysis of the data indicates a probability of randomness of < 0.001. As has been previously reported [1], large fluctuations of the event rate occur in connection with major meteor showers. Fig. 1 shows the event rates associated with eight major meteor showers. The points represent 4-day totals. A line graph with the point at the center of the 4-day period is used for clarity. There is a striking similarity, year to year, of the time of occurrence of peaks, as well as the height of the peaks; the differences in these is well within the fluctuations observed in ground-based data. The graphs in Fig. 1 thus strengthen the case for micrometer-sized lunar ejecta associated with major meteor showers. However, it should be emphasized that the 100 picogram sensor still shows no statistically meaningful enhancement during shower periods. This has been commented on in [1]. The new data have not changed the validity of the statement. This non-enhancement is important because, if the enhanced flux is caused by lunar ejecta related to showers, it establishes a mass cutoff for detected ejecta particles.

4. Discussion

Data from OGO 3 and Lunar Explorer 35 can be compared in a meaningful way with data from previous experiments in interplanetary, cislunar, and selenocentric space. The cumulative fluxes derived from Pioneer 1 [9], Mariners 2 and 4 [10, 11], Lunar Orbiter [12], Pioneers 8 and 9 [13], OGO 3, and Lunar Explorer 35 are given in Table 3. The mass sensitivities stated for Mariners 2 and 4 and Pioneer 1 have been revised since original publication because of the results of hypervelocity calibration studies reported by Alexander et al. [14].

Table 3
Dust particle experiments in interplanetary, cislunar and selenocentric space

	Mass sensitivity (g)	Cumulative flux ($pm^{-2}\,s^{-1}\,\pi^{-1}\,sr^{-1}$)
OGO 3	2.0×10^{-13}	5.0×10^{-3}
Pioneer 1	1.0×10^{-11}	3.0×10^{-4}
Mariner 2	1.0×10^{-10}	6.0×10^{-6}
Mariner 4	5.0×10^{-12}	7.3×10^{-5}
Lunar Orbiter	1.0×10^{-9}	2.0×10^{-6}
Pioneers 8, 9	1.67×10^{-11}	1.0×10^{-4}
Lunar Explorer 35 (mic.) (non-shower)	5.0×10^{-12}	2.97×10^{-4}
Lunar Explorer 35 (cap.-mic. coin.)	1.0×10^{-10}	7.4×10^{-6}

More recent findings by McDonnell [15] were used to determine mass sensitivity from a quoted momentum sensitivity for Pioneers 8 and 9 [13].

For the empirical equation $\overline{\Phi} = k\, m^{-\alpha}$, where $\overline{\Phi}$ is the flux ($pm^{-2}\,s^{-1}\,\pi^{-1}\,sr^{-1}$), m is the mass (g), and k and α are constants, the logarithmic equation

$$\log \overline{\Phi} = \log k - \alpha \log m$$

is linear with slope $-\alpha$ [1]. A linear curve was fitted to data in Table 3 using linear regression curve fitting techniques; excluding the Lunar Explorer 35 data, the slope of the curve is $\alpha = 0.83$. When the Lunar Explorer 35 non-shower time data and capacitor-microphone coincidence data are included, $\alpha = 0.97$ gives the best curve fit. The two curves are shown in Fig. 2. The data point for Lunar Explorer 35 shower time is shown for comparison.

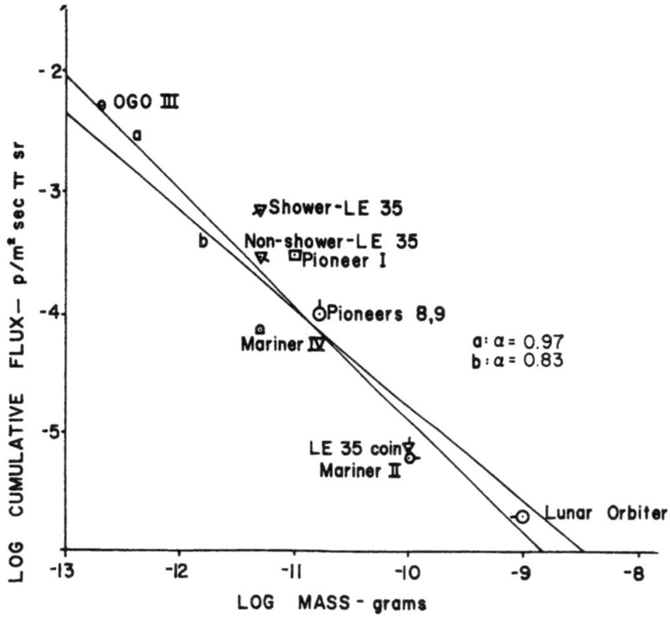

Fig. 2. Cumulative flux-mass distribution model.

At the present time, the physics of microparticle hypervelocity impact is only partially understood. The mass threshold sensitivities of the various sensors will be modified as more laboratory data on the nature of the impact phenomenon become available. Thus, the cumulative flux-mass distribution models will also reflect these changes. However, there is a high degree of consistency in the data from the various experiments listed in Table 3. The new data reported in this paper are in strong agreement not only with those previously reported for the two experiments but also with those from other experiments in interplanetary, cislunar, and selenocentric space.

Acknowledgments

The authors wish to acknowledge the cooperation of the OGO 3 and Lunar Explorer 35 project offices at Goddard Space Flight Center in supplying data tapes and information to the experimenters. The work was supported under NASA Research Grant NGR 39–012–001.

References

[1] W. M. ALEXANDER, C. W. ARTHUR, J. D. CORBIN and J. L. BOHN, in: Space Research X, North-Holland, Amsterdam 1970 (p. 252).
[2] W. M. ALEXANDER and J. L. BOHN, in: Space Research VIII, North-Holland, Amsterdam 1968 (p. 489).
[3] STAFF, Oklahoma State University Research Foundation Electronics Laboratory, Quarterly Status Reports Nos. 1 to 11 on Contract No. 19 (605) 1908.
[4] C. S. NILSSON, Science **153**, 1242 (1966).
[5] J. L. BOHN, W. M. ALEXANDER and W. F. SIMMONS, in: Space Research VIII, North-Holland, Amsterdam 1968 (p. 588).
[6] T. N. NAZAROVA, A. K. REJBAKOV and G. D. KOMISSAROV, in: Space Research VIII, North-Holland, Amsterdam 1968 (p. 496).
[7] B. P. KONSTANTINOV et al., Physical-Technical Institute of the Order of Lenin, 1968.
[8] W. M. ALEXANDER, C. W. ARTHUR and J. L. BOHN, Rev. Scient. Instrum., submitted July 1970.
[9] M. DUBIN, in: Proceedings of the First International Space Science Symposium, Nice 1960.
[10] W. M. ALEXANDER, Science **138**, No. 3545 (1962).
[11] W. M. ALEXANDER and J. L. BOHN, Science **149**, 1240 (1965).
[12] C. A. GURTLER and G. W. GREW, Science **161**, 462 (1968).
[13] O. E. BERG, H. BURTON and U. GERLOFF, Trans. Am. Geophys. Un. **51**, 339 (1970).
[14] J. L. BOHN, W. M. ALEXANDER and A. WEVER, in: Space Research VIII, North-Holland, Amsterdam 1968 (p. 596).
[15] J. A. M. MCDONNELL, private communication.

Space Research XI — Akademie-Verlag, Berlin 1971

METEOR STREAMS

B. A. LINDBLAD

Lund Observatory, Lund, Sweden

Computer techniques for the recognition of meteor streams are discussed and new photographic streams discovered by this technique are presented. Several new comet–meteor associations are presented. The possibility of extending the stream concept to groupings of asteroid orbits is discussed.

1. Introduction

A meteor stream may be defined as an assembly of interplanetary particles moving in similar heliocentric orbits. If near identity exists in all five orbital elements, the stream members cross the earth's orbit in essentially parallel paths. Owing to the perspective effect the luminous meteor trails produced by the stream members will appear to radiate from one particular point in the sky, the so-called radiant. The celestial coordinates of a stream radiant may be obtained from the intersection of a group of meteor paths plotted on a star map.

2. Visual Observations

Extensive catalogues of stream radiants have been published by Denning [1], Olivier [2], Astapovich [3], McIntosh [4], Hoffmeister [5] and others. Although most radiants listed in these catalogues probably exist, or have existed, a number of the minor stream radiants may be spurious and only represent chance intersections of unrelated meteor paths. Particularly Denning's catalogue has been criticized [6]. The visual observations have provided valuable information about radiants, hourly rates and dates of occurrence, but have yielded only limited information about the heliocentric orbits of meteors.

3. Radio Observations

Multi-station radar programs for the study of sub-visual meteors have been reported by Davies and Gill [7], Hawkins [8], Lebedinec [9] and Nilsson [10]. Although the multi-station radar studies give radiants and orbits of individual meteors, the individual orbits have, as a rule, not been published.

4. Photographic Observations

During the last 20–30 years photographic double-station programmes have provided detailed information about radiants and orbits of individual meteors. The main published catalogues are listed in Table 1. Additional, but smaller, data samples have been reported by Ceplecha [11] and others. The number of individual orbits in the photographic catalogues [12–20] is about 3600, or, if only precisely reduced orbits are considered, about 1800.

Table 1
Photographic meteor catalogues

Type of data	Station	No. of orbits	No. used in search	Reference
Small camera (bright meteors)	Mass NM	144	136	[12]
Super Schmidt	NM	413	411	[13]
Super Schmidt	NM	313*	311	[14]
Wide angle camera (very bright meteors)	Praire	100	100	[15]
Super Schmidt	NM	353	346	[16]
Small camera (bright meteors)	D	49	49	[17]
Small camera (bright meteors)	O	225	221	[18, 19]
Small camera (bright meteors)	D	253	253	[18, 19]
Total		1845	1827	
Super Schmidt	NM	1790**	–	[20]

* Meteors common to reference [13] have been exluded.
** Graphical reduction procedure. Meteors common to references [13] and [14] have been excluded.
Mass = Massachusetts; NM = New Mexico; Praire = Praire network; D = Dushanbe (Stalinabad); O = Odessa.

5. Meteor Streams

The division into stream meteors and sporadic meteors is to a certain extent arbitrary. The classifications hitherto used are based primarily on geocentric quantities such as radiants, velocities and dates of occurrence. The orbital elements are used as a secondary criterion for stream identification. The photographic studies have confirmed the major visual showers as well as several of the minor visual showers. The meteor streams detected in the Harvard programme have been summarized by Jacchia [21].

If precise orbital information is available for individual meteors, stream classifications may be based directly on the orbital elements. The use of the orbital elements has the advantage that these parameters are more fundamental, having regard to the initial development of a stream. This approach is particularly useful for detecting minor streams as only a few, well-determined, orbits are necessary for delineating a stream. Extensive stream searches based on inspection of the photographic orbit catalogues have been made by Terent'eva [22–24]. Similar searches in radio data have been reported by Nilsson [10].

6. Computer Techniques for Stream Detection

An attempt to introduce an objective method of classification made on the basis of the orbital elements and suitable for computer analysis was made by Southworth and Hawkins [25]. Applications of the Southworth criterion to photographic orbits have been reported by Lindblad [26, 27], to radio orbits by Hawkins et al. [28] and Elford et al. [29].

The stream detection method used by the Harvard group is based on the intercomparison of different sets of orbital elements. A mathematical term $D(a, b)$ is computed, which expresses the difference in orbital elements of any two given meteors A and B. These two meteors are then said to be associated if $D(a, b)$ does not exceed a stipulated value D_s. A stream is then defined as a group of meteors in which every member is associated with at least one of the other members. The rejection level D_s to be adopted in a given study varies with sample size N and the reductional errors in the sample. For photographic meteor samples Lindblad [27] recommends

$$D_s = 0.80 \times N^{-1/4} .$$

The definition of stream membership as given by the D criterion is impersonal. For well studied photographic streams the membership classification as obtained in the computer searches agrees well with that obtained by conventional stream classification techniques. The Southworth technique is, however, much more efficient than the classical method in detecting extended or dispersed streams. In addition, minor streams which may have been overlooked in previous studies, because of too few members, are effectively detected by the programme.

7. Photographic Radiants

The author has recently made a computer search in a sample of 1827 precisely reduced photographic orbits (Table 1). The purpose was twofold, firstly to obtain a list of currently active meteor stream radiants, secondly to study the orbital characteristics of the streams. The search, at $D_s = 0.13$, divided the sample into 828 sporadic meteors and 999 stream meteors; of the latter, 543 belonged to the well studied major streams and 456 belonged to various minor streams. Thus, approximately 55% of the meteor population was in streams.

Table 2 lists all those photographic radiants detected, in the search for which an identification with a previously studied visual, photographic or radio radiant appeared reasonably well established. The radiants are listed in order of right ascension. Columns 1–5 give the equatorial coordinates (epoch 1950), geocentric

Table 2
List of photographic meteor stream radiants

α_R	δ_R	V_G	Stream name	Duration	1827 precise str. no.	2401 graphical str. no.
8	63	52	Cassiopeids	8–9 Aug.	147	
9	6	24	Piscids	3 Sept.–14 Nov.	73	92
12	8	34	δ Piscids	16 Sept.–4 Oct.	68	(31)
15	– 6	18	θ Cetids 1	8–22 Oct.	42	126
28	1	18	θ Cetids 2	9–12 Nov.	61	
43	57	68	Perseids (Hyperb.)	8–12 Aug.	154	220
45	58	60	Perseids	23 July–23 Aug.	152	221
47	15	31	N Taurids S Taurids	10 Sept.–21 Dec.	52	61
51	16	17	S δ Arietids	3–19 Dec.	13	(42)
55	18	22	14 Taurids	25 Nov.–4 Dec.	39	133
72	20	12	Cyclids	Whole year	3	
73	33	69	ι Aurigids 1	27–29 Sept.	164	
91	16	44	ξ Orionids	25 Nov.–3 Dec.	126	
94	16	68	Orionids	2–29 Oct.	167	228
95	33	69	ι Aurigids 2	15–24 Oct.	165	
101	27	71	ε Geminids	16–23 Oct.	169	231
102	8	44	Monocerotids	10–15 Dec.	133	
102	29	22	ϱ Geminids	15–27 Jan.	2	90
103	28	25	N χ Orionids	20–28 Dec.	48, 64	(76)
103	41	66	Aurigids	9–14 Sept.	160	
103	41	14	ξ Orionids (Febr.)	11–24 Febr.	56	
107	14	40	S Geminids	4–13 Dec.	124	186
108	48	67	Lyncids	27–28 Sept.	158	225
111	32	37	Geminids	25 Nov.–15 Dec.	118	186
115	10	23	χ Geminids	21–23 Jan.	65	(102, 137)
126	2	60	σ Hydrids	3–15 Dec.	157	223
147	15	29	ψ Leonids	5–11 Febr.	14	
152	22	72	Leonids	15–20 Nov.	166	
160	38	62	Leo Minorids	23–24 Oct.	156	
163	1	73	τ Leonids	1–11 Dec.	168	
171	26	21	δ Leonids	1–23 March	66	21
173	60	14	ψ Ursae Majorids	9–21 April	91	160
176	24	65	Coma Berenicids	3–4 Jan.	159	
188	3	28	Virginids	5 March–1 April	60	(52, 62, 121)
192	–26	30	Hydraids	1–11 April	113	
197	– 6	20	N σ Leonids	31 March–13 May	21	8
203	–15	34	S λ Virginids	23 March–5 April	69	85
205	67	18	ω Ursae Majorids	14–22 May	122	152
207	–14	23	S σ Leonids	3–6 May	18	8
212	– 9	32	N λ Virginids	4–22 April	37	83
215	77	35	Ursids	17–22 Dec.	140	
216	36	25	γ Bootids	12–15 April	129	203
219	– 3	29	μ Virginids	13–26 April	89	144
230	49	43	Quadrantids	2–4 Jan.	144	216
231	45	17	φ Bootids	7–9 May	112	188
234	42	19	τ Herculids	1–9 June	123	168
237	–26	30	π Scorpiids	5–15 May	62	(59)
237	–14	22	θ Librids	23–31 May	34	
254	50	24	μ Draconids	11–12 June	131	
256	– 9	21	χ Scorpiids	5–23 June	51	119

α_R	δ_R	V_G	Stream name	Duration	1827 precise str. no.	2401 graphical str. no.
260	63	22	ζ Draconids	20–25 Aug.	130a	207
261	36	18	θ Herculids	18 July–8 Aug.	99	167
267	−28	29	θ Ophiuchids	4–16 June	53	98
270	−18	15	μ Ophiuchids	5–10 Aug.	31	45
273	34	48	Lyrids	21–23 April	145	217
274	−12	29	o Serpentids	11–25 June	79	146
275	0	24	η Serpentids	29 June–2 July	100	174
276	48	20	Draconids	9 Oct.	127	202
279	49	28	γ Draconids 1	23–27 July	134	
280	−19	24	μ Sagittariids	25 June–16 July	6	73
280	44	20	α Lyrids	4–13 Aug.	130	204
281	50	31	γ Draconids 2	20–27 July	136	
289	55	27	$\varkappa$ Cygnids	19–25 Aug.	132	
307	− 9	25	α Capricornids	17 July–10 Aug.	55	108
318	−17	35	S ι Aquarids 1	19–26 July	27	60
318	−10	36	N ι Aquarids 1	23–25 July	80	
327	−16	31	S ι Aquarids 2	21 July–5 Aug.	40	60
331	−15	37	S ι Aquarids 3	21 July–8 Aug.	43	(110)
335	− 5	14	Capricornids	16 Sept.–2 Oct.	28	
342	−16	42	S δ Aquarids 1	21 July–16 Aug.	119	196
344	1	19	β Piscids	10 Aug.–15 Dec.	10	(140)
344	46	47	Lacertids	6–12 Aug.	146	
346	1	41	N δ Aquarids	5–25 Aug.	108	171
351	− 1	33	N ι Aquarids 2	18 Aug.–4 Sept.	17	78
357	− 8	42	S δ Aquarids 2	12–29 Aug.	116	196

velocities (km s^{-1}), names and (extreme) dates of activity of the streams, column 6 lists the arbitrary stream number used in the present search and column 7 gives the corresponding number in a previous search [27]. It is believed that most of the streams listed are of a recurrent nature, but it should be remembered that the observational data available at present are far from comprehensive, in particular, most of the published Harvard data refer to the years 1952–1954.

Inspection of Table 2 shows that a number of streams are split up into two components symmetrical with respect to the ecliptic plane. It was somewhat unexpectedly found that the intense Geminid meteor stream has an associated southern branch (radiant 1827–124).

8. Stream Orbits

Mean orbital elements are available for the streams listed in Table 2, but are not given here. The major meteor streams are described extensively in the literature and mean orbital elements may be found in any standard textbook on meteors. Mean elements for most of the minor streams in Table 2 may be found by reference to the corresponding stream number in the 2401 meteor search [27]. These elements differ only slightly from those computed in the present sample.

A stream search divides the data sample under study into two groups: stream meteors and non-stream meteors. It is of interest to look for systematic differences in the orbital characteristics of the two populations. In this study it is expedient to exclude the major streams, since their orbital characteristics may be determined by the chance occurrence of a few, large comets. Accordingly the Taurids, α Capricornids, Andromedids, Geminids, δ Aquarids, Quadrantids, Perseids and Orionids were excluded. Figs. 1–3 show, respectively for minor streams and sporadics, the distribution of the longitude of the ascending node Ω, the argument of perihelion ω and the longitude of perihelion π. Shaded parts of the histograms represent low inclination objects ($i < 30°$ or $i \geqq 150°$). The data shown have not been corrected for observational selection.

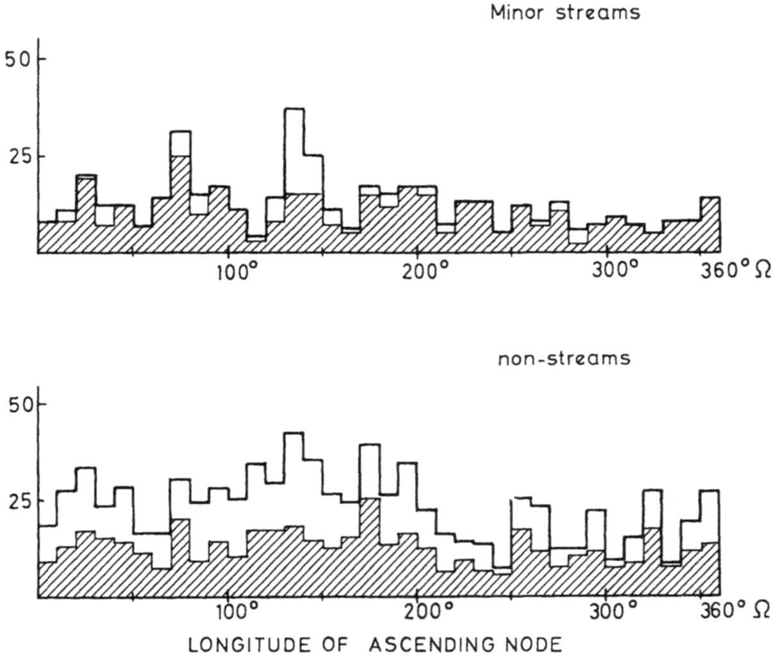

Fig. 1. Distribution of longitude of ascending node Ω in photographic meteor orbits. Shaded areas represent low inclination objects.

The distribution of Ω is similar for both minor streams and sporadics, with largest numbers occurring in the interval 0–180°. This predominance is probably caused by a combination of a seasonal observational selection and a slight tendency for the nodes to be aligned with Jupiter's orbit. The distribution of ω is similar for both streams and sporadics, with minimum number of meteors recorded at $\omega = 0°$ (or 360°). The maximum near 180° appears to be a selection effect peculiar to the northern hemisphere, in particular the non-stream sample includes many high inclination meteors observed near perihelion at their descending node.

The distribution of π appears to be random for both stream and non-stream members and for both high and low inclination objects. This is in contrast with the corresponding distribution amongst the asteroids. The distribution of π in meteor orbits has been discussed by McCrosky and Posen [20] who found

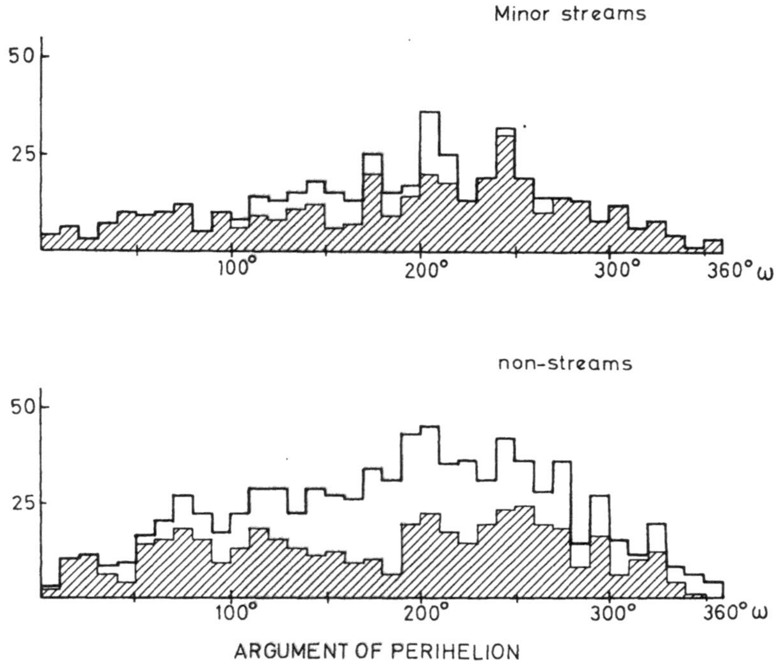

Fig. 2. Distribution of argument of perihelion ω in photographic meteor orbits.

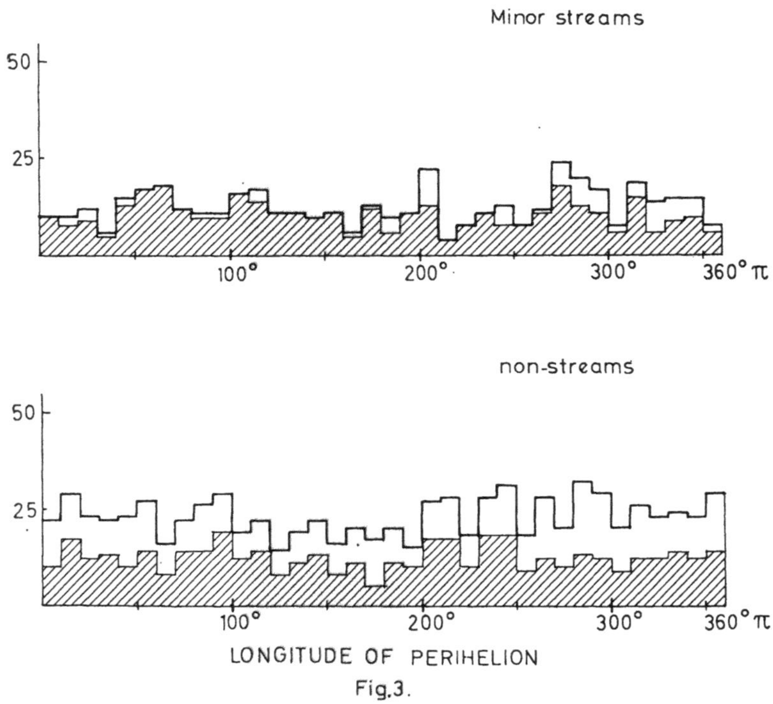

Fig. 3. Distribution of longitude of perihelion π in photographic meteor orbits.

a double-humped curve in the distribution of π. Their study differs from ours in that the major meteor streams were included. Their sample was also restricted to inclinations $i < 30°$ or $i \geqq 150°$. Our data, Fig. 3, do not show the double-humped feature reported by McCrosky and Posen, and we therefore conclude that this feature arises from the inclusion of the major meteor streams.

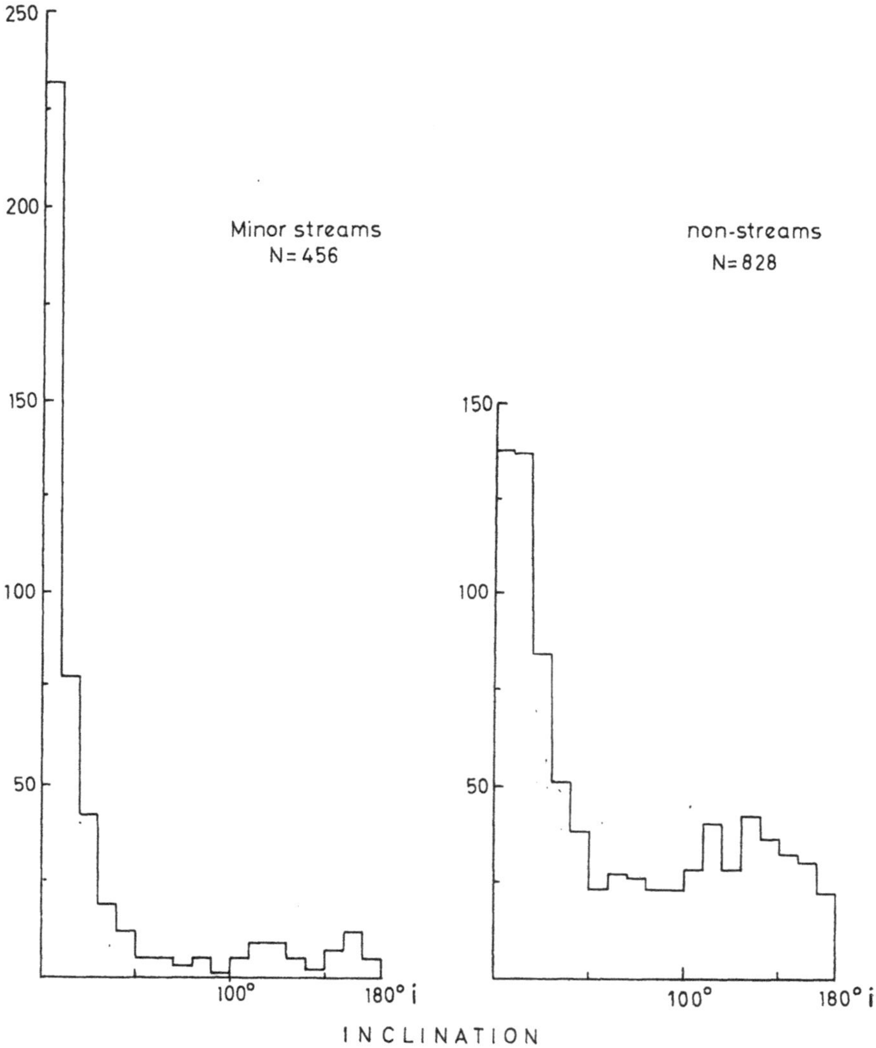

Fig. 4. Distribution of inclination i in photographic meteor orbits.

The distribution of inclination i is shown in Fig. 4. There is a significant difference between stream and non-stream members. It is seen that the inclination of the orbit plane is usually less than 30° for minor streams. In general, the minor stream orbits are similar to those of the short-period comets. Very few minor streams have retrograde orbits, which are, however, quite common amongst the non-stream meteors.

9. Comet–Meteor Associations

It is well known that when a comet orbit is interrupted by the earth, meteors have been detected which follow very nearly the same orbit as the comet. Porter [30] lists all elliptical comet orbits observed after 1700 which approach the earth's orbit to within 0.10 AU. Of eleven of these with radiants easily observable in the northern hemisphere, Porter noted that meteor showers were established in eight cases.

The streams detected in the present search were investigated for comet–meteor associations. Table 3 lists six possible associations. It is encouraging that in all cases of comet–meteor associations detected by us the distance of nearest approach Δ is small and less than 0.03 AU. If the comet–meteor associations, Table 3, are accepted, it is evident that all theoretical radiants listed by Porter as observable in the northern hemisphere have now been detected.

It is generally believed that meteor streams are formed as the result of comet disintegration. In support of this viewpoint it may be advanced that comet–meteor associations have been established for all known short-period comet orbits with $\Delta \leq 0.10$ AU. However, none of the three most intense meteor streams: the Southern δ Aquarids, the Geminids and the Daytime Arietids has an associated comet. The degree of orbital similarity within the δ Aquarid and Geminid streams speaks against an appreciable age of these streams. It is difficult to explain, firstly, how a parent comet could have moved in such a short-period orbit, and, secondly, how it could have completely disintegrated and at the same time left behind a meteor stream with very little internal dispersion among the members.

10. Asteroidal Streams

It is well known that the distribution of orbital elements among the minor planets is non-random. Hirayama [31, 32] showed the existence of families, i.e. groups of asteroids with almost similar values of the orbital elements a, e and i. Brouwer [33] later restudied the problem using the proper-elements and added several new families. Alfvén [34, 35] has drawn attention to the fact that within one of the asteroid families (Flora) there exist groups that exhibit similarity also in the orbital elements ω and Ω. Arnold [36] has recently used computer methods to detect families and streams amongst the asteroids. Danielsson [37] has studied the statistical significance of asteroidal streams.

From a purely geometrical point of view, the asteroidal streams detected by Alfvén are analogous to the meteor streams. On the time scale of the solar system the asteroidal streams must be of a recent origin. It is therefore difficult to explain them as the result of a disintegration process, such as the explosion or fragmentation of a primeval planet. Alfvén has therefore proposed that the asteroidal streams may be an intermediate product in the formation of planets. It will be interesting to see whether this accretional approach to the problem

Table 3
New comet–meteor associations

Stream/comet name	Observed/predicted date	q	a	e	i (°)	ω (°)	Ω (°)	π (°)	α (°)	δ (°)	V_G	Δ
μ Sagittarids	16 June–16 July	0.616	2.72	0.768	3.1	262.0	96.6	358.6	273	−18	22	—
Lexell (1770 I)	5 July	0.674	3.15	0.786	1.6	224.3	132.0	356.3	272	−21	21	0.015
τ Herculids	19 May–14 June	0.971	2.79	0.646	18.8	205.3	74.7	279.9	231	37	15	—
Schwassman-Wachmann (3) (1930 VI)	8 June	1.011	3.09	0.672	17.4	192.3	76.8	269.1	218	45	14	0.006
Monocerotids	10–17 Dec.	0.178	56.75	0.998	34.3	129.7	80.6	210.3	102	9	42	—
Mellish (1917 I)	15 Dec.	0.190	27.64	0.993	32.7	121.3	87.5	208.8	103	9	40	0.002
ζ Arietids	13–25 Aug.	0.973	17.90	0.945	172.5	19.5	326.0	345.5	49	14	71	—
Schmidt-Temple (1862 II)	21 Aug.	0.981	—	1.000	172.1	27.2	327.8	355.0	49	13	72	0.028
Southern α Triangulids	24 Oct.–7 Nov.	0.739	2.76	0.727	2.3	66.9	39.5	106.4	27	7	18	—
Grischow (1743 I)	12 Nov.	0.862	3.09	0.721	1.9	6.4	86.9	93.3	20	9	14	0.021
Aurigids	9–14 Sept.	0.705	−103.1	1.007	143.6	113.0	169.0	282.0	104	41	66	—
Kiess (1911 II)	31 Aug.	0.684	153.8	0.996	148.4	110.3	158.0	268.3	90	39	67	0.006

Note. Table 3 includes data from references [12]–[20]. The listed radiants therefore differ slightly from those of Table 2.

of asteroidal streams is of importance also for the formation of certain meteor streams.

Acknowledgments

I am indebted to Dr. R. Southworth for kind permission to use his computer program for stream selection and to Mrs. S. Rosenthal for help at various stages in the data reduction. This work has been partly supported by grants 204—43 and 204—44 from the Swedish Natural Science Research Council.

References

[1] W. F. Denning, Mem. Roy. Astron. Soc. **53**, 203 (1899).
[2] C. Olivier, Trans. Am. Phil. Soc., N. S. **22**, Part 1.
[3] I. S. Astapovich, Main Catalogue of 20th Century Meteor Radiants, Ashabad, Akad. Nauk Turkman SSR (1956).
[4] R. A. McIntosh, Mon. Not. Roy. Astron. Soc. **95**, 709 (1935).
[5] C. Hoffmeister, Meteorströme, Verlag Johann Ambrosius Barth, Leipzig 1948.
[6] C. Olivier, Publ. Leander McCormick Obs., II, Part 4 (1913).
[7] J. G. Davies and J. C. Gill, Mon. Not. Roy. Astron. Soc. **121**, 437 (1960).
[8] G. S. Hawkins, Smithson. Contr. Astrophys. **7**, 53 (1963).
[9] V. N. Lebedinec, Physics and Dynamics of Meteors, Reidel Publ. Co., Dordrecht 1968 (p. 241).
[10] C. S. Nilsson, Austral. J. Phys. **17**, 205 (1964).
[11] D. Ceplecha, Bull. Astron. Inst. Czech. **15**, 144 (1964).
[12] F. L. Whipple, Astron. J. **59**, 201 (1954).
[13] L. G. Jacchia and F. L. Whipple, Smithson. Contr. Astrophys. **4**, 97 (1961).
[14] G. S. Hawkins and R. B. Southworth, Smithson. Contr. Astrophys. **4**, 85 (1961).
[15] R. E. McCrosky, Physics and Dynamics of Meteors, Reidel Publ. Co., Dordrecht 1968 (p. 265).
[16] A. Posen and R. E. McCrosky, NASA Contr. Rep. CR-862 (1967).
[17] L. A. Katasev, Photographic Methods in Meteor Astronomy, Gosud. Izd. Tekhn. Teor. Lit., Moscow 1957. (Engl. trans. Monson Press, Jerusalem 1964.)
[18] P. B. Babadjanov and E. N. Kramer, Ionosphere and Meteors. Section V of IGY Program, No. 12, Moscow 1963.
[19] P. B. Babadjanov and E. N. Kramer, Smithson. Contr. Astrophys. **11**, 67 (1967).
[20] R. E. McCrosky and A. Posen, Smithson. Contr. Astrophys. **4**, 15 (1961).
[21] L. G. Jacchia, in: The Moon, Meteorites and Comets, The Solar System, Vol. 4, Univ. of Chicago Press, Chicago 1963 (p. 774).
[22] A. K. Terentéva, Rezult. Issled. MGP, Issled. Meteorov, No. 1, 62 (1966).
[23] A. K. Terentéva, Smithson. Contr. Astrophys. **11**, 109 (1967).
[24] A. K. Terentéva, Physics and Dynamics of Meteors, Reidel Publ. Co., Dordrecht 1968 (p. 408).
[25] R. B. Southworth and G. S. Hawkins, Smithson. Contr. Astrophys. **7**, 261 (1963).
[26] B. A. Lindblad, Smithson. Contr. Astrophys. **12** (in press).
[27] B. A. Lindblad, Smithson. Contr. Astrophys. **12** (in press).
[28] G. S. Hawkins, R. B. Southworth and S. Rosenthal, Harvard College Obs. Radio Meteor Project, Res. Rep. 7 (1964).
[29] W. G. Elford, G. S. Hawkins and R. B. Southworth, Harvard College Obs. Radio Meteor Project, Res. Rep. 11 (1964).
[30] J. G. Porter, Comets and Meteor Streams, Chapman Hall, London 1952 (p. 92).
[31] K. Hirayama, Japan. J. Astron. and Geophys. **5**, 137 (1928).
[32] K. Hirayama, Proc. Imp. Acad. Japan **9**, 482 (1933).
[33] D. Brouwer, Astron. J. **56**, 9 (1951).
[34] H. Alfvén, Astrophys. and Space Sci. **4**, 84 (1969).
[35] H. Alfvén, Astrophys. and Space Sci. **6**, 161 (1970).
[36] J. R. Arnold, Astron. J. **74**, 1235 (1969).
[37] L. Danielsson, Astrophys. and Space Sci. **5**, 53 (1969).

Space Research XI — Akademie-Verlag, Berlin 1971

INVESTIGATIONS OF THE METEOR INCIDENT FLUX DENSITY BY RADIO METHODS

N. S. Andrianov, O. I. Belkovich, L. B. Gussakovskaya, K. V. Kostylyov, V. V. Sidorov, D. I. Stepanov and Y. A. Pupyshev

Kazan University, Kazan, USSR

An investigation of the incident meteor flux has been carried out at Kazan University from radar and oblique scattering observations. Radar observations for a six-year period were used to obtain the seasonal variations of the incident flux over the celestial hemisphere. It was found that these were nearly the same every year. The mass exponent S was measured by three methods, the relative error being within 10% for all of them. The variations of S over the celestial hemisphere were obtained: the mean value was $S = 2.5$.

The values of the incident flux obtained from both radar and oblique scattering methods are in a good agreement, indicating a correct choice of the corresponding selecting factors.

The orbits, for meteors brighter than $+8^m$, were measured by the oblique scattering method.

The values of radii of the dust particles obtained from photoelectric twilight observations: are 1.2×10^{-5} cm at 60–70 km and 0.7×10^{-5} cm at 70–100 km. The volume density of the dust is 5×10^{-3} cm^{-3} at 70–80 km.

To investigate properties of the incident meteor flux we utilized radio reflections from meteor trains and the scattering of light from micrometeoroids in the atmosphere and from the products of meteoroid ablation during the photometric study of twilight.

During radio investigations we strove to obtain sufficient statistical data for the study of seasonal and long-term variations of the incident meteor flux. Therefore in order to investigate the density of meteor substance we employed a radar with an azimuthally rotating aerial and the statistical method of calculating the distribution of meteor density over the celestial sphere [1, 2].

Characteristics of the equipment are as follows:

pulse power 100 km,
pulse duration 10 μs,
aerial gain 32,
receiver sensitivity 17 μV.

The minimum electron line density is 1.3×10^{11} cm^{-1} which corresponds to meteors brighter than $+ 8^m$; there was a 30° aerial rotation in discrete steps every 5 minutes. The data are averaged over period of between 10 and 13 days.

To calculate the incident meteor flux a set of equations was used connecting the observed number of reflections N_n under certain observation conditions n with the meteor flux $\theta_{jn}(M_0)$ above a given mass M_0 incident in a given direction j.

$$N_n = \sum_j \theta_{jn}(M_0)\, F_{jn}\, k_n \qquad (n = 1, 2, \ldots, m) \tag{1}$$

k_n is the apparatus factor which depends on the aerial diagram and orientation; F_{jn} is the selecting factor which depends on the ionizing capability of the meteor, the initial radius and diffusion velocity and the parameter S for the distribution of meteor particle masses. We have more equations than observations made

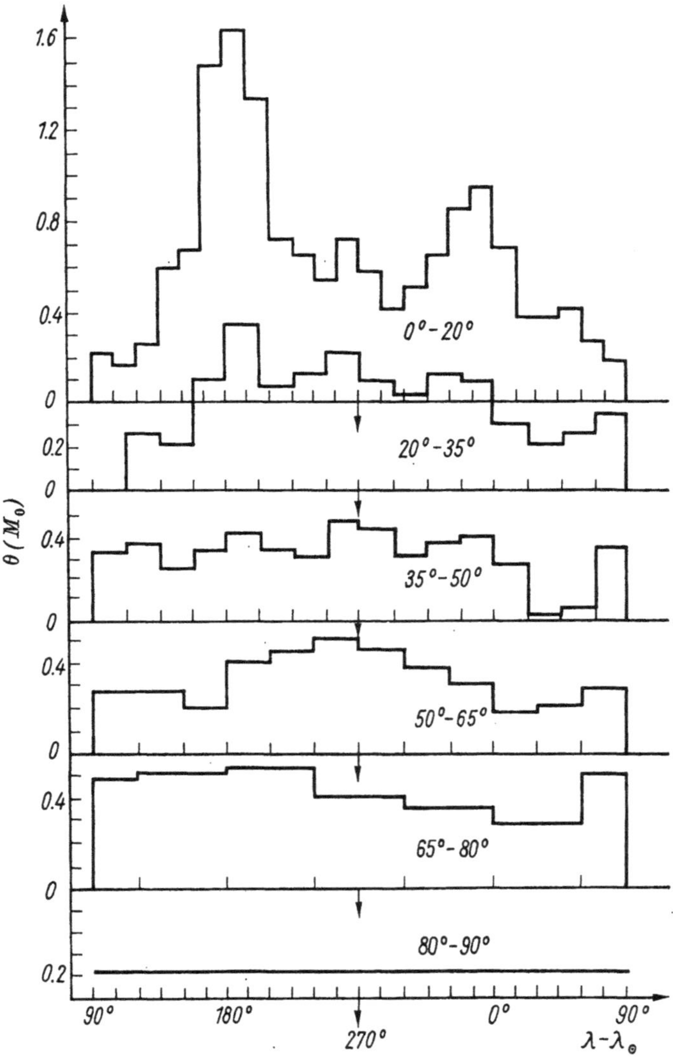

Fig. 1. The distribution of the mean annual flux $\bar{\theta}(M_0)$ over the celestial sphere for various intervals of ecliptic latitude, β. λ is the ecliptic longitude (the arrow indicates the longitude of the apex).

under different conditions (at different times of the day and different azimuths), in our case $m = 288$. The equations (1) were solved by a computer in respect to the flux terms $\theta_{jn}(M_0)$, the number of which was determined by the number of sections over the celestial sphere.

The mean annual distribution for the northern hemisphere of incident meteor flux with mass greater than $M_0 = 7 \times 10^{-5}$ g for various ecliptic latitudes β is given in Fig. 1. The flux distribution $\theta(M_0)$ is uneven and is approximately twice as great in the area with $\varepsilon < 90°$ than in the anti-apex. In the region with β from 0 to 35° solar, apex and anti-apex maxima appear. In latitudes $\beta > 35°$ the distribution $\theta(M_0)$ is more even. When comparing this distribution with similar ones for individual months it can be seen that the average annual distribution does not reflect the monthly variations. The flux density gradient $\theta(M_0)$ for various regions of the celestial sphere can reach 10 during some months, particularly in the ecliptic plane during the summer months.

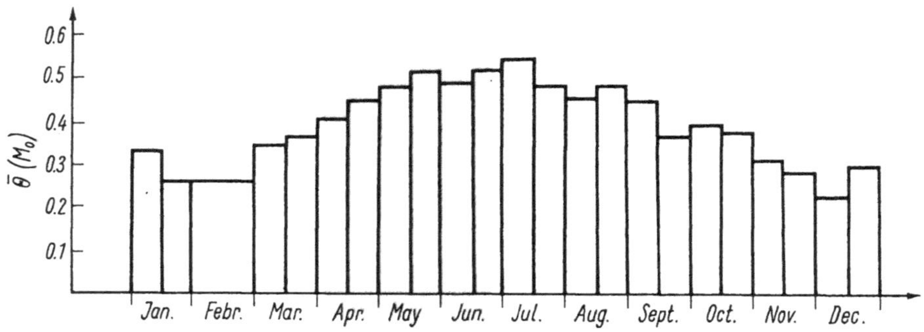

Fig. 2. The seasonal variation of the mean flux $\bar{\theta}(M_0)$ km^{-2} h^{-1} sr^{-1} (1964–1966).

The seasonal variations in the density of the meteor flux $\theta(M_0)$ calculated by integrating fortnightly distributions over the northern hemisphere are shown in Fig. 2. It is noteworthy that the seasonal flux variation $\theta(M_0)$ is observed at all latitudes, although in the celestial pole direction it is somewhat less than in the ecliptic plane. Seasonal variations and certain details of incident meteor flux recur annually with a mean correlation coefficient for fortnightly distributions of 0.86.

Calculation of the dispersion of the observed meteor numbers shows that it is roughly equal to 1.5 times the mean hourly number of meteors in March, and 3 times in September. The mean incident meteor flux density with $M_0 \geqq 7 \times \times 10^{-5}$ in the northern hemisphere varies from 0.9×10^{-10} m^{-2} s^{-1} sr^{-1} to $1.4 \times \times 10^{-10}$ m^{-2} s^{-1} sr^{-1}, the mean annual flux magnitude being 1.25×10^{-10} m^{-2} s^{-1} sr^{-1}.

To compare the experimental results $\theta(M_0)$ for different mass ranges one must know with sufficient accuracy the parameter S of meteor mass distributions: $\theta(M) \sim c\, M_0^{1-S}$.

The relative error in the determination of the flux of meteors with a mass greater than M, from the flux with a mass greater than M_0 is

$$\frac{\sigma\,\theta\,(M_1)}{\theta\,(M_0)} = 2.3 \left|\lg \frac{M_1}{M_0}\right| \sigma(S) \tag{2}$$

proves to be 100% at $(M_0/M_1) = 100$ and $\sigma(S) = 0.2$. To make the measurement of S more precise we made use of radar data for meteor showers and determined S for one and the same observation period by three different methods: (i) by the distribution of the reflection amplitudes of under-dense meteor trains

$$F_1(A) \approx A^{-k_1}, \qquad k_1 = \frac{s-1}{\gamma}. \tag{3}$$

(γ is the function of the initial train radius at a characteristic altitude) and delays in the formation and detection of the amplitude maximum; (ii) by the distribution of the reflection amplitudes of over-dense meteor trains

$$F_2(A) \sim A^{-k_2}, \qquad k_2 = 3.5\,(S-1) + 0.08\ ; \tag{4}$$

(iii) by the distributions of the reflection duration of over-dense trains

$$F(T) \sim T^{\frac{3}{4}(S-1)}. \tag{5}$$

Meteor showers were used in this analysis so that we could be sure of the constancy of S for a wide mass range (the three methods differ in threshold sensitivity). We must also distinguish between meteor showers and the sporadic background. The comparison of S calculated by the three methods for Geminids 1964 is given in Table 1. The mean values of the ratios S_1/S_2 and S_3/S_2 are equal

Table 1

Date	Observ. time	S_1	S_2	S_3
11–12 Dec.	21–24 h	2.11	2.10	2.16
	00–04 h	2.04	2.01	1.93
	04–06 h	1.60	1.63	1.87
12–13 Dec.	21–24 h	2.11	1.81	1.80
	00–04 h	1.91	1.79	1.76
	04–06 h	1.93	1.90	1.60
r.m.s.deviation		0.03	0.03	0.04

to 1.04 and 0.99 and the g.m.s. deviations of these ratios are 0.07 and 0.10, hence the radio echo methods of the determination of S have about 10% accuracy.

If we assume that the sporadic background is the sum of a great number of elementary showers we can expect variation in the value of S for different regions of the sky. Using a method similar to the one employed for the determination of the incident meteor flux, we succeeded in obtaining the distribution of S over the celestial sphere [2]. While calculating S, corrections for γ (see [3]) were taken into account. The tendency of S to increase in the apex direction is evident. The mean value of S in the meteor mass range from 10^{-3} to 10^{-5} g was $S = 2.5$. The same value of $S = 2.57$ was calculated from the photometric mass distribution obtained by photoelectric observations. It is this value that was used in the calculation of F_{in} in [1].

S tends to increase with decrease of M_0, which in accordance with Whipple [3] can be approximated by the formula

$$S = -\,0.05 \lg M_0 + 2.25\,, \tag{6}$$

where M_0 is given in grammes.

We obtain $\theta(M_0) = 3.6 \times 10^{-4}\,\mathrm{m}^{-2}\,\mathrm{s}^{-1}/2\pi$, from our data for $S = 2.6$ and $M_0 = 10^{-8}$, whereas the Explorer 8 data yield $7 \times 10^{-4}\,\mathrm{m}^{-2}\,\mathrm{s}^{-1}/2\pi$.

The lack of certainty about the reliability of the data concerning faint fast meteors under conditions of high velocity selectivity during radio echo observations made us turn to the oblique scattering method for the measurement of meteor orbits.

The increase in the effective wavelength resulted in the fact that the maximum of meteor detectability in terms of velocity shifted from 34 km s^{-1} to 70 km s^{-1} [4] and the sensitivity of the radio echo method to faint fast meteors increased.

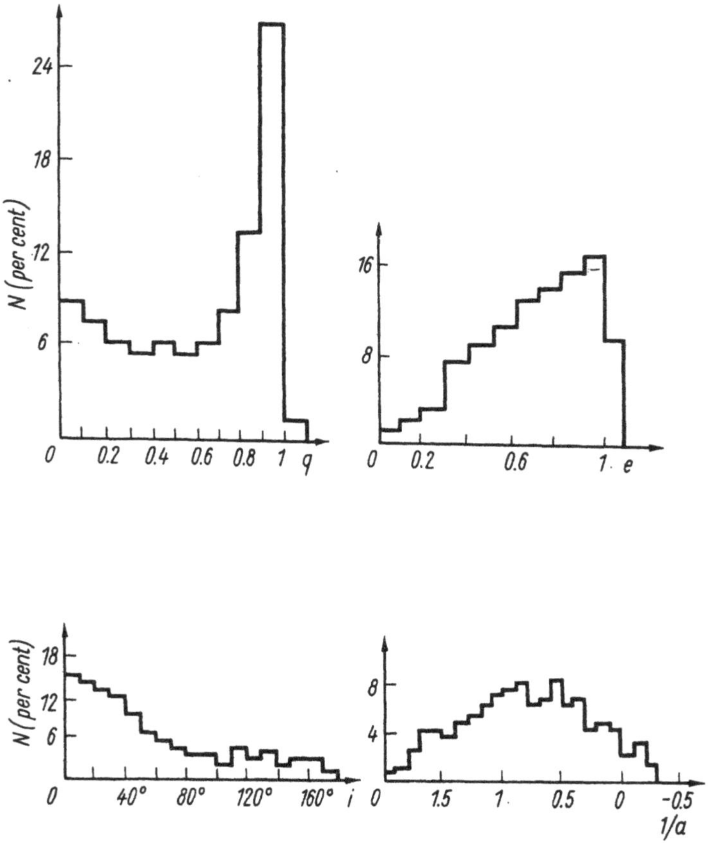

Fig. 3. The distribution of q, e, i and $1/a$.

In 1965 applying the oblique scattering to the measurement of various parameter of individual meteors' orbits, we also hoped to verify the selectivity corrections F_{jn} used in the determination of the incident meteor flux.

Our velocity distributions for fixed elongation angles close to the apex have two maxima whose positions are displaced towards lower velocities with the increase of elongation angle.

The analysis of meteor orbital element distributions (see Fig. 3) obtained by this method showed that the distribution of major semi-axes is characterized by a greater number of short period orbits, as compared with other authors' data (39% orbits with $a \leqq 1$ AU). A considerable number of orbits is not situated in the plane of the ecliptic; only 46% of orbits have inclinations $i \leqq 30°$. Orbits with eccentricity $e < 0.6$ accounted for 48%.

It can be seen that the correction for velocity selection increased the number of orbits in the interval $0.6 \leqq e \leqq 0.8$ whose role in the corresponding distributions of photometeors is important. The distribution of perihelion distances has a maximum in the value interval $0.9 \leqq q \leqq 1$ and a small increase in the region of small q which distinguishes it from a similar one found by Kasceev and Lebedinets [5] and a monotonically increasing one found by McCrosky and Posen [6].

On the whole our orbital studies confirm the previously established [5] general regularities in the distribution of orbital parameters of meteors of different mass ranges. The fact that there were again limitations in sensitivity towards faint fast meteors did not result in considerable variations in the character of the distributions of radiometeor orbit parameters. Good observations of the sky characteristic of oblique scattering permitted us to compare the results of the distribution variations of incident meteor fluxes over the celestial sphere obtained by the two methods. Table 2 gives, for different ecliptic latitude

Table 2

β(°)	65–80	50–65	35–50	20–35	0–20
Azimuthal radiants (%) (Kazan)	100	56	63	49	60
Individual radiants					
Oblique scatter (%) (Kazan)	110	47	46	40	47
Back scatter (%) (Kharkov)	60	34	20	20	15

intervals, the ratio n (in per cent) of mean flux values $\theta(M_0)$ in the anti-apex direction ($\varepsilon \geqq 90°$) and corresponding values in the apex direction ($\varepsilon < 90°$) for "azimuthal" and "individual" radiants in Kazan and "individual" radiants in Kharkov [5]. We can see from the table that there is good agreement between the results obtained by two methods used in Kazan.

The consideration of physical processes taking place during the entry of a meteor particle into the atmosphere [7] shows that, depending on the initial velocities and masses of the meteor particles, they can be either: (i) decelerated in the atmosphere without ablation, or (ii) ablated completely, leaving the products of ablation in the atmosphere, or (iii) partially ablated leaving decelerated bodies in the atmosphere.

In view of this one can expect the presence of three corresponding aerosol layers. In order to establish the way in which these aerosols are distributed with respect to altitude particle size we undertook an investigation of the dust component of the atmosphere by the twilight method [8]. Colour (in three spectral sections 4700 Å, 5220 Å and 5800 Å) and polarization electrophotometric observations of twilight were carried out simultaneously in the zenith, for the celestial pole and in the solar vertical. Having distinguished the first-order twilight brightness we employing dispersion coefficients obtained by the V. G. Fessenkov method, and using the Mie theory. Assuming the particles to be spherical we conclude that aerosol size is weakly affected by altitude, ranging from 2.2 to 2.5×10^{-5} cm. Twilight observations in the presence of strong meteor showers detected a depolarization effect at altitudes of 70–80 km, increasing with increase of the mean meteor number.

The general analysis of the polarization index distribution with altitude (240 polarization curves) revealed three well-defined layers in the altitude intervals 40–60 km, 60–90 km and 90–120 km.

The minima are stable and recur for different observation dates, the decrease in polarization being accompanied by the "reddening" of twilight.

During the activity of the Lyrids shower we succeeded in determining the rate of precipitation of the aerosols introduced by the meteor shower from the variations in the optical density gradient calculated from the twilight brightness curves. In the altitude interval 100–70 km it corresponds to precipitation of particles of radius 7×10^{-6} cm, and in the interval 70–60 km to particles of 1.2×10^{-5} cm. These estimates are close to those obtained from the dispersion coefficient.

The small variability of the mean particle size made it possible to estimate the density distribution of particles with altitude. Three highly dusty layers were discovered of which the 70–80 km layer was most clearly defined, where particle density reaches 5×10^{-3} cm^{-3}.

References

[1] Yu. A. Pupyshev, Meteorn. Rasprostr. Radiovoln **3–4**, 5 (Kazan 1966).
[2] O. I. Belkovich, Yu. A. Pupyshev and V. S. Tereskina, Meteorn. Rasprostr. Radiovoln **5–6**, 20 (Kazan 1969).
[3] F. L. Whipple, On maintaining the meteoritic complex, Smithson. Astrophys. Obs. Spec. Rep. (1967).
[4] N. S. Andrianov, Meteorn. Rasprostr. Radiovoln. **5–6**, 41 (Kazan 1968).
[5] V. N. Lebedinets and B. L. Kaseev, Astron. Zh. **4–3**, 854 (1966).
[6] R. E. McCrosky and A. Posen, Smithson. Contr. Astron. **1**, 4 (1961).
[7] K. V. Kostylyov, Meteorn. Rasprostr. Radiovoln **5–6**, 138 (Kazan 1969).
[8] D. I. Stepanov and L. B. Gussakovskaya, Meteorn. Rasprostr. Radiovoln **2**, 146 (Kazan 1964).

Space Research XI — Akademie-Verlag, Berlin 1971

RADAR METEOR INFLUX AND ITS COMPARISON WITH DIRECT COSMIC DUST MEASUREMENT DATA

V. N. Lebedinets

USSR

Meteoroid space density in the vicinity of the earth's orbit is determined from results of radar meteor observations. Various radar observation selectivity factors are taken into account: influence of diffusion and the initial radius of the meteor trail on the radio echo amplitude, velocity dependence of ionization probability, deviation of ionization curves and faint meteor heights from the values given by the simpliest physical theory of meteors.

Values of meteoroid space density obtained are in good agreement with direct cosmic dust measurements, if exponents 2–2.3 are adopted for the power of the meteoroid mass distribution.

1. Introduction

In recent years widely disparate data on the density of cosmic dust in the vicinity of the earth have been obtained by different authors from artificial satellite measurements [1–6]. This enhances the role of indirect methods of investigation of solid particles in the interplanetary medium, and specifically the radar meteor observation method. Radar meteor observations from the standpoint of space research have great interest because particles with masses $M \gtrsim 10^{-4}$ g can be really dangerous for spaceships [7].

Radar meteor investigation results as well as results of observations made by other methods are distorted by observational selectivity. In order to convert measured parameters to real ones for the whole complex of meteoroids (with masses M greater than some minimum value M_0) which met the earth it is necessary to take into account the different detectability of meteors with various velocities and radiants. In [8,9] it was shown that meteor detectability may be represented as a product of a geometry factor $1/P_1$ and a physical factor $1/P_2$. The geometry factor defines the relative detectability of meteors for varying radiant position; the physical factor defines the relative detectability of meteors for varying velocity.

The calculation of the geometry factor is comparatively simple — it is enough to know the antenna pattern. Various methods have been proposed for calculating P_1 [9–11] and for its determination from the observational data [8] which lead to a satisfactory agreement between the results obtained.

The calculation of the physical factor is more complicated. Different detectability of meteors with various velocities is defined mainly by the ionization probability dependence β on v, by the influence of the ionized meteor trail of initial radius r_0 on the radio echo amplitude, by the dependence of r_0 on the height and meteor velocity, by the dependence of the height and form of the meteor ionization curve on M and v, by the influence of meteor trail diffusion on the radio echo amplitude. In [9] we proposed a method of calculating P_2 taking into account all these factors and also a method of calculating meteoroid influx from the data of the radar meteor shower rate observations.

2. Meteor Showers

The flux of meteoroids with masses $M > M_0$ can be found from the formula [9, 12]:

$$\Phi(M_0) = \frac{N}{S_{\text{eff}}(A, z, v_0) \cos z}, \tag{1}$$

where N is the rate of the shower radiometeors, A, z are the azimuth and zenith angle of the shower radiant, v_0 is the extra-atmospheric velocity; S_{eff} is the effective sampling surface of the radar for shower meteors with given A, z, v_0. The meteoroid space density can be found from the formula:

$$D(M_0) = \Phi(M_0)\, v_g^{-1} \left(1 - \frac{g(R_0 + h)}{v_g^2}\right]^{-1}, \tag{2}$$

where v_g is the geocentric velocity, g the gravitational acceleration, R_0 the earth's radius, h the average meteor height.

To determine $\Phi(M_0)$ we use the radar meteor rate data obtained in Kharkov [13]. The radar parameters were: wavelength 8 m, transmitter power 75 kW, pulse duration 10 s, pulse frequency 500 s^{-1}, receiver sensitivity 2×10^{-14} W;

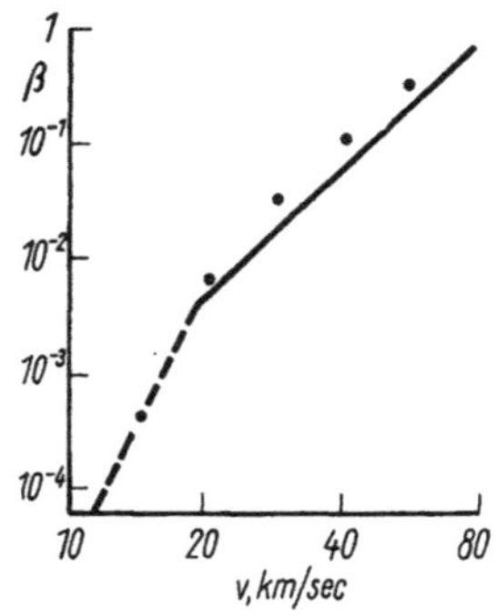

Fig. 1

transmitting and receiving antenna gains are 16. For this equipment we calculated S_{eff} and also the effective limit of meteoroid mass M_0 at various values of A, z and v_0.

The dependence of ionization probability β on the velocity assumed by us is shown as a curve in Fig. 1. Analysing all earlier published data on the effective ionization cross section for collision of different pairs of atoms and molecules we determined in [12] that in the velocity interval 20–80 km s^{-1} the dependence of β on v can be represented by:

$$\beta = \beta_0 v^{7/2}, \tag{3}$$

where $\beta_0 = 4\times10^{-25}$ s$^{7/2}$ cm$^{-7/2}$. An average atom weight for meteoric matter is assumed to be 23. At $v < 20$ km s^{-1} β must decrease rapidly with decreasing v since for the greater number of evaporated meteor atoms their energy of motion relative to atmospheric molecules becomes less than the ionization potential. Practically nobody has carried out ionization cross section measurements at $v < 20$ km s^{-1}. On the basis of his semi-empirical theory Öpik [14, 15] calculated values of β for stone meteors through the whole interval of meteor velocities; values of β for stone meteors are marked by dots in Fig. 1. Since in the interval v 20–80 km s^{-1} Öpik's data agree well with our results we used the value $\beta = 4\times10^{-4}$ which was found by Öpik at $v = 14.8$ km s^{-1} for the extrapolation of our curve $\beta(v)$ to $v < 20$ km s^{-1}.

From the data of simultaneous optical and radar meteor observations [16–18] the relation between the linear trail electron density α and the absolute visual magnitude of the meteor m is:

$$m = 35.10 - 2.5 \lg \alpha, \tag{4}$$

where α is in cm^{-1}.

The heights and ionization curves of faint meteors were determined by Lebedinets and Shushkova [19] taking into account the loss energy on thermal radiation. It was assumed that the initial radius of the ionized meteor trail r_0 is proportional to the meteor velocity and inversely proportional to the atmospheric density; at the height of 96 km and $v = 40$ km s^{-1}, $r_0 = 1$ m [12, 20, 21]. The coefficient of radio reflection from meteor trails obtained by Lebedinets and Sosnova was used [22]. The power meteoroid mass distribution was taken as

$$dN \sim M^{-S}\, dM \tag{5}$$

with $s = 2$.

Taking into account the sporadic background we succeeded in determining with great precision the radar meteor rate for Quadrantids, Arietids, Southern δ-Aquarids, Geminids and association No. 16 [23]; a very high activity of the Lyrid shower was observed in 1958. Table 1 shows values $\Phi(M_0)$ and $D(M_0)$ at $M_0 = 2\times10^{-4}$ g for maximum periods of these showers and also the number of particles falling to the earth for a day $\theta(M_0)$.

Let us compare the meteoroid space density obtained by us with other authors' radar results (Tables 2—5). The meteoroid space density in showers over a wide

Table 1

Shower	v_0 km s^{-1}	$\Phi\,(2\times10^{-4}$ g) km^{-2} s^{-1}	$D\,(2\times10^{-4}$ g) km^{-3}	$\theta\,(2\times10^{-4}$ g) day^{-1}
Quadrantids	43	$2.6 \cdot 10^{-5}$	$6.0 \cdot 10^{-7}$	$3.2 \cdot 10^{8}$
No. 16	37	$2.3 \cdot 10^{-5}$	$6.1 \cdot 10^{-7}$	$2.9 \cdot 10^{8}$
Lyrids	48	$4.8 \cdot 10^{-5}$	$1.0 \cdot 10^{-6}$	$5.7 \cdot 10^{8}$
Arietids	39	$5.4 \cdot 10^{-5}$	$1.4 \cdot 10^{-6}$	$6.7 \cdot 10^{8}$
Southern δ-Aquarids	42	$7.5 \cdot 10^{-5}$	$1.8 \cdot 10^{-6}$	$9.2 \cdot 10^{8}$
Geminids	36	$3.4 \cdot 10^{-5}$	$9.2 \cdot 10^{-7}$	$4.3 \cdot 10^{8}$

interval of radar meteor magnitude is given in [24–26]. Weiss [27] indicates value of parameters s accepted by him for different showers and this allow us to reduce his data to a limiting value $M_0 = 2\times10^{-4}$ g. The results of all authors are reduced on the meteor mass scale accepted by us.

Table 2
Quadrantids

Author	Method	$\Phi\,(2\times10^{-4}$ g) ($km^{-2}\,s^{-1}$)	$D\,(2\times10^{-4}$ g) (km^{-3})
Lebedinets	radio	$2.6\cdot10^{-5}$	$6.0\cdot10^{7}$
Weiss [27]	radio	$1.9\cdot10^{-6}$	
Levin [28]	vis		$3.7\cdot10^{-7}$

Table 3
Arietids

Author	Method	$\Phi\,(2\times10^{-5}$ g) ($km^{-2}\,s^{-1}$)
Lebedinets	radio	$5.4\cdot10^{-5}$
Weiss [27]	radio	$9.3\cdot10^{-6}$
Kaiser [25]	radio	$7.5\cdot10^{-7}$
Davies [24]	radio	$7.0\cdot10^{-6}$

Table 4
Geminids

Author	Method	$\Phi\,(2\times10^{-4}$ g) ($km^{-2}\,s^{-1}$)	$D\,(2\times10^{-4}$ g) (km^{-3})
Lebedinets	radio	$3.4\cdot10^{-5}$	$9.2\cdot10^{-7}$
Weiss [27]	radio	$3.8\cdot10^{-6}$	
Kaiser [25]	radio	$4.0\cdot10^{-6}$	
Davies [24]	radio	$3.6\cdot10^{-6}$	
Webster et al. [26]	radio	$2.0\cdot10^{-6}$	
Levin [28]	vis		$1.1\cdot10^{-6}$

Table 5
Southern δ-Aquarids

Author	Method	$\Phi\,(2\times10^{-4}$ g) ($km^{-2}\,s^{-1}$)
Lebedinets	radio	$7.5\cdot10^{-5}$
Weiss [27]	radio	$5.0\cdot10^{-6}$

From Tables 2–5 one can see that the meteoroid space density obtained by us is about one order higher than other authors' data based on radar meteor observations. Such great divergence can be explained apparently only by the influence of the initial radius of the ionized meteor trail on meteor detectability which was taken into account only by us. To define the accuracy of estimation of this factor in our work we will compare our results with visual observation data (the initial trail radius does not influence meteor detectability in optical observations).

Taking into account the meteor detectability at different magnitude and the dependence of luminosity coefficient on velocity Levin [28] from visual observation data estimated the space density of meteoroids with masses $M > M_{4.3}$ (where $M_{4.3}$ is the initial mass of meteor $+ 4.3^m$ with velocity 40 km s^{-1}) for different showers: Quadrantids $D(M_{4.3}) = 4.4 \times 10^{-8}$ km^{-3}, Geminids $D(M_{4.3}) = 1.32 \times 10^{-7}$ km^{-3}. In the mass scale accepted by us $M_{4.3} = 1.7 \times 10^{-3}$ g. In Tables 2 and 4 values of $D(M_0)$ obtained by Levin are reduced on the scale $M_0 = 2 \times 10^{-4}$ g; for both showers s is assumed to be 2. Values of space density of meteor showers obtained by us are in good agreement with visual observation data.

3. Sporadic Meteors

Now let us estimate the mean integrated space density of all meteoroids (sporadic and shower) in the vicinity of the earth's orbit. For this in the addition to the numbers of radio meteors it is necessary to know radiant and velocity distribution of the meteors. For this purpose we used the results of radiant and velocity radar measurements in Kharkov in 1960 on which the catalogue of 12500 radar meteor orbits were calculated. For all these meteors with radiants in the northern hemisphere we calculated the mean values of S_{eff} and M_0. The yearly average diurnal number of radar meteors was 8400 in 1960; knowing the radiant distribution we found 7250 meteor radiants to be situated to the north from the ecliptic and 1150 meteor radiants to the south. Assuming that during a year both hemispheres have the same number of meteors we deduced that $\theta = 5 \times 10^9$ particles with masses $M > 2 \times 10^{-4}$ g fall to the earth in a day. Knowing meteor velocity distribution we found $\Phi\,(2 \times 10^{-4}$ g$) = 1.1 \times 10^{-4}$ km^{-2} s^{-1} and $D\,(2 \times 10^{-4}$ g$) = 10^{-5}$ km^{-3}. By comparison with Table 1 it is seen that

Table 6
Sporadic meteors

Author	Method	$\Phi\,(2 \times 10^{-4}$ g) (km^{-2} s^{-1})	$D\,(2 \times 10^{-4}$ g) (km^{-3})	$\theta\,(2 \times 10^{-4})$ (day)
Lebedinets	radio	$1.1 \cdot 10^{-4}$	$1.0 \cdot 10^{-5}$	$5 \cdot 10^9$
Kaiser [25]	radio	$1.4 \cdot 10^{-6}$		
Davies [24]	radio	$1.4 \cdot 10^{-6}$		
Weiss [27]	radio	$1.9 \cdot 10^{-6}$		
Kruchinenko et al. [29]	radio			$1 \cdot 10^8$
Baldwin [31]	radio	$2.6 \cdot 10^{-6}$		
Elford [30]				
$S = 2.0$	radio	$2.5 \cdot 10^{-5}$		
$S = 2.34$	radio	$3 \cdot 10^{-6}$		
Nilsson [32]	radio	$3.5 \cdot 10^{-5}$		
Levin [28]				
$S = 2.0$	vis		$9.4 \cdot 10^{-6}$	
$S = 2.34$	vis		$1.9 \cdot 10^{-5}$	
Baharev [33]	tele			$2 \cdot 10^9$
Hawkins [34]	photo	$1.4 \cdot 10^{-4}$		

the meteoroid space density in meteor streams giving the greatest radar meteor activity is an order less than the mean density of sporadic particles.

Let us compare space meteoroid density with other authors' data (Table 6). Besides our evaluation [9] the meteoroid space density was determined from radar observational data by a number of other authors [24, 25, 27, 29–31]. In Table 6 all the data were reduced to our mass scale with limiting mass value $M_0 = 2 \times 10^{-4}$ g; values of s accepted by authors themselves at the processing of their observations were used (two limiting values $s = 2.0$ and 2.3 for sporadic meteors are pointed out in [30]; while calculating detectability it was pointed out that s is assumed equal to 2.34; in Table 6 the reduction results on $M_0 = 2 \times 10^{-4}$ g at two values $s = 2.0$ and 2.34 are shown).

The evaluation of meteoroid influx from optical meteor observations is also shown in Table 6. Analysing the results of numerous visual meteor observations Levin [28] obtained for sporadic meteors $D(M_{4.3}) = 1.1 \times 10^{-6}$ km^{-3}; Table 6 presents the extrapolation of Levin's results for $s = 2.34$. From telescopic observations of meteors $+\,7^{m} - +\,11^{m}$ Bakharev [33] determined the number of meteors brighter than the limiting magnitude m: $\lg \theta(m_0) = 6.066 + 0.434\, m_0$ reaching the earth during a day. Hawkins and Upton [34] estimated the meteoroid space density from Super-Schmidt meteor observation data. The dependence $\Phi(M_0)$ obtained by them is in our mass scale: $\lg \Phi(M_0) = -\,8.81 - 1.34 \lg M_0$. In Table 6 this dependence is extrapolated by us for $M_0 = 2 \times 10^{-4}$ g.

From Table 6 it is seen that the sporadic meteoroid space density obtained by us on radar observational data is in satisfactory agreement with the results of photographic, visual and telescopic meteor observations; the evaluations made by almost all the authors on radar meteor observations are less than our evaluations by one to two orders. As was indicated above, this fact is mainly explained by the influence of the initial radius on meteor detectability which has been taken into account only by us. In this connection it is surprising that there is a small discrepancy between our evaluation and the results of Nilsson and Southworth's extrapolation to large meteoroid masses if in their data reduction one takes the value $s = 2.05$ obtained by them. We will return to this result later.

4. Discussion

Both for sporadic meteors and for showers we have the same results: our evaluation of the radar meteor flux is in satisfactory agreement with optical meteor observations and is essentially larger than the radar data of other authors who do not take into account the influence of the initial radius of ionized meteor trails on radar meteor detectability. Therefore apparently one can deduce that on the basis of the correctness of our mass scale the evaluation of particle flux with masses $M > 2 \times 10^{-4}$ g, $\Phi\,(2 \times 10^{-4}\,\text{g}) = 1.1 \times 10^{-4}$ km^{-2} s^{-1} is trustworthy. However, two questions are unsettled: (i) is the mass scale assumed by us correct?; (ii) what value of the parameter s should we use in our data reduction for small and large masses? Let us discuss these questions in detail.

4.1. Meteor Mass Scale

Verniani and Hawkins [35] present a summary of ionization probability values (i.e. radar meteor mass scale) accepted by different authors. Table 7 shows this summary with the addition of data from other authors. Values of β at

Table 7

Authors	Year	β ($v = 40$ km s^{-1})	k
Herlofson [36]	1948	0.01	
McKinley [37]	1951		6
Greenhow and Hawkins [38]	1952	0.2	
Kaiser [39]	1953	$0.1 + 1.0$	0 ± 0.6
Evans and Hall [40]	1955	0.1	0.5 ± 0.5
Whipple [41]	1955		5
Hawkins [42]	1956		5.6
Millman [43]	1956		4
Opik [15] $20 < v < 80$ km s^{-1}	1958	0.095	3.6
Davies et al. [44]	1959		3 ± 1
Davies and Gill [8]	1960		0
McKinley [45]	1961		5
Davies and Hall [46]	1963	0.03 $v = 32$ km s^{-1}	
Lazarus and Hawkins [47]	1963	0.0004	3.4
Lebedinets [9] $20 < v < 70$ km s^{-1}	1963	0.08	3
Verniani [48]	1964	0.01 $v = 32$ km s^{-1}	
Verniani and Hawkins [35]	1964	0.026	4
Frichtenicht et al. [49]	1968	0.36	4
Lebedinets [12] $20 < v < 80$ km s^{-1}	1966	0.05	3.5

$v = 40$ km s^{-1} accepted by different authors, and the exponent k in the power dependence $\beta = \beta_0 v^k$ accepted by the majority of authors are presented. From Table 7 it is seen that both the value of β at the average meteor velocity $v = 40$ km s^{-1} and the exponent k vary over a very wide interval: β ($v = 40$ km s^{-1}) from 1 [39] to 0.0004 [47]; k 0 [39] to 6 [37].

However, in discussion of the results shown in Table 7 one must remember that any detail calculations of β based either on the theory of ionization by collision of atoms and molecules or on results of laboratory experiments were carried out only by Öpik [15], Lazarus and Hawkins [47], Frichtenicht et al. [49] and Lebedinets [12]. The works of Öpik [15] and Lebedinets [12] have been discussed earlier. Let us consider the other two.

Lazarus and Hawkins from theoretical calculations obtained the dependence of the effective ionization cross section Q_i on the velocity for Na and O collisions which is in satisfactory agreement with the results obtained by Bydin and Buhteev [50] for Na and O_2 collisions. However, the effective cross section of diffusion Q_d calculated by them is incorrect. Calculating Q_d Lazarus and Hawkins used the first Born's approximation, although in the same work they showed that Born's approximation is inapplicable in the meteor velocity interval. As a result of this inaccuracy they obtained, at $v = 40$ km s^{-1}, $Q_d = 10^{-14}$ cm^2, which is even larger than the cross section in gas-kinetic conditions when Q_d must decrease in inverse proportion to the velocity of the colliding particles [12, 23]. The value of Q_d is over-estimated by Lazarus and Hawkins by more than an order; just this error leads to an under-estimate of β for Na of more than an order. For estimating the value of β for other meteor atoms Lazarus and Hawkins used an unsubstantiated formula:

$$\beta \sim \frac{n\,\mu}{E_i}, \tag{6}$$

where n is the valency of the atom, μ is the molecular weight and E_i the ionization potential. This formula contradicts the results of laboratory measurements of the effective ionization cross section for different pairs of atoms and molecules [50–53] and is apparently incorrect. The formula (6) gives under-estimated values of β for typical meteor atoms. Thus the results of Lazarus and Hawkins should not be taken into account.

The work of Frichtenicht et al. [49] gives the most direct value of β since they used the results of laboratory measurements of ionization caused by very small iron particles speeded up to meteor velocities 20–60 km s^{-1} by a linear electrostatic accelerator and then evaporated in rarefied air. However, experimental conditions are not the complete models of the evaporation conditions for real meteoroids since the air density used was much greater than in the upper atmosphere meteor region. As the authors themselves pointed out, it could lead to over-estimation of β due to ionization process at collision of excited atoms and molecules (and simultaneously to the luminosity coefficient under-estimation which was detected). Thus value β obtained by Frichtenicht et al. [48] one can accept only as the upper limit.

Above mentioned consideration shows that at the present time the following assumption on value β seems to be trustworthy. As the upper limit β (and the lower mass scale limit) through all the intervals of meteor velocity one may accept the dependence obtained by Frichtenicht et al:

$$\beta = 1.4 \cdot 10^{-27}\, v^4\,, \tag{7}$$

where v is in cm s^{-1}. As the lower limit β (and the upper mass scale limit) for the velocity interval of 20—80 km s^{-1}, deduced in 4 times our evaluation β given by the formula (3) is accepted:

$$\beta = 10^{-25}\, v^{7/2}\,, \tag{8}$$

The lower limit β with $v < 20$ km s^{-1} is quite uncertain. As the most probable value β we will accept the value shown as curve in Fig. 1.

4.2. Parameter S

The parameter S for sporadic meteors and showers based on radar meteor observational data has been determined by many different authors. Values of S close to 2 were determined. However in the case of faint meteors in these works the influence of initial radius of the meteor trail on meteor detectability was not taken into account and this led to an under-estimate of S. Thus the value $S = 2$ can be accepted as the lower limit. As the upper limit we will accept $S = 2.34$ obtained by Hawkins and Upton [34] according photo observations. It is seen from Table 6 that extrapolation of the Hawkins and Upton data when $S = 2.34$ to $M_0 = 2 \times 10^{-4}$ g leads to a satisfactory agreement with our results. With $S = 2.34$ we have a larger discrepancy between our meteor influx estimate and that from the Nilsson and Southworth data which is due to the influence of the initial radius of the meteor trail influence on meteor detectability.

5. Comparison with Direct Measurement Data

Fig. 2 shows three evaluations of meteoroid influx obtained by us for a broad mass interval: the most probable evaluation with β according to Fig. 1 and $S = 2.2$ (solid curve); minimal evaluation with β according to (7) and $S = 2$

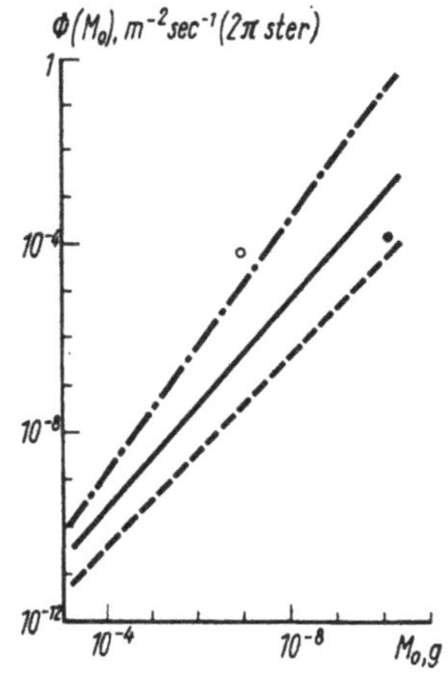

Fig. 2

(broken line); maximal evaluation with β according to (8) and $S = 2.34$ (chain line). Fig. 2 shows the results of micrometeoroid impact counts on a space vehicle far from the earth: Zond 3 and Venera 2 (circle) [54] and Mariner 4 (solid point) [55].

From Fig. 2 it is seen that the radar meteor influx evaluation obtained by us is in satisfactory agreement with direct measurements on space vehicles carried out far from the earth. Comparison with recordings of micrometeoroids in the earth's dust cloud makes no sense as particles in dense dust clouds either have to move along geocentric orbits or be kept in weighted conditions by some forces – such particles can not be recorded by radar and optical observation methods.

References

[1] T. N. Nasarova, Space Sci. Rev. 8, 455 (1968).
[2] C. W. McCracken and W. M. Alexander, Smithson. Contr. Astrophys. 7, 71 (1963).

[3] R. K. SOBERMAN and C. L. HEMENWAY, Smithson. Contr. Astrophys. **7**, 99 (1963).
[4] C. T. D'AIUTOLO, W. H. KINARD and R. J. NAUMAN, Smithson. Contr. Astrophys. **11**, 239 (1967).
[5] C. NILSSON, Private communication.
[6] B. P. KONSTANTINOV et al., Kosm. Issled. **6**, 738 (1968).
[7] F. L. WHIPPLE, J. Geophys. Res. **68**, 4929 (1963).
[8] J. G. DAVIES and J. C. GILL, Mon. Not. Roy. Astron. Soc. **121**, 437 (1960).
[9] V. N. LEBEDINETS, Astron. Zh. **40**, 719 (1963).
[10] T. R. KAISER, Mon. Not. Roy. Astron. Soc. **114**, 39 (1954).
[11] W. G. ELFORD, Harvard Radio Meteor Project Res. Report No. 8 (1964).
[12] V. N. LEBEDINETS, doctor's thesis Obninsk, USSR (1966).
[13] B. L. KASCHEYEV and V. N. LEBEDINETS, Radar Investigation of Meteor Phenomena, Publ. House Acad. Sci. USSR 1961.
[14] E. J. ÖPIK, Proc. Roy. Soc. A **230**, 463 (1955).
[15] E. J. ÖPIK, Physics of Meteor Flight in the Atmosphere, New York and London 1958.
[16] B. A. LINDBLAD, Smithson. Contr. Astrophys. **7**, 27 (1963).
[17] P. M. MILLMAN and D. W. R. MCKINLEY, Canad. J. Phys. **34**, 50 (1956).
[18] J. G. DAVIES and J. E. HALL, Proc. Roy. Soc. A **271**, 120 (1963).
[19] V. N. LEBEDINETS and V. B. SHUSHKOVA, in: Physics and Dynamics of Meteors, D. Reidel Publ. Co., Dordrecht, Holland 1968.
[20] V. N. LEBEDINETS and J. I. PORTNYAGIN, Geomagn. i Aeronom. **6**, 712 (1966).
[21] R. I. MOISYA, Astron. Tsirk. USSR, No. 409 (1967).
[22] V. N. LEBEDINETS and A. K. SOSNOVA, in: Physics and Dynamics of Meteors, D. Reidel Publ. Co., Dordrecht, Holland 1968.
[23] B. L. KASHCHEYEV, V. N. LEBEDINETS and M. F. LAGUTIN, Meteoric Phenomena in the Earth's Atmosphere, Izd. Nauka, Moscow 1967.
[24] J. G. DAVIES, Nature **179**, 121 (1957).
[25] T. R. KAISER, J. Atmos. Terr. Phys., Spec. Suppl. **2**, 119 (1955).
[26] A. R. WEBSTER, T. R. KAISER and L. M. C. POOLE, Mon. Not. Roy. Astron. Soc. **133**, 309 (9661).
[27] A. A. WEISS, Austral. J. Phys. **10**, 397 (1957).
[28] B. J. LEVIN, Bull. Astron. Inst. Czech. **7**, 45 (1956).
[29] V. G. KRUCHINENKO, R. I. MOISYA and I. V. BAIRACHENKO, Geophys. and Aeronom. Inform. bull., No. 5, 293, Kiev, USSR (1963).
[30] W. G. ELFORD and G. S. HAWKINS, NASA Research Report No. 9, Nov. 1964.
[31] J. P. BALDWIN and T. R. KAISER, Mon. Not. Roy. Astron. Soc. **129**, 85 (1965).
[32] C. S. NILSSON and R. B. SOUTHWORTH, Smithson. Astrophys. Obs. Spec. Rep., No. 263, Dec. 1967.
[33] A. M. BAHAREV, Bull. Astron. and Geodes. Soc. USSR, No. 16, 37 (1957).
[34] G. S. HAWKINS and E. K. UPTON, Astrophys. J. **128**, 727 (1958).
[35] F. VERNIANI and G. S. HAWKINS, Astrophys. J. **140**, 1590 (1964).
[36] N. HERLOFSON, Rep. Progr. Phys. **11**, 444 (1948).
[37] D. W. R. MCKINLEY, J. Appl. Phys. **22**, 202 (1951).
[38] J. S. GREENHOW and G. S. HAWKINS, Nature **170**, 355 (1952).
[39] T. R. KAISER, Adv. Phys. **9**, 495 (1953).
[40] S. EVANS and J. E. HALL, Astr. Contr. Univ. Manchester, ser. 2, No. 126, 1955.
[41] F. L. WHIPPLE, Astrophys. J. **121**, 241 (1955).
[42] G. S. HAWKINS, Astrophys. J. **124**, 311 (1956).
[43] P. M. MILLMAN, J. Roy. Astron. Soc. Canad. **44**, 209 (1950).
[44] J. DAVIES, J. S. GREENHOW and J. E. HALL, Proc. Roy. Soc. A **253**, 121 (1959).
[45] D. W. R. MCKINLEY, Meteor Science and Engineering, New York 1961.
[46] J. DAVIES and J. E. HALL, Proc. Roy. Soc. A **271**, 120 (1963).
[47] D. M. LAZARUS and G. S. HAWKINS, Smithson. Contr. Astrophys. **7**, 221 (1963).
[48] F. VERNIANI, Astrophys. J. **69**, 561 (1964).

[49] J. F. Frichtenicht, J. C. Slattery, E. Tagliaferri and D. O. Hansen, Astrophys. J. **74**, 321 (1969).
[50] J. F. Bydin and A. M. Buhteev, J. Techn. Phys. USSR **30**, 546 (1960).
[51] G. Messi and E. Burhop, Electron and Ionic Collision, Publ. House Foreign Literature, Moscow, No. 958.
[52] N. G. Utterback, Phys. Rev. **129**, 219 (1963).
[53] H. Fetz and H. Bussman, Z. Naturf. **A 192**, 1475 (1964).
[54] T. N. Nazarova, Space Res. **7**, 1439 (1967).
[55] W. M. Alexander et al., Smithson. Contr. Astrophys. **11**, 227 (1967).

Space Research XI — Akademie-Verlag, Berlin 1971

TEMPORAL VARIATIONS IN THE MASS DISTRIBUTION OF PARTICLES IN METEOR STREAMS

DAVID W. HUGHES

University of Sheffield, England

This paper reviews the results of visual and radio echo measurements yielding the size distribution of sporadic and shower meteorids as a function of magnitude and solar longitude. The values of the exponent s in the inverse power law describing the differential distribution of meteoroid mass, for the various meteor showers, are considered in relation to the observed diameter of the stream and the lifetime of meteoroids in the solar system. It is shown that the size dependent velocity of meteoroids as they leave the comet nucleus and the erosion of meteoroids by solar wind and interplanetary matter are competitive processes in the evolution of meteor showers.

1. Introduction

The mass distribution index s of sporadic meteors has an average value of 2.30 for magnitudes brighter than +3 and 2.01 for magnitudes fainter than +3. For the faint meteors in streams, s varies as a function of age from a value of 1.8 for young streams to 1.5 for much older more diffuse streams. This paper reviews methods of measuring s and the results obtained. The variation of s with stream diameter and age is discussed with reference to the evolution of the stream as debris from a comet nucleus and to the perturbing effects acting on a stream in the solar system.

2. The Definition of the Mass Distribution Index s

A meteor train is both luminous and ionized. The luminosity can be assigned a visual magnitude M_V. The ionization can be measured by radio techniques, and the line density α (the number of free electrons per centimetre along the train) obtained. The way in which the number of meteors observed varies as a function of magnitude and electron density can be represented by the following equations:

$$A(M) = C_1\, r^M \qquad (1)$$

and

$$\Phi(\alpha_Z) = C_2\, \alpha_Z^{-S} \qquad (2)$$

M is the zenithal magnitude, α_Z the zenithal line density (the maximum number of electrons produced per unit height), C_1 and C_2 are constants and $s = 1 + 2.5 \log_{10} r$. $A(M)\,\mathrm{d}M$ and $\Phi(\alpha_Z)\,\mathrm{d}\alpha_Z$ are respectively the fluxes of meteors with magnitudes between M and $M + \mathrm{d}M$ and electron line densities between α_Z and $\alpha_Z + \mathrm{d}\alpha_Z$. If $\theta(\alpha_Z)$ is the flux of meteors having a line density greater than α_Z, integration of Eq. (2) gives

$$\theta(\alpha_Z) = C_3 \alpha_Z^{(1-s)} \tag{3}$$

Eq. (2) may be derived from Eq. (1) by assuming that the point of maximum ionization along a train is the same as the point of maximum luminosity. A radio magnitude scale may be defined by the equation

$$M_R = 35 - 2.5 \log_{10} \alpha_Z \, .$$

The classical theory of meteoroid ablation assumes that individual meteoric atoms leave the surface of the meteor and produce light and ionization by collision processes. The luminosity I and ionization α produced by the meteoric atoms are proportional to the rate of ablation of the meteoroid and are related to meteoroid mass m by $I_{\max} \propto K\, m$ and $\alpha_{\max} \propto J\, m$ where K and J are luminous and ionizing efficiencies respectively, assumed to be independent of mass. The magnitude of a meteor is then given by

$$M = \text{const.} - 2.5 \log_{10} m \, .$$

From a study of the magnitude distribution and electron line density distribution the mass distribution of the associated meteoroids can be measured, subject to the condition that the classical ablation theory holds over the magnitude range in question. If the classical theory breaks down, as would be the case when meteoroids fragment on entering the atmosphere, then $x \propto m^d$ where d is a constant. The apparent mass distribution index s_f measured in this case is related to the actual mass distribution index s by the equation

$$s_f = 1 + \frac{s-1}{d} \, .$$

3. Methods of Measuring s

3.1. Visual, Telescopic and Photographic Methods

If there were no observational selection favouring brighter meteors these techniques would detect, in unit time, n_0 zero magnitude meteors, $n_1 = r\, n_0$ meteors of first magnitude, $n_2 = r^2\, n_0$ meteors of second magnitude and so on. Problems arise because the effective field of view for bright meteors is much greater than for faint ones. As one tries to observe fainter meteors a higher percentage are missed. The observations have to be corrected for this effect by assuming a value of s. Thus visual techniques can only be used to measure s down to magnitude -2, telescopic and photographic techniques down to $+2$ [1].

3.2. Radio Echo Amplitude

Radio echo techniques measure the electron line density in the vicinity of the specular reflection point. In general this is not the same as the point of maximum ionization. For under dense-trails, $\alpha < 2\times10^{12}$ ($M_R > 5$) and the echo amplitude is directly proportional to α. Measurements of the number of meteors producing amplitudes greater than certain fixed values enables s to be measured in the magnitude range $5 \leqq M_R \leqq 12$. Problem arise when the initial radius of the meteor train has to be taken into account. If this effect is present the return echo from meteor trains formed above a certain height are strongly attenuated. Thus the smaller meteors will not be observed and a low value of s will be obtained. The effect of the initial radius is still under debate. It is dependent upon wavelength and becomes much more important for short wavelengths.

3.3. Echo Height Distribution

For under-dense trains the echo duration τ is proportional to $\lambda^2 \varrho$ where λ is the radio wavelength and ϱ is the atmospheric density at the reflection point. Under-dense trains produce decay type echoes and by measuring the duration of these echoes the height of the reflection point can be calculated. Kaiser [2] has shown that the width of the height distribution for a group of meteors having the same velocity depends significantly only on the atmospheric scale height and the mass distribution of the incoming meteoroids.

At great heights the echo duration is very small and echoes may be missed due to radar systems having a finite pulse repetition frequency; also the removal of sporadic meteors from a shower meteor flux is a problem. This method is applicable to meteors in the magnitude range $5 \leqq M_R \leqq 12$.

3.4. Echo Duration Distribution

For over-dense trains, $\alpha > 2\times10^{12}$ cm, $M_R < 5$, the number of echoes $N(\tau)$ having a duration τ and greater is given by

$$\log_{10} N(\tau) \propto 0.75\,(1 - s)\log_{10}\tau\,,$$

assuming that diffusion alone is reducing the electron line density. If attachment is dominant the duration T is given by

$$\log_{10} N(T) \propto 4.5\,(1 - s)\log_{10} T\,.$$

Attachment becomes dominant for meteors of duration greater than 10 seconds formed at heights below about 90 km [3].

Under-dense trains can only be observed by backscatter if the train is perpendicular to the line of sight. Over-dense trains with durations longer than a few seconds can be distorted by atmospheric turbulence and wind shear so that echoes can be produced by trains which were originally formed well off the perpendicular condition [4]. This tends to reduce the measured value of s. This technique is used in the range $0 \leqq M_R \leqq 5$.

4. Measurements of s

Recent measurements of s for sporadic meteors are given in Table 1. The distribution shows a sharp change around magnitude $+3$ (meteoroid radius

Table 1
Observed values of the mass distribution index s for sporadic meteors

s	Mean mag.	Mag range	Ref. Method	s	Mean mag.	Mag range	Ref. Method
2.30	−5.5	$-8 \leqq M_V \leqq -3$	[5] A	2.26±0.02	4.0	$1 \leqq M_R \leqq 5$	[16] D
2.12[a]	−5.5	$-8 \leqq M_V \leqq -3$	[5] A	2.00	6.2	$5 \leqq M_R \leqq 7.5$	[9] B
2.39[b]	−5.5	$-8 \leqq M_V \leqq -3$	[5] A	2.00	6.5	$3 \leqq M_V \leqq 10$	[6] A
2.42	−5.0	$-10 \leqq M_V \leqq 0$	[6] A	2.014±0.003	6.6	$5 \leqq M_R \leqq 7.6$	[17] B
1.99	−1.2	$-4.5 \leqq M_V \leqq 2$	[7] A	2.02	7.2	$5 \leqq M_V \leqq 9.5$	[14] A
2.34	−0.5	$-2.5 \leqq M_{Ph} \leqq 1.5$	[8] A	1.98±0.02	7.5	$5 \leqq M_R \leqq 10$	[13] B
2.5	−0.2	$-3 \leqq M_R \leqq 2.5$	[9] D	2.45[c]	7.5	$5 \leqq M_R \leqq 10$	[18] B
2.27	0	$-2 \leqq M_V \leqq 2$	[10] A	2.00±0.02	7.5	$5 \leqq M_R \leqq 10$	[19] B
2.25	1.5	$0 \leqq M_V \leqq 3$	[6] A	2.35	7.5	$5 \leqq M_R \leqq 10$	[20] B
2.33	1.9	$0.4 \leqq M_R \leqq 3.4$	[11] D	2.17±0.02[d]	7.9	$5 \leqq M_R \leqq 10.8$	[21] B
2.35	2.0	$0 \leqq M_V \leqq 4$	[12] A	2.02±0.08[e]	7.9	$5 \leqq M_R \leqq 10.8$	[21] B
2.4±0.6	2.5	$0 \leqq M_R \leqq 5$	[13] D	1.97±0.02	8.5	$6.5 \leqq M_R \leqq 10.5$	[16] B
2.16	2.7	$0.5 \leqq M_V \leqq 5$	[14] A	2.12	9.2	$8 \leqq M_R \leqq 10.5$	[22] B
2.00	3.5	$-3 \leqq M_V \leqq 10$	[15] A	2.00±0.05	12.5	$11 \leqq M_R \leqq 14$	[23] B

Methods of observation: A, visual magnitude distribution; B, radio echo amplitude distribution; D, radio echo duration distribution.
[a] retrograde orbits, [b] direct orbits, [c] correction made for the effect of initial radius to the results of [19], [d] 00–12 h, [e] 12–24 h.

8×10^{-2} cm): Fig. 1. For brighter magnitudes the mean value of s is 2.30. This stronly suggest that fragmentation is important for M brighter than +3 and is unimportant for fainter magnitudes. Jones and Kaiser [24] find that solid brittle meteoroids having a mass greater than a certain critical value will

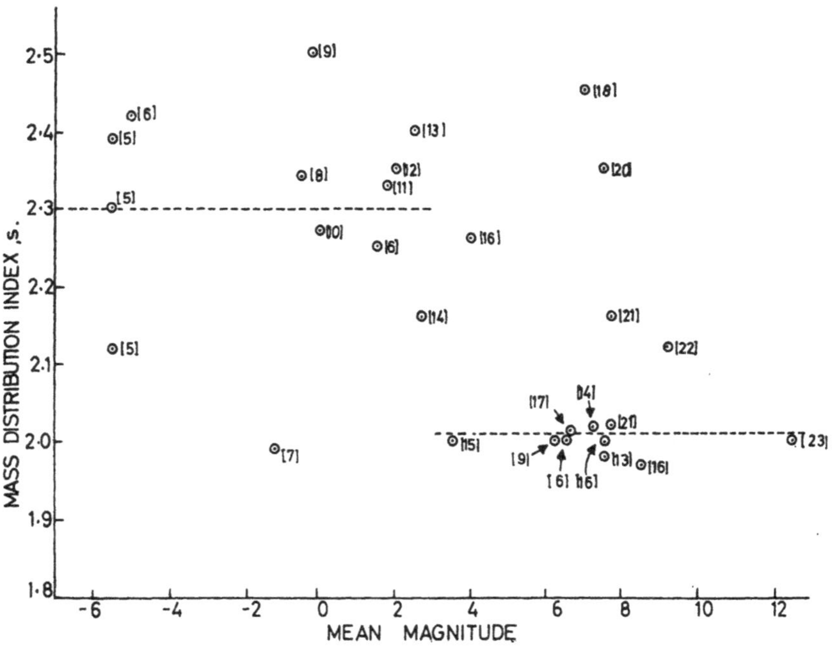

Fig. 1. The mass distribution index s for sporadic meteors as a function of magnitude. - - - - - mean value of s.

fragment. This critical mass is found to be equivalent to $M = +3$. Poole and Kaiser [25] find evidence from observations of radio meteors that fragmentation is unimportant for meteoroids producing under-dense trains, i.e. $M > +5$. Meteors fainter than $M + 3$ have a mean s of 2.01 and in this region the classical theory of ablation is applicable. (The result obtained by Greenhow and Hall [18] is ignored as subsequent work [26] has shown that the initial radius of a meteor train has a negligible effect on the amplitude of the returned radio echo at 17 MHz.) On these assumptions, using the above values for s we obtain $\alpha \propto m^{0.78}$ for meteors brighter than $M + 3$ (compared with $d = 0.73$ [27]) and $\alpha \propto m$ for fainter ones. Values of s for shower meteors are given in Table 2; again a transition appears to occur at $M = +3$.

Table 2

Observed values of the mass distribution index s of shower meteors

s	Mean Mag.	Mag. Range	Ref. Method
Quadrantids			
1.8	1.0	$0 \leq M_R \leq 2$	[13] D
2.15	2.0	overdense	[28] D
2.0	6.0	underdense	[28] B
1.78 ± 0.06	6.0	$5 \leq M_R \leq 7$	[13] C
1.55 ± 0.02	7.1	$6 \leq M_R \leq 8.1$	[29] B
η Aquarids			
2.0	−0.6	$-1.7 \leq M_R \leq 0.5$	[9] D
1.4	1.1	$0.5 \leq M_R \leq 1.75$	[9] D
1.61 ± 0.1	6.2	$5 \leq M_R \leq 7.5$	[9] B
Arietids			
1.8	1.0	$0 \leq M_R \leq 2$	[13] D
2.7 ± 0.3	6.0	$5 \leq M_R \leq 7$	[13] D
2.14 ± 0.10	6.0	$5 \leq M_R \leq 7$	[29] B
1.65 ± 0.06	6.6	$5.8 \leq M_R \leq 7.5$	[30] B
see also			[31]
Arietids plus ζ Perseids			
2.3 ± 0.2	2.5	$0 \leq M_R \leq 5$	[16] D
2.5 ± 0.3	6.0	$5 \leq M_R \leq 7$	[13] C
2.04	7.2	$6.4 \leq M_R \leq 8$	[16] B
1.95	8.6	$8 \leq M_R \leq 9.2$	[16] B
2.14 ± 0.2	8.6	$8 \leq M_R \leq 9.3$	[16] B
1.51	9.8	$9.2 \leq M_R \leq 10.5$	[16] B
1.55 ± 0.03	10.0	$9.3 \leq M_R \leq 10.8$	[16] B
δ Aquarids			
2.7	0.5	$0 \leq M_R \leq 1$	[9] D
2.70	1.0	$0 \leq M_R \leq 2$	[10] A
1.9	1.7	$1 \leq M_R \leq 2.5$	[9] D
1.54 ± 0.1	6.2	$5 \leq M_R \leq 7.5$	[9] B
1.50 ± 0.05	7.0	$4 \leq M_R \leq 10$	[31] E
Perseids			
2.40	−2	$-4 \leq M_V \leq 0$	[10] A
2.1	1	$M_V \leq 3$	[32] A
2.00 ± 0.02	1.5	$0 \leq M_R \leq 3$	[13] D
1.96 ± 0.17	2	$0 \leq M_R \leq 4$	[13] D
2.22 ± 0.03	3.2	$1.5 \leq M_R \leq 5$	D
1.59 ± 0.03	6	$5 \leq M_R \leq 7$	[13] C
1.55 ± 0.05^{a}	7	$6 \leq M_R \leq 8$	[13] B
1.27 ± 0.02^{b}	7	$6 \leq M_R \leq 8$	[13] B
1.82 ± 0.05^{a}	7.1	$6.0 \leq M_R \leq 8.3$	[29] B
1.93 ± 0.05^{b}	7.1	$6.0 \leq M_R \leq 8.3$	[29] B
Draconids			
1.73	−1.5	$-3 \leq M_{Ph} \leq 0$	[33] A
3.33	1	$0 \leq M_V \leq 2$	[34] A
2.07	4	$2 \leq M_V \leq 8$	[34] A
Leonids			
1.35 ± 0.02	2.5	$0 \leq M_R \leq 5$	[35] D
1.48 ± 0.05	2.5	$0 \leq M_R \leq 5$	[35] D
2.2 ± 0.2	2.5	$0 \leq M_R \leq 5$	[36] D
1.73 ± 0.08	7	$M_R \approx 7$	[36] B
Geminids			
2.3	0	$M_V \leq 3$	[37] A
2.7 ± 0.3	0.5	$0 \leq M_R \leq 1$	[9] D
2.3	1	$M_R \leq 3$	[13] D
2.24 ± 0.04	1.5	$0 \leq M_R \leq 3$	[13] D
1.7 ± 0.1	1.7	$1 \leq M_R \leq 2.5$	[9] D
1.62 ± 0.04	6.0	$5 \leq M_R \leq 7$	[13] C
1.50 ± 1	6.2	$5 \leq M_R \leq 7.5$	[9] B
1.45 ± 0.01	6.5	$6 \leq M_R \leq 7$	[13] B
1.72 ± 0.07	6.7	$5.8 \leq M_R \leq 7.5$	[30] B
1.70 ± 0.15	6.7	$5.8 \leq M_R \leq 7.5$	[30] B
1.75 ± 0.02	7.1	$6.0 \leq M_R \leq 8.3$	[29] B

Methods of observation: A, visual magnitude distribution: B, radio echo amplitude distribution: C, radio echo height distribution: D, radio echo duration distribution: E, radio echo rate curves for fixed aerial.

[a] discrete component, [b] diffuse component.

5. Variation of s with Observed Stream Diameter

The mean values of s have been calculated for each shower for the magnitude range $5 \leqq M \leqq 10$ (see Table 3). If $\Delta\lambda$ is the change in solar longitude during which a shower has an activity in excess of half its maximum value and ε is

Table 3
Observed stream diameter of major meteor streams

	$\Delta\lambda$ (°)	ε (°)	Observed stream diameter (10^5 km)	Mean s values
Quadrantid	0.5	37	7.9	1.78 ± 0.18
η Aquarid	10	71	250	1.61 ± 0.10
Arietid	20	60	460	1.90 ± 0.20
Arietid + ζ Perseid	20	67	480	1.84 ± 0.26
δ Aquarid	20	60	450	1.52 ± 0.02
Perseid	5	6	137	1.63 ± 0.22
Draconid	0.5	23	2.0	2.07 ± 0.15
Leonid	4	11	20	1.73 ± 0.08
Geminid	6	25	129	1.62 ± 0.11

the angle between the apex of the earth's way and the shower radiant then the observed stream diameter D is given by

$$D = \frac{2\,\pi\,\lambda\,\Delta\,\sin\varepsilon}{360}(149 \times 10^6)\ \text{km}\,.$$

D will only be the actual stream diameter if the earth passes through the centre. Fig. 2 shows the way in which s varies with the observed stream diameter.

6. The Evolution of Meteor Streams

Assume that meteor streams are the debris left behind by decaying comets. Most comets have a coma of diameter about 10^5 km and it can be seen from Fig. 2 that the Draconid, Quadrantid and Leonid meteor streams have diameters of the same order of magnitude. These streams must therefore be young. As a stream ages its diameter increases and the meteoroids become more evenly spread around the orbit. The intense narrow cores observed in the young streams are not evident in older ones. Fig. 2 shows that ageing leads to a steady decrease in s from 1.8 (which may be the mass distribution index of the particles in the original comet) to 1.5.

6.1. The Ice Conglomerate Comet Model

Consider Whipple's icy conglomerate comet model [38, 39]. An average comet has a nucleus of mass 10^{18} g and radius 10 km. Particles are ejected from the nucleus into the coma and finally into the meteor stream with a velocity Δu given by

$$\Delta u = \left(\frac{r_c}{L\,r}\right)^{1/2} \frac{3.28}{q^{9/8}}\ \text{m s}^{-1} \tag{4}$$

where r_c (km) is the radius of the nucleus, $1/L$ is the heat of absorption coefficient, r (cm) is the radius of the meteoroid and q (AU) is the perihelion distance of the cometary orbit.

Two important facts follow from Eq. (4).

(i) The velocity at which the particle leaves the comet is proportional to $r^{-1/2}$, i.e. the smaller particles move away from the nucleus more quickly. A stream

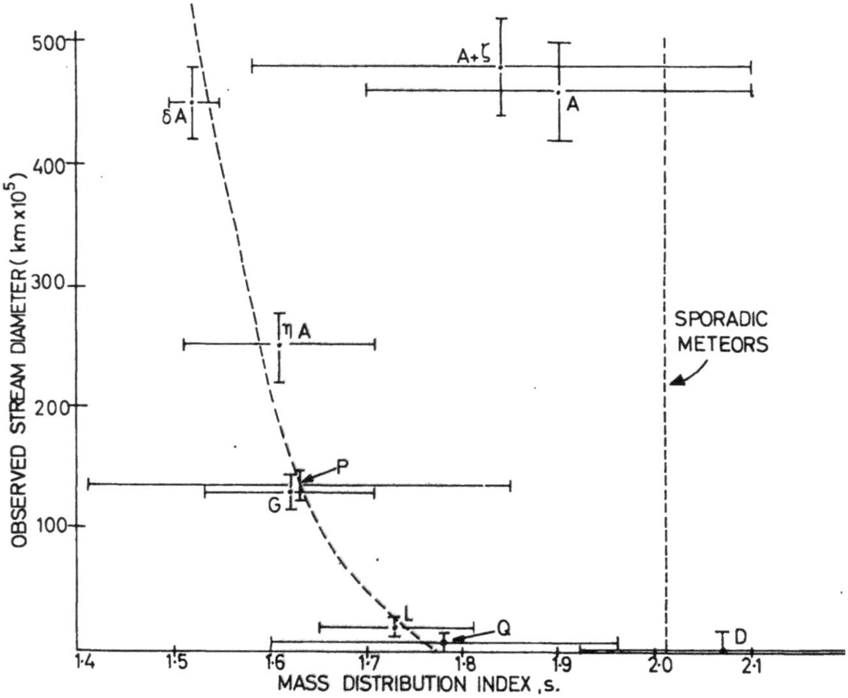

Fig. 2. The mass distribution index s for shower meteors in the magnitude range $5 \leqq M \leqq 1$ as a function of the observed stream diameter.
D, Draconids; Q, Quadrantids; L, Leonids; G, Geminids; P, Perseids; ηA, Eta Aquarids; δA, Delta Aquarids; A, Arietids; A + ζ, Arietids plus Zeta Perseids.

which is formed with a given value of s will evolve until the outer portion has a larger s value due to an enhancement of small particles and the inner region has a lower s value due to small particles being depleted. The decrease in s with age is due to the smaller particles escaping from the stream more quickly. The Geminid stream shows this well as it has a large s in the outer regions (especially on the antisolar side of the stream) and a smaller s in the centre. For the Perseids the diffuse outer region has $s = 1.93$ and the discrete centre $s = 1.82$ [29]. The escaping small particles add to the sporadic background.

(ii) Magnitude +3 meteoroids have a radius of about 8×10^{-2} cm and, from Eq. (4), a velocity with respect to the nucleus of 20 m s^{-1}. This is approximately equal to the escape velocity from an average cometary nucleus and would imply that meteor streams should be relatively deficient in bright meteors. Sporadic meteors may have been produced by the perturbation and decay of smaller comets but there could be two mechanisms producing the s change at $M = +3$, the onset of fragmentation and the effect of the finite escape velocity.

6.2. Meteor Stream Age

The Leonid stream was first observed in AD 1833, the Perseids in AD 36 and the Delta Aquarids in AD 401 [40]. These figures give minimum ages and, assuming that the stream diameter increases at a constant rate, they give the minimum age of the older Arietid shower as about 4000 y. As this shower is now 20° wide in solar longitude it is reasonable to assume that its total possible age cannot be more than a factor of ten greater. If we take the above minimum ages as actual ages then meteor streams will have a maximum age of between 10^4 and 10^5 years. For the variation of s with age to be due to perturbations and not simply to the different velocities of the particles with respect to the original nucleus, these perturbations have to produce a significant effect in a period of 10^4 y.

6.3. Perturbing Effects Acting on Streams

6.3.1. Radiation pressure

If a meteoroid of radius a, density ϱ has $a\varrho < 0.6 \times 10^{-4}$ g cm^{-2} it will be blown out of the solar system [41]. The smallest radio meteor is 500 times this size so this effect need not be considered.

6.3.2. Poynting–Robertson effect

Meteoroids lose angular momentum due to re-radiation of adsorbed solar radiation and spiral into the sun. The life-time t of a meteoroid in the solar system is given by $t = a\varrho k \times 10^7$ y where k depends on the initial eccentricity and perihelion distance (k varies between 0.1 and 20 for the showers considered in Table 3) [42]. According to Kresák [43] a change in stream s value from 1.8 to 1.5 due to the Poynting–Robertson effect would take about 10^7 y. Poole [29] states that the variation in s across the Geminid stream (inclination 23.5°) is due to this effect. Jones [30] finds a similar variation in the Perseid stream (inclination 113.6°) and, as the Poynting–Robertson effect can only change s in the plane of the ecliptic, it is thought that the mechanism discussed in § 6.1 must be the cause of this variation. As the Poynting–Robertson effect is so slow most showers would have dispersed before it becomes noticeable.

6.3.3. Planetary collisions

According to Öpik [44] planetary collisions take between 10^7 and 10^8 y to remove appreciable numbers of meteoroids. Planetary perturbations decrease the semi-major axis of stream orbits but do not affect the mass distribution.

6.3.4. Solar wind

Meteoroid retardation due to solar wind increases the Poynting–Robertson effect by 22% [45]. Interactions between the energetic hydrogen ions in the solar wind and the meteoroids leads to a cathode sputtering type of erosion which reduces the meteoroid size and thus alters the mass distribution causing s to decrease with time because, in a given magnitude range, particles are eroded and are only partially replaced by larger particles being reduced by ablation to the required size. Kresák [43] calculates that a change in s from 1.8 to 1.5 would take 10^4–10^5 y, the change being quicker for particles which spend their life closer to the sun. This perturbation seems to be about a factor of ten slower than the effect discussed in § 6.1.

6.3.5. Erosion by interplanetary dust

Whipple [46] has considered this effect coupled with sputtering by solar protons; they combine to give a meteoroid a life-time $t = a\,\varrho\,4.3 \times 10^4$ y. This erosion is rapid and as the relative mass loss accelerates with decreasing particle size the effect is compatible with a stream s-value decreasing with age and also with the lack of streams over 10^4 years old.

6.3.6. Mutually destructive collisions

This process requires a considerable interparticle velocity or very fragile low density particles. The majority of collisions will occur at the nodes of the orbit and, as velocity is inversely proportional to perihelion distance q, this effect is more important for orbits with small q. The destructive collision life-time is proportional to mass and again s decreases with age.

According to Jones [47] a grinding process where two meteoroids on collision break up into many smaller ones will lead to an increase in s with age. As this effect is not observed it can be assumed to be negligible.

7. The Sporadic Background

The meteoroid influx is divided between streams (27%) and sporadic background (73%) [48]. Sporadic meteoroids could be primeval, originating when the solar system was formed, the meteors which are observed now having been perturbed towards the earth's orbit from the outer regions of the solar system by the Poynting–Robertson effect.

Stream meteoroids might disperse into the sporadic background. This definitely happens for M fainter than $+3$. Because of this one would expect the ratio between sporadic and shower meteor fluxes to be increasing, unless the rate of removal of particles from the sporadic background is equal to the incoming rate from streams.

8. Conclusions

The mass distribution index of stream meteoroids decreases with age. For meteors fainter than $M + 3$ this is due to the size dependent velocities of the particles as they leave the comet nucleus and also to erosion and sputtering of meteoroids by the solar wind and interplanetary matter. The Poynting–

Robertson effect, planetary perturbations, interparticle collisions and radiation pressure have a negligible effect over the 10^4 y life-time of the average meteor stream.

References

[1] D. W. R. McKINLEY, Meteor Science and Engineering, McGraw-Hill, New York 1961 (p. 106).
[2] T. R. KAISER, Mon. Not. Roy. Astron. Soc. **114**, 39, 52 (1954).
[3] T. R. KAISER, Smithson. Contr. Astrophys. **7**, 175 (1963).
[4] E. R. BILHAM and I. C. BROWNE, Proc. Phys. Soc. B **69**, 98 (1956).
[5] E. ÖPIK, Contr. Armagh Obs. No. 26 (1958).
[6] P. M. MILLMAN, J. Roy. Astron. Soc. Can. **51**, 113 (1957).
[7] M. KRESÁKOVÁ and L. KRESÁK, Contr. Astr. Obs. Skalnaté Pleso **1**, 40 (1955).
[8] G. S. HAWKINS and E. K. L. UPTON, Astrophys. J. **128**, 727 (1958).
[9] A. A. WEISS, Aust. J. Phys. **14**, 102 (1961).
[10] B. A. LINDBLAD, Smithson. Contr. Astrophys. **11**, 171 (1965).
[11] J. CRYGAR, L. KOHOUTEK and Z. PLACOVÁ, in: Physics and Dynamics of Meteors, D. Reidel Publ. Co., Dordrecht 1968 (p. 63).
[12] M. KRESÁKOVÁ, Contr. Astr. Obs. Skalnaté Pleso **3**, 75 (1966).
[13] I. C. BROWNE, K. BULLOUGH, S. EVANS and T. R. KAISER, Proc. Phys. Soc. B **69**, 83 (1956).
[14] P. M. MILLMAN, Canad. Aeronaut. J. **5**, 358 (1959).
[15] F. G. WATSON, Between and Planets, Harvard Univ. Press 1956.
[16] D. BROOKS, Ph. D. Thesis, Sheffield University (1963).
[17] D. W. HUGHES and D. G. STEPHENSON, to be published.
[18] J. S. GREENHOW and J. E. HALL, Mon. Not. Roy. Astron. Soc. **121**, 183 (1960).
[19] D. W. McKINLEY, Canad. J. Phys. **29**, 403 (1951).
[20] M. ŠIMEK and B. A. McINTOSH, in: Physics and Dynamics Meteors (p. 362).
[21] J. P. BALDWIN and T. R. KAISER, Mon. Not. Roy. Astron. Soc. **129**, 85 (1965).
[22] T. R. KAISER, Annls Géophys. **17**, 50 (1961).
[23] W. G. ELFORD and G.S. HAWKINS, Harvard Radio Meteor Project, Res. Rep. No. 9 (1964).
[24] J. JONES and T. R. KAISER, Mon. Not. Roy. Astron. Soc. **133**, 41 (1966).
[25] L. M. G. POOLE and T. R. KAISER, Planet. Space Sci. **15**, 1131 (1967).
[26] L. M. G. POOLE, Ph. D. Thesis, Sheffield University (1965).
[27] B. J. LEVIN and S. V. MAJERA, Smithson. Contr. Astrophys. **7**, 229 (1961).
[28] B. A. McINTOSH, Canad. J. Phys. **44**, 2729 (1966).
[29] L. M. G. POOLE, Smithson. Contr. Astrophys. **11**, 157 (1965).
[30] J. JONES, Ph. D. Thesis, Sheffield University (1966).
[31] W. G. ELFORD, in: Physics and Dynamics of Meteors (p. 352).
[32] A. HRUŠKA, Bull. Astr. Insts. Czech. **5**, 13 (1954).
[33] L. G. JACCHIA, Z. KOPAL and P. M. MILLMAN, Astrophys. J. **111**, 104 (1950).
[34] F. G. WATSON, Bull. Harv. Coll. Obs. No. 895 (1934).
[35] Z. PLACOVÁ, in: Physics and Dynamics of Meteors (p. 432).
[36] B. A. McINTOSH, in: Physics and Dynamics of Meteors (p. 343).
[37] B. J. LEVIN, in: Meteors, Pergamon Press, London 1955 (p. 131).
[38] F. L. WHIPPLE, Astrophys. J. **111**, 375 (1950).
[39] F. L. WHIPPLE, Astrophys. J. **113**, 464 (1951).
[40] S. IMOTO and I. HASEGAWA, Smithson. Contr. Astrophys. **2**, 131 (1958).
[41] T. R. KAISER, Space Sci. Rev. **1**, 554 (1962).
[42] F. L. WHIPPLE, Astrophys. J. **111**, 134 (1950).
[43] L. KRESÁK, Bull. Astr. Insts. Czech. **11**, 1 (1960).
[44] E. ÖPIK, Adv. Astron. Astrophys. **2**, 219 (1963).
[45] F. L. WHIPPLE, NASA SP 150, p. 409 (1967).
[46] F. L. WHIPPLE, Smithson. Contr. Astrophys. **7**, 239 (1963).
[47] J. JONES, Canad. J. Phys. **46**, 1101 (1968).
[48] L. KRESÁK, in: Physics and Dynamics of Meteors (p. 391).

Space Research XI — Akademie-Verlag, Berlin 1971

THE FALL OF COSMIC MATERIAL IN ITALY

F. Di Benedetto

Laboratory of Space Astrophysics, National Research Council and University of Rome, Italy

An investigation of particles of cosmic origin in Italian soil reveals that there is evidence of a stratum exceptionally rich in cosmic particles in the Sicilian levels of the Pontina plain. Various characteristics of these particles are described and the Italian research programme on dust, consequent to this discovery, as planned through the Italian National Research Council is indicated. Emphasis is placed upon the space aspects of the subject.

1. Introduction

Three basic problems are of interest to space research: (i) the origin and evolution of our solar system, (ii) the origin and evolution of life and (iii) the dynamic processes which model the environment in which we live.

A relevant contribution to the first problem is given by the fact that it was discovered almost 100 years ago that primordial material of extra-terrestrial origin exists in exactly identical conditions on the earth, in its atmosphere and in interplanetary space, in the form of dust, rocks and spherules, and on the moon. The importance of this lies in the fact that it contains the basic data necessary to understand the processes which have acted in forming this material and, consequently, the planets of our solar system and, presumably, the universe.

2. Discovery of Spherules in Italy

The discovery of spherules in Italian soil is due to B. Accordi in a survey of the Po river plain. Some levels in the Pleistocene sediments were rich in micrometeorite particles. This discovery led to active research on micrometeorites in central Italy (Apennines and Pontina Plain). The main results reached so far, although of a preliminary nature, have been established [1]. The present author has also reviewed the subject, in particular the importance of space research activities in relation to cosmic dust and micrometeorites [2].

One can now establish that from the Trias (225 ± 5 million years ago) onwards, magnetic spherules are continuously present. In addition, the cores extracted from the Pontina plain and the samples of clay of the Quaternary, near Ascoli Piceno, show a significant increase in the concentration of spherules per kilogram of sediment. The increase of 35 spherules per gramme of sediment is very significant in the layers of the lower Sicilian.

The majority of particles are less than 5 μm in diameter and less than five particles have a diameter greater than 5 μm. These particles can be hollow or solid and may have any shape, although the spherical shape is most frequent (particles whose diameter is less than 30 μm are all spherical).

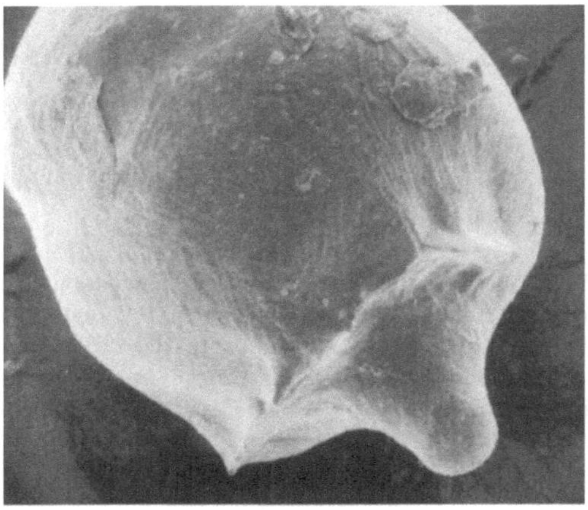

Fig. 1. Drop-shaped spherule.

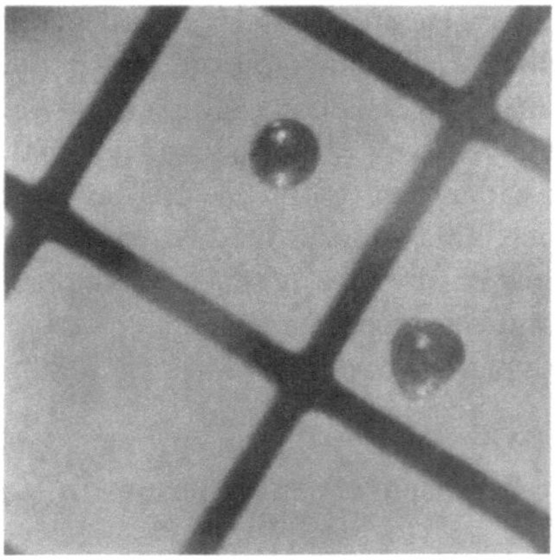

Fig. 2. Opaque black microspherule (above) and oxidized spherule (below).

In Fig. 1 is a drop-shaped micrometeorite. Note the swollen upper part and the "nipple" on the bottom. The streaks on the surface are also probably related to a dendroid cooling structure (Fig. 1 is magnified 200 times) [1].

In Fig. 2 is an opaque black microspherule (above) of diameter 113.4 μm whose surface is rugged and void of metallic glow. An oxidized spherule, diameter 126 μm, is shown (below). This spherule is partially covered by a reddish crust [2].

The structure of the inner surface of a hollow micrometeorite (Fig. 3) is typical of a melt which reaches the solidification point in a very short time.

The analysis of the Italian sediments is only preliminary and quite recently a comparison of this material has been made with micrometeorites of similar nature found on the moon which shows a great morphological affinity of the lunar material with that of the earth.

As this research must be developed through the collaboration of scientists from different branches of science including astrophysics, geochemistry and geology, a meeting was held on 12 June 1968 at the Institute of Geochemistry of the University of Rome in order to establish the initial steps necessary for the examination of this problem. Professor Accordi emphasized the value of an exchange of information between Italy and NASA, and as a preliminary

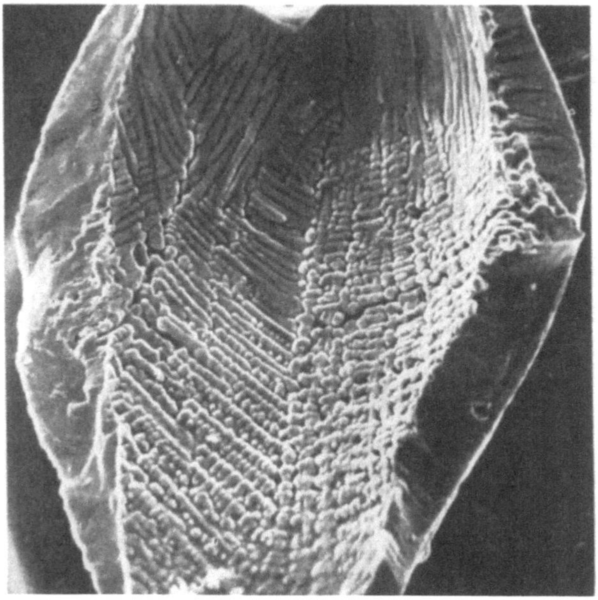

Fig. 3. Structure of inner surface of a hollow micrometeorite.

step to tackle this long-term problem, the CNR and the Ministry of Foreign Affairs have taken the initiative of proposing to send two experts of the Research Group on meteorites to NASA and to send the author to Washington.

3. Italian Research Programme for Astrophysics

The space research programme for astrophysics concerns the hypothesis of the origin of meteorites and cosmic dust, research on the interaction with the atmosphere (formation of shock waves at temperatures of some tens of thousands of degrees); ionization phenomena, interaction of shock waves with the parent meteorite, meteoritic ablation phenomena, research of eventual correlations with the geomagnetic field, paleomagnetic correlations, analysis of experimental data obtained by means of rockets, satellites and space probes and determination of the flux of meteorites (numerical and total), participation in space experiments for the direct collection of cosmic dust (balloons, rockets, satellites and space probes) and exchange of material and data with other organizations, conferences and congresses and an exchange of scientists with other appropriate institutions.

4. Italian Research Programme for Geochemistry

The geochemical interest is mainly in the laboratory analysis developed by means of modern techniques at the Institute of Geochemistry of the University of Rome, directed by Professor M. Fornaseri, and applied in the USA, mainly at the Astronomical Observatory, Dudley, directed by Professor C. Hemenway.

The study of the composition of the cosmic spherules requires the measurements of the specific weight and the semi-quantitative or quantitative determination of its components by the application of ultramicroanalysis and spectral analysis in the optical and X-ray fields, as well as the use of electronic microanalysis and of analysis by nuclear reactivation.

There should be collaboration with the work on the first item of the research programme for astrophysics, and conferences and scientific publications should be in common with astrophysics, remembering the necessity to extend collaboration and participation to the meetings of other institutions interested in geochemistry.

5. Italian Research Programme for Geology

The most compelling task, from a practical point of view, for geology is in the operations of perforation and extraction and in the classification of the material collected, both in Italy and elsewhere, through campaigns established by common agreement with other geological institutions.

The extraction requires the following operations: (i) perforation at continuous coring, with a sonde already available, of sedimentary series from the Miocene to the Quaternary, (ii) calibration and use of a serial system of extraction of cosmic material, (iii) classification of spherules and particles extracted (granulometry and diagram of number of particles as a function of depth), (iv) optical analysis (traditional microscope, scanning microscope and magnetic balance), and (v) study of the isotropic structures discovered by us on the surfaces of the spherules.

The Italian Research Group should participate, with men and means, in the campaign of extraction and perforation of the sea beds (not only in the Mediterranean) established by other institutions, with the collaboration of other international organizations such as UNESCO, etc. This collaboration would ensure that the study of the oceanographic aspects of the problem be carried out.

As for geochemistry, one must deal with the bibliography and all the questions related to international conferences, as well as the publication of scientific papers as publications of the CNR.

In addition the research personnel, together with the personnel for geochemistry and astrophysics, should be able to commence laboratory experiments for simulating impacts, on surfaces made of different material and of various thicknesses of micrometeoritic charged spherules, in combination with the experiments at present being carried out at the Physics Laboratory of the University of Heidelberg.

The collaboration of Italian and American astrophysicists, physicists, metallurgists, geologists and oceanographists, indispensable for the realization of this project, should constitute a fundamental aspect of the Italian Group.

References

[1] R. Funiciello and M. Fulchignoni, Cosmic spherules in Central Italy Sediments, Geologica Romana, 8 (1969), p. 117.

[2] F. di Benedetto, La caduta sul globo terrestre di materiale di origine cosmica nel passato e nel presente, Rapporto Interno 69/18 (1969), Lab. Astrofisica Frascati (Roma).

Space Research XI — Akademie-Verlag, Berlin 1971

A SIMULTANEOUS COLLECTION AND DETECTION EXPERIMENT FOR COSMIC DUST

H. FECHTIG, M. FEUERSTEIN and P. RAUSER

Max-Planck-Institut für Kernphysik, Heidelberg, FRG

In the micrometeoroid collection experiment copper and aluminium plates covered by a nitrocellulose foil (0.1 μm thick) were exposed. Micrometeoroids produce characteristic penetration holes in the foil and are analysed on the metal collectors below.

The detection method is based on the plasma emission at the impact of dust particles on a tungsten target. From the measured ion charge and its rise time the mass and velocity of dust particles can be determined. The detector has been calibrated with a 2 MV Van de Graaff accelerator using iron particles with velocities from 0.7 to 12 km s^{-1} and masses from 8×10^{-14} to 2×10^{-10} g.

The combined experiments were flown on 14 August 1969 at 0201 UT, from ESRange, Kiruna, apogee 112 km, in the maximum of the sporadic meteors. From the 127 ± 33 irregular particles (diameter $\sim$0.1 μm) found on 8 mm^2 of the foil a flux $\varphi = (28 \pm 7)$ $m^{-2} s^{-1}$ is obtained for masses $m > 1.6 \times 10^{-15}$ g, above 60 km. On 9 cm^2 84-fringed penetration holes ($d \sim 2$ μm) and on the collectors below, 20–50 fragmented particles ($d \sim 0.2$ to 0.3 μm) were counted. From the penetration holes a flux $\varphi = 0.3$ $m^{-2} s^{-1}$ for masses $m > 1.3 \times 10^{-12}$ g was derived Signals from the detector were registered at 105.5, 81.9, 80 and 76.8 km with attributed masses of meteoroids from 1.6×10^{-13} g to 2.7×10^{-11} g. Above 83.4 km an integral flux of $\varphi < 6$ $m^{-2} s^{-1}$ has been achieved for masses $m > 1.6 \times 10^{-13}$ g. The measured signals are compatible with porous and fragile meteoroids ($\varrho = 0.3$ g cm^{-3}). A mass between 25 t and 170 t is estimated to be collected per day by earth.

1. Introduction

A simultaneous micrometeoroid collection and detection experiment has been flown a Centaure rocket, C. 58/A, from ESRange, Kiruna, on 14 August, 1969, at 0201UT, during the maximum of the Perseid meteor shower and near the daily maximum of the sporadic meteors. The apogee was 112 km. The influx rate, masses and velocities of cosmic dust particles near the earth were investigated. The combined experiments made possible a comparison of the flux values obtained by different methods. The trays on which the experiments were mounted were exposed at 61.1 km, on upleg, and closed at 97 km, on downleg, with a total exposure time of 159.4 seconds. The payload was successfully recovered.

2. Experiments

2.1. Collection Experiment

In the collection experiment polished plates of aluminium and copper covered by a thin aluminized nitrocellulose foil (0.1 μm thick) were exposed. Micrometeoroids are partly collected on the foil; bigger particles penetrate the foil and produce characteristic penetration holes. These particles can be analysed on the metal collectors for shape and chemical composition using a Stereoscan analyser and an electromicroscope [1].

2.2. Detection Experiment

The detection method is based on the plasma emission of dust particles at the impact on a tungsten target. At velocities $\geqq 0.8$ km s^{-1} a varying fraction of the incident particle mass is vaporized and ionized, depending on the mass and velocity of the particle [2, 3].

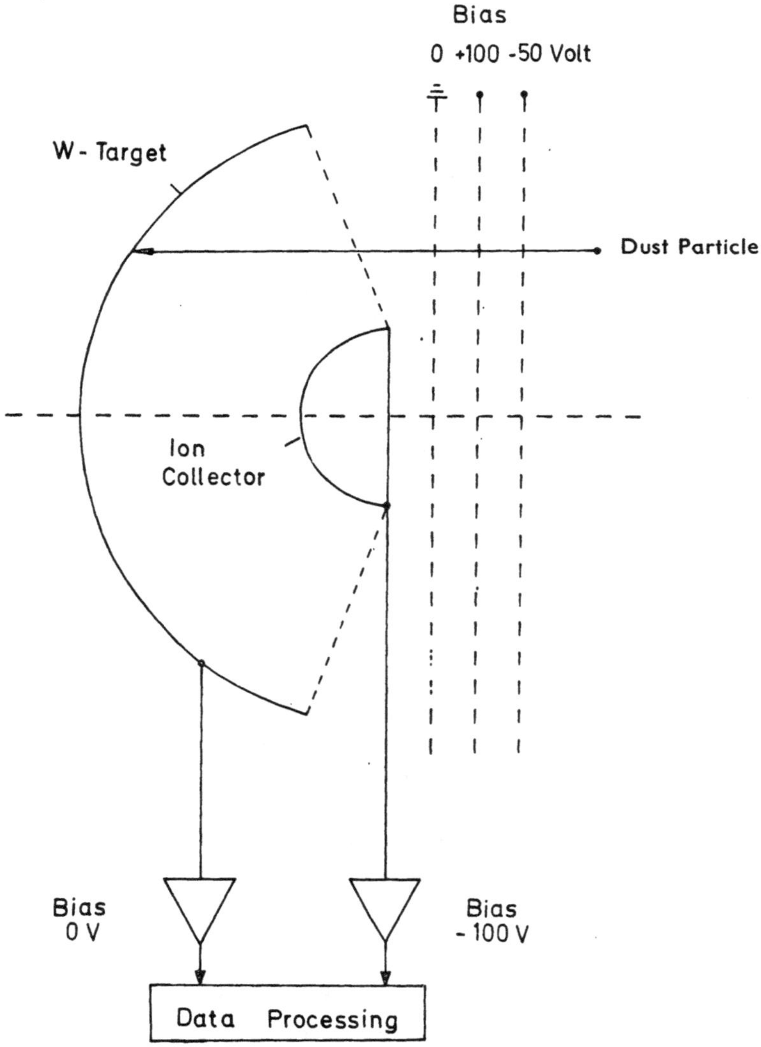

Fig. 1. Principle of the dust particle detector.

The principle of the detector is shown in Fig. 1. The target consists of a grounded tungsten hemisphere in the centre of which a negatively biased ion collector is located. When dust particles hit the target electrons and ions of the plasma cloud are separated in the electric field, the ions being detected at the collector. The charges measured with pre-amplifiers (sensitivity $\geqq 1 \times 10^{-15}$ coulomb) at the target and collector were found to be equal in shape, only differing in the polarity. In the experiment the amplitude and the rise time of the charge pulses are measured. From these quantities the mass and velocity of incident particles can be determined because the ion yield is a function of both mass and velocity of the particle whereas the rise time depends only on the velocity. To protect the thermal plasma of the ionosphere from entering the detector two grids were used biased with −50 V, +100 V. The area of the sensitive detector surface was 18.5 cm².

2.3. Cali bration

The detector has been calibrated with a 2 MV dust particle accelerator using micrometre sized iron spheres. Both telemetry channels were sensitive to particles in the velocity range 0.7–12 km s^{-1} and with masses from 8×10^{-14} to 2×10^{-10} g. At the ion collector charges from 1.4×10^{-15} coul and from 3 to 590×10^{-15} coul respectively could be recorded. The ranges of values given are valid together with a coincident measurement of the rise time. The electronic threshold of the noise detector was 3.5×10^{-15} coul.

The ion yield at the collector can be described by the relation $Q^+/m = 0.55\, v^3$ ($v \geqq 1$ km s^{-1}) which is plotted in Fig. 2 (v = particle velocity in km s^{-1}; m = particle mass in 10^{-12} g; Q^+ = ion charge in 10^{-15} coul). The yield agrees well with measurements using a plane tungsten target [2]. In Fig. 3 the rise time of the charge pulses is given as a function of the particle velocity showing a rather high variance in the low velocity region. Spectral measurements of the ion yield from the impact of iron particles showed that from 2 to 4 km s^{-1} only a small content of Fe ions is present. About 90% of the total charge consists of K and Ca ions and 10% of Na ions resulting from a known contamination of the Fe projectiles [4].

We assume that the flight signals were produced by dust particles with the cosmic composition [5] for alkalis of:

K, 0.09%; Na, 0.68%; Ca, 1.39%

or the lunar alkali composition [6]:

K, 0.12%; Na, 0.32%; Ca, 8.3% .

If we take it for granted that the individual ion yields are proportional to the partial masses we obtain a correction factor ranging from 26 to 60. The particle masses which were attributed to the flight signals can be smaller by the same factor compared with the laboratory calibrations.

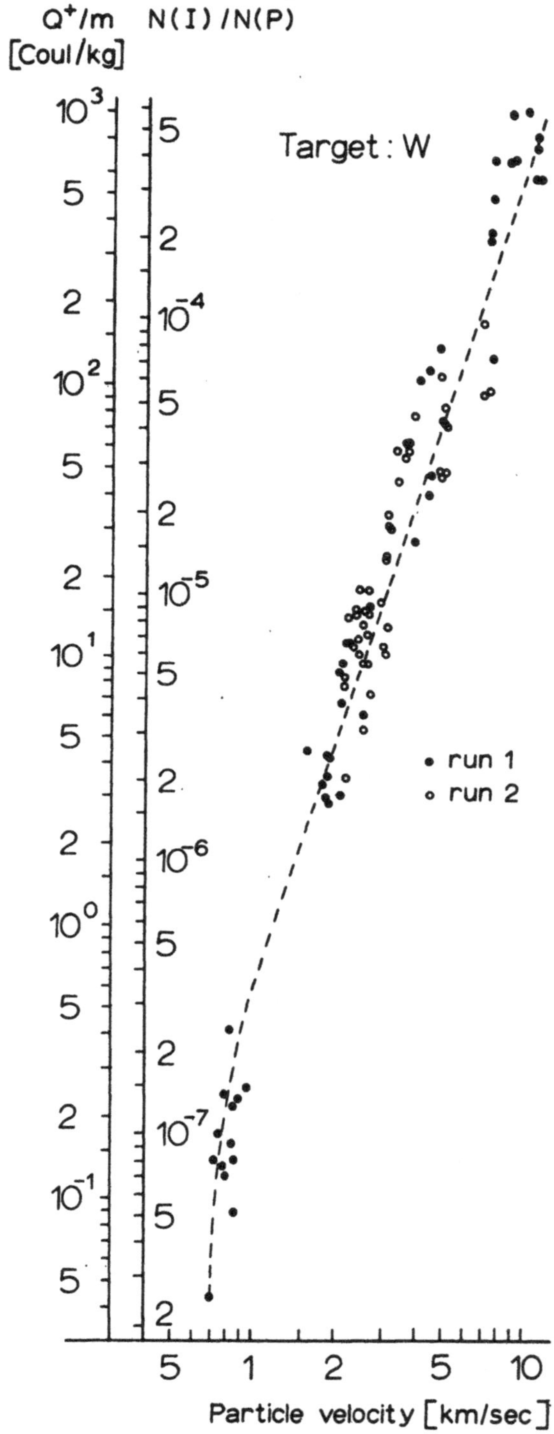

Fig. 2. Specific ion charge Q^{+}/m released at the impact of iron particles on a tungsten target as a function of the particle velocity v. $N(\mathrm{I})/N(\mathrm{P})$ is the ratio of the number of ionized atoms to the number of atoms of the incident particle.

3. Results

From an inspection of the experimental data about comets, radio and visual meteors Whipple [7], Jacchia [8, 9] and Verniani [10] conclude that meteoroids, on the average, have a bulk density of 0.3 g cm^{-3}. Some constituents are believed to consist of more dense grains ($\varrho \sim 3$ g cm^{-3}). The meteoroids have a porous and fragile structure leading to abrupt fragmentation during entry into the earth's atmosphere.

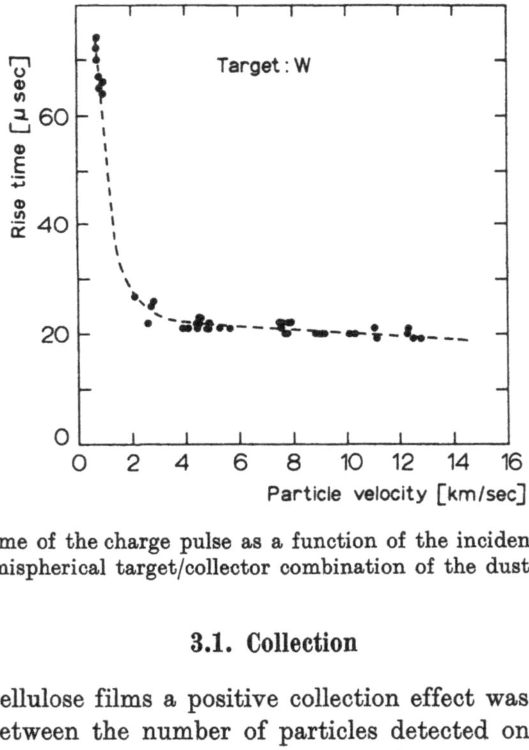

Fig. 3. The rise time of the charge pulse as a function of the incident iron particle velocity, for the hemispherical target/collector combination of the dust particle detector.

3.1. Collection

On the nitrocellulose films a positive collection effect was found. In Fig. 4 a comparison between the number of particles detected on flight and spare samples is given. On an area of 8 mm^2 77 ± 9 particles were seen. Approximately 170 ± 13 particles of an irregular shape are present compared which 120 ± 11 on spare samples. In total 127 ± 33 irregular particles with a mean diameter of $d \sim 0.1$ μm were collected above 60 km. The round particles found have very probably been produced during the evaporation process of the foils.

The particle influx φ has been calculated using the formula $\varphi = N/F\,t$; N is the number of collected particles, F the collection area, t the sum of exposure time and falling time of the particle. Assuming a density of 3 g cm^{-3} and a calculated falling time of 5.6×10^5 s down to 60 km an influx of

$$\varphi = (28 \pm 7)\ \mathrm{m}^{-2}\,\mathrm{s}^{-1}$$

with masses $m > 1.6 \times 10^{-15}$ g is obtained. The error is due to statistics and background.

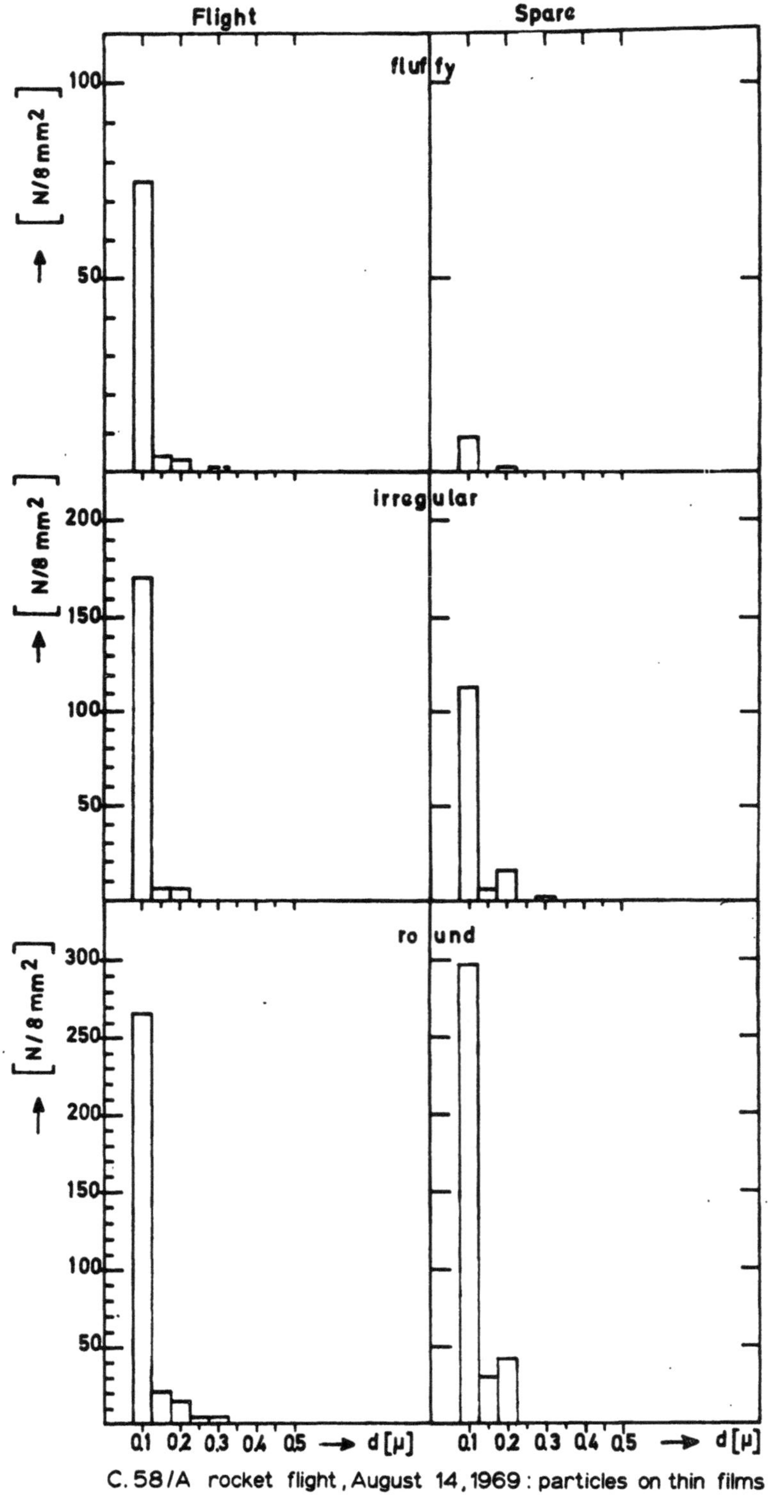

Fig. 4. Number of particles on flight and spare collection samples dependent on the particle diameter d.

Bigger particles were able to penetrate the nitrocellulose foils producing irregular fringed holes which are characteristic for impact velocities below 1 km s^{-1} [11]. Over a scanned area of 9 cm^2 the foils showed about 84 penetration holes of diameter $d \sim 2\,\mu m$. During penetration the loosely conglomerated particle easily breaks into its components. Below each hole there were counted 20 to 50 small fragmented particles with a mean diameter $d = 0.2$–$0.3\,\mu m$. A typical penetration hole is shown in Fig. 5 together with the particle debris on the collector. Assuming a bulk density of 0.3 g cm^{-3} for the primary particle, on the average, 30 to 100 fragmented particles of a density 3 g cm^{-3} with diameters from 0.2 to 0.3 μm can be expected on the metal collectors. A flux

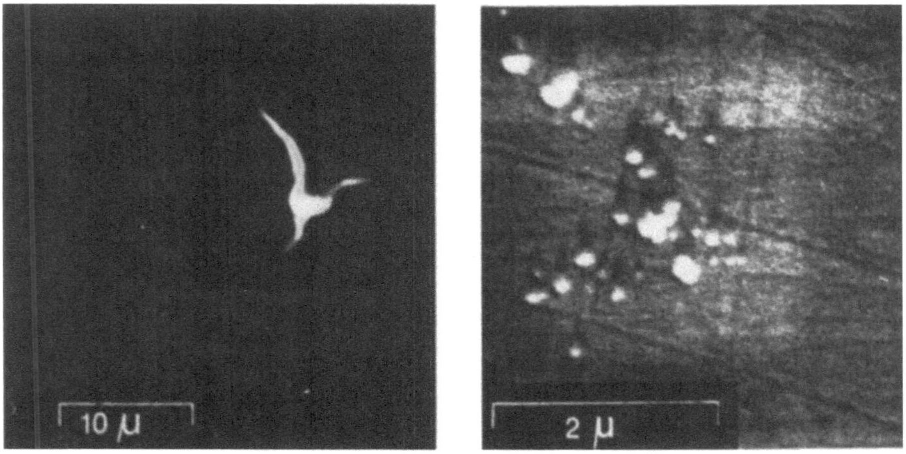

Fig. 5. Low velocity penetration hole in the foil and particle debris on the metal collector

value of $\varphi = 0.3\ m^{-2}\,s^{-1}$ is derived for particles with masses $m > 1.3 \times 10^{-12}$ g or diameters $d > 2\,\mu m$.

3.2. Detector

In Fig. 6 the trajectory and the velocity of the rocket are plotted. Below 105 km the angle of incidence was $<30°$, on upleg. Fig. 7 shows the ion charge collected over periods of 0.5 s as a function of the height or flight time. After the tray extension at 61.1 km both telemetry channels were overloaded. Within 10 s they recovered reaching zero level at 75 km. Between 63 km and 76 km the collector bias showed a 20% depression for an unknown reason. Stable flight conditions were found from 76 km on upleg.

Characteristic groups of pulses show up in the heights of 105.5, 82.3, 80 and 77.2 km. Three of these pulse distributions rise very sharply to high amplitudes at the low altitude end and decay rather slowly over an interval of 0.7–1.5 km. At 80 km a fairly symmetrical pulse group is observed. There were found only few ion pulses of low amplitude ($<3.5 \times 10^{-15}$ coul) till the end of the exposure. The coherent groups of signals concentrate in narrow layers <1.5 km. These signals could have been produced by porous dust particles of low density

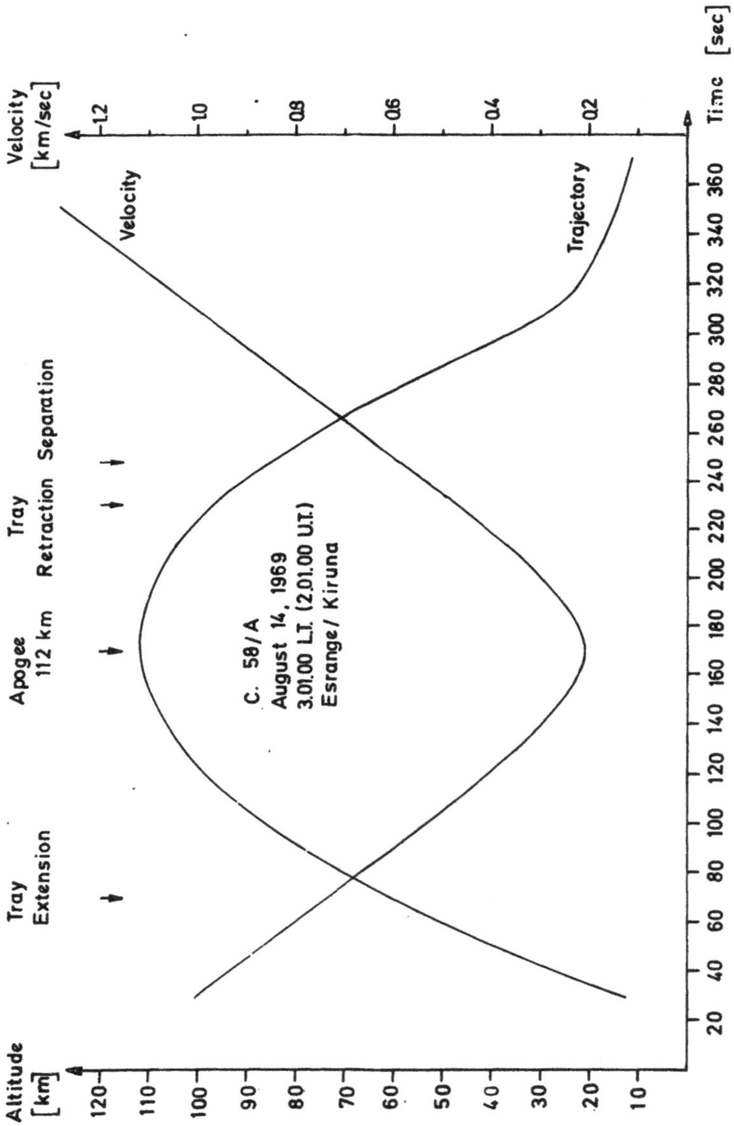

Fig. 6. Trajectory and velocity plot of the rocket.

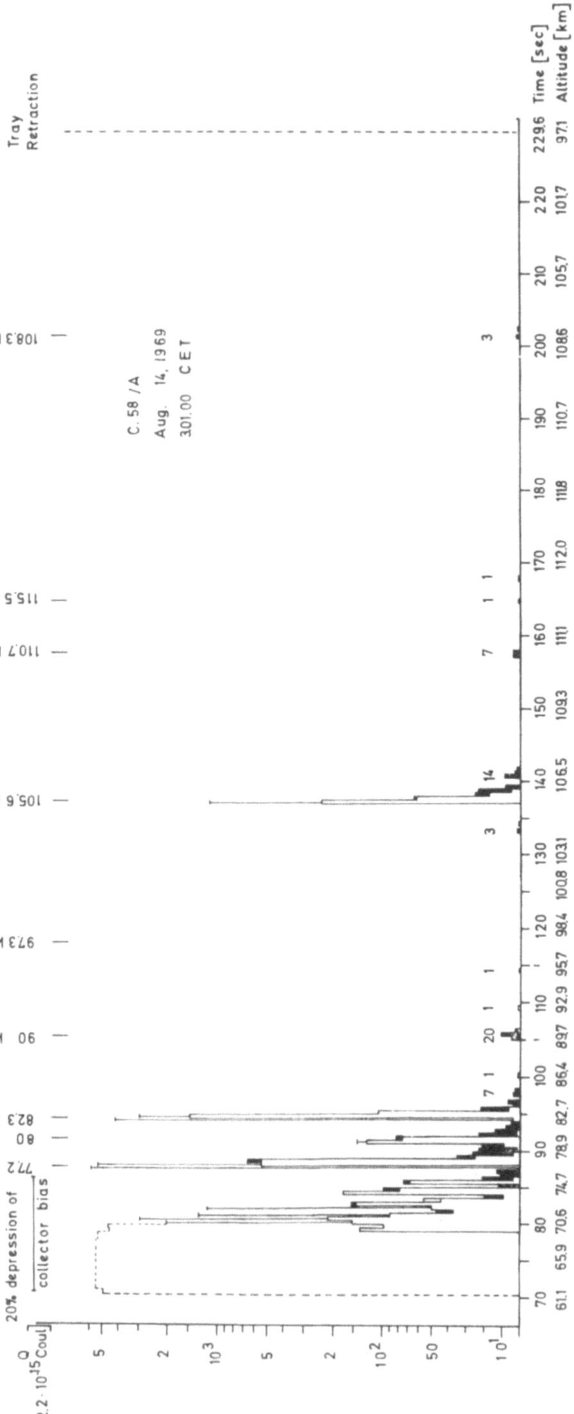

Fig. 7. Charge detected at the ion collector during the exposure time.

($\varrho \sim 0.3 \text{ g cm}^{-3}$) which had been severely decelerated during their entry into the earth's atmosphere. As a consequence of this shock the particles broke to pieces that, below 110 km, make impact on the detector with velocities only slightly above the rocket velocity. To a first approximation the breaking should begin when the area density (in g cm^{-2}) of the air is comparable with that of the particle. Such a behaviour leads to the observed asymmetrical distributions of the signals with a sharp rise at low altitudes and a slow decay.

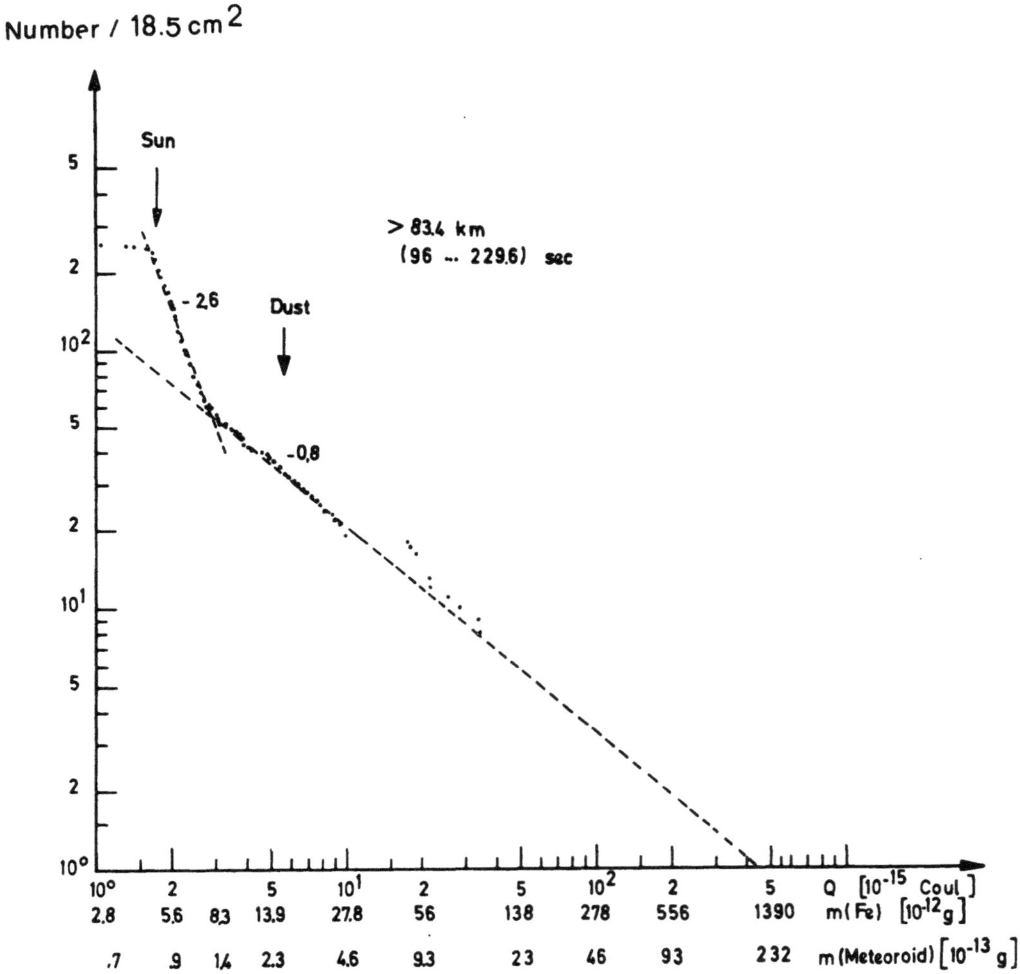

Fig. 8. Cumulative number of impact events above 83.4 km as a function of the released ion charge and meteoroid mass.

Some hints that the measured signals could have been produced by the ablation products of cosmic dust particles are given by mass spectrometric measurements [12, 13]. During the Leonid meteor shower Narcisi found a clear enhancement of positive metal ions (Fe, Mg, Al, Na, Ca) compared with non-shower conditions [14]: Iron-enriched layers have been found at 98 and 105 km and a silicon-enriched layer at 111 km. New rocket-borne cryogenic mass spectrometers of high resolution confirm the earlier measurements of metallic layers above 90 km [15, 16]. Because of the water-clusters found below it is possible that the measured pulse distribution around 80 km has been produced by the impact of ice crystals. Metallic ions of cosmic ablation products are believed to produce the condensation nuclei for the crystals.

Fig. 8 gives the cumulative number of events for altitudes above 83.4 km as a function of the collected ion charge Q^+. Above 97 km 53 periodic pulses

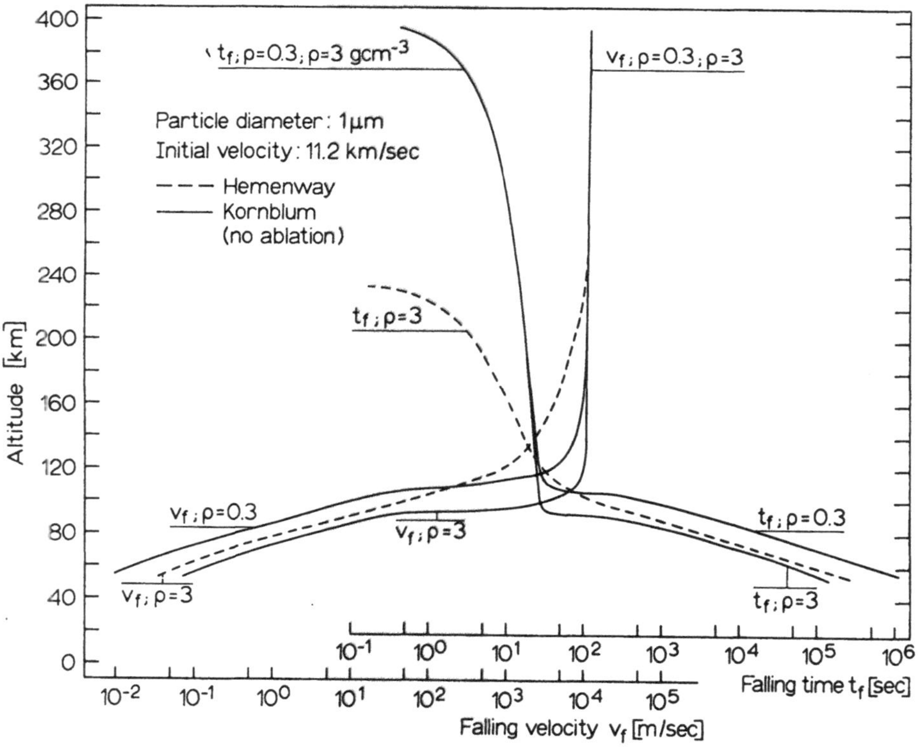

Fig. 9. Falling time t_f and falling velocity v_f in the atmosphere calculated for a dust particle ($d = 1$ μm) of density 0.3 and 3 g cm^{-3}.

($\sim 1.7 \times 10^{-15}$ coul) which were released by the sun hitting the grid system under a grazing angle of 0.9° have been eliminated.

Under the assumption of an enhanced plasma yield at the impact of cosmic or lunar dust particles meteoroid masses were attributed to the flight signals ranging from 1.6×10^{-13} to 2.7×10^{-11} g, or from 9.5×10^{-12} to 1.6×10^{-9} g when the calibration was made with Fe particles. Velocities measured at individual particles varied from 0.7 to 1.7 km s^{-1} with a mean value of 0.9 km s^{-1} (rocket velocity).

In the log–log plot the slope of the cumulative number against particle mass was found to be -0.8, or -2.4 if the particle diameter was plotted.

The influx of meteoroids has been derived from the detector signals. The falling time t_f and velocity v_f of the particles has been calculated [17,18). In this work at 400 km an initial velocity of 11.2 km s^{-1} has been taken. In Fig. 9

values of t_f and v_f are given for a 1-μm diameter particle with density 0.3 and 3 g cm^{-3} [19]. Mass loss by ablation, however, has not been included. The cumulative influx of dust particles above 83.4 km has been derived to be $\varphi \leqq 6\,\mathrm{m}^{-2}\,\mathrm{s}^{-1}$ for meteoroid masses $m > 1.6 \times 10^{-13}$ g. This value is very close to previous measurements [20] made near the earth. These measurements can be described in the flux diagram by the relation

$$\log \varphi = p + q \log m \quad \text{with} \quad q = -1.13 \quad \text{and} \quad p = -14.3\,.$$

By integration from 10^{10} g to 10^{-15} g the total mass influx to the earth can be estimated as 170 t/day. This value reduces to 25 t/day when we integrate below the curve ($q = -1$; $p = -14$) which is drawn through the Pegasus–Mariner–Explorer–OGO data points measured in outer space.

Acknowledgments

The authors would like to thank Prof. Dr. J. Zähringer for this encouraging support of this project. They are indebted to Mr. W. Weiss for preparation and to Miss H. Schröder for preparation and electronmicroscope-scanning of the collection samples; the help of Mr. D. Linkert and Mr. G. Matt in building the electronics is also acknowledged.

This work was supported by the Bundesministerium für wissenschaftliche Forschung.

References

[1] H. Fechtig and M. Feuerstein, J. Geophys. Res. **75**, 6751 (1970).
[2] S. Auer and K. Sitte, Earth Planet. Sci. Lett. **4**, 178 (1968).
[3] H. Dietzel, NASA Report N 69-32778, 1969.
[4] E. Grün and H. Dietzel, private communication.
[5] B. Mason, in: Meteorites, John Wiley, New York (1962) p. 151.
[6] T. Kirsten, O. Müller, F. Steinbrunn and J. Zähringer, Proc. Apollo 11 Sci. Conf. Geochim. Cosmochim. Acta, Suppl. 1, 1331 (1970).
[7] F. L. Whipple, in: The Zodiacal Light and the Interplanetary Medium, NASA SP. 150, 409 (1967).
[8] L. G. Jacchia, F. Verniani and R. E. Briggs, Smithson Contr. Astrophys. **10**, 1 (1967).
[9] L. G. Jacchia, Astrophys. J. **121**, 521 (1955).
[10] F. Verniani, Space Sc. Rev. **10**, 230 (1969).
[11] E. Grün and P. Rauser, Space Research IX, North-Holland Publ. Co., Amsterdam 1969 (p. 147).
[12] V.G. Istomin, Space Research III, North-Holland Publ. Co., Amsterdam 1963 (p. 209).
[13] R. S. Narcisi and A. D. Bailey, J. Geophys. Res. **70**, 3687 (1965).
[14] R. S. Narcisi, Space Research VIII, North-Holland Publ. Co., Amsterdam 1968 (p. 360).
[15] H. Wieder, J. Kissel, P. Lämmerzahl and J. Zähringer, International Symposium on Solar-Terrestrial Physics, Leningrad, USSR, 1970.
[16] F. Arnold, J. Kissel, D. Krankowsky, H. Wieder and J. Zähringer, paper presented at COSPAR XIIIth Plenary Meeting, Leningrad, 1970.
[17] J. J. Kornblum, J. Geophys. Res. **74**, 1893, 1908 (1969).
[18] U. Shafrir and G. J. Dittberner, Space Research IX, North-Holland Publ. Co., Amsterdam 1969 (p. 120).
[19] R. K. Soberman and C. L. Hemenway, J. Geophys. Res. **70**, 4943 (1965).
[20] S. Auer et al., Space Research X, North-Holland Publ. Co., Amsterdam 1970 (p. 287).

Space Research XI — Akademie-Verlag, Berlin 1971

ON THE NATURE OF THE CIRCUMTERRESTRIAL DUST CLOUD

V. G. Fessenkov

USSR

Photometric observations in two symmetrical points with respect to the solar vertical at $z = 70°$ at solar zenithal distances equal to 110° or more, made before dawn in the Astrophysical Institute, Alma-Ata, and also in Abastumani Observatory, have shown that the brightness of the nocturnal sky in the first point is always greater than in the symmetrical point. This difference has a very definite seasonal change: in the beginning of summer it is more prominent, being smaller in spring, sensibly low in autumn and nearly disappearing in winter. This phenomenon cannot be represented by any combination of all well known brightness components of the nocturnal sky, namely by integral brilliance of stars and interstellar medium, by zodiacal light and by ionospheric emission in a continuous region of spectrum (separate emission lines can be excluded by special interference filters).

The observed brightness difference in symmetrical points of the solar vertical can be explained only by the presence of some new component, the dust cloud around the earth. The detailed examination of possible models of such a cloud shows that it must have the shape of an ellipsoid of rotation elongated toward the poles of the ecliptic with sufficiently rapid decrease in density with distance from the earth's centre; the light scattered by dust particles of such a cloud has no noticeable polarization. The resulting increase in the nocturnal sky brightness is considerable near the horizon in the sun's direction but rapidly diminishing towards the zenith.

It is desirable to obtain more observational data from countries with suitable atmospheric conditions and especially to determine with the appropriate automatic camera raised to the height of some 30–40 km, the real indication of scattering, typical for the dust cloud near the earth.

The circumterrestrial dust cloud might be considered as a certain prolongation of the high atmosphere, so that its optical properties must be connected with twilight phenomena. After the sunset the twilight segment is at first very diffuse, then subsequently decreases to the horizon, being transformed into a thin band with a sharp falling-off of brightness with angular height, and finally vanishes. Before the morning twilight the course of events will be opposite. But the fact that just before the morning twilight the east part of the sky begins to be brighter is strange and not connected with the mechanism of light scattering in the high atmospheric layers. That fact shows that the scattering of light take place not in the atmospheric layers, but in the most expanded medium with a smaller density gradient with distance from the earth; in other words, this phenomenon corresponds to the presence of the circumterrestrial dust cloud.

This phenomenon has been observed by us many times, in expeditions, for example, in the Sary Ishik Otrau desert, to the south of the Balkhash lake, in conditions of high purity, homogeneity and transparency of the air. But for any precise conclusions quantitative observations are desirable.

For this purpose we have used photoelectric photometers for observations of the brightness at two symmetrical points of the solar vertical at zenithal distances $z = 70°$ at different negative heights of the sun (below the horizon), starting from 6°, up to the beginning of full night. The observations were carried out either before sunrise, or after sunset. From the twilight curves $\log J$, $\log J'$ obtained for these points it is possible, as has been shown by the author, to obtain data on the optical properties of the atmosphere up to 150 km in altitude, differentiating at the same time several layers of cosmic origin, situated presumably at an altitude of 80–90 km.

In all cases it is found that at night, when the twilight curves are parallel to the abscissa axis, i.e. do not show any decrease of brightness with zenithal distance of the sun (ζ), the curve of $\log J$ at symmetrical points to the solar vertical in the direction of the sun will always be higher than the curve of $\log J'$.

For the analysis of this phenomenon measurements carried out at morning twilight on Kamensky Plateau, near Alma-Ata, and partly in Abastumani, at high altitudes, were compared. The results of these observations show that $\Delta \log J$ reaches its maximum in the summer when the north declination of the sun is greatest, and then decreases to the autumn and to the beginning of winter. For different $\delta_{\odot}$ we have following ratios J/J' at symmetrical points:

$\delta_{\odot}$	$+23.5°$	0° (spring)	0° (autumn)	$-23.5°$
J/J'	1.60	1.45	1.41	1.09

To analyse this variation in the values of J/J' and to determine from these data the presence and general properties of the circumterrestrial dust cloud we must evaluate all the factors influencing the brightness of the night sky. These factors are: (i) the integral luminosity of stars and of the interstellar medium in the directions of selected points; (ii) the zodiacal component of the night sky airglow, depending on the position relative to the ecliptic; (iii) the ionospheric component, characterizing the properties of the upper atmosphere and consisting in the main part of the continuous spectrum (several emission lines could be easily eliminated with the aid of special filters); (iv) the luminosity of the circumterrestrial dust cloud, which needs to be evaluated.

The common brightness of the stars for both symmetrical points can be easily found for each day of the year from, for example, Allen's handbook. It can be shown that their value varies from about 0.5 to 1.5 of the luminosity of a star of 5^m per square degree. This unit will be employed in further analysis.

The evaluation of the influence of the zodiacal component, originated by scattering of light in the circumsolar dust cloud, is a more complicated problem. This component is large near the ecliptic at angular distances from the sun of 30–40°, where it exceeds in brightness even the brightest parts of the Milky Way. But the density of cosmic matter falls rapidly with distance from the ecliptic and the brightness of the zodiacal component accordingly decreases.

For evaluating this one can utilize the absolute isophotes of the zodiacal light, from which atmospheric extinction is excluded, adding to them our special observations made on the Ilijskaya desert north of Alma-Ata.

Having determined the ecliptical coordinates (latitude and longitude) for the first symmetrical point about the solar vertical one can derive the zodiacal component of the night sky brightness for each date of observation. The second symmetrical point with the given conditions ($z = 70°$, $\zeta = 110°$) is always situated accurately in the opposite direction in relation to the sun, i.e. it coincides with the "Gegenschein" (brightness of about 0.5 in our units).

A comparison of the variation of the stellar component at the two symmetrical points and that of the zodiacal component shows that the stellar component is much smaller and can be neglected. The zodiacal component reaches its maximum during the autumn months, especially in October, when there is a decrease in the observed values of $\Delta \log J$.

It is difficult to say anything about the ionospheric component and we are obliged to assume some average value obtained from other observations, equal in our units to 1.3 (luminosity of a star of 5^m per square degree). Then the observed ratio of brightness at symmetrical points would be

$$\frac{J}{J'} = \frac{A + B + C + x}{A' + B' + C'}, \tag{1}$$

where A, A', B, B', C,C' are respectively the stellar, zodiacal and ionospheric components for the two points and x is the (unknown) brightness of the circumterrestrial dust cloud (at the first symmetrical point): since the second symmetrical point of the sky at given values of z and ζ falls into the earth's shadow, the corresponding value of x' is zero). Then we have

$$x = \frac{J}{J'}(A' + B' + C') - (A + B + C)\,. \tag{2}$$

Calculations by the formula (2) give the following values for x (in our units):

summer	2.32–2.44
spring	1.25–1.35
autumn	0.37–0.47

The winter values of J/J' are very doubtful and for this reason they were not taken into account.

We have made estimates of the contribution of the circumterrestrial dust cloud to the brightness of the night sky at different seasons of the year and calculated the distribution of the brightness of this cloud at the first symmetrical point of the solar vertical: (i) on the assumption that the cloud has the form of a rotating ellipsoid enlarged in the plane of the earth's equator with $e = 0.8$ (it is clear that such a model is not real); (ii) and (iii) when the cloud is lengthened in the direction of the earth's axis to an ellipsoid of rotation with $e = 0.8$ and 0.9 respectively and with scale height of density $H = 2.5\ R_{\oplus}$. The last model is closer to reality but it cannot be adopted because it assumes full symmetry for the solar declination.

We may assume then that the circumterrestrial cloud is significantly longer in the direction perpendicular to the ecliptic plane. The calculations show that with this model one can obtain good agreement with observations which im proves as the eccentricity of the cloud towards the pole of the ecliptic is increased and as the gradient of density (or of scattering power) of the cloud is also increased.

We can conclude that the circumterrestrial dust cloud does exist and that it is elongated perpendicularly to the plane of the ecliptic and has a steep density gradient. Further estimates of its photometric properties by means of special observations from high-altitude balloons would be valuable.

One can suggest that since the polarization of the circumterrestrial dust is smaller than that of the zodiacal light, its particles must be larger than the interplanetary dust particles in the Solar System.

Photometric scanning following the solar vertical at zenithal distances of the Sun ζ not less than 110° must show a sharp decrease of brightness of the circumterrestrial cloud. component towards the zenith caused by a scattering index very elongated at small angles. This decrease of brightness will be greater in the autumn and less in spring and summer, but since the real form of the index is unknown, it would be reasonable to scan the night sky in different months at $\zeta = 110°$ following the almucantarat, for example at $z = 70°$.

It is evident that optical methods of measuring scattering of the light of the sky during the dark twilight time could be very effective for detection of the circumterrestrial dust cloud and for determination of its properties.

Space Research XI — Akademie-Verlag, Berlin 1971

THE RESULTS OF INVESTIGATION OF DUST IN THE UPPER ATMOSPHERE BY THE TWILIGHT METHOD

N. B. DIVARI

Odessa Polytechnical Institute, Odessa, USSR

Determination of primary twilight intensities is made at wavelengths 0.37 and 0.58 μm and the variations with scattering angle (relative phase function) of the sunlight scattered by the atmosphere are found for heights from 20 to 100 km. The scattering coefficients are defined for heights from 30 to 135 km at wavelength 0.37 μm. They disclose two maxima at heights near 50 and 100 km.

Since the use of exact methods for separating the primary twilight from the observed intensity of light of the twilight sky presents considerable difficulties, attempts were made to use different approximate methods for determining the scattering properties of the upper atmosphere. In this respect, some results were achieved by using Fessenkov's method of separating the primary twilight brightness from the twilight sky brightness observed at two points of the solar meridian with zenith angle $z = 70°$ [1]. In particular, the dust density at heights from 40 to 135 km was found [2]. In this way however, only the primary radiation intensity was determined from the twilight observations, the remaining scattering parameters being taken from the results of other investigations. Nevertheless, if the observations are not restricted to one single point, but involve the largest part of the sky, it is possible to determine some scattering properties of the upper atmosphere by means of twilight observations, in particular the relative phase function. This problem may be solved by the method of successive approximations.

Using the Fessenkov method of observing two points of the solar meridian, the dust diffusion coefficient has been determined as a function of height above the earth's surface. On the basis of these data, using the atmospheric aerosol phase function found by E. V. Pyaskovskaya-Fessenkova [3] and the known molecular scattering parameters, the intensity of primary and secondary twilight for the atmospheric model CIRA 1961 was calculated with a general ozone content of 0.30 cm [4]. The calculations were carried out for four wavelengths and four values of the atmospheric transmission coefficient. The results of these computations may be used to deduce the primary twilight from the brightnesses of the twilight sky observed at different points of the sky hemisphere. For this, on the sun side of the hemisphere near the horizon, it is convenient to use Fessenkov's method, that is, to find the ratios $k = B_2/B_2'$ of the intensity B_2 of the secondary twilight at a given point at the zenith distance z to the intensity B_2' of the secondary twilight at the same zenith distance z, but at a point lying in the earth's shadow ($A = 180°$) where the primary twilight intensity is equal to zero. Then, knowing the value of the factor k, it is possible to determine the secondary twilight intensity B_2 and after this to find the primary twilight intensity B_1 from the formula

$$B_1 = B_{obs} - k\, B_2' , \tag{1}$$

where B_{obs} is the intensity measured at the point on the sun side of the solar meridian. At large sun depressions, it is necessary to eliminate the night sky radiation. This method is described in detail by Fessenkov [1]. However, for points remote from the sun, where the primary twilight contributes only a small part to the observed radiation flux, such a method is useless, since B_{obs} is very near to $k\, B_2'$, and there will be a large error in computation by formula (1). In this case it is better to define the primary twilight by multiplying the observed intensity B_{obs} by the intensity ratio $B_1/(B_1 + B_2) = k$ which may be determined for the standard atmosphere [4] for corresponding wavelengths and a given value of atmospheric transmission coefficient.

Observations of the twilight were made by J. I. Zaginaylo in 1968–1969 at the observatory Mayaky near Odessa (USSR). The measured brightnesses of the primary twilight were used for the determination of the atmospheric phase function at different heights above the earth's surface.

The intensity B_1 of the primary twilight is the sum of the molecular B_m and dust B_d components. The brightnesses B_m and B_d are defined by the following formulae:

$$B_m = I_0 \frac{\beta}{n(0)} 3.046 \times 10^{-4} \frac{3}{16\pi} (1 + \cos^2\theta) \times$$
$$\times \int_0^\infty n(h) \exp(-\beta\chi - \alpha\chi_0 - \beta\chi' - \alpha\chi_0) \sec\gamma\, dh ,$$

$$B = I_0 f_1(\theta)\, (q\pi a^2) \int_0^\infty n_1(h) \exp(-\beta\chi - \alpha\chi_0 - \beta\chi' - \alpha\chi_0') \sec\gamma\, dh .$$

Here, I_0 is the flux of the sun radiation outside the atmosphere, $n(h)$ and $n_1(h)$ are respectively the densities of air molecules and dust particles at height h above the earth's surface, β is the absorption coefficient of pure air at height $h = 0$, α is the ozone absorption coefficient at height h, θ is the scattering angle, γ is the angle between the line normal to the earth's surface and the direction of vision at the scattering point; χ and χ_0 are the optical paths of the sun's ray on the way to the scattering point; χ' and χ_0' are the same paths for the ray from the scattering point to the observer, $f_1(\theta)$ is the phase function of the dust particles, $q\pi a^2$ is the average value of the effective scattering cross section of the dust particles. The factor 3.046×10^{-4} is introduced into the phase function to normalize the intensities of the scattering radiation to 1 square degree.

Introducing the designations

$$\mu = \beta \cdot 3.046 \times 10^{-4} \frac{3}{16\pi} (1 + \cos^2\theta) ,$$
$$\nu = f_1(\theta)\, (q\pi a^2) ,$$
$$\tau = \beta\chi + \alpha\chi_0 ,$$
$$\tau' = \beta\chi' + \alpha\chi_0' ,$$

we get

$$B_1 = I_0 \mu \int_0^\infty n(h) \exp(-\tau) \exp(-\tau') \sec\gamma \, dh +$$
$$+ I_0 \nu \int_0^\infty n_1(h) \exp(-\tau) \exp(-\tau') \sec\gamma \, dh . \quad (2)$$

The integrand of the first integral reaches its maximum value at some height h, which is called the effective height of the twilight ray. The values of the height h have been calculated by us in [5]. The plots of the integrand versus the height h show the profile of the twilight ray. The half-width of this profile, that is, the width at the height where the intensity of the scattered light is equal to half the maximum intensity, depends on the sun depression and on the wavelength of the scattered light. The half-width Δh may be computed if one plots the graph of the scattered light intensity against the heights h above the earth's surface. Such graphs have been plotted for different points of the sky at sun depressions $g_\odot = 2, 4, 6, 10, 14$ and $16°$ for four wavelengths 0.37, 0.50, 0.58 and 0.70 μm on the basis of the calculated data described in [4].

Values of the ratios $\Delta h/h$ of the half-widths Δh of the twilight ray to its height h above the earth's surface, taking into account ozone absorption (0.30 cm) have been computed for a standard atmosphere, CIRA 1961, and are given in Table 1. As seen from the table, these ratios depend on the sun depression and on the radiation wavelength.

Table 1
The values of the ratio $\Delta h/h$ of the half-width Δh to the height h of the twilight ray

$g_\odot$ (°)	λ (μm) 0.37	0.50	0.58	0.70
2	0.50	0.74	0.87	1.75
4	0.43	0.58	0.64	0.96
6	0.29	0.39	0.45	0.51
10	0.22	0.23	0.30	0.30
14	0.21	0.23	0.25	0.27
16	0.21	0.22	0.22	0.25

Using these values of Δh it is possible to write the integrals (2) in the following approximate form:

$$B_1 = I_0 [\mu \, n(h) + \nu \, n_1(h)] \exp(-\tau) \exp(-\tau') \sec\gamma \, \Delta h . \quad (3)$$

The values of τ may be calculated for every wavelength according to the formula

$$\tau = \tau_m + \tau_0 + \tau_d ,$$

where τ_m, τ_0 and τ_d are the optical paths along the tangent trajectory for air, ozone and dust respectively:

$$\tau_m = \beta \int_s n(h)/n(0) \, ds ,$$
$$\tau_0 = \alpha \int_s n_0(h) \, ds ,$$
$$\tau_d = \xi \int_s n_1(h) \, ds .$$

Here, β is the absorption coefficient of pure air, α is the ozone absorption coefficient, ξ is the average effective cross section of the dust particles. The angle γ in the formula (3) may be computed from the formula

$$\sin \gamma = h \sin z/(R + h) ,$$

where R is the earth's radius and z is the zenith distance of the observed point.

For τ' we have

$$\tau' = \tau'_m + \tau'_0 + \tau'_d$$

and

$$\tau'_m = \beta \int_0^h n(h)/n(0) \sec \gamma \, dh ,$$

$$\tau_0 = \alpha \int_0^h n_0(h) \sec \gamma \, dh ,$$

$$\tau_d = \xi \int_0^h n_1(h) \sec \gamma \, dh .$$

Knowing the values I_0, $\exp(-\tau)$, $\exp(-\tau')$, $\sec \gamma$, Δh and the observed brightness B_1, the scattering coefficient

$$\sigma(\theta, h) = \mu \, n(h) + \nu \, n_1(h) = B_1 \, [I_0 \exp(-\tau) \exp(-\tau') \sec \gamma \, \Delta h]^{-1}$$

may be determined as a function of the scattering angle and the height h above the earth's surface.

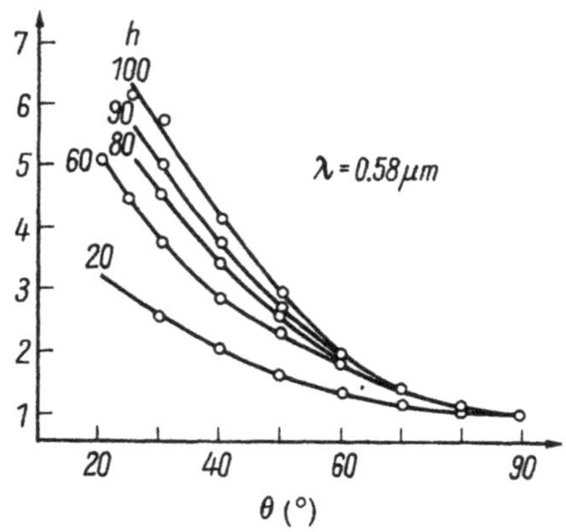

Fig. 1. The relative phase function for scattering at heights 20, 60, 80, 90 and 100 km for $\lambda = 0.58$ μm.

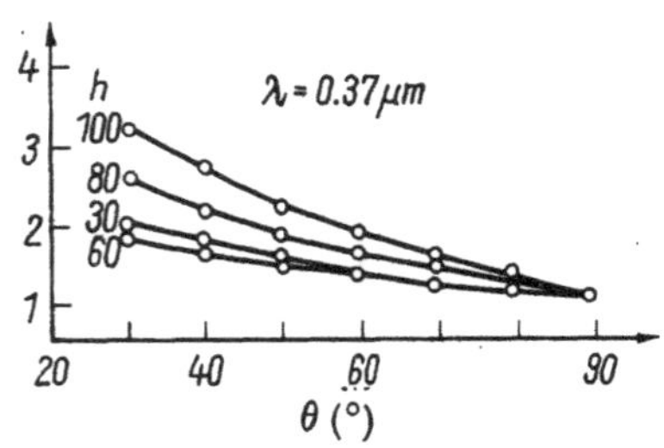

Fig. 2. The relative phase function for scattering at heights 30, 60, 80 and 100 km for $\lambda = 0.37$ μm.

In Figs. 1 and 2, the values of the relative phase function $f(\theta)$ are shown for two wavelengths 0.37 and 0.58 μm and for different heights $h = 20$, 30, 60, 80, 90 and 100 km. These relative phase functions are normalized to unity at $\theta = 90°$ $(f(\theta) = \sigma(\theta, h)/\sigma(90°, h))$.

As is seen from the Figs. 1 and 2, the phase function grows steeper with decreasing θ as h becomes greater. This fact indicates the increasing role of aerosols in scattering the sunlight in the upper atmosphere.

In the region of $\lambda = 0.37$ μm, the increase of scattering with decrease of scattering angle θ is flatter than in the region of $\lambda = 0.58$ μm.

It is of interest to compare the observed scattering coefficient $\sigma(\theta, h) = \mu\, n(h) + \nu\, n(h)$ with the scattering coefficient of pure air $\sigma_R(\theta, h) = \mu\, n(h)$. In Fig. 3, the values $\sigma(90°, h)/\sigma_R(90°, h)$ are shown for 0.37 μm in the interval of heights from $h = 30$ km to $h = 135$ km for four sets of observation data. The values $\sigma_R(90°, h)$ have been computed for the standard atmosphere model CIRA 1961. As seen from these graphs, the scattering maxima are at the heights of approximately 50 km and 100 km. It may be pointed out that the maximum at the height $h \approx 50$ km was found by Dave and Mateer [6].

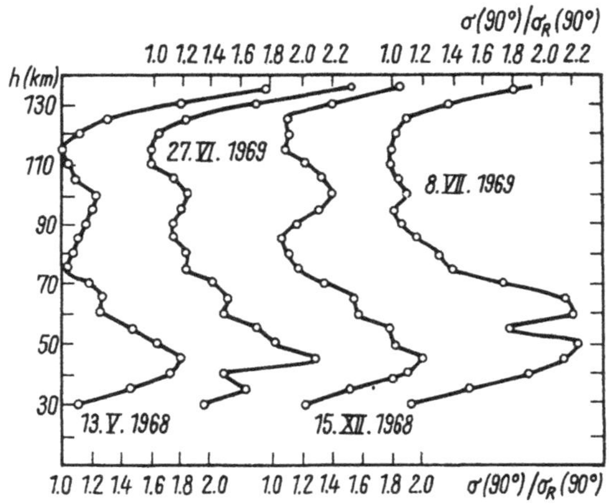

Fig. 3. The ratio $\sigma(90°, h)/\sigma_R(90°, h)$ of the observed coefficient of scattering to Rayleigh's coefficient for different data of observations at the scattering angle $\theta = 90°$.

An anomalous increase in solar visible light scattering between 45 and 55 km was found by Rossler [7] in rocket measurements. The maximum near the 100 km height lies in the meteoric zone and corresponds to the dust layer found by Kohl [8] in the polarization measurements of twilight. This maximum is near to the maximum found by Mikirov [9] in rocket photoelectric measurements.

References

[1] V. G. Fessenkov, Izv. Astrophys. Inst. Acad. Nauk Kaz. SSR (Alma-Ata) **12**, 3 (1961) (in Russian).
[2] N. B. Divari, Geomag. i Aeron. **4**, 886 (1964) (in Russian).
[3] E. V. Pyaskovskaya-Fessenkova, Issled. Rass. Sveta v Zemnoi Atmos., Akad. Nauk SSSR, Moskow 1957 (in Russian).
[4] N. B. Divari and L. I. Plotnikova, Astron. J. (USSR) **42**, 1090 (1965) (in Russian).
[5] N. B. Divari, in: Atmosphernaya Optika, Moscow, 1968 (p. 105).
[6] J. V. Dave and C. L. Mateer, J. Geophys. Res. **73**, 6897 (1968).
[7] F. Rossler, in: Space Research VIII, North-Holland, Amsterdam 1968 (p. 633).
[8] G. Kohl, Z. Met. **21**, 165 (1969).
[9] A. E. Mikirov, in: Space Research III, North-Holland, Amsterdam 1963 (p. 155).

Space Research XI — Akademie-Verlag, Berlin 1971

METEOR PARTICLE STUDIES FROM SPACE VEHICLES

T. N. NAZAROVA and A. K. RYBAKOV

V. I. Vernadsky Institute of Geochemistry and Analytical Chemistry, Moscow, USSR

Some problems related to methods of investigation of meteor particles by means of space vehicles and to the measuring instruments utilized by the authors of this paper are considered.

New data, obtained by means of detectors not sensitive to acoustic and vibrational noise of the vehicle are taken into account.

Discussion of the data on sporadic particles has been made on the basis of a comparison between the results of such experiments during which the same characteristics of particles were recorded in the same space region by different types of sensor.

Experimental data selected according to the above principle are in agreement (within the possible fluctuation range) and show that the spatial density of sporadic meteor particles in the vicinity of the earth at distances over 200 km from its surface exceeds the density of material in the zodiacal cloud.

Investigations of meteoric matter from rockets and satellites have been conducting for about two decades. For a long period of time these investigations were made with the help of piezoelectric sensors of different design, although from the very beginning researchers knew that the sensitivity of such detectors to acoustic and vibrational noise and hence the threshold sensitivity of instrumentation was limited due to the noises of the space vehicle.

Simultaneously with the accumulation of experimental data piezoelectric detectors were improved and detectors of another type were developed which were insensitive to vibrational and acoustic noise. Hence, with time scientists were provided with detectors for recording of various physical phenomena accompanying the impact of a fast flying meteoric body against the barrier.

Measuring results obtained by different methods defy simple comparison from a viewpoint of data about spatial density and the spectrum of particle masses. However, the use of instrumentation differing in principle makes it possible to obtain information about different properties of meteoric particles: energy characteristics, penetration, etc.

Along with merits each type of sensor has its demerits. For instance, piezoelectric sensors of different design have reliable calibration and large sensitive areas together with low weight. However, their sensitivity depends strongly on the noise level of the space vehicle. Other sensors, recording barrier penetration by meteor particles, can also have large sensitive areas, but their calibration is complicated and rather uncertain. Nevertheless sensors recording penetration of the barrier can be successfully used for studying meteoric hazard. Other types of sensors also have their strong and weak points. Thus, depending on the objective of the experiment one should choose the type of the detector most suitable for the specific case.

It is very desirable to use sensors based on various principles but recording the same physical phenomena accompanying the impact of particles against the barrier (for instance, their kinetic energy), since in this case easily comparable experimental data are obtained. However, while analysing the data one should not forget the existence of spatial and time fluctuations of the density of meteoric particles which may cause dispersion of experimental data.

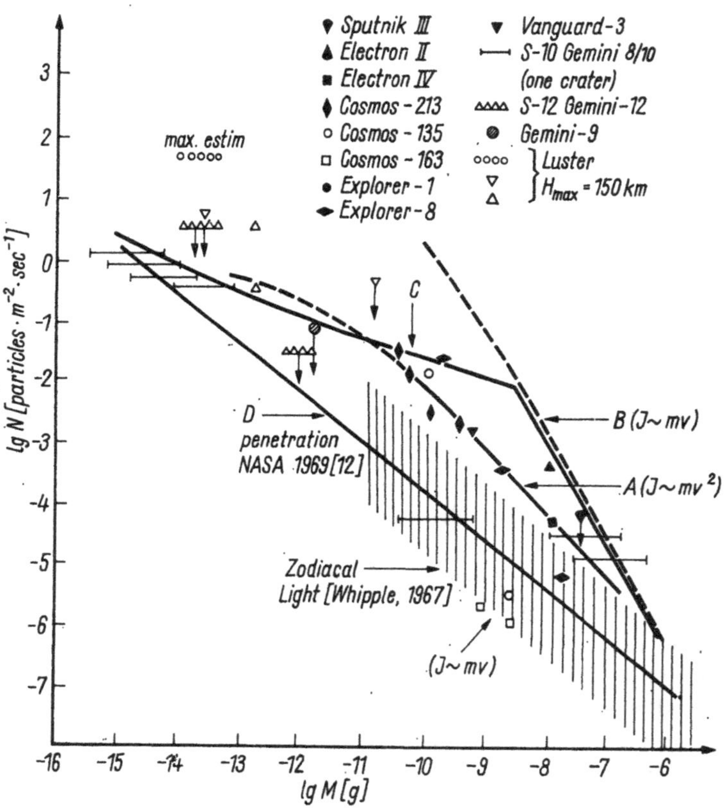

Fig. 1. Mean value of sporadic meteor particle flux in the vicinity of the earth at distances of over 200 km from its surface, according to data of experiments with sensors recording kinetis energy of particles (curve A). The data are recalculated for particles with $v = 20$ km s^{-1}. Corrections in keeping with a new calibration $I \sim mv^2$ [6] have been introduced into measured results from Explorer 1, Explorer 8 and Vanguard 3. Curve B ($I \sim mv$ [2]) is given for comparison. Curve C is from Hemenway [7]. Curve D is adopted on the basis of penetration detectors data by NASA the Smithsonian Astrophysical Observatory and Washington University [11].

Let us discuss the experimental data obtained from Soviet and foreign experiments concerning sporadic meteoric particles in the vicinity of the earth at distances of over 200 km from its surface. As pointed out above, we shall take into consideration the results obtained by the sensors recording the same parameter, namely, the energy of the particle.

Curve A in Fig. 1 is plotted from the results of Soviet and American measurements [1, 2] performed with the aid of acoustic sensors of different design and re-checked for the effect of temperatures [3, 4] as well as results of Soviet measurements by luminescent sensors [5]. We have introduced corrections to the data of the American measurements [2] using their new calibration by high-speed particles [6] ($I \sim mv^2$ instead of $I \sim mv$) which has led to a decrease of the particle flux.

On the same graph we have plotted experimental data from the instrument S-10 installed on Gemini 8/10 which made it possible to see on the earth the trails from the impacts of meteoric particles from circumterrestrial space in the form of craters on a polished surface which are also the result of a display of particle energy [7].

The S-10 instrument was exposed for four months. The survey of the instrument's receiving area has shown the presence of three large craters on it (two from particles with $m \sim 10^{-8}$–10^{-7} g and one from $m \sim 10^{-10}$ g). As far as small craters are concerned, due to very difficult labour of decoding, the results of processing only 1 mm² of the surface (six small craters with $d \sim 0.4$–1 μm) are included in [5]. Therefore, to obtain the final result about the number of small particles from the data of the experiment with S-10 it is necessary to obtain data from additionally processed areas.

Fig. 1 gives also results of measurements obtained with instrument S-12 installed on the Gemini 12 satellite to record craters from the impacts of meteoric particles [8, 9].

On the graph we have plotted Soviet data [10, 11] obtained with acoustic sensors calibrated for the pulse of the impacting particle and installed on Cosmos 135 and Cosmos 163. These data are in good agreement with the data obtained by the sensors recording barrier penetration (curve D) and approximately two to three orders of magnitude different from other data. If these sensors have been calibrated by particle energy, the latter discrepancy would have, naturally, been increased.

Curve A indicates the mean value of the particle flow per square metre per second of a surface arbitrarily oriented in space. This curve makes the curve C suggested earlier by Hemenway [7] more precise. For the sake of comparison we have plotted curve B from [2] (by calibration used earlier by the US scientists $I \sim mv$).

Over the last three years some scientists have considered the results obtained by means of acoustic sensors, including the enhanced density of matter near the earth, to be erroneous. NASA, the Smithsonian Astrophysical Observatory and Washington University [12] have adopted curve D as a model of the distribution of meteoric particle flux by masses near the earth obtained from recording the penetration of thin barriers.

As evident from Fig. 1, the discrepancy between curves A and D is on the average about 1.5 orders of magnitude. We do not cast doubt on curve D. However, new data obtained by us with luminescent sensors insensitive to vibrational and acoustic noise of the space vehicle are of a single character, but agree with the results of measurements performed with other detectors recording the energy of particles meeting a barrier of infinite thickness and indicate the excess of the spatial density of interplanetary matter in the vicinity of the earth over the density of this matter in the zodiacal cloud. The coincidence of data makes us continue to experiment in space with sensors of different types and to accumulate experimental data.

According to McDonnell [13] there is no appreciable dependence of the depth of penetration of a fast moving body into the barrier on the density of this body up to $\varrho = 0.4$ g cm^{-3}. (But porous aluminium used for this apparently cannot serve as a model of a conglomerate). In this case only one curve should have existed characterizing a spatial density of interplanetary particles and not two curves. One of the curves is incorrect due to some reason. However, there is a methodological difference between the experiments on the basis of which curves A and D are plotted and if the discrepancy in the results of corresponding measurements had continued, it would have become a subject of special consideration since it would have been a source of additional information on meteoric bodies, namely on their structure.

We must stress that the object studied by us has the specific property of low spatial density. Therefore, naturally the degree of reliability of the result of the experiment depends on the size of the surface exposed which is sensitive to impacts of meteoric particles S, exposure time t and St. Unfortunately, there are not a few experiments which, owing to the smallness of the above-mentioned values, should not have been used for quantitative estimates.

Fig. 1 shows data of some experiments on rockets Luster [8, 9, 14, 15] which, according to the authors, refute the results obtained by acoustic detectors though, as evident from the graph, the data of [13–15] lie above curve A. Hence, we believe that it is premature to draw final conclusions about the spatial density of interplanetary matter in the vicinity of the earth.

As for possible mechanisms leading to increase of spatial density of interplanetary matter near the earth, besides the opinion of Shapiro et al. [16–19], who have replied to this question in the negative, there are other points of view, among them the calculation of the motion of charged particles in the magnetic field of the earth [20–22].

References

[1] T. N. NAZAROVA, Cosmic Research 4, No. 6 (1966).
[2] W. M. ALEXANDER, C. W. MCCRACKEN, L. SECRETAN and O. E. BERG, Space Research III, North-Holland Publ. Co., Amsterdam 1963 (p. 891).
[3] T. N. NAZAROVA, G. D. KOMISSAROV and Z. V. VASUCOVA, Cosmic Research 6, No. 5, 797 (1968).
[4] J. L. BOHN, W. M. ALEXANDER and W. F. SIMMONS, 10th COSPAR meeting, London 1967.
[5] T. N. NAZAROVA, A. K. RYBAKOV, G. D. KOMISSAROV, Z. V. VASUKOVA and I. N. ORLOV, Cosmic Research 6, No. 6, 795 (1969).

[6] J. L. Bohn, W. M. Alexander and A. Wever, 10th COSPAR meeting, London 1967.
[7] C. L. Hemenway, D. S. Hallgren and I. F. Kerridge, 10th COSPAR meeting, London 1967.
[8] J. Weihrauch, U. Gerloff and H. Fechtig, Space Research VIII, North-Holland Publ. Co., Amsterdam 1968 (p. 566).
[9] D. E. Brownlee, P. W. Hodge and F. W. Wright, J. Geophys. Res. **74**, 1 (1969).
[10] B. P. Konstantinov et al., Cosmic Research **6**, 738 (1968).
[11] B. P. Konstantinov et al., Cosmic Research **7**, No. 6 (1969).
[12] NASA, NGR 48-002-033 Progress Report.
[13] J. A. M. McDonnell, Space Research X, North-Holland Publ. Co., Amsterdam 1970 (p. 314).
[14] M. H. Carr and H. J. Gabe, J. Geophys. Res. **72**, 4007 (1967).
[15] H. Fechtig, U. Gerloff und J. H. Weichrauch, J. Geophys. Res. **73**, 5029 (1968).
[16] I. I. Shapiro, D. A. Lautman and G. Colombo, J. Geophys. Res. **71**, 5695 (1966).
[17] G. Colombo, D. A. Lautman and I. I. Shapiro, J. Geophys. Res. **71**, 5705 (1966).
[18] G. Colombo, I. I. Shapiro and D. A. Lautman, J. Geophys. Res. **71**, 5719 (1966).
[19] D. A. Lautman, I. I. Shapiro and G. Colombo, J. Geophys. Res. **71**, 5733 (1966).
[20] N. B. Divari, Geomagn. i Aeron. **4**, 886 (1964).
[21] V. G. Fessenkov, Space Research XI, Akademie-Verlag, Berlin (p. 347).
[22] A. V. Artemjev, Cosmic Research, in press.

Space Research XI — Akademie-Verlag, Berlin 1971

COSMIC DUST AND METEOR SHOWERS

E. P. Mazets

Academy of Sciences of USSR, Moscow, USSR

A study of the cosmic dust in recent years showed that the spatial density of sporadic micrometeoroids in the vicinity of the earth and the moon and in interplanetary space do not differ appreciably. Consideration of the data from ground-based optical and radar observations leads to the conclusion that the probability of detection of small particles in meteor showers by direct measurement in space should be small. Measurements carried out on the Cosmos 135 and Cosmos 163 satellites in 1966–1967 extending over most of the major annual meteor showers failed to show any manifestation of their activity in the mass range of 10^{-9} g. Doubt is cast on reports of registering meteor showers wth microphone sensors, the high densities and pronounced variations of the micrometeoroid flux in space reported earlier obviously being simulated by noise and interference.

At early stages of the experimental study of micrometeoroids in space, most direct measurements on satellites and rockets employed the acoustic method using piezoelectric detectors. The unexpectedly high values of the cosmic dust flux in the vicinity of the earth obtained in these measurements [1] contradicted strongly, first, estimates of the spatial density of dust particles based upon zodiacal light studies and upon extrapolation of ground-based meteor observations and, second, the penetration rate measurements on Explorer 16 [2].

The first contradiction was removed temporarily by assuming the cosmic dust to concentrate near the earth in the form of a dust cloud, the second having been resolved by assuming a low density, and hence low penetrating power, of the major part of the dust particles.

Experiments performed in recent years on satellites and rockets (Fig. 1) removed the discrepancy between impact and penetration rate measurements while showing definitely that the concentration of micrometeoroids near the earth and the moon does not differ markedly from that of the cosmic dust particles in interplanetary space. Evidence was obtained that early microphone data were erroneous as a result of inadequate protection of equipment against noise of mechanical, electrical and thermal origin [3–5].

Acoustic measurements of the micrometeoroid flux near the earth carried out during a prolonged period in 1966–1967 on the Cosmos 135 and Cosmos 163 satellites with a noise-protected equipment [4–6] have shown the micrometeoroid impact rate to be 10^3–10^4 times lower than the values accepted earlier, and to agree well with penetration rate measurements on the Explorer 16 and 23, Pegasus 1 and 2 [7] and Ariel 2 [8] satellites. Later experiments on the Lunar Orbiter [9] and Explorer 35 [10] have yielded a value of the micrometeoroid flux in the cislunar space half that near the earth.

Despite methodological differences and uncertainties in instrument calibration, all these measurements covering a particle mass range of 10^{-7}–10^{-11} g agree well, on the whole, with one another and with estimates [11, 12] of dust particle concentration in the zodiacal cloud. The data thus obtained disprove the concept of the existence of a dust cloud around the earth.

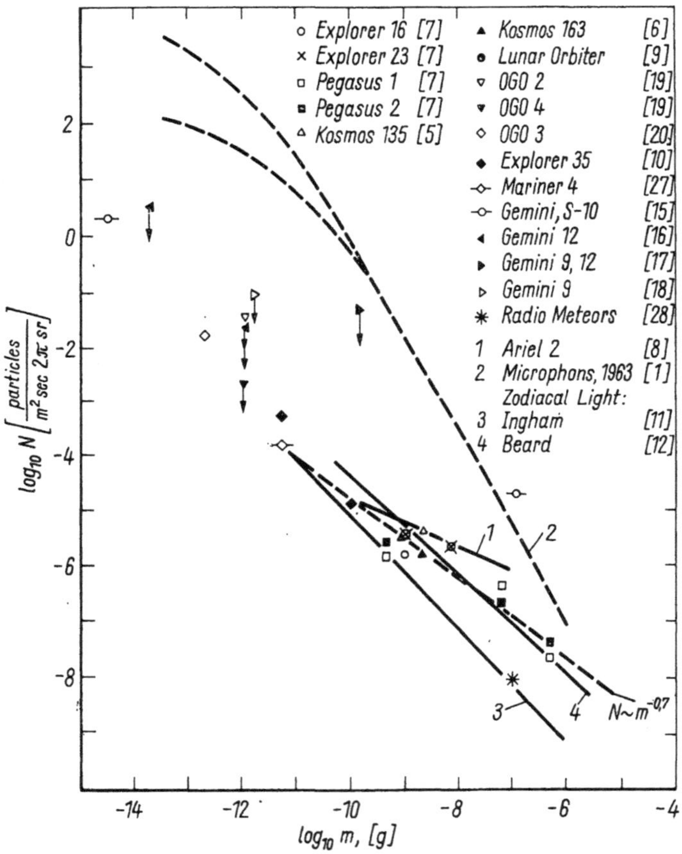

Fig. 1. Cumulative micrometeoroid flux.

The dust cloud hypothesis also received no support from a theoretical consideration of possible mechanisms of cosmic dust concentration in the vicinity of the earth [13]. Also it does not agree with observations of the distribution of zodiacal light brightness [14].

Equipment for the detection of very small particles with masses of 10^{-11} to 10^{-14} g was installed on the Gemini manned spacecraft [15–18] and OGO 2, OGO 4 [19] and OGO 3 [20] satellites. In most of these experiments such particles have not been reliably detected, the data obtained permitting one to establish only an upper limit for the flux of such small dust particles. Taken in their totality, these data apparently reveal a trend towards a decrease of the upper limit for the flux of particles of 10^{-11}–10^{-14} g down to $\approx 10^{-2}$ to 10^{-4} m^{-2} s^{-1}.

Results of cosmic dust collection in the upper atmosphere carried out by the Luster project have clearly shown the seriousness of the contamination problem in such experiments [21, 22]. The high degree of contamination of the collecting surfaces made it possible to obtain only upper limits for the number of collected particles of supposedly cosmic origin which correspond to essentially lower particle flux values than those obtained in similar experiments earlier [23].

Thus experiments on the detection of micrometre and submicrometre size particles likewise do not support the hypothesis of dust concentration near the earth.

The experimentally established low background of sporadic micrometeoroids in space might seem to favour detection of micrometeoroids belonging to strong annual meteor showers. However an extrapolation of ground-based optical and radar observational data on sporadic meteors and meteor showers leads one to a conclusion that the probability of detection in meteor showers of small particles of 10^{-8}–10^{-11} g should be low.

Erickson [24] considered optical observational data on meteors and, assuming the mass distribution law for sporadic and shower meteoroids to be the same and to remain unchanged also for small particles, evaluated the mean annual mass influx to the earth due to strong annual showers to be $\approx 1\%$ of that for the sporadic meteoroids. The increase of the micrometeoroid count rate by satellite detectors expected from these estimates at the maximum of the shower does not exceed the count rate due to sporadic meteoroids, and with detector sensitive areas available it should lie within statistical errors.

It should be noted that the assumption of the applicability of a common mass distribution law for both sporadic micrometeoroids and particles of the showers is hardly justified. In connection with this problem, Millman [25] has analysed radar data on meteors. He has emphasized that the cumulative mass distribution law $N \propto m^{-s}$ for sporadic meteors has the mean value $s \approx 1.3$, whereas the slope for annual showers is considerably less and lies in the 0.7 to 1 range. Extrapolation of such distributions to the region of small particles of 10^{-7}–10^{-11} g shows that the contribution of meteor showers to the total micrometeoroid count rate should be very small. Moreover, with a decrease of particle size there should be an increase in the role of nongravitational forces acting on such particles and restricting essentially their lifetime in the Solar System. From this viewpoint, the decrease of the slope of the cumulative distribution for micrometeoroids of 10^{-7}–10^{-11} g down to 0.6–0.8 (Fig. 1) as compared with $s = 1.3$ for meteoroids of the optical and radar range is quite significant. The action of such forces on particles in meteor streams could result in selective removal of small particles from the stream. In this case, detection of small particles in showers should become still more difficult.

At present there is apparently no reliable evidence on direct detection of meteor shower particles by satellite-borne instruments. Prolonged measurements on the Cosmos 135 and Cosmos 163 satellites in 1966–1967 during the epochs of the major meteor showers Geminids, Ursids, Quadrantids, Arietids, ζ-Perseids, β-Taurids, δ-Aquarids and Perseids failed to reveal any activity due to these showers in the particle mass range 10^{-10}–10^{-9} g above the sporadic micrometeoroid background of 10^{-5}–10^{-6} m^{-2} s^{-1} [6]. However in some works variations in the count rate of the piezoelectric microphones are considered as evidence for the detection of showers, both known and unknown, from ground-based observations [26].

We shall present some data on the time variations of noise obtained in measurements on the Cosmos 135 and 163 satellites. These data illustrate clearly the sources of possible errors in the acoustic method of micrometeoroid detection.

Two identical instruments were installed on Cosmos 135 to detect impact

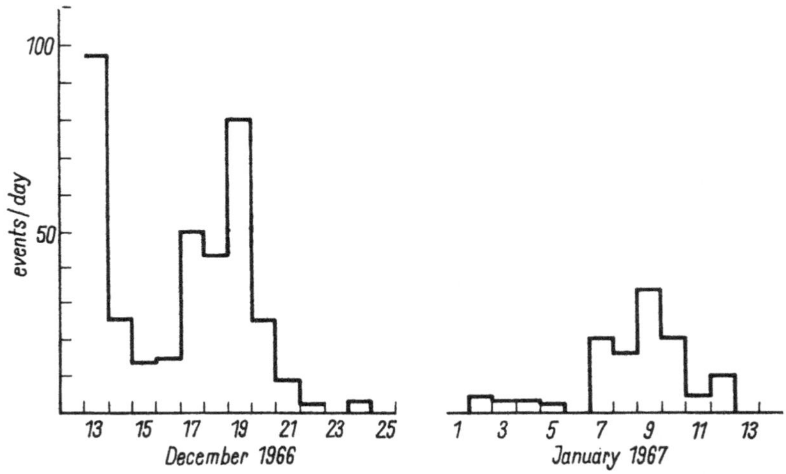

Fig. 2. Noise counted by detector mounted on satellite body.

with a sensitivity of 2×10^{-9} g [4, 5]. The sensors of the first instrument were fixed to a metal panel with a thorough acoustic isolation from the satellite body, the sensors of the second instrument being mounted directly on the body, as was usual practice earlier.

Fig. 2 presents the count rate of the second instrument vs. time which exhibits strong variations at a mean level of 2×10^{-4} m^{-2} s^{-1}. The count rate variations in Fig. 2 may look like a manifestation of a meteor shower, however the second instrument has been shown [5] to detect mechanical noise produced in the satellite body by temperature variations. During the same time the sensors mounted on the panel protected against mechanical noise have detected only three micrometeoroid impacts, which corresponds to a sporadic dust flux of about 5×10^{-6} m^{-2} s^{-1}.

The equipment on board Cosmos 163 consisted in measuring the micrometeoroid flux with two fully identical instruments also provided with multichannel pulse-height analysis. The sensitivity of the equipment was set at $\approx 2\times10^{-10}$ g; however, because of a weak noise of thermal origin only data corresponding to a threshold of 7×10^{-10} g were used in determining the dust particle flux [6].

During the period June to October 1967 eleven impacts have been detected at this sensitivity threshold by the two instruments in the absence of noise. The average particle flux constituted $3.5 \times 10^{-6}\ m^{-2}\ s^{-1}$.

Fig. 3 presents data illustrating the intensity of low-amplitude noise in the highest sensitivity channels of the instruments ($<7 \times 10^{-10}$ g). The period chosen for illustration is 7 August to 9 September 1967. The count rates of both instruments are shown in the diagrams. The readings of the second instrument lie, on the average, somewhat above those of the first; however, at the end of August one sees a considerable increase in the count rate of the second instrument well beyond the limits of statistical error. A conclusion that

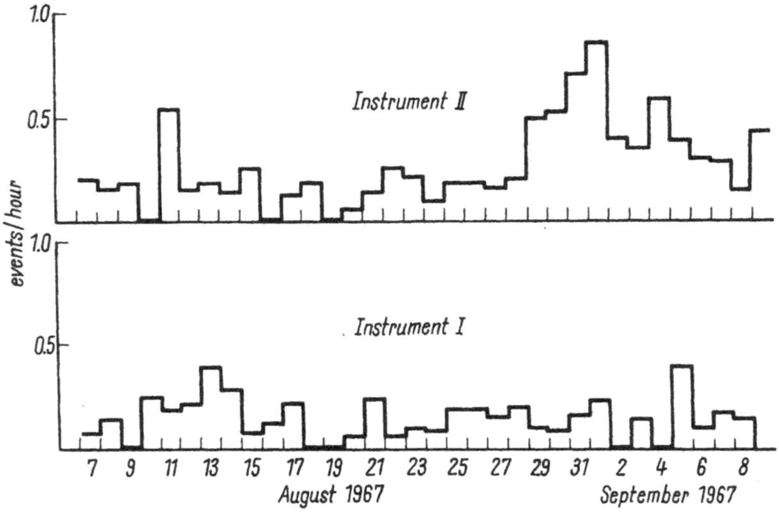

Fig. 3. Variations of thermal noise intensity.

we detected here some meteor shower would be erroneous. Let us consider some characteristic features in the instruments' readings.

Variations in the low-amplitude signal count rate are observed during the whole period of measurements extending from June to October 1967. The intensity of these signals drops sharply with increasing amplitude. Approximating the integral pulse-height spectrum by a power law yields a slope s of 2.5–3 in strong disagreement with any reasonable micrometeoroid mass distribution. A discrepancy in the readings of two identical detectors persisting for many days could not exist in the case of micrometeoroid registration, since the detector sensitivity should be isotropic due to averaging over a long period as a result of rotation of the non-oriented satellite.

Also the count rate variations cannot be explained as due to variations in the equipment sensitivity, whose stability was regularly checked in flight.

An analysis of signal distribution over the satellite orbit reveals a strong correlation of the count rate with detector temperature variations. Fig. 4 presents the distribution of the number of events in time after the satellite's emergence from the earth shadow. Time is measured here in fractions of the satellite's orbiting period T_s and is reckoned from the moment of crossing the shadow boundary. Also shown in the diagram is a typical variation of detector temperature along one turn. The measurement period in question was divided into two equal parts, one of them including a period of increased count rate. The total number of detected signals being small, the readings of the two instruments are summarized. A comparison of the two distributions shows them to be practically the same for the first and second periods of observations. The highest signal count rate is observed at detector temperatures below 40–35 °C. The corresponding parts of the precessing satellite orbit lie during this measure-

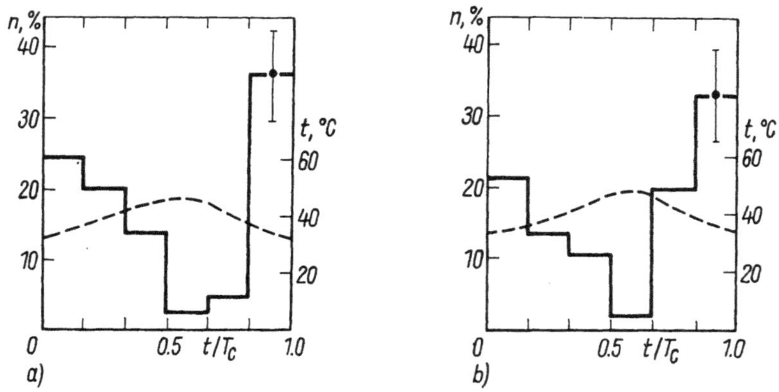

Fig. 4. Noise distribution in time after emergence of satellite from earth shadow. Solid line, relative number of signals n; broken line, variation of detector temperature: a, measurement period 7–24 Aug. 1967; b, measurement period 24 Aug.–9 Sept. 1967.

ment period near the earth's shadow boundary above the northern hemisphere.

The temperature dependence, pulse-height spectrum and average signal intensity agree with the results of pre-flight study of thermal noise in these detectors [5, 6]. It should be noted that the intensity of such noise in our equipment is very low. Even assuming all noise signals to be due to micrometeoroid impacts, the average count rate at a sensitivity level of $\approx 2 \times 10^{-10}$ g will be $\approx 1.0 \times 10^{-4}\ m^{-2}\ s^{-1}$ which is 10^3 times less than the cosmic dust fluxes derived from previous microphone data. However such an intensity is more than an order of magnitude higher than the real values of the micrometeoroid flux, and assuming the above time variations as representing real fluctuations of the dust particle flux would lead one to utterly wrong conclusions.

It can be certain that statements on the alleged detection of meteor shower particles with microphone sensors still appearing in literature are unfounded since the high densities and pronounced variations of the micrometeoroid flux in space reported earlier have undoubtedly been simulated by noise and interference.

Detection of small particles even in the strongest annual meteor showers represents a very difficult experimental problem which apparently cannot be solved by a statistical analysis of the detector count rate variations, and will require a development of methods involving determination of individual particle orbits.

References

[1] W. M. Alexander, L. Secretan, C. W. McCracken and O. E. Berg, in: Space Research III, North-Holland, Amsterdam 1963 (p. 891).
[2] C. T. D'Aiutolo, in: Space Research IV, North-Holland, Amsterdam 1964 (p. 858).
[3] C. Nilsson, Science **153**, 1242 (1966).
[4] B. P. Konstantinov, M. M. Bredov and E. P. Mazets, Dokl. Akad. Nauk SSSR **174**, 580 (1967).
[5] B. P. Konstantinov et al., Kosm. Issled. **6**, 738 (1968).
[6] B. P. Konstantinov et al., Kosm. Issled. **7**, 911 (1969).
[7] C. T. D'Aiutolo, W. N. Kinard and R. J. Naumann, Smithson. Contr. Astrophys. **11**, 239 (1967).
[8] R. C. Jennison, J. A. M. McDonnell and I. Rodger, Proc. Roy. Soc. **A 300**, 251 (1967).
[9] C. A. Gurtler and G. W. Grew, Science **161**, 462 (1968).
[10] W. M. Alexander, J. D. Corbin, C. W. Arthur and J. L. Bohn, Space Research X, North-Holland Publ. Co., Amsterdam 1970 (p. 252).
[11] M. F. Ingham, Space Sci. Rev. **1**, 576 (1962).
[12] D. B. Beard, Astrophys. J. **129**, 496 (1959).
[13] I. I. Shapiro, D. A. Lautman and G. Colombo, J. Geophys. Res. **71**, 5695 (1966).
[14] S. J. Peale, J. Geophys. Res. **73**, 3025 (1968).
[15] C. L. Hemenway, D. S. Hallgren and J. F. Kerridge, in: Space Research VIII, North-Holland, Amsterdam 1968 (p. 521).
[16] D. E. Brownlee, P. W. Hodge and F. W. Wright, J. Geophys. Res. **74**, 876 (1969).
[17] G. V. Ferry, N. H. Farlow and M. B. Blanchard, J. Geophys. Res. **73**, 3035 (1968).
[18] J. H. Weihrauch, U. Gerloff and H. Fechtig, in: Space Research VIII, North-Holland, Amsterdam 1968 (p. 566).
[19] C. S. Nilsson, F. W. Wright and D. Wilson, J. Geophys. Res. **74**, 5268 (1969).
[20] W. M. Alexander, J. D. Corbin, C. W. Arthur and J. L. Bohn, Space Research X. North-Holland Publ. Co., Amsterdam 1970 (p. 252).
[21] N. H. Farlow, J. Geophys. Res. **73**, 4363 (1968).
[22] H. Fechtig, U. Gerloff and J. H. Weihrauch, J. Geophys. Res. **73**, 5029 (1968).
[23] C. L. Hemenway and R. K. Soberman, Astron. J. **67**, 256 (1962).
[24] J. E. Erickson, J. Geophys. Res. **74**, 576 (1969).
[25] P. M. Millman, Space Research X, North-Holland Publ. Co., Amsterdam 1970 (p. 260).
[26] R. D. Ducus, Observatory **89**, 60 (1969).

Space Research XI — Akademie-Verlag, Berlin 1971

ON THE CONCENTRATION OF METEOR PARTICLES IN THE VICINITY OF THE EARTH'S ORBIT

E. N. Kramer

The Odessa State University, Odessa, USSR

A meteor activity prognosis in the vicinity of the earth's orbit based on meteor astronomy data can be used for comparison with rocket, satellite and interplanetary probe measurements.

Sources of meteor matter in the region of the earth's orbit are considered as follows: (i) A background of sporadic meteors. Results of optical investigations show a rather high concentration of meteor' particles near the earth's orbit; the perihelion distribution function increases at $q = 1$ AU. (ii) The cometary meteors: The method and results of cometary radiant calculations are described; they can be used for computing the coordinates and velocity of meteor particles in regions of probable increased meteor activity.

The concentration of meteor particles in different points of interplanetary space depends on the distribution of meteor orbits. The terrestrial methods of meteor investigations give data on orbits with perihelion distances less than 1.017 AU. The observed distribution with respect to perihelia is distorted by factors which depend on meteoric mass and velocity, on orbital elements and other parameters. It is usually possible to eliminate the influence of all these factors and obtain a "true" ("weighted") distribution of the orbital elements for various meteor gropus. For example, a "cosmic weight" factor is used to exclude the orientation and size of orbits. The "cosmic weight" was calculated for each orbit [1, 2] and used for statistical investigations. While processing the photographic and radio observations, it is necessary to take into account geometrical, physical and other probability parameters, as well as cosmic weight. When the true distribution is calculated, each of the meteors is considered with its total weight.

Table 1 contains the following data:

In the first column, the intervals of perihelion distances q are given; the table is then divided into two sections: the first section, I, contains the results obtained in Odessa and the second section, II, those obtained by American observers [3]. The first data are relating to "bright" meteors, the second to the "faint" meteors photographed by Super-Schmidt cameras. The columns n and g contain the observed and true distributions of perihelion distances; n is the meteor number, and g is the sum of corresponding weights.

As a rule, the interval length Δq was not taken into account when the data in the two last lines were calculated. For example, supposing that $0.9 < q < 1.0$, it was not taken into account that the observer cannot observe orbits with $q > r_0$, where r_0 is the heliocentric distance of the earth. Over a period of

Table 1
Distribution of perihelion distances (sporadic meteors)

q	I n	I g	I $G \times 10^{-3}$	II n	II g	II $G \times 10^{-3}$
0.0–0.1	9	50.0	12	78	80	51
0.1–0.2	12	78.9	3	275	388	36
0.2–0.3	9	35.5	0.45	125	112	3
0.3–0.4	18	79.4	0.50	150	128	1.8
0.4–0.5	10	40.2	0.17	150	160	1.3
0.5–0.6	14	32.4	0.08	250	252	1.4
0.6–0.7	16	89.8	0.17	220	224	0.9
0.7–0.8	16	93.2	0.13	240	268	0.8
0.8–0.9	17	78.2	0.09	315	352	0.8
0.9–1.0	45	≈160		550	≈356	
>1.0	20	≈ 60		130	≈ 68	

about six months, r_0 changes from 0.983 to 1.000 AU, therefore the true number of orbits in the interval of perihelion distance $0.9 < q < 1.0$ is greater than the observed number.

If $q > 1.0$ and the cosmic weight be calculated, as indicated in [1, 2], the formula

$$\text{C. W.} = \frac{V_g \sin i}{V_\infty^4}\left(2 - \frac{1}{a} - p\right)^{1/2}, \tag{1}$$

where V_g and V_∞ are the geocentric and outside-atmosphere velocities of the meteoric body, loses its meaning. Then the cosmic weight must be calculated by means of a corrected formula in which heliocentric distance of the observer

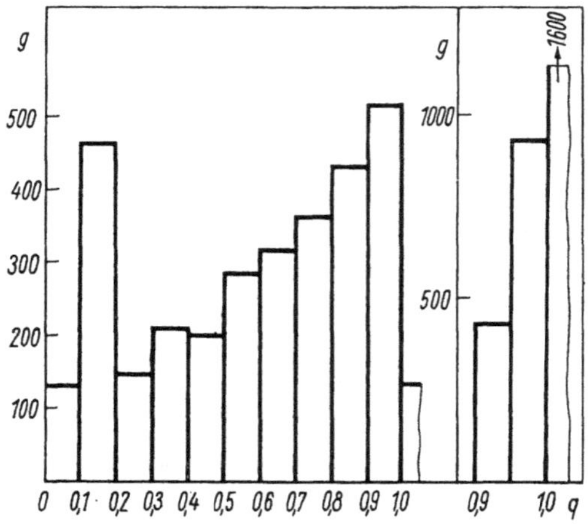

Fig. 1. Distribution of perihelion distances for "bright" and "faint" meteors (I + II).

differs from 1 AU. Moreover, it is necessary to remember that: (a) meteor orbits with $q > 1.017$ cannot be observed from the earth, and (b) during the six months mentioned above, orbits with $q > 1.00$ are not observed, since, during that time, the heliocentric distance of the earth is less than 1 AU.

The figures in the last line of Table 1, $n = 20$ and $n = 130$, correspond to an interval the length of which $\Delta q \ll 0.1$ AU. In order to reduce these figures to a standard interval $\Delta q = 0.1$ AU, it is necessary to multiply them by a correction coefficient which depends on the true interval of observational perihelion distances. All this has been used in compiling Table 2, in which the observed and true distributions at $q = 1$ AU are brought to standard conditions.

Table 2
Corrected distribution of perihelion distances

q	I			II		
	n	g	$G \times 10^{-3}$	n	g	$G \times 10^{-3}$
0.900–0.950	16	53	0.05	420	376	0.4
0.950–0.975	64	260	0.22	810	400	0.4
0.975–1.000	124	272	0.31			
1.000–1.017	250	750	0.60	1600	850	0.7

In the last columns of Tables 1 and 2, the distributions of average density G of perihelia are given. The distribution in the perihelion distance both for bright and faint meteors shows a sharp maximum at $q = 1$ AU (Fig. 1). The distribution of average density G decreases almost strictly monotonically with perihelion increase, and only at q near 1.0 AU does G increase slightly, but does not reach the value corresponding to that near the sun (Fig. 2).

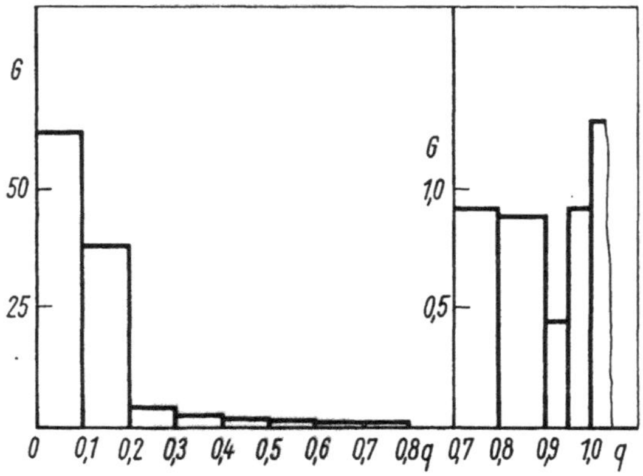

Fig. 2. Average density distribution of perihelion distances (I + II).

The decrease rate of G in the distribution curve at $q = 0.1$ AU for bright meteors is greater than for faint ones. The increase of G at $q = 1.0$ AU is greater for bright than for faint meteors.

The radio observations give only one maximum in the perihelia distribution for small values of q [4]. The number of orbits decreases almost monotonically with increase in perihelion. The decreasing tendency is somewhat slowly rectified when the figures in the range $0.9 < q < 1.017$ are brought to a standard interval.

We have applied all the correlations considered above to sporadic meteors. Analogous results may be obtained for shower meteors. From the point of view of meteoric activity prognosis, it is, however, more interesting to investigate the associations between meteor streams and comets. The hypothesis commonly admitted is that meteor streams are originated by comet disintegration. The evolution of meteor streams comes from gravitational perturbations. Moreover, meteoric dust is dispersed as a result of the Poynting–Robertson effect and other causes. After a comparatively short time, the meteor particles are gradually spread along the orbit of the parental comet. The width of the stream grows with time and reaches many millions of kilometres (e.g. about 30 million for the Perseid stream).

The associations between meteor showers and individual comets can be studied by means of calculations of the meteoric orbits and their comparison with the cometary ones. This method is too complicated, and its application is not always possible. Therefore it is better to calculate the theoretical cometary radiants and compare them with the observed meteor radiants. The method of cometary radiants can be successfully used for the planning of observations [5–8].

Table 3

Cometary radiants (a part of a cosmic radiant list)

No.		l	r	A	B	C	T	α	δ	V_g
735	1950 II	324	1.38	−0.04	−1.03	0.33				
740	1950 VII	125	1.48	−0.23	−0.96	−0.30				
742	1951 II	311	1.06	0.78	−0.04	−1.13	VIII−3	22	−37	50
744	1951 IV	166	1.22	0.29	−1.08	−0.26	III−6	138	−27	12
748	1951 VIII	119	1.40	−0.07	−1.02	−0.22				
751	1952 I	76	1.49	−0.83	0.72	−0.37				
		256	1.46	0.83	0.73	0.38				
753	1952 III	86	1.42	0.38	−0.99	−0.21				
755	1952 V	302	1.41	−0.36	0.42	1.05				
759	1953 II	163	1.22	0.77	0.13	1.01	III−3	229	+20	51
761	1953 IV	76	1.46	−0.06	−0.89	−0.54				
764	1953 VII	45	1.15	−0.32	−1.16	−0.07	XI−8	248	−43	9
771	1954 VII	75	0.88	0.25	−0.38	1.35	XII−7	198	+68	45
774	1954 X	182	0.98	0.18	−0.85	1.14	III−22	254	+57	34
776	1954 XII	255	1.18	0.79	−0.02	1.03	VI−6	299	+20	49
781	1955 IV	304	1.45	0.13	−0.74	−0.89				
782	1955 V	339	0.88	−0.15	0.45	−1.44	IX−2	78	−20	58
784	1955 VII	63	1.16	−0.10	−1.15	0.33	XI−26	297	+32	12
793	1957 II	299	1.38	0.07	−1.02	0.23				
798	1957 VII	125	1.48	−0.23	−0.96	−0.30				
800	1957 IX	210	0.95	−0.95	1.00	−0.43	IV−20	332	−23	67
		30	1.24	0.95	0.77	0.33				
801	1958 I	301	1.54	−0.11	−1.00	0.08				
803	1958 III	151	1.35	0.17	−1.16	−0.31				
809	1959 III	285	1.25	0.07	−1.23	0.28	VII−7	228	+34	11
810	1959 IV	339	1.59	−0.54	−0.64	0.72				
812	1959 VI	45	0.68	1.48	0.26	0.81				
814	1959 VIII	16	0.94	−0.07	−1.16	0.69	X−9	267	+52	21
819	1960 III	86	1.42	0.39	−0.99	−0.21				
821	1960 V	76	1.46	−0.06	−0.90	−0.54				
824	1960 VIII	42	1.18	−0.32	−1.15	−0.07	XI−4	243	−32	11
827	1961 II	357	1.06	−0.47	1.20	0.66	IX−19	101	+39	70

With the development of direct meteor investigation, the problem has arisen of the distribution of meteor particles in interplanetary space, and first of all in the Venus–Mars interval. Since the cosmic vehicle orbits lie near the plane of the ecliptic (the orbit inclination is less than 1°), it is expected that meteor particles of cometary origin would be recorded at the points of intersection of comet orbits with the ecliptic plane [9]. It is assumed that recording apparatus and methods will be improved and their efficiency raised.

Table 3 contains a part of the cosmic radiant list. In the first columns of the table, are given the serial numbers and official comet designations [10]. Using the Porter catalogue, the following data have been calculated:

(i) The ecliptical longitudes of the centre with higher concentration of meteor particles of cometary origin

$$l = \Omega \quad \text{or} \quad l = \Omega + 180° \quad (1950.0) \tag{2}$$

(ii) The corresponding heliocentric distances

$$r = \frac{a(1 - e^2)}{1 \pm e \cos \omega} . \tag{3}$$

(The table contains the data in the interval 0.7–1.6 AU only.)

(iii) The rectangular coordinates of true (heliocentric) radiants

$$A \equiv V_h \cos(L - l) \cos B = -K\sqrt{p}\, e \sin\theta , \tag{4}$$

$$B \equiv V_h \sin(L - l) \cos B = -K\sqrt{p}\, r^{-1} \cos i , \tag{5}$$

$$C \equiv V_h \sin B = -K\sqrt{p}\, r^{-1} \sin i . \tag{6}$$

The axis of abscissae is directed to the corresponding node; that of ordinates lies in the ecliptic plane, and the axis Z is directed to the pole of the ecliptic; the true radiant is directed away from the velocity vector of the meteor particle at the time of nodal passage; $\theta = -\omega$ for the ascending node, and

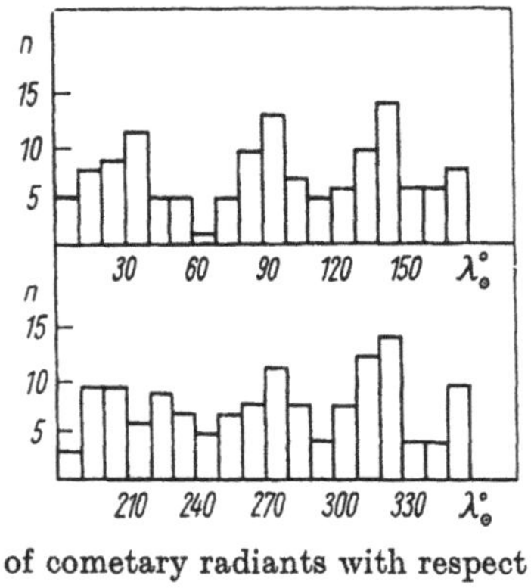

Fig. 3. Distribution of cometary radiants with respect to the sun's longitude.

$i = -i$, $\theta = 180° - \omega$ for the descending node; L and B are the ecliptic longitude and latitude of the true radiant.

The eighth column of Table 3 contains the time T of minimum distance between the earth and cometary orbits. The last three columns contain the right ascension α and declination δ of the geocentric radiant (equinox 1950.0) and V_g velocity with respect to the centre of the earth in km s^{-1}. Fig. 3 depicts the distribution of cometary radiants with respect to the sun's longitude. Table 3 and Fig. 3 can be used to estimate the closest approach between a space vehicle and meteoric particles of cometary origin.

In each separate case, the circumstances of the approach can be defined more precisely on the basis of the method proposed by Dubiago [11]. It is not possible, however, to expect an exact coincidence of the calculated and observed place of approach and the radiant data, because perturbations lead to dissipation and destruction of meteor streams. Separate clouds can be moved away from the stream axis.

The uttermost interest is excited by the prognosis of meteoric activity for hypothetical streams, the origin of which is associated with short-period comets. The permanent influence of solar radiation leads to rapid disintegration of a comet and to the formation of meteor streams; planetary perturbations, especially at close approaches, when the planetary attraction transfers the comet and the meteor particle clouds to different orbits and renders the prognosis too unreliable.

References

[1] E. Öpik, Proc. Roy. Irish Acad. Sect. **A 54**, No. 12 (1951).
[2] L. Jacchia and F. Whipple, Smithson. Contr. Astrophys. **4**, No. 4 (1961).
[3] R. McCrosky and A. Posen, Smithson. Contr. Astrophys. **4**, No. 2 (1961).
[4] B. L. Kascheev, V. N. Lebedinets and M. F. Lagutin, Meteoric Phenomena in the Earth's Atmosphere, Nauka, Moscow 1967.
[5] K. D. Pokrovsky and G. A. Shain, Sur les radiants cométariens, Travaux de l'Université de Perm **1** (1918).
[6] I. G. Porter, J. Brit. Astron. Ass., **60**, 1 (1949).
[7] E. N. Kramer, Izv. Astron. Obs. Odessa State University, **3**, 2 (1953).
[8] I. Hasegawa, General Index—List Theoret. Radiant, Documentation des Obs. Paris **11**, 1 (1958).
[9] E. N. Kramer, The Physics of Comets and Meteors, Naukova Dumka, Kiev 1965.
[10] I. G. Porter, Catalogue of Cometary Orbits, Mem. Roy. Astron. Soc. **39**, 3 (1961).
[11] A. D. Dubiago, The Determination of Orbits, Moscow-Leningrad 1949.

Space Research XI — Akademie-Verlag, Berlin 1971

SOUNDING ROCKET SAMPLINGS OF COSMIC DUST

D. S. Hallgren and C. L. Hemenway

Dudley Observatory, Albany, New York, and the State University of New York at Albany, New York, USA

Four new sounding rocket collections of cosmic dust in the upper atmosphere were made in 1969 as part of a continuing program to sample the dust concentration at various times of the year. The first two samplings were made during and just after the Zeta Perseid and Arietid meteor showers. The second pair of samplings was made during and after the Perseid shower as a comparison with the 1968 samplings and in general do confirm the concentrations observed at that time. Particle concentrations as a function of altitude and typical electron micrographs are shown.

In the past year four more payloads have been successfully flown using the Pandora 2 micrometeorite collection system. The rockets were launched in pairs, on 28 May 1969 and 9 June 1969, to sample before and during the Zeta-Perseid and Arietid showers and on 13 August 1969 and 22 August 1969 to sample during and after the Perseid shower. The second pair was a repeat of the 1968 sampling. The particle concentrations observed were similar to those obtained in comparable experiments in August and December 1968.

Table 1 shows a summary of the particle collection made in 1968 and 1969. Flights H, J, K and L are the new flights. Flight H, launched just prior to the Zeta-Perseid and Arietid showers indicates an altitude dependence for the particle concentration similar to that observed on flights F and G in December 1968, namely that event III, the region between 100–120 km, was nearly devoid of particles.

In both 1968 and 1969 a decrease in particle concentration with altitude was observed at the time of the peak of the Perseids and one week later. The particle concentration per unit volume from these flights is in general agreement with the Venus Fly Trap [1] as shown in Table 2. It is interesting to note that the particle concentration observed from the noctilucent cloud control flight is about a factor of 10^2 higher than most of the Pandora results. It is hoped that an opportunity to remeasure the high latitude dust concentration will result from a campaign in Kiruna in the summer of 1970.

On the Pandora 1 [2] flight a large number of particles were collected, mostly of a single type, rounded and low density. If this unique group of particles is subtracted from the total count the remaining particles represent a concentration about equal to the results from the Pandora 2 flights.

The in-flight shadowing technique used on the Pandora 2 payload has proved to be a very powerful device in acquiring reliable data from sounding rockets. It has completely eliminated any concern for post-flight contamination as

Table 1
1968 Pandora 2 rocket collections

Flight	No. of Particles	Range (km)	Particles/m³
D	3	87–102	1.62×10^2
12 Aug. 1968	7	102–124	1.15×10^2
	4	124–153	0.56×10^2
E	11	103–124	0.92×10^2
20 Aug. 1968	2	124–152	0.65×10^2
F	35	88–100	6.46×10^2
3 Dec. 1968	1	102–124	0.07×10^2
	20	124–153	1.72×10^2
G	40	85–100	18.9×10^2
14 Dec. 1968	3	100–121	0.34×10^2
	11	121–148	2.35×10^2
H	9	85–100	1.63×10^2
28 May 1969	0	100–122	0
	8	124–155	1.52×10^2
J			
9 June 1969	2	90–104	0.70×10^2
K	20	84–98	2.28×10^2
13 Aug. 1969	6	99–121	0.91×10^2
	2	121–143	0.48×10^2
L	2	75–84	5.84×10^2
27 Aug. 1969	27	84–98	1.51×10^2
	11	99–121	0.29×10^2
	9	121–143	0.32×10^2

Table 2
Sounding rocket collections

Flight	No. of Particles	Range (km)	Particles/m³
Venus Flytrap 6 June 1961	254	68–168	1.77×10^2
Noctilucent Cloud Control 7 Aug. 1967	354	75–98	1.76×10^4
Noctilucent Cloud 11 Aug. 1962	301	75–98	1.11×10^6
Pandora 1	44	53–62	1.14×10^4
18. Nov. 1965	824	62–71	2.77×10^5
	295	71–88	1.02×10^5
	272	88–148	2.43×10^4

evidenced by the extreme case of Flight H on 28 May 1969 where the payload was recovered lying on the desert with most of the doors open. Despite the fact that the grids were badly contaminated with non-shadowed particles it was still possible with patience to observe the single and double shadowed particles and thus to obtain reliable data.

The in-flight performance of this shadowing technique has been shown to be very good. Performance in this case is measured by the ability of the system to shadow faithfully all the particles all the time and not produce any false particles in the process. In order to monitor the quality of the shadows over all areas, the grids are seeded with tungsten oxide particles prior to flight. Seed particles can be found in almost every grid opening and provide an assurance that both first and second shadows can be detected in that area. Laboratory tests have shown that the seed particles must not be more than about a factor of ten larger than the cosmic dust particle or it is possible that reliable shadows will appear for the large particle but not for the small one. As is now well known, shadowing can produce spherical contaminants. These particles when found are readily identified, even when they are as small as 0.05 μm, by the fact that the particle impacts on the film with sufficient velocity to remove the evaporated material from the nitrocellulose film, leaving a particle with a gouge such as seen on the left side of Fig. 1. A laboratory test of the in-flight shadowing unit using seeded grids has shown that all of

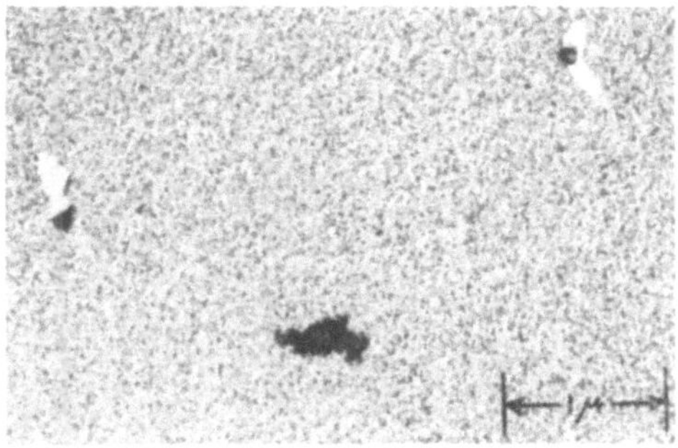

Fig. 1. Particle impact gouge on test film.

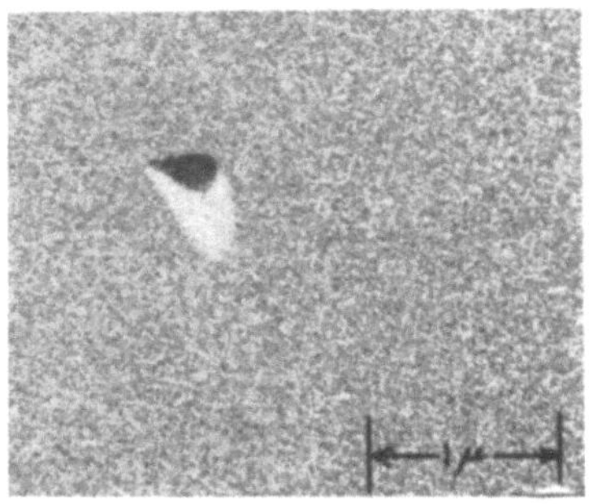

Fig. 2. Micrometeorite particle.

the particles observed are either double shadowed or are of the type with a gouge caused by the impact of a particle.

On flights prior to K the collector doors were opened in sequence and a fraction of a second was allowed for any contaminants to be swept away before the shadowing was initiated. Beginning with flights K and L all of the door openings were completed one half second prior to the first shadow event to allow more time for outgassing and for contaminants to be swept away in the subsequent events. It will be noted that there were no significant differences between the K and L flights and D and E both of which sampled the Perseids. There is no evidence that contamination played a significant role in either technique. Also there is no indication that the particles collected in either year were debris from the Perseid shower. On the Pandora 1 [2] experiment graphite was painted on the nose cone to test whether particles can originate from the outer surfaces of the rocket forward of the collector. In addition downward-looking grids monitored possible backstreaming from sources aft of the collection surfaces. In both cases negative results were obtained. In summary, it is not possible to attribute the collected particles to contamination sources.

It appears that only sub-micrometer particles have sufficient area-to-mass ratio to be slowed down sufficiently with respect to the rocket by outgassing

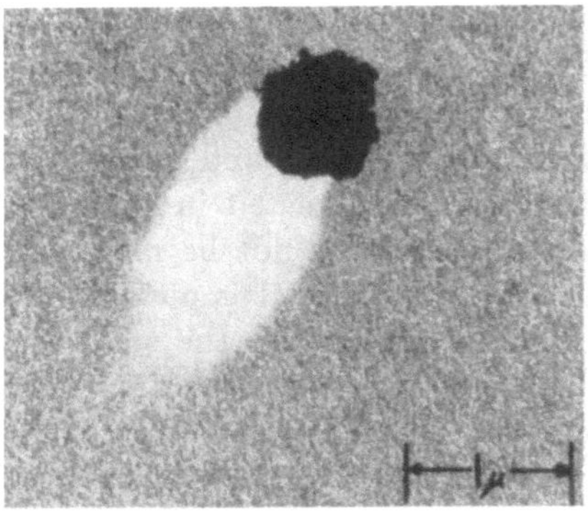

Fig. 3. Micrometeorite particle.

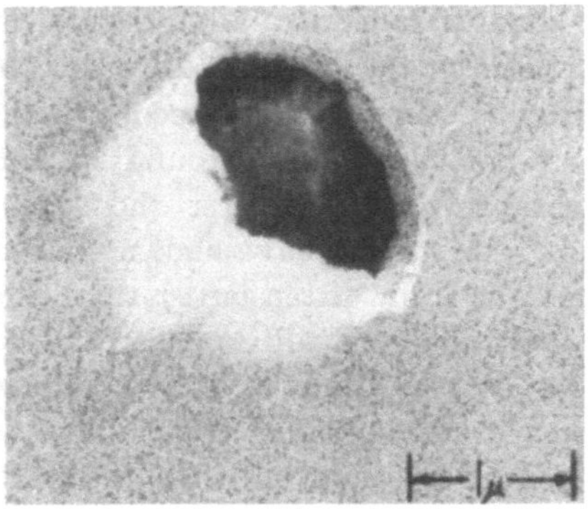

Fig. 4. Micrometeorite particle.

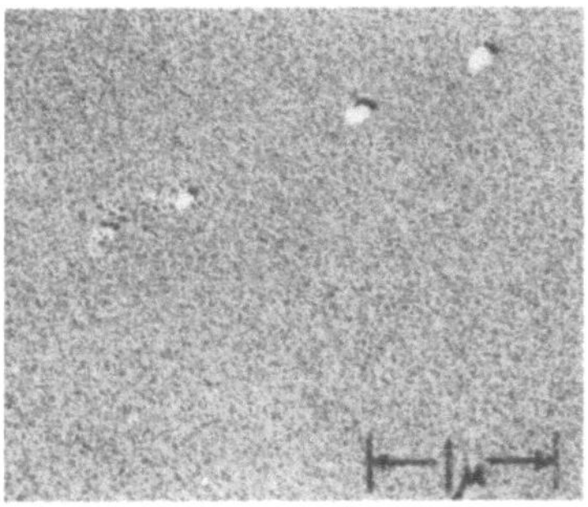

Fig. 5. Cluster of single shadowed particles.

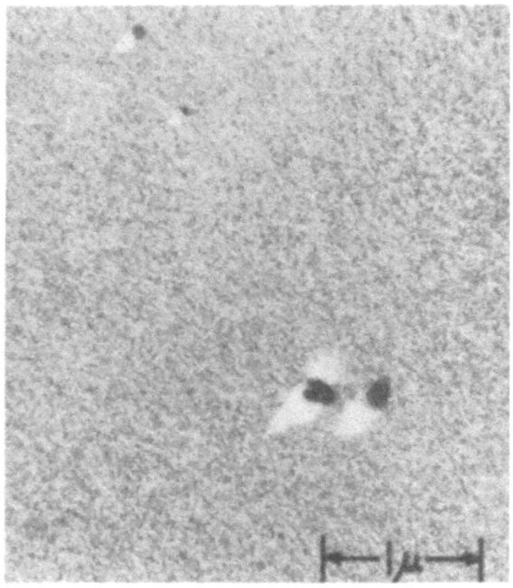

Fig. 7. Cluster of single shadowed particles.

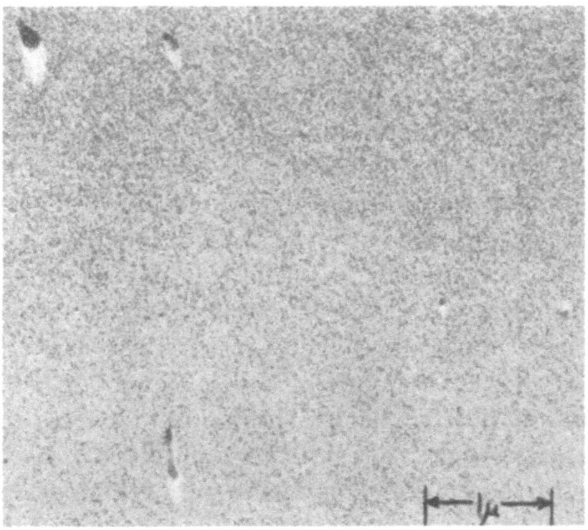

Fig. 6. Cluster of single shadowed particles.

and ram pressure effects to allow these particles to be collected on thin films. Penetration holes from larger particles were first observed in the Venus Flytrap experiment [3].

In general the collected particles are typical of those found in the past, that is irregular particles with a medium to low electron optical density. Fig. 2, 3 and 4 show several typical particles.

On the K and L flights in August two unusual observations were made. In event III on flight K and in event II on flight L an area no greater than two or three specimen grids (3 mm diameter) received a much higher concentration of particles than the rest of the grids in that event. The region of highest concentration of flight L averaged 30–40 particles in each 0.01 mm^2 grid opening. On flight K the concentration was approximately 75 particles per grid rather than the usual 3 to 5 particles. There is no evidence to indicate that either of these events was caused by contamination. Figures 5, 6 and 7 show some of the areas of high particle concentration. One possible explanation for the large number of small particles is that a larger particle has fragmented as it interacted with the rocket "atmosphere" and the debris was collected before it had time to be spread out. The particle concentrations in Table 1 do not include the large number of particles found on the grids mentioned above.

This fragmentation appears to be a similar phenomenon to that observed in the Venus Flytrap and to the clusters of penetration holes observed in the Gemini S-012 experiment [4]. Thus a significant fraction of the dust population appears to be a fragile conglomerate of small particles which would be unlikely to register on penetration sensors but could be detected by microphone and plasma sensors.

Acknowledgments

The authors wish to express their thanks to Amy Jettner, Seth-Ann Hayes, Anthony Laudate, and David Wachtel for many busy days spent in preparing and scanning samples, and to Donald Tackett for the operational support for these rocket flights. Our thanks also are given to the Sounding Rocket Branch at Goddard for their support.

The financial support for these experiments was given under NASA Grant NGL-33-011-001 and Air Force Cambridge Research Laboratory Contract F 19628-67-C-0191.

References

[1] C. L. Hemenway and D. S. Hallgren, in: Space Research X, North Holland Publ. Co., Amsterdam 1970, (p. 272).
[2] C. L. Hemenway et al., Smithson. Contr. Astrophys. 7, 93 (1963).
[3] C. L. Hemenway and D. S. Hallgren, in: Space Research IX, North Holland Publ. Co., Amsterdam 1969 (p. 140).
[4] C. L. Hemenway, D. S. Hallgren and J. F. Kerridge, in: Space Research VIII, North Holland Publ. Co., Amsterdam 1968 (p. 521).

Space Research XI — Akademie-Verlag, Berlin 1971

ELECTRON MICROPROBE STUDIES OF COSMIC DUST IMPACT CRATERS

O. K. Griffith, T. S. Renzema, D. S. Hallgren and C. L. Hemenway

State University of New York at Albany, New York, and Dudley Observatory, Albany, New York, USA

An electron microprobe analysis of the cavity surface and lip of each of three impact craters found in stainless steel plates which were exposed for four months at an altitude of approximately 400 km in the Gemini S-010 experiment is described. Specimen current images and X-ray images are presented showing the distributions of the principal elements in these locations. Estimates of the compositions of the impacting particles are made on the basis of sample-to-standard count ratios obtained in a dispersive analysis of the X-rays from selected spots within the craters.

1. Experiment

The Gemini S-010 experiment provided cosmic dust impact data from four highly polished stainless steel plates. These plates were exposed for four months at an altitude of approximately 400 km before being returned to earth. A technical description of the S-010 experiment and the results of a Stereoscan electron microscope study of surface features have been previously reported [1, 2]. Three craters larger than 10 μm were found on these plates. These craters are identified by their longest dimension, namely 290 μm, 160 μm and 17 μm.

Laboratory studies of hyperballistic impacts have shown that the walls of the crater produced may contain material left by the impacting particle [3]. A qualitative chemical analysis of the S-010 impact craters has been carried out.

The craters were analyzed using a CEC-CAMECA electron microprobe. The analysis was performed at an accelerating potential of 20 kV. An anti-contamination device and low beam current were used to avoid depositing carbon on the samples. The CAMECA probe has an X-ray take-off angle of 18°. In order to detect material in the deepest crater a modified sample holder was constructed which tilted the specimen so as to obtain an effective take-off angle of 54°. Point analyses were made at selected spots on the bottom, sides, and lip of each crater. The probe was used to detect elements with atomic number 6 and above. The highly irregular surface of the crater walls precludes any absolute determination of the amounts of material remaining in the craters. This is especially true of elements emitting extremely soft X-rays. Relative concentrations have been estimated for each crater by adjusting count rates for the detector sensitivity as determined by pure elemental standards. Scanning photographs were taken for the X-rays lines of several elements to determine their relative distribution. The arrow near the edge of each figure points toward the location of the spectrometer.

290 μm Crater

The 290 μm crater from the Gemini S-010 experiment is a large multiple crater that is shallow over most of its area. Its dimensions are 290 μm × 180 μm × 32 μm. Fig. 1 shows a stereoscan picture of the crater. There

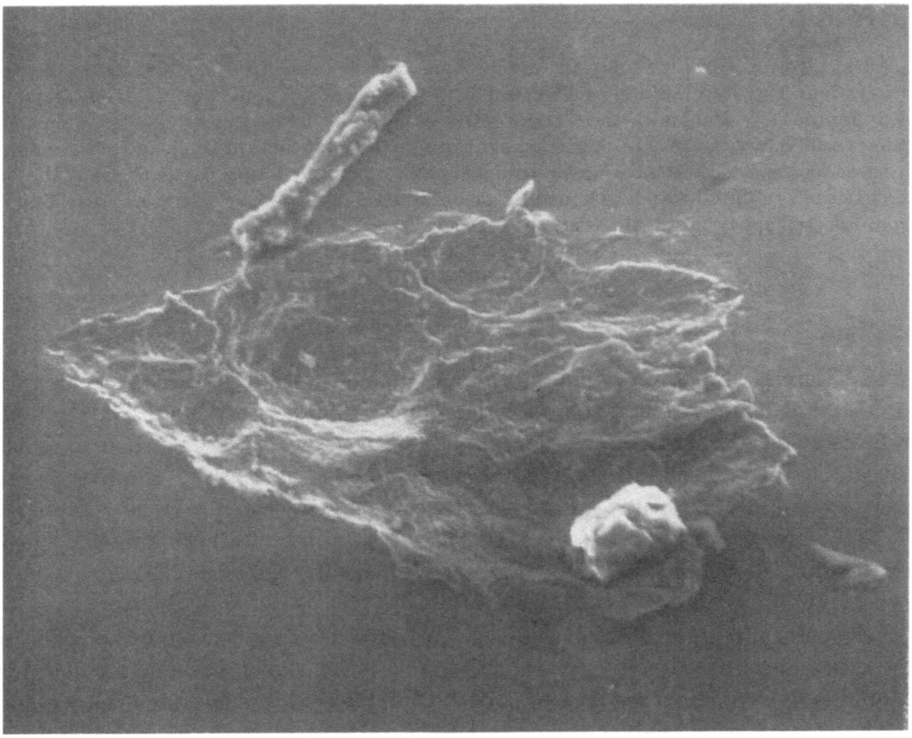

Fig. 1. 290 μm crater — scanning electron microscope image.

are two prominent particles in the figure which appear not to have been associated with the particle or cluster of particles which formed the crater. There is a long particle outside the crater to the right of the narrow tip and a roughly spherical particle inside the crater opposite the first. This second particle extends above the lip of the crater by 15 to 20 μm. Probe analysis of these particles shows them to differ significantly from the crater surface. Fig. 2 shows the scanning image produced by carbon Kα radiation. Fig. 3 is a similar scan for the silicon Kα line. Carbon and silicon are the most abundant elements in the particles and crater surface debris respectively. Further analysis showed a third particle located just below the large spherical particle in Fig. 1. A scan

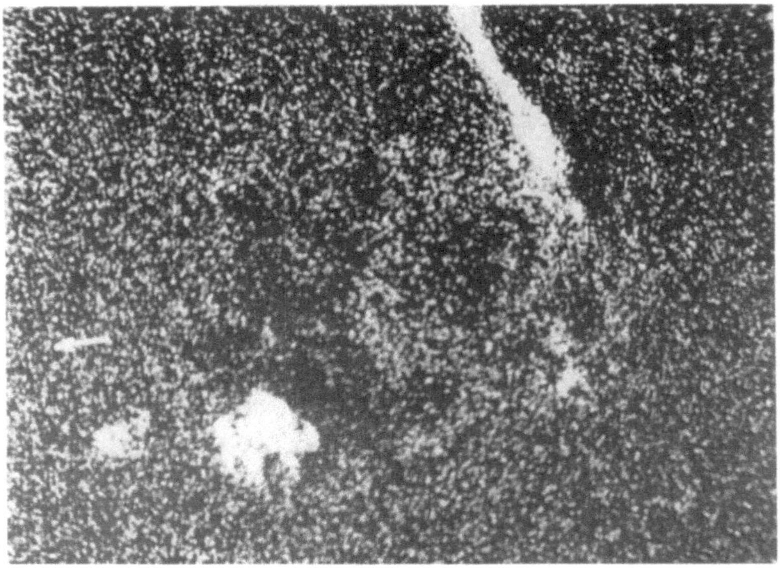

Fig. 2. 290 μm crater — carbon Kα image.

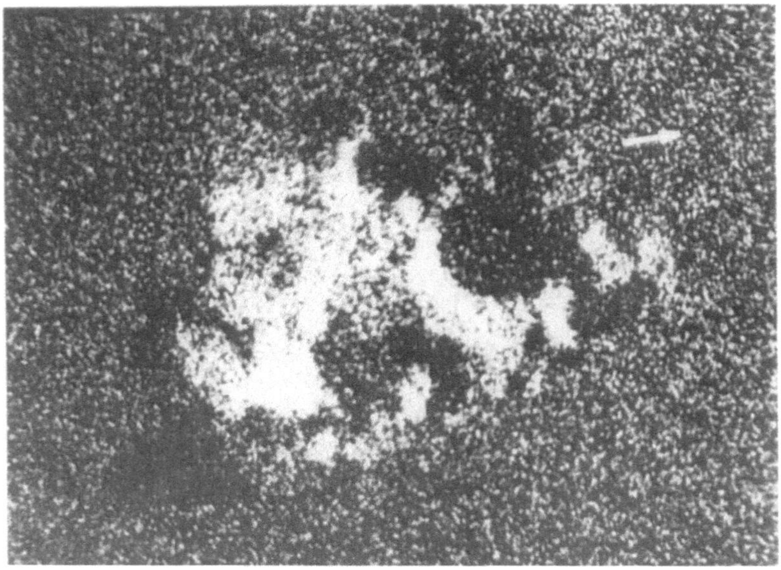

Fig. 3. 290 μm crater — silicon Kα image.

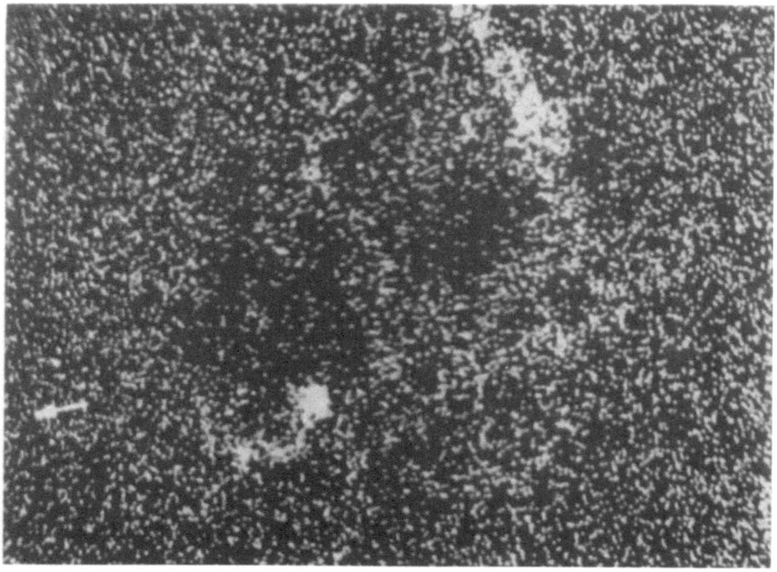

Fig. 4. 290 μm crater – zinc Kα image.

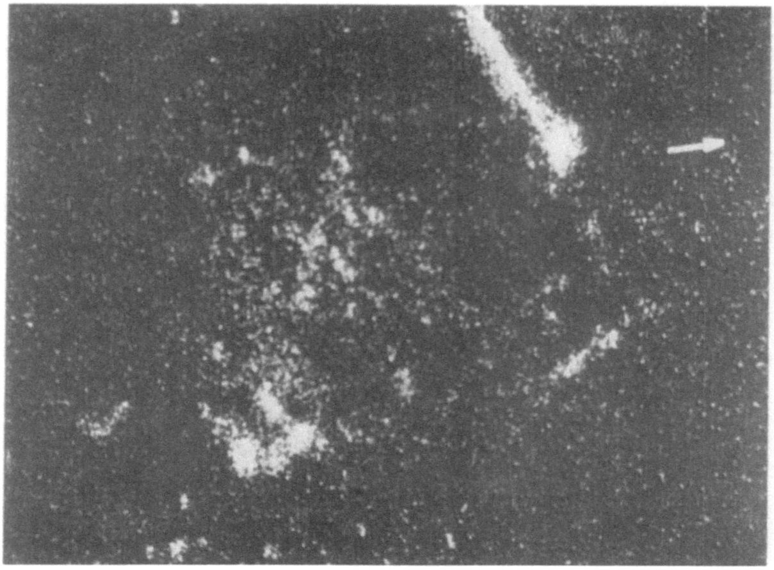

Fig. 5. 290 μm crater – sodium Kα image.

in zinc Kα radiation shows this small particle (Fig. 4). Copper Kα gives a similar photograph. Aluminium radiation was detected from the small particle but not from the other two particles. From their appearance and the probe analysis it is quite clear that these particles are contaminants. The stainless steel surfaces were contaminated in flight and during the recovery from the Agena spacecraft by the astronauts [4].

Analysis of the interior surface of the crater shows oxygen evenly distributed while silicon and magnesium, always in the same ratio and over the entire crater, vary greatly in intensity from point to point (Fig. 3). While sodium is found over much of the crater (Fig. 5), its distribution is quite different

Table 1.
Analysis of craters

Crater	Element and estimated relative abundance* (%)							
	C	O	Si	Mg	Na	S	Ti	P	Al
290 μm	10	25	35	20	10				
160 μm	9	24	45	4		8	4	2	
17 μm	5	20		5					30

* Based on normalized sample to standard count ratios for elements not components of the stainless steel.

from the other elements. A trace of sulphur was found uniformly distributed and a few spots indicated carbon. In Table 1 an estimate has been given of the percentages of the various elements.

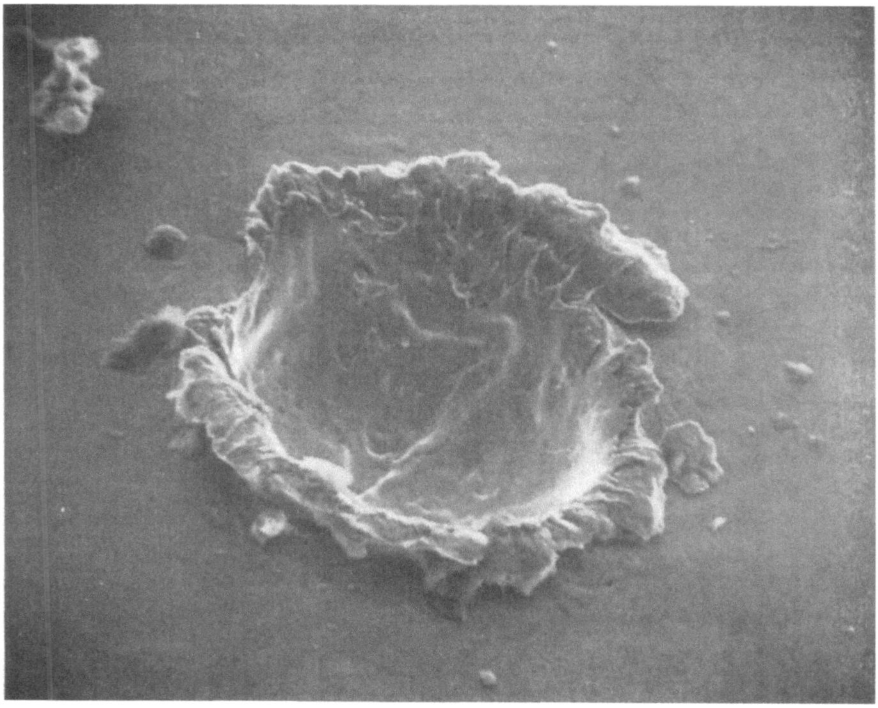

Fig. 6. 160 μm crater – scanning electron microscope image.

3. 160 μm Crater

This is a deep (70 μm) circular crater, 160 μm in diameter having very steep sides and a prominent raised lip. Fig. 6 shows a stereoscan picture. The dark irregular area around the crater is the lip. Due to the steepness of the sides of this crater and the probe's low X-ray take-off angle much of the

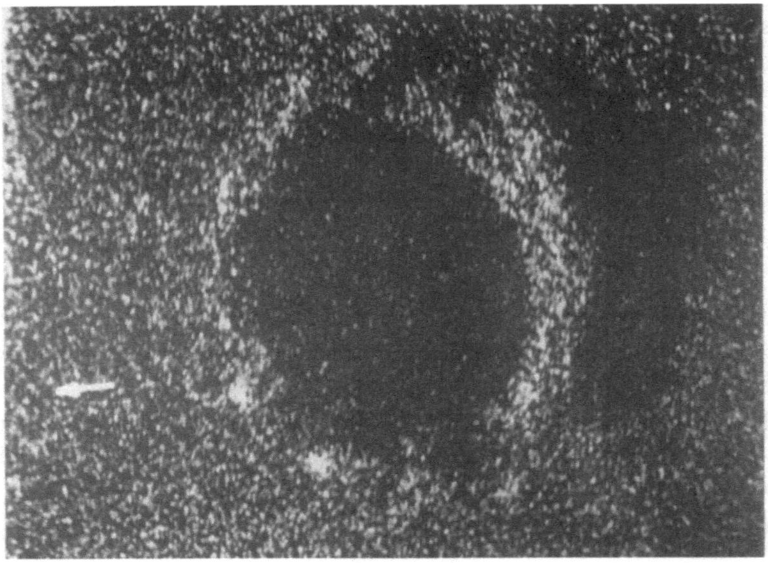

Fig. 7. 160 μm crater — oxygen Kα image.

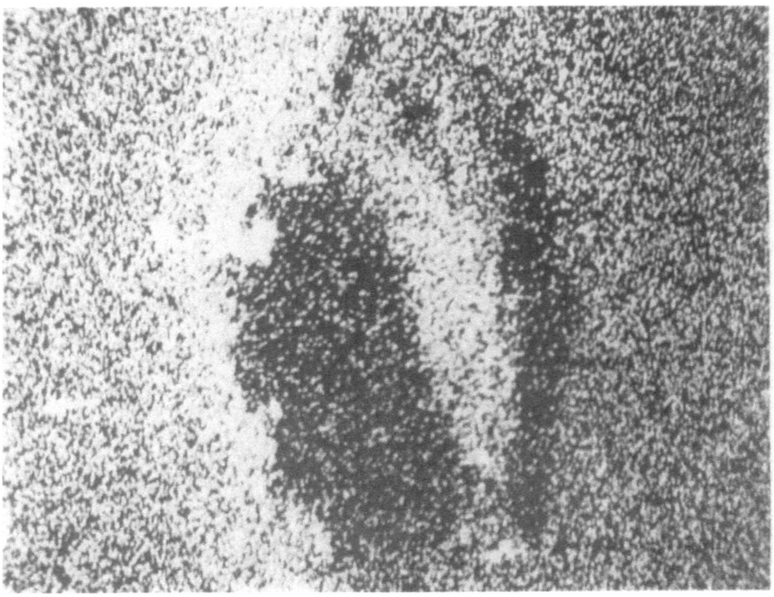

Fig. 8. 160 μm crater — silicon Kα image.

analysis of this specimen was carried out with the stainless steel plate tilted in the direction of the spectrometers.

Point analysis of the crater's interior showed only oxygen and silicon to be widely distributed. Analysis of the plate in the vicinity of the crater showed both oxygen and silicon traces. Fig. 7 was taken of the oxygen radiation with the plate level. It shows some oxygen distributed over the crater lip, with a few spots rich on oxygen. Fig. 8 is a similar picture of the silicon radiation but with the plate tilted so that the inner part of the lip in the upper part

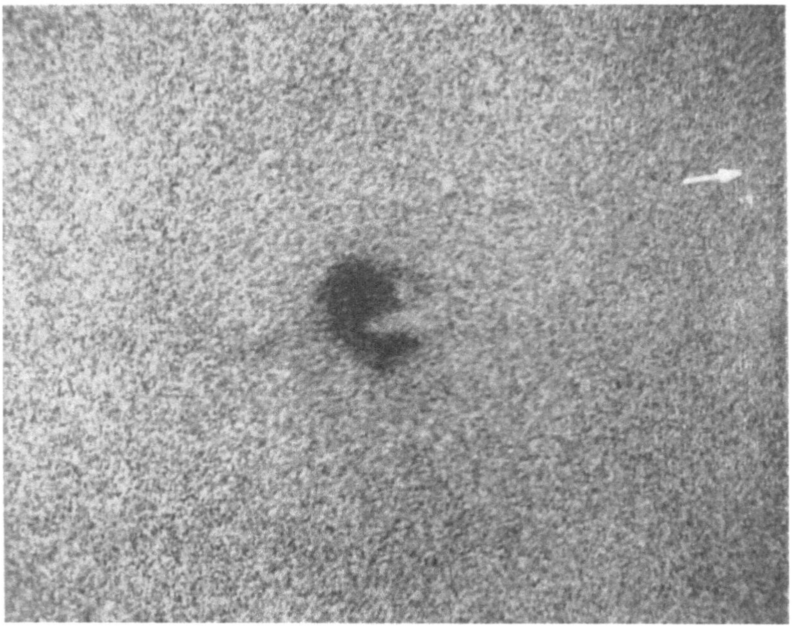

Fig. 9. 17 μm crater – iron Kα image.

of the photograph is exposed to the spectrometer. The outline of the crater and its lip are distinguishable in the photograph and clearly there are areas outside the crater which are rich in silicon. The larger bright area inside the upper lip and just to the right of center in Fig. 8 showed in addition to oxygen and silicon: titanium, magnesium, sulphur, phosphorus and aluminum. These elements appear in differing proportions at other locations within the crater. Estimates of the percentages are given in Table 1.

4. 17 μm Crater

The smallest of the craters is about 7 μm deep. Unlike the larger craters there was material within this crater which greatly reduced the iron Kα radiation (see Fig. 9). Both carbon and oxygen were found concentrated at the location of the circular dark area covering about one third of the crater. Fig. 10 shows the location of oxygen. Silicon and aluminum are found in roughly equal

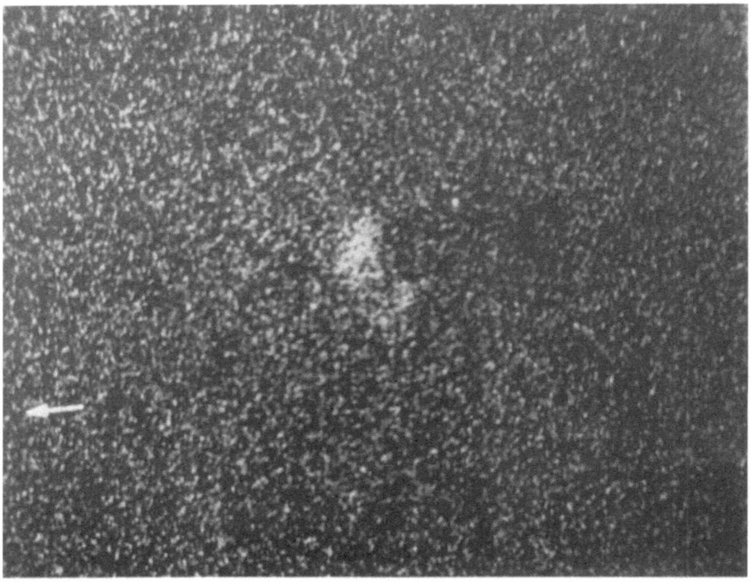

Fig. 10. 17 μm crater — oxygen Kα image.

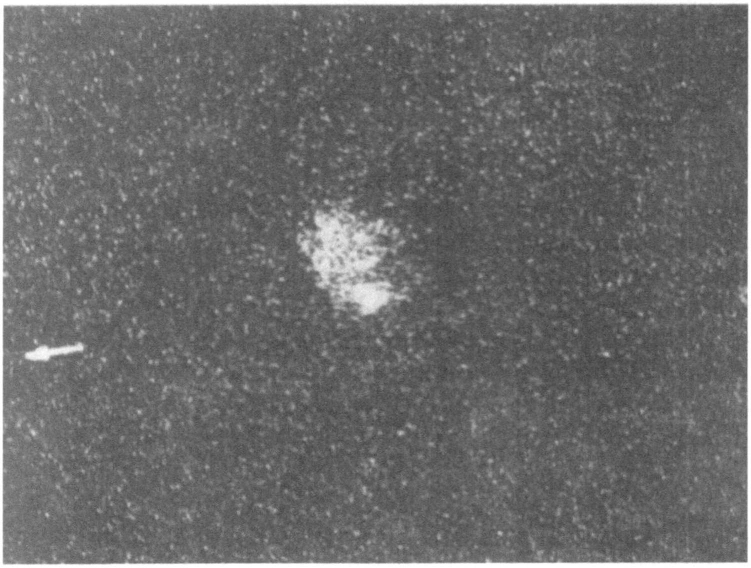

Fig. 11. 17 μm crater — silicon Kα image.

amounts both at the location of the oxygen response and more concentrated in adjacent area. Traces of sulphur and of magnesium were found but were not as definitely localized. Figs. 11 and 12 show the silicon and aluminum locations respectively.

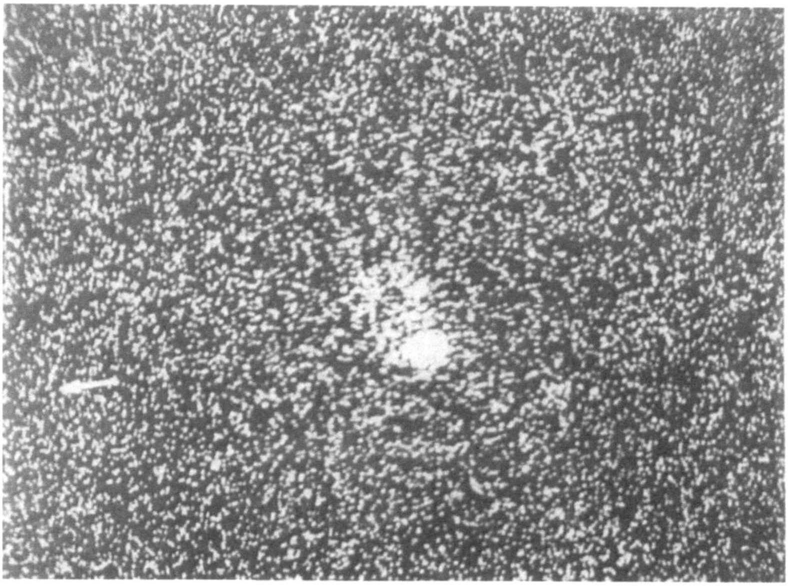

Fig. 12. 17 μm crater — aluminum Kα image.

Summary

Although the S-010 plates do show a general detectable level of surface contamination, the interior of each crater gives a unique analysis. One can feel confident, then, that there is material in the crater which was left by the impacting cosmic dust particle. The results of the microprobe study are summarized in Table 1. It should be pointed out again that these craters are in a stainless steel plate so that it is not possible to say if any of the elements Fe, Cr, Ni, or Mn were brought to the crater by the impacting particles.

It is possible that elements were present in the cosmic dust particles involved which were not found in this analysis because they were completely volatilized during the impact.

Acknowledgments

This work was carried out under NASA Grant NGL 33-011-001 and AF Contract F1962-67-C-0191.

References

[1] C. L. Hemenway, D. S. Hallgren, R. E. Coon and L. A. Bourdillon, in: Space Research VIII, North-Holland, Amsterdam 1968 (p. 510).
[2] C. L. Hemenway and D. S. Hallgren, in: Space Research VIII, North-Holland, Amsterdam 1968 (p. 521).
[2] K. Sitte, Micrometeoroids and interplanetary dust, Trieste meeting of the Italian Physical Society (1966).
[4] D. S. Hallgren and C. L. Hemenway in: Space Research IX, North-Holland, Amsterdam 1969 (p. 102).

Space Research XI — Akademie-Verlag, Berlin 1971

A NEW HIGH ALTITUDE BALLOON-TOP COSMIC DUST COLLECTION TECHNIQUE

C. L. Hemenway, D. S. Hallgren, A. T. Laudate, H. Patashnick, T. S. Renzema and O. K. Griffith

Dudley Observatory, Albany, New York, and the State University of New York at Albany, New York, USA

A new technique for the collection of cosmic dust utilizing high altitude balloons is described. The fluxes obtained by this method are compared with fluxes obtained by a technique previously described. Direct measurements of individual particle composition and masses have been made. Evidence for particle ablation has been sought using the scanning electron microscope. Typical electron micrographs are shown.

A summary of eight recent balloon-top cosmic dust collections is given. Four of these collections utilized a sweep-up technique while the remaining four were collections using floating balloons at an altitude of 110,000 feet as described at COSPAR in Vienna [1]. In the sweep-up technique the balloon-top collector is opened at about 65,000 feet and remains open until just before float altitude is reached. The actual open time for the sweep-up collections is about 45 minutes. In computing the flux for the sweep-up collections, the effective collection time is computed in the same way as in rocket flights [2]. The effective collection time is equal to the actual open time plus the time for a particle of a given size to fall through the collection interval.

Table 1

Flight	Date	Technique	No. of particles collected	Particle masses (g) $>2.0\times10^{-10}$	$>1.6\times10^{-9}$	$>1.3\times10^{-8}$	$>1.0\times10^{-7}$
				Approximate fluxes (particles $m^{-2}\,s^{-1}$)			
25	18 Nov. 1967	Float	19	4.2×10^{-2}	2.2×10^{-2}	1.3×10^{-2}	–
27	11 June 1968	Sweep-up	12	6.2×10^{-4}	5.0×10^{-4}	–	–
28	13 Aug. 1968	Sweep-up	15	1.0×10^{-3}	8.6×10^{-4}	6.1×10^{-4}	–
29	9 June 1969	Float	4	5.7×10^{-3}	1.4×10^{-3}	–	–
30	10 June 1969	Sweep-up	6	3.0×10^{-4}	1.8×10^{-4}	–	–
31	12 Aug. 1969	Float	13	2.9×10^{-2}	1.8×10^{-2}	4.5×10^{-3}	–
32	26 Feb. 1970	Float	7	1.4×10^{-2}	3.9×10^{-3}	–	–
33	27 Feb. 1970	Sweep-up	6	1.3×10^{-3}	1.2×10^{-3}	1.1×10^{-3}	9.1×10^{-4}

Table 1 lists the flight numbers and times, identifies the technique used, indicates the number of particles collected, and gives the approximate fluxes as a function of particle mass intervals. It will be noted that the fluxes obtained during these recent collections are lower by about a factor of ten from those reported earlier [1]. These lower values appear in part to be due to three reasons.

First, a direct weighing technique [3] has been developed which has shown that the particles collected were on the average about a factor 3 lighter than estimated and a number of particles were then too small to be counted. Second, better clean laboratory facilities and techniques probably reduce the chance of unrecognized contaminants being counted. Finally, microprobe techniques are now available to analyze chemical composition of particles and these techniques have revealed a previously unrecognized source of contamination in the collector.

On two occasions the float and sweep-up techniques were employed in pairs to sample debris from the Zeta-Perseid and Arietid meteor showers in June

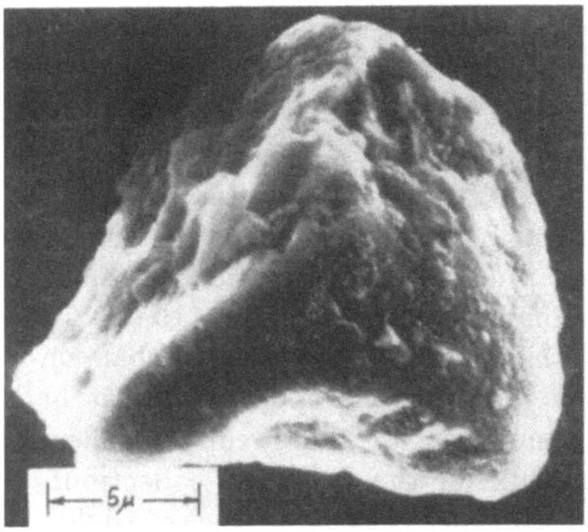

Fig. 1. Stereoscan picture of cosmic dust particle.

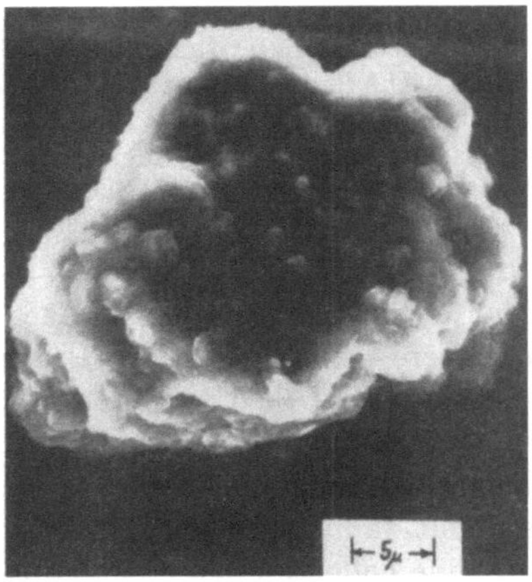

Fig. 2. Stereoscan picture of cosmic dust particle showing possible surface ablation.

1969 (flights 29 and 30) and to sample at the time of the transit of the earth through the plane of Encke's comet (flights 32 and 33) in February 1970. It will be noted that the sweep-up technique followed the float technique by one day. For particles about 30 μm in diameter, these techniques should, in principle, give equivalent results and appear within statistical limits to do so. For smaller particles the sweep-up techniques appears to give systematically lower fluxes.

During three flights (24, 25 and 26), in an attempt to determine whether the balloons are contributing contaminating particles to the balloon-top collections, the tops of the balloons in the vicinity of the collector were each coated with approximately 10^9 aluminum microspheres (10–80 μm in diameter). Telemetry from flight 24 showed it to be unusually violent in its up and down motion during float due to a balloon malfunction, and nine microspheres entered the collector. Consequently no flux was determined. Flights 25 and 28 were normal and no contamination test spheres were found in the collector. It is not considered that the balloons used are a significant source of contamination for large cosmic dust particles, >5 μm in diameter.

Figs. 1–4 show Steroscan electron microscope pictures of particles collected during the Zeta-Perseid–Arietid and the Leonid meteor showers. All of the collected particles have been examined closely in the Stereoscan microscope to look for evidence of surface melting. Few particles show any evidence of ablation; Fig. 2 is an example of a particle which may show such evidence. Most of the particles appear to be irregular; some are very weak mechanically and break up when touched with a 10 μm diameter quartz fiber: an example of a broken particle is shown in Fig. 4.

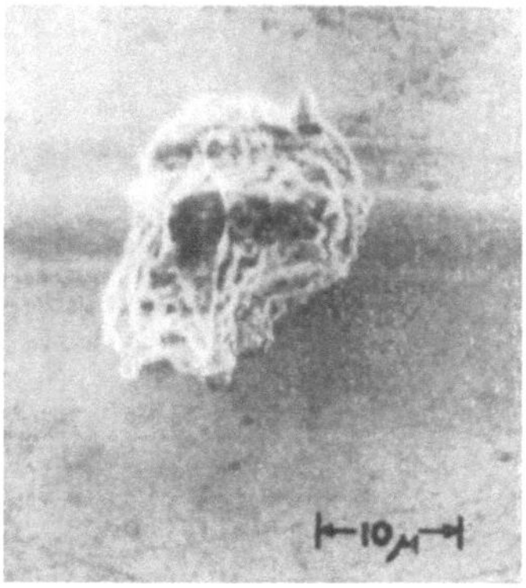

Fig. 3. Stereoscan picture of cosmic dust particle.

Fig. 4. Stereoscan picture of broken cosmic dust particle.

Forty of the collected particles for which no contamination source could be found were analyzed using an electron beam probe which can detect the presence of all elements heavier than boron. Table 2 shows the number of the forty particles in which an element was detected. Silicon was the most frequently occurring element with carbon, oxygen and aluminum almost as abundant.

Table 2

C	29	Al	28	Cl	1	Mn	2
O	29	Si	35	K	5	Fe	14
Na	11	Ca	9	Ca	14	Co	1
Mg	18	S	20	Ti	4	Zn	3

We note with interest the absence of nickel even though iron was found in 14 particles.

References

The authors wish to express their thanks to Amy Jettner and David Wachtel for their help in scanning the slides, and to the crew at the NCAR balloon base at Palestine, Texas, for launching the balloons and recovering the collectors.

This work was carried out with support from NASA Grant NGL 33-011-001 and Air Force Cambridge Research Laboratory Contract F 19628-67-C-0191.

References

[1] C. L. Hemenway, D. S. Hallgren and R. E. Coon, in: Space Research VII, North-Holland Publ. Co., Amsterdam 1967 (p. 1423).

[2] R. K. Soberman and C. L. Hemenway, J. Geophys. Res. **70**, 4943 (1965).

[3] H. Patashnick and C. L. Hemenway, Rev. Scient. Instrum. **40**, 1008 (1969).

Space Research XI — Akademie-Verlag, Berlin 1971

A MODEL FOR PREDICTING THE RESULTS OF IN SITU METEOROID EXPERIMENTS: PIONEER 8 AND 9 RESULTS AND PHENOMENOLOGICAL EVIDENCE

U. Gerloff and O. E. Berg

NASA Goddard Space Flight Center, Greenbelt, Md, USA

The paper summarizes the theoretical and astronomical basis of *in situ* micrometeoroid experiments. It is shown that recent micrometeoroid results obtained by Pioneer 8 and 9 reveal strong correlations with wellknown astronomical evidence such as zodiacal light and the comet Encke. A full understanding of these accurate data, however, demands a quantitative meteoroid model that is consistent both with photometric zodiacal light observations and ground-based meteor data. Such a model is also urgently needed to improve and optimize the design of cosmic dust experiments. The main elements of the micrometeoroid environment have now been established: flux, cut-off and mass distribution of both sporadic and shower meteoroids; cometary origin of the bulk of micrometeoroids; concentration of cometary dust in the orbital planes of comets.

1. Introduction

The present state of *in situ* meteoroid experiments can be described as the transition period from qualitative to quantitative research. More specifically, in order to be feasible and interpretable future meteoroid experiments will have to be based on a much more sophisticated theory than have been earlier and contemporary approaches to the problem. The development of such a theory has not been possible in the past, mainly because the amount of good data available was not sufficient fully to characterize the meteoroid ensemble in interplanetary space. For instance, photometric observations of the zodiacal light do not determine enough parameters over sufficiently large intervals to define both the size and the orbit distribution of meteoroids. *In situ* meteoroid experiments, because of initial difficulties of the new technique, have not yielded reliable results until very recently.

At the present, however, both the astronomical and technological knowledge exist to design meteoroid experiments that are highly reliable and sophisticated. As, for instance, the results of Pioneer 8 and 9 show, it is possible with the help of these experiments to ask relevant questions and to obtain revealing answers with regard to the nature and history of interplanetary dust. However, there appears a danger that may severely hamper, if not jeopoardize, the success of *in situ* meteoroid experiments: the danger that there is no adequate theoretical basis for *in situ* experiments, and the consequence might be that even the most reliable and promising techniques cannot be properly applied.

In this paper, some facts are compiled that, it is hoped, will stimulate some theoretical work in the field of cosmic dust research. First, a general sketch is given of what a comprehensive theory would have to deal with. Second, the role of a COSPAR meteoroid is anticipated. Third, the Pioneer 8 and 9 data are discussed. Fourth, those features of the interplanetary meteoroid ensemble are described that most probably will have a sensible impact on future efforts in the field.

2. Working Program to Establish a Theoretical Basis for in situ Meteoroid Experiments

A theory of *in situ* meteoroid experiments, in this context, may be characterized by two features:

(i) it must provide for quantitative evaluations of *in situ* meteoroid experiments in order to optimize their design and orbit,

(ii) it must help interpret the actually measured results in terms of astronomical significance.

The first feature demands that a model of the interplanetary dust ensemble be designed. The second feature requires that this model be based on physical parameters rather than on a purely mathematical fit of the measured data.

It is the opinion of the authors that such an amibitious theory can be established first, because there are more independent data available now than in the past, second, by considering both ground-based and *in situ* meteoroid data, it should be possible to improve simultaneously both experiment and theory, which will eventually lead to a reasonably consistent model of the interplanetary meteoroid ensemble.

In order to achieve its twofold purpose of predicting and interpreting the meteoroid impact rate encountered by a given meteoroid satellite, the model must be heavily based on orbits. The flux measured by satellite experiments is critically dependent on the aberration of the apparent meteoroid radiants due to the satellite's velocity, and only by explicitly taking into account the orbits of both the meteoroids and the satellite can these aberration effects be reflected in a model. A successful flux model for deep space meteoroid experiments, then, must accommodate two rather different theories: (i) a dynamic orbit model for the zodiacal dust cloud that includes such features as dust concentrations in the orbital planes of comets, and (ii) a kinematic theory of selection effects, such as aberration, that influence the results of satellite meteoroid experiments. Neither of these theories is well developed at the present time, nor has either been applied to satellite meteoroid experiments more than occasionally. One reason for this negligence may be seen in the fact that the results of earlier flux experiments indicated so many dust particles that a loss by a factor 10 (ignoring aberration effects) did not seem significant. Obviously, this reason does not apply now, and further, we know that the flux is very low indeed, and that a kinematic factor 10 may well decide the feasibility of any one of the current meteoroid experiments. Thus, the demand of the hour is to optimize the flux sensitivity of meteoroid experiments.

A kinematic theory that may provide the missing factor 10 in the impact rate of satellite flux experiments should enable us to change properly from heliocentric coordinates to the spacecraft's moving frame of reference, and vice versa. It should provide the means to convert the heliocentric orbit distribution, given, for instance, by an astronomical model of the zodiacal dust cloud, into the apparent radiant and velocity distribution measured by a satellite experiment. Once we have learned how to estimate the radiant and velocity distribution to be measured, optimizing the orientation and response function of the meteoroid detector (as well as the altitude and orbit of the spacecraft) begins to become feasible.

Let us now consider meteoroid measurements in the near-earth environment and the atmosphere, i.e. rocket and balloon techniques. We first notice that the above-described deep space meteoroid model has a crucial impact also on this type of experiment. The flux variations and orbit characteristics of the deep space meteoroid ensemble that consist of sporadic meteoroids, meteor streams, asteroidal meteoroids, and, most important, cometary dust concentrations, define the supply of meteoroid particles entering the earth's atmosphere. In the earth's atmosphere, the distribution of primary meteoroids is mixed up with aerosol particles, moon ejecta, ablation products and fragments of larger meteoroids. In addition, the velocity spectrum is considerably modified by atmospheric deceleration. As a result, a near-earth meteoroid ensemble is finally created that varies strongly with altitude and time.

Again, in order theoretically to control the various parameters of this ensemble, a model must be developed that, probably, is not less complex than the deep space meteoroid model of the zodiacal dust cloud. Obviously, the influence of the measurement techniques must again be considered quantitatively in order to optimize near-earth and atmospheric meteoroid experiments. Selective effects may, in this case, be due to contamination of the sample, identification of meteoroids, particle losses in the shock front of a rocket nose cone or of a balloon-borne impactor.

In conclusion, activity in four fields of theoretical research is suggested in order to provide a theoretical basis for and to utilize fully the scientific potential of *in situ* meteoroid experiments:

1. Orbit model of the deep space meteoroid ensemble including cometary dust
2. Kinematics of deep space meteoroid experiments including selection effects
3. Model of the near-earth meteoroid ensemble including moon dust trajectories
4. Theory of atmospheric meteoroid experiments including aerodynamic effects.

This paper aims toward clarifying the theoretical and empirical situation of today's *in situ* meteoroid experiments. A later paper will discuss the kinematic effects influencing satellite meteoroid experiments to provide a quantitative means for optimizing the design of future experiments. The above four theories provide a working program for interested scientists that should result in a substantial enhancement of the scientific return of *in situ* meteoroid experiments.

3. The COSPAR Meteoroid Model

The COSPAR Flux Revision Committee has asked for suggestions for a better representation of flux results; this means the development of a flux model for the interplanetary space and the upper atmosphere that provides a proper basis upon which to interpret and estimate the flux results of *in situ* and ground-based meteoroid experiments. Such a model, however, can direct and guide future efforts in this field only if it provides quantitative solutions to each of the four theories listed above.

The COSPAR meteoroid model, if it is developed comprehensively, will be of historical significance. From astronomical observations of the zodiacal light and meteors, no consistent and generally accepted model of the interplanetary dust ensemble could be deduced in the past 50 years. Since these ground-based observations were limited to the earth's orbit and restricted to remote sensing techniques, their results did not allow adequately provided data to specify the many features of a realistic model of the interplanetary dust ensemble. *In situ* experiments, on the other hand, have yielded erroneous data for some length of time. As a consequence of this situation, the scientific communication between astronomers and space physicists has gradually deteriorated as the lack of success on both sides has persisted.

The COSPAR meteoroid model provides an ideal subject to reopen the dialogue between astronomers and space physicists. By combining the theoretical and statistical results of century-long astronomical studies with the detailed data about individual meteoroids gathered now by *in situ* measurements, a joint effort should bring about a highly sophisticated and reliable model of the interplanetary dust ensemble. The desperate need for, and potential impact of, such a model on the design and results of space experiments have been indicated in the first section of this paper.

4. The Pioneer 8 and 9 Results until August 1969

The Pioneer 8 and 9 cosmic dust experiments have revealed clear-cut information about meteoroids and meteoroid orbits [1–4] that defines some main features of the meteoroid model to be designed. The two spacecrafts are in direct heliocentric orbits with perihelia between 0.7 and 1.0 AU and aphelia between 1.0 and 1.1 AU respectively. Their orbit velocity is approximately 30 km s^{-1}. Both satellites are spinning at 1 rev s^{-1} around an axis vertical to the ecliptic plane and carry a micrometeoroid detector that faces in the ecliptic plane [5]. The detector consists essentially of two thin film-grid detectors separated by a distance of 5 cm, that produce an electric signal when the film is penetrated by a micrometeoroid. Each film has a sensitive area of 100 cm^2 and is composed of 16 segments, that provide both direction and the time-of-flight needed for the meteoroid to traverse the 5 cm distance between the front-film and the rear-film sensor.

Fig. 1 gives the measured azimuth a) and pulse height b) distribution for time-of-flight signals and front-film signals. The detector, that "looks" into the sun once per revolution, "sees" a noise pulse at the front sensor only at any 500000th revolution. The few noise pulses accumulated during 20 months give rise to the dominant peak at the sun's direction in the azimuth diagram. However, since the noise rate is extremely low, a smaller peak, pointing to the apex direction, can also be resolved in thefront-signal distribution. This peak indicates true meteoroid impacts, as is concluded from eight clearly identified events recorded by the time-of-flight sensor. The concentration of true events around the apex direction of the satellite corresponds directly to the well-known enhancement of meteor activity at the morning side of the earth and has been seen in a satellite meteoroid experiment for the first time.

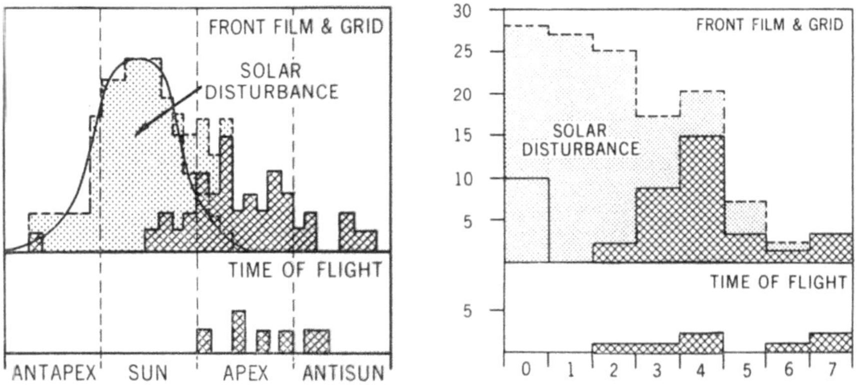

Fig. 1. Meteoroid impacts as measured by Pioneer 8 during the first 20 months in orbit: a) azimuth distribution; b) pulse-height distribution.

The pulse height distribution, at the right-hand side of Fig. 1, indicates a cut-off in the particle energy. Since it appears in both the time-of-flight and front-film distribution, it is not caused by meteoroid fragmentation on the front film, but constitutes a general feature of the interplanetary dust environment.

Fig. 2 gives the mass distribution of eight time-of-flight meteoroids which was evaluated by considering the measured velocities of these particles. For each individual time-of-flight event, the actually detected particle mass is, in the average, 100 times larger than the threshold sensitivity of the dector. The cut-off mass of 0.8×10^{-11} g is in good agreement with a "flat" size distribution of zodiacal dust particles [6]. However, no statement can be made about the existence or abundance of meteoroids having masses smaller than the detection limit of the Pioneer 8 and 9 detector of 10^{-13} g.

In Fig. 3, the velocity distribution measured by the time-of-flight sensor, is shown as folded in the ecliptic plane. We see that all but one of the detected meteoroids are in direct heliocentric orbits. Since meteoroids in retrograde orbits have a larger relative velocity and therefore a higher probability of being detected by the satellite, the predominance of direct orbits is actually somewhat more pronounced than indicated in this diagram. Fig. 4 gives an idea of the

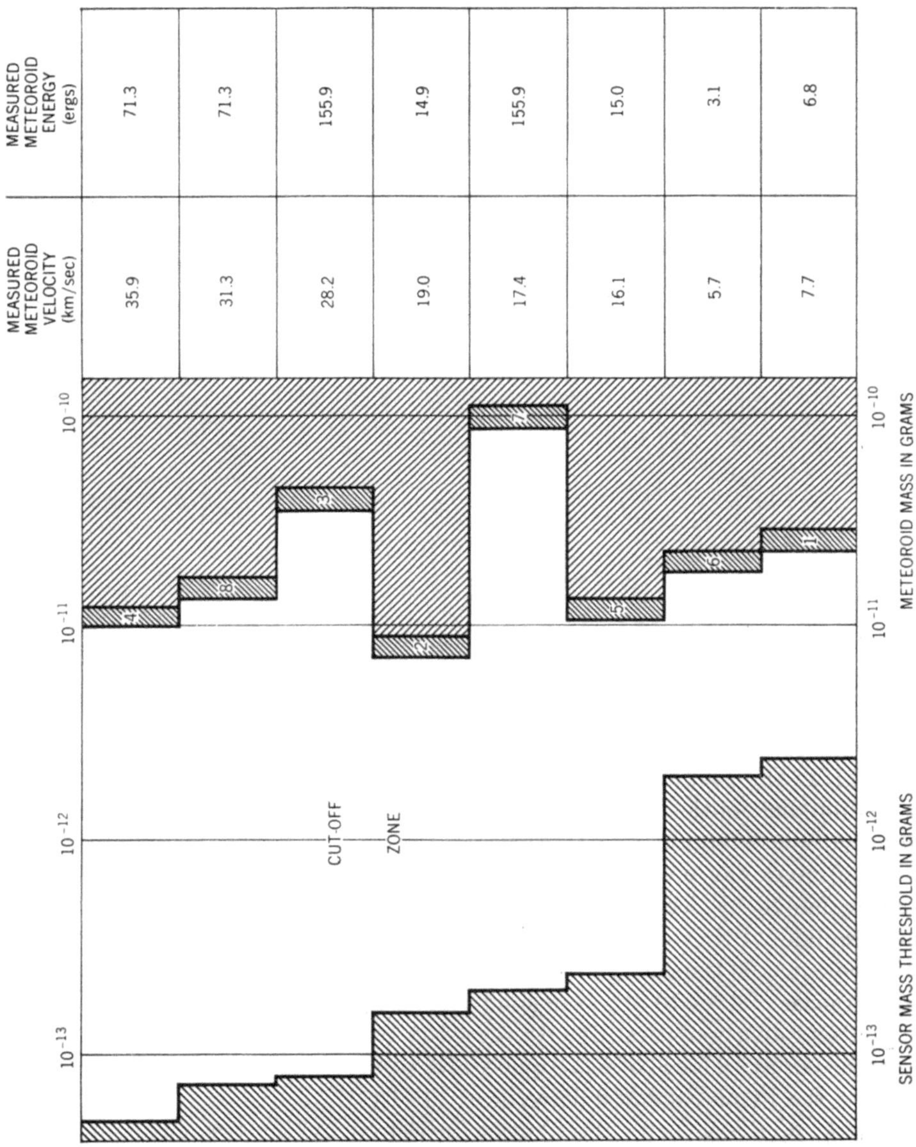

Fig. 2. Meteor orbits determined by Pioneer 8 and 9. Masses of 8 time-of-flight meteoroids indicating a cut-off.

apparent and heliocentric inclinations measured for the eight recorded time-of-flight meteoroids. The roundly shaped curves simulate approximately the response function of the detector. We find a strong predominance of low-inclination orbits, which of course, is partly imposed by the low-inclination orbit of the Pioneer satellites.

Fig. 5 shows the dependence of the measured meteoroid orbits on the meteoroid's mass-density ϱ due to the radial component of the solar radiation pressure. In the $(1/a, e)$ diagram, (a = semi-major axis, e = eccentricity), stable elliptic

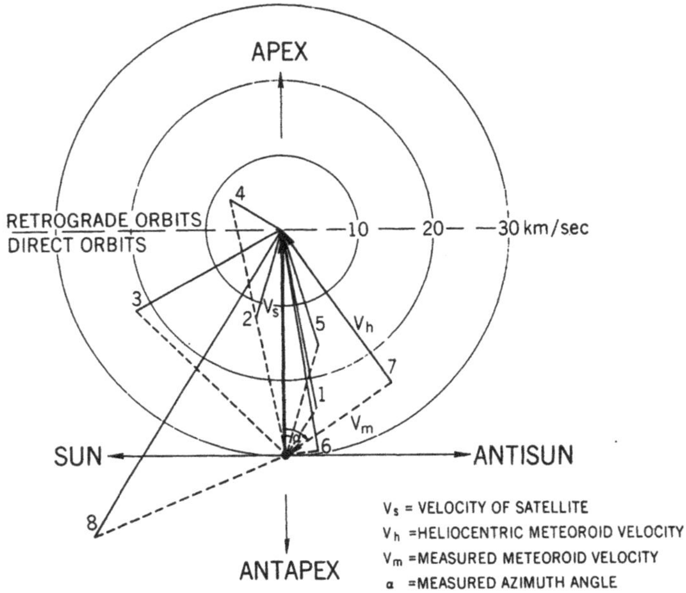

Fig. 3. Meteoroid orbits determined by Pioneer 8 and 9 (velocity distribution).

orbits are located in the triangles in the top half of the diagram, and unstable hyperbolic orbits are located in the rectangular regions beginning at the lower right-hand corner of each triangle. Among eight orbits determined by Pioneer 8 and 9, seven are elliptic and one is hyperbolic, assuming a meteoroid's mass-density not lower than 1 g cm^{-3}. With decreasing mass-density ϱ the orbits move slightly to the lower right, since the solar radiation pressure tends to increase the orbit's semi-major axis and eccentricity. Finally, at $\varrho = 0.3$ g cm^{-3}, the upper triangle is almost depleted of meteoroid orbits and most of the elliptic orbits have been converted into hyperbolas. Now, the probability of detecting particles having elliptic orbits is approximately 1000 times that of detecting particles having hyperbolic orbits due to repeated intersections with the satellite's orbit. Elliptic rather than hyperbolic orbits must be expected. We conclude, therefore, that a lower limit for the average mass-density of micrometeoroids exists between 1 and 0.3 g cm^{-3}. This limit does not apply to meteoroid No. 8, which is in a hyperbolic orbit for any mass-density assumed. Hyperbolic particles, because of their highly unfavourable encountering probability, should be detected only at regions in space where they are selectively enhanced, as, for example, in the orbital plane of comets.

Fig. 6 shows how the orbits of micron-sized meteoroids (mass-density $\varrho = 3\ \text{g cm}^{-3}$) compare with the orbits of larger bodies that produce meteor trails when entering the earth's atmosphere. The abscissae give the absolute brightness of the meteors and the corresponding mass of the meteoroids. Three orbital

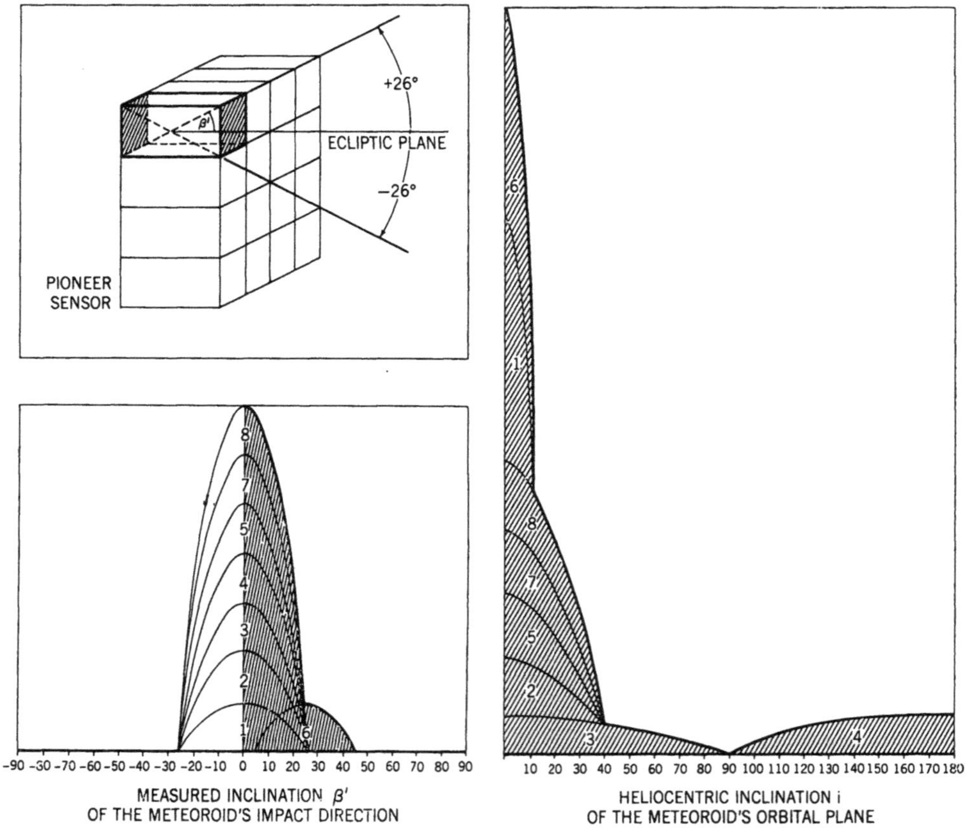

Fig. 4. Meteoroid orbits determined by Pioneer 8 and 9 (inclination distribution).

parameters are shown as ordinates: the semi-major axis a, the eccentricity e, and the perihelion distance q. The results of three major meteor surveys [7–9] are compiled in such a manner that the diameter of the circles shown is proportional to the number of meteors per orbital parameter increment. At the very left, the orbits of the eight micrometeoroids are shown as detected by Pioneers 8 and 9. Although the Pioneer data could not yet be corrected for selection effects, as have been the meteor data in this diagram, the following conclusions can safely be drawn: The trend toward smaller semi-major axis a, slightly reduced eccentricity e, and smaller perihelion distances q, indicated by the successive meteor measurements, when going to smaller masses, is clearly confirmed by the Pioneer results. The orbits of micrometeoroids, thus,

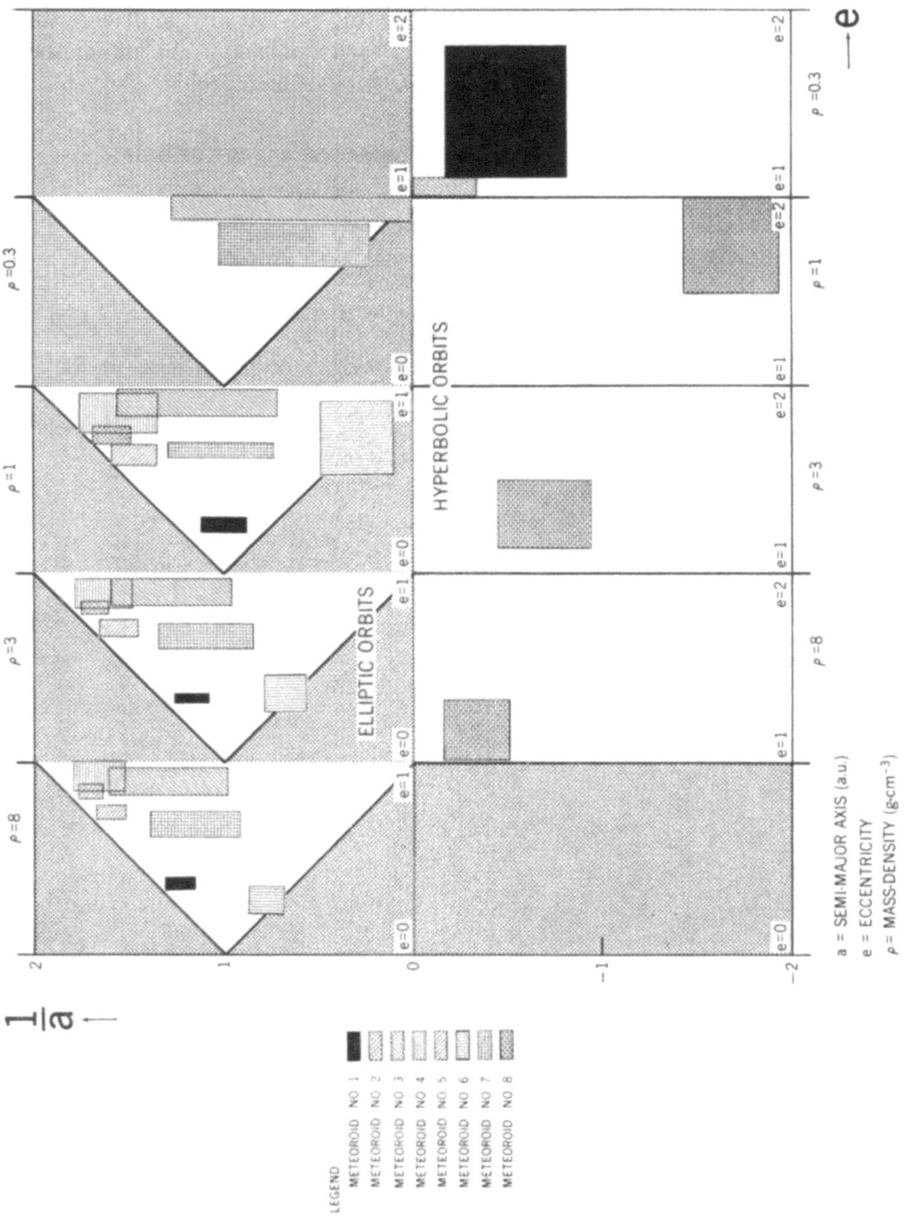

Fig. 5. Meteoroid orbits determined by Pioneer 8 and 9 (influence of radiation pressure).

differ markedly from the orbits of meteor-producing bodies. They are neither similar to short period comets nor circular. Both types of orbit similarities have been assumed for dynamical studies of the zodiacal dust cloud. The measured meteoroid orbits with a mean eccentricity of 0.7 are readily interpretable as the result of the Poynting–Robertson effect acting on initially high eccentric meteoroids of cometary origin. They cannot easily, however, be understood on the basis of an asteroidal origin for the majority of meteoroids.

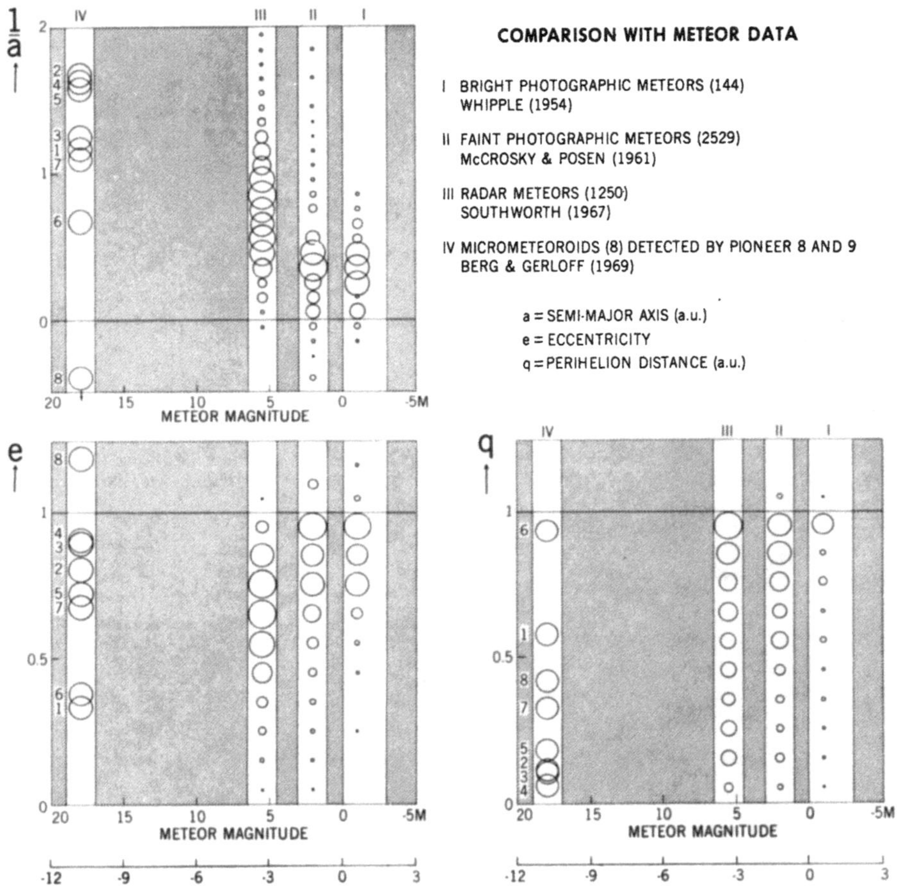

Fig. 6. Meteoroid orbits determined by Pioneer 8 and 9 (comparison with meteor data).

Table 1 shows that two of the eight meteoroids with measured orbits are directly related to comets. As indicated by the first three columns of the table, these particles move outside the comet's orbit but within the comet's orbital plane. This behavior is expected on the basis of radiation pressure considerations. The parameter β (the ratio between radiation pressure and solar gravitation experienced by a meteoroid) is given by Eq. (1):

$$\beta = \frac{5.78 \times 10^{-5}}{r \varrho} (\text{g cm}^{-2}) \qquad (1)$$

Table 1
Meteoroid orbits determined by Pioneer 8 and 9
— correlation to comets —

	Solar distance of node (AU) $R_M < R_C$	Longitude of node (deg.) $\Omega_M \approx \Omega_C$	Inclination of orbit (deg.) $i_M \approx i_C$	Eccentricity of meteoroid orbit	Radiation parameter of meteoroid as function of eccentricity e and mass-density ϱ (g cm^{-3})	Characteristic feature of meteoroid
Meteoroid No. 3	$R_3 = 1.05$	$\Omega_3 = 208$	$i_3 = 0-54$	$e_3 < 1$	$0.13 < \beta_3 < 0.27$ for $3 < \varrho_3 < 1$	Fragmentation
Comet Grigg-Skellerup	$R_{GS} = 0.856$	$\Omega_{GS} = 215$	$i_{GS} = 17.6$	–	$0.15 < \beta < 0.25$ for $e < 1$	–
Meteoroid No. 8	$R_8 = 0.9$	$\Omega_8 = 143$	$i_8 = 0-14$	$e_8 > 1$	$0.4 < \beta_8 < 0.9$ for $1 > \varrho_8 > 0.3$	Fragmentation
Comet Encke	$R_E = 0.34$	$\Omega_E = 155$	$i_E = 12.4$	–	$0.6 < \beta$ for $e > 1$	–

where r = meteoroid radius (cm), ϱ = meteoroid mass-density (g cm^{-3}). The relations for β given in the table must be fulfilled if the solar radiation pressure "felt" by the meteoroid should be sufficient to bring the particle out from its place of origin at the comet's orbit to a radial distance from the sun of about 1 AU where it actually has been detected by the Pioneer satellite. We see that the postulated parameters β and the actually measured parameters β_3 and β_8 can be matched by assuming a reasonable mass-density between 3 and 0.3 g cm^{-3} for both meteoroids. It is intriguing, further, that the Encke meteoroid, which is in a hyperbolic orbit and therefore only a hundred days old, has a lower mass-density than any one of the other seven meteoroids investigated. Both comet-related particles, and only those, showed fragmentation when being recorded.

On the basis of seven time-of-flight particles and $40 \pm {}^{0}_{10}$ front-film impacts recorded by Pioneer 8 during the first 20 months in orbit, a flux has been evaluated that is represented in Fig. 7. This flux diagram differs from earlier versions in two respects: First, it contains no data from atmospheric experiments because, as was pointed out earlier, the theoretical and practical difficulties of these techniques may impose systematical errors with respect to deep space satellite measuremets. Second, since this paper aims toward providing a guide line for future experiments, any satellite data of merely historical interest, such as upper limit measurements exceeding the lowest known flux value, have been omitted from the diagram.

The Pioneer flux result may be stated in the following ways:

(i) A cosmic dust detector that (a) is sensitive to 1 μ meteoroids, (b) has a cross section area of 100 cm^2, (c) has an angle of acceptance of π sterad ($\pm 60°$ half-angle), (d) is mounted on the circumference of a spin-stabilized satellite (vertical axis) so as to face into the ecliptic plane, and (e) is orbiting with a heliocentric velocity of 30 km s^{-1} at approximately 1 AU around the sun, measures an impact rate of (2 ± 0.5) per month.

(ii) A flux of $(2 \pm 0.5) \times 10^{-4}$ m^{-2} s^{-1} $(2\pi$ sr$)^{-1}$ is encountered by a spacecraft under the above-described conditions.

(iii) Stating the flux result independently of spacecraft parameters: the space density of micrometeoroids in the ecliptic plane at approximately 1 AU is $(2 \pm 0.5) \times 10^{-14}$ cm^{-3}.

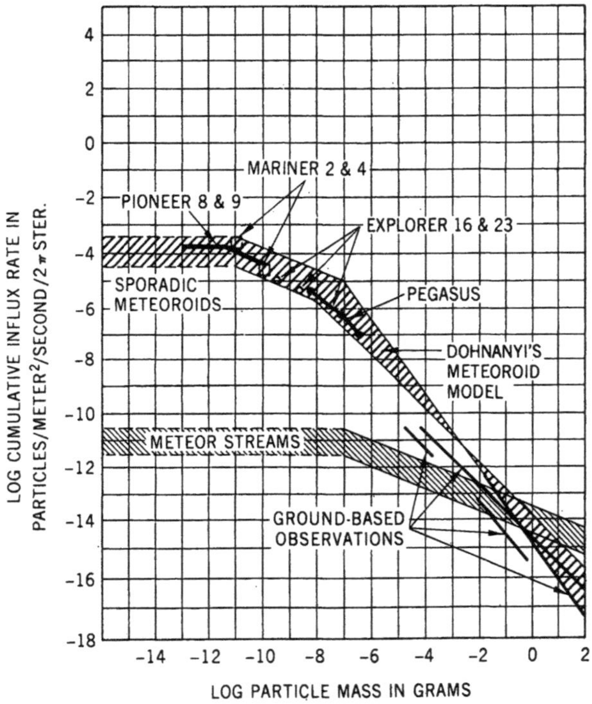

Fig. 7. Flux of meteoroids (recent satellite measurements and theoretical predictions on the basis of a cometary origin of meteoroids).

The only feasible way for a micrometeoroid satellite to encounter a space-density in excess of this value, is by intersecting the orbital plane of some comets at a time related to the comet nucleus and at a distance from the comet's node related to the particle size to be investigated.

Thus the Pioneer 8 and 9 Cosmic Dust Experiments mark a turning point in the history of cosmic dust research. They have removed the threat of an undetectably low flux of micron-sized meteoroids by revealing evidences of unique clarity. The result, mainly achieved by measuring *orbits* of meteoroids, is equally significant for astronomy, space technology, and the development of future meteoroid experiments.

5. Phenomenological Description of the Interplanetary Dust Ensemble near 1 AU

The findings of Pioneers 8 and 9 are consistent with, and lend strong support to, the hypothesis that the bulk of meteorite dust is of cometary origin. A cometary origin of meteoroids has the consequence, that the dust particles are given off along the comet's orbit and are injected with initial velocities similar to the comet's velocity at the injection point. Many meteoroids are injected with high initial velocity at the perihelion regions of the comet's orbit, and few meteoroids are injected with low initial velocity at the aphelion portion of the comet's orbit. Since comet orbits are highly eccentric, and therefore close to the parabolic limit, a slight change in the injection velocity caused by the evaporation or collision process that produced the particle or a slight weakening of the central force caused by the solar radiation pressure which reduces the gravity of small dust particles, has a considerable impact on the young meteoroid's orbit.

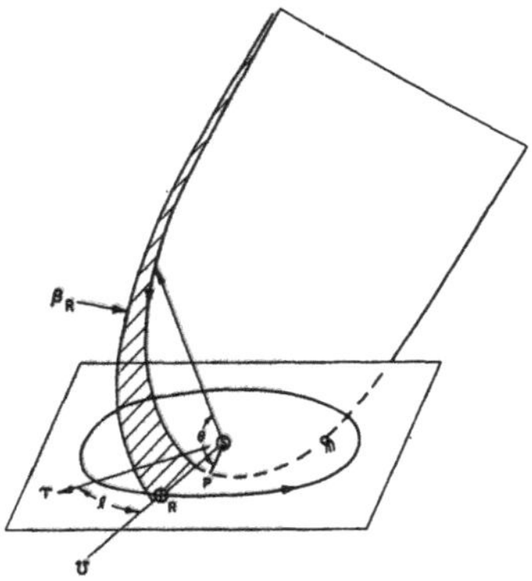

ONLY METEOROIDS WITH $\beta > \beta_R$ CAN REACH A DISTANCE R FROM THE SUN.

$$\beta = \frac{5.78 \cdot 10^{-5}}{r \rho} \ (\mathrm{g \cdot cm^{-2}})$$

β = METEOROID RADIATION PARAMETER
r = METEOROID RADIUS (cm)
ρ = METEOROID MASS-DENSITY ($\mathrm{g \cdot cm^{-3}}$)

Fig. 8. Meteoroid orbits determined by Pioneer 8 and 9 (geometry of dust related to comets).

For the micron-sized meteoroids considered here, the radiation pressure effect predominates. Small particles are forced into orbits that have a larger eccentricity and aphelion distance than the parent comet's orbit. They therefore move outside the original orbit, but within the plane of the parent comet's orbit, as is shown in Fig. 8.

For any further consideration, now, we should know two things: (i) in what type of orbit (elliptic or hyperbolic) the young meteoroid was born, and (ii) in what relative abundance ratio the two types of meteoroid orbits are produced. If, for instance, the ratio of hyperbolic to elliptic orbits is less than say 1000, meteoroids in elliptic orbits contribute most to the flux encountered by space experiments when intersecting the comet's orbital plane, because they stay at least 1000 times longer within satellite distances from the earth than particles in hyperbolic orbits. On the other hand, if the ratio greatly exceeds 1000 then the advantage in life-time of the elliptic particle is overcompensated, and a search for hyperbolic particles in the comet's orbital plane seems more promising.

At the present time, no quantitative theory about the exact size and orbit distribution of dust given off by comets is established. The only semi-quantitative approach known to the authors are the calculations by Silverberg [10] as represented in Fig. 9. Silverberg predicts the radiation parameter β for dust particles that can be found at a certain distance from the comet's node. But

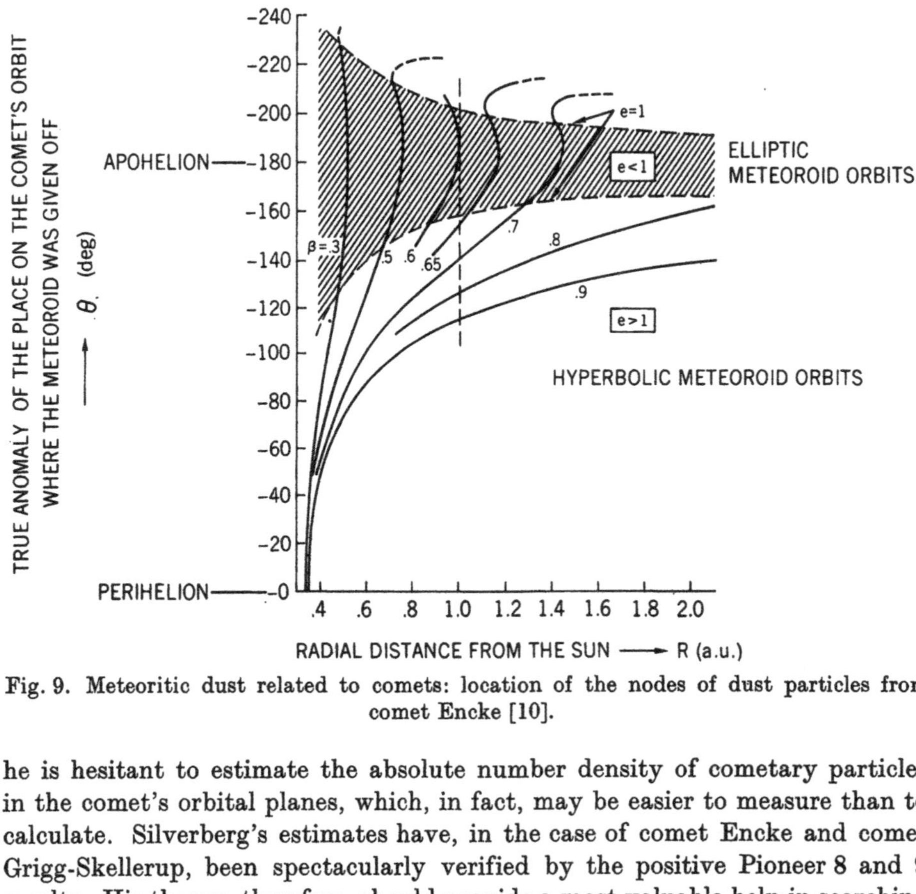

Fig. 9. Meteoritic dust related to comets: location of the nodes of dust particles from comet Encke [10].

he is hesitant to estimate the absolute number density of cometary particles in the comet's orbital planes, which, in fact, may be easier to measure than to calculate. Silverberg's estimates have, in the case of comet Encke and comet Grigg-Skellerup, been spectacularly verified by the positive Pioneer 8 and 9 results. His theory, therefore, should provide a most valuable help in searching for further comet-related dust particles by means of space probe meteoroid experiments. The flux rate encountered by a deep space meteoroid experiment can, most probably, be considerably enhanced by selecting the spacecraft's orbit with respect to a certain comet (for instance Encke) and by further adapting the orientation and response function of the meteoroid detector to the velocity and size of the cometary particles expected. Besides discussing several comet-related peaks in the flux measured by microphone satellite experiments, he also mentions an evidence of a completely different nature: "Gegenschein observations at the University of Texas by Rob Roosen [11], have always failed to show the earth's shadow in the backscattering of the zodiacal light with the exception of one observation. The one exception was near the plane of comet Encke on 21 February 1969. This indicates that the dust density near the earth was higher on 21 February 1969 than on the other dates."

In view of the strong evidence, both in experiment and theory, supporting a cometary origin of small meteoroids, the meteoroid model by Dohnanyi [12],

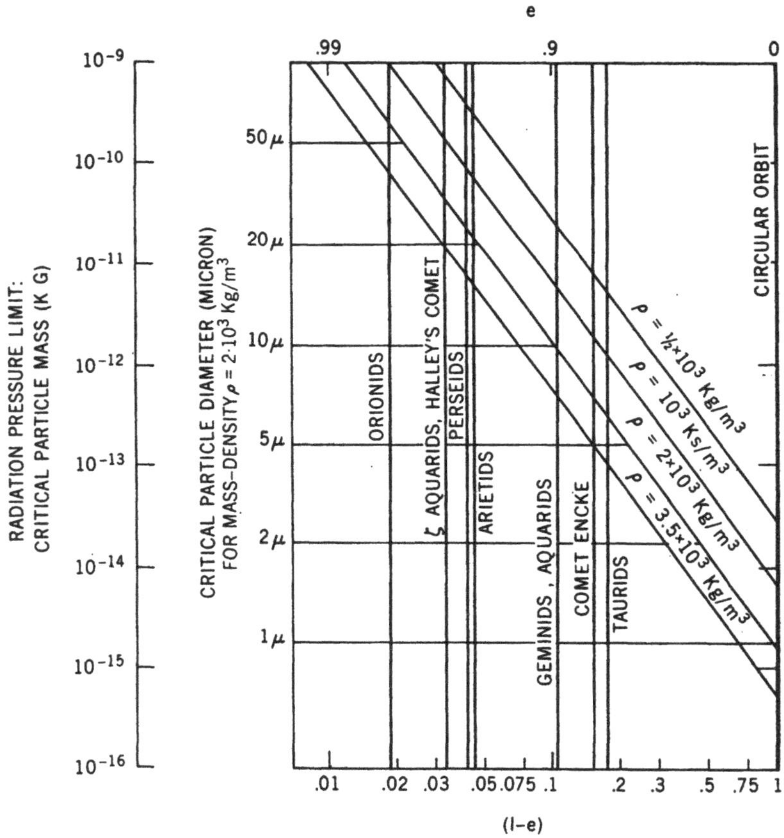

Fig. 10. Radiation pressure limit for sveral streams and comets as a function of particle densities and eccentricities.

which is based on just this hypothesis, deserves particular attention. After calculating the dynamical radiation pressure limits for certain meteor streams and comets (Fig. 10) he discusses the collision between those meteoroids that stay in the solar system by considering their orbital statistics. The upper and lower limit of the mass-distribution of meteoroids as obtained by Dohnanyi is represented by the shaded area labeled Sporadic Meteoroids in Fig. 7. The significant reduction in the slope of this distribution at about 10^{-7} g is due to the fact that, below that mass, a large portion of stream meteoroids is continuously lost by the action of the radiation pressure. Only the marginal number of meteoroids that stays in elliptic orbits can further contribute to the (therefore reduced) slope of the mass distribution at smaller meteoroid masses. The ultimate cut-off at $\sim 10^{-12}$ g shown in the diagram is justified by the twofold evidence of the Pioneer cut-off-measurement (§ 4), and the fact that the radiation pressure limit for circular injection orbits, and hence the most favourable case for small particles, is not below 10^{-12} g according to Fig. 10.

The discussion of the radiation pressure limit is based on the geometrical cross section of the meteoroid. However, even the assumption of "exotic" indices of refraction for the meteoroid material probably cannot counterbalance the enormous loss of particles due to their highly unstable initial orbits based on a cometary origin. An interesting discussion of many related facts is reviewed by Bandermann [13].

6. Conclusions

The new field of *in situ* meteoroid research is now on the verge of contributing considerably to the astronomical knowledge of meteoroids. The Pioneer 8 and 9 results prove that satellite meteoroid experiments are both scientifically worthwhile and technologically feasible. The well-established flux result now justifies the design, and guarantees the success of further meteoroid space experiments.

However, many interesting facts about the interplanetary meteoroid environment will be discovered only if we first estimate how they manifest themselves during the spacecraft misssion. To initiate action, we make the following suggestions:

(i) Dohnanyi's model [12] modified by the Pioneer 8 and 9 cut-off results, as given in Fig. 7, provides a safe basis for estimating the flux result of future satellite meteoroid experiments. It should be adopted as a preliminary standard until a more detailed orbit model can be developed, possibly by the COSPAR Flux Revision Committee.

(ii) Silverberg's work [10] on dust concentrations in the orbital planes of comets, supplemented by the present and future Pioneer 8 and 9 *in situ* measurements of the exact orbits of comet-related particles, provides a first guide line to direct spacecraft meteoroid experiments to regions of enhanced meteoroid number density in space.

(iii) Theoretical studies should be undertaken in the four areas suggested in § 2: particularly, the orbit optimization of space probe meteoroid experiments should be investigated with regard to cometary dust concentrations.

Acknowledgment

This research was carried out while one of the authors (U.G.) held a National Research Council Postdoctoral Resident Research Associateship supported by NASA.

References

[1] O. E. Berg and U. Gerloff, in: Space Research XI, Akademie-Verlag, Berlin 1971 (p. 225).
[2] U. Gerloff, O. E. Berg and F. F. Richardson, Paper presented at AGU Conference, Washington 1970.
[3] O. E. Berg, U. Gerloff and Burton, Paper presented at AGU Conference, Washington 1970.
[4] O. E. Berg and U. Gerloff, J. Geophys. Res., in press.
[5] O. E. Berg and F. F. Richardson, Rev. Scient. Instrum. **40**, 1333 (1969).
[6] H. Elsässer and T. Schmidt, Z. Naturforsch. **21a**, 1116 (1966).
[7] F. L. Whipple, Astron. J., **59**, 501 (1954).
[8] R. E. McCrosky and A. Posen, Smithson. Contr. Astrophys. **4**, 15 (1961).
[9] R. B. Southworth, NASA Sp-150, Conf. on The Zodiacal Light and the Interplanetary Medium, Hawaii 1967.
[10] E. C. Silverberg, Univ. of Md, Tech. Rep. No. 70-066, 1970.
[11] R. G. Roosen, An independent investigation of the Gegenschein thesis, University of Texas 1969.
[12] J. S. Dohnanyi, Bellcomm. Inc. TR 69-105-3-2, 1969.
[13] L. W. Bandermann, Univ. of Md, Tech. Rep. No. 771, 1967.

Space Research XI — Akademie-Verlag, Berlin 1971

REVIEW OF IN SITU MEASUREMENTS OF COSMIC DUST PARTICLES IN SPACE

J. A. M. McDonnell

University of Kent at Canterbury, Canterbury, England

Direct measurements of cosmic dust particles in the near-earth environment and interplanetary space are considered and aspects of reliability and calibration discussed. Improved scaling factors for impulse sensors and penetration sensors are introduced and the measurements used to consider momentum distribution and penetration models. Good agreement exists between the penetration data and the controlled momentum measurements in the vicinity of the earth. These indicate that the near-earth flux is enhanced over interplanetary values by only a small factor which is no greater than two for masses 10^{-7} to 10^{-11} g. For masses less than 10^{-11} g the cislunar data of OGO 3 shows an enhancement over the interplanetary flux measured by Pioneer 8 and Pioneer 9.

1. Introduction

Although *in situ* measurements on the meteoric flux in space have been well supported over the last 20 years since the advent of rockets and satellites, some doubt still exists on the magnitude of the flux to the extent of three orders of agnitude. Sophistication has developed from the earliest piezoelectric transducer mounted on a V-2 rocket skin by Bohn [1] in 1949 to triple coincidence time-of-flight and momentum measurements carried to interplanetary space by Pioneer 8 and 9 [2]. Increasing complexity has not always been followed by improved reliability in the measurement, and the problems which face measurements of this type have been well described by Nilsson [3] analysing the OGO 2 and OGO 4 data. Tight schedules for experimental design construction and integration have not left sufficient time for reliability and control studies, and unfortunately this has accentuated the problem, because well conceived experiments of the OGO 'time-of-flight' series were designed to measure the high flux rates indicated by early measurements of perhaps questionable reliability.

This review looks at the data obtained from different measurement techniques separately; most of the data separate into two main groups: (i) momentum (either with or without associated measurements) and (ii) penetration. Data are approached initially with no astronomical preconceptions; they are considered

Note. Revised values for several key experiments referred to in the preprint of this review were presented during the 1970 Leningrad meeting. This review now includes revised and updated data from OGO 3, Lunar Explorer 35, Pioneer 8 and 9; minor improvements to the model have resulted.

and compared on statistical significance and the reliability of the measurement. The possibility of temporal variations is not fully considered; more weight is attached to extended observations, and where possible, a model is then evaluated for the two main techniques representing an average value of (i) the momentum distribution and (ii) the penetrating power of particles in space.

The models derived from these separate analyses are then compared and their significance and correlation discussed.

2. Particle Momentum Measurements

2.1. Microphone Measurements without Control—Near Earth

Table 1 lists satellite data from most significant microphone sensors and also includes the luminescent sensor results of Cosmos 213. The data have been corrected for earth shielding; revised data are used whenever possible, sometimes from author's reviews, recent literature or personal correspondence in reply to the Cospar Cosmic Dust panel request for information [4]. Microphone

Table 1
Near earth satellite measurements from microphone sensors without control or coincidence measurements (Fig. 1).
Mass sensitivity is calculated at 20 km s^{-1} impact velocity

Satellite	Orbit (km)	Date of observations	No. of events	Mass sensitivity (g)	Flux rate ($m^{-2} s^{-1}$ $2\pi sr^{-1}$)	Data point (Fig. 1)
Explorer 1 [19, 4]			145	$3 \cdot 10^{-10}$	$2 \cdot 10^{-2}$	7
Sputnik 3 [7]	400/700	18–25 May 1958	0	$3 \cdot 10^{-8}$	$<10^{-4}$	10
Vanguard 3 [19]			~3500	$5 \cdot 10^{-10}$	$2 \cdot 10^{-3}$	6
Midas 2 [4]	500/500	24–25 May 1960	67	$2 \cdot 10^{-11}$	10^{-1}	
Explorer 8 [8, 50]		1961	3600 75 1 or 2	$5 \cdot 10^{-11}$ $5 \cdot 10^{-10}$ $5 \cdot 10^{-9}$	$3 \cdot 10^{-2}$ $6 \cdot 10^{-4}$ 10^{-5}	5
Electron 2 [4, 7, 51]	400/7130	1 Jan.– 10 Mar 1964	38	10^{-8}	$6 \cdot 10^{-4}$	9
Electron 4 [4, 7]	400/7130	2 Jul. to 9 Sept. 1964	9	10^{-8}	$6 \cdot 10^{-5}$	8
Cosmos 213 [4] (Luminescent sensor)	205/291	15–20 Sept. 1968	19	$3 \cdot 10^{-11}$ $3 \cdot 10^{-10}$	$4 \cdot 10^{-2}$ $2 \cdot 10^{-3}$	12

experiments calibrated by beads have been increased in sensitivity by a factor of 3 in line with recent calibration studies [5] at 6 km s^{-1}. An energy scaling law (based on mv^2) is used from 6 km s^{-1} to an assumed cosmic impact velocity of 20 km s^{-1}. This scaling is in line with other reviews [6, 7]; the apparent sensitivity of the Explorer 8 data, for example, is an order of magnitude greater than the first published data [8] (see Appendix for discussion).

The reliability of single parameter piezoelectric sensor measurements has been questioned by Nilsson [9] but since refuted by Bohn et el. [10] experimenting on flight 'back-ups' and associated sensors. It is a highly questionable matter whether such measurements can now be used to improve the weighting of experiments which have already been flown, because: (i) laboratory tests have not taken place in a simulated thermal vacuum environment, (ii) they did not have the full complement of on-board equipment and experiments operating, and (iii) the noise characteristics of two apparently similar structures may differ widely due to such trivial factors as the tension on a mounting bolt. Certainly the Explorer 8 satellite did not have opportunity to prove itself on a thermal gradient run before launch, and many other experiments must share this deficiency. Little weight can be given to any particular microphone data point, therefore, but it should be noted that a correlation between several results could lead to an improved collective weight. The data points are displayed in Fig. 1.

Interplanetary and Lunar Space

The various measurements of particle momentum and also the luminescent sensors of Zond 5 and Zond 6 are tabulated in Table 2 and illustrated in Fig. 1. The interplanetary environment usually offers a less critical environment for piezoelectric sensors, where thermal drifts are less severe and skin temperatures are not subjected to orbital cycling. A higher weighting can be given to these than to the near earth microphone measurements.

2.2. Controlled Measurements

An improvement of the simple microphone has been developed in the USA for impact measurements: a thin capacitor is deposited on the surface of the piezoelectric sensing diaphragm. A discharge pulse of the capacitor caused by a penetration is then registered in conjunction with the momentum pulse. This was flown successfully on Lunar Explorer 35 [11] and also attempted on Cosmos 163 [12]. The OGO series employed further control by using a thin 'front' film sensor prior to a combined microphone and plasma (or capacitor) unit placed a short distance away. Three events were first recorded by these experiments on OGO 1 [13] but later analyses of OGO 2 and 4 [3] have indicated that these might have been spurious. Certainly these early results from four hours' observation are not typical of the later OGO measurements which serve as significant upper limits. OGO 3 (14) has also measured relatively low values of the cislunar flux. These data have been revised [15, 54].

Pioneer 8 and 9 instrumentation [16], a development from the OGO units, provides a very wide acceptance angle and promises to provide the most significant controlled flux measurements to date. Measurements taken over two years [16] in interplanetary space show a flux of $2 \times 10^{-4}\ m^{-2}\ s^{-1}$ at 10^{-11} g and an absence of particles of lower mass (Table 3).

Even controlled sensors must be treated in perspective, however. Apart from the possibility of extraneous stimulation, the control also can affect the measurement. The front film of a time-of-flight sensor will require a certain

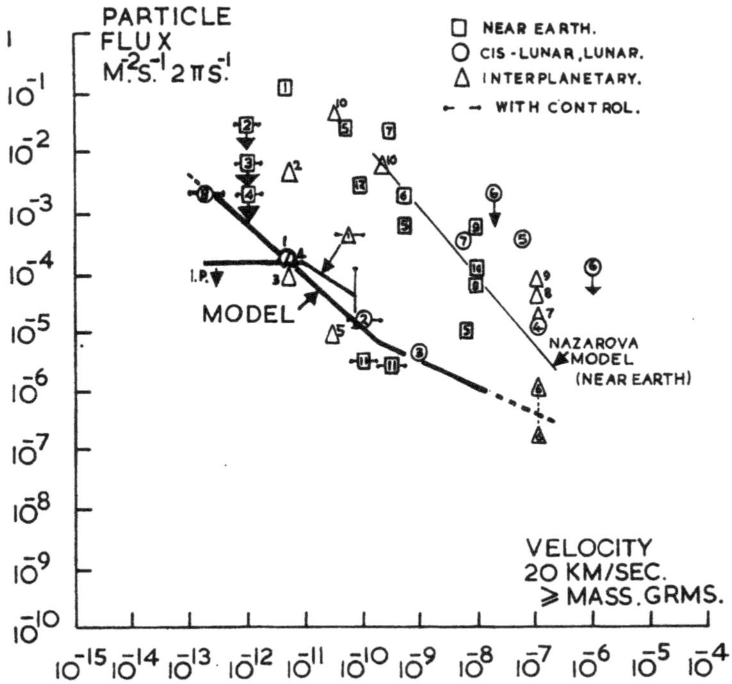

Fig. 1. Microphone measurements. Experiments with secondary coincidence detection are indicated thus: ●— —●

□ near earth

1. Midas 2
2. OGO 2
3. OGO 4 — A Sensors
4. OGO 4 — A & B Sensors
5. Explorer 8
6. Vanguard 3
7. Explorer 1
8. Electron 4
9. Electron 2
10. Sputnik 3
11. Cosmos 135/163
12. Cosmos 213 (luminescent)

○ cislunar/selenocentric

1. Lunar Explorer 35 (mic.)
2. Lunar Explorer 35 (mic/cap)
3. Lunar Orbiter
4. Lunar 2
5. Lunar 3 I.P. Station
6. Lunar 1
7. Lunar 3
8. OGO 3

△ interplanetary

1. Pioneer 8
2. Pioneer 1
3. Mariner 4 (1966)
4. Mariner 4 (1967)
5. Mariner 2
6. I. P. Background [7]
7. Mars probe
8. Venus 2
9. Zond 3
10. Zond 5 and 6 (luminescent)

Table 2

Distant geocentric, lunar and interplanetary measurements from microphone sensors without control or coincidence measurements (Fig. 1)

Satellite or probe	Region of space	Date	No. of events	Mass sensitivity (g)	Flux rate ($m^{-2} s^{-1}$ $2\pi sr^{-1}$)	Data point (Fig. 1)
Lunar 1 [7]	cislunar	2 Jan. 1959		$2 \cdot 10^{-8}$–10^{-7}	$\leqq 2 \cdot 10^{-3}$	⑥
				10^{-7}–10^{-6}	$\leqq 5 \cdot 10^{-4}$	
				$>10^{-6}$	$\leqq 10^{-4}$	
Lunar 2 [7]	cislunar	12 Sept. 1959	0	10^{-8}	$<2 \cdot 10^{-5}$	④
Pioneer 1 [19]	inter-planetary			$6 \cdot 10^{-12}$	$4 \cdot 10^{-3}$	2*
Mariner 2 [52]	inter-planetary (Venus)	1962	1	10^{-10}	10^{-5}	5*
Mars [7] Probe	inter-planetary	31 Dec. 1962 – 30 Jan. 1963		10^{-7}	$5 \cdot 10^{-3}$	7*
		2 Nov.– 30 Dec. 1962		10^{-7}	$2 \cdot 10^{-5}$	7*
Zond 3 [7]	inter-planetary	18 Jul 1965– 21 Jan. 1966		10^{-7}	$7 \cdot 10^{-5}$	9*
Venus 2 [7]	inter-planetary (Venus)	12 Dec. 1965 –Jan. 1966		10^{-7}	$3 \cdot 10^{-5}$	8*
Mariner 4 [53]	inter-planetary (Mars)	1966 and 1967		$6 \cdot 10^{-12}$	10^{-4}	3*
					$1.4 \cdot 10^{-4}$	4*
Lunar Orbiter [11]	Lunar Orbit			10^{-9}	$4 \cdot 10^{-6}$	③
Lunar Explorer 35 [11, 15, 54]	Lunar Orbit	1967 and 1968		$5 \cdot 10^{-2}$	$2 \cdot 10^{-4}$	①
Lunar 3 [7]	cislunar	4–8 Oct. 1969		$.7 \cdot 10^{-9}$–$2 \cdot 10^{-8}$	$4 \cdot 10^{-4}$	⑤
				$2 \cdot 10^{-8}$–$6 \cdot 10^{-8}$	$2 \cdot 10^{-3}$	
				$> 6 \cdot 10^{-8}$	$4 \cdot 10^{-4}$	
Zond 5 [55] (Luminescenr sensor)	inter-planetary	15 Sept. 1968		$3 \cdot 10^{-11}$	$4.8 \cdot 10^{-2}$	10*
Zond 6 [55] (Luminescent sensor)	inter-planetary	10 Nov. 1968		$3 \cdot 10^{-10}$	$4.4 \cdot 10^{-3}$	10*

penetrating power and will also modify the particle before impact on the sensor. Capacitor discharge sensors are also subject to a critical penetrating power of the particle.

2.3. Model for the Particle Flux Based on Momentum Measurements and Combined Sensors

For measurements in the vicinity of the earth the first striking feature is the large variability amongst early microphone data. There are several examples

Figures with asterisks are those in triangles.

Table 3

Controlled sensor measurements (Fig. 1)

Satellite or Probe	Control	Mass sensitivity (g)	Flux rate (m^{-2} sec^{-1} 2π sr^{-1})	Data point Fig. 1
Lunar Explorer 35 [11, 15, 54] (Lunar, 1967 and 1968)	Coincidence of momentum and capacitor discharge	10^{-10} g	$1.4 \cdot 10^{-5}$	②
OGO 2 and 4 [3] (near earth)	Time of flight. Front film (A) sensor and plasma microphone (B)	10^{-12} g		
	OGO 2.		$3 \cdot 10^{-2}$	
	OGO 4. Based on A events (49)		$\sim 7 \cdot 10^{-3}$	[3]
	OGO 4. Based on A and B events (0)		$< 2 \cdot 10^{-3}$	[4]
OGO 3 Cislunar [11, 15, 54] 1966 and 1967	Time of flight. Front film (A) sensor and plasma microphone. Based on A events ($\sim$70)	$2 \cdot 10^{-13}$	$2.5 \cdot 10^{-3}$	⑧
Cosmos 135 [18] Cosmos 163 [12] (near earth 1966 and 1967)	Coincidence of momentum and capacitor discharge			
	Cosmos 135	10^{-10}	$< 10^{-6}$ (mean)	[11]
		$3 \cdot 10^{-10}$	$4.5 \cdot 10^{-6}$ (mean)	[11]
	Cosmos 163	10^{-10}	$3 \cdot 5 \cdot 10^{-6}$ (mean)	[11]
		$3 \cdot 10^{-10}$	$1.6 \cdot 10^{-6}$ (mean)	[11]
Pioneer 8 and 9 [16] (interplanetary)	Time of flight. Front film sensor (A) and plasma/microphone (B) Position identification on A and B	10^{-13} g to 10^{-11} g	$2 \cdot 10^{-4}$ at 10^{-11} g. Flux distribution plotted (Fig. 1)	1*

of upper limits lying 'below' flux measurements, and there is difficulty in ascribing an average value to the measurements. The Explorer 8 microphone data, which has formed the basis for a model for the near-earth flux, do show some correlation with other microphone data, however, to within something like an order of magnitude. But several of the high flux rate experiments have been short lived, and must be subject to statistical fluctuations as well as the acknowledged extra risk of extraneous stimulation due to the proximity to launch. Nazarova [17] has recently supported the high flux of the microphone data and the model from that review is shown in Fig. 1.

There is a more significant distinction between the microphone measurements and the controlled measurements of Cosmos 135 [18], Cosmos 163 and OGO 4. Since the reliability of piezoelectric sensors has been questioned [9] it would be a simple matter to discount any measurements without control, but there is still some correlation amongst the high microphone data and in fairness it must be asked if there is any feature which could explain the low rate of controlled experiments. One feature of some controlled experiments must be examined: the penetration resistance of any thin film or a capacitor which might require penetration for coincidence measurement. It has been suggested that the particles may be fluffy in order to explain the low rate of penetration experiments many times, but the effect of low particle densities is not nearly so drastic as previously anticipated (see Appendix). The penetration at hypervelocities is largely determined by the particle mass and velocity, and for the density range of 0.4 g cm^{-3} to 19 g cm^{-3} the minimum particle *mass* able to penetrate a given thickness of target varies only by a factor of 3 (see Appendix). If the low flux rate of OGO 2 and 4 is to be explained by weakly penetrating particles, then exceptionally diffuse or elongated particles are called for — even the loose agglomerates typical of many faint radar meteors of density 0.2 to 0.4 g cm^{-3} will not explain this discrepancy. On Cosmos 135 and 163 penetration was not required and these low fluxes conflict directly with the early microphone data.

No positive measurements have taken place near the earth for particle masses $<10^{-10}$ g with control but the OGO 4 data give an upper limit at mass $\sim 10^{-12}$ g. Looking however at more massive particles from 10^{-11} to 10^{-8} g the lunar measurements of Explorer 35 and Lunar Orbiter, and the extended interplanetary measurements of Mariner 2 and 4 and Pioneers 8 and 9 a clearer picture emerges. Good correlation is shown between these and they also substantially agree with the near-earth Cosmos 135 and 163 data. Significance is attached to the background interplanetary level from Soviet microphones [7], where no events were measured for periods of weeks to a month for masses $\geqq 10^{-7}$ g. Little significant difference can be discerned from these measurements between interplanetary, selenocentric far or near geocentric measurements for masses $\geqq 10^{-11}$ g. A model is shown in Fig. 1 representing these measurements. The slightly lower slope of the Pioneer data in this region is compatible and within the statistical uncertainties determined by the eight events. For masses $<10^{-11}$ g an interesting situation arises. Pioneer 8 and 9 indicate a 'cut-off', but OGO 3 in cislunar space detects a higher flux. This may indicate that in the vicinity of the earth–moon system, the submicrometre flux is enhanced, but it must be emphasized that if this does exist it is limited to mass $\leqq 5 \times 10^{-12}$ g and that the OGO 4 data preclude this from being closely associated with the earth in the way the earth's 'dust belt' was proposed. It may be tentatively ascribed to the earth–moon system and might originate from submicrometre lunar ejecta.

3. Penetration Measurements

3.1. Near Earth Measurements

Most available telemetered measurements and recovered sample studies cover the near-earth environment. Telemetered data are summarized in Table 4 and certain recovered sample studies in Table 5; only data which seek evidence of high velocity impact craters are listed here. The penetration of various metals is referred to the equivalent thickness of aluminium. Different experimental calibration factors are still to be expected after this scaling, however; for example the Pegasus data refer to a particle of diameter perhaps 20–40% greater

Table 4
Penetration data from near-earth satellites (Fig. 2).
The critical penetration thickness is scaled to the equivalent thickness of aluminium (appendix)

Satellite	Experiment	Penetration thickness (μm)	Equivalent thickness Al (μm)	Date	Events	Fluxrate ($m^{-2} s^{-1}$ 2π sr^{-1})	Data point (Fig. 2)
Explorer 1 [19]	Wire grids	17 copper	39		0	$<3 \cdot 10^{-4}$	11
Explorer 3 [19]	Wire grids	17 copper	39	–	2	$8 \cdot 10^{-3}$	13
Midas 2 [4]	Wire grids	20 copper	46	25 May 1960	0	$<3 \cdot 10^{-4}$	10
Samos 2 [4]	Wire grids	20 copper	46	31. Jan.– 1 Feb. 1961	8	$4 \cdot 10^{-4}$	12
Various near-earth particle counters	Photomultiplier cover penetration (Van Allen)	~3 mica	~4	various	0	$<2 \cdot 10^{-2}$	3
Vanguard 3 [19] Explorer 7 [19] Explorer 16 [20]	Aluminiumised 5 μm Mylar covering CdS cell	5 Mylar. From the total eroded area, the flux rate is calculated assuming holes are ~5 μm diameter	$\simeq 5$			$\leqq 7.5 \cdot 10^{-2}$ $3.3 \cdot 10^{-1}$ 0.86	4 5 6
Explorer 16 [56, 57]	Pressurised cells	25 BeCu 51 BeCu 127 BeCu	58 118 292	Dec. 62– Jul. 63	44 11 0	$5.6 \cdot 10^{-6}$ $2.9 \cdot 10^{-6}$ $<1 \cdot 4 \cdot 10^{-6}$	15
Explorer 16 [56, 57]	Wire grids	51 copper 76 copper	116 176	1962 & 1963	1 1	7.10^{-6} $3.5 \cdot 10^{-6}$	
Explorer 16 [56, 57]	Stainless steel over grids	25 stainless steel 76 152	64 194 390	Dec. 62– May. 63	6 1 0	$1.3 \cdot 10^{-6}$ $8.7 \cdot 10^{-7}$	17
Explorer 23 [58]	Pressurised cells	25 stainless steel 51	65 130	Nov. 64– Jul. 65	43 54	$5.4 \cdot 10^{-6}$ $2.7 \cdot 10^{-6}$	16
Ariel 2 [21]	Aluminium covering photocell	9 Al 16 Al	9 16	April– Sept. 64	0 0	$\leqq 10^{-3}$ $\leqq 1.5 \cdot 10^{-5}$	7 8
Pegasus 1 [57, 59]	Aluminium covering capacitor discharge sensor	40 Al 200 Al 400 Al	40 200 400	Feb.– May 65	121 9 4	2.10^{-6} $6.3 \cdot 10^{-7}$ $3.6 \cdot 10^{-8}$	14
Pegasus 2 [57, 59]		40 Al 200 Al 400 Al	40 200 400	Jun.– Aug. 65	121 18 58	$3.2 \cdot 10^{-6}$ $2.5 \cdot 10^{-7}$ $7.0 \cdot 10^{-8}$	14
Pegasus 3 [57, 59]		40 Al 200 Al 400 Al	40 200 400	Aug. 65	23 4 11	$3.6 \cdot 10^{-6}$ $3.0 \cdot 10^{-7}$ $5.2 \cdot 10^{-8}$	14

Table 5
Data from recovered near-earth satellite experiments showing evidence of high speed impacts (Fig. 2)

Vehicle	Experiment	Sensitivity	Events	Flux rate (m^{-2} s^{-1} $2\pi sr^{-1}$)	Data point (Fig. 2)
Gemini 9 [34] May 1966	Recovery of polished Cu and Al Flight surfaces show no significant difference compared to control	~0.5 μm crater diam.	0	$1.4 \cdot 10^{-1}$	2
Gemini 8 [37]	Recovery of polished stainless steel. 3 craters observed indicating high speed impact. Surface contaminated 7 μm crater depth in steel ≃ 30 μm marginal penetration of aluminium	7 μm depth 32 μm 70 μm	1 1 1	10^{-4} fpr penetration ±30 μm in thin Al	9
Gemini 12 [48]	Recovering of polished surfaces optical and electron microscope scanning	3 μm 0.75 μm		$3 \cdot 10^{-2}$ $\leqq 4 \cdot 3$	1

than a marginally penetrating particle, and the pressurized cells of Explorers 16 and 23 may refer to a particle which makes only a fine fissure: perhaps 10% smaller than marginal penetration sensed by a photocell. There is a severe lack of reliable penetration data more sensitive than the Ariel 2 data. The aluminized Mylar sensors, Explorer 3 [19], Vanguard 3 [19] and Explorer 16 [20] might have provided information in this region, but the data conflict with other upper limits and they must be considered to be a little less reliable than metallic foil experiments. Aluminized Mylar is not an ideal material for such a demanding use as a near-earth satellite and its durability has not been proved.

3.2. Model for the Flux of Penetrating Particles

Good agreement is obtained in many cases when a similar experiment is repeated; the Explorer 16 and 23 pressurized cells are an example and also the Pegasus series. These data in particular show an important feature of the flux: for particles penetrating less than approximately 100 μm of aluminium, the flux slope decreases markedly. This is indicated by Explorer 16 and 23 and also Pegasus 1, 2 and 3. Statistical uncertainties are not likely to cause this apparent change of slope, and the Ariel 2 data also bear this out: they also show that this low slope must extend to the smaller particles capable of penetrating ~16 μm of aluminium [21]. More sensitive reliable measurements are scarce, however, and only an upper limit can be indicated for aluminium ≦16 μm. This penetration model is in excellent agreement with the NASA meteoroid environmental Model [22].

4. Impact Measurements from Sounding Rockets

Impact measurements yielding telemetered or recorded in-flight data are considered separately from collection data at this stage because: (1) they

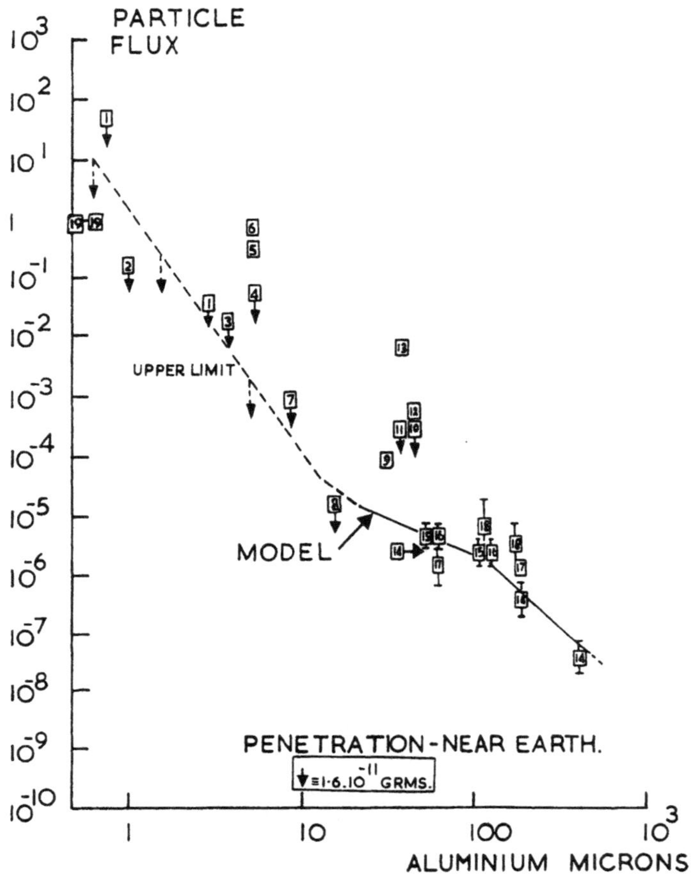

Fig. 2. Near-earth penetration measurements referred to the equivalent thickness of aluminium.

1. Gemini 12 [48]
2. Gemini 9 [34]
3. Van Allen (P. M.)
4. Vanguard 3 (CdS cell)
5. Explorer 7 (CdS cell)
6. Explorer 16 (CdS cell)
7. Ariel 2 (DROD)
8. Ariel 2 (IROD)
9. Gemini 8 [37]
10. Midas 2
11. Explorer 1
12. Samos 2
13. Explorer 3 (grids)
14. Pegasus 1, 2, 3 (mean value)
15. Explorer 16 (BeCu cans)
16. Explorer 23 (S.S. cans)
17. Explorer 16 (S.S. plate)
18. Explorer 16 (grids)
19. Gemini 8 (small holes) [37]

might be generally less immune to some of the contamination problems which face collection experiments and (ii) the time history of the events is usually known which may improve the measure of confidence.

The environment of such flights is far from ideal, however, since aerodynamic heating is very significant and experiments can reach excess temperatures of several hundred degrees within a matter of minutes. Momentum measurements are particularly difficult, therefore, and since early instrumentation of rocket skins with piezoelectric sensors [1, 19], controlled measurements have been attempted. Wlochowicz [23] used three piezoelectric sensors on a rocket skin, but was still troubled by extraneous mechanical noise. McDonnell [24] used two identical momentum sensors, in the same housing pointing to the zenith and earthwards, combined with an impact plasma measurement. The plasma emission experiment suffered interference during flight but the microphones yielded an upper limit well below existing collection data at that time. Since

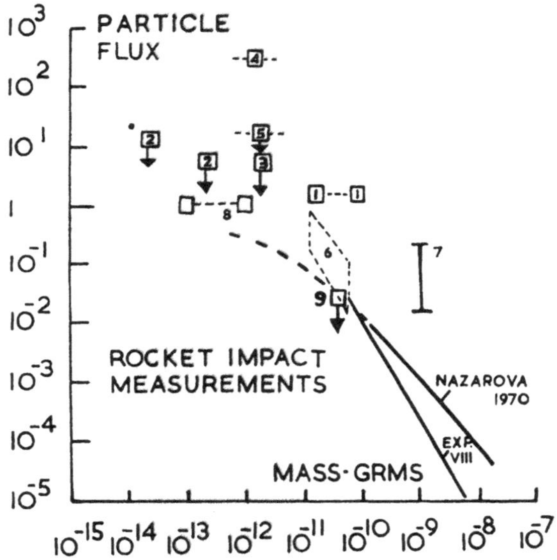

Fig. 3. Impact measurements from sounding rockets.

1. Wlochowicz (mic.) [23]
2. M.P.I. charge sensing [25]
3. McDonnell (mic.) [24]
4. Berg and Meredith (P.M.) [27]
5. Lovering (P.M.) [28]
6. OSO (mic.) [19]
7. Geophysical rockets (mic.) [7]
8. Lindblad (mic.) [26]
9. McDonnell (pen.) [49]

that date Auer et al. [25] have improved the plasma emission technique in rockets using a thin covering (0.3 μm) to exclude ambient plasma. A microphone sensor was also included in this experiment. The results of two flights have yielded several plasma emission pulses, but the authors conclude that these could prove extraneous in origin. Lindblad et al. [26] have flown several rockets each instrumented with microphone sensors in a controlled array, either hidden in the rocket body or mounted vertically and horizontally in close proximity. Several events were detected which could be impacts. The control offered by this will undoubtedly prove a valuable asset particularly if multiple detection can be included (Fig. 3).

Flights by Berg and Meredith [27] using a photomultiplier to detect the impact flash have yielded high rates but in a similar version of the experiment by Lovering [28] no events were detected. Little weight can be attached to these two experiments because of the lack of laboratory impact calibration and in-flight control. The technique is interesting, however, and could be usefully developed to yield particle composition measurements.

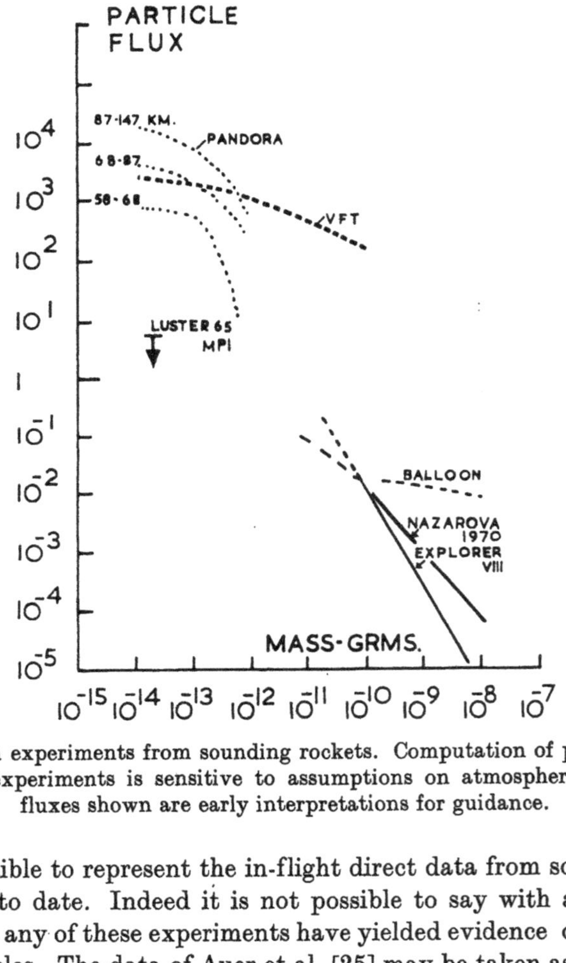

Fig. 4. Collection experiments from sounding rockets. Computation of particle flux rates from collection experiments is sensitive to assumptions on atmospheric processes, and fluxes shown are early interpretations for guidance.

It is not possible to represent the in-flight direct data from sounding rockets by any model to date. Indeed it is not possible to say with any measure of confidence that any of these experiments have yielded evidence of the existence of cosmic particles. The data of Auer et al. [25] may be taken as possible upper limits of the flux for the mass range 10^{-13} to 10^{-12} g in the range of 100–200 km; these masses could be enhanced over the flux at higher altitudes.

5. Collection Measurements

Although the study of particles in the upper atmosphere is closely related to direct measurements of the flux of cosmic particles in space, collection experiments cannot always be directly compared with space flux measurements without a number of assumptions on the properties of the atmosphere. A review of those measurements is not attempted in this paper, but significant results of the measurements are presented for comparison with space measurements.

The first reports of the possible recovery of extraterrestrial particles were received with great interest from the Venus Flytrap experiment [29]. The ease at which so many particles could be accumulated stimulated many further experiments, and has led to improvement in the sampling and control techniques. Sampling techniques aimed at recovering noctilucent cloud particles have also been reported [30].

The problems which face experimenters and theoreticians in the obtaining and interpretation of collection data are not insignificant. Physical processes in the atmosphere are not fully understood, but it is well known that noctilucent clouds can form in the time-scale of one day and show particle concentrations enhanced by orders of magnitude. It is not known if these are condensations on extraterrestrial nuclei at the inversion layer or if particles may be transported up to this layer by atmospheric currents. Even above this level of 85 km other stratified concentrations have been reported [31, 32] and, of course, meteor fragmentation could add to the population. Interpretation of collection data, therefore, must be cautions.

Further doubts have been raised on experimental problems, namely contamination. On later flights using improved experimental conditions, particle concentrations several orders of magnitude lower than the Venus Flytrap data have been reported ([33, 34, 35] and Fig. 4). Sources of contaminations have not always been possible to isolate or explain; even perfectly clean laboratory conditions cannot be guaranteed to yield unambiguous results. On a rocket flight experiment containing a solar coronagraph, for example, Tousey et al. [36] observed trajectories of some 30 particles of 10 to 20 μm travelling vertically upwards with the rocket to apogee. The equipment was sensitive only to a narrow range of relative velocities and could not detect submicrometre particles. Tousey did not attribute the source of these to the rocket, but an alternative origin is very difficult to explain. Nevertheless, residues from expended motors and explosive bolts, and the outgassing of surfaces may contribute significantly to the atmosphere around a short-duration rocket flight. Manned space flights also have possibly more severe problems: thruster motors for stabilization and dumps of waste led to very dirty samples being recovered from Gemini surfaces [37].

The Venus Flytrap data also are in disagreement with later rocket measurements using microphones and plasma sensors shown in Fig. 3 [24–26], and it is likely that typical concentrations of particles in the region 80–150 km are several orders of magnitude below the Venus Flytrap data. The S-10 small crater measurements of Hemenway et al. [37] are in good agreement with an extension of the satellite penetration data to masses of 10^{-14}–10^{-15} g.

Balloon data are even more difficult to relate to extraterrestrial fluxes, but for comparison representative measurements from Hemenway [38] are illustrated with other data in Fig. 4.

6. Current Knowledge of the Flux of Particles in Space from Direct Measurements

Momentum and Combined Sensor Measurement

A review of the data leads to several conclusions on the average flux distribution: (i) In the mass range 2×10^{-10}–10^{-11} g a flux distribution of slope -0.85 (log mass scale) characterizes within experimental and statistical uncertainties the lunar, cislunar and interplanetary fluxes and is also compatible with the near-earth measurements from two controlled microphone experiments. (ii) The Pioneer interplanetary data show a "cut off" for particles $<10^{-11}$ g, but the OGO 3 distant geocentric measurements indicate an increasing flux of smaller particles to 10^{-13} g.

Penetration data

A review of the near earth data permits the definition of an average penetrating flux referred to aluminium from masses 10^{-6}–10^{-10} g. For particles of mass $\sim 10^{-6}$–$\sim 10^{-8}$g the slope may be characterized by a value of -1 which fits well the faint radar meteor data [39]. For particles of mass 10^{-8}–3×10^{-11} g the penetration flux shows a lower slope of ~ 0.5 and on the assumption of a constant impact velocity this would indicate a relative decrease of particles in this mass range. For masses below 10^{-10} g few reliable penetration data have been obtained, but upper limits indicate that the maximum slope of the flux is in the region of -1.

Comparison of Penetration and Momentum Measurements

Although all data have been scaled to 20 km s^{-1} impact velocity, and hence may only be referred to particle mass subject to this condition, the scaling of both momentum and penetration measurements is characterized by a similar velocity exponent, and assumptions on impact velocity lead only to second-order changes in the relative displacements of these two measurements. Also, uncertainty of the impacting particle density in both momentum and penetration measurements is not expected to cause significant differences in the two measurements techniques unless the particles are of *extreme* shape factor (Fig. 5).

The difficulty in explaining the near-earth enhancement inferred from some microphones has been discussed [40, 41] but no satisfactory mechanism has been proposed. We must now add the restriction that if such an enhancement were to exist, then to reconcile this with the vast amount of penetration data, the particles must be of extreme shape factor. But this hypothesis leads to other facts which are not borne out by measurements. These particles would cause a shallow depth of penetration at the expense of an increased area of surface damage, and therefore possibly explain the low rate of some controlled momentum measurements and penetration measurements, but recovered samples of Gemini flights would conversely have been expected to yield very much higher numbers of shallow hypervelocity craters. This evidence, together

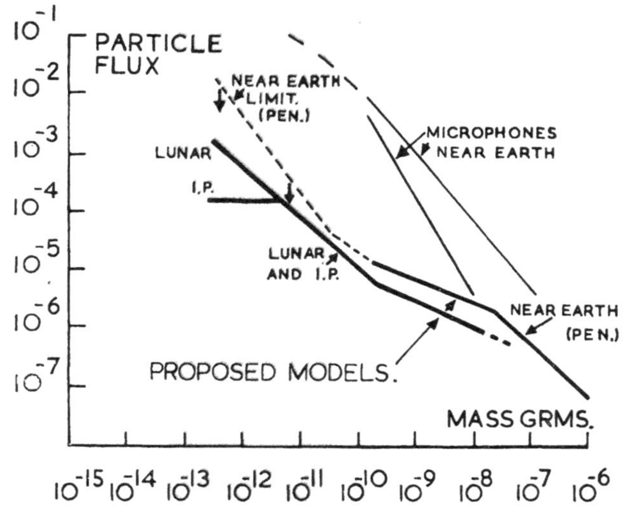

Fig. 5. Comparison of flux rate models. The lunar and interplanetary model is based on controlled experiments.

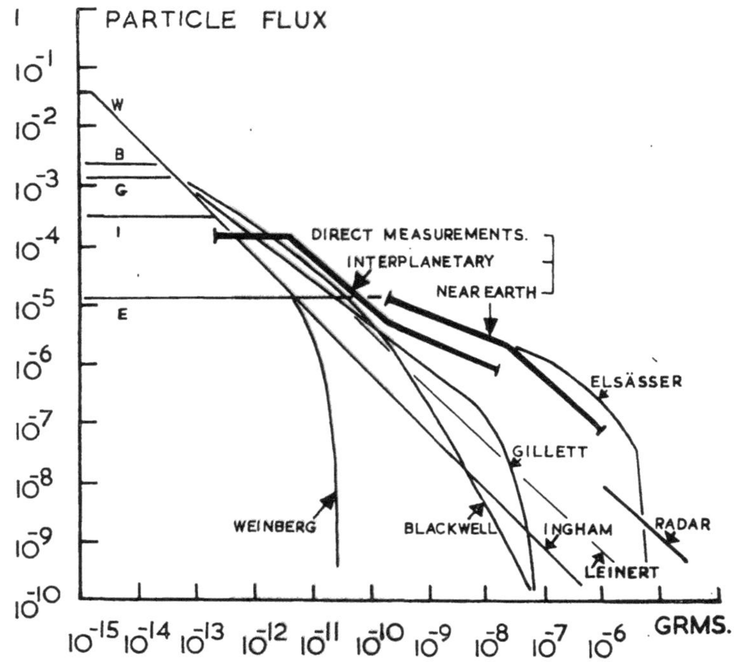

Fig. 6. Comparison of the proposed flux model with zodiacal light measurements and interpretations.

Table 6

Impact measurements from sounding rockets (Fig. 3)

Rocket	Experiment	Results	Sensitivity (g)	Flux Rate ($m^{-2} s^{-1} 2\pi sr^{-1}$)	Data point (Fig. 3)
OSO [19] Rockets Aerobee–Nike	Microphone sensors using rocket skin	events recorded	10^{-11}–10^{-10}	$10^{-12} \rightarrow 1$	[6]
Geophysical rockets [7] 1957–1958	Microphone sensors		10^{-8}	$5 \cdot 10^{-2}$–$7 \cdot 10^{-1}$	[7]
Aerobee 25 [27] White Sands 103 km apogee. 1955	Photomultiplier. Luminescent Sensor	$\sim$100 events between 85 and 103 km. No calibration at high velocities	$\sim 10^{-12}$ –10^{-10}	$2 \cdot 10^{-2}$	[4]
Long Tom Woomera [28] 135 km apogee. 1959	Photomultiplier. Luminescent Sensor (as Aerobee 25, above)	No impacts recorded. No calibration at high velocities	$\sim 10^{-12}$–10^{-10}	<2	[5]
Canadian [23] Churchill Range. Apogee $\sim$150 km 1963, 1964	Three microphone sensors using rocket skin	Many events detected. Idenfication of possible impacts not definite due to extraneous noise	10^{-11}–10^{-10}	$\lesssim 2.8$	[1]
Aerobee [24] 4.195 White Sands 180 km apogee. 1966	Two microphone sensors control by geometry (one up, one down)	Several pulses detected – shown unlikely to be true impacts	$3 \cdot 10^{-12}$	$\lesssim 6$	[3]
Skylark 115[49] Woomera	Penetration of aluminium foil	No penetrations recorded	12 μm Al ($3 \cdot 10^{-11}$ g)	$2.4 \cdot 10^{-2}$	[8]
Nike–Apache [25] Kiruna KNA 12 and KNA 13 180 km apogee. 1968	Impact plasma and microphone sensor	Several possible (plasma) impacts recorded. Not definitely identifiable	$2 \cdot 10^{-13}$ $2 \cdot 10^{-14}$	$\lesssim 4$ $\lesssim 11$	[2]
Centaure [26] Kiruna. Apogee 140 km. 1967, 1968	Microphone, control by geometry. Shielded sensor (1967) and vertical/horizontal sensors (1968)	Several events recorded. Possible impacts	10^{-13}–10^{-12}	1.4	[8]

with the direct contradiction of the data by OGO 4, Cosmos 135 and Cosmos 163 must surely lead to a rejection of the 'high' microphone model.

Referring to the proposed models in Fig. 5 it may be noted that the penetration data define the near-earth flux for masses above 10^{-10} g. Controlled momentum measurements define for masses 10^{-10}–10^{-11} g the flux common to cislunar, lunar and interplanetary space. Enhancement in the near-earth region above the interplanetary level is approximately by a factor of two for the mass range 10^{-8}–10^{-10} g and is unlikely to be more than an order of magnitude for the mass range 10^{-6}–10^{-12} g.

Radar meteor data [39] and several zodiacal light interpretations [42–47] are shown in Fig. 6 for comparison with the proposed models. Agreement with the zodiacal light curves for the smaller masses is good but perhaps little more than fortuitious in view of the latitude in the interpretation of these data [42]. Agreement with the radar meteor data for masses $\sim 10^{-6}$ g is significant and indicates good continuity of the cometary population to masses as low as 10^{-8} g. Both the near-earth and interplanetary measurements show evidence of a change of the particle flux distribution between 10^{-8} and 10^{-10} g. The decrease in the flux of smaller particles could be attributed to the removal of the less dense cometary particles by the Poynting–Robertson effect and solar radiation pressure. It would be informative to compute the flux rate inferred from zodiacal light measurements based on a particle distribution model indicated by this interplanetary flux model.

Acknowledgments

The author is indebted to R. C. Jennison for helpful discussions on the subject and for making available the material gathered for the Cosmic Dust Panel. He thanks Miss D. Paine for preprint and manuscript preparation.

References

[1] J. L. Bohn and F. H. Nadig, Research Institute of Temple University Res. Report No. 8, p. 1–26 (1950).
[2] O. E. Berg and F. F. Richardson, Rev. Scient. Instrum. **40**, 1333 (1969).
[3] C. S. Nilsson, F. W. Wright and D. Wilson, Smithson. Inst. Rep. 902-8 (1969).
[4] Personal Correspondence – Information supplied for COSPAR Cosmic Dust Panel in response to questionaire distributed by R. C. Jennison.
[5] J. A. M. McDonnell, J. Scient. Instrum. (J. Phys. E), Ser. 2, **2**, 1026 (1969).
[6] T. N. Nazarova, Paper presented at Int. Symp. Optical Characteristics of the Upper Atmosphere and Space, October 1969. Preprint.
[7] T. N. Nazarova, Space Sci. Rev. **8**, 455 (1968).
[8] C. W. McCracken, W. M. Alexander and M. Dubin, Nature **192**, 441 (1961).
[9] C. S. Nilsson, Science **153**, 1242 (1966).
[10] J. L. Bohn, W. M. Alexander and W. F. Simmons, Paper presented at Xth COSPAR Meeting, London 1965.
[11] W. M. Alexander, J. D. Corbin and C. W. Arthur, Paper presented at XIIth COSPAR Meeting, Prague 1969.
[12] B. P. Konstantinov et al., Cosmic Research **7**, 911 (1969).
[13] C. S. Nilsson et al., Nature **208**, 673 (1965).
[14] W. M. Alexander, J. D. Corbin, C. W. Arthur and J. L. Bohn, Paper presented at XIIth COSPAR Meeting, Prague 1969. Preprint.

[15] W. M. Alexander, C. W. Arthur and J. L. Bohn, in: Space Research XI, Akademie-Verlag Berlin 1971 (p. 279).
[16] O. E. Berg and U. Gerloff, in: Space Research XI, Akademie-Verlag, Berlin 1971 (p. 225).
[17] T. N. Nazarova and A. K. Rybakov, in: Space Research XI, Akademie-Verlag, Berlin 1971. (p. 357).
[18] B. P. Konstantinov et al., Cosmic Research **6**, 738 (1968).
[19] W. M. Alexander, C. W. McCracken, L. Secretan and O. E. Berg, COSPAR Symposium (1962) NASA TN-D-1669 (1962).
[20] L. Secretan, in: Symp. "Meteor Orbits and Dust" NASA SP-135, 1965 (p. 307).
[21] R. C. Jennison, J. A. M. McDonnell and I. Rodger, Proc. Roy. Soc. A **300**, 251 (1967).
[22] Meteor Environment Model, 1969, NASA SP. 8013, March 1969.
[23] R. Wlochowicz, Canad. J. Phys. **44**, 1 (1966).
[24] J. A. M. McDonnell, J. Geophys. Res. **72**, 6110 (1967).
[25] S. Auer et al., Preprint of COSPAR paper.
[26] B. A. Lindblad, G. Arinder and T. Wiesel, in: Space Research X, North-Holland Publ. Co., Amsterdam 1970 (p. 295).
[27] O. E. Berg and L. H. Meredith, J. Geophys. Res. **61**, 751 (1966).
[28] J. F. Lovering, Planet. Space Sci. **2**, 75 (1959).
[29] C. L. Hemenway and R. K. Soberman, Astron. J. **67**, 255 (1962).
[30] C. L. Hemenway et al., Environmental Research Paper Number 83; AFCRL-65-122 (1965).
[31] G. Fiocco and L. D. Smallin, Nature **199**, 1275 (1963).
[32] G. Fiocco and G. Colombo, J. Geophys. Res. **69**, 1795 (1964).
[33] U. Gerloff, J. H. Weihrauch and H. Fechtig, in: Space Research VII, North-Holland Publ. Co., Amsterdam 1967 (p. 1412).
[34] J. H. Weihrauch and U. Gerloff and H. Fechtig, in: Space Research VIII, North-Holland Publ. Co., Amsterdam 1968 (p. 566).
[35] N. M. Farlow, M. B. Blanchard and G. V. Ferry, Paper presented at Xth COSPAR Meeting, London 1967.
[36] R. Tousey, M. J. Koomen, R. E. McCullough and R. T. Seal, in: Proc. Symp. Meteor Orbits and Dust, NASA SP-135, 1965 (p. 333).
[37] C. L. Hemenway, D. S. Hallgren and J. F. Kerridge, Results from Gemini S 10 and S 12 micrometeorite experiments. Preprint.
[38] C. L. Hemenway, D. S. Hallgren and R. Coon, Paper presented at COSPAR Meeting, Vienna 1966.
[39] C. S. Nilsson and R. B. Southworth, in: Physics and Dynamics of Meteors, D. Reidel Publ. Co., Dordrecht, Holland 1968 (p. 280).
[40] C. W. McCracken, in: Proc. Symp. Meteor Orbits and Dust NASA SP-135, 1965 (p. 213).
[41] I. I. Shapiro, D. A. Lautman and G. Colombo, J. Geophys. Res. **71**, 5695 (1966).
[42] C. Leinert, in: Space Research XI, Akademie-Verlag, Berlin 1971 (p. 249).
[43] D. E. Blackwell, M. F. Ingham and A. D. Petford, Mon. Not. Roy. Astron. Soc. **136**, 325 (1967).
[44] F. C. Gillett, Thesis, University of Minnesota (1966).
[45] M. F. Ingham, Mon. Not. Roy. Astron. Soc. **122**, 171 (1961).
[46] J. L. Weinberg, Annls d'Astrophys. **27**, 718 (1968).
[47] H. Elsasser, Space Research X, North-Holland Publ. Co., Amsterdam 1970 (p. 244).
[48] D. E. Brownlee, P. W. Hodge and F. Wright, J. Geophys. Res. **74**, 876 (1969).
[49] J. A. M. McDonnell, Ph. D. Thesis, University of Manchester, England (1964).
[50] J. L. Bohn, W. M. Alexander and A. Wever, Paper presented at Xth COSPAR Meeting, London 1967.
[51] T. N. Nazarova and A. K. Rybakov, Paper presented at VIIIth COSPAR Meeting, Buenes Aires 1965. Preprint.
[52] W. M. Alexander, Science **138**, 3545 (1962); Science **139**, 3549 (1963).

[53] W. M. ALEXANDER, C. W. McCRACKEN and J. L. BOHN, Science **149**, 1240 (1965).
[54] Communication to Open Cospar Dust Panel Meeting, Leningrad 1970.
[55] T. N. NAZAROVA et al., Cosmic Research **7**, 759 (1969).
[56] E. C. HASTINGS, NASA TMX-810 (1963).
[57] C. T. D'AIUTOLO, W. H. KINARD and R. J. NAUMAN, Proc. Symp. Meteor Orbits and Dust, NASA SP-135, 1965 (p. 239).
[58] R. L. O'NEAL, NASA TN-D-4284, 1968.
[59] J. B. DOZIER, in: The Meteoroid Satellite Project Pegasus. NASA TN-D-3505 (p. 65).
[60] K. P. STANYUKOVICZ, in: Artificial Earth Satellites, USSR Academy of Sciences, Moscow 1966, **4**, 292.
[61] M. A. LAVRENTYEV, in: Artificial Earth Satellites, USSR Academy of Sciences, Moscow 1959, **3**, 85.
[62] J. A. M. McDONNELL, Paper presented at XIIth COSPAR Meeting, Prague 1969.
[63] M. H. WAGNER, N. B. BROOKS and R. L. BJÖRK, Proc. VIIth Hypervelocity Impact Symposium **3** (1965).
[64] R. L. BJÖRK, Proc. VIth Hypervelocity Impact Symposium **2**, 1 (1963).
[65] T. D. RINEY and J. F. HEYDA, NASA CR 1196 (1958).

Appendix: Sensor Calibration Factors

1. Momentum Measurements

Space measurements using microphone sensors have been characterized very often by a thin plate acting as a sounding board for one or more piezoelectric transducers. Early laboratory studies showed that such a plate yields a signal of amplitude proportional to the particle momentum for a wide range of particle masses and velocities up to several metres per second (L. Secretan et al. personal communications). This is a general characteristic expected of most linear elastic systems, where the maximum displacement is proportional to the initial momentum given to the system. The crystal output is a voltage of amplitude proportional to this displacement. Other measurements by Wlochowicz [23] have found a dependence of $m^{0.8}\, v^{1.1}$ for low-speed impacts, but this result is liable to be influenced by the excessive contact time of the polystyrene spheres which were used.

At high impact velocities doubt also exists of the role caused by ejected matter. An enhancement is undoubtedly expected, and a dependence of mv^2 has been proposed by Stanyukovicz [60] and $m\,v^{1.6}$ by Lavrentyev [61]. Experimentally, the response to hypervelocity particles on a diaphragm has been measured for several satellite sensors [50] to 7 km s^{-1}. A dependence of mv^{α} where $\alpha \geqq 2$ is indicated for the region 5 –7 km s^{-1}. More recently this high velocity dependence has not been confirmed in similar measurements by McDonnell [5] but a calibration error of approximately 3 was shown to arise from bead calibrations compared with hypervelocity impacts because of the extended contact time of beads. More important, this work has shown that the momentum from the impact region is transmitted to the crystal by the bending wave. This type of system is therefore ballistic and hence yields an output proportional to the true resolved momentum exchange at impact. Calibrations of sensors in the micrometre region do not extend yet above 7 km s^{-1} to the 'cosmic' velocity of 20 km s^{-1} taken for this analysis. Fortunately, choice between the various proposed velocity scaling laws lead only to a factor of 3 in the calculated impacting particle mass if laboratory measurements are extrapolated from 7 km s^{-1}. A scaling proportional to mv^2 which is chosen for this review is thus liable to be correct within a factor of $\leqq 2$ at 20 km s^{-1}. Where microphone measurements were calibrated by bead dropping, a factor of 3 is used to give a revised value of the increased sensitivity.

2. Penetration Measurements

Most penetration measurements in space have used marginal penetration sensing, and since the flux of particles generally increases for the smaller particles, marginal penetration will represent the majority of recorded penetrations. This region has been studied for various metallic thin foils and for aluminized Mylar [62]. Data extend to 7 km s^{-1} for most metals and to 16 km s^{-1} for aluminium. Marginal penetration data for various target metals are used in Table 7 to scale commonly used space materials to the equivalent thickness of aluminium. Again, small errors in this scaling due to uncertainties of impact-

Table 7
Thin foil marginal penetration factors used to scale commonly used materials to the equivalent thickness of aluminium (appendix)

Material	Equivalent thickness Al	Minimum hole size*	Marginal penetration limit**
Aluminium	1.0	0.6	8 ± 1
Beryllium copper	2.3	$\lesssim$1.5	
Copper	2.3	1.5	
Stainless steel	2.6	1.4	
Mylar (aluminized)	1.0	$\leqq$0.3	

* at 5–10 km s^{-1} (× foil thickness)
** at 20 km s^{-1} (× particle diameter)

Table 8
Estimated minimum penetrating particle penetration limits for differing particle densities
The data scaled from the measured impacts of iron particles (appendix)

Impacting particle	Density (g cm^{-3})	Marginal penetration limit*	Minimum penetrating particle mass**
Iron	7.8	1.00	1.00
Tungsten	19.3	1.50	0.77
Titanium	4.5	0.68	1.90
Aluminium	2.7	0.53	2.30
Porous Aluminium	0.4	~0.25	~2.3

* Compared with iron particles (× particle diameter)
** Compared with iron particles.

ing particle density or velocity will not lead to noticeable effects, compared with other uncertainties. Scaling for impacting particle density is shown to be weaker than previously anticipated in a study based on values of the equation of state for porous aluminium [63]. Table 8 summarizes the estimated marginally penetrating masses calculated from hypervelocity impact work [63, 64] for particle densities varying from 0.4 g cm^{-3} to 19 g cm^{-3} referred to the mass of an iron particle which was used in the measurements. Agglomerates of mean density $\sim$0.4 g cm^{-3} have been shown to possess similar penetrating power [65]. Experimental work at up to 16 km s^{-1} is extrapolated to yield a penetration ratio of 8 ± 1 for aluminium at 20 km s^{-1}; expected values for other densities are tabulated. Conversion of penetration thickness to particle mass is made for a marginally penetrating particle of nominal density $\sim$3 g cm^{-3} (Fig. 2).

More important effects are expected from the particle shape factor. Although a factor of 2 to 3 would have little effect at 20 km s^{-1} a shape factor of >10 would have a drastic effect on the penetrating power for randomly orientated particles. A mass error of some 3 orders of magnitude could arise from such an extreme shape factor.

In addition to the scaling of the sensor materials to aluminium, various experimental factors must be considered. For example the pressurized sensors of Explorer 16 [57] could detect a fine crack, although this mode of failure is not expected at hypervelocities [62]. In contrast the capacitor discharge sensors protected by a plate as in the Pegasus series [57] may require considerably more than a marginal penetration, perhaps some 20% to 50% greater than the minimum particle dimensions.

Solid Earth Physics

Space Research XI — Akademie-Verlag, Berlin 1971

APPLICATION OF SPACE TECHNIQUES TO SOLID-EARTH AND OCEAN PHYSICS

C. A. LUNDQUIST

Smithsonian Astrophysical Observatory, Cambridge, Mass., USA

Typical spacecraft-tracking data support simultaneous refinement of ground-station coordinates and the parameters that influence spacecraft positions. The geodetic parameters reported by Gaposchkin and Lambeck for the 1969 Smithsonian Standard Earth (II) are products of this procedure. As the results from the refinement process become more detailed, especially as a consequence of improved observation techniques, dynamical phenomena within the earth become accessible to study. Both Newton and Kozai have studied the tidal mass displacements in the earth through the effects of this displacement on the orbits of satellites. Subsequently, Kozai isolated a seasonal variation in J_2. Anderle and Beuglass have shown that it is possible to detect changes in position of the earth's pole from analysis of satellite orbits. Still further instrumentation advances, such as laser ranging to decimeter accuracy, very long baseline radio interferometry, satellite-to-ocean altimeters, satellite surface-force compensation, and satellite-to-satellite tracking, allow investigation of other topics in solid-earth and ocean physics.

1. Initial Considerations for Applications of Space Techniques to Earth Physics

Some of the earliest scientifically useful data from space activities were tracking observations. Their subsequent analysis has revealed many details of the geometry and mass distribution of the earth. Improved spacecraft and ground instrumentation continues to support such analyses, which now include features in the physics of the solid earth and the oceans.

The information about the earth derived to date from tracking data is only a beginning toward the understanding of earth processes that seems possible with observation techniques and accuracies now achievable. In August 1969, some 65 specialists assembled in Williamstown, Massachusetts,* to examine the current and future promise of the application of space and astronomic techniques to solid-earth and ocean physics [1]. They reported an impressive array of prospects in instrumentation and for new investigations of the earth. The following discussion draws liberally from their conclusions.

As a background for this review, Fig. 1 illustrates very schematically how progress unfolds in the topics of interest here. The process is cyclic; hence, instrumentation is as satisfactory as any other place to begin the discussion. The instruments of initial concern typically measure some quantity that is

* Sponsored by the National Aeronautics and Space Administration.

a function of the position and motion at some time of both a spacecraft and an instrument on earth. Hence, these data carry information about the motion of each end of the station-to-spacecraft vector. This dual character deserves emphasis in a discussion of solid-earth and ocean physics. It is just as valid to consider that the earth is being tracked relative to the spacecraft as it is to consider that the spacecraft is being tracked from ground stations.

The positions of spacecraft and tracking stations each carry different information about the earth. For example, the spacecraft motions depend upon

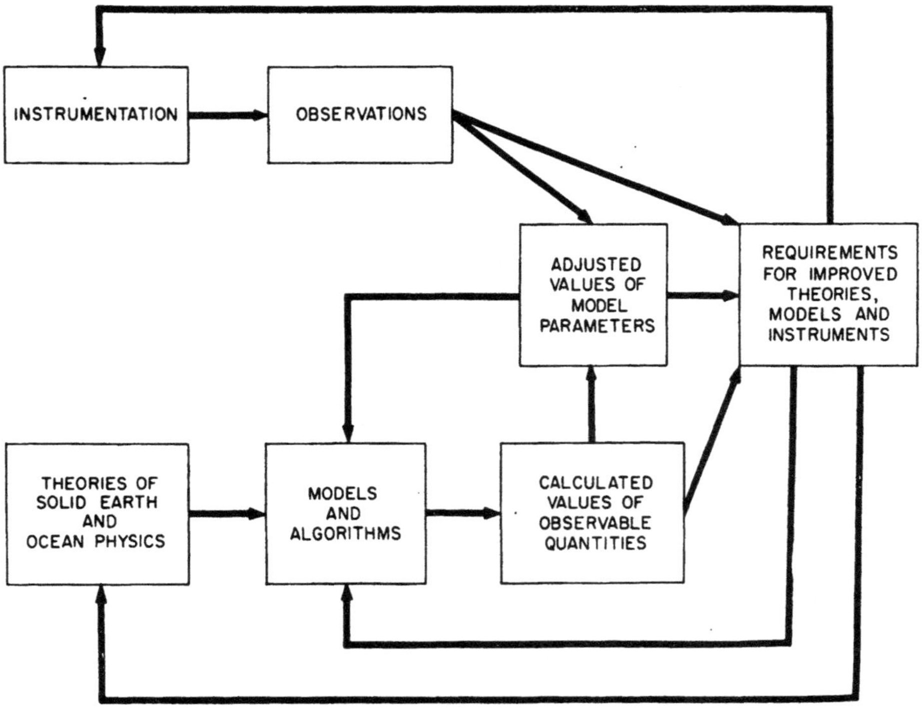

Fig. 1. Schematic information cycle for application of space techniques to solid-earth and ocean physics.

the mass distribution of the earth, including any time dependence. The collective and relative station motions are influenced by many dynamical phenomena of the earth, ranging from the daily rotation to continental drift. To describe the state and dynamics of the earth and other natural or artificial bodies at any instant there are current concepts and theories. These theories yield mathematical models or algorithms that can generate expected positions and motions for the spacecraft and stations as a function of time. The calculated positions or observable functions of them depend on the values of parameters that enter the models or algorithms. The conventional data treatment proceeds with an adjustment of these parameter values to minimize the differences between observed and calculated positional quantities. By this general procedure, the parameters involved in motions of the earth and the spacecraft are refined together. As the refinement progresses, the calculated positions of the spacecraft and the ground stations must maintain commensurate accuracies. In an ideal situation, the models would be refined until they generate values of observable quantities that agree with observed values to within the uncertainties of the observations. Sooner or later, requirements emerge for improved theories, models, and instruments, and so the cycle continues.

Historically, the observation techniques available in 1957 had accuracies far surpassing those of the best earth models and celestial-mechanics algorithms. During the following decade, the theoretical models progressed through many stages of refinement. Until the models began to approach the observations in accuracy, there was little impetus to sharpen observation techniques. Late in 1960s, when theory reached about the same state of refinement and observation, new instrumentation emerged with vastly improved accuracy. The resulting opportunity for further refined theory now leads to many topics in solid-earth and ocean physics.

2. Status of Earth Geometry and Kinematics

A necessary first step toward analysis of tracking data is selection of a coordinate system or systems in which to express the mathematical formalism [2]. Whatever system is chosen for convenience, it should be related in some well-defined way to an inertial coordinate system or to a sufficiently accurate approximation of an inertial system. A fundamental time system is also necessary; one based on atomic clocks is the best choice. Distant astronomical objects that appear as stationary point sources provide a usable means of defining directions for coordinate axes in an inertial system. This choice leads in § 4 to a consideration of the techniques by which other directions can be measured with respect to directions to the distant sources.

The simplest model for the position of a point on the surface of the earth assumes first that a coordinate system can be defined that is fixed in the solid earth. (This assumption will have to be reconsidered in § 6.) The coordinates of a reference point on a tracking instrument can then be expressed in this terrestrial system. In this simplest model, the time-dependent orientation of the terrestrial system is specified relative to astronomical right ascension and declination. The direction of the rotational axis of the earth and the angular position about this axis are measured and tabulated by international services (Bureau International de l'Heure, BIH; and International Polar Motion Service, IPMS) using reference stars and classical astronomical methods.

When the early satellites were launched, the orientation as a function of time for terrestrial coordinate axes was available from the astronomical services with far better accuracy than the coordinates of a tracking instrument could be found globally in the terrestrial system by conventional surveying procedures. Thus, one early requirement from satellite tracking was a refinement of instrument coordinates. To refine instrument coordinates, two principal methods have been widely used. One, geometric in character, depends on simultaneous satellite observations from two or more sites. A second depends on many individual observations of one or more satellites whose orbits are accurately represented by some orbit-determination procedure. By using a combination of both methods, Gaposchkin and Lambeck in 1969 obtained coordinates with an uncertainty of between 5 and 10 m for the principal camera and laser satellite-tracking sites [3, 4]. Other investigators have produced station

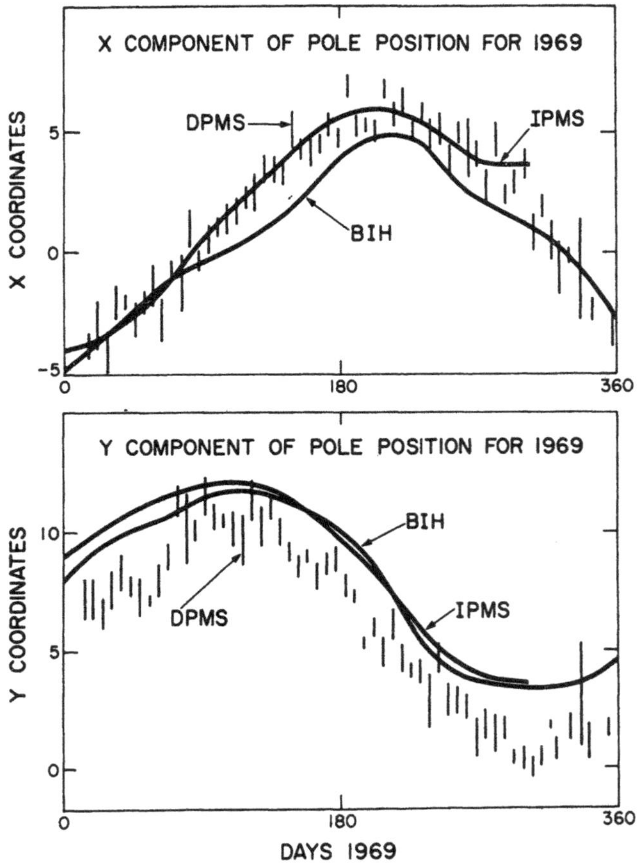

Fig. 2. Components of the pole position for 1969 (in meters) relative to the mean pole of 1900–1905, from Anderle and Beuglass [8].

coordinates of similar accuracy by using Tranet Doppler tracking data [5] and BC-4 camera observations of balloon satellites [6].

The uncertainties in the pole position and the rotational position of the earth as tabulated by the international services are roughly comparable with these uncertainties in station position. To improve the position model for ground stations by, say, an order of magnitude, terrestrial station coordinates, polar position, and earth-rotation angles all need to be improved. Anderle and Beuglass in 1969 first used Doppler satellite-tracking data for US Navy navigation satellites to determine the pole position independently of the conventional astronomical determinations [7]. In this approach, the orbits of one or more satellites essentially define the reference to which the pole position is measured. This calculation was extended in 1970 to cover the years 1967, 1968 and 1969 [8]. For 1969, they report random errors in the computed pole position of 1 m on the basis of 48 h of observations of one satellite. However, a bias of 1 to 2 m exists with respect to values from the BIH or IPMS. Fig. 2, adapted from their paper, illustrates this agreement.

A calculation similar to that of Anderle and Beuglass could probably extract from satellite observations the angular position of the earth about its axis

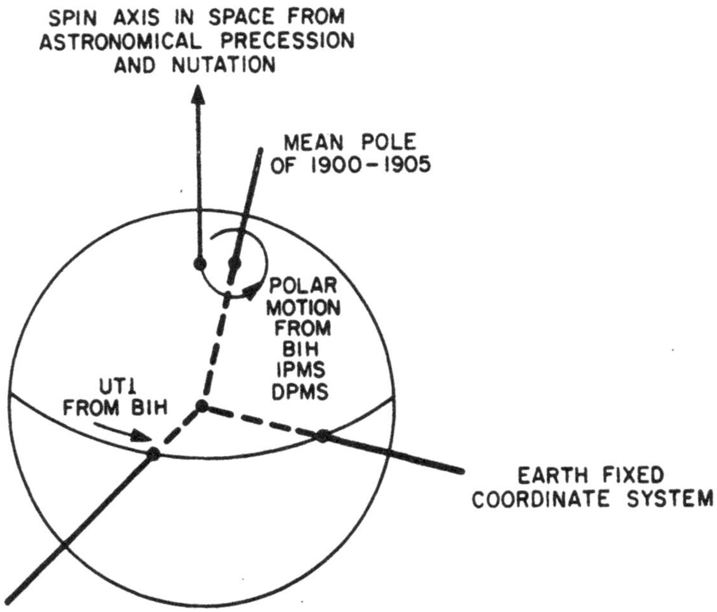

Fig. 3. Current model for calculating the position of a ground station as a function of time.

(UTI). At the equator, a point on the earth moves almost 0.5 millimeter per second. An examination of differences between UTI tabulations from different time services reveals differences corresponding to a displacement of almost 5 m. Gaposchkin and Lambeck [4] feel that the uncertainty in UTI limits the accuracy of their station-coordinate determinations.

In summary, the model (illustrated in Fig. 3) now in general use for station positions as a function of atomic time adopts a fixed geocentric coordinate system in the earth with its Z axis through the conventional mean pole of 1900–1905 and its XZ plane defined by adopted longitudes for the observing sites of the BIH. Station coordinates in this global system are most accurately obtained from satellite observations. The models adopt a rotation direction in space provided by formulas for astronomical precession and nutation, and instantaneous pole position in the earth as tabulated by BIH, IPMS, or the Dahlgren Polar Monitoring Service (DPMS). The angular rotation about this axis is obtained from tables of UTI provided by the BIH. It is worth emphasizing that in this model space techniques already play an essential or important role in locating the center of mass, refining station coordinates, and measuring polar motion.

3. Status of Geopotential Models

The mass distribution of the earth has many irregularities that generate a correspondingly irregular gravitational field, which in turn induces many measurable perturbations in the orbits of artificial satellites. If the geopotential is represented as a series of spherical harmonics, the coefficients in this series can be treated as parameters whose values can be determined by differential adjustment, as sketched in § 1. As more and more tracking data accumulated on a variety of satellites, more terms could be included in the solutions for geopotential coefficients. The solution by Gaposchkin and Lambeck [3, 4] for the 1969 Smithsonian Standard Earth (II) contains 296 coefficients for tesseral harmonics. Twenty zonal-harmonic coefficients derived by Kozai [9] are also part of this geopotential model. Orbits of 21 satellites contributed to this solution. French and United States laser ranging data with an accuracy of a few meters were available for six of the satellites. Baker-Nunn photographs provided the bulk of the data; they have an accuracy of a few arcsec. The geopotential solution also incorporated surface-gravity measurements. Fig. 4 is a map of the geoid calculated from the 1969 Smithsonian Standard Earth (II). When this geopotential was used in determining several-week-long arcs for Geos 1 and Geos 2, the laser data always had an rms deviation of less than 10 m, with the mean at 7 m. Thus, the station coordinates and satellite positions associated with the 1969 Smithsonian Standard Earth (II) have comparable accuracies.

Efforts to analyze satellite orbits to detect time-dependent terms in the geopotential due to tidal deformations began in 1965 [10]. The tidal-disturbing potential due to the sun and moon can be represented to a first approximation in terms of spherical harmonics of the second degree. This potential and the external potential due to the resulting deformations of the earth are related by a constant k, the ratio of the additional potential produced by deformation to the deforming potential. The value of this Love's number k depends on the elastic and other properties of the earth. In subsequent analyses, Newton [11] used four satellites in polar orbits for which Doppler tracking data from the Tranet system were available over intervals of 6 to 18 months. He obtained the following Love's numbers:

$$k_s = 0.359 \pm 0.042 \text{ for the solar tides,}$$
$$k_m = 0.314 \pm 0.036 \text{ for the lunar tides,}$$
$$k = 0.336 \pm 0.028 \text{ as a combined value.}$$

In an independent analysis, Kozai [12], using Baker-Nunn data covering intervals from 2 to 5 y for three satellites at lower inclinations, obtained the value

$$k = 0.29 \pm 0.03 .$$

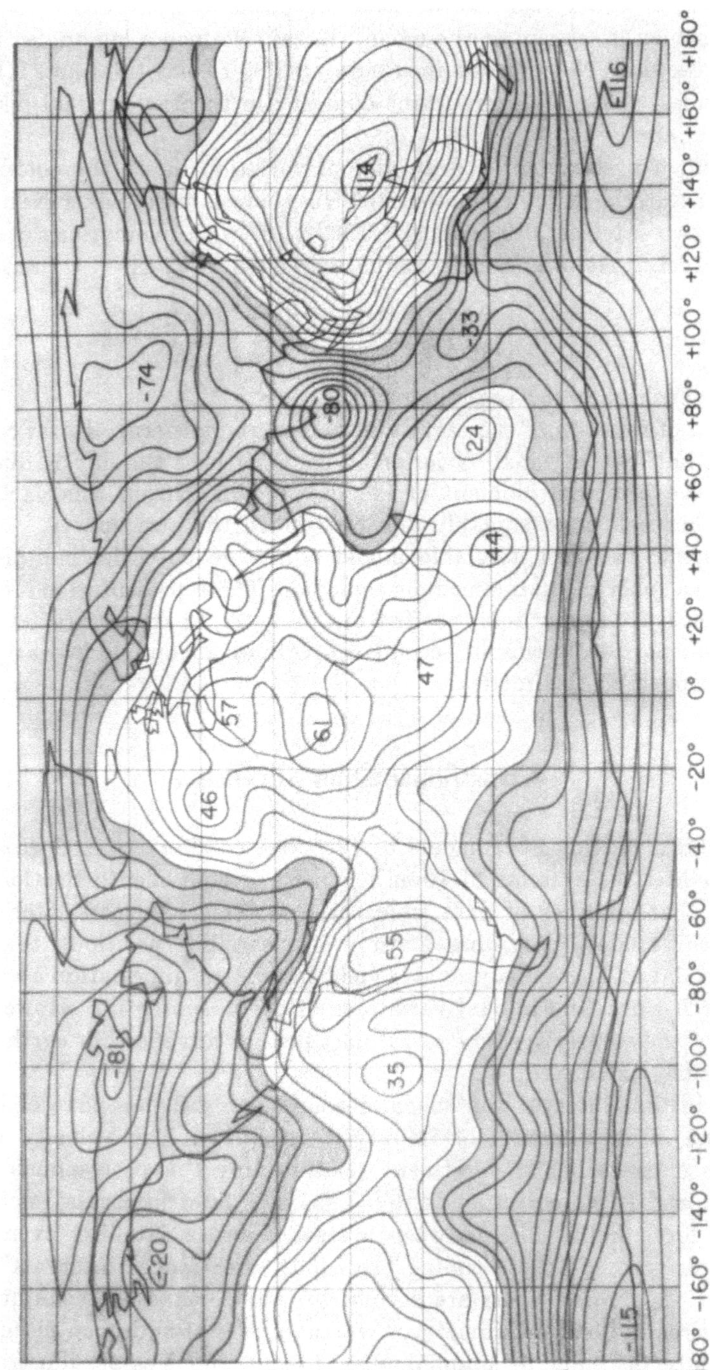

Fig. 4. Geoid heights in meters corresponding to a reference ellipsoid of flattening $f = 1/299.67$, from Gaposchkin and Lambeck [4].

Both Newton and Kozai also derived the phase of the earth tides relative to the longitude of the sun or the moon. Kozai obtained a lag angle of $5 \pm 3°$ for solar tides, and Newton obtained $0.8 \pm 0.3°$ for solar tides and $2.1 \pm 0.2°$ for lunar tides. This is an important quantity in theories of tidal dissipation of rotational energy.

Kozai [13] has also identified an annual variation in J_2, the coefficient of the oblateness term in the geopotential. His analysis used Baker-Nunn observations of two satellites, 1960 ι 2 and 1962 $\beta\mu$ 1, with inclinations of 47° and 50° respectively. He obtained an annual variation of J_2 as

$$\begin{aligned} \delta J_2 = {} & 1.3 \times 10^{-9} \cos (2 \pi t + 160°) \\ & \pm 0.2 \qquad\qquad\qquad \pm 10° , \end{aligned}$$

where the epoch is at the beginning of the year.

The speed of rotation of the earth also has an annual term. Kozai concludes that about half the seasonal variation in the length of the day is due to the change in the principal moment of inertia calculated from this variation in J_2. The other half is presumably due to wind in the atmosphere.

In summary, the best available model of the geopotential contains some 300 terms in a spherical-harmonic representation. This representation is complete through indices 16, 16, but some coefficients above 12, 12 are weakly determined. Time-dependent tidal terms of second degree and an annual variation of J_2 can be appended to this model.

4. Instrumentation Advances

The satellite-tracking instruments of 1960 can be characterized grossly, for typical satellites, as systems of 10-m accuracy or equivalent. Sections 2 and 3 conclude that analyses of data from these systems ten years later support mathematical models providing station and satellite positions to roughly 10-m uncertainty. These analyses also yield significant information about polar motion, earth tides, and annual variations of the geopotential. However, the capability of these early systems to support further advances in earth physics has been nearly exhausted.

Progress in time-interval and epoch measurement was the spearhead for the development of instruments of improved accuracy. The velocity of light implies the elementary but pertinent relationship: 1 m corresponds to 3 ns flight time, and 10 cm corresponds to 0.3 ns. For brief intervals, such as the round trip for a light pulse between a station and a satellite, even crystal oscillators can provide these timing precisions. For longer intervals at least rubidium or cesium oscillators are necessary. For demanding techniques such as long baseline radio interferometry, the capabilities of hydrogen-maser clocks are called for [14]. Fig. 5, adapted from Levine and Vessot, illustrates the typical stability of these alternative standards as a function of averaging time. The curve for hydrogen-maser clocks is obtained by comparing two such systems.

Ruby-laser ranging instruments introduced on an experimental basis in 1964 were the first tracking systems to demonstrate an accuracy approaching 1 m [15]. The range is calculated from the time required for a light pulse to reach the satellite and return. A useful block of data from five laser systems tracking six satellites with retroreflectors entered the 1969 Smithsonian Standard Earth (II). In 1969, range measurements were initiated to the retroreflector array placed on the moon by Apollo 11 [16, 17]. The International Satellite Geodesy Experiment (ISAGEX), scheduled to begin in the latter part of 1970, anticipates participation by more than ten laser systems with global distribution [18]. Even these few unrelated remarks about laser tracking indicate that this activity has achieved a substantial level of operational maturity. However, the full potential of the laser technique has yet to be realized. Both existing transmitters and the satellite retroreflectors now have the ability to provide 0.1 ns ranging. Currently, receivers are able to achieve an accuracy of only about 1 ns, so some improvement in photodetectors would be needed to reach 0.1 ns precision.

It seems, however, that tropospheric-propagation uncertainties will be a basic limitation in laser ranging from the earth's surface. Using surface measurements of refractive index to estimate the necessary corrections, Bean and Thayer [19] show that the ultimate accuracy of single-wavelength optical ranging to satellites or the moon appears to be limited to about 6 cm, or 2.5% of the total correction at the zenith. Radio systems are worse by a factor of about 2 because of the increased influence of water vapor in the radio spectral region. Of the radio frequency systems with prospects of 1 m or better accuracy, very long baseline interferometry (commonly abbreviated VLBI) seems to offer the greatest versatility. This technique depends upon local frequency standards of high quality; preferably hydrogen masers, at two or more radio telescopes separated by distances on earth as great as allowed by common visibility of a radio source [20]. The frequency standards provide time references for magnetic tape recording of signals from the same space source. These tapes are later correlated at a central computing facility, and the time difference for arrival of the same wavefront is determined.

One application of this technique uses natural radio sources of small apparent size. For example, a number of compact, distant radio sources remain unresolved at 6 cm wavelength over a 6319 km baseline between the United States and Sweden [21]. With data from a suitable observation campaign on several such point sources, the baseline distance and direction can be determined. The directions are relative to directions defined by the astronomical sources. With due refinement of the technique, there seems to be no obstacle in principle to achieving 0.001 arcsec accuracy.

Another application of VLBI is to spacecraft tracking. Several groups have demonstrated this possibility for interplanetary spacecraft [22] or for satellites in earth-synchronous orbits [23, 24]. If both natural and artificial sources figure in an observing campaign, the satellite directions can be referenced to directions defined by the natural sources.

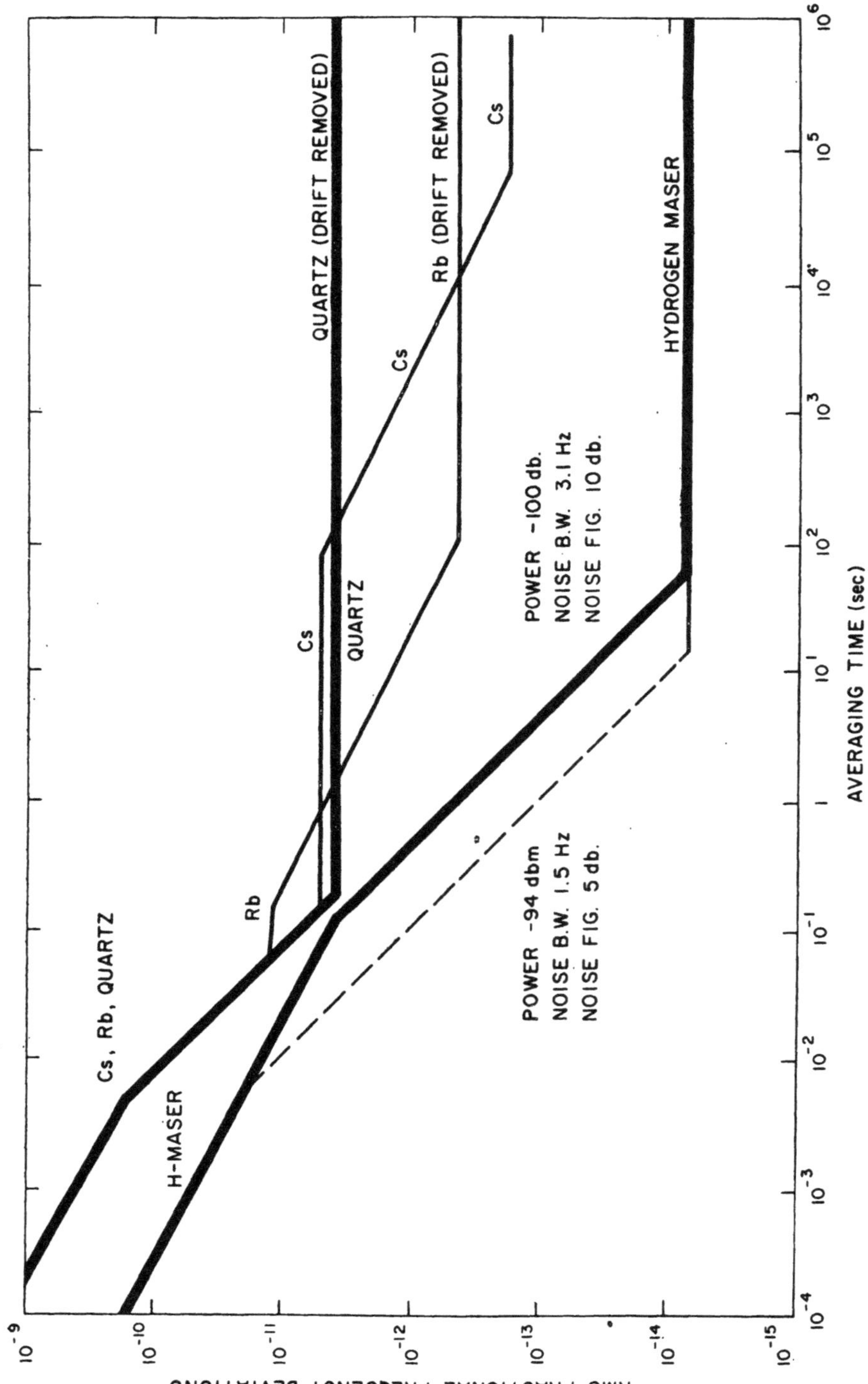

Fig. 5. Stability of frequency standards, from Levine and Vessot [14].

Satellite-to-satellite range-rate tracking offers substantial advantages. A very high satellite can track a low satellite continuously for about half the latter's orbit. This continuity should allow application of the same data-analysis techniques used so successfully by Muller and Sjogren [25] to determine the moon's potential from range-rate tracking of lunar orbiters. Satellite-to-satellite tracking also has the advantage that the tropospheric propagation errors unavoidable from ground-station tracking are eliminated. A suitable choice of frequencies can circumvent problems associated with propagation through the ionosphere. With instrumentation components and concepts already identifiable, it seems possible in a few years to achieve accuracies of 0.03 millimeter per second. A first experiment with satellite-to-satellite tracking is scheduled between ATS-F, in a synchronous orbit, and Nimbus-E [26].

To sense as many features of the gravity field as possible, ultimately the tracked satellite should be as low as practical. A satellite in a low orbit but free of atmospheric drag complications is possible with a surface-force compensation system [27, 28]. The essential element of such a system is an unsupported reference mass contained in a cavity at the mass center of the satellite. A control system in the satellite senses motions of the satellite relative to the reference mass and actuates small thrusters that force the satellite to follow the reference mass without touching it. Hence, the satellite follows an orbit influenced only by gravitational forces. Such a system is scheduled for flight, perhaps in 1971, on an experimental navigation satellite.

Satellite-to-ocean radio altimeters have been recognized as a possibility for many years [29–31]. An experimental instrument flew on a Saturn rocket in 1963 [32]. It seems possible with available components to build a space-qualified altimeter in which the error contributed by the altimeter does not exceed 1 m for averaging times of 10 s. To achieve an accuracy of 10 cm, it will be necessary to develop sophisticated satellite hardware, to find a way to make the propagation corrections, and to cope with the uncertainties in sea-surface scattering. As later sections will indicate, a 1 m instrument will be useful for geopotential refinement and for tidal measurements on the continental shelves. The 10 cm capability is eventually needed for physical oceanography.

Certainly the instruments discussed above will not be the only ones that contribute to future space programs in earth physics. They are probably the instruments that now seem most likely to provide major advances.

5. Prospects for Short-Term Dynamics of the Earth

The short-term dynamical phenomena now recognized as significant for the earth fall into three broad groupings: rotational motions, tides, and temporal variations of the geopotential. The measurement techniques discussed in the previous section promise data for substantial advances in each of these topics. A judicious choice of fundamental coordinate systems becomes increasingly important, and difficult, as the treatments of these earth phenomena attain ever greater precision. The reference systems must be defined at least to an accuracy comparable with that required to express and to analyze the phenomena under investigation. Both inertial and terrestrial systems are needed.

A catalog of positions for astronomical objects seems still to be the most promising tool to define an inertial reference system. The minimum requirement should be 0.01 arcsec uncertainty for star positions and proper motions. It is questionable whether ground techniques could provide this accuracy in optical frequencies, mainly because of atmospheric propagation effects. Perhaps an orbiting astrographic telescope could generate the data for an improved catalog of optical stars. In principle, a more precise inertial system could be established with a catalog of very distant point radio sources generated by VLBI techniques. An accuracy of 0.001 arcsec is expected with this method.

The required terrestrial coordinate system should have its origin at the earth's center of mass and in principle could have its Z axis defined by the principal axis of inertia. Satellite dynamics already provide the best location of the center of mass and can determine the location of the principal axes of inertia. The geopotential terms with coefficients C_{21} and S_{21} are usually taken to be zero because they are rigorously zero for an expansion about the principal axis of inertia. For an expansion about the adopted geographic pole, C_{21} and S_{21} will be very small but not zero. Their values change as the principal axis moves around the earth. With enough refinement of the geopotential representation and tracking accuracy, values for C_{21} and S_{21} can be determined as a function of time. Because the principal axes may not be determined to the required accuracy for some time, arbitrary geographic axes could be used in the interim. The relative positions of the geographic Z axis, the instantaneous spin axis, the axis of inertia, and the angular-momentum axis would be tabulated as functions of atomic time.

In practice, the terrestrial system would be realized by a model assigning coordinates and their time variations to a number of fundamental observing stations. Some of the stations would have optical or VLBI instruments monitoring the motions of the earth relative to the catalog stars. Some stations would have systems tracking the spacecraft employed to determine the center of mass and principal axes of the earth. The fundamental station-coordinate model would be refined occasionally on the basis of all the observations by using a procedure along the lines of Fig. 1. An accuracy of 1 m is expected to be achieved in the next few years, and of 10 cm later in the 1970s.

The procedures that relate the inertial and the terrestrial coordinate systems as a function of time can equivalently be viewed as the means of monitoring the rotational motion of the earth. The earth, of course, rotates about an instantaneous axis that continuously changes. Its direction in space has a 25800 year cycle (astronomical precession and nutation), caused mainly by torques from lunar and solar gravitational interactions with the oblateness of the earth. The instantaneous axis of rotation relative to the arbitrary geographical axes in current use appears to have a secular motion of the order of 0.003 arcsec y^{-1} [33]. The axis also performs an irregular precession or wobble with an amplitude of roughly 0.15 arcsec and with two main periods, of 12 and about 14 months. The 14-month component is the Chandler wobble.

There is some indication of longer period terms, but no noticeable feature at frequencies higher than 1 cycle y^{-1} in the spectrum derived from current measurements.

Comparison of pole paths generated by IPMS and BIH show differences as great as 0.1 arcsec ($\sim$3 m). This is presumably some indication of the accuracy of past methods. For 1969, Anderle and Beuglass [8] report differences of about 1 m between their satellite-derived pole position and values published by BIH or IPMS.

Unresolved questions about the Chandler wobble include the excitation and damping mechanisms that control its character. Proposed models of excitation mechanisms resort to changes in mass distribution of the atmosphere, electromagnetic core-mantle coupling, and extensive mass shifts accompanying earthquakes, but all seem to fall short of predicting the observed amplitude. Some observational support for the earthquake-excitation hypothesis has been found in a reported correlation between times of occurrence of major earthquakes and changes in the BIH pole path [34]. More accurate determinations of the pole path should help resolve the theory of the Chandler wobble and other features of polar motion. Using either precise satellite ranging or VLBI techniques, measurement of pole positions to 0.01 arcsec (0.3 m) at 2-day intervals would seem to be a useful and achievable objective.

The rate of rotation of the earth about its instantaneous axis shows secular accelerations ($\sim 2\times10^{-10}$ y^{-1}), irregular accelerations ($\sim 10^{-9}$ y^{-1}), seasonal fluctuations (0.5 ms in length of day), and irregularities of frequency higher than 1 cycle y^{-1}. Explanations of the secular accelerations involve tidal dissipation or perhaps angular momentum exchange between the mantle and the core. Seasonal fluctuations are thought to be due to winds and tides. Irregular changes are poorly understood and may have their origins in atmospheric interactions with the sea and earth or within the earth. A reasonable accuracy goal for measurements of the rotation about the instantaneous axis would seem to be 0.01 arcsec on a daily basis. This would not only support a more penetrating search for the origins of changes in the rotation rate but would also provide the necessary station-position accuracy for orbital analyses.

The sun and the moon generate tidal forces that produce low-frequency oscillations in the solid body of the earth, with a maximum surface-elevation excursion of about 50 cm near the equator. Two principal aspects of these earth tides can be investigated by space techniques. One apparent possibility is the measurement of all three spatial components of the actual motion executed by a ground-based instrument. This motion can be extracted from laser-range and VLBI data when they attain decimeter accuracies. Another measurement, already demonstrated, concerns the total mass displacement associated with tidal deformation. Section 3 discussed the status of this measurement. As further refinements of the gravity field evolve, particularly from accurate long-term tracking, the tidal mass displacements should exhibit details that will be useful in separating the earth and ocean tides and in identifying the various spectral components of the tides. Section 7 discusses the direct measurement of ocean tides by a satellite altimeter. The most important contribution from space techniques to an understanding of solid-earth tides may come indirectly from the ocean measurements. This seems likely because the ocean-tide loading on continental margins is an important factor in the analysis of earth tides, and this effect is currently very difficult to treat because the ocean tides are poorly known.

Section 3 considered the successful detection of seasonally varying geopotential terms, as well as tidally induced time-dependent effects. Non-tidal time dependence in the geopotential is a fruitful topic for more detailed examination as satellite tracking and orbital analyses evolve to greater accuracy. There will be many correlations of these effects to be made with polar motion and variations in the rotation of the earth.

6. Prospects for Long-Term Dynamics of the Earth

An impressive accumulation of evidence now indicates that the earth's surface is extremely mobile. Large surface blocks or plates 50 to 100 km thick and thousands of kilometers in horizontal extent seem to be moving relative to each other at average long-term rates from 1 to 15 cm y^{-1} [35, 36]. The interactions of these plates appear to be responsible for a wide variety of effects, including large earthquakes, mountain building, generation of tsunamis, and confinement of nearly all active volcanoes to only a few narrow belts. Almost all large-scale geological and geophysical phenomena occurring in the outer 600 km of the earth appear to be intimately related to this global pattern of motions.

Rudimentary models of crustal motions based on existing geophysical data can evaluate the anticipated relative motion of arbitrary points on the earth's surface [35]. However, continental drift and other large-scale earth deformations have not yet been detected conclusively by direct astronomical or geodetic techniques. The possibility of such measurements by the techniques discussed in § 4 is an exciting prospect to promote further understanding of the long-term dynamics of the earth. Laser ranging to the moon or artificial satellites in stable orbits, and VLBI observations of natural and artificial sources all promise 10 cm accuracy once the instrument refinements already identified have been made. Appropriate global observation campaigns that integrate these several techniques must be planned and executed in due course to realize the desired measurements of crustal motions. Such campaigns should evolve naturally as successors to cooperative international efforts such as ISAGEX.

It is important not only to measure the relative motions of the surface plates but also to find their absolute motions with respect to well-defined fundamental coordinate systems. Clearly, if the plates and sites on them are moving about as they seem to be, a coordinate system anchored to geographical sites is not a well-defined one. Space and astronomical techniques offer some possible approaches to this problem, as discussed in § 5.

The relationships between space capabilities and conventional or innovative surface techniques must also receive due attention in plans for future observation programs. The space techniques make their greatest contribution when applied on a global or at least a continental scale. But there is limited value in knowing that a radio telescope site in Europe has a certain motion relative to a site in North America if it is not known whether the sites are moving locally relative to some mean coordinates for their crustal block. Local surface geophysical measurements must relate fundamental instrument sites to their surroundings. Local measurements across fault zones, island arcs, and in general across the contacts of surface plates are necessary to a comprehensive model for motions of points on the earth's surface.

A complicating aspect to this whole picture is the recognition that the motions, at least locally, are not uniform. Across a feature such as the San Andreas fault, there can be both sudden motions accompanying earthquakes, and gradual deformations as strain accumulates. Whether something analogous is true for the motion of whole crustal blocks is not known, and the answer must be sought eventually in the observation programs.

It is not scientifically sufficient only to measure surface motions; it is also important to ask what can be learned about the mechanisms that drive the motions. Clearly, the plate motions themselves are related to the driving forces, but in a complicated way because they can transmit forces over distances of thousands of miles. Another relevant factor is the gravity field determined by satellite techniques. Kaula [37] has discussed some relationships between geopotential models and the patterns of tectonic activity.

Present published geopotential determinations are limited to spherical harmonics of degree 16, plus assorted higher terms related to orbital resonance. Yet simple calculations show that the gravity field will not be dominated by the strength of the lithosphere until degree 50. It therefore appears that extended representations of the geopotential can supply much more information about the driving mechanisms. Fortunately, both satellite-to-ocean altimetry and satellite-to-satellite tracking can provide data for substantial enlargement of geopotential models.

7. Prospects for Ocean Physics

A satellite-to-ocean altimeter is a natural tool to apply to problems of ocean physics [29, 38]. The information to be derived from the altitude data depends strongly on the accuracy of the measurement. A resolution of ± 5 m should provide usefully detailed information concerning the shape of the geoid, but very little oceanographic information. A resolution of ± 1 m would permit the detection of tides on continental shelves, storm surges, and possibly the surface elevations associated with some currents. A resolution of ± 0.1 m would permit detection of the general ocean circulation, tides, and other dynamical processes affecting sea level.

This progression of objectives corresponds to the difference between an equipotential surface and the ocean profile. At 5 m resolution, the two surfaces are indistinguishable, and a measurement of the profile is essentially a measurement of the geoid. Objectives of physical oceanography require a measurement of the difference between the topography of the sea surface and the geoid, and this difference is typically in the decimeter range. An altimeter with an uncertainty of a few meters poses no severe design or construction problems [29–31]. Such an instrument is being seriously considered for flight on the next GEOS satellite. Besides demonstrating the altimeter concept, the objective will be refinement of the geopotential through measurement of the ocean geoid.

Acquisition of altitudes of decimeter accuracy is a more difficult undertaking, as discussed in § 4. Further, the satellite-orbit calculations must achieve an uncertainty in the decimeter range for the fullest use of the data in mapping the global ocean surface. The best geopotential models from which to derive the geoid for the ocean will probably come from satellite-to-satellite tracking. However, this technique may not be accurate enough to reveal geoid variations of less than 100 km wavelength. Thus, the ultimate objectives of ocean physics pose measurement problems that remain to be fully resolved. Measurements of tides with useful accuracy for the deep ocean also demand the ultimate in altimeter refinement and sophisticated data processing.

The general circulation pattern of the oceans is a fundamental problem in physical oceanography that can be attacked effectively by space techniques. The solution could be realized from a spacecraft by monitoring the positions of free-drifting tracking beacons. The beacons could be carried either by surface buoys or in subsurface floats that periodically rise to the surface and are there located by the satellite. The daily position should be obtained to an accuracy of 1 km or better.

Another application of space techniques to oceanography concerns the adequate horizontal positioning of research vessels. A useful system should provide the position of a moving ship to ± 100 m in a global coordinate system.

8. Technological Integration and International Cooperation

The global nature of most significant problems in earth physics dictates that their solutions require global distributions of observations. The interaction between the several dynamical subsystems of the earth demands coordination of the observation campaigns mounted for various objectives. Hence, for maximum effectiveness, technological integration and international cooperation are essential to progressive investigation of topics in earth physics. The instrumentation panel of the Williamstown study [1] offered one example of how a unified system of ground and spacecraft instruments might be assembled to meet most requirements of a vigorous program in earth physics. With respect to international coordination, there is fortunately a long tradition of productive global cooperation on these topics. The BIH, the IPMS, the Smithsonian tracking network, and the ISAGEX are but a few examples.

All these considerations, and the substantial scientific results to date, support the conclusion that the new decade promises to be a golden age for the application of space and astronomical techniques to earth physics.

Acknowledgment

This work was supported in part by grant NGR 09-015-002 from the National Aeronautics and Space Administration.

References

[1] W. M. KAULA, Chairman, NASA Contractor Report NASA CR-1579 (1970).
[2] G. VEIS, in: The Use of Artificial Satellites for Geodesy, North-Holland, Amsterdam 1964 (p. 201).
[3] E. M. GAPOSCHKIN and K. LAMBECK, Trans. Am. Geophys. Un. **50**, 603 (1969).
[4] E. M. GAPOSCHKIN and K. LAMBECK, Smithson. Astrophys. Obs. Spec. Rep. No. 315 (1970); also submitted to J. Geophys. Res.
[5] W. H. GUIER, R. R. NEWTON and G. C. WEIFFENBACH, The Johns Hopkins University, Applied Physics Laboratory Technical Memorandum TG-653 (1965).
[6] H. H. SCHMID, in: The Use of Artificial Satellites for Geodesy, vol. II, National Technical University, Athens 1967 (p. 247).
[7] R. J. ANDERLE and L. K. BEUGLASS, Trans. Am. Geophys. Un. **50**, 602 (1969).
[8] R. J. ANDERLE and L. K. BEUGLASS, Trans. Am. Geophys. Un. **51**, 266 (1970).
[9] Y. KOZAI, Smithson. Astrophys. Obs. Spec. Rep. No. 295 (1969).
[10] R. R. NEWTON, J. Geophys. Res. **70**, 5983 (1965).
[11] R. R. NEWTON, Geophys. J. Roy. Astron. Soc. **14**, 505 (1968).
[12] Y. KOZAI, Publ. Astron. Soc. Japan **20**, 24 (1968).
[13] Y. KOZAI, Smithson. Astrophys. Obs. Spec. Rep. No. 312 (1970).
[14] M. W. LEVINE and R. F. C. VESSOT, presented at Symposium on Very Long-Baseline Interferometry, Charlottesville, Va. (1970); Radio Sci. **5**, 1287 (1970).
[15] H. H. PLOTKIN, COSPAR Information Bull. No. 29, 18 (1965).
[16] C. O. ALLEY et al., Science **167**, 368 (1970).
[17] C. O. ALLEY et al., Science **167**, 458 (1970).
[18] Centre National d'Etudes Spatiales, International Satellite Geodesy Experiment, Preliminary Document, March (1970).
[19] B. R. BEAN and G. D. THAYER, J. Res. NBS **67D**, 273 (1963).
[20] B. F. BURKE, Phys. Today **22**, 54 (1969).
[21] M. H. COHEN, D. L. JAUNCEY, K. I. KELLERMANN and B. G. CLARK, Science **162**, 88 (1968).
[22] J. GUBBAY, A. J. LEGG, D. S. ROBERTSON, A. T. MOFFET and B. SEIDEL, Nature **222**, 730 (1969).
[23] C. A. KNIGHT et al., Trans. Am. Geophys. Un. **51**, 267 (1970).
[24] S. CRISWELL, G. VEIS, S. ROSS and J. SHAW, Paper presented at Symposium on Very Long-Baseline Interferometry, Charlottesville, Va. (1970).
[25] P. M. MULLER and W. L. SJOGREN, Science **161**, 680 (1968).
[26] T. L. FELSENTREGER, T. J. GRENCHIK and P. E. SCHMID, NASA-GSFC preprint X-552-70-96 (1970).
[27] B. LANGE, AIAA J. **2**, 1590 (1964).
[28] Stanford University, Final Technical Report, NASA Contract NAS 12-695 (1969).
[29] E. J. FREY, J. V. HARRINGTON and W. S. VON ARX, in: Meteorological and Communication Satellites, Proc. 16th Int. Astronaut. Congr., vol. 4, Gauthier-Villars, Paris 1965 (p. 53).
[30] T. W. GODBEY, in: Oceanography from Space, Woods Hole Oceanographic Institution Ref. No. 65-10 (1965) 21.

[31] M. KOLKER and E. WEISS, NASA Contractor Report NASA CR-1298 (1969).
[32] O.T. DUGAN, Paper presented at AIAA Guidance and Control Conference, Cambridge, Mass., AIAA paper 63-352 (1963).
[33] W. MARKOWITZ, in: Continental Drift, Secular Motion of the Pole, and Rotation of the Earth, Int. Ass. Geod. Symp. No. 32, 25 (1968).
[34] L. MANSINHA and D. E. SMYLIE, J. Geophys. Res. **72**, 4731 (1967).
[35] X. LE PICHON, J. Geophys. Res. **73**, 3661 (1968).
[36] W. J. MORGAN, J. Geophys. Res. **73**, 1959 (1968).
[37] W. M. KAULA, J. Geophys. Res. **74**, 4807 (1969).
[38] J. A. GREENWOOD et al., in: Remote Sensing of Environment, American Elsevier, New York 1969 (p. 71).

Space Research XI — Akademie-Verlag, Berlin 1971

GEODETIC SATELLITE MISSION AND GEOS-C SPACECRAFT

F. O. VONBUN

NASA–Goddard Space Flight Center, Greenbelt, Md, USA

The purpose of the GEOS-C mission is: (i) to complete the gravimetric objectives of the National Geodetic Satellite Program (NGSP) and to extend our knowledge of the earth's gravitational field, (ii) to establish a common world datum for the participating tracking stations, (iii) to demonstrate the feasibility of a radar altimeter, (iv) to use satellite-to-satellite tracking for geodetic purposes, (v) to intercompare the accuracy of the participating tracking systems and thus their value for geodesy and earth physics.

In order to accomplish these tasks, the spacecraft will be equipped with a radar altimeter, laser corner reflectors, C-band Van Atta array, unified S-band system (Apollo), satellite-to-satellite tracking, C-band transponder, Goddard range and range rate system, Minitrack, flashing lights and Tranet system. Systems characteristics and their usefulness for geodetic measurements, as well as their capabilities, are discussed. Accuracy figures and their influence on the spacecraft orbit, the key element for a dynamic gravity field analysis, are given.

1. Introduction

The GEOS-C spacecraft and its main objectives are described in this paper [1]. Particular attention is paid to two new and powerful tracking systems. These are: (1) a radar altimeter [2–4] and (2) a satellite-to-satellite range and range rate tracking system [5, 6]. Other tracking systems mentioned in this paper are those already flown on GEOS-1 and 2. These are needed to fulfil one of the major goals of this spacecraft, that is the completion of the National Geodetic Satellite Program (NGSP).

The radar altimeter will be used, as shown, not only to improve our orbit determination capability by providing a completely new measurement (the spacecraft height), but also to determine ocean surface anomalies. The determination of ocean trenches [7], for instance is as yet only possible via direct ship measurements, a rather cumbersome and expensive task to survey the many existing ones. Satellite-to-satellite range rate measurements [5, 8] will, for the first time, make it possible to determine harmonic variations in the spacecraft velocity via "direct" measurements. Further discussed in some detail are the errors of tracking systems [5, 9–14] used on board the spacecraft as well as on the ground. These errors are important quantities since they are one of the main limitations for our major goal, namely the accurate determination of the earth's gravity field and the establishment of a world datum.

2. Mission Objectives

The GEOS-C spacecraft will be the third to be launched in this series of the National Geodetic Satellite Program (NGSP). Its intent is to contribute to its completion and further to the development and applications of new techniques and systems to be flown by future spacecrafts. Precise measurements of position, velocity and accelerations are needed to perform this task [8]. Completion means the determination of the gravity harmonic coefficients to an order and degree of at least 12 to 15 so that their correlation coefficient is of the order of 0.2 to 0.4. The coefficients should have a quoted error so that they contribute not more than, say, ± 3 mgal error to the mean anomalies Δ_g at the earth's surface over an area of, say, $(10 \times 10)^\circ$. Closely related to the above is the determination of the geocentric location of the participating tracking stations on a worldwide basis or the establishment of a common world datum.

Besides the dynamic requirement of the NGSP, the static or geometric objective is to develop an earth reference system of global-distributed points to an accuracy of approximately 10 m.

In addition to the above, the GEOS-C spacecraft will, for the first time, carry a radar altimeter on board, which will be able determine the spacecraft height (distance between orbit and sea level) to an accuracy of 3–5 m. This will not only improve our capability for more accurate orbit determination, a key element of dynamic geodesy, but will also make it possible to measure anomalies [7] of the ocean's surface. Ocean surface trench measurements will thus be possible for the first time on a global basis between latitudes $\pm 21^\circ$, the planned inclination of GEOS-C. A direct acceptance of the ocean surface topography as equipotential would be a good and accurate map of geoidal inclinations. For this purpose it would actually be advantageous to fly the GEOS-C in a polar, or near polar, orbit which would make coverage of all the oceans possible [8]. Other constraints, however, as discussed later led to the present selection of an orbit with 21° inclination.

GEOS-C will also be equipped with a satellite-to-satellite range and range rate tracking system. Tracking accuracies of the order of 0.7–0.1 mm s^{-1} over a time interval of 40–100 seconds should be realized. The "tracking" spacecraft, being ATS-F, can "see" GEOS-C for half of its orbit around the earth.

3. Orbit Design Considerations

A large number of factors, ranging from geodetic usefulness and the completion of NGSP to network coverage and consideration, have to be taken into account in a selecting the orbit. In the following a brief account of some of the major factors for orbit selection are presented.

3.1. Inclination

Almost all spacecraft launched so far, have orbit inclinations above 28°. Considering this fact, improvement in the determination of harmonic coefficients

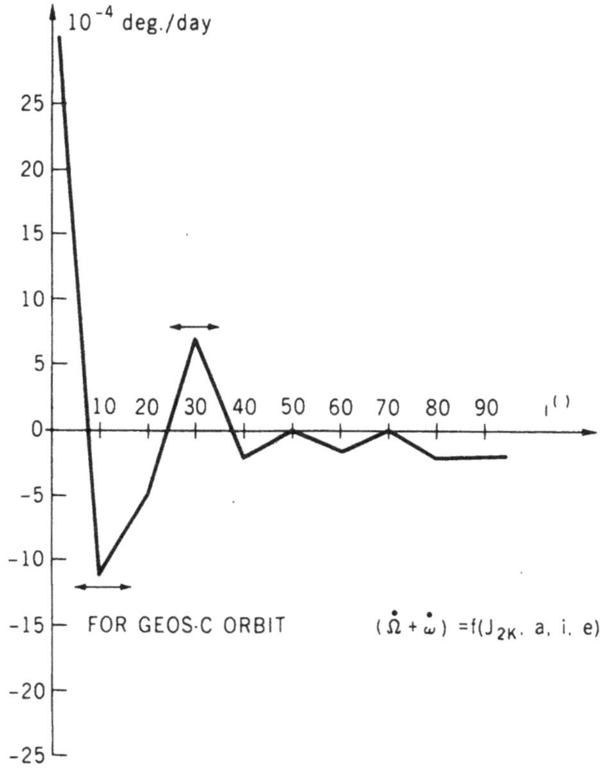

Fig. 1. Difference of $\dot{\Omega} + \dot{\omega}$ between two geopotential fields (Kozai–Anderle).

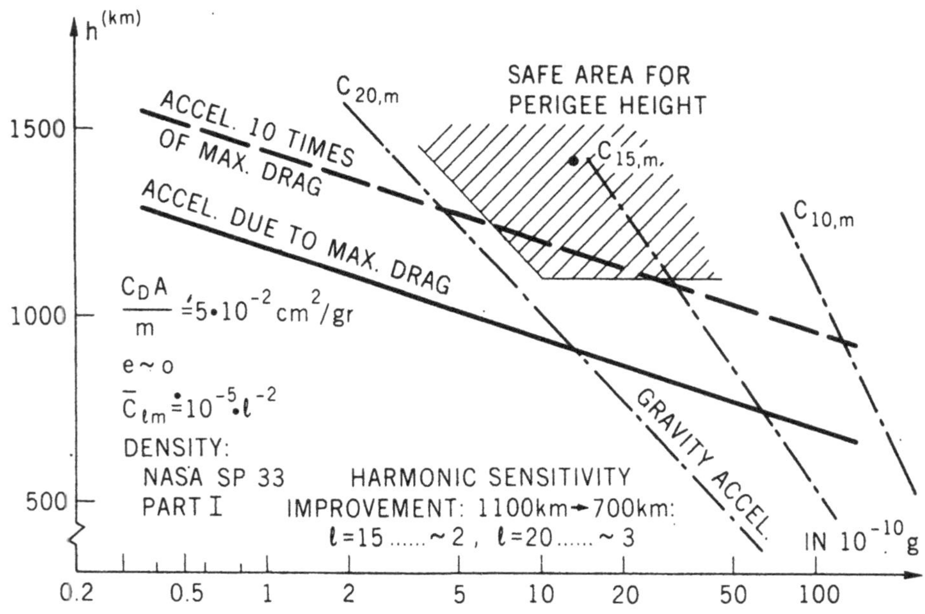

Fig. 2. Accelerations on GEOS-C atmosphere and gravity harmonics.

(perturbations of the orbit) requires a variety of inclinations and, in particular, some below 28° in order to obtain a more uniform coverage; a low-inclined orbit would be useful in the determination of the even zonal harmonics [15–17]. Fig. 1 depicts the difference between the sum of the time derivatives of the argument of the node and perigee of a GEOS-C type orbit as a function of the inclination for two earth models [18]. The discrepancies are, as expected, due to lack of available orbits as the inclination decreases. Thus, originally the GEOS-C orbit was chosen to be between, say, 15 and 25°.

3.2. Height and Eccentricity

The perigee of this orbit was chosen using Fig. 2 which shows drag accelerations undergone by GEOS spacecraft ($C_D A/m = 5 \times 10^{-2}$ cm^2 g^{-1}) as function of its height. Accelerations due to tesseral harmonics are also shown for an eccentricity $e \leqq 0.04$, assuming that $\overline{C}_{lm} \approx 10^{-5} l^{-2}$. A perigee height of 1100 km was chosen.

The eccentricity should be small in order to avoid large height changes for the onboard altimeter ($\dot{H}_{max} = 210$ m s^{-1}, radar bandwidth). But e should be large enough to allow good and easy determination of the argument of perigee. Thus, $0.01 \leqq e \leqq 0.04$ seems to be a good eccentricity range. Only a moderate improvement, by a factor of 2, in the sensitivity to high-order harmonics of the earth potential could be obtained by lowering the orbit to, say, 800 km, which does not seem to warrant the risk of drag [19] interference. (Radiation pressure acceleration is small compared with those considered before.)

4. Radar Altimeter

A radar altimeter [2–4] to measure the distance between the spacecraft and the sea surface has been proposed for GEOS-C and is now being incorporated into the spacecraft. This will be the first time that height variations of large ocean areas can be observed.

For this spacecraft it is planned to turn on the radar altimeter only over the international zones of the oceans. Hopefully, for future spacecraft, an international agreement can be reached for inland water and even land operations.

4.1. Altimeter Objective

The main objective of the altimeter is to demonstrate its operation and utility to measure variations in the ocean heights to about ± 3 to 5 meters. Geoidal anomalies having wavelengths of say 100 km (400,400 field) to 2000 km (20,20 field) cannot be obtained from conventional satellite geodesy and fields to terms higher than 20,20 may never be obtained via straight satellite ground tracking as has been done in the past.

Further, the height measurement will be used to test the satellite-to-satellite tracking loop thus closing the electronic tracking path as shown in Fig. 3.

The height measurements will further be used as an additional and independent set of measurements for exact orbit determination so essential for any dynamic geodetic work. A substantial improvement in orbit determination accuracy is expected because never before has it been possible to track continuously a near-earth spacecraft over such large areas.

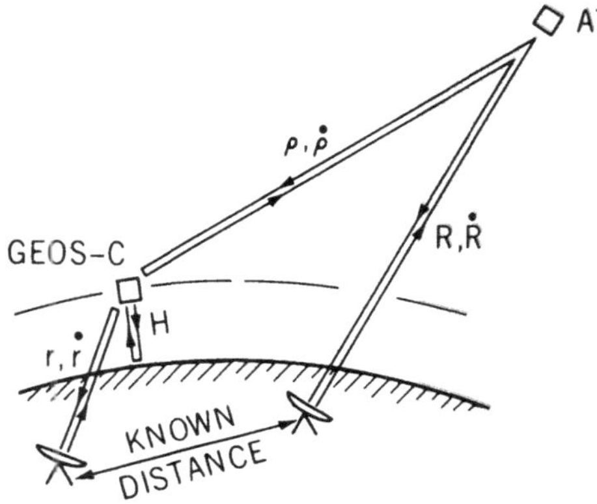

RANGE m		AVERAGE RANGE RATE cm/s		INTEGRATION TIME
BIAS	NOISE	BIAS	NOISE	SECONDS
4	2	0.0	0.60	1
4	2	0.0	0.07	10
4	2	0.0	0.02	40
4	2	0.0	0.01	100

Fig. 3. Range R and range rate $\dot{R}$ satellite-to-satellite tracking system errors (ATS-F and GEOS-C).

4.2. Determination of Ocean Trenches

Oceanographers [4, 7, 20] have pointed out that undulations of the sea surface are of importance; for instance it is shown [7] that the sea surface

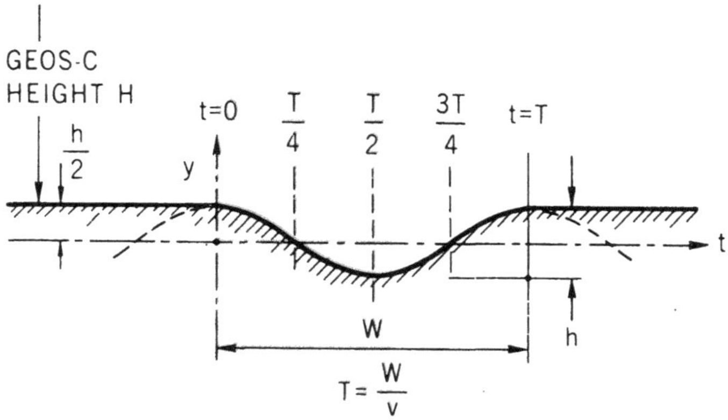

Fig. 4. Radar height changes over a trench.

$\dot{H} \approx v\, e \sin \eta$ GEOS-C height velocity $e \ll 1$

$$y = \frac{h}{2} \sin\left(\frac{2\pi}{T} t + \frac{\pi}{2}\right) \text{ Trench profile (Approx.)}$$

$$\dot{y} = \frac{h}{2}\left(\frac{2\pi}{T}\right) \cos\left(\frac{2\pi}{T} t + \frac{\pi}{2}\right) \text{ Trench height velocity}$$

$$\ddot{y} = -\left(\frac{2\pi}{T}\right)^2 \frac{h}{2} \sin\left(\frac{2\pi}{T} t + \frac{\pi}{2}\right) \text{ Trench height acceleration}$$

for GEOS-C:

with $h = 15$ m, $w = 210$ km, $T = 30$ s, $v \approx 7$ km s^{-1}, $e = 0.0322$

$\dot{H}_{max} \approx 230$ m s^{-1}, $\dot{y}_{max} \approx 1.6$ m s^{-1}, $\ddot{y}_{max} \approx 0.34$ m s^{-2} $= 3.4 \times 10^{-2}$ g boundaries approximately follow the trench surface. Satellites equipped with altimeters will be able to contribute much information on height variations. Fig. 4 shows the height variations expected to be seen by a radar when the spacecraft passes over a 15 m trench having a width of $\sim$200 km [7]. This kind of survey is much simpler to perform than to send a surface ship expedition.

5. Satellite-to-Satellite Tracking

Tracking of a near-earth spacecraft from an "Orbiting Tracking Station" has been proposed [5] in order to enhance the capability of NASA's ground tracking networks. It is now planned to perform a first test of this technique using the ATS-F and the Nimbus-G spacecraft. The purpose here is to demonstrate the technique and test the expected improvement of the orbit determination capability.

This system will also be incorporated into the GEOS-C [6]; the use and exploitations differ, however. Nimbus will be in a near-polar orbit while GEOS-C will be in a low-inclination orbit. Comparison of real harmonic velocity variations will be made using these two difference spacecraft orbits. Since the low-inclined spacecraft does not experience as "many" harmonic terms as a highly-inclined orbit, it may actually be easier to "see" direct variations in the range rate measurements.

5.1. ATS-F and GEOS-C Geometry

Fig. 5 depicts the geometry of the ATS-F and GEOS-C spacecrafts and gives an estimate of the velocity variations due to gravitation irregularities which can be expected to be measured. A good approximation of $\delta\dot{\varrho}$ is $\delta V_G \times \cos(\delta \boldsymbol{V}_G, \boldsymbol{\rho})$ and a precision of 0.02 to 0.01 cm s^{-1} can be expected in its determination. The advantage here is that the spacecraft can be "seen" over half

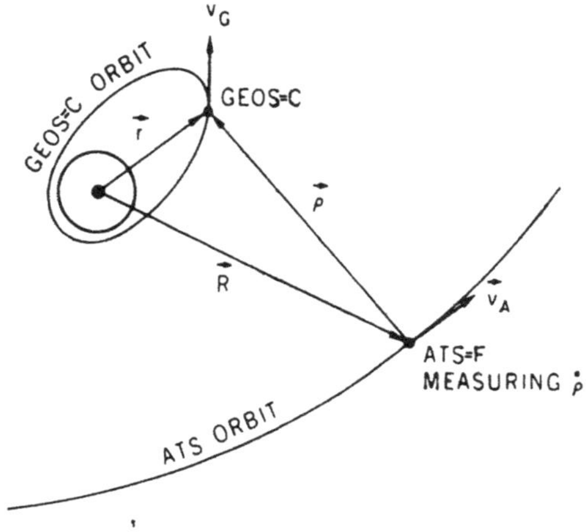

Fig. 5. Geometry ATS-F and GEOS-C.

the earth and thus we will be able to test the altimeter and measure accelerations due to gravity terms.

5.2. Direct "Acceleration" Measurements

Fig. 6 depicts the rms velocity variations, after Kaula, as accumulated over an approximate orbital time interval of 1000 seconds [8]. As can be seen from this graph, a "direct" harmonic term measurement to a degree $l = 28$ can be

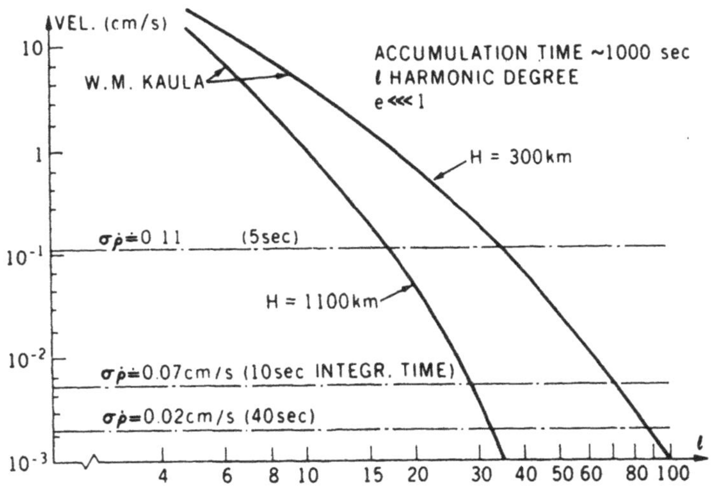

Fig. 6. Velocity variations (rms) plotted against l, and satellite-to-satellite tracking sensitivity.

expected from a satellite of a height of 1100 km and a small eccentricity, using an integration time of 10 sec.

It should be pointed out here that this is only so if the synchronous satellite velocity variation over the 1000 s time interval is negligible, as is the ionospheric noise. These assumptions will be checked with GEOS-C and the outcome will determine the usefulness of the orbiting tracking station's concept for geodetic work. Using present day techniques of harmonic analyses, one may be able to solve for a 20,20 gravity field, but this seems to be the limit; an accuray to 0.005 cm s^{-1} will be required to determine high-order gravity field variations [8], equivalent to, say, a 80,80 or 100,100 field. Only orbiting tracking station techniques may give the resolution needed for geodetic work in the future.

5.3. ATS-F and GEOS-C Orbit Errors

The GEOS-C thus equipped to be able to be tracked by an "orbiting station" (the ATS-F in this case) can be "fixed" in space with great accuracy [5, 9]. Fig. 7 depicts the result of error analysis carried out by J. L. Cooley for this spacecraft combination. Assuming conservative bias and noise error values as shown in this graph, the GEOS-C position error, the square root of the sum of the squares of the components, may be determined to, say, 15–20 m. But the height error is only of the order of 1 m or less, which constitutes an important factor for the altimeter calibration.

Since as shown in Fig. 7, the expected errors are rather small, one can be encouraged to exploit the radar altimeter system as well as dynamic geodesy techniques requiring extremely accurate orbit determination.

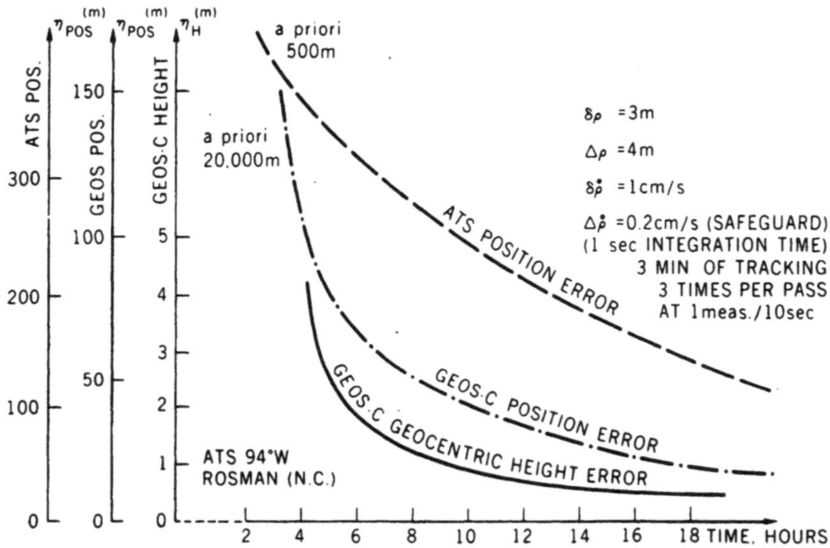

Fig. 7. Orbit errors ATS-F and GEOS-C.

6. GEOS-C Tracking Instrumentation

In general, the GEOS-C spacecraft structure will be almost the same as that of GEOS-1 and 2 in order to prevent a major and costly structural re-design. This, in turn, will, to a certain extent, limit the tracking instrumentation systems to be carried onboard the spacecraft.

6.1. GEOS-C

Fig. 8 shows an overall view of the GEOS-C spacecraft at this stage of the development phase. We are considering the suggestion of D. E. Smith to add a corner reflector system on the gravity gradient boom weight, as shown in Fig. 8, which may give an opportunity to determine the spacecraft attitude using laser ranging. If it is possible, this would be the first time that such a system was utilized. Table 1 shows the planned electronic tracking systems and stations. In addition, the spacecraft will be tracked by the two Goddard laser systems, the five Smithsonian Astrophysical Observatory (SAO) laser systems and possibly by lasers of other nations [21]. This would considerably increase the number of tracking stations and their distribution around the globe and thus contribute to a better gravity solution.

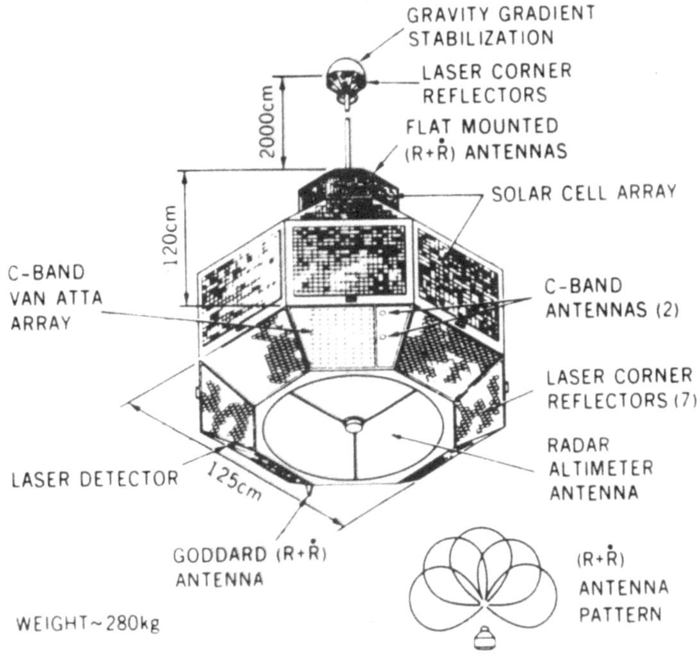

Fig. 8. GEOS-C.

Table 1
NASA Radio Tracking Stations for GEOS-C

	C-Band	Unified S-band	$R + \dot{R}$	Minitrack
Cape Kennedy	×	×		
Grand Bahama	×	×		
Bermuda	×	×		
Antigua	×	×		
Grand Canary*	×	×		
Ascension	×	×		
Madrid		×		
Tananarive	×		×	×
Carnarvon+	×	×	×	
Canberra+		×		×
Guam		×		
Kauai	×	×		
Goldstone+		×		
Guaymas		×		
Corpus Christi		×		
Goddard		×		
Quito				×
Rosman			×	
Santiago			×	

+ 85′ antennas.
* MPS-26 Radar

6.2. Ground Tracking System Errors

Table 2 shows a listing of the errors of the electronic ground systems to be used for the orbit determination of GEOS-C. These systems, together with the satellite-to-satellite tracking systems, the radar altimeter, and the laser

Table 2
GEOS-C Electronic Ground Tracking Systems Errors

Tracking systems	Range (m) Bias	Range (m) Noise	Average range rate ($cm\ s^{-1}$) Bias	Average range rate ($cm\ s^{-1}$) Noise	Integration time (s)
C-Band transponder	3	4	—	—	10 meas./s
Apollo USB	15	5	0	0.60	1
	15	5	0	0.07	10
	15	5	0	0.02	40
Goddard ($R+\dot{R}$)	5	2	0	0.60	1
	5	2	0	0.07	10
	5	2	0	0.02	40

Minitrack system	Angles (m rad) Bias	Angles (m rad) Noise
	0.1	0.1

Angular errors: USB $\sim$ 0.1–0.5 mrad
($R+\dot{R}$) $\sim$ 0.1–0.5 mrad

systems will ultimately be used for the determination of definitive orbits for GEOS-C which at the same time will, it is hoped, furnish the information on the gravity harmonics and station locations needed to finish the NSGP. The influence of ground tracking system errors and station location errors on spacecraft orbits is discussed in [22–24].

References

[1] Geonautics, Inc., NASA-Contract Techn. Rep. No. 184-711, December 1968.
[2] A. SHAPIRO and B. S. YAPLEE, NRL Rep. No. 7018, Naval Research Laboratory, Washington, D.C. 1970.
[3] M. I. SKOLNICK, NRL Rep. No. 2025, Naval Research Laboratory, Washington, D.C. 1969.
[4] A. J. GREENWOOD et al., in: Remote Sensing of Environment **1**, Elsevier Publ. Co., Amsterdam 1969 (p. 59).
[5] F. O. VONBUN and J. T. MENGEL, Spacecraft and Rockets **5**, 863 (1968).
[6] T. L. FELSENTREGER, T. J. GRENCHIK and P. E. SCHMID, Goddard Space Flight Center Rep. X-552-70-96, 1970.
[7] W. S. VON ARX, Science **154**, 1651 (1966).
[8] "Solid-Earth and Ocean Physics", Report of a Study at Williamstown, Mass., August 1969, sponsored by NASA.
[9] J. L. COOLEY and A. MARLOW, Goddard Space Flight Center Rep. X-551-69-1, January 1969.
[10] R. H. GREEN, IBM Rep. TM 68-40, 1968, NASA Contract NAS5-10022.
[11] P. E. SCHMID, Goddard Space Flight Center Rep. X-507-65-398, 1968.
[12] F. O. VONBUN, NASA TN D-4121, 1968.

[13] NASA/Wallops Rep. X-16-68-1, 1968.
[14] J. Berbert and H. Parker, Goddard Space Flight Center Rep. X-514-68-458, 1968.
[15] D. G. King-Hele et al., Planet. Space Sci. **14**, 49 (1966).
[16] S. Ross, Smithson. Astrophys. Obs. Contract SPC-9-5001, 1968.
[17] W. M. Kaula, Theory of Satellite Geodesy, Blaisdell Publ. Co., Waltham, Mass. (1966).
[18] R. J. Anderle and Smith, Naval Weapons Lab. Rep. No. 2106, July 1967.
[19] F. O. Vonbun, Paper presented at XII COSPAR Plenary Meeting, Prague 1969.
[20] W. S. von Arx, in: 1st Marine Geodesy Symp. 1966.
[21] Centre National D'Etudes Spatiales, "International Satellite Geodesy Experiment", Preliminary Document, 10 March 1970.
[22] W. D. Kahn and F. O. Vonbun, NASA TN D-3776, 1966.
[23] NASA/ANWG Rep. No. AN-2.1, 1966.
[24] J. L. Cooley, Goddard Space Flight Center Rep. X-507-67-60, 1967.

Space Research XI — Akademie-Verlag, Berlin 1971

TEMPORAL VARIATIONS OF THE GEOPOTENTIAL DERIVED FROM SATELLITE OBSERVATIONS

Y. Kozai

Tokyo Astronomical Observatory, Mitaka, Tokyo, Japan

By analysing Baker-Nunn observations of artificial earth satellites it is found that there are annual terms in the longitudes of the ascending nodes for two satellites, 1960 ι 2 and 1962 $\beta\mu$ 1, possibly due to annual variation of J_2, the coefficient of the dominant term in the expression of the geopotential. The amplitude of the variation in J_2 is of order of 10^{-9} and the phase is almost identical with that of the annual variation of the rotation rate of the earth. It is concluded from these data that a half of the annual variation of the length of day is due to the change of the principal moment of inertia and the other half is possibly due to atmospheric conditions.

Methods to derive both Love's number of the earth and the time lag of the earth tides from satellite observations are also discussed, and their recent results are summarized.

1. Introduction

For the past decade coefficients in the expressions of the geopotential expanded into spherical harmonics have been evaluated by many investigators by analysing satellite observations under the assumption that the geopotential is time-independent, i.e. that only combinations of observations made for different periods and for different satellites could determine the geopotential. The same assumption was made in the present work and all the available observations were used together in the determination of orbital elements and parameters of the geopotential, since there were too many unknowns and too few observations well distributed in space and time to allow a search for time dependence in the geopotential.

However, we know that the geopotential is changing due to tides, atmospheric motions, earthquakes and other mass motions inside and on the earth: the length of day is changing annually, secularly and irregularly, and the geomagnetic field is drifting westwards.

Since satellite tracking has the reputation of having provided the most powerful technique for measuring the geopotential and the gravity field, the next target for satellite geodesy should be the detection of temporal variations of the geopotential. For this purpose satellite geodesy has a particular advantage over the time-consuming gravity measurements of classical geodesy.

There is another type of temporal variation of the geopotential, not due to variation of the mass distribution on or inside the earth, but due to the motion of the equator and the axis of the earth referred to an inertial frame caused by precession, nutation and polar motion. The motion of the satellite in Newtonian dynamics must be specified in the inertial coordinate system, and not referred to the equator of the earth. The expression of the geopotential cannot be time-independent with respect to the inertial system. However, effects on satellite motion due to precession and nutation can be evaluated with sufficient accuracy with existing knowledge, and dynamical effects on satellite motion due to polar motion are extremely small.

The geocentric coordinates of observing stations cannot be fixed even if they are referred to the equator because of continental drift, latitude variations and so on. In future satellite geodesy, such variations of the station coordinates can be derived from time to time by determining from satellite observations not only orbital elements but also the station coordinates. Anderle and Beuglass [1] have already reported that they have successfully determined polar motions from latitude variations of the tracking stations by analysing Doppler tracking data of the Tranet system. Instantaneous positions of the pole based on two day's data are now regularly derived and published by Anderle and his group as the "Dahlgren Polar Monitoring Service".

2. Method of Analysis

Attempts to detect temporal variations of the geopotential by analysing precisely reduced Baker-Nunn observations for three satellites, 1959 α 1, 1960 ι 2 and 1962 $\beta\mu$ 1 have been in progress for two years. These satellites were selected because precise orbital elements are available for long intervals of time. In Table 1 the adopted time interval, orbital elements and the area-to-

Table 1
Details of satellites used in investigation

Satellite	1959 α 1 Vanguard 2	1960 ι 2 Rocket of Echo 1	1962 $\beta\mu$ 1 ANNA Geodetic satellite
Period	Feb. 1959–Dec. 1964	Sept. 1960–Dec. 1964	Nov. 1962–Dec. 1964
Mean motion n (day^{-1})	4119–4133°	4391°	4804°
Inclination i	32.88°	47.23°	50.14°
Eccentricity e	0.1660–0.1642	0.0114	0.0070
Secular motion of node $\dot{\Omega}$ (day^{-1})	−3.498–3.521°	−3.101°	−3.609°
Secular motion of perigee $\dot{\omega}$ (day^{-1})	5.258–5.293°	2.978°	2.963°
A/M (cgs)	0.21	0.21	0.07

mass ratio (A/M) are given for each satellite. For 1959 α 1 some of the orbital elements changed considerably due to the air-drag as indicated in the Table.

The orbital elements every two days are derived by using the Smithsonian Astrophysical Observatory's Differential Orbit Improvement (DOI) program [2] from precisely reduced Baker-Nunn observations over periods of four days (eight day observations were used in one set when enough observations are not available in four days). To derive the orbital elements, the effects of first-order short-period perturbations due to the oblateness of the earth, perturbations due to non-zonal harmonics depending on longitude in the geopotential, and lunar perturbations which depend on the lunar mean longitude are taken into account. The orbital elements for each epoch are then corrected by subtracting from them lunisolar long-periodic perturbations [3] which are not included in the DOI and solar radiation pressure perturbations [4] which are computed using the constant value of A/M given in Table 1. The amplitudes of long-period perturbations with arguments ω, 2ω and 3ω are evaluated using constant values for the coefficients of odd-order zonal harmonics, and are subtracted from the orbital elements to give mean orbital elements at each epoch.

The secular motion of the longitude of the ascending node and of the argument of perigee can be computed from the mean orbital elements thus derived by using constant values for the coefficients of even-order zonal harmonics. Using these values for the secular motion we can predict the argument of perigee and the longitude of the ascending node at the next epoch. Therefore, by assuming that (O — C)'s* at the initial epoch for the two angular elements are zero, (O — C)'s at each epoch can be computed in a uniform system even if the time interval is more than a year. Since the instantaneous values of the eccentricity and the mean motion are used to compute the perturbations, we can suppose that the effects of air-drag have been eliminated from O — C for the two angular elements.

From these values of O — C thus derived corrections to the secular motion and the amplitude of the long-period perturbations due to the zonal harmonics in the geopotential can be derived by the method of least squares and by using these data the coefficients of the zonal harmonics can be improved [5].

For 1959 α 1 the semi-major axis was decreased by 20 km due to air-drag; the perigee distance q (km) and the inclination i (°) are expressed as functions of a — 8305 km, where 8305 is the value of the semi-major axis at the mean epoch, as follows:

$$\left.\begin{aligned} q &= \underset{\pm 7}{6936.567} + \underset{\pm 1}{0.041}\,(a - 8305)\,, \\ i &= \underset{\pm 2}{32.87948} + (57.3 \times \underset{\pm 6}{0.0072})\,\frac{a - 8305}{8305}\,. \end{aligned}\right\} \qquad (1)$$

For the other two satellites, for which the air-drag effects are extremely small because of the large perigee height, the eccentricities, the inclinations and the mean motions are fitted by linear functions of time.

Then the residuals of the orbital elements for each epoch are derived by subtracting the equations (1), or the linear expressions plus the corrections to the long-period terms derived by the method of least squares, from O — C.

---

* O — C, here currently used, is the difference between observed and computed values of a parameter.

3. Determinations of Love's Number and Time Lag of Tides

For these three satellites the inclination is the most accurately determined element of the six orbital parameters, since most of the observations were made near apices in their orbits by virtue of the geographical distribution of the Smithsonian astrophysical observing stations. In the inclinations there are perturbations due to the tidal deformation of the earth caused by the sun and the moon. The most important of these perturbations for the three satellites is expressed [6] by:

$$\frac{-49^\circ\, k_2 \cos i}{n\,(1-e^2)^2\, a^3\, \dot{\Omega}} \cos\,(\Omega - \nu\, \Delta t)\,, \tag{2}$$

where k_2, Δt and ν are, respectively, Love's number for the earth, time lag of the tide due to friction, and the angular rotational velocity of the earth, a is the semi-major axis of the satellite orbit expressed in the earth equatorial radii, and the angular velocities are expressed in degrees per day. Since the secular motion is expressed as

$$\dot{\Omega} = -\, J_2\, n \cos i/p^2\,, \tag{3}$$

where $p = a\,(1 - e^2)$, and since $n^2\, a^3$ is constant because of Kepler's third law, it is found that the amplitude of the inclination perturbation given in Eq. (2) has a value which does not depend on the orbital elements and is equal to $0.80 \times 10^{-3}\, k_2$.

Perturbations with argument Ω, that is, with a similar argument, are produced by lunisolar gravitational attractions and by the motion of the equator of the earth due to precession [7]. The amplitude of the term with argument Ω due to precession in the inclination is $-0.87 \times 10^{-3}/\dot{\Omega}$. Table 2 lists the amplitude of these terms for the inclination of the three satellites. Roughly speaking, the amplitudes of the lunisolar terms with argument Ω are ten times larger than those due to the precession; the latter are of the same order of magnitude as those due to tidal deformation.

Table 2
Amplitudes (in degrees) of the perturbation terms with argument Ω

Satellite	Direct Lunisolar	Precession
1959 α 1	2.82×10^{-3}	-0.25×10^{-3}
1960 ι 2	2.43×10^{-3}	-0.28×10^{-3}
1962 $\beta\,\mu$ 1	1.80×10^{-3}	-0.24×10^{-3}

After subtracting perturbations due to precession and nutation, the following periodic variations of the inclinations with argument Ω are derived by the method of least squares from the residuals every two days [8]:

$$\left.\begin{array}{ll} 0.25 \times 10^{-3} \cos\,(\Omega - 0^\circ)\,, & \text{for } 1959\,\alpha\,1\,, \\ \pm 2 \qquad\qquad\quad \pm 5 & \\ 0.26 \times 10^{-3} \cos\,(\Omega - 6^\circ)\,, & \text{for } 1960\,\iota\,2\,, \\ \pm 2 \qquad\qquad\quad \pm 5 & \\ 0.18 \times 10^{-3} \cos\,(\Omega - 9^\circ)\,, & \text{for } 1962\,\beta\,\mu\,1\,. \\ \pm 2 \qquad\qquad\quad \pm 7 & \end{array}\right\} \tag{4}$$

From these data the value of Love's number is derived as

$$k_2 = 0.29 \pm 0.03 \tag{5}$$

and the time lag of the tide is computed from $\nu\, \Delta t = 5 \pm 3°$ as

$$\Delta t = 20 + 12 \text{ minutes in time.} \tag{6}$$

Newton [9] used four polar satellites, for which Doppler tracking data from the Tranet system were available over intervals of 6 to 18 months. He computed perturbations due to every source more carefully than Kozai. To derive tidal parameters he used terms

$$\genfrac{}{}{0pt}{}{\sin}{\cos}\, \Omega\,, \quad \genfrac{}{}{0pt}{}{\sin}{\cos}\, 2\,\Omega \quad \text{and} \quad \genfrac{}{}{0pt}{}{\sin}{\cos}\,(2\,\lambda - \Omega)\,,$$

where λ is the mean longitude of the sun or the moon, both for the inclination and for the longitude of the ascending node, since the eccentricity cannot be disturbed by tidal deformation. Therefore he could derive Love's number and the time lag from the lunar and solar tides separately, although from a theoretical point of view they cannot differ appreciably. The results are:

$$k_2 = 0.359 \pm 0.042\,, \quad \nu\, \Delta t = 0.8 \pm 0.2\,, \quad \text{for solar tides} \tag{7}$$

$$k_2 = 0.314 \pm 0.036\,, \quad \nu\, \Delta t = 2.1 \pm 0.2\,, \quad \text{for lunar tides.} \tag{8}$$

Although the differences of the tidal parameters derived by Newton and by Kozai are not very significant, Kaula [10] tried to explain them by assuming that the tidal parameters depended on latitude since different satellites with different inclinations were used by the two authors.

4. Annual Variation of J_2

Efforts have been made to identify the effects due to a seasonal variation of J_2, the coefficient of the dominant term in the geopotential, from the residuals of the longitudes of the ascending nodes. For this case the data for $1959\,\alpha\,1$ has not been used, since the longitude of the ascending node is not so accurately determined for this satellite because the inclination is low. Thus the following terms with annual period are derived:

$$\left.\begin{array}{ll} \delta\Omega = \underset{\pm 4}{0.22} \times 10^{-3} \sin\,(2\,\pi\, t + \underset{\pm 10}{350°})\,, & \text{for } 1960\,\iota\,2\,, \\ \delta\Omega = \underset{\pm 4}{0.30} \times 10^{-3} \sin\,(2\,\pi\, t + \underset{\pm 10}{330°})\,, & \text{for } 1962\,\beta\,\mu\,1\,, \end{array}\right\} \tag{9}$$

where $t = 0$ corresponds to the beginning of the year.

The longitude of the ascending node is approximately expressed as

$$\Omega = -\int (J_2/p^2)\; n \cos i \; dt\,. \tag{10}$$

If it can be assumed that there is an annual variation in J_2, that is, if J_2 in (10) can be replaced by

$$J_2 + A \cos\,(2\,\pi\, t + \beta)\,, \tag{11}$$

the following expression is derived from (10):

$$\Omega = \dot{\Omega}\,\{t + (A/2\,\pi\,J_2)\sin(2\,\pi\,t + \beta)\}\,. \tag{12}$$

By assuming that the annual variation in the longitude of the ascending node is due to that in J_2, we get

$$\left.\begin{aligned} \delta J_2 &= \underset{\pm 2}{1.2}\times 10^{-9}\cos(2\,\pi\,t + \underset{\pm 10}{170^\circ})\,, \quad \text{for } 1960\,\iota\,2\,,\\ \delta J_2 &= \underset{\pm 2}{1.4}\times 10^{-9}\cos(2\,\pi\,t + \underset{\pm 10}{150^\circ})\,, \quad \text{for } 1962\,\beta\,\mu\,1\,. \end{aligned}\right\} \tag{13}$$

The results derived from the two satellites agree with each other within the errors estimated and, if their mean is taken, the annual variation of J_2 is expressed as

$$\delta J_2 = \underset{\pm 2}{1.3}\times 10^{-9}\cos(2\,\pi\,t + \underset{\pm 10}{160^\circ})\,, \tag{14}$$

where the epoch is at the beginning of the year. Because of this term the geoid oscillates around its mean position, by 8 mm at most, with a one year period, and J_2 has its maximum value around 20 July and minimum around 20 January.

It is known that the rotation speed of the earth changes annually, and that the expression of UT2−UT1 (in seconds) as used by the Bureau International de l'Heure [11] has the following terms:

$$0.025\sin(2\,\pi\,t + 331^\circ) + 0.009\sin(4\,\pi\,t + 130^\circ)\,. \tag{15}$$

Therefore, the angular velocity of the earth ν changes as

$$(\delta\nu/\nu)\times 10^9 = 5.0\cos(2\,\pi\,t + 331^\circ) + 3.6\cos(4\,\pi\,t + 130^\circ)\,. \tag{16}$$

If we assume that the angular momentum of the earth $C\,\nu$ is conserved, we must have the following relation:

$$\delta C/C + \delta\nu/\nu = 0\,. \tag{17}$$

Since the expression for J_2 takes the form

$$J_2 = \left\{C - \frac{1}{2}(A + B)\right\}\Big/ M\,a_e^2\,, \tag{18}$$

and assuming that the mass M is constant we get

$$\delta J_2 = \delta C/2\,C = -\,\delta\,\nu/2\,\nu\,, \tag{19}$$

as $\delta A + \delta B + \delta C = 0$ and $C = M\,a_e^2/3$. Here a_e is the equatorial radius of the earth and A, B and C are the principial moments of inertia of the earth. In Eq. (19) a term allowing for the effect of the small variation of a_e, of second order, has been neglected.

Therefore, under the assumption that the angular momentum of the earth is conserved, the anticipated seasonal variation of J_2 becomes

$$\delta J_2\times 10^9 = 2.5\cos(2\,\pi\,t + 151^\circ) + 1.8\cos(4\,\pi\,t + 310^\circ)\,. \tag{20}$$

By comparing expression (20) with (14) derived from satellite observations, we see that the angular momentum is not conserved. However, since the amplitude of the annual term in (20) is twice that in (14) and the phase almost identical, about half of the annual variation of the rotation velocity may be due to the mass displacement expressed by (14), that is, the seasonal variation in C, the largest principal moment of inertia; the other half, as well as the semi-annual variation, is presumably due to the effect of winds in the atmosphere, since the semi-annual variation cannot be found in J_2 from the satellite observations in this analysis.

5. Discussion

There is also a possibility that other coefficients of the zonal harmonics are changing in a similar way. If J_3, the coefficient of the third zonal harmonics,

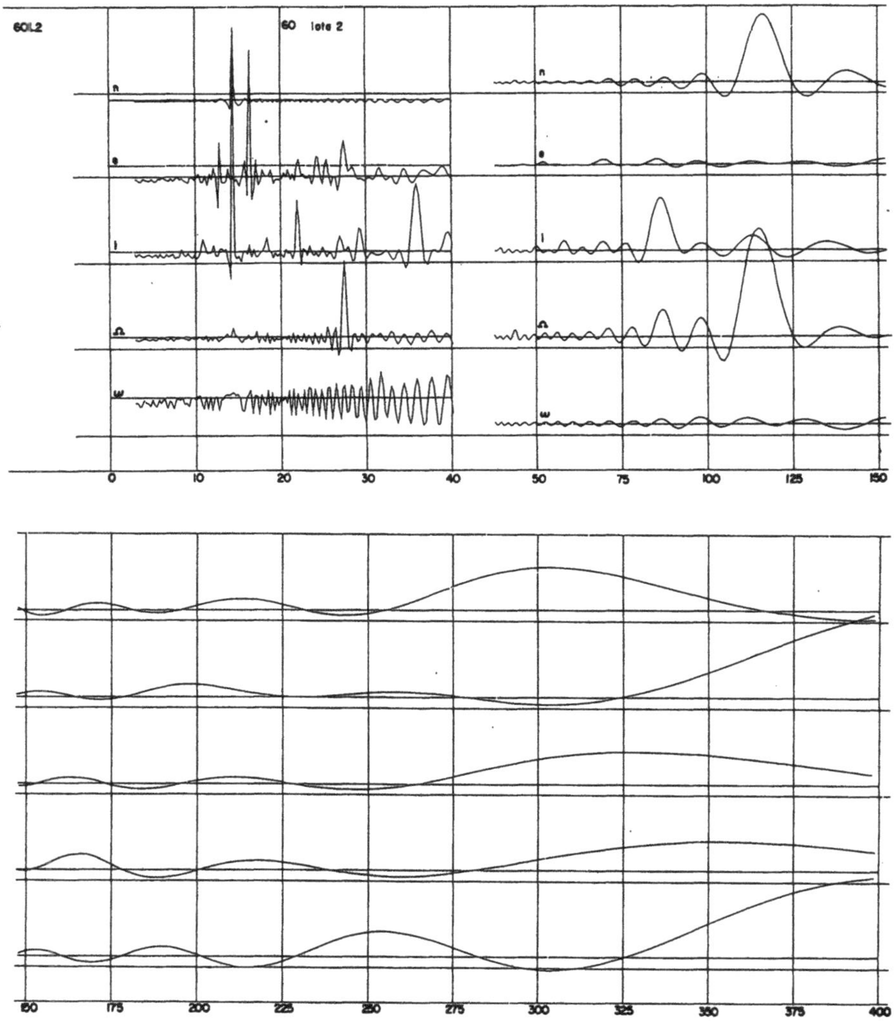

Fig. 1. Power spectra of five orbital elements for 1960 ι 2. Units of abscissa are days (note that the scale changes between 40 and 50 days).

changes periodically with argument $2\pi t + \beta$, long-period perturbations with $2\pi t \pm \omega + \beta$ as arguments must appear in the orbital elements. Efforts have been made to identify such perturbations from the residuals in some of the orbital elements for the two satellites, but no definitive results have yet been

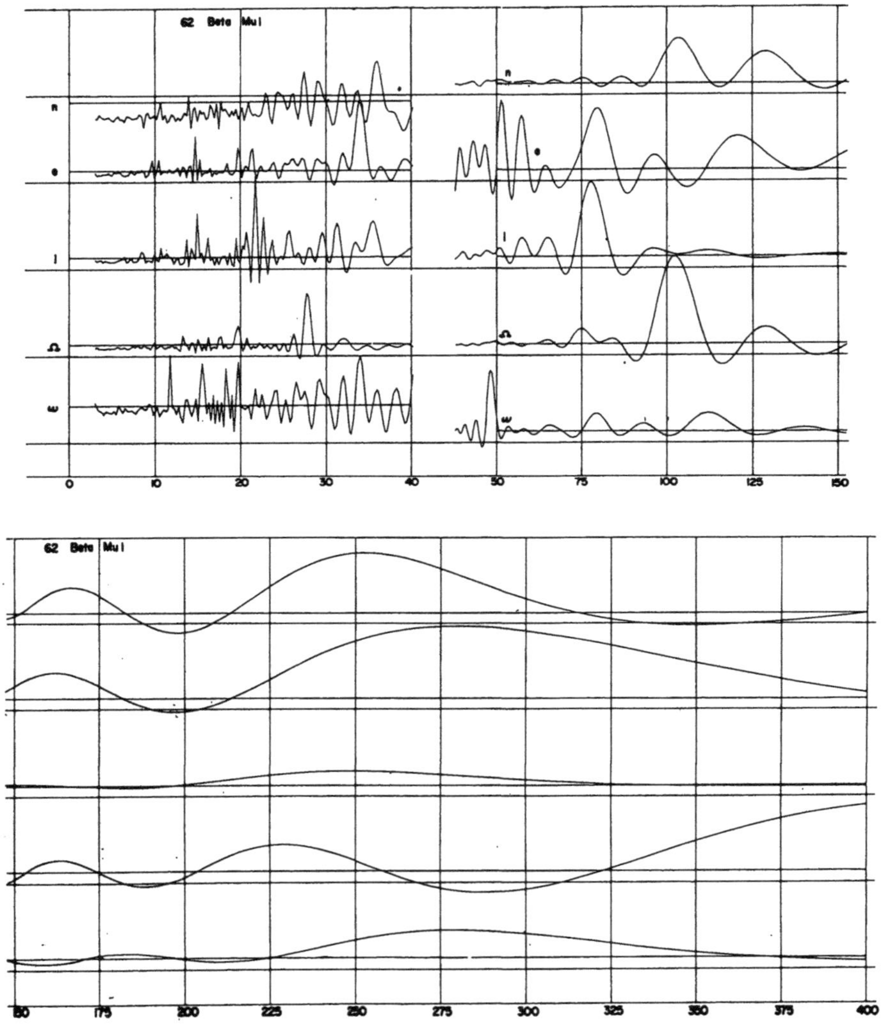

Fig. 2. Power spectra of five orbital elements for 1962 $\beta\mu$ 1. Units of abscissa are days (note that the scale changes between 40 and 50 days).

obtained. This would suggest that the amplitude of the variation is less than 10^{-9} if it exists.

In Figs. 1 and 2 are plotted power spectra computed from the final residuals of five orbital elements for the two satellites adopted to identify the J_2 variation. (Note that the scale is larger for the first forty days.) Some of the steep peaks in the spectra are connected with periods of resonance perturbations due to non-zonal harmonics in the geopotential, although most of them are very difficult to identify with periods having physical meaning. It is probable that most of the peaks in Figs. 1 and 2 will disappear if the perturbations due to the solar radiation pressure are computed more rigorously by including earth-reflected radiations and variables values of A/M. If this is done it should be possible to determine not only periodic variations but also secular and irregular variations of J_2 and other harmonic coefficients as well as those of G, the gravitation constant, from satellite tracking data for many years. Also it is expected that terms with a period of 19 years associated with lunar tides will be detected in the angular elements of satellites and another kind of Love's number of the earth derived [7] when observations become available extending over more than 19 years.

The problem whether or not non-zonal terms depending on longitude in the geopotential are time-dependent is interesting in connection with the westward drift of the geomagnetic field, and will be treated by comparing data derived from satellite observations made at different epochs.

References

[1] R. J. Anderle and L. K. Beuglass, Bull. Geodésique (1970) in press.
[2] E. M. Gaposchkin, Smithson. Astrophys. Obs. Spec. Rep. No. 161 (1964).
[3] Y. Kozai, Smithson. Astrophys. Obs. Spec. Rep. No. 22, 7 (1959).
[4] Y. Kozai, Smithson. Astrophys. Obs. Spec. Rep. No. 56, 25 (1961).
[5] Y. Kozai, Publ. Astron. Soc. Japan **16**, 263 (1964).
[6] Y. Kozai, Publ. Astron. Soc. Japan **17**, 395 (1965).
[7] Y. Kozai, Astron. J. **65**, 621 (1960).
[8] Y. Kozai, Publ. Astron. Soc. Japan **20**, 24 (1968).
[9] R. R. Newton, Geophys. J. Roy. Astron. Soc. **14**, 505 (1968).
[10] W. M. Kaula, Astron. J. **74**, 1108 (1969).
[11] Bureau International de l'Heure, Rapport annuel pour 1967, Observatoire de Paris 1968.

Space Research XI — Akademie-Verlag, Berlin 1971

NEW GEODETIC PARAMETERS FOR A STANDARD EARTH

K. Lambeck[a] and E. M. Gaposchkin[b]

[a] Observatoire de Paris, Meudon, France, and Smithsonian Astrophysical Observatory, Cambridge, Mass., USA
[b] Smithsonian Astrophysical Observatory, Cambridge, Mass., USA

Geodetic parameters describing the earth's gravity field and the positions of satellite tracking stations in a geocentric reference frame have been computed. These parameters were estimated by means of a combination of four different types of data, routine and simultaneous satellite observations, observations of deep space probes and surface gravity measurements. This combination solution gave better results than any subset of data types. The total gravity field is represented by spherical harmonics complete to degree and order 16 plus a number of higher order terms. The accuracy of the global field is about ±3 meters in geoid height or ±9 mgal. Station coordinates of many of the stations are determined with an accuracy to 10 m or better.

1. Introduction

In 1966 the Smithsonian Astrophysical Observatory (SAO) published numerica parameters for the earth's gravity field and the coordinates of the Baker-Nunn satellite tracking cameras [1]. In 1967 SAO published further refinements to this solution [2]. Subsequently several advances have been made which resulted in a new solution that is superior to any previous set of geodetic parameters. These improvements are due to a significant amount of new Baker-Nunn data, to increased cooperation with other observatories, to the use of precise laser range data and to a better geographical distribution of the tracking stations and of the observations. The improvements are also due to a major revision of the analytical techniques, computer programs and theories.

The new geodetic parameters have been estimated from a combination of results obtained from the dynamic and geometric methods of satellite geodesy, station position information determined by the Jet Propulsion Laboratory for its Deep Space Tracking Network (DSN) and surface gravity data in the form of 300-n-mi square free-air anomalies. The final solution yields harmonic coefficients in the geopotential expansion complete to degree and order 16, plus 14 pairs of higher coefficients in addition to the geocentric coordinates of 39 stations (or groups of co-located stations).

A solution of this kind is the result of several iterations and of the testing of the intermediate results obtained from the four component data sets against each other and against external standards. In the final dynamic solution about 60000 observed quantities of 21 satellites were used. About an equal number of data from various optical stations in addition to those from 19 Baker-Nunn stations were used in the geometric solution. Data from the tracking of deep space probes were provided by Mottinger [3] in the form of relative longitudes and the distances to the earth's axis of rotation for five stations. The free-air anomalies were provided by Kaula [4]. In addition to these four data sets some recently acquired surface gravity data not used in the Kaula compilation and some astro-geoid data were employed for independent tests of the solution.

The methods and results have been discussed in some detail by the authors [5] and in the present paper only the final results together with a brief summary of some of the results of the various comparisons made are presented.

2. Results

The final results for the station geocentric cartesian coordinates are given in Table 1 and for the tesseral harmonics in Table 2. Fig. 1 gives the geoid heights and Fig. 2 the free air gravity anomalies corresponding to the final combination solution. Table 1 contains the precision estimates of the station coordinates and Fig. 3 the precision of the geoid heights as a function of latitude. These estimates have been made from the statistics of the final iteration but subsequent comparisons indicate that they are quite realistic. The geoid height estimates refer to the generalized geoid only.

3. Comparison of Results

3.1. Comparison with Satellite Orbits

Each iteration resulted in improved orbital residuals when satellite orbits computed from the combination solution were compared with observations. For the final solution the orbital residuals for satellites such as GEOS 1 and GEOS 2 are less than 10 m. These orbits are computed from a combination of laser and Baker-Nunn data for 30 day arcs. The optical data residuals are less than 2 arcsec and the data residuals have an rms value of about 7 m.

3.2. Comparison of Geometric and Dynamic Satellite Solutions

Fig. 4 shows some typical results for comparisons of the directions between stations computed by the geometric and dynamic satellite solutions. The difference in the positions derived from the individual solutions is a good indication of their relative accuracies and of the accuracy of the combination solution. These differences are the result of uncertainties in the coordinates at both stations, and at each station a number of such comparisons can usually be made. Thus the accuracy of the station positions of the combination solution, relative to the earth's center of mass should be somewhat better than these figures indicate.

Table 1

Geocentric coordinates (in mm) of the stations determined in the final combination solution

The fifth column gives the formal precision estimates of the coordinates in meters

Station	X	Y	Z	σ	Station Name
1021	1.118029	−4.876316	3.942984	7	Blossom Point, Md.
1034	−0.521702	−4.242049	4.718731	7	Grand Forks, Minn.
1042	0.647515	−5.177924	3.656707	7	Rosman, N.C.
7036	−0.828496	−5.657458	2.816812	7	Edinburg, Tex.
7037	−0.191286	−4.967280	3.983262	7	Columbia, Mo.
7039	2.308239	−4.873597	3.394580	10	Bermuda
7040	2.465067	−5.534924	1.985510	10	Puerto Rico
7045	−1.240479	−4.760229	4.048995	9	Denver, Col.
7050	1.130674	−4.831368	3.994111	7	Goddard Space Flight Center
7075	0.692628	−4.347059	4.600483	9	Sudbury, Ont.
7076	1.384174	−5.905685	1.966533	10	Jamaica
7815	4.578370	0.457951	4.403134	5	Haute Provence, France
7816	4.654337	1.959134	3.884366	5	Stephanion, Greece
7818	5.426329	−0.229330	3.334608	15	Colomb-Bechar, Algeria
7901	−1.535757	−5.166996	3.401042	5	Organ Pass, N.M.
8015	4.578328	0.457966	4.403179	5	Haute Provence, France
8019	4.579466	0.586599	4.386408	5	Nice, France
9001	−1.535757	−5.166996	3.401042	5	Organ Pass, N.M.
9002	5.056125	2.716511	−2.775784	7	Pretoria, S. Africa
9003	−3.983776	3.743087	−3.275566	6	Woomera, Australia
9004	5.105588	−0.555228	3.769667	5	San Fernando, Spain
9005	−3.946693	3.366299	3.698832	10	Tokyo, Japan
9006	1.018203	5.471103	3.109623	9	Naini Tal, India
9007	1.942775	−5.804081	−1.796933	7	Arequipa, Peru
9008	3.376893	4.403976	3.136250	9	Shiraz, Iran
9009	2.251829	−5.816919	1.327160	7	Curaçao, Antilles
9010	0.976291	−5.601398	2.880240	5	Jupiter, Fla.
9011	2.280589	−4.914573	−3.355426	9	Villa Dolores, Argentina
9012	−5.466053	−2.404282	2.242171	7	Maui, Hawaii
9021	−1.936782	−5.077704	3.331916	15	Mt. Hopkins, Ariz.
9023	−3.977766	3.725102	−3.303035	6	Island Lagoon, Australia
9025	−3.910437	3.376361	3.729217	10	Dodaira, Japan
9028	4.903750	3.965201	0.963872	12	Addis Ababa, Ethiopia
9029	5.186461	−3.653856	−0.654325	12	Natal, Brazil
9031	1.693803	−4.112328	−4.556649	15	Comodoro Rivadavia, Argentina
9050	1.489753	−4.467478	4.287304	14	Harvard, Mass.
9065	3.923411	0.299882	5.002945	12	Delft, Holland
9066	4.331310	0.567511	4.633093	7	Zimmerwald, Switzerland
9074	3.183901	1.421448	5.322772	10	Riga, Latvia
9077	3.907421	1.602397	4.763890	10	Uzhgorod, USSR
9080	3.920178	−0.134738	5.012708	9	Malvern, England
9091	4.595157	2.039425	3.912650	5	Dionysos, Greece
9113	−2.450011	−4.624421	3.635035	7	Rosamund, Cal.
9114	−1.264838	−3.466884	5.185467	12	Cold Lake, Canada
9115	3.121280	0.592643	5.512701	17	Harvestua, Norway
9117	−6.007402	−1.111859	1.825730	15	Johnston Isl., Pacific

Table 2

Fully normalized coefficients of the spherical harmonic expansion of the geopotential obtained in the final iteration of the combination solution

C_{lm} are the cosine terms of degree l and order m and S_{lm} are the sine terms

l	m	C_{lm}	S_{lm}	l	m	C_{lm}	S_{lm}
2	2	2.4129E−06	−1.3641E−06	3	1	1.9698E−06	2.6015E−07
3	2	8.9204E−07	−6.3468E−07	3	3	6.8630E−07	1.4304E−06
4	1	−5.2989E−07	−4.8765E−07	4	2	3.3024E−07	7.0633E−07
4	3	9.8943E−07	−1.5467E−07	4	4	−7.9692E−08	3.3928E−07
5	1	−5.3816E−08	−9.7905E−08	5	2	6.1286E−07	−3.5087E−07
5	3	−4.3083E−07	−8.6663E−08	5	4	−2.6693E−07	8.3010E−08
5	5	1.2593E−07	−5.9910E−07	6	1	−9.8984E−08	3.7652E−08
6	2	5.4825E−08	−3.5175E−07	6	3	2.7873E−08	4.4626E−08
6	4	−4.0342E−10	−4.0388E−07	6	5	−2.1143E−07	−5.2264E−07
6	6	8.8693E−08	−7.4756E−08	7	1	2.4142E−07	1.1567E−07
7	2	2.8306E−07	1.5645E−07	7	3	2.0285E−07	−2.3448E−07
7	4	−1.9727E−07	−1.1390E−07	7	5	−8.7024E−10	9.8461E−08
7	6	−2.5847E−07	1.0209E−07	7	7	1.5916E−07	−6.7710E−08
8	1	3.1254E−08	2.5696E−08	8	2	4.8161E−08	8.4140E−08
8	3	−5.7444E−08	1.8086E−08	8	4	−1.5378E−07	7.5264E−08
8	5	−5.6733E−08	6.1636E−08	8	6	−5.3903E−08	2.5930E−07
8	7	3.4390E−08	8.9168E−08	8	8	−7.7364E−08	6.7607E−08
9	1	1.3823E−07	−1.6100E−08	9	2	6.6741E−09	−8.1733E−08
9	3	−9.6463E−08	−1.1817E−07	9	4	5.7125E−08	1.1183E−07
9	5	−6.1435E−09	3.3551E−09	9	6	2.4186E−08	2.2028E−07
9	7	−5.0450E−08	−1.2700E−07	9	8	2.3359E−07	5.7239E−08
9	9	−8.2490E−08	9.2326E−08	10	1	1.1251E−07	−1.0167E−07
10	2	−3.1225E−08	−1.0450E−07	10	3	−2.3346E−08	−1.4137E−08
10	4	−4.8185E−08	−4.3248E−08	10	5	−8.0004E−08	−1.4279E−07
10	6	−3.2486E−08	−2.0153E−07	10	7	5.4961E−08	3.2003E−08
10	8	7.3957E−08	−7.9706E−08	10	9	−6.8563E−09	6.2498E−09
10	10	1.2377E−07	−2.5885E−08	11	1	4.3900E−09	2.9751E−08
11	2	4.8900E−08	−9.1994E−08	11	3	−6.3247E−08	−1.3109E−07
11	4	−3.0193E−08	5.4317E−08	11	5	3.2523E−08	1.3215E−07
11	6	3.7517E−08	6.9005E−09	11	7	4.5726E−08	−1.3862E−07
11	8	6.4546E−08	−1.6993E−08	11	9	1.1750E−07	−9.9451E−09
11	10	−1.1736E−07	−1.8900E−08	11	11	1.1785E−07	−4.0688E−08
12	1	−4.5955E−08	−3.1000E−08	12	2	2.7481E−08	7.5986E−08
12	3	5.8386E−08	5.4784E−08	12	4	−4.3649E−08	−2.2262E−08
12	5	2.3375E−08	4.2637E−08	12	6	−2.3868E−08	−6.6770E−10
12	7	1.4507E−08	9.9784E−08	12	8	−5.7854E−09	3.3752E−08
12	9	−3.2232E−08	4.2858E−08	12	10	−1.8590E−08	4.8382E−09
12	11	−4.4921E−08	−4.8206E−08	12	12	−1.9407E−08	−5.7771E−08
13	1	−5.6042E−08	2.6288E−08	13	2	−4.7456E−08	1.7367E−08
13	3	2.3833E−08	−2.8930E−08	13	4	−1.9980E−08	5.7030E−09
13	5	9.6637E−08	−4.7760E−08	13	6	−8.3417E−08	5.9782E−08
13	7	−5.2217E−08	−3.2562E−09	13	8	−4.1759E−08	−2.0231E−08
13	9	−2.5623E−08	1.0767E−07	13	10	8.6589E−08	−1.0528E−08
13	11	−3.3749E−08	5.8541E−08	13	12	−1.3229E−09	8.2192E−08
13	13	−7.0288E−08	7.4643E−08	14	1	−2.3090E−08	4.9664E−08
14	2	3.2120E−08	−4.5289E−08	14	3	1.9042E−08	1.1919E−09
14	4	7.8017E−09	−3.7527E−08	14	5	−2.5958E−08	−2.3344E−08
14	6	1.9140E−08	−5.8721E−08	14	7	1.1061E−08	8.4132E−09
14	8	−3.0273E−08	−6.0838E−08	14	9	4.9539E−08	9.2345E−08
14	10	5.3732E−08	−4.3168E−08	14	11	2.7833E−08	−8.1637E−08

l	m	C_{lm}	S_{lm}	l	m	C_{lm}	S_{lm}
14	12	1.2481E−08	−5.7314E−08	14	13	5.1554E−08	4.5453E−08
14	14	−5.2082E−08	−1.2840E−08	15	1	−3.5971E−09	4.0142E−08
15	2	−4.4833E−08	−1.6056E−08	15	3	8.3016E−09	−5.7218E−09
15	4	1.3916E−08	6.6644E−08	15	5	3.1684E−08	1.8250E−09
15	6	7.0020E−08	−1.1872E−07	15	7	1.1856E−07	4.2690E−08
15	8	−9.7657E−08	−3.5710E−08	15	9	2.2064E−08	2.6632E−08
15	10	−2.0648E−08	5.3724E−10	15	11	−3.2585E−08	9.4052E−08
15	12	1.0524E−08	6.8726E−09	15	13	−3.7348E−08	4.0249E−09
15	14	1.2193E−08	−2.6786E−08	15	15	1.4515E−09	−1.4802E−08
16	1	−2.3789E−08	7.6413E−08	16	2	2.1327E−08	3.0669E−08
16	3	−4.7358E−08	3.2610E−08	16	4	−1.1591E−08	4.3001E−08
16	5	−4.4201E−08	3.2230E−08	16	6	−5.8439E−08	−4.2809E−08
16	7	1.0591E−07	8.1008E−09	16	8	−8.4738E−08	−2.4677E−09
16	9	9.0001E−09	−1.0628E−07	16	10	−2.9849E−08	−5.2467E−10
16	11	6.8502E−09	−7.0765E−08	16	12	2.2834E−08	−3.4087E−08
16	13	3.5475E−08	2.0683E−08	16	14	−7.3590E−09	−2.2626E−08
16	15	−3.5485E−08	8.4126E−10	16	16	−2.9522E−08	8.6217E−09
17	12	8.3097E−08	3.5424E−09	17	13	3.2749E−08	4.2920E−10
17	14	−1.6058E−08	2.7286E−08	18	12	1.1662E−08	8.4724E−09
18	13	4.6903E−09	−3.5547E−08	18	14	−2.7446E−08	−4.8376E−08
19	12	6.7115E−08	−8.2623E−09	19	13	3.3201E−08	−6.3128E−08
19	14	−3.9779E−09	−2.3817E−08	20	13	5.8374E−08	3.3320E−08
20	14	1.1130E−08	−1.6183E−08	21	13	3.6928E−09	−1.6288E−08
21	14	5.2067E−08	3.0801E−10	22	14	−8.0549E−09	2.6440E−08

3.3. Comparison of Satellite and DSN Solutions

The DSN sites can be related to Baker-Nunn positions by ground survey. For three station groups, 9002-4751, 9003-4741, 9113-4712, the distance between the two instrumentation types does not exceed 200 km. For the other two stations groups, 9003-4742, 9004-4761, the distances are larger, about 500 and 1000 km respectively, and the ground survey information uncertainties may influence the comparison results. Table 3 gives the results in the form of differences in longitude $\Delta\lambda_i$ and in distance to the earth's axis of rotation Δr_i. The table also gives the accuracy estimates deduced from the statistics of the two solutions and the ground survey data. The comparisons reflect a systematic longitude difference $\overline{\Delta\lambda}$ of -3.2 μrad, the DSN longitudes being to the east of the SAO longitudes. When this systematic part is removed the residuals

Table 3
Results of SAO–JPL stations comparison

Station	$\lambda_{SAO} - \lambda_{JPL} = \Delta\lambda_i$ (μrad)	$\overline{\Delta\lambda} - \Delta\lambda_i$ (μrad)	(m)	$\sigma(\Delta\lambda_i)$ (m)	$r_{SAO} - r_{JPL} = \Delta r_i$ (m)	$\sigma(\Delta r_i)$ (m)
4751−9002	−3.5	+0.3	+ 1.9	7.7	+5.9	4.9
4741−9003	−2.2	−0.9	− 5.2	6.8	−7.3	4.5
4742−9003	−1.2	−2.0	−10.4	9.0	−6.5	4.5
4761−9004	−4.5	+1.4	+ 6.9	6.6	−1.2	4.5
4712−9113	−4.9	+1.7	+ 9.2	12.4	+7.6	5.5

$\Delta\lambda_i$ is the longitude difference and Δr_i the difference in distance to the earth's axis of rotation for the two solutions. $\overline{\Delta\lambda}$ is the weight mean longitude difference.

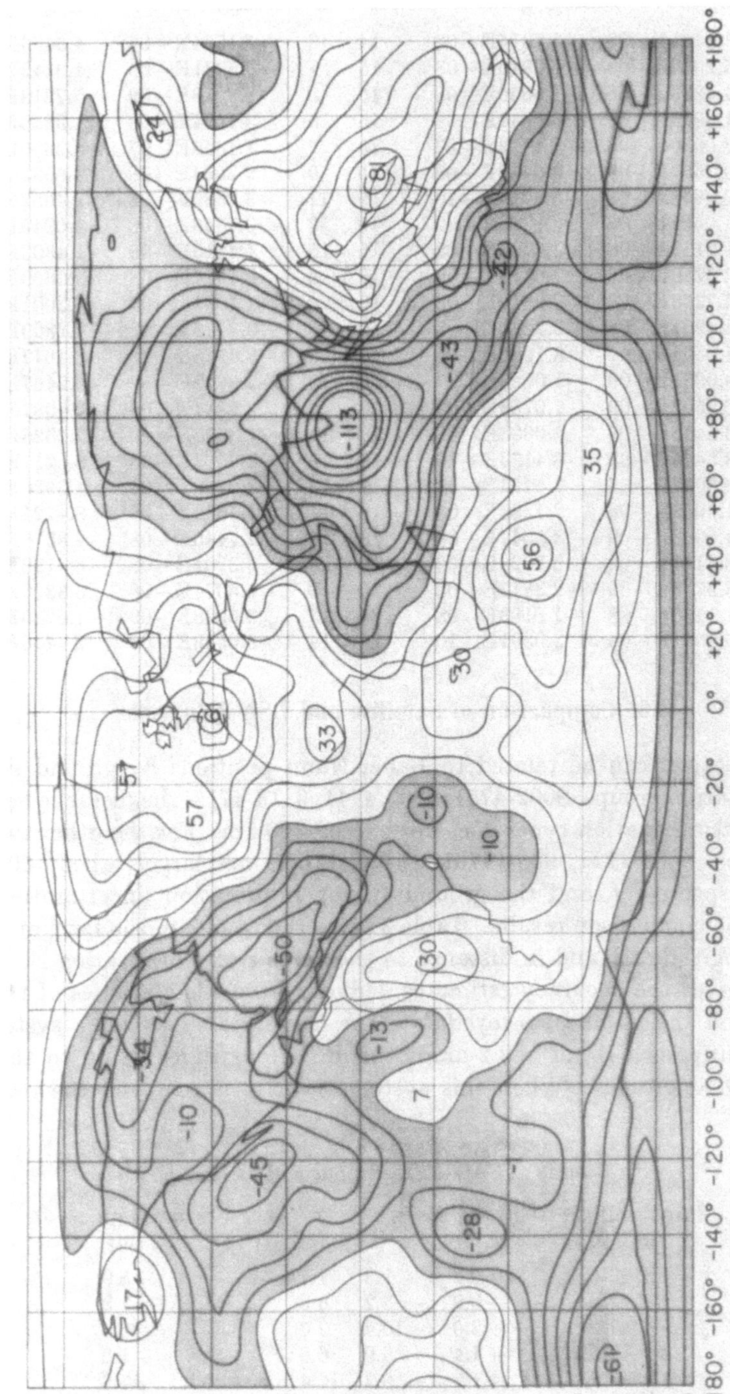

Fig. 1. Geoid heights in meters of the new combination solution corresponding to a reference ellipsoid of flattening $f = 1/298.255$. (10-m-contours.). Shaded areas are regions of negative geoid heights.

Fig. 2. Gravity anomalies with respect to the best fitting ellipsoid $f = 1/298.255$ from the adopted solution (10-mgal contours). Shaded areas are regions of negative gravity anomalies.

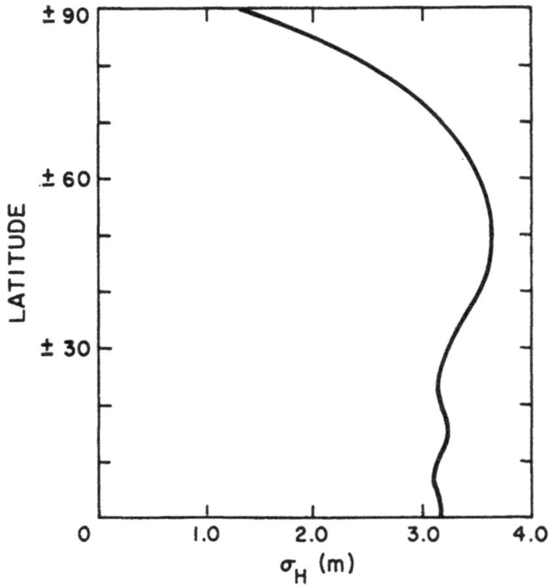

Fig. 3. Precision estimates of geoid heights determined from the harmonic coefficient precision estimates.

Table 4
Cartesian Geocentric coordinates for JPL-Stations

Station	X (mm)	Y (mm)	Z (mm)
4751	5.085451	2.668252	−2.768728
4741	−3.978706	3.724858	−3.302213
4742	−4.460972	2.682424	−3.674618
4761	4.849242	−0.360290	4.114869
4712	−2.350454	−4.651975	3.665631

are all less than about 10 m and support the accuracy estimates given in Table 1 for the Baker-Nunn stations 9002, 9003, 9004, 9113. Table 4 gives the cartesian geocentric coordinates for these stations.

3.4. Comparison with Surface Gravity Data

Table 5 summarizes the results obtained from a comparison of the geopotential derived from (i) the new satellite solution, (ii) the new combination solution presented here, and (iii) the 1966 Standard Earth solution (also known as the M1 solution), with the 300-n-mi squares for which at least 20 surveyed 60-n-mi squares were available for the compilation. The quantities are defined in [4] and [5]. Briefly, g_t is the terrestrial gravity anomaly and g_s is the satellite or combination solution anomaly. The ε_t and ε_s are the expected errors in the estimates g_t and g_s and δg is the expected truncation error. The estimate D of the gravity field contained by the spherical harmonic expansion is obtained

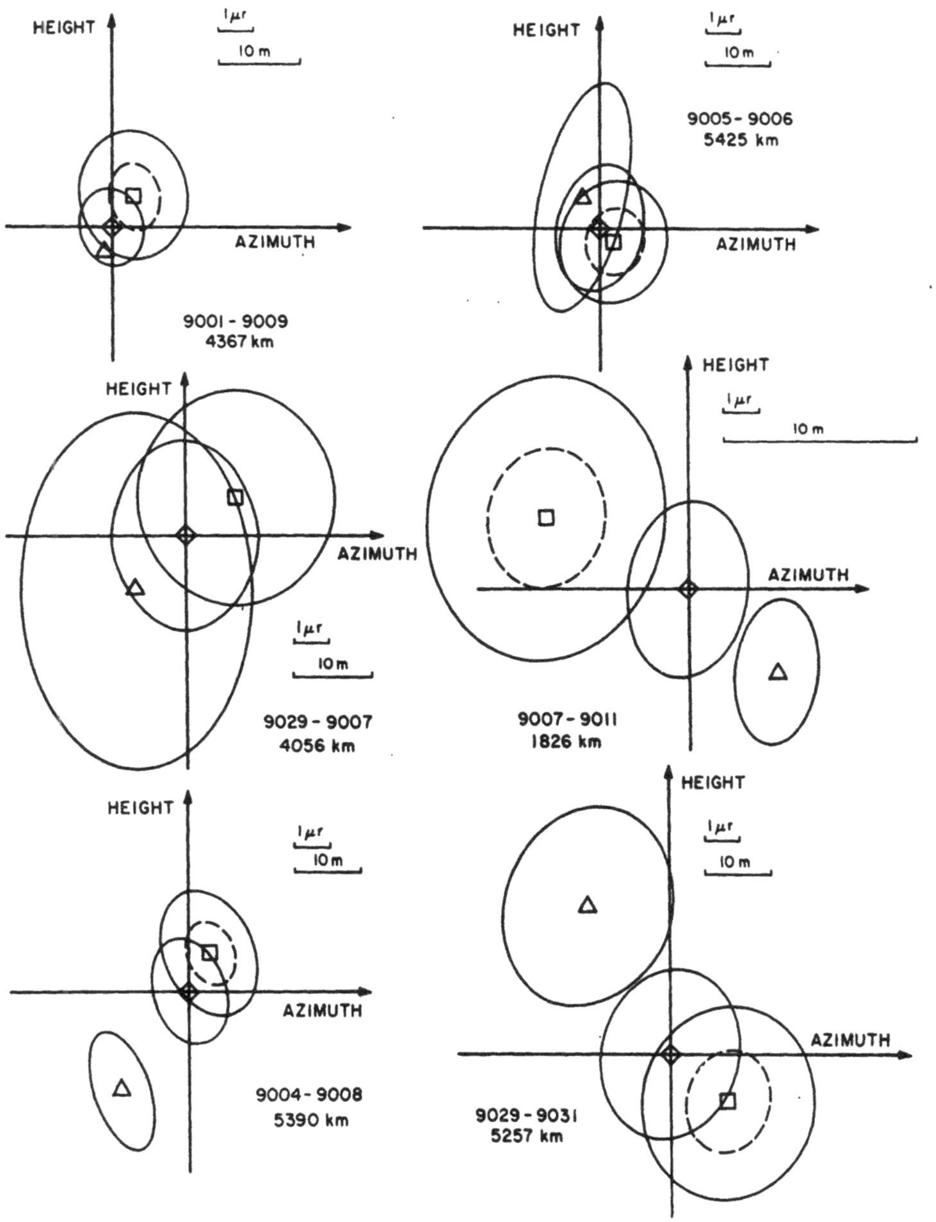

Fig. 4. Comparisons for station–station vectors computed from the geometric solution △, the dynamic solution □, and the combination solution ◇. The two error ellipses centered at □ refer to the formal statistics of the dynamic solution (the inner ellipse) and after the covariance matrix has been multiplied by the factor k_1^2 (outer ellipse).

Table 5

Comparison of satellite and combination solutions with surface gravity measurements (mgal²)

Solution		$\langle(g_T - g_S)^2\rangle$	$\langle g_T\, g_S\rangle$	$\langle g_S^2\rangle$	D	g_T^2	$E\{\varepsilon_S^2\}$	$E\{\varepsilon_T^2\}$	$E\}\delta g^2$
				$n \geqq 20$, $N = 136$,		300-n mi squares			
Combination solution									
$l \leqq 8$	$m \leqq 8$	165	90	92	102	253	2	11	152
$l \leqq 10$	$m \leqq 10$	132	119	116	120	253	−3	11	123
$l \leqq 11$	$m \leqq 11$	135	126	134	126	253	8	11	116
$l \leqq 12$	$m \leqq 12$	134	129	138	129	253	9	11	113
$l \leqq 14$	$m \leqq 14$	109	156	166	146	253	10	11	87
$l \leqq 16$	$m \leqq 16$	75	184	186	163	253	2	11	58
				$n \geqq 20$, $N = 136$,		300-n mi squares			
Satellite solution									
$l \leqq 8$	$m \leqq 8$	179	86	98	102	253	12	11	156
$l \leqq 10$	$m \leqq 10$	145	109	110	120	253	1	11	133
$l \leqq 11$	$m \leqq 11$	151	115	126	126	253	11	11	127
$l \leqq 12$	$m \leqq 12$	163	111	128	129	253	17	11	131
$l \leqq 14$	$m \leqq 14$	173	117	150	146	253	33	11	125
Total field		177	118	161	143	253	43	11	124
				$n \geqq 20$, $N = 136$,		300-n mi squares			
M1 solution									
$l \leqq 8$	$m \leqq 8$	168	85	85	102	253	0	11	157
Total field		168	93	101	108	253	7	11	148

from the surface gravity. If both the satellite solution and the surface gravity give "perfect" results for all the terms up to a certain degree, then

$$\langle g_S^2\rangle = \langle g_T g_S\rangle = D$$

and

$$E\{\varepsilon_s^2\} = E\{\varepsilon_t^2\} = 0\,.$$

The comparisons are made for the three fields truncated at different degrees as well as for the total fields.

The combination solution, as may be expected since it contains the data against which the tests are made, gives the best results in that there is good agreement between the three estimates $\langle g_s^2\rangle$, $\langle g_t g_s\rangle$, D of the true value of the contribution to the gravity anomaly from the geopotential coefficients. Also the $E\{\varepsilon_t^2\}$ and $E\{\varepsilon_s^2\}$ are small. The negative value for $E\{\varepsilon_s^2\}$ when $l = 10$ is because the combination solution already contains the surface gravity information. The estimates of the errors of omission $E\{\delta g^2\}$ are still quite large when compared with the estimates of ε_2^s and ε_t^2, indicating that the surface gravity data contain additional information that has not been extracted in this solution.

The results obtained from the satellite solution alone are not in as good agreement with the surface gravity data as the combination solution. For the solution complete to 8,8 there is no significant difference between the two satellite solutions and both contain most of the information in a "correct" 8,8 field. The combination solution at 8,8 is also not significantly different from either satellite solution indicating that the surface gravity does not contribute significantly to the lower order harmonics. At 10,10 the new solution shows a marked improvement in the comparison $\langle (g_s - g_t)^2 \rangle$ and this field also appears to be about as good as can be expected, but beyond about the 11 th order the comparisons for the satellite solution deteriorate and the surface gravity contribution to the combination solution begins to dominate.

Table 6

Summary of comparisons between surface-gravity measurements and gravity anomalies gr_s

Profile		$\langle (g_s - g_t)^2 \rangle$* (mgal²)	$[\sigma(g_t)]^2$ (mgal²)	$[\sigma(g_s)]^2$** (mgal²)
$\varphi = 32.5°$	North Atlantic	84	25	59
NW–SE	North Atlantic	68	25	43
NW–SE	South Atlantic	222	25	197
$\varphi = 0°$	Indian Ocean	80	25	55
$\varphi = -25°$	Indian Ocean	166	25	144
				$[\sigma(g_s)]^2_{av} = 99$ mgal² ≈ 10 m²

* g_t from Talwani and Le Pichon; g_s computed from the combination solution for selected profiles.

** $[\sigma(g_s)]^2 = \langle (g_s - g_t)^2 \rangle - [\sigma(g_t)]^2$.

Further tests with recently acquired and independent gravity data have also been made and some typical results are shown in Table 6 for profiles in the Atlantic and Indian oceans (see also Fig. 5). The results are in good agreement with the earlier indicated estimates.

3.5. Comparison with Astro-geoid Data

Geoid heights obtained from astro-geodetic levelling can also be used for comparison purposes once these profiles are transformed into a geocentric reference system. Such transformations can be established through a comparison of geodetic datum coordinates and coordinates derived from satellite analyses. A summary of the geoid comparisons is given in Table 7 and Fig. 6. The results are again in good agreement with the earlier accuracy estimates.

Table 7

Summary of the comparison between the geoid profiles obtained from the combination solution and the astrogeoids referred to the geocentric system*.

Datum	Profile	$\overline{\Delta h}$	$\sigma^2_{\delta h}$	σ^2_a	$\sigma^2_s = \sigma^2_{\delta h} - \sigma^2_a$
NAD	$\varphi = 35°$	−15	8	1.5	6.5
NAD	$\lambda = 260°$	−16	6	1.5	4.5
AGD	$\varphi = -28.75°$	−12	10	1.5	8.5
AGD	$\lambda = 136.25°$	−12	12	1.5	10.5
IND	$\lambda = 75°$	−36	30	1.5	28.5
EUR	$\lambda = 16°$	−42	6	1.5	4.5
					$(\sigma^2_s)_{av} = 10.5$ m²

* $\overline{\Delta h}$ is the systematic height difference between the profiles, $\sigma^2_{\delta h}$ the variance of the difference between the two profiles, σ^2_a the variation of the astrogeoid heights, and σ^2_S the contribution of the combination solution to $\sigma^2_{\delta h}$.

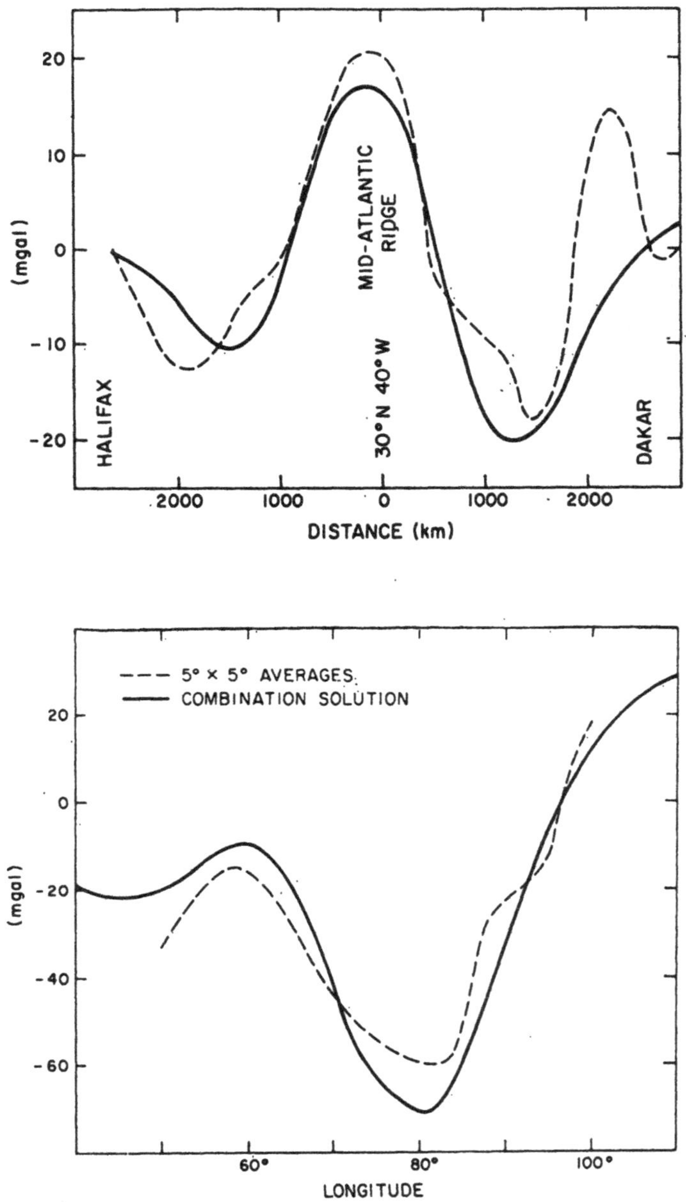

Fig. 5. Comparisons of continuous gravity profiles from shipboard measurements compiled by Talwani and Le Pichon (broken lines) with profiles computed from the combination solution (solid lines). Above: North Atlantic, NW–SE, profile from Halifax to Dakar. Below: Indian Ocean $\varphi = 0°$.

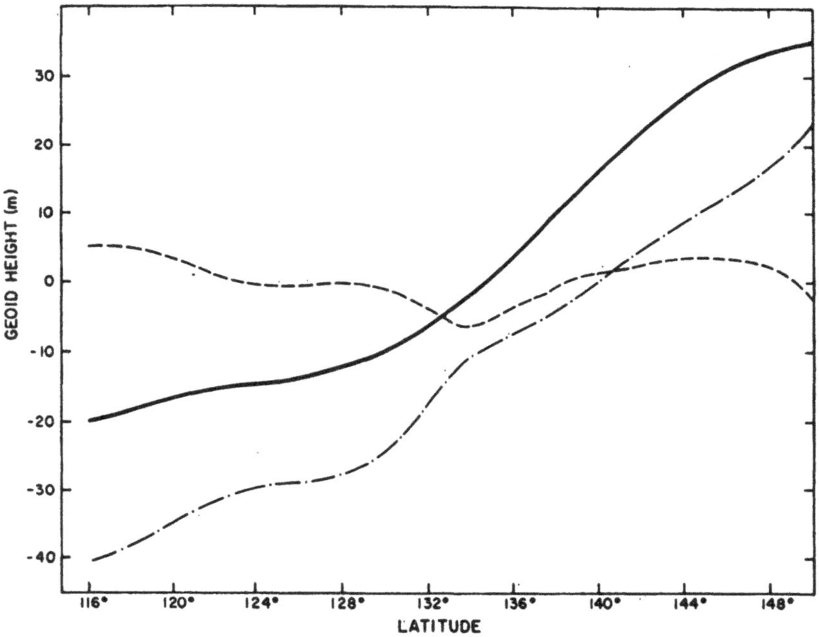

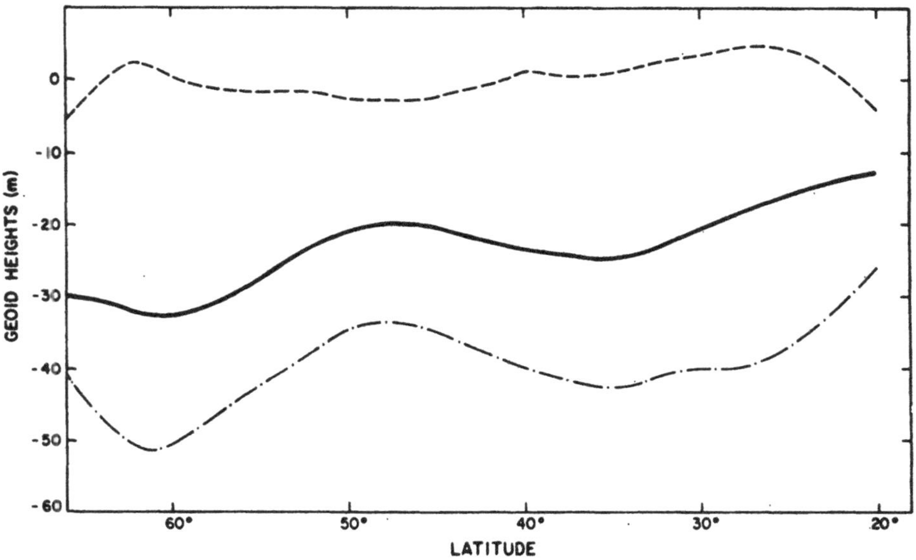

Fig. 6. Comparisons between geoid profiles obtained from the combination solution (solid lines) with profiles obtained from astro-geodetic measurements transformed into the global reference system (chain lines). The difference between the two profiles, after the systematic part has been subtracted, is indicated by the broken line. Top: Australian Geodetic Datum $\varphi = -28.75°$. Bottom: North American Datum $\lambda = 260°$.

4. Conclusions

1. A combination of the four methods of estimating parameters used in this analysis gives better results than any subset of these methods.

2. The geocentric station coordinates of the 12 fundamental Baker-Nunn sites and the laser-optical sites at Haute Provence (8015-7815) and Athens (9091-7816) are determined to an accuracy between 5 and 10 m.

3. The geopotential has been determined complete up to $l = 16$, $m = 16$. Comparisons with surface gravity data indicate that up to 10,10 the satellite solution is about as good as can be expected but some of the higher terms are poorly determined. The terms between 11 and 16 are largely determined from the surface gravity data. The M1 8,8 solution is not improved upon by the new solution truncated at 8,8.

4. Comparisons with independent data sets indicate that the generalized geoid is reliable to about 3 m.

Two important further conclusions based on the experience gained in this solution are:

5. To improve the satellite gravity field determination for terms beyond $l = 11$, more satellites in lower orbits, at different inclinations, and tracked with greater accuracy and uniformity are required.

6. To improve upon the accuracy of the station position determination, more laser data for the dynamic solution and laser-optical simultaneous data for the geometric solution are required.

References

[1] C. A. LUNDQUIST and G. VEIS, Smithson. Astrophys. Obs. Spec. Rep. 200 (3 vols.), (1966).
[2] C. A. LUNDQUIST, Smithson. Astrophys. Obs. Spec. Rep. 264 (1967).
[3] N. MOTTINGER, Space Programs Summary No. 37-60, 2, 77, Jet Propulsion Laboratory, Pasadena, California (1969).
[4] W. M. KAULA, J. Geophys. Res. 71, 5303 (1966).
[5] E. M. GAPOSCHKIN and K. LAMBECK, Smithson. Astrophys. Obs. Spec. Rep. 315 (1970). Also submitted to J. Geophys. Res.
[6] K. LAMBECK, Bull. Géodésique, No. 98 (1970).
[7] K. LAMBECK, Bull. Géodésique, No. 99 (1971).

Space Research XI — Akademie-Verlag, Berlin 1971

THE DETERMINATION OF THE FINE STRUCTURE OF THE EARTH'S GRAVITATIONAL FIELD

L. P. PELLINEN

Central Research Institute of Geodesy, Moscow, USSR

The feasibility of determining the fine structure of the earth's gravitational field by the use of satellites is considered. The estmation of the variances and autovariances of different geopotential derivatives at different altitudes above the earth's surface is deduced from the statistical analysis of gravity. The expected variances of satellite coordinates and rates are calculated for the case of a circular orbit.

1. The Fine Structure of the Gravity Field on the Earth's Surface

The fine structure of the earth's gravitational field is defined as that part represented by spherical harmonics of high degree. Recently all harmonic gravity coefficients up to the eighth degree have been sufficiently well determined from the analysis of satellite orbit perturbations. Therefore we shall treat the effect of harmonics of degrees higher than eighth.

It will be noted that it is hardly reasonable to represent the fine structure of the gravitational field by spherical harmonics because the statistical analysis of gravity anomalies on the earth's surface shows that their distribution is very irregular and the harmonic series which represent them converge very slowly. Pellinen [1] has therefore obtained an empirical formula for high degree variances of gravity anomalies Δg_n^2, if $n \geqq 15$

$$\Delta g_n^2 = 166\, n^{-1.12}\ \text{mgal}^2\,. \tag{1}$$

According to [1] the following simple formula may be used instead of (1) if $4 < n < 100$

$$\Delta g_n^2 = \frac{A}{n} = 120\, n^{-1}\ \text{mgal}^2\,. \tag{2}$$

The estimates obtained from (2) are slightly smaller than those which may be obtained with the aid of Kaula's empirical formula [2] for a normalized geopotential harmonic coefficient

$$\mathrm{O}(\overline{C}_{nk}, \overline{S}_{nk}) = 10^{-5}\, n^{-2}\,.$$

For the whole of the earth's surface variances of residual gravity anomalies caused by geopotential harmonics of degrees higher than the 15th are either 1007 mgal² [1] or 946 mgal² [2]. Yet for areas with different relief the following values were obtained by us:

areas of cenozoic orogenesis both on land and oceans	2860 mgal^2
mountainous areas of moderate altitudes and transition areas	
from land to ocean	995 mgal^2
flat land areas	469 mgal^2
ocean platform areas	201 mgal^2.

It is evident that the estimates differ quite noticeably. High degree variances Δg_n^2 are mainly due to some large gravity anomalies in relatively small areas. They will no doubt worsen the approximation of anomalies by spherical harmonics in relevant areas of the earth.

Thus, after extracting some planetary part out of the earth's gravity field it will be reasonable to use mean values of gravity anomalies for certain standard areas instead of spherical harmonics in representing the residual gravity field. Such area representation is consistently applied by Arnold and Stange (e.g. [3]). Other area geopotential characteristics (quasigeoidal heights, densities of an auxiliary simple layer etc.) may also be used.

2. Methods of Determining the Fine Structure of the Earth's Gravitational Field

Classical methods for determining the earth's figure — astrogeodetic and gravimetric — enable local features of the gravity field for some areas to be obtained directly. These methods will no doubt preserve their importance in future used both singly and, especially, in combination with satellite methods.

Up to now only one of the satellite methods has the same advantage as the classical ones. It is the geometrical method of determining with the aid of a radar altimeter the geoidal figure in the oceans by measuring the height above the water surface of a satellite whose orbit is known [4].

The feasibilities of orbital (and dynamic) methods are more problematical. Nowadays perturbations accumulated during one or more revolution of a satellite around the earth are used. Those perturbations are functions of mean gravity anomalies for standard areas either on the whole of the earth's surface or on a considerable part of it. To definite the effect of each standard area we have to develop a uniform and dense world net of tracking stations. The lack of stations may partly be compensated by a great number of observed satellites whose trajectories pass over the areas under investigation in various directions.

In principle a more definite result may be obtained by analysing the orbit perturbations within arcs shorter than one revolution of a satellite around the earth. Such a method was successfully applied investigating the details of the moon's gravitational field by using measured range rates of the "lunar orbiter" spacecrafts [5]. But it was found that this method was less effective in studying the earth's gravity field because lunar satellites may be closer to the planet surface than earth satellites and lunar harmonic gravitational coefficients greater than terrestrial. Sheffield [5] has estimated this by using some typical models of disturbing topographical masses. We are able now to get a more reasonable estimate of perturbations of the earth's satellite orbits by using the results of our statistical analysis of surface gravity anomalies (see § 1).

3. Statistical Estimates of the Disturbing Potential and its Derivatives at Different Heights above the Earth's Surface

Proceeding from Stokes' relations between various characteristics of the earth's gravitational field let us write the formulae for degree variances of the disturbing potential and its derivatives on a sphere with a radius ϱ which is larger than the mean earth's radius R (see [1]):

the disturbing potential

$$T_n^2(\varrho) = \frac{R^2}{(n-1)^2} \cdot \Delta g_n^2 \, e^{-(n+1)\alpha}, \tag{3}$$

the radial derivative

$$T_{\varrho n}^2(\varrho) = \left(\frac{n+1}{n-1}\right) \cdot \Delta g_n^2 \, e^{-(n+2)\alpha}, \tag{4}$$

the horizontal derivative

$$T_{tn}^2(\varrho) = \frac{n(n+1)}{2(n-1)^2} \cdot \Delta g_n^2 \, e^{-(n+2)\alpha}, \tag{5}$$

where $\alpha = 2 \ln (\varrho/R)$.

The following relation exists between degree variances of any function v and its variance $\varepsilon(v)$:

$$\varepsilon(v) = \sum_{n=1}^{\infty} v_n^2 . \tag{6}$$

The next formula may be used in estimating the effect of high degree harmonics

$$\varepsilon_{\bar{n}}(v) = \sum_{\bar{n}+1}^{\infty} v_n^2 \approx \int_{\bar{n}}^{\infty} v_n^2 \, dn . \tag{7}$$

Inserting in (7) successively (3), (4) and (5) and using (2) we get, after integration:

$$\varepsilon_n(T) = R^2 A \, e^{-\alpha} \left[\frac{e^{-n\alpha}}{n-1} + (1-\alpha) \, e^{-\alpha} \, \mathrm{Ei}\,(-n\alpha+\alpha) - \mathrm{Ei}\,(-n\alpha)\right], \tag{8}$$

$$\varepsilon_n(T_\varrho) = A \, e^{-2\alpha} \left[\frac{4 \, e^{-n\alpha}}{n-1} + 4\alpha \, e^{-2\alpha} \mathrm{Ei}\,(-n\alpha+\alpha) - \mathrm{Ei}\,(-n\alpha)\right], \tag{9}$$

$$\varepsilon_n(T_t) = \frac{A \, e^{-2\alpha}}{2} \left[\frac{e^{-n\alpha}}{n-2} + (\alpha-2) \, e^{-2\alpha} \mathrm{Ei}\,(-n\alpha+\alpha)\right], \tag{10}$$

$$\mathrm{Ei}\,(-t) = -\int_t^{\infty} \frac{e^{-x}}{x} \, dx .$$

Let us consider a case of a circular satellite orbit and find the autocovariance functions of radial, along-track and cross-track accelerations of a satellite. According to Kaula [6] we have

$$C(T_\varrho, \psi) = \sum_{\bar{n}}^{\infty} T_{\varrho n}^2 \, P_n(\cos\psi) \tag{11}$$

$$C(T_u, \psi) = \sum_{\bar{n}}^{\infty} T_{tn}^2 \left[P_n(\cos\psi) - \frac{P_{n,2}(\cos\psi)}{n(n+1)} \right] \quad (12)$$

$$C(T_n, \psi) = \sum_{\bar{n}}^{\infty} T_{tn}^2 \left[P_{n-1}(\cos\psi) + \frac{P_{n-1,2}(\cos\psi)}{n(n+1)} \right] \quad (13)$$

where $\bar{n}$ denotes the lowest degree of harmonics representing the fine structure of the gravity field and ψ is the spherical distance. Figs. 1 and 2 give the results of the calculation based on the above formulae for $\bar{n} = 8$. To find $T_{\varrho n}^2$ and T_{tn}^2

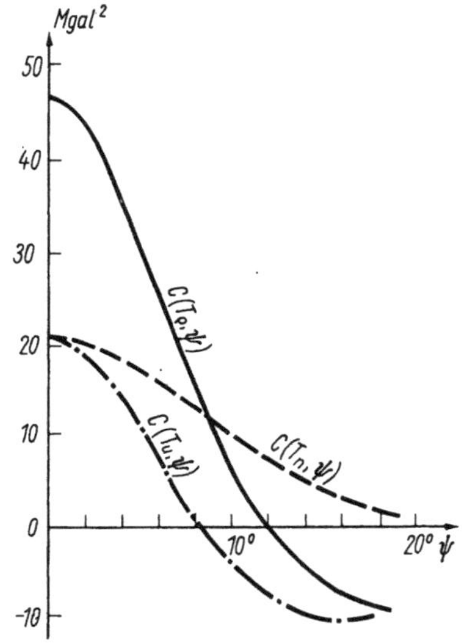

Fig. 1. Autocovariances of the derivatives of the disturbing potential. $\alpha = 0.1$ ($H = 327$ km), $\bar{n} = 8$.

we used the formulae (2), (4) and (5). Having obtained the estimates according to formulae (9) and (10) we restricted in all cases the summation up to $n = 30$. The most interesting results are those for an orbit at a height of 327 km ($\alpha = 0.1$). Radial accelerations are greater than both the along-track and cross-track ones.

4. The Estimation of Perturbations of Satellite Rate and Position

To give a tentative prediction of satellite perturbations for short arcs we may neglect secondary effects due to the earth's rotation and the curvature of satellite trajectories. The change of any component v_x of satellite rate in a direction x along an arc ψ_0 is

$$v_x = \frac{1}{\bar{n}} \int_0^{\psi_0} T_x \, d\psi \, . \quad (14)$$

The related perturbation of the x coordinate is

$$\Delta x = \frac{1}{\bar{n}^2} \int_0^{\psi_0} T_x (\psi_0 - \psi) \, d\psi \, . \tag{15}$$

In (14) and (15) $\bar{n}$ denotes the mean motion.

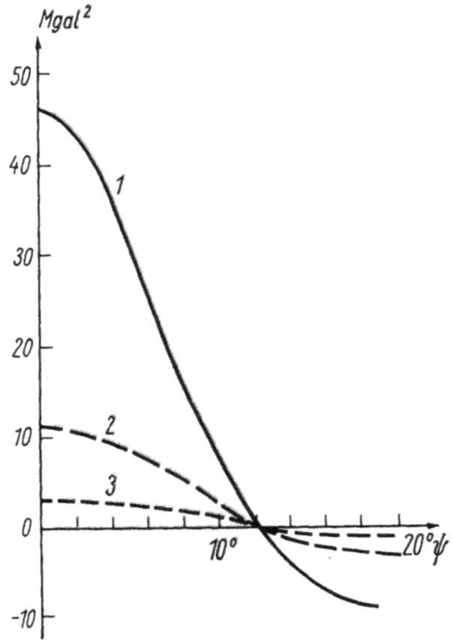

Fig. 2. Autocovariances of the vertical derivative of the disturbing potential, $\bar{n} = 8$.
1. $\alpha = 0.1$ ($H =$ 327 km),
2. $\alpha = 0.2$ ($H =$ 670 km),
3. $\alpha = 0.3$ ($H =$ 1030 km).

Variances of v_x and Δx are

$$\varepsilon_x(v) = \frac{2}{\bar{n}^2} \int_0^{\psi_0} (\psi_0 - \psi) \, C(T_x, \psi) \, d\psi \, , \tag{16}$$

$$\varepsilon(\Delta x) = \frac{1}{3 \, \bar{n}^4} \int_0^{\psi_0} (2 \, \psi_0^3 - 3 \, \psi \, \psi_0^2 + \psi^3) \, C(T_x, \psi) \, d\psi \, . \tag{17}$$

Variances of satellite rates and geocentric distances were computed for ψ up to 16° based on data in Fig. 1 and formulae (16) and (17). Then rms perturbations were determined which are given in Table 1.

Table 1

ψ (°)	2	4	6	8	10	12	14	16
$m(V_\varrho)$ (cm s^{-1})	0.2	0.4	0.6	0.8	0.9	1.1	1.2	1.3
$m(\delta_\varrho)$ (m)	0.04	0.13	0.3	0.5	0.7	1.0	1.3	1.5

According to our data in § 1 the perturbations over the most anomalous areas are about twice, and over the extensive ocean platform areas about half, those given in Table 1. Perturbations of horizontal coordinates and rates are approximately 1.5 times smaller than $\Delta\varrho$ and v_p.

As may be seen, position perturbations do not exceed 0.5 m if $\psi < 10°$. They are much smaller than errors of modern angular and linear satellite tracking equipments. Rate perturbations are greater than the errors of the most precise Doppler systems used in tracking deep space probes [7], but those perturbations are smaller than errors of the equipment used by tracking close satellites (for instance, the accuracy of the Doppler system Tranet is ± 1–$2\,\mathrm{cm\,s^{-1}}$ [8]).

These results are disappointing as regards the feasibility of determining the fine structure of the earth's gravitational field; but they have also a positive aspect. They show we should not be afraid of errors caused by local gravity perturbations within short orbital arcs in solving the geometrical problems of satellite geodesy with the aid of dynamic or semi-dynamic methods. These errors are small even for a low orbit at an altitude of 327 km ($\alpha = 0.1$). Fig. 2 shows that they are much smaller still at a height of 1030 km ($\alpha = 0.3$) which is usual for geodetic satellites.

Acknowledgment

I thank G. V. Demjanov for his help in computations made for this paper.

References

[1] L. P. Pellinen, Studia geophys. et geod. **14**, 168 (1970).
[2] W. M. Kaula, Theory of Satellite Geodesy. Applications of Satellites to Geodesy, Blaisdell Publ. Co, Waltham. Mass., Toronto, London 1966.
[3] K. Arnold, Veröff. Geod. Inst. Potsdam **27**, 55 (1965).
[4] J. A. Greenwood et al., Remote. Sens. Environ **1**, 59 (1969).
[5] Ch. Sheffield, Ann. N.Y. Acad. Sci. **163**, 40 (1969).
[6] W. M. Kaula, Army Map Service USA Techn. Rep. No. 24 (1959).
[7] W. G. Melbourne, in: Dynamics of Satellites, Springer-Verlag, Berlin 1970 (p. 63).
[8] L. H. Solomon, Smithson. Astrophys. Obs. Spec. Rep. No. 264, 129 (1967).

Space Research XI — Akademie-Verlag, Berlin 1971

DETERMINATION D'UNE BASE GEODESIQUE A LONGUE DISTANCE

A. Cazenave[a] and O. Dargnies[b]

[a] Centre National d'Etudes Spatiales, Paris, France
[b] Observatoire de Meudon, Meudon, France

L'observation simultanée d'un satellite à partir de stations faisant soit des mesures de distance par laser, soit des photographies du satellite sur fond d'étoiles, permet de déterminer les vecteurs joignant ces différentes stations par une méthode purement géométrique.
Dans le cadre de la liaison géodésique Europe–Afrique il a été nécessaire de définir de façon précise un triangle géodésique de base en Europe par emploi simultané de télémètrie laser et d'observations optiques sur satellites munis de prismes réflecteurs cataphotes. Ce sont les résultats relatifs au triangle San Fernando (Espagne), Haute Provence (France), Dionysos (Grèce), que nous présentons ici.

1. Presentation

Nous présentons ici la réduction d'observations photographiques et de mesures de distance par laser, effectuées sur des satellites munis de réflecteurs laser depuis le 1er avril 1968 jusqu' à la fin de l'année 1968 à partir des stations suivantes:

San Fernando (Espagne)	Station optique et station laser
Dionysos (Grèce)	Station optique
Haute Provence (France)	Station optique et station laser
Nice (France)	Station optique
Meudon (France)	Station optique

L'ensemble des points sub-satellite utilisés dans ce traitement a été reporté sur la figure 1.

2. Les Données et leur Prétraitement

2.1.

Une mesure de distance se présente sous la forme d'un temps d'aller et retour station–satellite–station du faisceau laser. L'instant associé à la mesure est $t + \frac{1}{2}\,(t' - t)$ si t est l'instant de départ du faisceau et t' l'instant d'arrivée. La distance ayant pour valeur $c \cdot \frac{1}{2}(t' - t)$, où c est la vitesse de la lumière.
Une mesure photographique se présente sous la forme d'un cliché à partir duquel les coordonnées équatoriales (α, δ) du satellite sont déduites des coordonnées équatoriales (α, δ) des étoiles, ces dernières étant rapportées à l'équinoxe

moyen 1950,0. La direction station–satellite correspondante est la direction géométrique du vecteur joignant le satellite à l'instant de depart de la lumière et la position de la station à ce même instant.

L'élimination des mauvaises mesures est effectuée par un traitement en correction différentielle, les mauvaises mesures de distance ayant déjà été éliminées par lissage polynomial en fonction du temps.

2.2. Calcul des Distances Synthétiques

Mesures de distances et mesures photographiques ne sont pas en réalité exactement simultanées. Au cours d'un passage un ou plusieurs clichés peuvent être pris à partir d'une station optique. Si un tir au laser a lieu au cours de ce même passage à partir d'une station laser, la période des mesures laser qui s'étend en général sur plusieurs minutes encadre le plus souvent la période des mesures photos, cette dernière s'étendant seulement sur quelques secondes. Un éclair n'est jamais simultané avec un écho laser.

On se propose donc de calculer des distances synthétiques aux instants des éclairs (plus précisément aux instants de départ de la lumière du satellite). Pour cela, on détermine la fonction distance satellite–station en fonction du temps; on utilise un programme de correction différentielle qui ajuste les paramètres d'orbite du passage considéré à l'aide des vraies mesures laser, avec un modèle de potentiel simplifié. La position du satellite étant calculée à tout instant, on en déduit facilement la distance satellite–station laser à l'instant de la mesure photographique. Cette distance théorique étant obtenue, il est nécessaire de lui affecter une correction due au fait que le modèle de forces utilisé est imparfait; les vraies mesures laser du passage permettent de tracer la courbe des résidus (distance calculée moins distance mesurée), en fonction du temps. Il suffit alors d'interpoler pour déterminer le résidu (valeur calculée moins valeur observée, $c - o$) à l'instant de la mesure photographique, à affecter à la distance calculée; on a alors la distance synthétique définitive:

$$d = d_{\text{théor.}} - (c - o) .$$

Il arrive que l'instant de la mesure photographique soit extérieur au passage laser. Il faut alors extrapoler la courbe des résidus, ce qui peut entraîner une incertitude importante sur la valeur de la correction $c - o$. On affecte à la distance synthétique définitive une erreur moyenne quadratique δd rendant compte de la précision avec laquelle on estime avoir fait le calcul.

2.3.

Les coordonnées équatoriales (α, δ) du satellite sont transformées en cosinus directeurs dans un repère choisi pour référence et dans lequel sont effectués tous les calculs. Lé repère $\sum$ choisi est défini comme suit:

axe **Z** dirigé vers le pôle moyen 1900–1905,

axe **X** dans le plan du méridien moyen de Greenwich défini par le BIH,

axe **Y** dirigé vers l'est, perpendiculaire au plan (**X**, **Z**).

Les coordonnées équatoriales (α, δ) du satellite rapportées à l'équinoxe moyen 1950,0 sont corrigées de l'aberration annuelle puis transformées en cosinus directeurs dans le repère céleste moyen 1950,0; les corrections de précession de 1950 à l'époque t de la mesure puis de nutation sont effectuées. On se situe alors dans le repère céleste instantanné de la date. On effectue une rotation de temps sidéral vrai et enfin une correction de mouvement du pòle de t à 1900–1905.

Les (α, δ) sont affectés d'une erreur de l'ordre de 1'' à 2'' dues à la qualité des clichés et à la précision théorique des caméras. La correction de réfraction parallactique a été effectuée [1, 2].

3. Methode de Calcul

On définit un nouveau repère $\sum'$ parallèle à $\sum$ centré en l'une des stations optique. A chaque instant, la position du satellite est fixée par ses trois coordonnées dans $\sum'$. Si à un instant donné, on mesure à partir d'une station S_i, l'une des coordonnées topocentriques du satellite (distance ou direction), on obtient une relation liant les coordonnées de la station et celles du satellite. Si on mesure simultanément n grandeurs indépendantes ($n \geqq 4$) à partir de plusieurs stations, on peut alors éliminer les trois coordonnées du satellite entre les n relations obtenues et obtenir $n - 3$ relations indépendantes où ne figurent plus que les coordonnées des stations.

Ces relations sont linéarisées au voisinage d'une solution approchée connue (coordonnées du système Europe 50) puis traitées par la méthode des moindres carrés, avec d'autres relations obtenues de la même façon par l'observation simultanée d'autres positions du satellite à partir de différentes stations. Les solutions obtenues sont les composantes des vecteurs écarts qu'il faut ajouter aux vecteurs stations E 50 pour obtenir les vecteurs vrais.

Une mesure de distance fournit une relation indépendante alors qu'une photo, corrsepondant à deux informations α, δ, fournit deux relations indépendantes. Ainsi si N_P stations optiques font des mesures angulaires à un instant donné et si N_L stations laser font des mesures de distance à ce même instant, le nombre de relations indépendantes est $2\,N_P + N_L - 3$.

Une position S du satellite est observée par plusieurs stations photo P_i (la mesure de direction correspondante étant $\mathbf{U}_i$), ou par plusieurs stations laser L_j (la mesure de distance correspondante étant D_j) dont on connait les positions approchées P_{i0}, L_{j0}. L'élimination de S entre les relations d'observations permet d'écrire une ou plusieurs relations du type

$$f(\mathbf{P}_i, \mathbf{L}_j, \mathbf{U}_i, \mathrm{D}_j) = R \qquad i = 1, \ldots, N_P$$
$$j = 1, \ldots, N_L$$

où $\mathbf{P}_i$ est le vecteur station optique P_i, $\mathbf{L}_j$ le vecteur station laser L_j, $\mathbf{U}_i$ le vecteur unitaire de la direction observée en P_i. D_j est la distance mesurée en L_j. R est le résidu qui doit être nul pour les valeurs vraies. Chaque relation f est linéarisée de manière à faire apparaître les vecteurs écarts $\mathbf{dP}_i$, $\mathbf{dL}_j$.

3.1. Etablissement des relations $\bar{f}$

Nous disposons toujours d'au moins une relation photo qui permet d'orienter la figure 1.

3.1.1. Mesures de direction seules

Il n'y a pas de mesures de distances et l'échelle de la figure ne peut être déterminée. Les relations f expriment alors que les droites issues des stations et parallèles aux directions mesurées se coupent en un point que l'on ne cherche pas à déterminer.

3.1.2. Mesures de direction et mesures de distance

Si on dispose simultanément d'au moins une mesure laser et d'une mesure photo, on peut déterminer une position approchée S du satellite. On écrit alors que: pour les autres stations laser L_j, $\mathbf{SL}_j$ est égal à la distance mesurée D_j et que pour les autres stations P_i, le vecteur $\mathbf{SP}_i$ est parallèle à la direction mesurée $\mathbf{U}_i$. On obtient ainsi une méthode systématique permettant de traiter un nombre quelconque de mesures laser et de mesures photographiques [3, 4].

4. Resultats

La station San Fernando Optique (9004) a été choisie pour origine. Toutes les coordonnées initiales des stations sont les coordonnées Europe 50 les plus récentes fournies par l'Institut Géographique National.

Dans les différents cas traités, les petits vecteurs San Fernando optique–San Fernando laser, ainsi que Haute Provence optique–Haute Provence laser, ont été figés avec un poids très fort.

$$\overrightarrow{\text{SFO–SFL}} \begin{bmatrix} +20\text{ m} \\ -22\text{ m} \\ -31\text{ m} \end{bmatrix}, \qquad \overrightarrow{\text{HPO–HPL}} \begin{bmatrix} +44\text{ m} \\ -16\text{ m} \\ -42\text{ m} \end{bmatrix}.$$

Sur le tableau 1 ont été reportées les coordonnées initiales des stations avec leur code et leur numéro. Les Tableaux 2 et 3 présentent les résultats provenant des différents traitements effectués (traitements partiels puis traitement

Tableau 1
Coordonnées Initiales des Stations (Europe 50)

	Xm	Ym	Zm
SFO (9004)	5105681,0	−555102,9	3769799,3
HPO (8015)	4578413,0	458091,0	4403312,0
NIO (8019)	4579554,2	586729,1	4386535,6
DYO (9091)	4595251,6	2039577,3	3912795,0
MDO (8030)	4205717,6	163840,9	4776660,7
SFL (7804)	5105701,2	−555125,7	3768768,3
HPL (7815)	4578457,2	458075,2	4403270,5

global). Différents lots de mesures que nous avons traités ont fourni au total 1407 relations linéairement indépendantes, mais non statistiquement indépendantes, car tous les eclairs d'un même cliché ont une valeur géométrique équivalente:

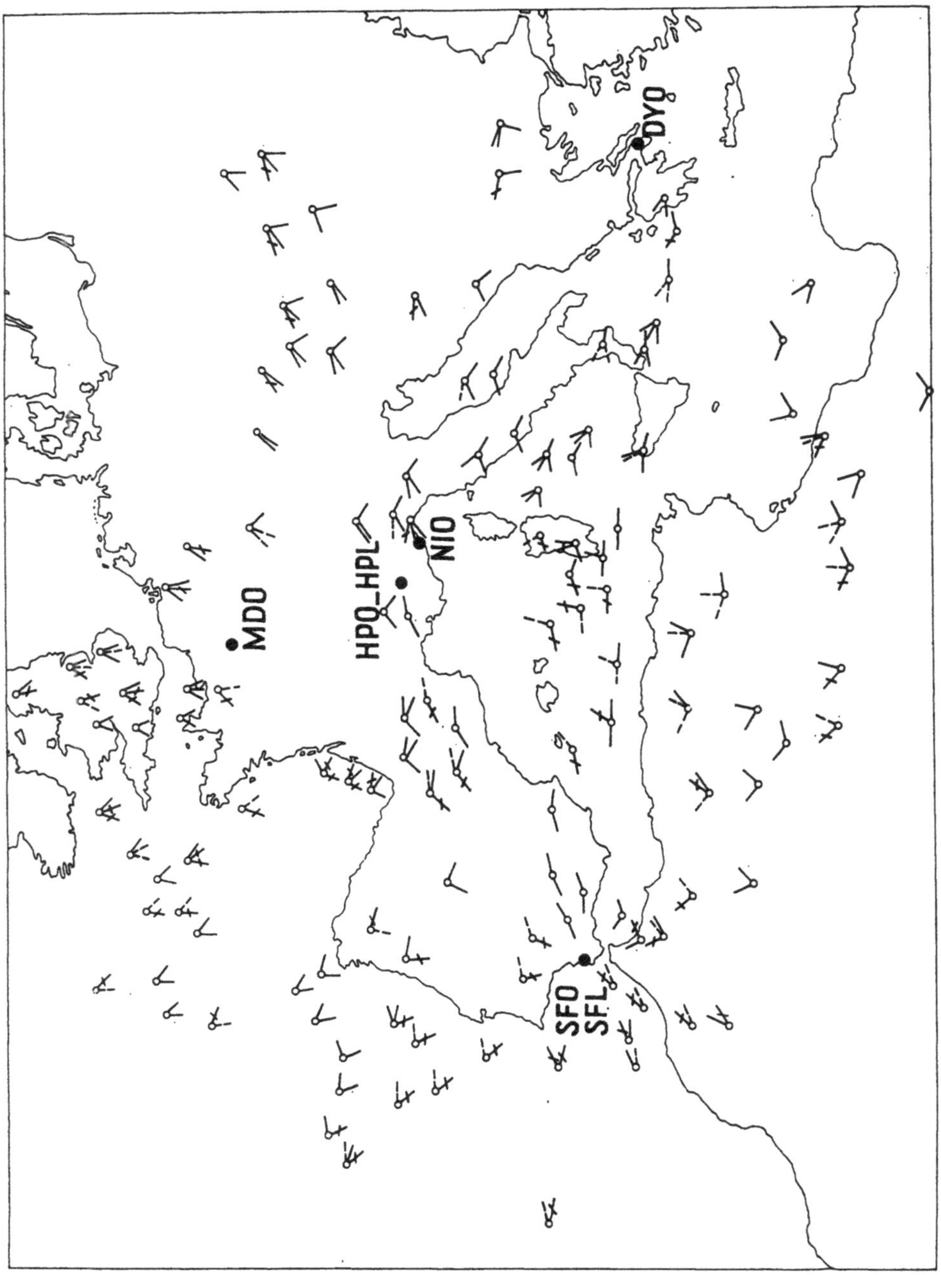

Fig. 1. Répartition géographique des points sub-satellite utilisés.
● Station; ○ point sub-satellite; ○— visée optique; ○- - visée laser; ○-/- visée optique + visée laser.

Tableau 2
Vecteur SFO–HPL (9004–7815)

	341 relations Mesures de juin et juillet 1968		846 relations Mesures d'août à décembre 1968		1407 relations Mesures de toute l'année 1968 Traitement global
	HPO-NIO libre	**HPO-NIO** figé	**HPO-NIO** libre	**HPO-NIO** figé	**HPO-NIO** libre
dX	9,6 (1,3)	8,8 (1,2)	8,9 (1,1)	5,9 (0,9)	8,3 (0,6)
dY	14,0 (1,1)	13,5 (1,1)	20,1 (1,6)	18,1 (1,3)	15,9 (0,7)
dZ	2,3 (1,3)	4,2 (1,2)	4,2 (1,7)	11,3 (1,4)	4,5 (0,7)
$d\varrho$	9,2 (1,4)	9,0 (1,4)	14,0 (3,0)	17,1 (2,5)	11,2 (1,0)
du 10^5	1,02 (0,12)	0,95 (0,11)	1,12 (0,13)	0,98 (0,10)	0,98 (0,06)
dv 10^5	0,52 (0,09)	0,50 (0,08)	0,71 (0,10)	0,36 (0,06)	0,55 (0,04)
dw 10^5	0,01 (0,11)	−0,01 (0,10)	−0,20 (0,14)	0,23 (0,11)	−0,07 (0,05)
x	−527214,2	−527215,0	−527214,9	−527217,9	−527215,5
y	1013192,1	1013191,6	1013198,1	1013196,2	1013194,0
z	633473,0	633474,9	633474,9	633482,0	633475,2
X	4578466,8	4578466,0	4578466,1	4578463,1	4578465,5
Y	458089,2	458088,7	458095,3	458093,3	458091,1
Z	4403272,3	4403274,2	4403274,2	4404281,3	4403274,5
$\frac{d\varrho}{\varrho}$	$+7,0\times10^{-6}$	$+6,8\times10^{-6}$	$+10,7\times10^{-6}$	$+13\times10^{-6}$	$+8,5\times10^{-6}$

Tableau 3

	Vecteur **SFO-HPL** (9004–7815)	Vecteur **SFO-NIO** (9004–8019)	Vecteur **SFO-DYO** (9004–9091)	Vecteur **HPL-DYO** (7815–9091)
dX	8,3 (0,6)	8,7 (1,0)	2,0 (2,2)	−6,3
dY	15,9 (0,7)	15,7 (0,9)	10,8 (2,3)	−5,1
dZ	4,5 (0,7)	13,2 (1,0)	−5,1 (2,7)	−9,6
dp	11,2 (1)	15,4 (1,1)	9,9 (3,2)	−2,08 (3,7)
du 10^5	0,98 (0,06)	1,03 (0,06)	−0,29 (0,03)	
dv 10^5	0,55 (0,04)	0,22 (0,04)	0,04 (0,00)	
dw 10^5	−0,07 (0,05)	0,46 (0,05)	−0,21 (0,03)	
x	−527215,5	−526118,1	−510427,4	+ 16788,1
y	1013194,0	1141847,7	2594691,0	+1581497,0
z	633475,2	616749,5	142990,6	490484,6
X (E.50)	4578465,5	4579562,9	4595253,6	
Y (E.50)	458091,1	586744,8	2039588,1	
Z (E.50)	4403274,5	4386548,8	3912789,9	
$\frac{\partial\varrho}{\varrho}$	$+8,5\times10^{-6}$	$+10,9\times10^{-6}$	$+3,7\times10^{-6}$	

toutes ces relations ont été mélangées lors du traitement global. Tableaux 2 et 3 de résultats comprend: les composantes du vecteur écart dX, dY, dZ (en mètres), l'écart sur la corde $d\varrho$ (en mètres), les ecarts sur les cosinus directeurs $du\times10^5$, $dv\times10^5$, $dw\times10^5$, les composantes finales du vecteur dans le systeme Europe 50 x, y, z (en mètres), les coordonnées finales de la station–non origine–dans le systeme E 50 X, Y, Z (en mètres), et enfin le facteur d'échelle $(d\varrho)/\varrho$ (tous les sigmas sont entre parenthèses). Au cours des différents traitements, le vecteur **HPO–NIO** (8015–8019) a été figé puis libéré.

5. Conclusion

Le but de la campagne, c'est-à-dire la détermination d'un triangle de base par géodésie géométrique, avec une précision de l'ordre de 10^{-6}, a été atteint. Une extension à l'ensemble des stations du réseau européen qui ont observé GEOS-B à la même époque est en cours de traitement et permettra de tirer des résultats concernant l'ensemble du réseau. Certes il est toujours possible que des erreurs systématiques demeurent; des tests indépendants ont cependant été effectués: en ce qui concerne les mesures, l'expérience „HOPLASI" réalisée fin 1969 et qui consistait à effectuer des tirs simultanés de deux stations laser implantées au même site [5], a montré que l'erreur systématique de la mesure laser était inférieure au mètre; des tirs terrestres à grande distance (150 km) entre deux stations ont confirmé les résultats.

En le qui concerne les résultats obtenus pour les positions de stations, l'accord avec des déterminations dynamiques utilisant des mesures Doppler et/ou laser, est excellent [6–9]. Ce qui demeure, en tout cas, c'est la répétabilité de l'expérience: la cohérence externe entre des résultats provenant de lots de mesures indépendants est très significative.

Il semble donc établi qu'on puisse envisager de dépasser ce stade et s'orienter vers un programme à buts géophysiques, specialement si des stations plus précises sont construites et si des satellites hauts munis de réflecteurs laser sont lancés. L'expérience Isagex constituera une première étape de ce nouveau type de recherches.

Bibliographie

[1] F. Barlier, Bull. Astron. **25**, 325 (1965).
[2] R. Bivas, Bull. Astron. [3e série] **3**, 517 (1968).
[3] M. Bivas, Méthode générale de géodésie géométrique, Symposium sur la triangulation spatiale Européenne. Paris, Février 1969.
[4] G. Balmino, Thèse de 3e cycle (Octobre 1969) non publiée.
[5] A. Roumiguier et J. Gaignebet, Doc. CNES, AR/AC/0.377/CB/MT.
[6] M. Lefèbvre, dans: Dynamics of Satellites, Springer Verlag, Berlin 1970 (p. 162).
[7] J. P. Chassaing, dans: Dynamics of Satellites, Springer Verlag, Berlin 1970 (p. 154).
[8] A. Roumiguier et J. P. Chassaing, Document CNES 9.481/CB/MT.
[9] M. Lefèbvre et J. P. Chassaing, Campagne Diadème — Résultats obtenus en géodésie semi-dynamique à l'aide de mesures Doppler, Symposium sur la triangulation spatiale européenne, Paris 1969.

Space Research XI — Akademie-Verlag, Berlin 1971

NASA STADAN, SPEOPT AND LASER TRACKING STATION POSITIONS DERIVED FROM GEOS 1 AND 2 OBSERVATIONS

J. G. Marsh[a], B. C. Douglas[b] and C. F. Martin[b]

[a] NASA Goddard Space Flight Center, Greenbelt, Md, USA
[b] Wolf Research and Development Corporation, Riverdale, Md, USA

Coordinates for 22 NASA Stadan and Speopt optical and two NASA laser tracking sites have been dynamically estimated from GEOS 1 and 2 data. To strengthen the solutions, Smithsonian Astrophysical Observatory (SAO) Baker-Nunn optical data were also used with the SAO station coordinates held fixed at their 1969 C6 values. Multiple 2-day arcs were used to minimize model error effects.

An analysis of model error effects and comparison of results with those of SAO suggest an uncertainty of perhaps 5 m north–south (latitude) and east–west (longitude) for most stations in North America. The height of the NAD (North American Datum) sites, and latitude, longitude, and height elsewhere, are uncertain by 10–15 m in most cases.

1. Introduction

The GEOS 1 and 2 satellites, extensively observed by a wide variety of tracking systems with good geographic coverage, provide an unusual opportunity for satellite geodesy. Literally tens of thousands of precision-reduced optical observations in the period 1966–1969 are available, along with many thousands of passes of laser and electronic data. Initial analyses were designed to provide accurate coordinates for NASA optical and laser sites.

The GEOS 1 and 2 satellite orbital specifications are listed in Table 1. The orbits are complementary to one another, in that combined GEOS 1 and 2 solutions provide excellent geometry for station recovery. Resonance is

Table 1
Orbital elements of GEOS 1 and 2

	GEOS 1	GEOS 2
Epoch	2 January 1966	28 April 1968
Apogee height	2273 km	1569 km
Perigee height	1116 km	1077 km
Eccentricity	0.07	0.03
Inclination	59.4°	105.8°
Anomalistic period	120.3 min	112.1 min

a problem for the GEOS satellites, but can be dealt with either by adjusting resonant geopotential coefficients [1] or by using multiple short orbital arcs.

Our approach to the station estimation problem was dynamical, using Cowell's method [2] to solve the equations of motion. The orbital solutions included an equal amount of SAO Baker-Nunn data in addition to the Stadan–Speopt Minitrack Optical Tracking System (MOTS) observations. These served greatly to strengthen the solutions, particularly for those stations with limited data.

The Baker-Nunn station positions were held fixed at their 1969 [3] values referred to the C6 Earth. Thus the SAO 1969 gravity [3] model was used for all analyses. This model appears to be the best available. However, unmodeled long-period pertubations approaching 50 m (due to resonance) and additional short periodic errors exist.

To estimate the errors of our positions, we analyzed the effects of gravity model error, SAO station position error, and error in *GM*. The position uncertainty produced by the assumed gravity model error was generally 5–15 m. Other effects, including data noise, account for 2–3 m position error. These good results, in some cases obtained with GEOS 1 data alone, are the direct result of the high quality data and good coverage afforded by the satellite-borne flashing lights.

2. Selection of Optical Data

Fig. 1 shows the geographic distribution of the SAO Baker-Nunn and Stadan/Speopt sites. Note the concentration of the Stadan/Speopt sites in North America, in contrast to the worldwide distribution of SAO positions.

In order to obtain suitable geometry at each observing station that minimizes model error effects passes on all sides of the station and in opposite directions are necessary. Unfortunately for the case of GEOS 2 data this latter requirement could not be met because the data were taken only on south–north

Table 2
Number of Optical Observations per station used in Dynamical Estimation

Station	Station number	No. of observations	Station	Station number	No. of observations
Blossom Pt., Md	1021	650	Tananarive, Madagascar	1043	339
Ft. Myers, Fla	1022	680	Edinburg, Texas	7036	1109
Woomera, Australia	1024	152	Columbia, Mo.	7037	1540
Santiago, Chile	1028	264	Bermuda	7039	448
Mojave, Cal.	1030	932	San Juan, P. R.	7040	475
Johannesburg, S. Africa	1031	623	Greenbelt, Md.	7043	158
St. Johns, Newfoundland	1032	179	Denver, Col.	7045	573
E. Grand Forks, Minn.	1034	542	Jupiter, Fla.	7072	516
Winkfield, UK	1035	400	Sudbury, Ontario	7075	699
Rosman, N.C.	1037, 1042	1015	Kingston, Jamaica	7076	388
Orroral, Australia	1038	482			

passes. Fig. 2 shows the geometry of the passes for an ideal case, the Edinburg, Texas, site. The length of the line indicates the pass length. Coverage of the extent shown in Fig. 2 was not available in all cases, but was sufficient in nearly every case to obtain an accurate position.

Table 2 presents the number of observations used for each site. 140 two-day arcs containing approximately 60000 observations were examined; about

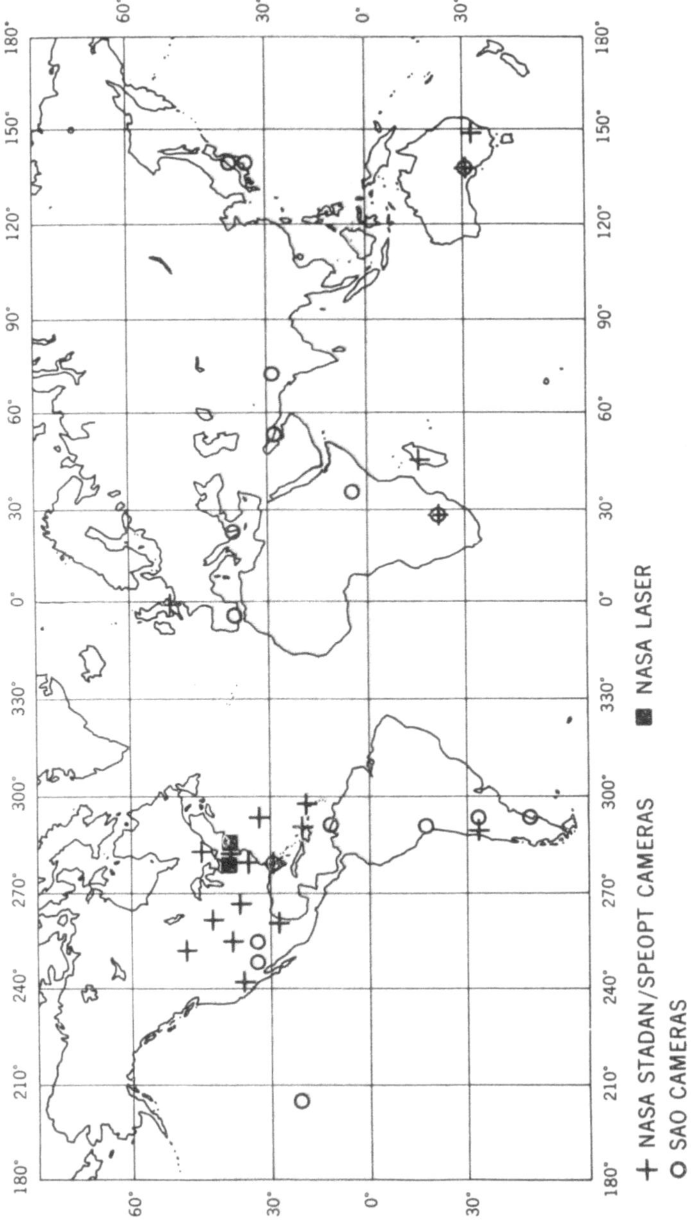

Fig. 1. Tracking station positions.

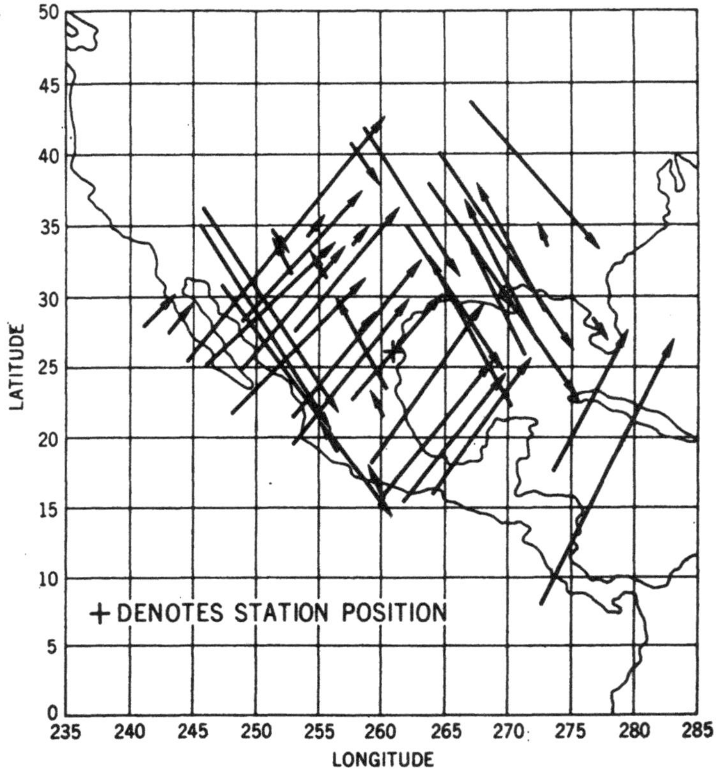

Fig. 2. Subsatellite plot of GEOS 1 and 2 passes, Edinburg, Texas.

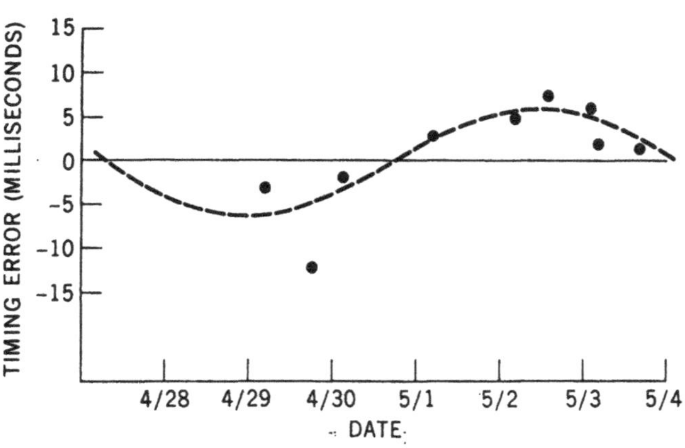

Fig. 3. Apparent timing errors for Rosman passes of range data based on GEOS 2 orbital arc, 28 April–4 May 1968.

13000 Stadan/Speopt observations and an equal number of SAO Baker-Nunn points in 60 two-day arcs were used in the determination.

The SAO 1969 [3, 4] gravity model and station coordinates were adopted in the work. An evaluation [5] of gravity models has demonstrated the general superiority of this model although significant errors still exist. Fig. 3 shows *apparent* timing errors for range data from the Rosman, North Carolina, Goddard Range and Range Rate (GRARR) site computed from a $5\frac{1}{2}$ day GEOS 2 optical reference orbit. It is seen that the satellite alternately leads and lags in its orbit by up to 6 or more milliseconds. The 6^{+} day period of this oscillation is of course the same as the beat period of GEOS 2 and is thus a result of inadequately modeled resonance. However, the choice of two-day arcs effectively eliminated unmodeled resonance as a significant factor: good solutions were obtained by the use of many of these relatively short arcs.

3. Results for Optical Stations

Table 3 presents the C-6 center of mass positions for 22 Stadan and Speopt sites.

The errors considered were those in the gravity model, the SAO station positions, and an error in the quantity GM (G is the universal gravitational

Table 3
Stadan/Speopt station coordinates*

Station Name	Number	Geodetic latitude	East longitude	Height (m) above ellipsoid
Blossom Pt, Md	1021	38°25′49.9″	282°54′48.5″	−49
Fort Myers, Fla	1022	26°32′53.2″	278° 8′ 4.1″	−45
Woomera, Aust.	1024	−31°23′26.2″	136°52′14.9″	123
Santiago, Chile	1028	−33° 8′59.0″	289°19′53.7″	711
Goldstone, Calif.	1030	35°19′47.9″	243° 5′58.8″	876
Johannesburg, S.Africa	1031	−25°53′ 1.7″	27°42′26.1″	1535
St. Johns, Newfoundland	1032	47°44′29.3″	307°16′46.1″	48
E.Grand Forks, Minn.	1034	48° 1′21.7″	262°59′19.3″	212
Winkfield, UK	1035	51°26′46.4″	359°18′ 7.9″	88
Rosman, N.C.	1037	35°12′ 7.4″	277° 7′41.4″	852
Orroral, Aust.	1038	−35°37′33.0″	148°57′14.7″	946
Rosman, N.C.	1042	35°12′ 7.3″	277° 7′40.7″	850
Tananarive	1043	−19° 0′32.6″	47°17′59.3″	1360
Edinburg, Texas	7036	26°22′46.5″	261°40′ 7.1″	5
Columbia, Mo.	7037	38°53′36.3″	267°47′40.8″	213
Bermuda	7039	32°21′50.0″	295°20′35.3″	−28
San Juan, P.R.	7040	18°15′28.6″	294° 0′23.5″	−21
Greenbelt, Md	7043	39° 1′15.8″	283°10′18.7″	−26
Denver, Col.	7045	39°38′48.1″	255°23′38.3″	1746
Jupiter, Fla	7072	27°01′14.3″	279°53′12.7″	−39
Sudbury, Canada	7075	46°27′21.6″	279° 3′10.3″	222
Kingston, Jamaica	7076	18° 4′34.5″	283°11′27.0″	403

* referred to an ellipsoid having a semi-major axis of 6378.155 km and a flattening of 1/298.25.

constant and M is the mass of the earth). The error in the gravity model was taken to be the difference between the SAO 66 and SAO 69 gravity models through (8,8) plus 20% error chosen for the resonant terms. This error of resonant terms is about twice that indicated in Fig. 3, but an error of this size proved to have little effect because of the short arcs (2 days) used. The error assumed for the low degree and order terms is also pessimistic, in that it implies that the SAO 1969 model is not an improvement over the SAO 1966 model for $(l, m) < (8,8)$.

The SAO station position error was taken to be 10 m in each coordinate, which is probably pessimistic. It will be seen that an error of this magnitude has little effect on estimates of station position. The gravity model error of low degree and order was by far the most important.

Table 4

Effects of assumed SAO station Position (S) and gravity model (G) error on station coordinates

Station Name	SN	Latitude (m) G	S	Longitude (m) G	S	Height (m) G	S
Blossom Pt, Md	1021	6	1	7	1	3	1
Fr. Myers, Fla	1022	10	2	3	3	8	1
Woomera, Aust.	1024	2	2	1	2	6	1
Santiago, Chile	1028	8	2	8	1	8	1
Goldstone, Calif.	1030	1	2	7	2	5	1
Johannesburg, S.Africa	1031	8	1	8	1	2	1
St. Johns, Newfoundland	1032	19	1	4	1	8	1
E. Grand Forks, Minn.	1034	2	2	3	1	4	1
Winkfield, UK	1035	12	1	15	1	3	1
Rosman, N.C.	1037	13	2	8	2	11	1
Orroral, Aust.	1038	9	5	7	1	3	1
Rosman, N.C.	1042	7	2	6	1	8	1
Tananarive, Madagascar	1043	13	1	9	1	12	1
Edinburg, Texas	7036	6	2	6	2	9	1
Columbia, Mo.	7037	7	1	2	2	9	1
Bermuda	7039	16	2	5	2	4	1
San Juan, P.R.	7040	16	2	11	2	5	1
Greenbelt, Md	7043	8	1	14	1	4	1
Denver, Col.	7045	1	2	9	2	7	1
Jupiter, Fla	7072	2	1	1	1	3	1
Sudbury, Ontario	7075	2	1	5	2	7	1
Kingston, Jamaica	7076	14	2	7	1	8	1

SN = station number; G = gravity model error; S = station position error.

Table 4 shows the relative contributions to error in latitude, longitude and height of the assumed gravity model error through (8,8) and the assumed 10 m (each coordinate) SAO station position error. The contribution of SAO station error is nearly constant and only a few meters. The gravity model error effects are particularly large for the GEOS 2 stations where inferior coverage was obtained.

In Table 5 the present results for some MOTS stations are compared with those obtained by Gaposchkin and Lambeck [3] by a different (geometrical) method. The agreement in latitude and longitude is very satisfying, and suggests that our estimate of gravity model error was indeed pessimistic. The systematically much larger difference in height was not expected from gravity model error, and hence requires further investigation.

Table 5
Differences between SAO and Goddard station coordinates on the SAO C-6 standard earth

Stations	Stn No.	Latitude ″ of arc	(m)	Longitude ″ of arc	(m)	Height (m)
Edinburg, Texas	7036	0.11	(3)	0.02	(0)	−17
Columbia, Mo	7037	0.25	(7)	0.05	(1)	−14
Bermuda	7039	0.24	(7)	−0.22	(−6)	−16
San Juan, P.R.	7040	0.21	(6)	−0.04	(−1)	−11
Denver, Col.	7045	0.02	(1)	−0.02	(−1)	−20
Sudbury, Ontario	7075	0.20	(6)	−0.32	(−7)	−2
Kingston, Jamaica	7076	0.43	(13)	−0.02	(0)	−27
E. Grand Forks, Minn.	1034	0.37	(11)	−0.16	(−3)	−5
Rosman, N.C.	1042	0.04	(1)	−0.05	(−1)	−5
Goddard	7050	−0.10	(−3)	0.20	(6)	−3
Average			5 m		−1 m	−12 m

4. Laser Station Estimation

The ultimate goal of our work is to obtain solutions from combined laser, optical, and electronic data. The combined solutions should be much stronger, and many more control points for datum orientations will be obtained.

As an initial attempt to obtain combined solutions, positions have been obtained for the Goddard Space Flight Center and Wallops Island laser tracking sites from combined optical and laser data. Five two-day arcs were used for the Wallops site and seven two-day arcs for Goddard Space Flight Center. The C6 positions derived are given in Table 6.

Table 6
Laser site locations

Station	Geodetic latitude	E. longitude	Height (m) above ellipsoid
Goddard (7050)	39°01′14.1″	283°10′18.4″	+2
Goddard (SAO det.) (7050)	39°01′14.2″	283°10′18.2″	+5
Wallops Is. (7052)	37°51′35.9″	284°29′24.1″	−65

The solution for the Wallops laser site contained about 2000 optical observations and 800 laser ranges. The a priori standard deviation assumed was two seconds for optical data and 3 m for range data. The rms fit for the right ascension and declination measurements in the converged solution was 1.9 seconds of arc. The rms of fit for range measurements was 2.7 m. Similar results were obtained for the Goddard laser solution. This latter solution contained 1800 optical observations and 1250 laser observations. In both cases, only about one third of the available laser data in each pass was used.

Table 6 also shows the SAO [1] result for the Goddard laser site.

5. The Goddard 1970 Directory of Tracking Station Coordinates

There is a great need for a unified set of station coordinates for the complete spectrum of tracking instruments. To this end, our positions and the SAO

Baker-Nunn positions have been used as control points to obtain shifts from local to center-of-mass coordinates (SAO C-6) for more than 100 sites including minitrack, C-band, unified S-band, laser, and many others. This directory, to be published by NASA in the summer of 1970, will in most cases provide positions accurate to 10 m or better. These coordinates will provide a highly accurate means for many orbit analyses and will serve as an accurate starting point for future investigations.

6. Conclusions

It is concluded that the use of multiple short (2 day) orbit arcs is a valid means of determining accurate station positions. Gravity model error effects, particularly resonance, are minimized by the use of short arcs. Also, the solution is less complex because periods where observations are sparse or absent are excluded. The strength of the solution lies in the use of multiple arcs with passes on all sides of the station.

Comparison of SAO optical and laser results with the present optical and laser results shows systematic differences in station height, but very close agreement in latitude and longitude. Thus, the latitude and longitude uncertainties estimated from gravity model error are pessimistic, being larger than indicated by the agreement between present results and SAO. The existence of the large systematic difference in height between Goddard and SAO optical solutions indicates that further analysis is required.

Acknowledgments

We wish to acknowledge the important contributions to this work of P. Ruttkay and S. Klosko.

References

[1] B. C. DOUGLAS and J. G. MARSH, Celest. Mech. 1, 479 (1970).
[2] D. BROUWER and G. M. CLEMENCE, Methods of Celestial Mechanics, Academic Press, New York and London 1961 (p. 169).
[3] E. M. GAPOSCHKIN and K. LAMBECK, New geodetic parameters for a standard earth, presented at the Fall Meeting of the American Geophysical Union, San Francisco, California, December 1969.
[4] Y. KOZAI, Smithson. Astrophys. Obs. Spec. Rep. No. 295 (1969).
[5] J. G. MARSH, B. C. DOUGLAS and M. DUTCHER, NASA Document No. X-552-70-48, February 1970.

Space Research XI — Akademie-Verlag, Berlin 1971

CONDENSATION DES DONNEES D'OBSERVATION EN VUE D'UNE UTILISATION GEODESIQUE

B. LAGO et A. M. MAINGUY

Centre National d'Etudes Spatiales, Brétigny, France

An original method is proposed for condensing data obtained from geodetic satellites. Generally, a typical Doppler pass contains 100–150 individual data points and a laser pass 200–250 points; the method consists in substituting for all the measurements of a pass above a station, one pseudo-measurement, expressed by a vector involving at most six components.

Using a good reference orbit the experimental data of a pass are treated by differential correction using a least squares fit. At the time of closest approach t_M, we obtain displacements $\Delta P(t_M)$, $i = 1.6$ (position–velocity) which are not significant. As we require the difference between the arc of the reference orbit and the corrected arc, we normalize the displacements and the covariance matrix and obtain linear combinations of these displacements. These combinations are then projected on the bases of the eigen-vectors of the covariance matrix, keeping only the combinations considered significant. There are three significant combinations for laser passes and either two or three for Doppler passes.

This method has been applied to laser and Doppler experimental data of the Diademe satellite; the results obtained after a semi-dynamical geodesy treatment using the pseudo-measurements are compared with the results obtained during the same period using real measurements. It is seen that these are equivalent, and we can hope that these pseudo-measurements will be useful in dynamical geodesy.

1. Définition

La géodésie dynamique permet, en poursuivant plusieurs satellites par plusieurs stations durant de nombreux mois et grâce à la diversité des mesures, de remettre en cause les caractéristiques du modèle de potentiel terrestre choisi.

Mais comme nous n'avons à notre disposition qu'un nombre faible de stations (groupées autour de la Méditerranée) nous n'avons pas pu procéder à ce traitement. A partir d'un bon modèle de potentiel terrestre, en étudiant les équations dynamiques nous avons restitué une trajectoire (dite par la suite trajectoire de référence) et précisé la position des stations par rapport à celle-ci. Cette méthode a reçu le nom de géodésie semi-dynamique.

Notons cependant que le but poursuivi ici n'est pas d'étudier la valeur du traitement de semi-dynamique dont nous avons rappelé le principe, mais d'étudier lors de celui-ci le comportement des mesures condensées par rapport aux mesures réelles.

2. Methode: Données Condensées

Tous les traitements qui vont suivre ont en commun qu'ils utilisent la méthode des corrections différentielles basée sur le critère des moindres carrés. Pour obtenir ces données condensées nous avons procédé par étapes

a, détermination d'une bonne orbite de référence (c'est-à-dire les six paramètres d'orbite pour une date initiale donnée);

b, construction des mesures condensées: pour cela nous avons traité les mesures arc par arc (c'est-à-dire nous avons groupé les mesures par passage au dessus de chaque station). Ce qui suit, est valable pour un arc.

2.1. Par la méthode des moindres carrés nous établissons les corrections différentielles à l'instant de passage au plus près t_M (avec une seule itération). Cet instant a été choisi car il peut être considéré comme caractéristique du passage. Nous obtenons les corrections aux paramètres d'orbites, $\Delta P_i(t_M)$ $i = 1{,}6$ (que nous pouvons représenter par le vecteur $\Delta \mathbf{P}(t_M)$ auxquelles est associée une matrice de covariance C^{-1}, (si C est la matrice des moindres carrés).

2.2. Comme nous désirons mettre en évidence l'écart entre l'arc de référence et l'arc corrigé, et comme les $\Delta P_i(t_M)$ obtenus n'ont pas eu eux-mêmes de valeurs significatives, nous normalisons ces déplacements.

La norme choisie est

$$d_2^s = [t_f - t_0]^{-1} \cdot \int_{t_0}^{t_f} \delta M^2(t)\, dt \text{ avec } \delta M(t) = \begin{pmatrix} \delta X(t) \\ \delta Y(t) \\ \delta Z(t) \end{pmatrix}$$

permet de déterminer les combinaisons linéaires des $\Delta P_i(t_M)$ intéressantes pour la suite du traitement.

Faire cette normalisation revient à définir le produit scalaire associé à la norme, et à projeter sur la base orthonormale pour ce produit scalaire. Nous obtenons $\Delta P_M(t_M)$, C_N^{-1}.

2.3. Nous projetons enfin sur la base des vecteurs propres de la matrice de covariance, ce qui revient à étudier les corrections sur les axes de l'ellipsoïde d'erreurs (donné par la matrice de covariance). Nous obtenons un vecteur V dont les composantes sont statistiquement indépendantes. La matrice de covariance associée est diagonale [1, 2].

2.4. C'est ce vecteur à six composantes auquel est associée une matrice diagonale que l'on peut considérer comme donnée condensée. Nous allons utiliser ces données dans un programme de correction différentielle comme pseudo-mesures. Pour les utiliser, nous aurons besoin de connaître l'instant où elles ont été définies et le jeu de paramètres à partir duquel elles ont été construites, c'est-à-dire, *le vecteur station satellite* $\boldsymbol{R}_{\text{réf.}}$ et sa *dérivée* $\boldsymbol{V}_{\text{réf.}}$ exprimés dans un repère géocentrique inertiel.

Les données, ou maintenant les pseudo-mesures, sont exprimées dans une base dépendant de chaque arc et qui n'est pas le repère géocentrique inertiel. Pour pouvoir les traiter il nous faut donc fournir la matrice de *changement de base M* qui permet de passer du repère géocentrique inertiel à la base de la pseudo-mesure $\bar{V} = M\,\Delta P(t_M)$; $\Delta P(t_M)$ est le vecteur correction suivant $(X\ Y\ Z - \dot{X}\ \dot{Y}\ \dot{Z})$. Il faut associer également à chaque composante V_i son poids π_i qui est calculé à portée de la matrice de covariance $\pi_i = s/\Delta_i$ où Δ_i est le terme correspondant de la matrice diagonale.

3. Généralités d'Utilisation

Le programme GIPSE qui est un programme de correction différentielle avec critère des moindres carrés, utilise les pseudo-mesures regroupées dans un traitement de géodésie semi-dynamique.

Les conditions initiales sont l'orbite de référence et les positions des stations. Ce sont ces mêmes paramètres que nous laisserons libres.

L'évalaution théorique correspondant aux pseudo-mesures ainsi définies sera faite lors d'une intégration numérique par la méthode de Cowell: d'ordre 8. Les modèles de forces introduits dans les seconds membres sont: (i) le potentiel terrestre (modèle standard earth B 13.1); (ii) le frottement atmosphérique modèle empirique de Jacchia. Bien que le programme puisse en tenir compte, nous avons négligé la pression de radiation et les effets gravitationels luni-solaires. Remarquons que nous sommes en cela parfaitement cohérents avec les programmes précédents: détermination de l'orbite; construction des données condensées; et le programme qui traite les mesures réelles pour la semi-dynamique nous servant de comparaison.

4. Résultats

Nous proposons ici les résultats obtenus sur le satellite D1D-67.14 A. Nous avons travaillé avec les mesures obtenues à partir de trois stations: Haute-Provence, Stéphanion, Colomb-Bechar. Nous avons traité les pseudo-mesures obtenues à partir, les unes de mesures laser, les autres de mesures Doppler. Les périodes de traitement ont été de 1 jour à 1 jour $\frac{1}{4}$ (20 arcs en moyenne en Doppler, 11 en laser). Nous avons ajouté une condition supplémentaire qui a été de figer la longitude du centre de gravité du triangle formé par les trois stations.

Comme le but de cette étude est de comparer la qualité des pseudo-mesures par rapport à celle des mesures réelles: nous comparons nos résultats à ceux obtenus (pour les mêmes périodes avec des mesures réelles) par M. Chassaing avec des mesures Doppler, par M. Roumiguier avec des mesures laser [3, 4].

4.1. Doppler

Les périodes traitées sont

Cas A 1 mars 1967 (0 h) à 2 mars 1967 (4 h)
Cas B 2 mars 1967 (20 h) à 4 mars 1967 (4 h)
Cas C 5 mars 1967 (4 h) à 6 mars 1967 (4 h)
Cas D 6 mars 1967 (22 h) à 8 mars 1967 (2 h)

Tableau 1
Déplacement proposé au systeme SAO (Doppler)

	Cas A				Cas B				Cas C				Cas D			
	Mesures réelles	2 σ	Mesures condensées	2 σ	Mesures réelles	2 σ	Mesures condensées	2 σ	Mesures réelles	2 σ	Mesures condensées	2 σ	Mesures réelles	2 σ	Mesures condensées	2 σ
BCD 72	18	(10)	20	(10)	37	(13)	45	(11)	61	(11)	68	(8)	39	(12)	50	(9)
	−13	(6)	−2	(5)	−1	(9)	−1	(5)	−3	(9)	−5	(4)	−0,6	(8)	−5,5	(4)
	−36	(52)	−13	(43)	19	(65)	30	(39)	−102	(85)	−82	(50)	−25	(78)	14	(53)
STD 75	12	(9)	14	(11)	8	(12)	12	(11)	−2	(12)	9	(10)	9	(11)	22	(13)
	9	(6)	5	(5)	10	(10)	8	(5)	6	(10)	21	(5)	9	(8)	14	(5)
	−56	(46)	−22	(39)	61	(58)	67	(38)	33	(76)	−14	(44)	11	(70)	48	(46)
HPD 79	28	(9)	22	(9)	23	(11)	32	(10)	46	(11)	47	(9)	50	(9)	56	(9)
	12	(6)	6	(4)	10	(9)	6	(4)	13	(9)	2	(4)	5	(7)	9	(4)
	−68	(44)	−51	(37)	52	(56)	42	(37)	−73	(71)	−52	(41)	−6	(65)	34	(44)

Tableau 2
Déplacement proposé au systeme SAO (Laser)

	Cas A				Cas B			
	Mesures reelles	2 σ	Mesures condensées	2 σ	Mesures réelles	2 σ	Mesures condensées	2 σ
BCL 82	−54	(18)	36	(28)	30	(7)	47	(14)
	−14	(2)	−14	(4)	−10	(1)	−11	(2)
	9	(32)	−17	(48)	56	(15)	−7	(21)
STL 85	3	(10)	6.5	(19)	−13	(9)	−5	(8)
	6	(4)	6,6	(8)	−2	(2)	0,5	(3)
	16	(28)	0,2	(43)	63	(14)	20	(20)
HPL 89	4	(9)	24	(17)	1	(4)	21	(9)
	17	(3)	17	(5)	14	(2)	19	(3)
	50	(24)	48	(39)	78	(13)	40	(15)

Les déplacements sont proposés aux coordonnées géocentriques de référence définie par l'ellipsoïde SAO $R = 6375{,}155$ km, $\varepsilon = 1/298{,}25$, dont les valeurs sont données ci-dessous:

$$\text{Colomb-Bechar 72-BCD} \begin{cases} X = 5\,426\,229 \text{ m} \\ Y = -229\,424 \text{ m} \\ Z = 3\,334\,754 \text{ m} \end{cases}$$

$$\text{Stephanion 75-STD} \begin{cases} X = 4\,654\,351 \text{ m} \\ Y = 1\,959\,182 \text{ m} \\ Z = 3\,884\,355 \text{ m} \end{cases}$$

$$\text{Haute-Provence 79-HPD} \begin{cases} X = 4\,578\,447 \text{ m} \\ Y = 458\,083 \text{ m} \\ Z = 4\,403\,026 \text{ m} \end{cases}$$

voir les résultats, tableau 1. Ils sont en bon accord avec ceux obtenus à l'aide de mesures réelles, mais présentent des intervalles de confiance plus réduits surtout en ce qui concerne la détermination du Z.

4.2. Laser

Les périodes traitées sont

Cas A 8 mai 1967 (23 h) à 10 mai 1967 (2 h)

Cas B 9 mai 1967 (23 h) à 11 mai 1967 (2 h)

Le système de référence est le même

$$\text{Colomb-Bechar 82-BCL} \begin{cases} X = 5\,426\,324{,}29 \text{ m} \\ Y = 229\,305 \text{ m} \\ Z = 3\,334\,585{,}61 \text{ m} \end{cases}$$

$$\text{Stephanion 85-STL} \begin{cases} X = 4\,654\,341{,}9 \text{ m} \\ Y = 1\,959\,153{,}8 \text{ m} \\ Z = 3\,884\,370{,}7 \text{ m} \end{cases}$$

$$\text{Haute-Provence 89-HPL} \begin{cases} X = 4\,578\,365{,}2 \text{ m} \\ Y = 457\,942{,}23 \text{ m} \\ Z = 4\,403\,127{,}46 \text{ m} \end{cases}$$

Les résultats sont données dans le tableau 2. Les résultats semblent aussi en accord avec ceux obtenus avec mesures réelles. La légère différence qui apparaît sourtout sur Z est due au fait que les cas laser ne comportent qu'une dizaine d'arcs alors que nous en avons toujours une vingtaine en Doppler.

5. Conclusion

Les résultats que nous venons d'obtenir en géodésie semi-dynamique sont bons. Cette technique est, certes, plus longue car nous passons par une étape intermédiaire qui est la condensation. Mais celle-ci est particulièrement efficace et nous ne perdons rien sur les mesures réelles, ce qui permet de penser qu'il serait nécessaire de l'utiliser en géodésie dynamique.

Bibliographie

[1] B. Lago, Document CNES, BL/ZK/8-237/MT, Brétigny, juin 1968.
[2] A. M. Mainguy, Document CNES, AMM/GD/0-028/MT, Brétigny, mai 1970.
[3] B. Lago, in: Dynamics of Satellites, Springer-Verlag Berlin 1970 (p. 138).
[4] J. P. Chassaing, Bull. Géodésique No. 94, 403 (1969).

Space Research XI – Akademie-Verlag, Berlin 1971

DETERMINATION OF THE EQUATIONS OF CONDITION FOR THE ZONAL HARMONICS USING THE DIAL SATELLITE

A. CAZENAVE[a] and F. FORESTIER[b]

[a] Centre National d'Etudes Spatiales, Brétigny, France
[b] Observatiore de Meudon, Paris, France

Interferometric measurements have been obtained over a revolution of the perigee of the DIAL satellite, whose inclination is only 5°.
In spite of a low perigee height, we have obtained three equations of condition for the zonal harmonics by studying the long period variation of the eccentricity and the secular variation of the argument of perigee and of the ascending node. These equations are given only up to the fourteenth order.

1. Introduction

It is possible to draw a curve showing the different values of the secular motion of the argument of perigee, as a function of the inclination, with the semi-major axis and eccentricity given, using different sets of zonal harmonics. The differences which appear are especially large for satellites with low inclinations. These differences between the various sets of harmonics appear also on the curves showing the secular motion of the node or the amplitude of the long periodic term in the eccentricity, as a function of the inclination.

Hence one can expect much improvement in the knowledge of the zonal harmonics using a low inclination satellite. The DIAL satellite, in orbit since 10 March 1970, is the first low inclination satellite, with an inclination of 5°. A difficulty exists, however, as the height of the perigee is only 320 km; such a low altitude introduces an important air drag. In spite of this we have obtained three equations of condition.

2. Tracking System

Only interferometric tracking, from two CNES network stations, has been made for the DIAL satellite: Kourou 5° N 53° W and Pretoria 25° S 28° E. Four passes each day were visible from Kourou, and only one from Pretoria.

3. Procedure and Results

Measurements have been obtained for 30 days, over a revolution of the perigee, the secular motion for the argument of perigee being about 12° a day. Mean orbital elements of the satellite were obtained by treatment of four-day arcs, with two-day overlaps using a differential correction programme [1] which takes into account zonal harmonics only up to J_5. After elimination of air-drag, solar radiation and luni-solar attraction effects, based on Cook's formulae [2] and shown graphically in Fig. 1, the only remaining perturbation terms are those due to imperfections in the air-drag correction and imperfections in the set of zonal harmonics used in the differential correction programme; tesseral terms are not taken in account.

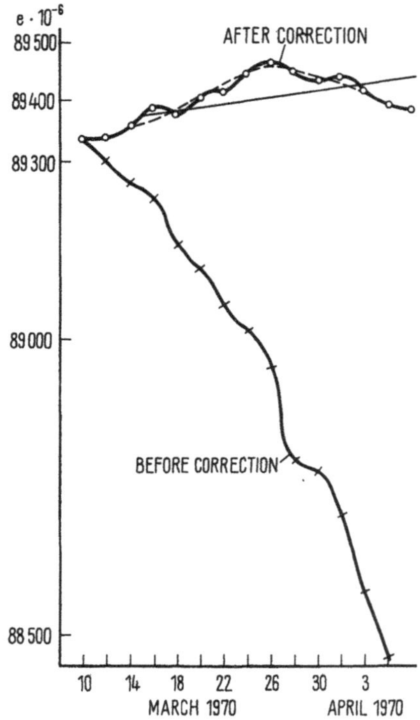

Fig. 1. Curves showing e before and after air-drag correction.
– – – long-period variation; ——— secular drift.

Equations of condition for harmonics of even order can be found from the secular motion of the perigee ω and the node Ω; odd harmonics can be found from the long-period variation of the eccentricity e and eventually from the long-period variations of ω and Ω. Orbital elements can be expressed as Fourier series; for example the relation for e is:

$$e = e_0 + A_{1e} \sin \omega + A_{2e} \cos 2\,\omega + A_{3e} \sin 3\,\omega + e_1 \cdot \Delta t\,, \qquad (1)$$

where Δt is the time since the mean initial elements, e_1 is an unknown quantity which is adjusted by a least squares method, the relations for Ω and ω are:

$$\Omega = \Omega_0 + A_{1\Omega} \cos \omega + A_{2\Omega} \sin 2\,\omega + A_{3\Omega} \cos 3\,\omega + \Omega_1\,\Delta t\,, \qquad (2)$$

$$\omega = \Omega_0 + A_{1\omega} \cos \omega + A_{2\omega} \sin 2\,\omega + A_{3\omega} \cos 3\,\omega + \omega_1\,\Delta t\,. \qquad (3)$$

A programme [3] using Kozai's set of zonal harmonics has been used to compute terms like A_{2e}, A_{3e}, . . . and other coefficients.

After having subtracted from the first members of Eqs. (1)–(3) terms such as $(A_{2e} \cos 2\,\omega + A_{3e} \sin 3\,\omega)$, a new system of equations is computed by a least squares method, each equation being weighted by a function of the deviations obtained on the mean orbital elements coming from the differential correction method.

We find, for example:

$$\left.\begin{aligned} A_{1e} &= -\,(3.9 \pm 0.5)\times 10^{-5} \\ A_{1\Omega} &= -\,(5.3 \pm 2.4)\times 10^{-4} \\ A_{1\omega} &= -\,(7.3 \pm 4.5)\times 10^{-4} \end{aligned}\right\}\text{(in radians)} \tag{4}$$

$$\left.\begin{aligned} \Omega_1 &= -\,1.2466\times 10^{-6} \pm 2.0\times 10^{-10} \\ \Omega_1 &= +\,2.4752+10^{-6} \pm 3.5\times 10^{-10} \end{aligned}\right\}\begin{array}{l}\text{(in radians} \\ \text{per second)}\,.\end{array} \tag{5}$$

Since the errors in $A_{1\Omega}$ and $A_{1\omega}$ in the least squares programme are of the same order of magnitude as $A_{1\Omega}$ and $A_{1\omega}$ themselves, we have given up any attempt to obtain equations of condition for $A_{1\Omega}$ and $A_{1\omega}$. We therefore decided to hold them fixed, and the equations of condition thus obtained for Ω_1 and ω_1 are given in Eq. (5).

The error in A_{1e} in the least squares programme is some 8 times smaller than A_{1e}. This shows that the equation of condition for e, although not as good as the equations for ω_1 and Ω_1 has some significance. The reason for this large uncertainty is that interferometric measurements are not very accurate. For example, for e, the differential correction programme gives $2\,\sigma = 0.2\times 10^{-4}$ and the amplitude of the long periodic term in e is about 0.2×10^{-4}.

We have tried to improve the equation in e, taking into account the day–night modifications of the atmosphere. For this purpose we write:

$$a(t) = a_0 + \int_{t_0}^{t} a \, \mathrm{d}t$$

with $a_1 = 140\,\frac{m}{j}\cos 2\,\pi\left(\frac{D\,J - 7394}{T}\right)$

in which: DJ is the date expressed in Julian days since 1 January 1950, T is the period (70 days) and a is the semi-major axis expressed in metres. It follows that:

$$e = e_0 + A_{1e}\sin\omega + A_{2e}\cos 2\,\omega + A_{3e}\sin 3\,\omega$$
$$-\,140\,\frac{m}{j}\left(\frac{1-e_0}{a_0}\right)\sin 2\,\pi\left(\frac{D\,J - 7394}{T}\right)\frac{T}{2\,\pi}\,.$$

Hence the new equations of condition for e are:

For the long period variation of eccentricity:

$$(-0.29 \pm 0.10)\times 10^{-4} = -\,37.369\,\Delta J_3 + 70.392\,\Delta J_5 - 92.692\,J_7 + $$
$$+\,104.432\,J_9 - 107.595\,J_{11} + 104.473\,J_{13}$$

J_3 and J_5 being those of Kozai's II set of zonal harmonics (1966).

For the secular variation of the argument of perigee

$$
\begin{aligned}
+0.24750\times10^{-5} \pm 0.27\times10^{-9} = {} & 0.227\times10^{-2} J_2 - 0.427\times10^{-2} J_4 + \\
& + 0.562\times10^{-2} J_6 - 0.632\times10^{-2} J_8 + \\
& + 0.648\times10^{-2} J_{10} - 0.626\times10^{-2} J_{12} + \\
& + 0.576\times10^{-2} J_{14}
\end{aligned}
$$

For the secular variation of the ascending node

$$
\begin{aligned}
-0.12490\times10^{-5} \pm 0.13\times10^{-9} = {} & -0.114\times10^{-2} J_2 + 0.217\times10^{-2} J_4 - \\
& - 0.292\times10^{-2} J_6 + 0.337\times10^{-2} J_8 \\
& - 0.358\times10^{-2} J_{10} + 0.361\times10^{-2} J_{12} - \\
& - 0.350\times10^{-2} J_{14}
\end{aligned}
$$

References

[1] Thèse de Mme Bauchereau, Paris 1969, Faculté des Sciences.
[2] G. E. COOK, Geophys. J. Roy. Astron. Soc. **6**, 271 (1962).
[3] A. CHALLE and F. J. LACLAVERIE, Astron. Astrophys. **3**, 15 (1969).

Space Research XI — Akademie-Verlag, Berlin 1971

USE OF SPACE TELEVISION AND PHOTO IMAGES FOR EARTH SCIENCES

B. V. VINOGRADOV

Institute of Physics, Leningrad State University, Leningrad, USSR

Numerous local television pictures of the earth from the meteorological satellites Nimbus, Essa and Cosmos as well as the local and global photographs from spacecraft Gemini, Soyuz, Zond and Apollo are applied in investigations of the earth, viz. soil, geobotanical, structural, geological, geomorphological and hydrological. They are used for the interpretation of the natural resources with scale details of 0.2×10^{-5}–0.2×10^{-7} and for the correction of the generalization schemes in the compilation of small-scale schematic maps. They are even more important for the study of the rhythms of natural processes of plant vegetation, soil moisture, the formation and melting of snow cover. The interpretation of the underlying surfaces is also of great auxiliary value for the meteorological analysis of satellite data. Some features of the soil, vegetation cover, geological–geomorphological structure and hydrographical land features are also interpreted from the infra-red images of the earth's surface received from the Nimbus and Cosmos satellites.

1. Introduction

The earth's imagery derived from artificial satellites, manned spacecraft and automated interplanetary stations is of great importance for a geographical study of the earth.

Space methods for earth science are determined as a complex of the apparatus and methodical developments aimed at the study of the composition, structure, dynamics and rhythmics of the geosphere by interpretation of the images and spectra of natural formations obtained from space vehicles [1].

Among the principal space methods for earth science the following are considered: visual, photographic, television, spectrometric (in the ultraviolet, visible and infrared spectral regions), microwave.

The achievements made in the individual fields of space method development are different. Here we shall confine ourselves to the discussion of the most informative recording methods, viz. photographic, television and spectrophotometric.

2. Space Imagery

Among the space images analysed three categories can be singled out according to the degree of integration which they afford [2, 3]: global, regional and local.

2.1. Global Imagery

Global imagery provides photos and spectra of the entire (or nearly so) sunlit portion of the Earth's disk from the altitudes higher than 2×10^4 (Fig. 1).

Fig. 1. Global earth image as obtained from the automated station Zond 5 from altitude 9×10^4 km, exhibiting landscape and climatic zones over Africa and Atlantic: 1, temporary humid; 2, subtropical with hard-leaved evergreen forest and shrub; 3, desert steppe and subdesert; 4, desert; 5, dry savannah; 6, typical savannah; 7, intertropical convergence with moist savannah; 8, equatorial rain forest.

Such imagery is strongly affected by the earth's sphericity, its real scales even in the central parts being smaller than 10^{-7}. The imagery referred to has been obtained from the altitudes higher than 2×10^4 km from the automated interplanetary stations Zond 5 and 7 (6×10^4 and 9×10^4 km), the artificial satellite Molniya (about 2×10^4 km), ATS ($3{,}6 \times 10^4$ km) and the manned spacecraft Apollo (up to 4×10^5 km). Were the latitudinal circulation, climatic [4] and geographical zones, both within the African continent and over the Atlantic ocean have been interpreted from the global imagery. By the use of global imagery an attempt has also been made to obtain regional information on the distribution of structural geological and geographical complexes within the Central and West Sahara [5].

2.2. Regional Imagery

Regional imagery embraces considerable areas more than 10^5 km^2 within geographical countries and regions (Fig. 2). The spherical character of the surface is pronounced, the image distortions being allowed for within the whole image. The real scale is smaller than 10^{-6}. The observations were carried out, as a rule, from the altitudes higher than 300–500 km. Such imagery is regularly obtained from the meteorological satellites Meteor, ESSA (regional photos have been obtained from the American manned spacecraft Gemini 11 from an altitude of about 1400 km). From the regional imagery a number of researches has been carried out on the study of the composition of underlying surfaces with synoptic interpretation of the cloud cover [6], schematic interpretation of the hydrographical network, classes of the vegetation formations, lithological composition of rocks and relief in the scale of about $0.5–0.2 \times 10^{-6}$ [7], investigation of the distribution of sea ice cover [8] etc.

2.3. Local Imagery

Local imagery represents individual regions and landscapes with area smaller than $10^5–10^4$ km^2. The spherical character of the surface is slightly pronounced, and for non-topographic interpretation the correction due to the earth's curvature is not required. Such imagery (considerable in amount) has been obtained in the USA from the manned spacecraft Mercury, Gemini and Apollo according to the programme on the study of the land surface and ocean, and in the USSR from the manned spacecraft Voskhod and Soyuz according to the programme on photography of geologo-geographical objects of the main types of the earth's surface in the scales of about 10^{-6} from rather low orbits (180–300 km). Interpretation of these images demonstrated the great scientific and practical importance of local imagery for various kinds of special mapping in the scales of 10^{-6} and larger [9–11].

3. Space Methods

Apart from their numerous technical advantages space methods for the investigation of the natural environment have a number of methodological peculiarities due to which perspective can be introduced into geographical research. Among the numerous advantages of space methods we shall point out only the main three [12]:

a) horizontal integration, i.e. the reproduction on one image of a large extent of the earth's surface; b) vertical integration, i.e. the reproduction on one image of various landscape components: atmosphere, lithosphere, hydrosphere,

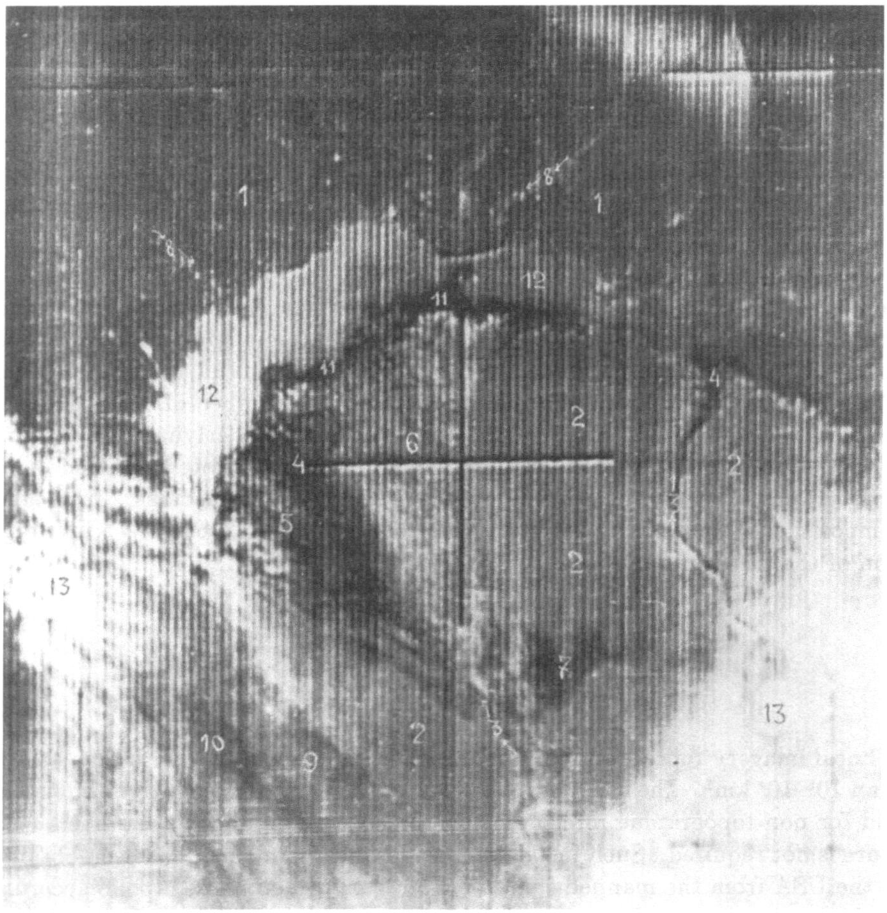

Fig. 2. Regional image of the Balkhash lake district as obtained from the meteorological satellite Cosmos 226, exhibiting landscape regions: 1, stony desert; 2, sand massif; 3, recent river valley with dense vegetation and cultivated lands; 4, recent low-level river delta with swamps; 5, recent middle-level river delta with dry meadows; 6, ancient river delta with salt lands; 7, ancient river plain with clay soils and dense shrubs; 8, dry valleys of temporary rivers; 9, Piedmont plain with grey loose soils; 10, mountains; 11, low land with swamps along lake foreshore; 12, Balkhash lake; 13, clouds.

biosphere and antroposphere; c) dynamical integration, i.e. the derivation with one recording system of subsequent images of the same territory at certain time intervals.

3.1. Study of Structure, of Natural Complexes

Most perspective applications of space methods are for studying the structure of natural complexes, territory integration being the principal advantage. Due to this it becomes possible to obtain the image of large territory complexes and to identify them directly by boundary interpretation from the super-small-scale earth's image instead of by subjective identification and multi-stage generalization from medium- and large-scale maps. The following scheme has been obtained of the relationship between different types of geographical territory complexes and space survey scales (the equivalent resolving power of the image being about 100 cm^{-1}) (see Table 1).

Table 1

Natural complexes	Equivalent scale of imagery	Representative area	Dimension of resolved details moderate contrast
Geographical zones, regions	10^{-7}	10^5 km^2	1 km^{-1}
Geographical areas, landscapes	10^{-6}	10^4 km^2	10 km^{-1}
Geographical land systems, land units	10^{-5}	10^3 km^2	100 km^{-1}

To detect natural complexes of different extents it is necessary to have space imagery of different picture areas. The area of the earth's surface fixed on one photograph should embrace the entire object, in one dimension at least, reveal the orientation of the object and reflect nearby objects in order to determine their interrelation, the shape of the boundaries and the magnitude of optical contrast. Hence, for the detection of global structure: geographical latitudinal zones, planetary tectonic lineaments and other macroscopic elements of the geosphere, pictures are needed which would embrace hundreds of thousands of square kilometres. For the detection of local structures: geographical land units and systems, tectonic features, various geomorphological elements pictures embracing thousands and a few tens of thousands of square kilometres are sufficient.

To obtain different kinds of information different resolving powers of space imagery are needed. The main advantage of space imagery is its large surveillance, which, in turn, is to a certain degree inversely dependent on detail. Attempts to improve the latter being natural, the priority remains in favour of surveillance.

Taking into account the above points and allwoing for the influence of atmospheric transparency and refraction as well as consideration of the legal order the limits of real resolution for a satellite are from 100 km^{-1} to 25 km^{-1} [13, 14].

Many authors, particularly specialists in agriculture and forestry have often pointed out that if space recording facilities did not possess a certain resolving power the interpretation of space images would lose its practical importance [15].

For interpretation to be effective, the resolution of space imagery is approximately established from the mean statistical dimensions of the objects to be recognized and interpreted:

Geologo-geomorphological elements	25–10 km^{-1}
Engineering-geological phenomena	50–25 km^{-1}
Forest and agricultural lands	100–25 km^{-1}

The detail required depends on the scale of the investigation, viz. a geological survey in the scale of 10^{-6} requires the resolution of 8–25 km^{-1}, whereas for a sruvey in the $2.5^{-1} \times 10^{-5}$ scale the required resolution is 50–25 km^{-1} [11].

3.2. Study of Interrelations of Natural Phenomena

A study of origin and inter-relations of the separate components of natural complexes is another direction in which space methods may develop. So far only methods demanding sufficiently narrow specialized interpretation of space pictures have been developed, viz. nephananalysis, ice reconnaissance, fire detection, etc. An idea has been put forward for the creation of numerous narrow by specialized satellites to survey natural resources [16], but wider interpretations, particularly with increased demands for details and reliability, necessitate the use of an integrated interpretation method. This comprises a combined interpretation of the geospheric component of interest with other related information; for example, the interpretation of cloud cover pictures must be combined with recognition of underlying surfaces [6, 17], that of frontal zones of oceanic currents must be combined with the recognition of the forms and boundaries of cloud fields [18, 19], etc.

3.3. Study of the Dynamics of the Natural Environment

A study of the rhythmics and dynamics of the natural environment is quite promising. The most reliable method of studying the dynamics is by repeated surveys by one recording space system of the same territories, which ensures a sufficient comparability of technical conditions for obtaining images [20, 21]. However, in view of the insufficient lifetime of space vehicles and, sometimes, the long-term character of the processes the dynamics of the natural environment can also be determined from single photographs.

A dynamic approach determines the frequency of observations required for studying the natural environment, which may be summarized as follows:

Meteorological situation	1–2 times per 24 hr period
Atmospheric contamination	1–2 times per week
Ice situation	1–2 times per week
Snow cover in the course of its formation and melting	1–2 times per decade

Soil cover in the period of drying, moistening, agricultural tillage	1–2 times per month
Land use and forest management	1–2 times per year
Vegetation cover in periods of intensive phenological changes	1–2 times per month
Ripening of crops	1–2 times per decade
Oceanic currents	1–2 times per month
Geological structure	once in a period of 5–10 years
Engineering-geological processes	1–2 times per quarter
for most active cases	1–2 times per decade

Thus, different intensities of single physico-geographical processes and different rhythms require different periodicities for the optimum terms of space survey.

At present, apart from the survey of the movement of cloud masses and ice fields (examples of which are well known) repeated surveys are being developed for studying other processes, viz. formation and melting of the snow cover, phenological vegetation development, movement of dust-sand storms and flows, local atmospheric and oceanic pollution, etc. [19–23].

Yet, from single images one can also derive satisfactory information, at a given stage, at least, on the extent of eroded soils and burned-out forests, directions of movement of coastal mantles, etc.

4. Conclusions

The principal methods for the development of the technique of space methods to study earth science are combined (space, aircraft and ground) integrated surveys of the key area and profile systems which characterize the main landscapes of the globe as well as their imagery in different spectral regions. In developing this programme, in the USA and outside since 1963 investigation has been undertaken of more than a hundred key areas to determine the interpretation parameters that could be used for the recognition of natural conditions from manned spacecraft images of various regions of the globe (including similar areas not visited) [24]. The method of combined satellite, aircraft and ground observations of the satellite trajectories has been developed and perfected jointly by the Laboratory of Aerial and Space Methods for Earth Sciences, University of Leningrad, and the Main Geophysical Observatory.

As a result, geographical and geophysical profiles of the atmosphere and the earth's surface have been obtained. This method which we have earlier called the method of landscape profiles, is most useful for studying the regularities of formation of space imagery of the earth's surface and determining the atmospheric optical transfer function.

References

[1] B. V. Vinogradov, K. Ya. Kondratyev and V. D. Stepanenko, Kosm. Issled. **6**, 4 (1968).
[2] C. Lescure and C. Watchel, Air et Cosmos, **6**, 35 (1969).

[3] B. V. VINOGRADOV and K. YA. KONDRATYEV, Izv. Akad. Nauk SSSR, Ser. Geogr. **2**, 88 (1970).
[4] V. A. BUGAYEV et al., Meteorol. i. Gidrol. **7**, 3 (1969).
[5] B. V. VINOGRADOV, Zemlya i Vselennaya **1**, 31 (1970).
[6] D. M. SONETCHKINE, Meteorol. i Gidrol. **9**, 3 (1968).
[7] B. V. VINOGRADOV, Izv. V. G. O., **101**, 210 (1969).
[8] K. P. VASILYEV, Zemlya i Vselennaya **1**, 40 (1970).
[9] P. D. LOWMAN, Photogrammetria **24**, 77 (1969).
[10] B. V. VINOGRADOV and A. A. GRIGORYEV, Izv. Akad. Nauk SSSR, Ser. Geol. **16** (1970).
[11] F. J. WOBBER, Photogrammetria, **24**, 107 (1969).
[12] B. V. VINOGRADOV, Izv. V. G. O. **98**, 101 (1966).
[13] B. P. MILLER and H. GURK, IAF Paper, TS 186, 1968 (p. 16).
[14] G. W. MORGENTHALER, IAF Paper, TS 193, 1968 (p. 18).
[15] R. C. WILSON, Photogramm. Engng. **33**, 483 (1967).
[16] W. A. FISHER, IAF Paper, AS 155, 1968 (p. 2).
[17] H. KAMINSKY, Weltraumfahrt und Raketentechnik **19**, 69 (1968).
[18] S. I. RASOOL and T. S. HOGAN, Bull. Am. Met. Soc. **50**, 130 (1969).
[19] J. C. WILKERSON, Oceanus **13**, 2 (1967).
[20] K. E. BERGSTEN, in: Televisionsystem i Naturgeografisk Forsking, Stockholm 1968 (p. 63) (in Swedish).
[21] R. N. COLWELL, IAF Paper, AD 163, 1968 (p. 10).
[22] A. R. BARRINGER, B. C. NEWBURY and A. T. MOFFAT, Proc. Fifth Symp. on Remote Sensing of Environment, University of Michigan 1968 (p. 123).
[23] J. W. SHERMAN, AAAS Paper, 69-062 (p. 20).
[24] R. C. BADGLEY, L. CHILDS and W. FISHER, Am. Astronaut. Soc., Sci. and Techn. Ser. **4**, 183 (1965).

Space Research XI — Akademie-Verlag, Berlin 1971

A PRELIMINARY REPORT ON THE REMOTE SENSING OF COCONUT TREES

P. R. PISHAROTY

Indian Space Research Organisation

A short note describing the possibility of using infrared photography to detect disease in coconut trees.

Under a USA–INDIA Joint Project, Coconut plantations with healthy and unhealthy trees were photographed from the air. A 70 mm Hasselblad camera and ektachrome infrared sensitive colour films as well as panchromatic colour films were used. A helicopter given by the Soviet Government to the Thumba range for surveillance, flying at an altitude of about 600 m, served as the aerial platform.

The higher the chlorophyll content of the leaves the brighter the red colour recorded on the films. The brightest red colour belongs to the Jack trees. (These trees yield fruits somewhat similar to the pineapple in appearance and flavour, but about a hundred times larger in mass.) The least bright ones are the cashew nut trees, which have poor chlorophyll content. The coconut trees range in their infra-red reflectivity, or chlorophyll content, between the Jack and cashew-nut trees. The bright red ones are the healthy coconut trees and the pale red ones are the unhealthy. A virus affects the roots; the leaves curl and become pale. The average yield of a tree is about 60 nuts per annum in Kerala, and about 100 elsewhere in India; this yield drops to less than ten nuts per annum when affected by the disease, called the Wilt (root) disease. The disease is widely prevalent in South Kerala, and leads to a loss of about ten million rupees per annum.

It is hoped that a significant fall in the infrared reflectivity would occur in the early phase of the disease and thereby enable suitable remedial action to be taken. Normally the agriculturist can detect the disease only when the only cure is to cut and remove the tree along with its roots full of the virus.

We have about 200 photographs similar to the coloured ones projected. Detailed plans have been drawn up to study the picture of each tree along with the case-history of the corresponding tree, maintained at the Coconut Research Centre, in whose farm the tree is located. The studies are being carried out in close cooperation with the Indian Council of Agricultural Research, in overall charge of the Coconut Research Centres.

Remote Sounding of the Atmosphere

Space Research XI — Akademie-Verlag, Berlin 1971

DEVELOPMENTS IN SPACE METEOROLOGY DURING 1969/1970

F. Möller

Meteorologisches Institut der Universität München, München, FRG

It is shown with what success measurements of the temperature as obtained from satellites can be used for improvement of numerical predictions of the hemispheric or global circulation. Methods are described by which vertical temperature profiles can be inferred from spectral radiance measurements of the instruments SIRS and IRIS in the satellite Nimbus 3. One of the three methods is described in some detail. Examples are given of the determination of vertical profiles of water vapour and ozone as well as of the total ozone amount from IRIS measurements. The use of microwaves to determine the total water vapour amount of the atmosphere is explained as it has been applied in the satellite Cosmos 243. Finally some procedures for determining the displacement of clouds or wind direction and speed from the cloud pictures of geostationary satellites are described.

1. Introduction

COSPAR is a scientific organization and its proposals concern scientific investigations of physical quantities, in our particular case those of the lower atmosphere. Some of these investigations remain pure research tasks, they serve to the widening of our knowledge, and to give a deeper insight into natural processes. Other results turn out to be of the utmost value for practical tasks. Investigations of these quantities cannot be stopped just because they can be applied. It is the aim of the Global Atmospheric Research Programme, GARP, to utilize measurements from satellites for the diagnosis and analysis of the atmosphere and thus for the prediction of the general circulation.

Modern weather prediction with large computers is based on the application of the differential equations of hydrodynamics and thermodynamics to the situation on a single particular day. The objects of calculating the general circulation are to determine the changes and transport of energy in all its forms, development of potential energy, transport and production of kinetic energy, transport and changes of thermal energy including transport and fluxes of radiative energy as well as transport of latent energy and its change. This is the transport of water vapour and the change of water vapour into liquid water or the origin of clouds and precipitation.

It is generally well known that the precalculation of motion and of the pressure patterns in the atmosphere has become accessible since large computers have been constructed.

2. Global Meteorological Observations

As a purely meteorological introduction to this survey on satellite techniques and results an example from an investigation of Miyakoda et al. [1] at the ESSA Geophysical Fluid Dynamics Laboratory in Princeton demonstrates the significance of global observations of the lower atmosphere for the basic problem of the general circulation. The initial values of Miyakoda's calculations were

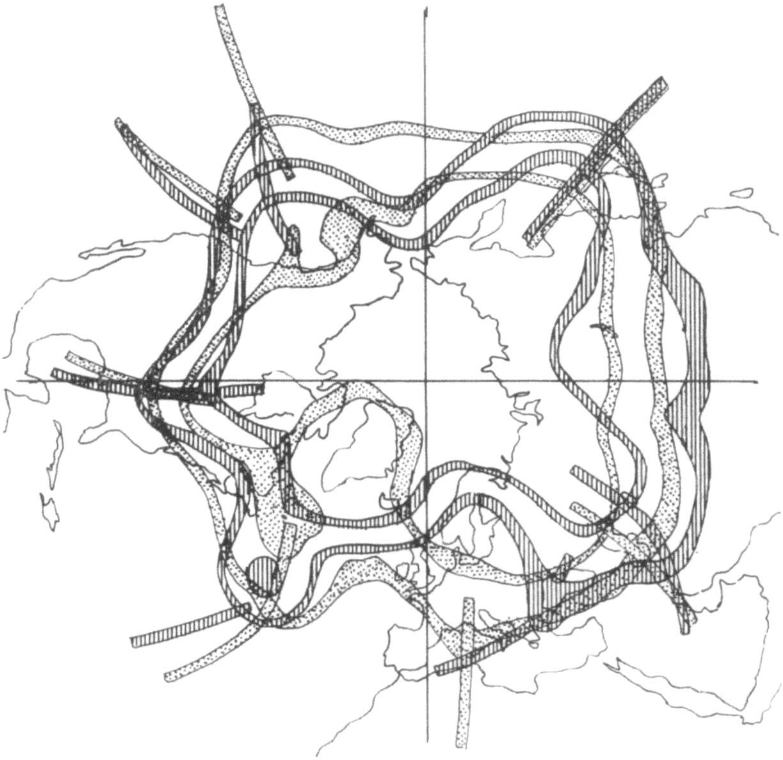

Fig. 1. 500-mb surface of 20 January 1964, indicated by the intervals between contour lines of 5220 and 5280 gpm and of 5460 and 5520 gpm. Hatched, observed; stippled, numerical prediction from observations 11 days before. Major troughs in the same marking. After Miykoda et al. (1969).

given by the distribution of pressure, temperature etc. according to the observations for 9 January 1964, 1200 GMT. The development of the circulation over the northern hemisphere was calculated over a time interval of 14 days. The results were compared with the real situation for each day. Fig. 1 shows in different marking the contour lines of the 500 mb surface 11 days later. Some contour lines of the observed situation at 20 January are hatched, the calculated values are stippled. The two wide stripes are the intervals between each two consecutive contour lines of the 500 mb surface. The crossing lines give the position of the major troughs in observation and computation. Coincidence or deviation of the stippled and hatched lines demonstrate the quality of the numerical prediction after 11 days. It is highly satisfactory to see that the large scale troughs and ridges are situated at the same places in both these maps. One recognizes a good correlation in many places. Two troughs west of North America, one over the east of North America, a closed upper low over the Atlantic Ocean, another trough over the Mediterranean and one more over the Caspian Sea show up very distinctly in both maps at the same places although somewhat deformed; finally a large trough over Japan has almost exactly the same position in prediction and observation.

If it is possible to recognize whether in a particular area a stationary blocking high will develop bringing a long drought or if it is possible to predict 14 days

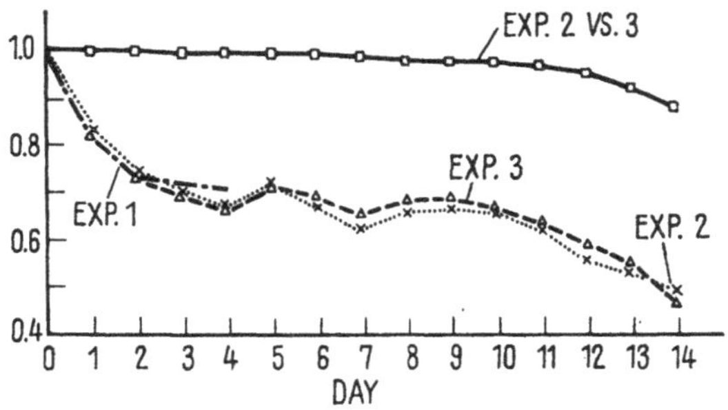

Fig. 2. Correlation coefficients between the observed and forecast pattern of the change in geopotential height of 500 mb from the initial time. Numbers refer to different refinements of the model. After Miyakoda et al. (1969).

ahead whether or not such an upper ridge or cold trough will stay, that will be a statement of immense economic value.

Fig. 2 gives another illustration of the same study: It shows the correlation coefficient between the observed and the predicted field of the change of the geopotential height of the 500 mb surface from the initial day to the forecast day. Experiments 1, 2 and 3 are different calculation experiments with different refinements of the model. Because only the deviations of the predicted field against the initial field have been used all stationary pressure patterns, as for example the Icelandic low, do not enter into the correlation. From the second to the tenth day the correlation remains at a value of about +0.7 and then decreases to +0.5 on the 14th day. The steep decrease from day zero to the second day is attributed by the authors to a strong cyclogenesis over the Pacific Ocean. That occurred just at the place where the fewest observations were available and where the initial situation was least clear. A much better prediction and a less steep decrease of the correlation coefficient would have been expected if reliable observations had been available in that area. This example demonstrates very clearly the significance of observations completely covering the total hemisphere. Such observations cannot be more easily obtained by other means than meteorological satellites.

3. The Use of Satellites for Prediction

In order to clarify whether and how satellite measurements of temperature, pressure and wind influence a prediction, Charney, Halem and Jastrow have made a very interesting theoretical experiment [2]. In its first part they simulated numerically the development of the general circulation over the northern hemisphere over a time interval of 155 days. This quasi-historical development considered the true or observed circulation with which all future computations would be compared. In following experiments it was to be investigated in what way observational errors could disturb the predictions and whether or

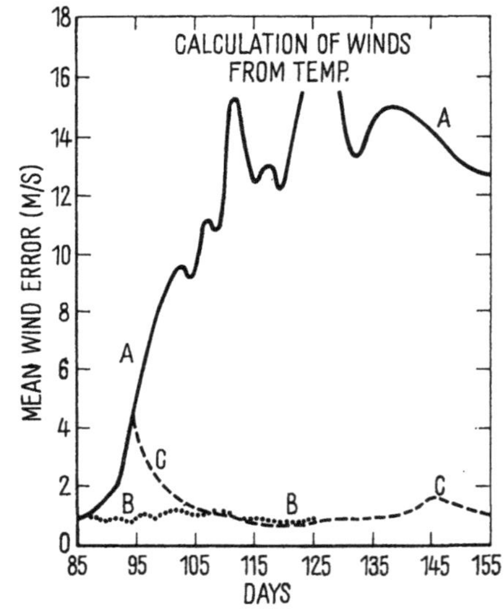

Fig. 3. Increase of wind error between the true and the perturbed situation (A); with insertion of satellite-observed temperatures beginning after 1 day (B), and after 10 days (C).

not future calculations could be improved by incorporation of new observations. Thus, in the next step of this experiment small errors were superimposed on the true field at the 85th day of the development; this particular date was chosen arbitrarily and had no particular meaning. The errors were randomly distributed, their size being between $+1°$ and $-1°$. For the disturbed situation the integration was carried out with that somewhat changed initial situation of the general circulation. To show the difference between the true and the disturbed situation the rms error in the wind difference between the two situations was chosen. Fig. 3 shows the increase of the error which was 1 m/s at day zero (the 85th day). After 3 days hardly any change can be recognized. At the 5th day the difference is 1.5 m/s, after 10 days 4.5 m/s and after 20 days it has increased to 10–15 m/s. With a rms error of 6 m/s one may call it a poor forecast. A difference of 15 m/s can be considered as a complete incoherence; there is no similarity between the two situaticns [3].

In the next step, beginning with the 86th day, temperatures of the true situation, i.e. of the original model, were taken and introduced into the disturbed situation. This was done with the arrangement in time and space of the temperatures observed by two polar orbiting satellites 90° apart. Temperatures with a mean error up to $\pm 1°$ randomly distributed were applied instead of the true temperatures from the original situation in order to imitate the measuring errors of a real satellite instrument. The results are shown in the dotted line *B* showing that the deviations between the original and the disturbed but again corrected situation remained at 1 m/s rms.

In a third experiment the disturbed situation was given 10 days to develop. The error increased to the high value of 4.5 m/s. At day 95, the same procedure of inserting correct temperatures was followed as described above. The errors decreased quickly and attained after 10 days again the small size of 1 m/s rms.

This picture shows convincingly that even a wrong prediction can be brought into order by feeding in correct temperatures even in the asynoptic way provided by polar orbiting satellites. Temperatures only were used for the corrections while wind was used as an indicator of errors. This shows at the same time the connection between temperature and pressure patterns and the wind field, and demonstrates that corrections in one element are sufficient for correcting the whole model.

At the Goddard Institute for Space Studies various similar experiments have been carried out: instead of the two polar orbiting satellites only one has been used; or measurements from four geostationary satellites have been imitated assuming that they could provide temperature profile measurements in their viewing field; smaller "measuring errors" have been applied to the observed temperatures before using them for the correction, e.g. errors of $\pm 1/2°$ or even the exact values; the rms errors for tropical and higher latitudes have been calculated separately. Finally it can be shown that the results become worse by taking measurements from four or six instead of two polar orbiting satellites. This is hard to understand at first. One reason may be that gravitational-inertial instabilities are caused which have no time to vanish or to be damped. These are interesting and important investigations per se, and progress in satellite meteorology should be seen in the light of these theoretical studies.

These investigations, in which simulated weather situations as well as simulated temperature measurements from satellites have been used, show that measurements from satellites are valuable. It is now considered whether measurements of this kind can be obtained at all or have been performed already. Temperature measurements of the required kind are already available and have been used increasingly since May 1969.

3.1. Temperature Measurement

3.1.1. SIRS

By the successful launching of Nimbus 3 immediately two instruments came into operation to measure atmospheric temperatures. These are the Satellite Infra Red Spectrometer SIRS, developed by Wark and coworkers at ESSA

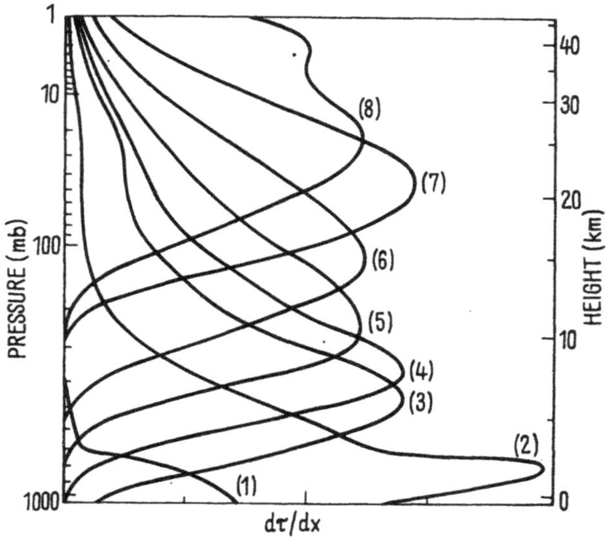

Fig. 4. Weighting functions of SIRS instrument in Nimbus 3.

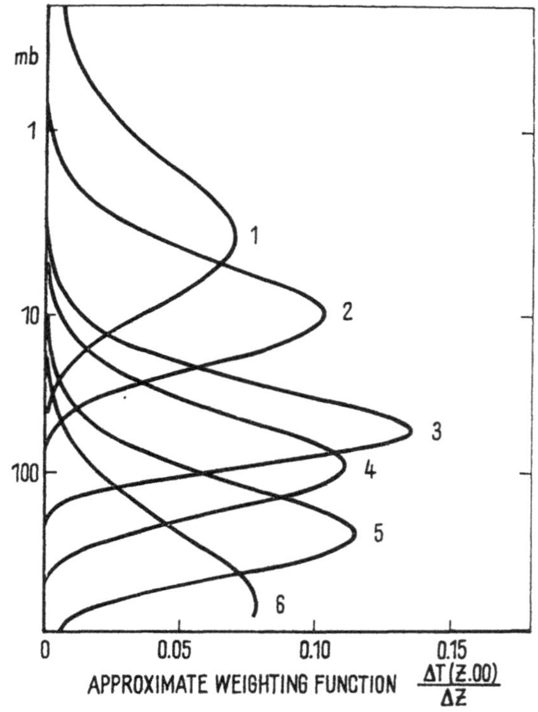

Fig. 5. Weighting function of SCR instrument in Nimbus 4.

and the InfraRed Interferometer Spectrometer IRIS designed by Hanel and coworkers at Goddard Space Flight Center [4]. Both of them measure the long wave radiance emitted by the earth + atmosphere at several frequencies. The first instrument is a grating spectrometer measuring the radiation emitted at seven frequencies of the 15 μm CO_2 band and in one channel at 11 μm in the window. Since the pioneering work of L. Kaplan it is well known that measurements inside an absorption band of an atmospheric gas at constant concentration permit the determination of the vertical temperature profile. Fig. 4 shows the weighting functions of the radiation measured with the SIRS instrument. At low absorption the radiation from the surface and the lowest layers emerges from the atmosphere. The stronger the absorption, the higher in the atmosphere is the emission layer. This fact enables us to derive the vertical temperature distribution from the radiation in the different frequency regions. There is, however, an overlapping of the weighting functions which makes simple calculations of the temperatures difficult. The peak of the uppermost weighting function, belonging to the Q-branch of the band, is at about 20 mb or 25 km. Fig. 5 shows a very similar experiment which was developed by Houghton and Smith at the Universities of Oxford and Reading and was launched in Nimbus 4 early in 1970. The weighting functions are somewhat different because of the different technique used. The uppermost peak is about 3 mb or 40 km; thus this instrument will give information up to the upper stratosphere.

Six weeks after the launching of Nimbus 3 the temperature profiles derived from the SIRS measurements were already being cautiously used in the daily weather analysis in USA, although Nimbus 3 must be considered an experimental and not an operational satellite. Cloud pictures from television which for years have been obtained by the US and Soviet Union satellites have been used in the weather services. There is, however, a difference. The traditional method of weather analysis by looking at the pressure systems in weather maps has not been changed in principle by viewing cloud pictures. These have been used as another means of recognizing pressure or frontal systems. There was no easy way of including the cloud pictures into numerical weather prediction because from computing procedure there are not calculated cloud distributions to compare with the real observed cloud distribution. This is quite different in the case of the temperature measurements: as described above temperature provides a direct initial value for numerical calculations of the circulation patterns.

For many years, mathematical procedures and experiments have been developed to convert radiation data to meaningful temperatures. Essentially three different methods have been tested and are now being utilized. The first method consists in developing a large number of measured vertical temperature profiles in series of empirical orthogonal functions, in which studies research workers from many nations have participated [5]. It has to be presupposed that a large quantity of temperature profiles have been measured by other methods, e.g. standard radiosondes, in order to derive the statistical functions. The constants of the mathematical series will be determined from the radiation measurements. A second procedure has been given by Chahine [6]; this does not need any presuppositions and is a pure, exact mathematical procedure without using statistical information. The third method [7] is a purely statistical procedure relying on the regression between radiation and temperature as measured by radiosondes in the main pressure levels of the atmosphere*. Figs. 4 and 5 showed the weighting functions of the different spectral regions with peaks at different altitudes and widths of about 6 km. Thus, other layers of the atmosphere also give contributions to the emerging radiation. Table 1

Table 1
Radiation–Temperature Correlation Matrix, May–June 1969

Pressure	Channel (Central Wave Number in cm^{-1})						
Level (mb)	669.3	677.8	692.3	699.3	706.3	714.3	750.0
1000	−0.69	−0.70	−0.74	−0.73	0.35	0.74	**0.91**
850	−0.73	−0.79	−0.82	−0.78	0.50	0.86	**0.91**
700	−0.80	−0.86	−0.90	−0.83	0.64	**0.93**	0.89
500	−0.81	**−0.87**	**−0.91**	−0.81	0.72	**0.92**	0.78
400	−0.78	−0.85	−0.89	−0.77	0.73	0.88	0.70
300	−0.74	−0.81	−0.82	−0.66	**0.76**	0.82	0.60
250	−0.44	−0.48	−0.39	−0.15	0.63	0.42	0.18
200	0.62	0.67	0.80	**0.87**	−0.33	−0.71	−0.71
150	0.75	0.82	**0.90**	**0.85**	−0.55	−0.82	−0.74
100	**0.88**	**0.94**	**0.95**	0.83	−0.66	**−0.89**	**−0.80**
50	**0.86**	0.89	0.85	0.77	−0.51	−0.78	−0.78
30	0.75	0.77	0.74	0.70	−0.32	−0.61	−0.65
10	0.14	0.12	0.10	0.15	0.13	−0.01	−0.14

shows a correlation table between the radiations of the seven channels and the temperatures at the main pressure levels. The first channel has its strongest absorption in the Q branch of the CO_2 band. This has the best correlation with the temperatures at 50 and 100 mb, the next with 100 mb, the third with 100 and 150 mb and so on downwards until the channels with the least absorption have their best positive correlation with the temperatures of the lower troposphere.

The best correlations are always fairly high. But there are also negative correlations attaining values of −0.8 or −0.9 with tropospheric temperatures or, for the stronger absorptions, with stratospheric temperatures. The reason for this is the well-known negative intercorrelation between tropospheric and stratospheric temperatures. One realizes that it will be possible by measurements in one spectral region to obtain information of stratospheric *and* tropospheric temperatures and vice versa. The regressions will, of course, depend somewhat on season and geographical latitude. Inversion methods have been founded on these multiple regression relations and have proved very successful in practice. (A first series of simultaneous radiation measurements and radiosoundings is of course needed to obtain the regression coefficients.)

The very good coupling between air temperature and pressure in the atmosphere has been known for many years. Nevertheless Table 2 is interesting.

* *Editor's note*: The fourth method has been developed in the USSR ('regularization technique'): see K. Ya. Kondratyev and Y. M. Timofeyev. Thermal sounding of the Atmosphere from Satellites, Gidrometeoizdat, Leningrad, 1970.

Table 2
Radiation–Geopotential Height Correlation Matrix, May–June 1969

Pressure Level (mb)	Channel (Central Wave Number in cm^{-1}) 669.3	677.8	692.3	699.3	706.3	714.3	750.0
1000	−0.09	−0.10	−0.16	−0.27	−0.28	−0.08	0.03
850	−0.54	−0.57	−0.64	−0.72	0.06	0.47	0.63
700	−0.72	−0.77	−0.83	−0.86	0.30	0.73	0.84
500	−0.81	−0.87	−0.92	**−0.90**	0.49	0.87	**0.89**
400	−0.83	−0.89	−0.94	**−0.91**	0.55	0.90	**0.88**
300	−0.84	−0.90	**−0.95**	−0.89	0.61	**0.92**	0.86
250	**−0.85**	**−0.91**	**−0.96**	−0.88	0.64	**0.92**	0.85
200	**−0.85**	**−0.92**	**−0.95**	−0.85	0.68	**0.93**	0.83
150	**−0.85**	**−0.92**	−0.92	−0.79	**0.71**	0.91	0.81
100	−0.82	−0.88	−0.86	−0.71	**0.71**	0.89	0.78
50	−0.32	−0.34	−0.28	−0.16	0.46	0.45	0.36
30	0.35	0.35	0.37	0.44	0.08	−0.15	−0.24
10	0.44	0.45	0.44	0.49	0.04	−0.26	−0.41

It shows the immediate correlations between radiation in the seven frequency ranges of SIRS and the heights of the standard pressure levels. As is to be expected, there is in this case a very high correlation, either positive or negative, with the pressure heights in the upper levels from 500 to 100 mb, while correlations at surface pressure are rather weak. Direct determination of pressure height from radiation measurements is possible as can be seen from the Table and from the examples in following figures.

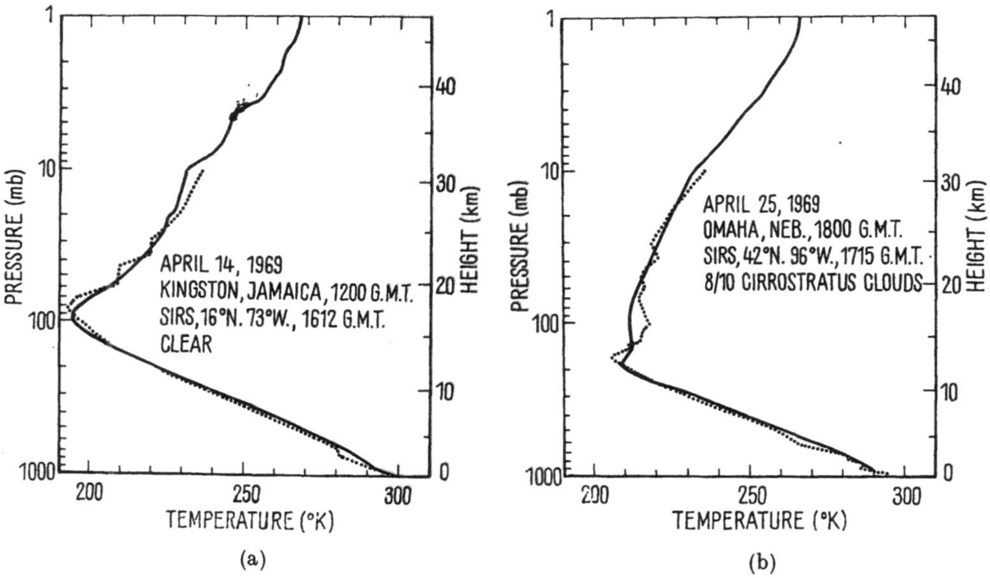

Fig. 6. Vertical temperature profile: full line, retrieval from SIRS measurements; dotted, radiosounding.
a) Kingston, Jamaica, 14 April 1969, 1200 GMT
SIRS, 16° N 73° W, 1612 GMT. Clear.
b) Omaha, Neb., 25 April 1969, 1800 GMT
SIRS, 42° N 96° W, 1715 GMT, 8/10 Cirrostratus clouds.

Fig. 6a shows the very first temperature inversion profile compared with the radiosounding over Jamaica on 14 April 1969. The dotted line is the radiosounding and the full line the measurement from the satellite. The agreement is generally very good. The details of the temperature structure at the tropopause cannot be resolved by the satellite; similarly the resolution of the small temperature inversion in the troposphere near the earth's surface is rarely possible. This is due to the width of the weighting function of the measuring channels. It is seen that apparently the radiation measurements give too high temperatures in the whole troposphere. The same effect can be seen in Fig. 6b.

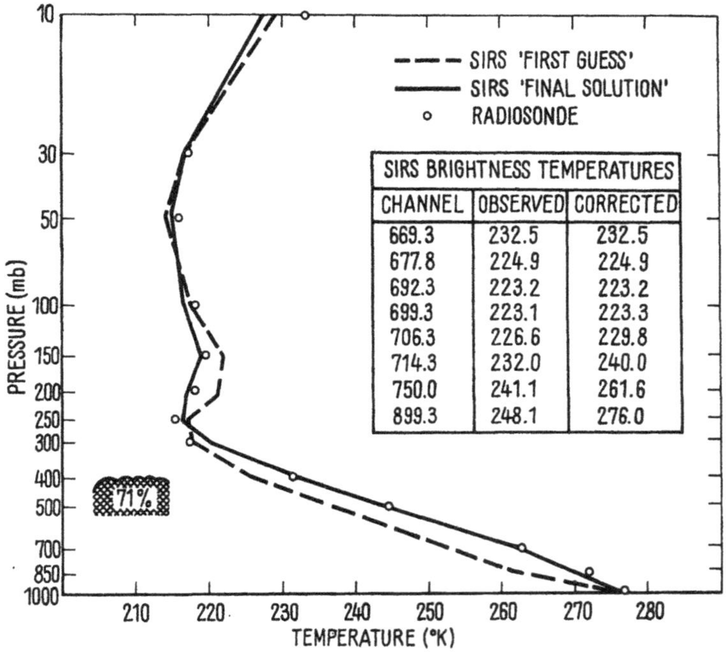

Fig. 7. Temperature profile retrieval of SIRS measurements under cloudy conditions. The table shows the radiances after correction for cloudless sky.

These inversions have been calculated by the direct method which is based essentially on theoretical transmission functions. Such a systematic error may hinder the application of the temperature inversions to operational analysis. The transmissions have therefore been corrected empirically.

Fig. 6b shows that there was 8/10 cirrostratus cloud. Evidently this had no influence on the evaluation. Originally, there had been doubts of the applicability of infrared sounding methods because of the prevention of infrared radiation by clouds, but it was found that even *with* cloudiness in about 90% of all cases reliable temperature profiles could be obtained.

Fig. 7 shows such an evalutaion. The broken line is the first guess, the full line the definite profile. The large points show the temperatures as measured at the nearest radiosonde station. The table in the right upper corner shows the measured radiances of the eight channels and the values corrected for cloudiness, all expressed in brightness temperatures. One sees that the corrections in the four opaque channels are negligible, while those in the four transparent channels attain values of 20 to 30°.

The possibility of determining directly the heights of the pressure surfaces by correlation (Table 2) is an essential advantage. The procedure is analogous to that of temperature inversion. Fig. 8 shows the contour lines of the 500 mb surface over the Pacific Ocean on 24 June 1969, 0000 GMT, where the network

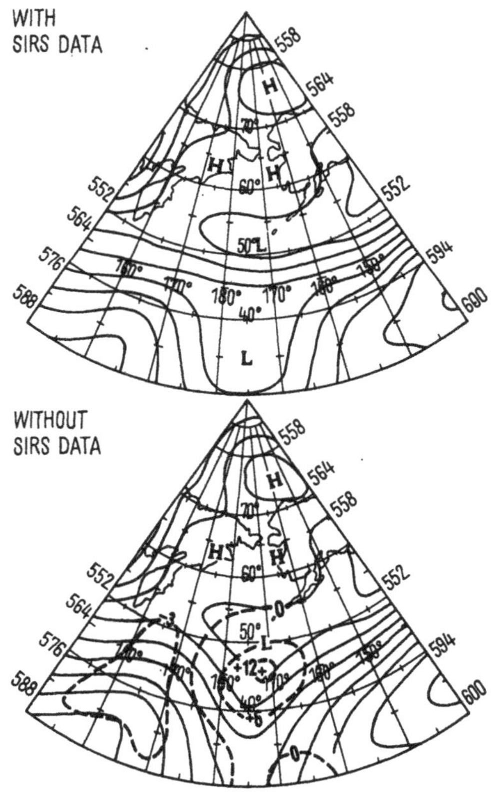

Fig. 8. Comparison of analysis of 500-mb surface obtained with and without SIRS soundings. 24 June 1969. The differences in geopotential decametres are shown by the dashed isolines.

of surface or radiosonde stations is very thin. The lower picture shows the analysis without, the upper with use of the SIRS data. The dashed lines give the corrections which have been applied and already incorporated in the upper picture by means of the SIRS measurements of the satellite. They amount to +120 geopotential metres and quite decisively change the trough which was found from routine methods into a separate low. It is to be expected that by this method even more dramatic changes will be found over the southern hemisphere and in winter [8].

The SIRS instrument has been developed for purpose of weather analysis. It supplies invaluable data for daily use. During 1969–1970 in Nimbus 3 and in other satellites some other instruments have been flown which are not yet so well adapted to operational use in the weather service. Although such an application might not be excluded the experiments first serve for pure research.

3.1.2. IRIS

The second instrument in Nimbus 3 built for a similar purpose is the IRIS of R. Hanel [4]. It gives a complete spectrum of the emerging long wave radiation from 5 to 25 μm or in wavenumbers from 2000 to 400 cm^{-1}, thus supplying about 300 measuring points in a range where SIRS gives eight data points. Fig. 9 shows such a record. The spectral resolution is 5 cm^{-1}. The envelope

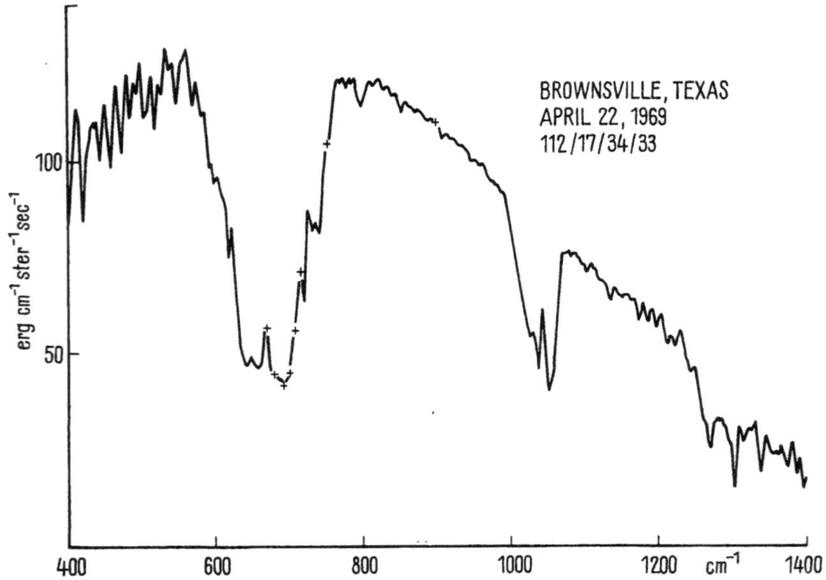

Fig. 9. Emission spectrum measured by IRIS instrument from Nimbus 3. Crosses indicate simultaneous measurements by SIRS.

to the uppermost parts of the record represents the Planck radiation curve of the ground temperature which in these parts is not disturbed by absorption of the atmospheric gases. The large dips are produced by the low radiation temperatures within the strong absorption bands. At the right side the vibration rotation band of H_2O at 6.3 μm (the centre of the band is somewhat outside the picture), at 1050 cm^{-1} the ozone absorption band with its two peaks, and between 600 and 800 cm^{-1} the CO_2 band. The crosses give the measurements in the eight channels of SIRS at the same time. The coincidence is excellent if one considers that both these instruments are designed on completely different principles and with independent calibration. The agreement was not in all cases as good. The reason may be that the aperture of the SIRS is somewhat larger, that its viewing point moves slightly with the motion of the satellite while the IRIS is allowed to swing in order to keep the same spot on the earth in the centre of the viewing angle. Therefore the two instruments may see different cloud systems. At the left side of the picture is the structure of the H_2O rotation band. It should be emphasized that the structures discernible outside the large bands are by no means instrumental noise. Many single lines can be identified by comparison with theoretical calculation or with laboratory measurements of higher resolutions as absorption lines of H_2O, CO_2 or other gases. They are real and could be evaluated.

The exhaustive evaluation of the information given by these spectra would need enormous effort, and therefore only a few evaluations have been made.

Fig. 10 shows the temperatures as calculated from the measurements in Fig. 9 together with the evaluations of the simultaneous SIRS measurements and a radiosonde of a station nearby. The agreement is very good.

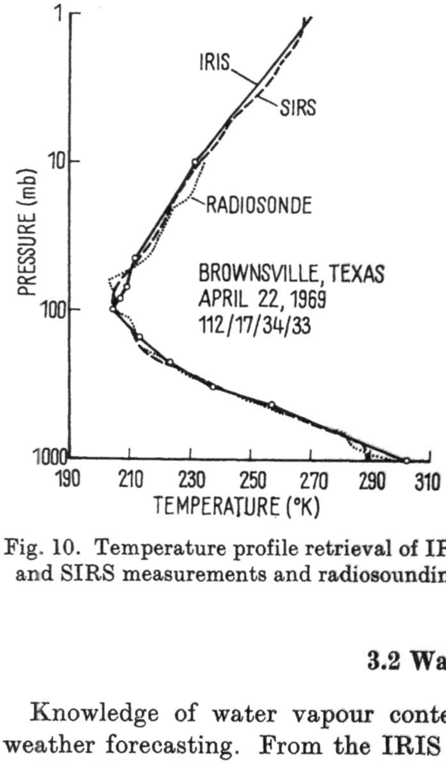

Fig. 10. Temperature profile retrieval of IRIS and SIRS measurements and radiosounding.

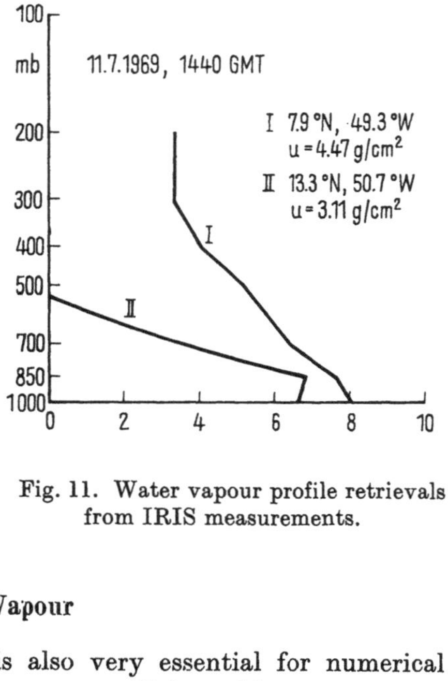

Fig. 11. Water vapour profile retrievals from IRIS measurements.

3.2 Water Vapour

Knowledge of water vapour content is also very essential for numerical weather forecasting. From the IRIS measurements H_2O profiles can be evaluated. Two which appear particularly interesting from a meteorological point of view are shown in Fig. 11. On 11 June 1969 Nimbus 3 crossed the intertropical convergence zone (ITC) over the Atlantic Ocean at about 10° N 50° W. Seven measurements south of the ITC showed a high water vapour content between 4.1 and 4.5 g cm^{-2}; north of it the vapour concentration close to the surface was almost the same, but above 600 mb the vapour density was so small that it could not be distinguished from zero. The total water vapour mass was between 3.1 and 3.4 g cm^{-2}. In the cloudy region of the ITC itself evaluation of the water vapour concentration is impossible.

Water vapour determinations can also be made by measurements in the 1.35 cm absorption band of the microwave region where there is much less prevention by clouds. The Soviet satellite Cosmos 243 carried an instrument to measure the radiation at four wavelengths in the microwave region, one in the centre of the 1.35 cm band, two others one on each side in order to obtain the undisturbed radiation of the sea surface. Obouchov and Tatarskaya [9] succeeded in drawing a map of the distribution of the total water vapour content (Fig. 12) for the conditions of 24 September 1968 over the Southern Pacific. In this picture the South Pole is at the top, Australia at the right and

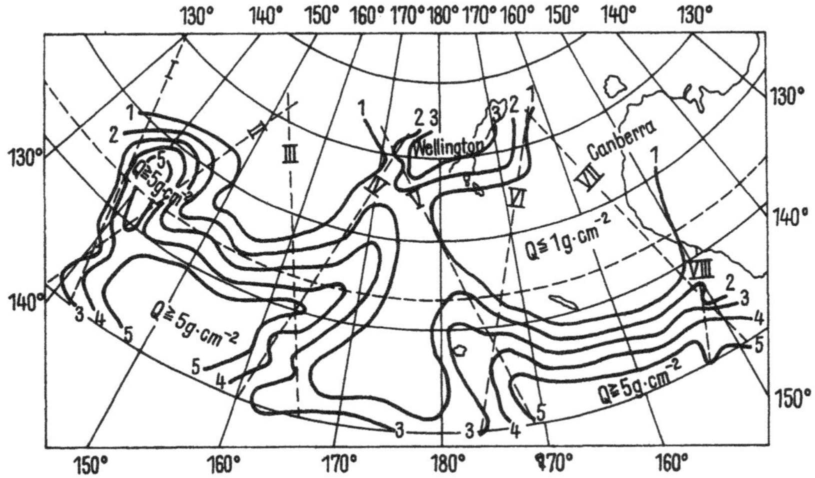

Fig. 12. Total water vapour content ($g\,cm^{-2}$), derived from microwave measurements from Cosmos 243. After Oboukhov and Tatarskaya [9].

New Zealand in the centre; the lowest latitude circle is 10° S. The dry areas of the horse latitudes with less than 1 g cm^{-2} water vapour content and the increase towards the equator are clearly discernible as well as the higher water vapour content in a low pressure area at New Zealand. In principle an O_2 line at 0.5 cm wavelength could be used to determine temperature profiles. It is, however, a weak line and there are disturbances due to the variable emissivity of the sea surface depending on the state of the sea and the amount of foam on it. These difficulties have not yet been overcome. The emissivity of water surfaces is about 0.4, thus the brightness temperature of the water is about 140 °K. Rain clouds over sea appear as warm spots.

Over land the emissivity of the ground strongly depends on the water content of the soil. One can therefore identify moist spots in a desert by microwaves because of their low brightness temperature. Differences between the Sahara and the tropical rain forest could be distinguished by the Cosmos 243 satellite.

3.3. Ozone

In Fig. 9 the 9.6 μm ozone absorption band is clearly seen. Prabhakara [4] has developed a two-parameter model for the vertical distribution of ozone because the resolution of the band does not allow the determination of more parameters or better resolution in the vertical profile. He applied the method of empirical orthogonal functions. Essentially only the total ozone amount can be found. Fig. 13 shows a profile in the tropics which gives a good coincidence between directly measured ozone concentrations and the IRIS evaluation. Even a mapping of the total ozone amount over the globe was possible, as shown in Fig. 14 for 24 April 1969. This is spring in the northern hemisphere.

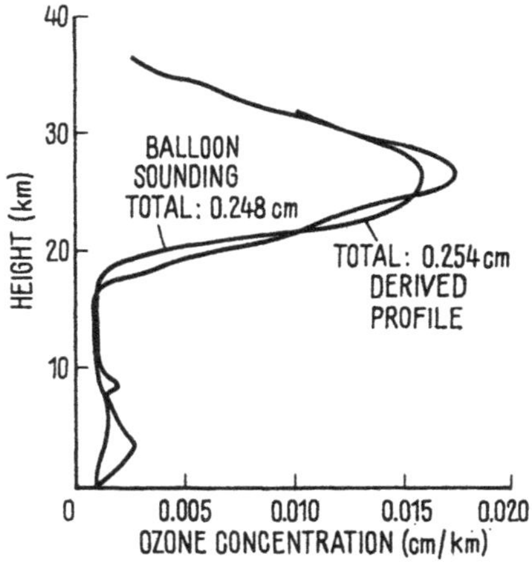

Fig. 13. Vertical ozone profile retrieval from IRIS measurements. Balboa, 18 April 1969.

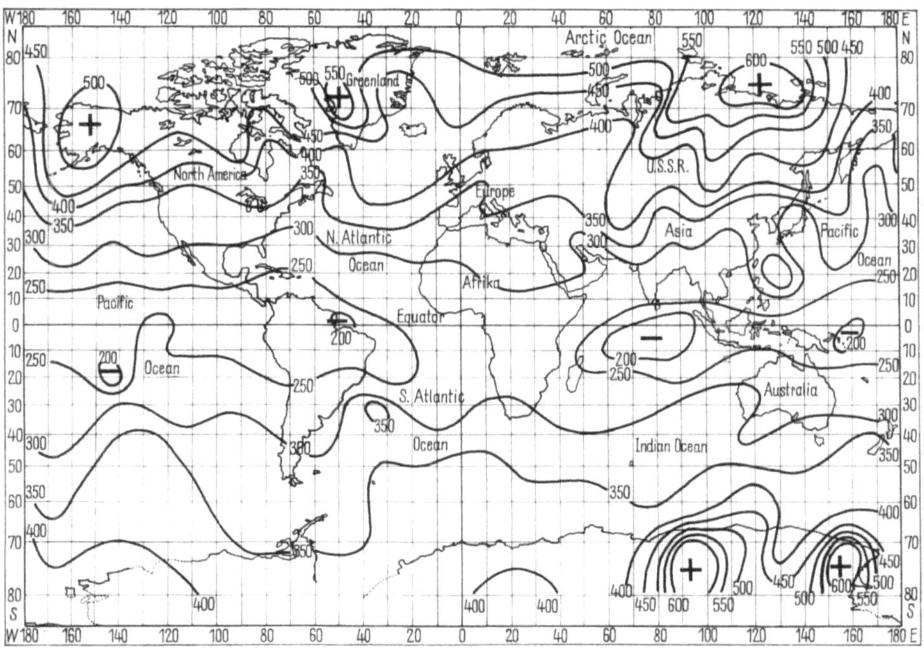

Fig. 14. Total ozone content (cm STP) derived from IRIS measurements taken from Nimbus 3. 24 April 1969.

A latitudinal zone of small values can be distinguished along the equator with values occasionally below 0.2 cm. In the north polar area the amounts are higher and reach 0.6 cm over North-East Siberia and 0.5 cm over Alaska. The amount over the southern hemisphere is generally a little lower. This map seems to be the first map which shows quantitatively the ozone distribution over the whole globe. Ozone is a good tracer of atmospheric motions in the lower stratosphere. When such a map can be drawn for every day of the year, incomparably more material will be available for ozone research than could ever be gained by a network of Dobson instruments.

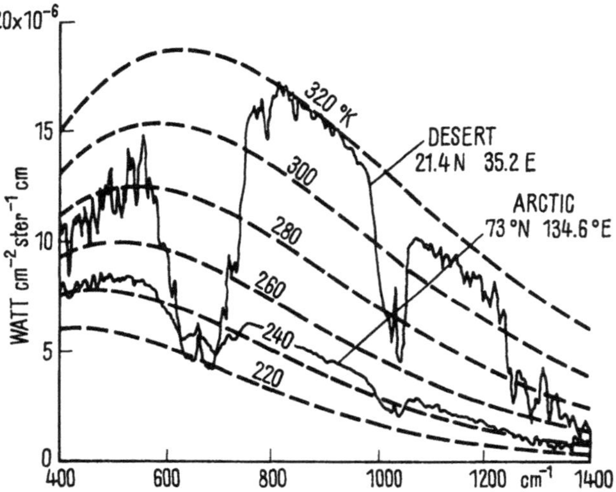

Fig. 15. Emission spectrum measured by IRIS from Nimbus 3.

Fig. 15 shows again a record with the IRIS instrument, in this case over a desert. The temperature envelope in the range of 800 to 900 cm^{-1} indicates a surface temperature of 320 °K, while on the other side of the ozone band the temperature is 10° lower. This is an effect of the Reststrahlen which are quite evident with the mineralic soil of the desert. Quartz and feldspar, the most important minerals of sand and all soils have just in that region strong Reststrahlen reflection bands and it appears therefore possible to distinguish by the diminished emission the nature of the surface minerals over a planet which has no atmospheric absorption, for instance Mars. The spectrum with low temperatures over the snow-covered Arctic in Fig. 15 does not show this effect.

3.4. Winds

I shall now come to the determination of winds in the atmosphere from satellites. Generally, wind can be determined by swimming and drifting bodies in the atmosphere. Such bodies will easily be seen in clouds which can be followed from geostationary satellites. This is hardly possible with cloud pictures taken from polar orbiting satellites. Clouds can, however, be used as tracers only with some restrictions because they are always changing. Cloud systems may originate over large areas without having drifted in, or they may disappear, "melt away", and by that simulate a huge travelling speed, or they may become stationary while the wind is blowing through. It is not always easy to overcome these difficulties. Four different methods have so far been developed to determine the cloud motion from two pictures following each other at an interval of 20 minutes, which is the time interval of the pictures taken by the US geostationary satellites.

One procedure is purely manual and rather slow. From the series of pictures taken on one day a photographic film is made, eventually showing only a section of the globe. The film is projected on a screen showing the geographical coordinates and the countries. A well trained meteorologist marks with a pointer a cloud whose motion can be followed and fixes the initial and the end position. If sufficient cloud structures are discernible he can cover the whole field with wind arrows. Another method uses only two consecutive pictures. The positions of a selected cloud spot in both pictures is accurately fixed by careful comparison of the isophots, the displacements taken into punchcards and the wind vector evaluated by a computer. A third method uses the apparent stereo effect. In stereographic viewing of two cloud pictures, clouds travelling from west to east are seen in a higher or lower position than clouds drifting in the opposite direction. The impression in a stereoscope is plainly fascinating and striking. One recognizes cloud layers high above others in apparently quite low layers. A well trained eye can immediately draw contour lines of the apparent altitude of the cloud layers which are not heights but isolines of wind speed in the west–east direction. Turning the pictures through 90° one can determine in the same way the north–south displacement of the clouds. Usage of a stereocomparator as in geodesy to fix the contour lines of the earth's surface should result in excellent wind maps. The true altitude of the clouds has however yet to be determined from their appearance or temperature, for the illusion is perfect that the apparently high cloud is really a high cloud.

The fourth method finally can be called a fully automatic one. In two consecutive pictures a computer is drawing very carefully lines of equal brightness. These pictures show many fine structures. Both pictures will be so displaced with respect to each other that a best fit of the shifted contours will be obtained. The method gives good results for a time interval of 20 minutes. In comparing two pictures one hour apart, i.e. omitting two pictures of the ATS satellite in between, the change of the individual cloud structures is too large and the best fit does not agree with the mean of the three shorter intervals. This may also be one reason why pictures from polar orbiting satellites cannot be used for wind determination from cloud displacements, for by this method it is not possible to determine cloud altitude. Whatever method is applied, it is applicable only when clouds are present. Experience has shown that it will hardly be possible to identify more than two cloud layers in a picture or to determine winds at more than two altitudes in the atmosphere.

4. Conclusion

In this survey many items have been omitted: e.g. the continuing success of the television pictures of clouds as provided by the Meteor and ITOS satellites and their statistical averaging, which opens new aspects of research in cloud climatology; the mapping of the annual course of temperatures in the highest layers of the stratosphere by which our climatology of the stratosphere receives a "new look"; the measurements of atmospheric elements from drifting balloons, the collection of their signals and the determination of their locations from satellites, which gives another way of determining the wind in the atmosphere. These measurements may, however, become very important in the coming Global Atmospheric Research Programme.

Acknowledgments

The author is deeply indebted to Messrs. B. J. Conrath, M. Halem, R. A. Hanel, W. C. Jacob, R. Jastrow, C. Prabhakara, W. L. Smith, D. Q. Wark and H. M. Woolf for making available valuable information and hitherto unpublished papers, of which he has made use in this article.

References

[1] K. MIYAKODA, J. SMAGORINSKY, R. F. STRICKLER and G. D. HEMBREE, Mon. Weath. Rev. **97**, 1 (1969).
[2] J. CHARNEY, M. HALEM, and R. JASTROW, J. Atmos. Sci. **26**, 1160 (1969).
[3] R. JASTROW and M. HALEM, Bull. Amer. Met. Soc. **51**, 490 (1970).
[4] B. J. CONRATH, R. A. HANEL, V. G. KUNDE and C. PRABHAKARA, J. Geophys. Res. **75**, 5831 (1970).
[5] D. Q. WARK and H. E. FLEMING, Mon. Weath. Rev. **94**, 351 (1966).
[6] M. T. CHAHINE, J. Opt. Soc. Amer. **58**, 1634 (1968).
[7] W. L. SMITH, H. M. WOOLF and W. C. JACOB, Mon. Weath, Rev. **98**, No. 8 (1970).
[8] W. L. SMITH, H. M. WOOLF and D. Q. WARK, WMO, WWW Planning Rep. No. 30, App. H (1969).
[9] A. M. OBOUKHOV and M. S. TATARSKAYA, Meteorol. i Gidrol., No. 11, 36 (1969).

Space Research XI — Akademie-Verlag, Berlin 1971

METEOROLOGICAL RESULTS FROM NIMBUS SIRS OBSERVATIONS

W. L. SMITH and D. Q. WARK

National Environmental Satellite Center
Environmental Science Services Administration
Washington, D.C., USA

The Nimbus 3 and Nimbus 4 meteorological satellites carry the first and second models of the Satellite Infra-Red Spectrometer (SIRS). The Nimbus 3 model, SIRS-A, measures the radiation emitted by the earth and atmosphere in seven spectral intervals of the 15 μm band of carbon dioxide and one spectral interval of the 11 μm window region. In addition to the window and carbon dioxide spectral observations, the Nimbus 4 model, SIRS-B, measures the earth's outgoing radiation in six spectral intervals of the rotational water vapor band. The absolute accuracy of the SIRS radiance observations is better than one per cent; the relative accuracy is better than one-half per cent. The SIRS radiance observations are being used to specify the three-dimensional distribution of temperature, geopotential height, and water vapor.

This paper presents various meteorological results obtained from the Nimbus SIRS experiments. Sample temperature and water vapor profiles calculated from SIRS radiances are given. A case is presented illustrating an improved weather analysis with SIRS data over the central Pacific Ocean which improved the 3-day forecast for the central and western part of North America. Tropospheric and stratospheric constant pressure analyses for the Southern Hemisphere are presented to show the usefulness of the SIRS data in regions largely devoid of conventional data.

1. Introduction

The Satellite Infra-Red Spectrometer A (SIRS-A) experiment on board the Nimbus 3 spacecraft measures the radiation emitted by the earth and atmosphere in seven narrow spectral intervals of the 15-μm carbon dioxide band and one narrow spectral interval of the 11-μm window regions. The radiation measured in the various spectral intervals corresponds to a vertically weighted mean temperature of different atmospheric layers.

In addition to the window and carbon dioxide spectral observations, SIRS-B, flying on Nimbus 4, measures the earth's outgoing radiation in six narrow (5 cm^{-1}) spectral intervals of the rotational water vapor absorption band. The radiation measured in these spectral intervals can be used to estimate the water vapor profile of the atmosphere.

Fig. 1 shows the weighting functions ($d\tau/d \ln p$ where τ is the atmospheric transmittance and p is atmospheric pressure) pertaining to the radiances measured in the eight SIRS-A channels. As indicated, radiances measured in the most absorptive spectral intervals of the 15-μm carbon dioxide band originate from the upper levels of the atmosphere. The radiance measured in the 11-μm

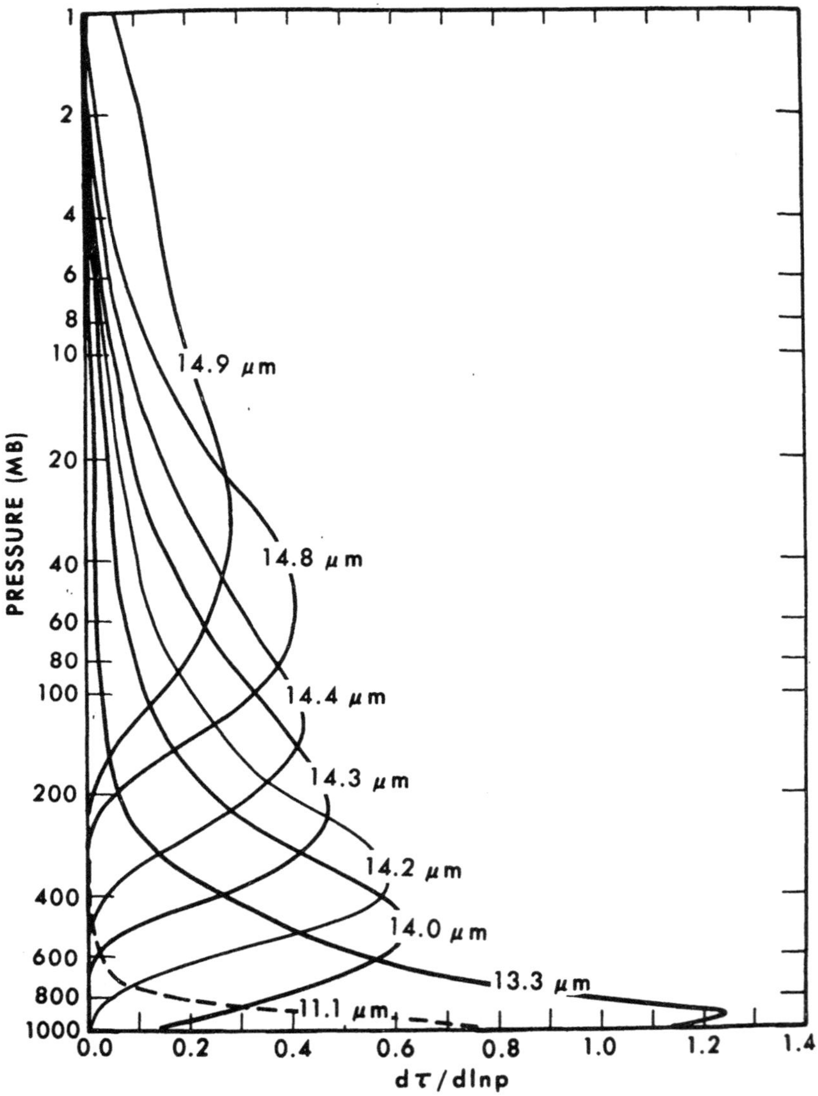

Fig. 1. The derivative of transmittance with respect to the logarithm of pressure. These functions approximately describe the relative sensitivity of the eight SIRS-A radiance observations to temperature variations in various altitude layers of the atmosphere.

window, where atmospheric absorption is very weak, originates mostly from the earth's surface. Radiances measured in other 15-μm spectral intervals originate from various intermediate layers of the atmosphere. Each spectral channel receives its radiation mostly from a layer about 6 km thick. The layers sensed by adjacent spectral channels partially overlap each other. The weighting functions pertaining to the SIRS-B window and carbon dioxide channels are similar to those of SIRS-A. Weighting functions of water vapor exist for the radiances measured in the SIRS-B water vapor channels, but are dependent upon the water vapor distribution. The information content of these measurements resides mainly in the troposphere where a significant amount of water vapor exists.

Table 1 shows the correlation of SIRS-A radiances with radiosonde temperature observations between 35° N and 55° N. As shown, the eight SIRS-A radiances are most highly correlated with the temperature at eight different

Table 1
Correlation of SIRS Radiances with Radiosonde Temperature Observations (700 Soundings: 35° N–55° N; 29 May–15 June 1969)

Pressure Level of Highest Individual Correlation (mb)	Highest Individual Correlation SIRS Channel	Correlation	Multiple Correlation With All SIRS Channels
1000	1 (11.1 μm)	0.97	0.98
850	2 (13.3 μm)	0.91	0.93
500	3 (14.0 μm)	0.92	0.96
250	4 (14.2 μm)	0.63	0.76
200	5 (14.3 μm)	0.87	0 92
100	6 (14.4 μm)	0.95	0.97
50	7 (14.8 μm)	0.89	0.91
30	8 (14.9 μm)	0.77	0.81

pressure levels between the earth's surface and the middle stratosphere. It is the vertical independence of the different spectral radiance observations which allows a temperature profile to be determined. The multiple correlation for almost all pressure levels below 10 mb is greater than 0.90. The high radiance-level temperature correlations are a result of the high interlevel correlation of atmospheric temperature. The detailed vertical temperature profile is determined from all eight spectral radiances.

There are numerous techniques for calculating temperature profiles from satellite-observed spectral radiances. The statistical regression method was the first technique employed for large scale reduction of the SIRS-A data in terms of atmospheric temperatures. The regression method, which makes use of the multiple regression relations for both temperature and geopotential height, is derived from coincident satellite and conventional observations. The advantage of this particular method is that the transmittance characteristics of the atmosphere need not be as precisely known, as required by all other techniques.

Temperature values may be calculated at any arbitrary number of pressure levels, although the results are only meaningful at levels which contribute significant amounts of energy to the measured radiances. Other types of statistical and non-statistical techniques, which have no requirement for current conventional data, are proving to be equally successful with adequate atmospheric transmission functions. An iterative method is being used to infer water vapor profiles from Nimbus 4 SIRS-B rotational water vapor band spectral radiance observations.

Retrieval procedures have been implemented with Northern Hemisphere SIRS-A observations on a routine operational basis since 20 May 1969. Southern Hemisphere implementation up to now has been on an experimental basis. This paper presents some of the meteorological results obtained from the Nimbus 3 SIRS-A and Nimbus 4 SIRS-B observations.

2. Meteorological Results

Fig. 2 compares a temperature profile calculated from SIRS-A radiances observed over Berlin wtih the Berlin radiosonde temperature profile. There is good agreement of the gross characteristics of the two soundings. However, it is shown that small-scale vertical features of the temperature profile were not retrieved from the infra-red radiance observations.

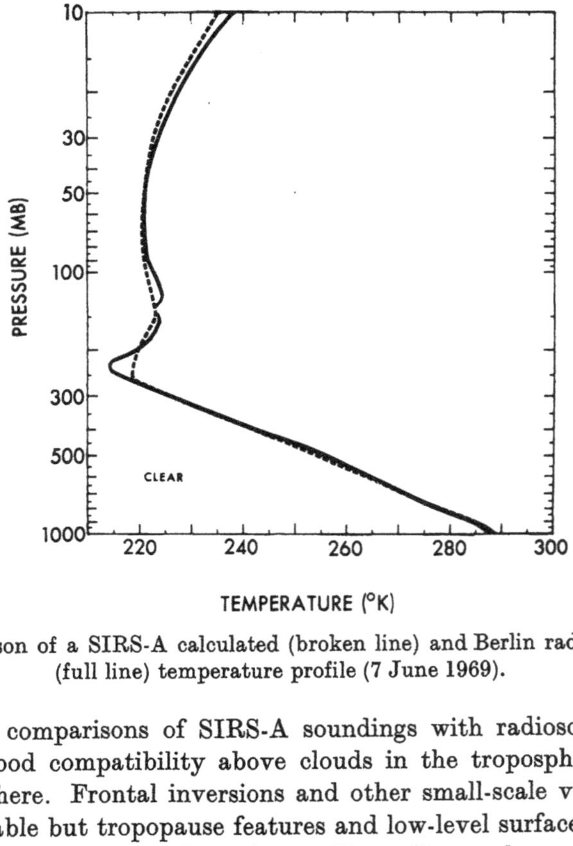

Fig. 2. Comparison of a SIRS-A calculated (broken line) and Berlin radiosonde observed (full line) temperature profile (7 June 1969).

Many other comparisons of SIRS-A soundings with radiosonde soundings have shown good compatibility above clouds in the troposphere and in the lower stratosphere. Frontal inversions and other small-scale vertical features are not detectable but tropopause features and low-level surface inversions are generally retrieved quite well from the satellite radiance observations.

Several experiments have been conducted to assess the possible improvement of weather forecasts by addition of satellite data to the existing conventional network of meteorological observations. The first experiment consisted of running forecasts up to 72 hours from two slightly different initial conditions:

(i) the objective analysis obtained from conventional data only, and (ii) the objective analysis obtained with SIRS-A and conventional data. The two initial conditions were obtained from observations on 24 June 1969.

Fig. 3 shows the differences in the 500-mb analysis constructed with and without SIRS-A data over the mid-Pacific. In this region the primary difference in the analysis produced by the SIRS-A soundings is the increase in the magnitude of the heights and the increase in the speed of the westerlies across the northern

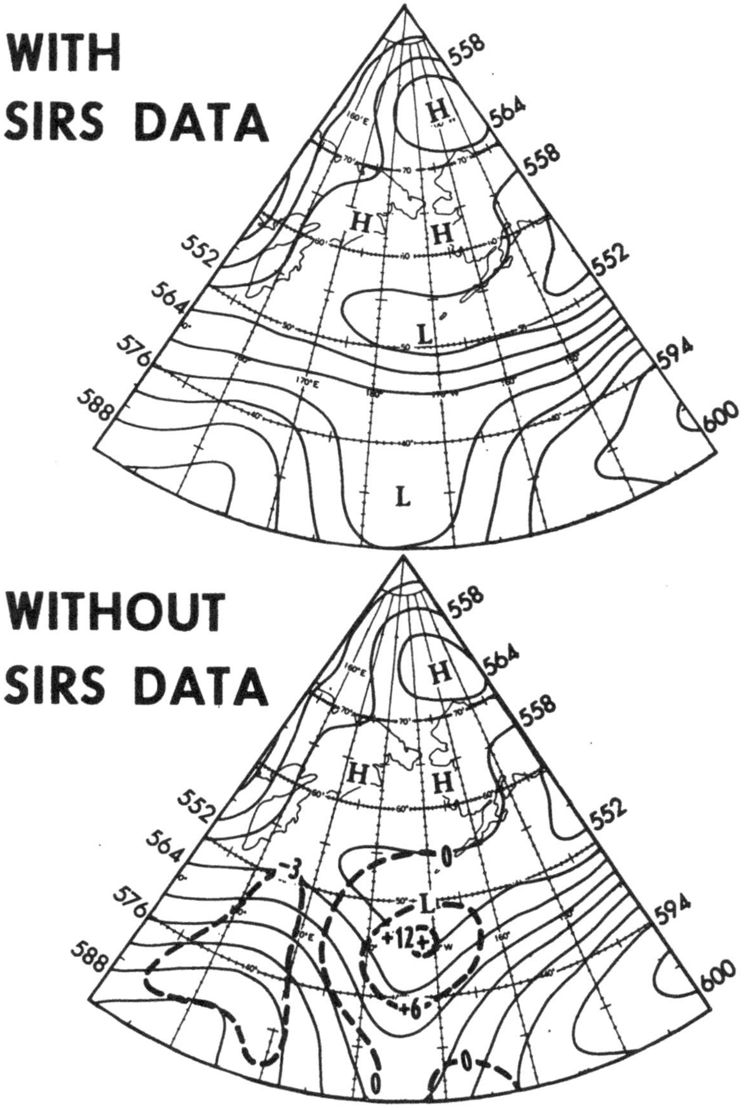

Fig. 3. Comparison of objective analyses of 500-mb height obtained with and without SIRS-A soundings (24 June 1969). The differences in decameters are shown by the dashed isolines.

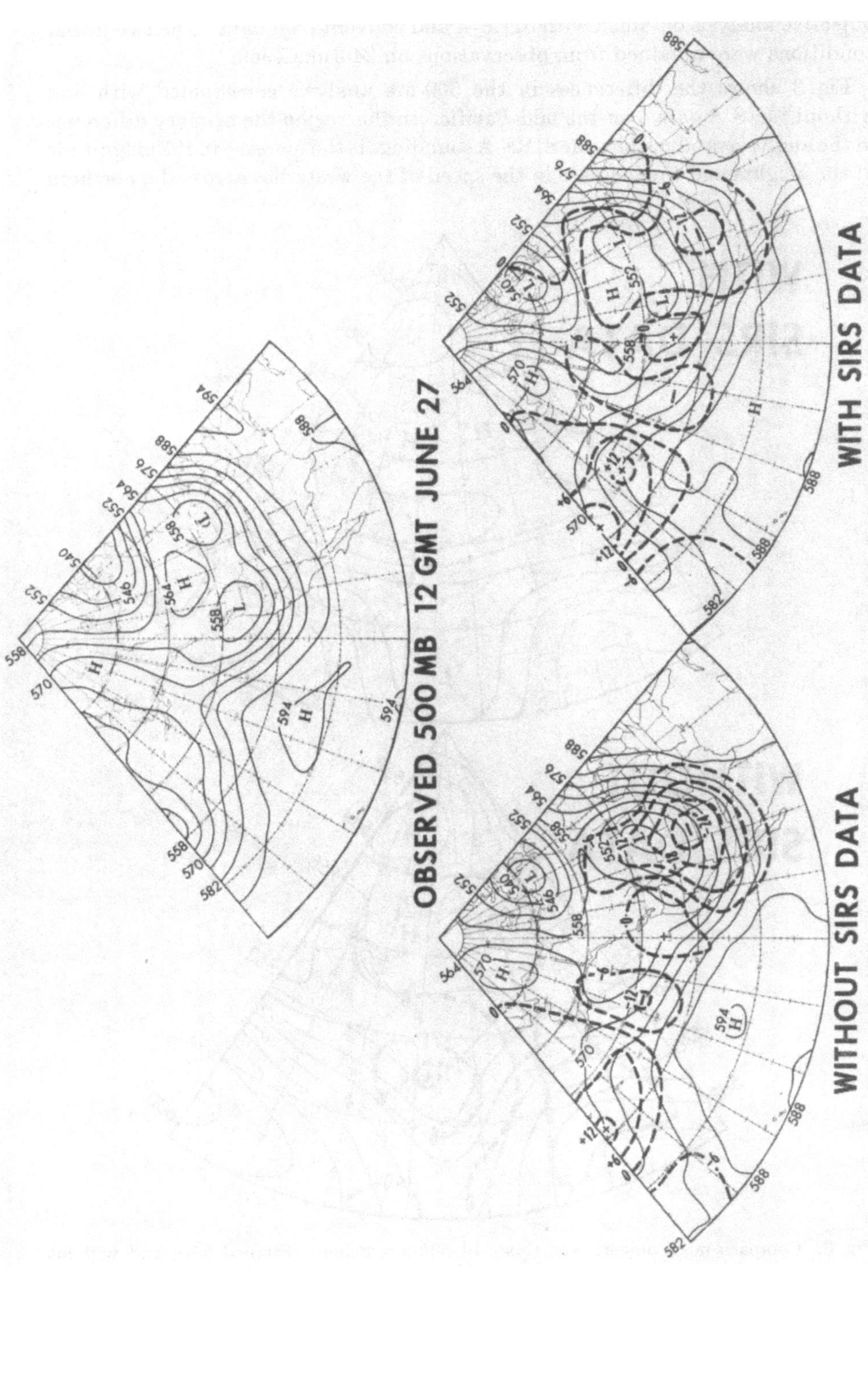

Fig. 4. Comparison of 72-hour 500-mb forecasts obtained from the initial condition specified with and without SIRS-A data (27 June 1969). Differences of each forecast with the observed 500 -mb distribution are indicated in decameters by the dashed isolines.

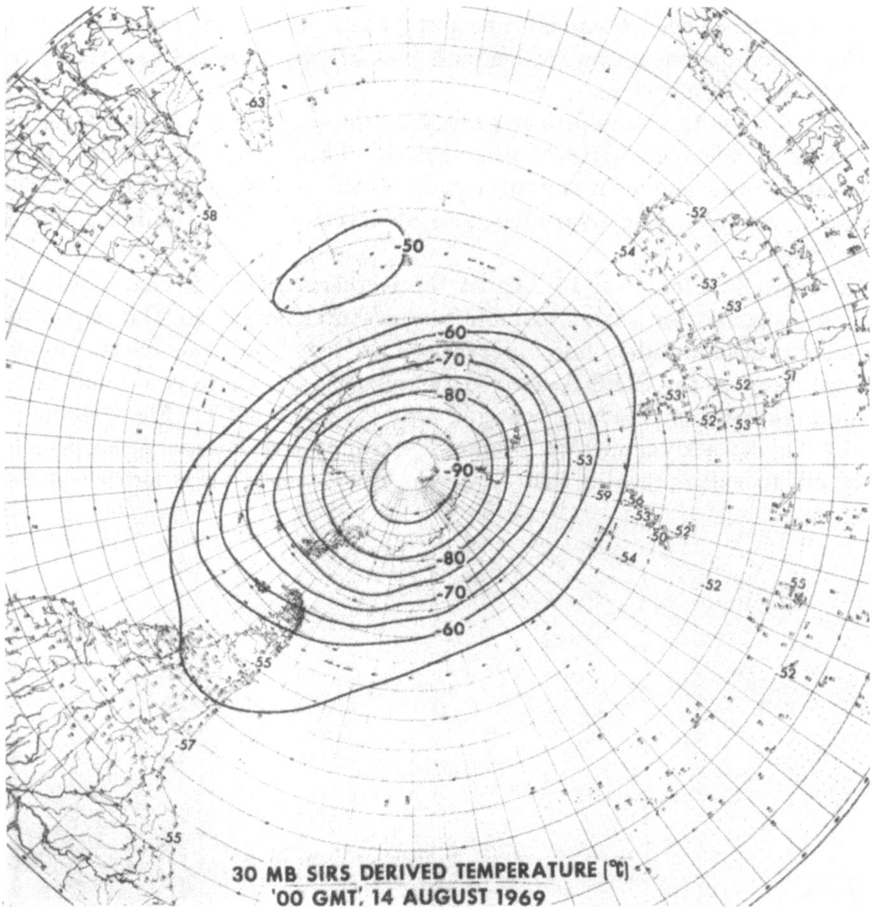

Fig. 5. Analysis of the Southern Hemisphere 30-mb temperature field obtained from SIRS-A radiance observations.

portion of a trough at 175° W, leaving a low center at 35° N. As shown, the largest absolute differences were 12 decameters.

Fig. 4 shows the differences in the two 3-day forecasts. Over the United States the forecast based on the satellite analysis closely resembles the observed situation. The main feature is the double low over the United States, separated by a short wave ridge. This feature was not forecast without the satellite data. The largest absolute errors of the SIRS-A forecast are only half of those resulting from initial conditions specified without the satellite data.

This example illustrates how a relatively small improvement in the analysis over the Pacific can produce a relatively large improvement downstream in a short-range prediction. Similar experiments have indicated some improvement in the forecast, although they are usually less dramatic than in the case illustrated above.

The Northern Hemisphere is rich with data in comparison with the Southern Hemisphere. For the first time, the temperature and mass distributions in the latter area have been obtained, using the SIRS-A spectral radiances, with a degree of completeness comparable with that usually accepted as a requisite in the Northern Hemisphere.

Fig. 5–7 show three temperature analyses obtained for the 30-mb level from Southern Hemisphere SIRS-A observations. This sequence of charts clearly illustrates the development and subsequent breakdown of a Southern Hemisphere "Sudden Stratospheric Warming" as observed by the Nimbus 3 SIRS-A instrument.

As shown in Fig. 5, on 14 August the isothermal field is characterized by a circumpolar vortex with maximum temperatures near −50 °C (temperatures from conventional radiosonde observations are given on these charts to verify the accuracy of the SIRS-A derived temperatures). There is a small area of slightly warmer temperatures centered near 47° S and 50° E. Fig. 6 shows that by 16 August a dynamically produced warming has increased temperatures at 30 mb to values greater than −40 °C. The warmest air is located at 57° S

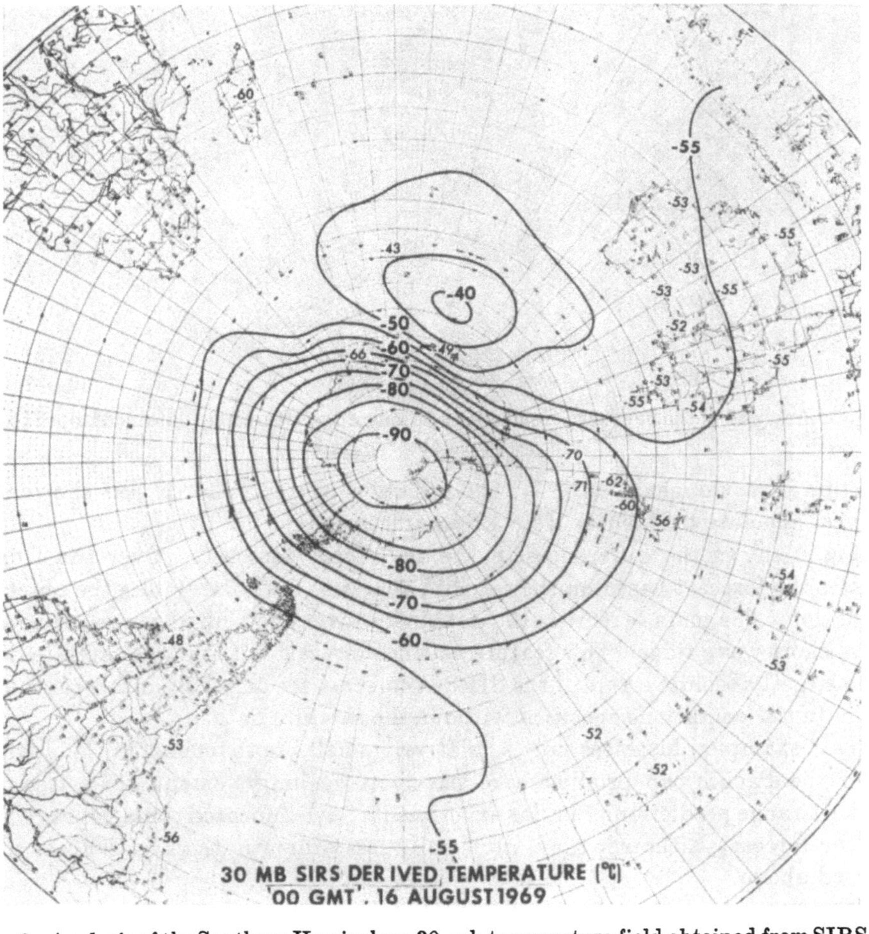

Fig. 6. Analysis of the Southern Hemisphere 30-mb temperature field obtained from SIRS-A radiance observations.

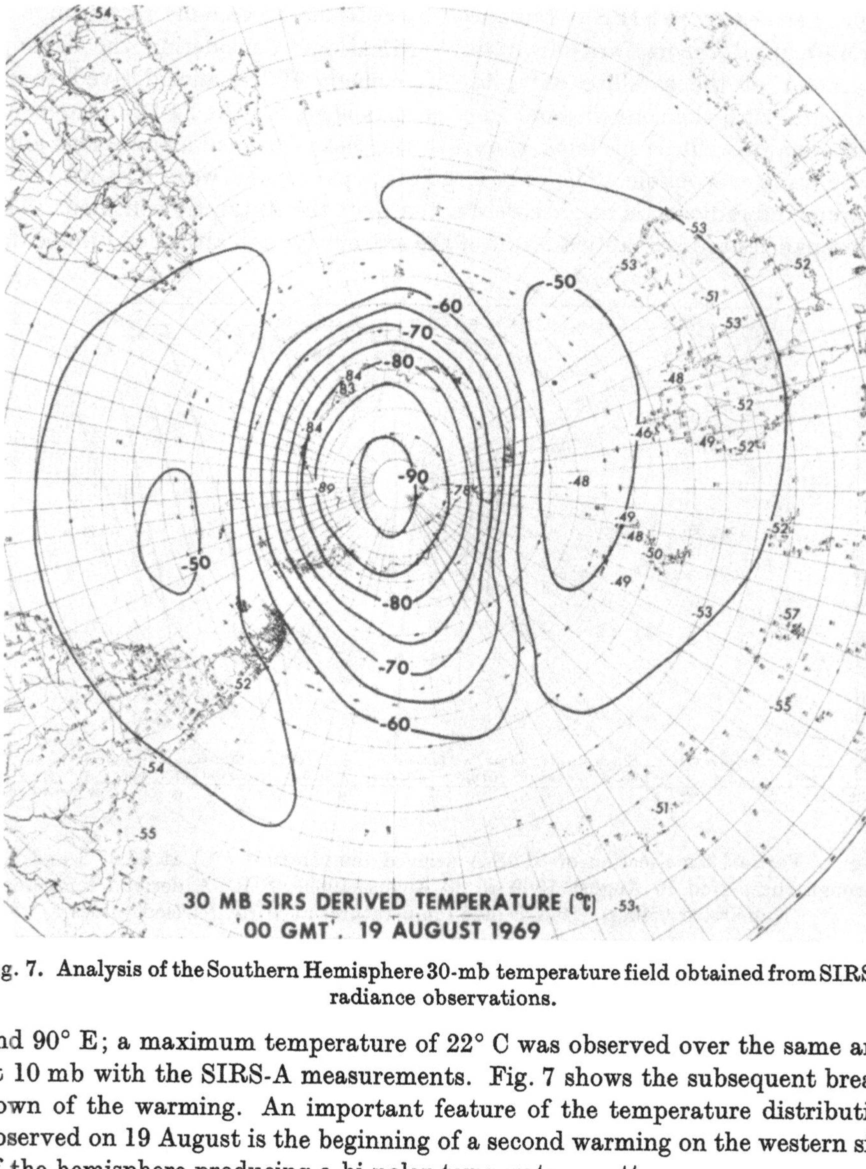

Fig. 7. Analysis of the Southern Hemisphere 30-mb temperature field obtained from SIRS-A radiance observations.

and 90° E; a maximum temperature of 22° C was observed over the same area at 10 mb with the SIRS-A measurements. Fig. 7 shows the subsequent breakdown of the warming. An important feature of the temperature distribution observed on 19 August is the beginning of a second warming on the western side of the hemisphere producing a bi-polar temperature pattern.

Figure 8 shows a vertical time-section constructed from temperature profiles derived from SIRS-A observations near the Mirny radiosonde station (66.5° S, 93° E). The dramatic temperature changes observed in the SIRS-A data near the area where the sudden stratospheric warming occurred are verified quite well by the radiosonde observations at Mirny (the circled temperature values). These results illustrate the ability of the SIRS-A instrument to detect, as well as the ability to retrieve, rapidly changing stratospheric temperatures.

Of more immediate interest to meteorology are the patterns found in the troposphere. Fig. 9 shows an example of the 300-mb geopotential height contours derived from SIRS-A radiance observations. Available rawinsonde and aircraft wind reports are given for verification. Considering the asynoptic character of the satellite data, the agreement of radiance-derived contour gradients with the conventional wind reports is good. The 24-hour movements of the major weather systems, shown by the heavy dashed arrows, also appear to be quite reasonable. The only large discrepancies between SIRS-A derived heights and radiosonde heights were noted over the Antarctic continent. These discrepancies are probably a result of the extremely high altitude of the surface.

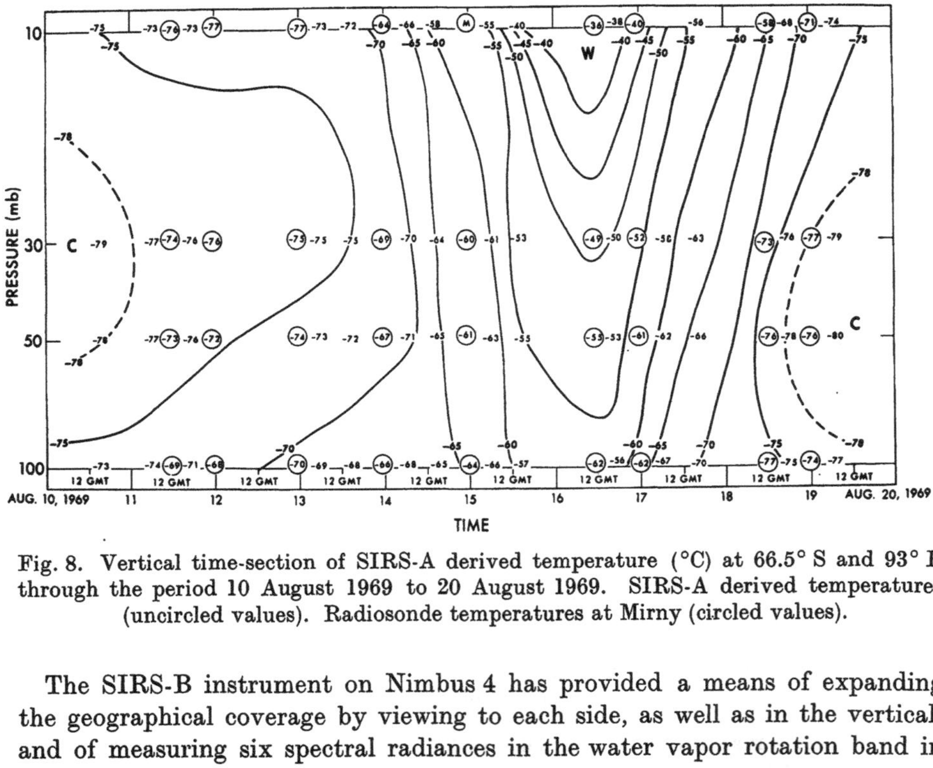

Fig. 8. Vertical time-section of SIRS-A derived temperature (°C) at 66.5° S and 93° E through the period 10 August 1969 to 20 August 1969. SIRS-A derived temperatures (uncircled values). Radiosonde temperatures at Mirny (circled values).

The SIRS-B instrument on Nimbus 4 has provided a means of expanding the geographical coverage by viewing to each side, as well as in the vertical, and of measuring six spectral radiances in the water vapor rotation band in order to deduce the abundance of that atmospheric constituent.

Soon after the launch of Nimbus 4 an opportunity arose to compare a SIRS-B vertical view with one viewing along a slant path, and to compare both of these with a SIRS-A sounding. Fig. 10 shows the area viewed by: I, SIRS-B in the vertical at 1532 GMT, 8 April 1970; II, SIRS-A 12 hours later; and III, SIRS-B, viewing at a local zenith angle of $\theta = 43.6°$ about 25 hours after the first observation. In Fig. 11 are the temperature profiles derived from the respective measurements. The SIRS-A sounding is clearly subtropical, demonstrating the higher temperatures in the upper troposphere and the lower stratosphere compared with a tropical sounding. This results from the higher latitude (21° N vs 13° N) of the SIRS-A measurements. The SIRS-B soundings indicate some differences near the tropopause which are comparable with changes found in the Manila (Clark A. F. B.) soundings during the same period. More striking, however, is the large difference in temperature (8–14 °K) in the levels above 20 mb. Although this phenomenon is not verified from direct observations, there are strong reasons to believe that it is genuine within the limits imposed by Fig. 1.

Finally, Fig. 12 shows a comparison of a water vapor profile calculated from SIRS-B radiances with those observed by the Key West radiosonde. The first-guess water vapor profile is climatological, assuming a constant relative humidity of 40%. As with the temperature profiles, the gross characteristics of the water vapor profile were calculated fairly well from the radiances observations. The radiosonde observation indicates much drier conditions than the first guess. The general shape of the curve follows the Key West sounding, rather than the first guess. This indicates that the shape of the moisture profile was sensed and retrieved satisfactorily from the SIRS-B water vapor channel radiances.

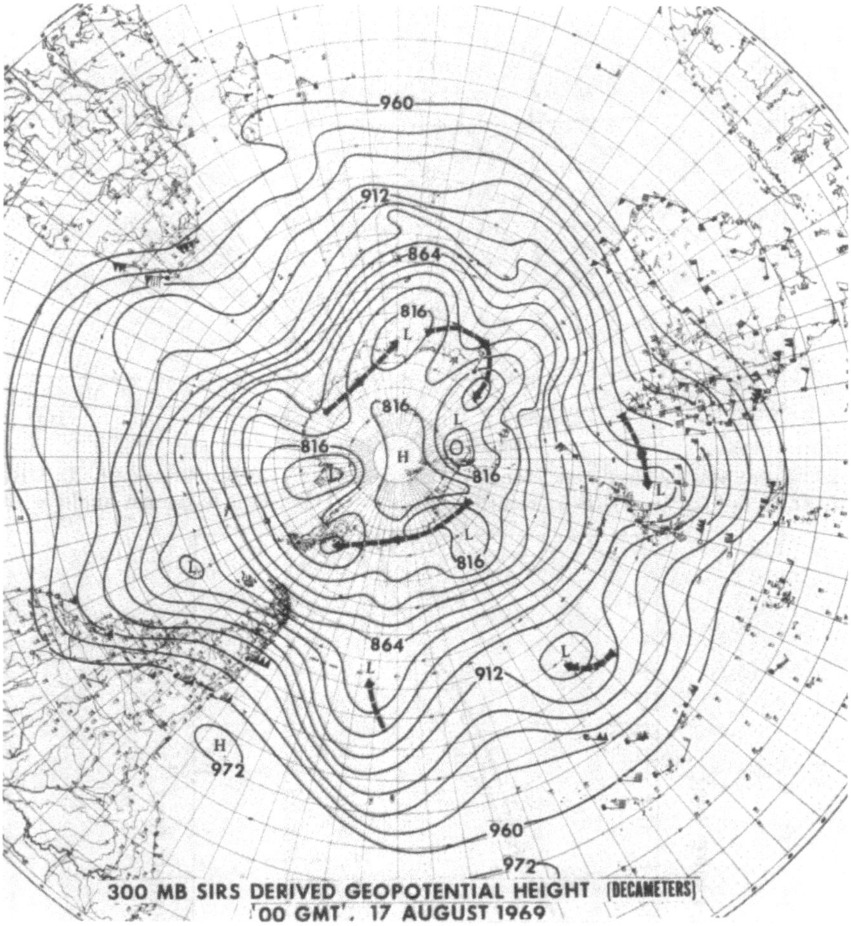

Fig. 9. Analysis of the Southern Hemisphere 300-mb geopotential height obtained from SIRS-A radiance observations. Radioseond observations, 00 GMT (filled circles), 12 GMT (open circles), aircraft wind observations (squares).

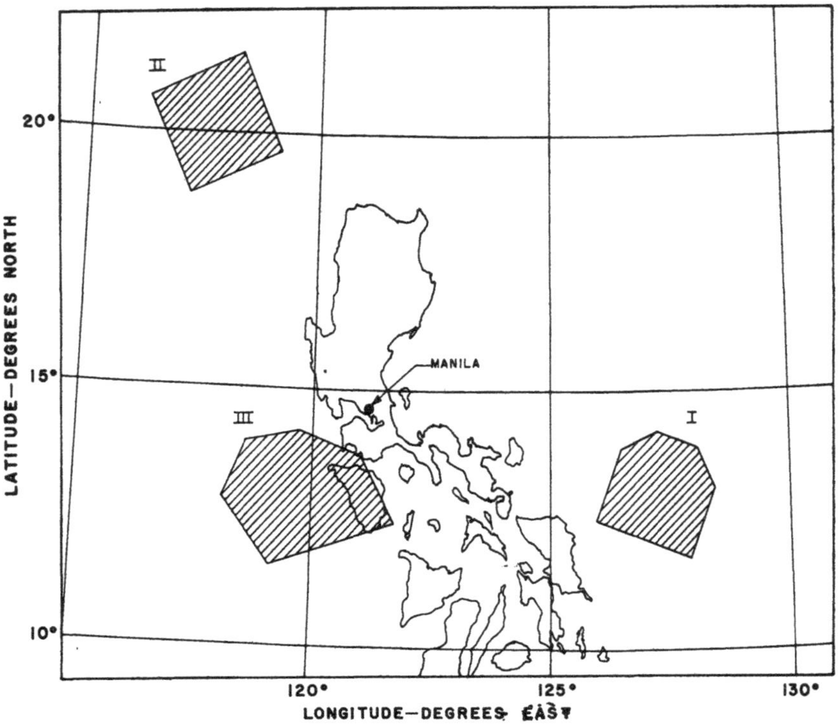

Fig. 10. Three areas viewed near Manila by SIRS-A and SIRS-B instruments. I: SIRS-B, zenith angle = 0°; II: SIRS-A, zenith angle = 0°, III: SIRS-B, zenith angle = 43.6°.

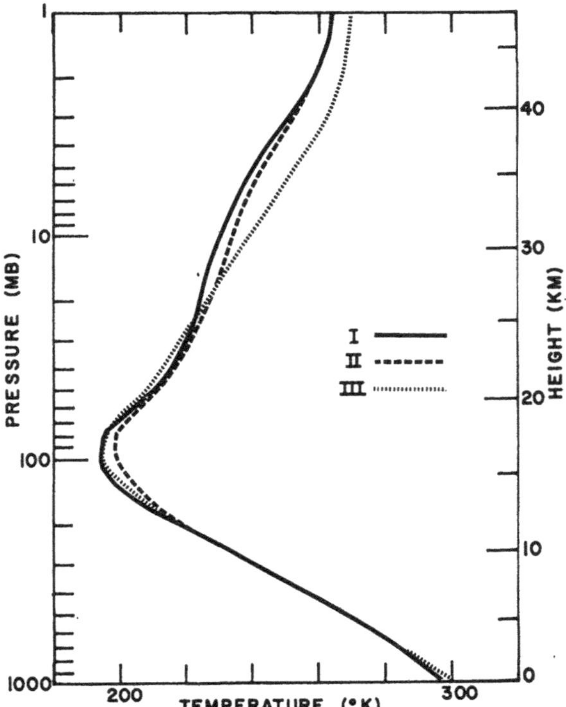

Fig. 11. SIRS-A and SIRS-B derived temperature profiles from radiances measured over the three areas as shown in Fig. 10.

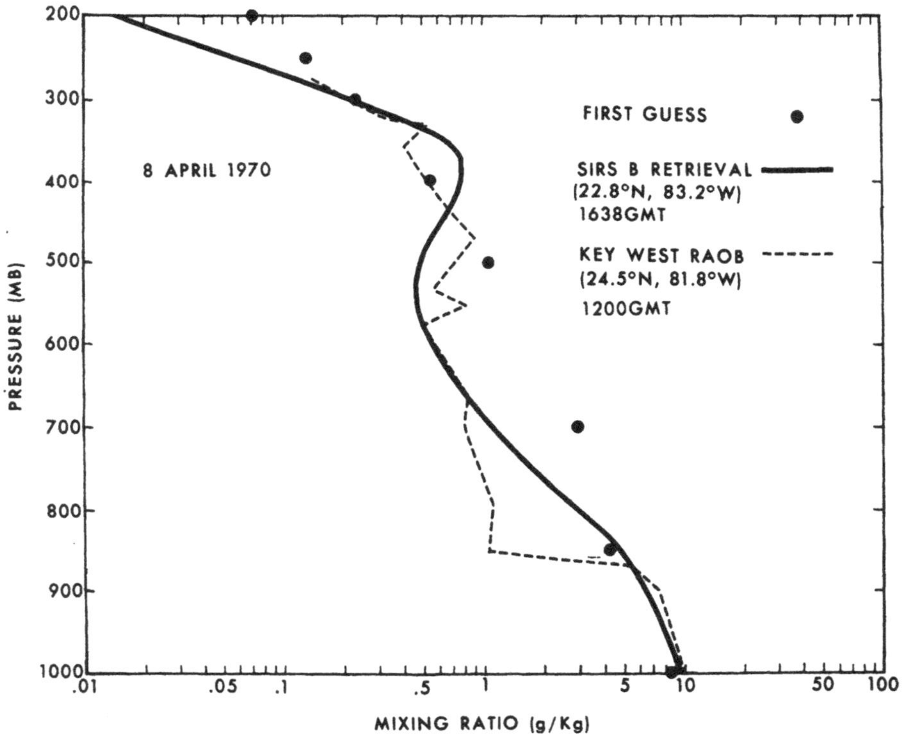

Fig. 12. Comparison between a SIRS-B calculated water vapor profile and a radiosonde-observed water vapor profile.

3. Conclusions

The Nimbus 3 and Nimbus 4 SIRS experiments have demonstrated that meaningful vertical soundings can be obtained from satellites on a global scale. Moreover, the accuracy of temperature and geopotential height profiles obtained from satellite radiometric measurements will be improved significantly over the next few years. Scanning radiometers with high spatial resolution, which are scheduled for flight in the latter part of 1971, will permit accurate temperature profiles to be calculated down to the earth's surface in all areas except those with extensive overcast cloud conditions. Futhermore, improved data reduction techniques are constantly being developed; these alleviate the dependence of the interpretive algorithm on current conventional data without significant degradation in the accuracy of the temperature determination. With the implementation of these instrumental and data reduction improvements, it is anticipated that temperature, geopotential height, and water vapor profiles of a quality sufficient for large scale numerical forecasting will soon be provided over the entire globe from satellite observations.

Space Research XI — Akademie-Verlag, Berlin 1971

GLOBAL DISTRIBUTION OF OZONE FROM NIMBUS 3

C. Prabhakara and B. J. Conrath

Goddard Space Flight Center, Greenbelt, Maryland, U.S.A.

A method to invert the radiance measurements made by the Infra-Red Interferometer Spectrometer (IRIS) on board the Nimbus 3 satellite is developed to deduce the vertical distribution and, thereby, the total amount of atmospheric ozone. The standard error of total ozone amount derived from this method is about 6% with respect to "Dobson" measurements.

The satellite-observed total ozone distribution on the globe shows a close relationship with the tropospheric weather systems. Generally low values of total ozone are found in the equatorial belt and the values increase towards the poles. Total ozone amounts well in excess of 0.5 cm (STP) at Arctic regions are observed during spring.

1. Introduction

The Infra-Red Interferometer Spectrometer (IRIS) on board the Nimbus 3 satellite yielded spectral measurements of the radiation emanating from the earth and its atmosphere in the region 7 to 25 μm, during the period 15 April to July 22 1969. This spectrometer had a resolution of 5 cm^{-1} which is adequate to derive useful meteorological information from the spectral measurements in the 15 μm CO_2 band, the 9.6 μm ozone band, and the water vapor bands in the 6.3 μm and the 20 μm regions.

In this paper a brief account is given of the method [1] developed to deduce the atmospheric ozone from the IRIS measurements. A map of the global ozone distribution derived in this manner is shown and its meteorological significance is discussed.

Atmospheric ozone is produced by solar near-ultraviolet radiation in the stratosphere but its geographic and seasonal variations cannot be explained by the corresponding variations in the incoming solar energy [2]. On the other hand, from the early findings of Dobson et al. [3], atmospheric ozone variations were found to have strong association with the tropospheric weather, but, as there are not more than 100 stations over the globe which report total ozone measurements, ozone has not yet become a useful parameter in day-to-day weather predictions. With the development of a satellite observing capability the present situation can be greatly improved.

2. Method

There is on the average about 0.35 cm (STP) of ozone per cm^2 on the earth, but it is distributed so that there is more ozone at low pressures in the stratosphere and less in the troposphere. Therefore, the atmosphere is optically thin in the 9.6 μm band and the absorption follows the "weak line" approximation. Thus all the weighting functions obtainable for this gas in the 9.6 μm absorption band have essentially the same shape and closely resemble the shape of the vertical ozone profile itself [4]. Hence detailed information on the vertical

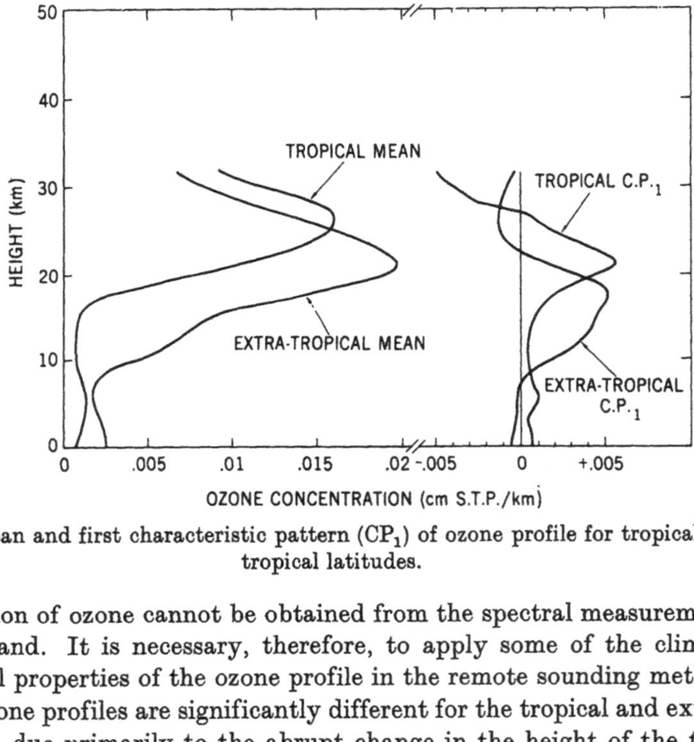

Fig. 1. Mean and first characteristic pattern (CP_1) of ozone profile for tropical and extratropical latitudes.

distribution of ozone cannot be obtained from the spectral measurements made in this band. It is necessary, therefore, to apply some of the climatological statistical properties of the ozone profile in the remote sounding method.

The ozone profiles are significantly different for the tropical and extratropical latitudes, due primarily to the abrupt change in the height of the tropopause at about 30° N and 30° S. The mean and the "characteristic patterns" [5] of the ozone profile for the tropical and extratropical latitudes have therefore been derived separately from the "ozone sonde observations over North America" [6]. Fig. 1 shows the mean profile and the first characteristic pattern (CP_1) for the tropical and extratropical latitudinal zones. These two climatological properties are used to represent the vertical profile in our remote ozone sounding method. Specifically the distribution of ozone with height Z in a given latitudinal zone is assumed to be given by

$$O_3(Z) = \overline{O}_3(Z) + \alpha_1 f_1(Z) , \tag{1}$$

where $\overline{O}_3(Z)$ and $f_1(Z)$ are the mean and the CP_1 for the latitudinal zone. The parameter α_1 in equation (1) characterizes the ozone profile. The objective of the remote ozone sounding method is thus to determine this parameter from the IRIS 9.6 μm spectral measurements by suitable application of the radiative transfer formalism.

The radiative transfer equation in a nonscattering atmosphere under local thermodynamic equilibrium may be written as

$$I_\nu = B_\nu[T(Z_0)]\,\tau_\nu(Z_0) + \int_{\tau_\nu(Z_0)}^{I} B_\nu[T(Z)]\,d\tau_\nu(Z)\,, \tag{2}$$

where ν is the wavenumber: B_ν is the Planck intensity; $T(Z_0)$ and $T(Z)$ are the absolute temperatures at the surface and any height Z in the atmosphere; and $\tau_\nu(Z_0)$ and $T_\nu(Z)$ are the transmissions from the surface and any level Z

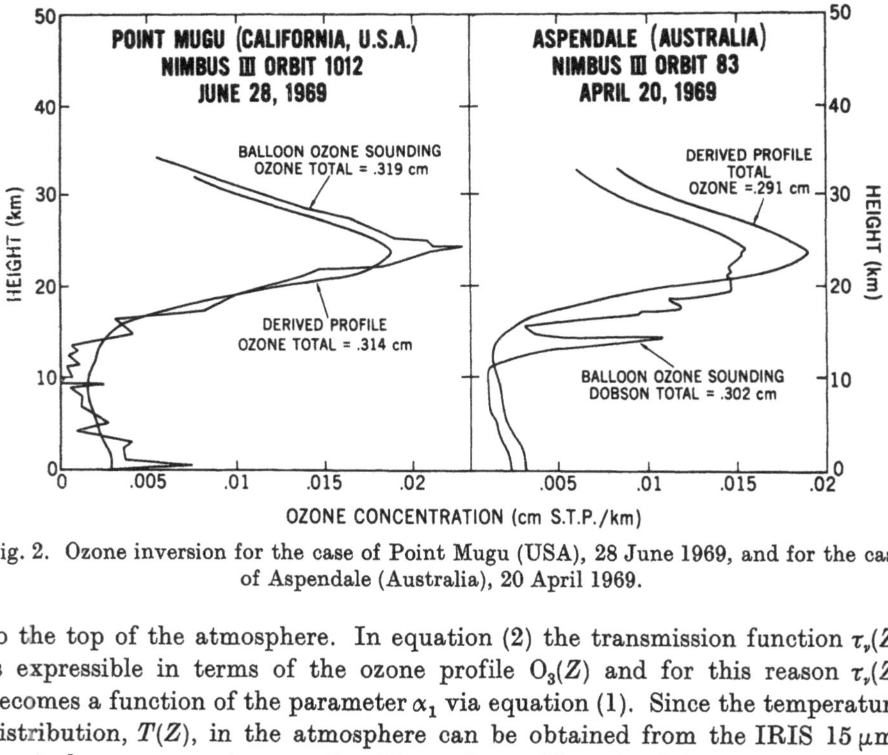

Fig. 2. Ozone inversion for the case of Point Mugu (USA), 28 June 1969, and for the case of Aspendale (Australia), 20 April 1969.

to the top of the atmosphere. In equation (2) the transmission function $\tau_\nu(Z)$ is expressible in terms of the ozone profile $O_3(Z)$ and for this reason $\tau_\nu(Z)$ becomes a function of the parameter α_1 via equation (1). Since the temperature distribution, $T(Z)$, in the atmosphere can be obtained from the IRIS 15 μm-spectral measurements, equation (2) can be readily solved for α_1. This briefly is our method [1] of remote sounding of ozone. The total ozone is then obtained by a simple integration of the ozone profile with respect to height.

3. Results and Discussion

From § 2 we see that the ozone profile is determined with the help of climatological statistics and with only one free parameter derived from the IRIS spectral measurements. Obviously such a procedure cannot always lead to good vertical profile determination. The point is illustrated in Fig. 2 where the IRIS-derived soundings and the balloon chemical ozone-sonde measured profiles for the two stations Point Mugu (USA) and Aspendale (Australia) are compared. There is good agreement between the IRIS-derived and balloon-measured profiles in the case of Point Mugu but not in the case of Aspendale; this is understandable because the statistics used in the IRIS method were typical of North America.

It must be emphasized, however, that for both the cases shown in Fig. 2, the IRIS-derived total ozone agrees fairly well with the observed total. To

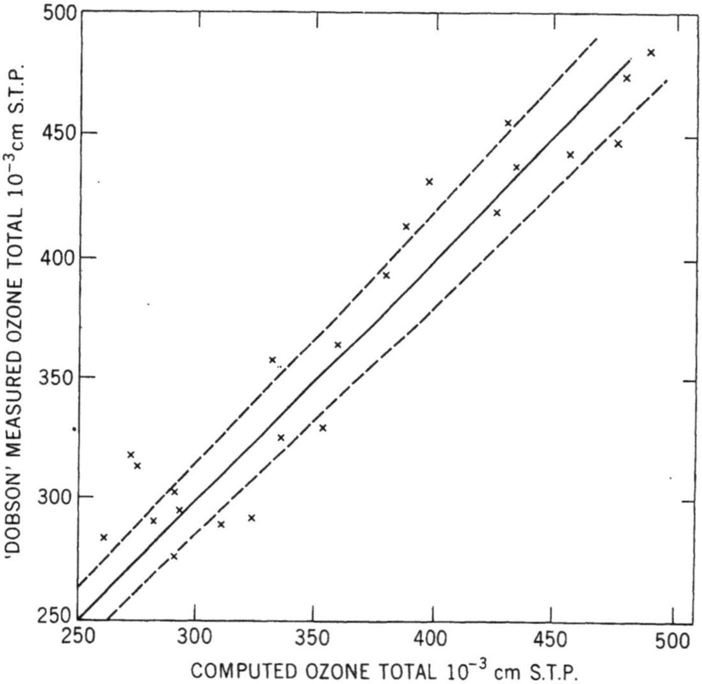

Fig. 3. Total ozone comparison between IRIS measurements and "Dobson" observations.

demonstrate this agreement further, we have plotted in Fig. 3 the IRIS-derived total ozone against the "Dobson" measured totals for 23 stations in Canada, Europe, Japan, India and Australia. All the points lying along the diagonal solid line represent an exact agreement, the broken lines on either side indicate $\pm 5\%$ deviation. The "Dobson" totals in this sample ranged from about 0.25 cm to 0.5 cm. An examination of the figure shows that the remote sounding method estimated the totals failry well over the entire range. The standard error of estimation is 6%. This kind of accuracy is necessary in order to make a study of the geographic and temporal variations of ozone.

Another important point is that in at least ten cases in the sample presented in Fig. 3 skies were cloudy; and so since the bulk of ozone is well above the tropospheric clouds it appears valid to conclude that the clouds do not have a significant effect on the method of remote sounding of ozone.

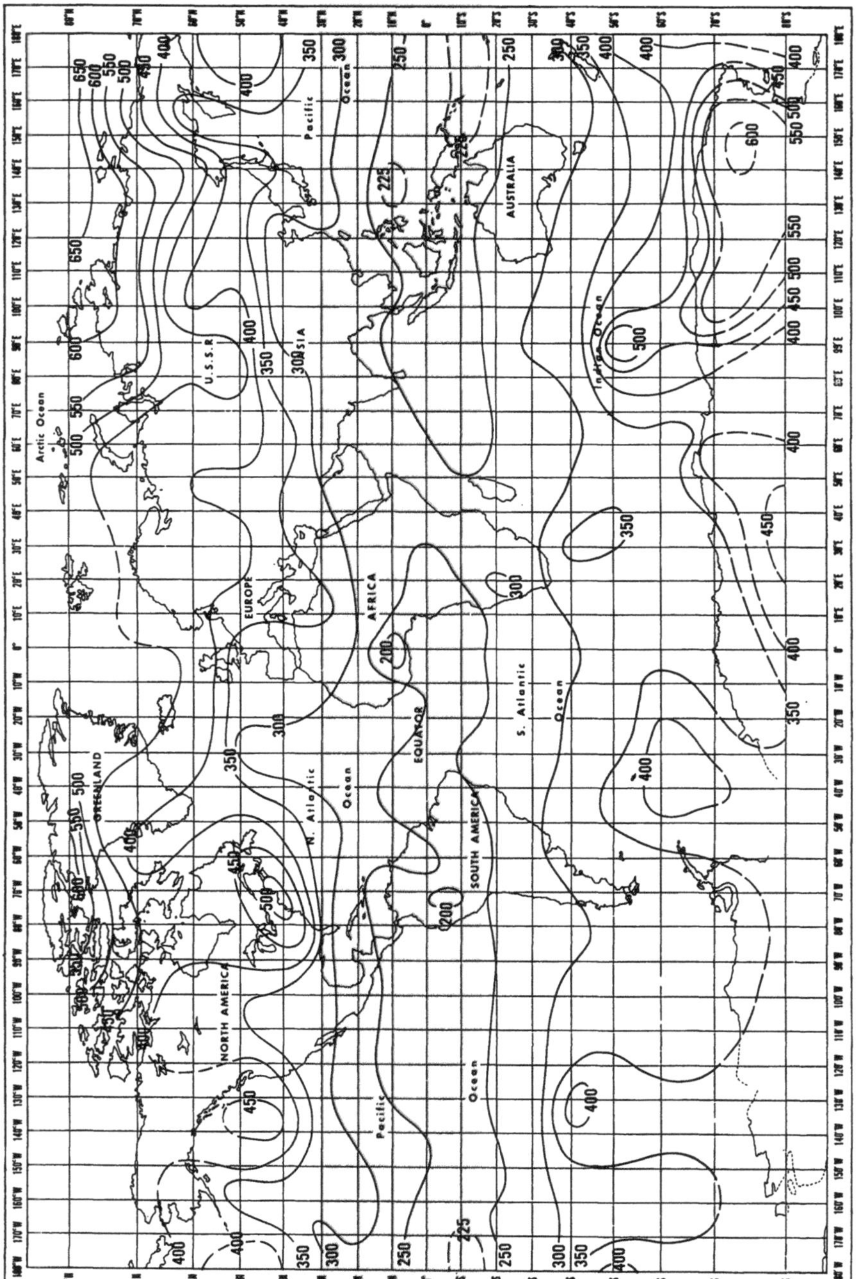

TOTAL OZONE CONTENT $(10^{-3}$ CM STP)
APRIL 22, 1969

Fig. 4. Global distribution of ozone determined from IRIS data for 22 April 1969.

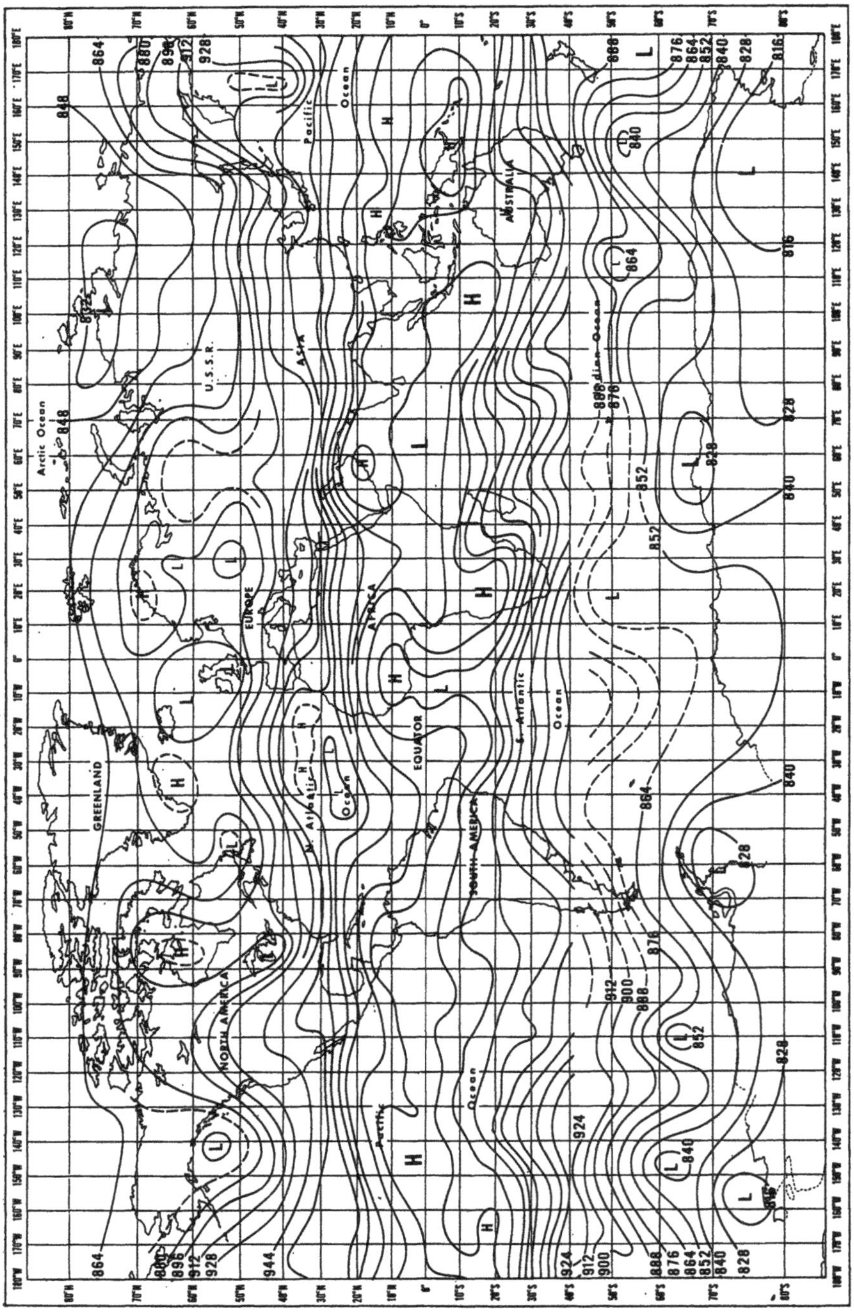

300 MB CONTOURS - 0000Z
APRIL 22, 1969

Fig. 5. 300-mb height field map of 22 April 1969.

By application of this technique to IRIS data the total ozone distribution over the globe between 80° N and 80° S has been derived. Fig. 4 shows the resulting map for the day of 22 April 1969. The isolines of total ozone drawn on this map are based on the data from successive Nimbus 3 orbits that are about 28° of longitude apart on the globe falling in the time interval 0000 UT to 2400 UT on 22 April 1969. Thus the ozone map is not a synoptic one and the data do not conform to the meteorological synoptic scale (approximately 5°). This map shows ozone values in excess of 0.6 cm in the Arctic region, and values as low as 0.2 cm over tropical Africa and South America. The equatorial region in general has low amounts of ozone with increasing amounts towards the North and South Poles.

Despite some shortcomings in the geographic distribution of data the ozone map shown in Fig. 4 reveals several "highs" and "lows" and a fairly well organized pattern of the ozone distribution over the globe. In order to bring out the meteorological significance of the ozone distribution Fig. 5 shows the 300-mb height field map of 22 April 1969. The close correspondence between the two maps is apparent. In general the high pressure systems on the 300-mb map are associated with minima of ozone and the low pressure systems correspond to high ozone centers, for example, the low pressure centers located over Siberia, Japan and west of the USA are all regions of ozone maxima. However, there is one large ozone maximum over Northeast USA which corresponds to two low pressure centers on the 300-mb map separated by a ridge. This discrepancy in the comparison is strictly due to the poor geographic resolution of the ozone data mentioned above.

There is similar association between the pressure and ozone patterns in the tropical regions of Brazil and Nigeria and also in the Southern Hemisphere. It is thought that the ozone amounts over Antarctica are not accurate because the vertical temperature profile there is often nearly isothermal, a condition under which the remote sounding methods do not perform well, but over the Arctic region in this season, April to July, a substantial variation in the vertical temperature profile does exist, and remote soundings should yield reliable data.

The strong association between the 300-mb height field and ozone over the globe suggests that ozone can be used as a meteorological variable. Further since the tropospheric clouds have a negligible effect on the remote sounding of ozone, such measurements are ideally suited for satellite techniques. In this sense the future weather prediction models based on satellite measured data should use ozone as a valuable predictor of weather.

Acknowledgments

The authors are grateful to J. Steranka for analyzing the ozone data on the global map and to Dr. A. B. Pittock of the CSIRO Laboratory, Aspendale, Australia, and W. S. Hering and T. R. Borden of the AFCRL, Bedford, Massachusetts, for making available data from many chemical soundings.

The services rendered by the personnel at Wallops Island, Virginia, Meteorological Station and at Point Mugu, California, in flying balloon-borne ozone sondes coincident with Nimbus 3 passes have enabled the authors to make this study quantitative.

References

[1] C. PRABHAKARA, B. J. CONRATH, R. HANEL, and W. J. WILLIAMS, J. Atmos. Sci. **27**, July (1970).
[2] R. A. CRAIG, Meteorological Monographs, Vol. 1, No. 2 (1950) American Meteorological Society, Boston.
[3] G. M. B. DOBSON, D. N. HARRISON and J. LAWRENCE, Proc. Roy. Soc. London A, **122**, 456 (1928).
[4] C. PRABHAKARA, Mon. Weath. Rev. **97**, 307 (1969).
[5] C. L. MATEER, J. Atmos. Sci. **22**, 370 (1965).
[6] W. S. HERING and T. R. BORDEN, Ozone sonde observations over North America, Vol. 4, AFCRL-64-30 (IV) Environmental Research papers, No. 279, Air Force Cambridge Research Laboratories, Bedford, Mass., USA (1967).

Space Research XI — Akademie-Verlag, Berlin 1971

AN ANALYSIS OF THERMAL SOUNDINGS OF THE ATMOSPHERE FROM SATELLITES

K. YA. KONDRATYEV[a], W. NORDBERG[b], B. J. CONRATH[b], O. M. POKROVSKY[a], YU. M. TIMOFEYEV[a] and R. HANEL[a]

[a] Leningrad University, Leningrad, USSR
[b] Goddard Space Flight Center, Greenbelt, Md, USA

Satellite data on the outgoing radiation in the 15 μm CO_2 band are used to determine the vertical temperature profile of the atmosphere. Two methods are used for interpreting the data of measurements: the regularization method and the Chahine method. The data characterizing the accuracy of temperature profile determination are represented, the possible causes of errors are analysed and a comparison of the two methods of interpretation of the radiation data is made.

Considerable progress has already been made towards the practical solution of the problem of the indirect determination of atmospheric temperature profiles from measurements of the spectral distribution of the outgoing thermal radiation [1]. This progress was promoted by the intensive investigations in various directions performed by scientists of various countries. The practical solution of the problem is based on work in the following three principal directions:

1. development of effective methods for the interpretation of thermal radiation measurements;
2. investigation of the absorption characteristics of the real atmosphere;
3. construction of instruments capable of measuring thermal radiation with high accuracy and resolution (both spectral and spatial).

More than a dozen techniques based on different principles have been proposed so far. A careful analysis of these techniques enables a formulation of the basic requirements for the correct solution of the problem. Practically, in all the techniques the following problems occur in some form or other:

a) introduction of a priori information on the solution sought for in the inversion algorithm;
b) choice of the optimum measuring conditions;
c) stabilization of the inverse operator of the problem.

The first problem results from the necessity of interpreting radiation data known to contain some error. Therefore, an infinity of solutions is available which satisfy the corresponding integral equation but differ widely between themselves because of the instability of the inverse operator. Different a priori assumptions are introduced in different techniques, viz. smoothness, forcing the solution to agree with a priori empirical information, etc.

The problem of the choice of optimum measuring conditions arises in the designing of a concrete spectral instrument. If there is a possibility of measuring the outgoing radiation with high spectral resolution over the entire absorption band of interest this problem is automatically solved. The use of 10–20 radiation values in the interpretation leads to complete exhaustion of the information on the atmospheric temperature profile.

Stabilization of the inverse operator of the problem is carried out based on two principles:

a) "narrowing" a given operator by a certain subspace of a smaller dimension;
b) "distortion" of the operator to permit the stabilization throughout the whole of space.

The main problem in the first approach is to find the subspace in which the operator (of the problem) is stable, that in the second to choose a stable operator with a minimum deviation from the data. Both problems are strictly solved with the presence of the empirical statistics on the solution sought for.

A practical solution of the problem is connected with a careful investigation of the absorption properties of the real atmosphere and with the development of techniques to correct for the errors introduced by incorrect knowledge of atmospheric absorption. Most promising solutions of this problem are the direct calculations of transmission function as well as the determination of the absorption characteristics based on simultaneous measurements of outgoing radiation and temperature profiles with the help of radiosondes.

So far, a number of spectral instruments designed for thermal atmospheric soundings have been constructed and solutions of inverse problems have been successfully tested.

In April 1969, the meteorological satellite Nimbus 3 was launched carrying two such instruments: an eight-channel infrared spectrometer (SIRS) and a Michelson's interferometer (IRIS). The interferometer measured the outgoing radiation over the range of 500–2000 cm^{-1} with a resolution of 5 cm^{-1}. The present paper reports the results of processing several spectra, by two methods employed for the inversion to temperature profiles: Chahine's iterative method and the regularization method. Comparison of the techniques is of great practical importance. For the operational processing of the enormous number of radiation data it is necessary to use an economical and stable technique which enables one to derive the temperature profile with the maximum accuracy.

To analyse the accuracy of restoration with the above-mentioned interpretation techniques radiosonde data were applied. By the use of identical initial data (radiation values, transmission functions) the same, on the average, restoration accuracy has been obtained by means of both techniques, mean-square errors being 2–4°, maximum error amounting to 8–9°. In the authors' opinion, the main sources of error are the errors in determination of the kernel of the corresponding integral equation. More detailed comparison of the techniques shows that the regularization method allows more accurate temperature determination in the region above 100 mb, but is less accurate, as compared with Chahine's technique, for the pressure range of 100–500 mb. It should be noted that these methods do not use any a priori statistical information about the real temperature profile of the atmosphere. A variety of spectral intervals, differing in numbers, was used for the solution by the regularization method: 20, 15, 10, 7 and 6 intervals were used on the short-wave side of the 15 μm band. The results of calculations indicate that in the case of a reasonable choice of spectral intervals there is no essential difference in the temperature profiles obtained with the variously chosen spectral intervals. This fact suggests that the number of independent radiation measurements is limited.

Determination of the integral equation kernel allowed one to increase the accuracy of temperature profile determination to mean-square errors of 2–3° and maximum errors of 4–6°. This once again testified to the importance of the accurate fixation of the absorption characteristics of the real atmosphere and the necessity of their specification in the process of the solution of the inverse problem.

References

[1] K. Ya. Kondratyev and Yu. M. Timofeyev, Thermal Sounding of the Atmosphere from Satellites, Gidrometeoizdat, Leningrad 1970, in Russian.

Space Research XI — Akademie-Verlag, Berlin 1971

ANALYSIS OF THE TECHNIQUES FOR THE SOLUTION OF THE PROBLEM OF THERMAL ATMOSPHERIC SOUNDING

O. M. Pokrovsky and Yu. M. Timofeyev

Leningrad University, Leningrad, USSR

The present state of the interpretation of radiation data in the problem of thermal atmospheric sounding is considered. The main stages of the correct solution of the integral Fredholm equation of the first kind are examined; viz., the introduction of a priori information on the function sought for in the transformation algorithm, choice of the optimum measuring conditions, stabilization of the algebraic system. A comparative analysis of some methods of interpretation is performed.

The question of the choice of optimum measuring conditions is studied for the solution of the problem of the thermal atmospheric sounding from measurements of the outgoing radiation in the 15 μm CO_2 band.

On the basis of numerical experiments for real temperature profiles the accuracy of the temperature profile restoration is investigated using a number of methods for the solution of the inverse problem, such as the regularization methods, that by Tourchin–Westwater–Strand and the method of eigenfunction decomposition. Data showing the influence of the error in determining the integral equation kernel on the restoration accuracy are given.

The possibilities of indirect determination of the real atmosphere absorption characteristics are discussed.

1. Introduction

Various methods have been proposed for the numerical solution of the problem of atmospheric thermal sounding [1]. (Some of these have been successfully tested in numerical experiments in processing of real data on thermal radiation measurements.) In developing the operational system for processing the data there is the problem of the choice of optimum interpretation techniques, a problem which is far from being solved. The present paper compares the practical application of a number of techniques based on numerical experiments. It is considered that this method of solution has a number of advantages over the interpretation of the data on real measurements by various methods. In the case of numerical experiments it is easy to analyse, separately, the influence of various sources of errors in temperature profile restoration, whereas in the interpretation of real data the reasons for discrepancies are difficult to determine, and therefore it is difficult to isolate the particular factors which need improvement.

2. Details of Calculations

Three methods of interpretation are analysed in the present paper:

(i) The regularization method without the use of statistical information.
(ii) The method of statistical regularization.
(iii) The method of expansion in empirically orthogonal functions.

Experiments were made for a great number of real temperature profiles taken from paper [2]. For the calculations of outgoing radiation use was made of the transmission function of a real atmosphere for spectral intervals in the 15 μm CO_2 band, corresponding to six- and eight-channel models of the instrument. On the calculated radiation values were superposed accidental perturbations that imitated accidental errors of measurements with a relative error of 0.5, 1.0 and 2.0%. To investigate the influence of the errors in determining the kernel of the corresponding integral equation on the restoration accuracy of the temperature profile, a solution of the inverse problem was deduced both with the unperturbed kernel $\Delta = 0$ and with a kernel carrying a systematic error caused by different physical reasons ($\Delta = \Delta_1$) (neglecting the influence of water-vapour absorption, temperature dependence of transmission etc.).

3. Results of Calculations and Conclusions

The techniques investigated in this paper differ in a number of features. First, they use different a priori information on the solution sought. The first method uses a sufficiently general assumption on its smoothness or constraints; the second and third methods use the assumption that it belongs to an empirical statistical set. The second difference is in the ways of stabilizing the inverse operator of the problem. The first two methods refer to α-parametric regularization, the latter referring to regularization based on the cut-off in the expansion of an empirically orthogonal function.

The results typical for five characteristic profiles are represented in Tables 1 and 2. The principal conclusions based on the numerical material obtained can be formulated as follows:

Table 1

Temperature profile restoration with unperturbed kernel ($\Delta = 0$)

Profiles		T_1			T_2			T_3			T_4			T_5		
Interpretation technique	error %	0.5	1.0	2.0	0.5	1.0	2.0	0.5	1.0	2.0	0.5	1.0	2.0	0.5	1.0	2.0
I	σ_T	2.1	2.8	3.5	1.8	2.2	3.0	1.5	1.8	2.9	1.4	2.1	3.1	1.1	1.6	2.6
	ΔT_m	3.3	5.8	7.1	3.3	4.6	5.6	2.9	3.0	5.0	2.0	5.0	7.9	2.8	3.2	4.2
II	σ_T	1.3	1.7	2.1	0.9	1.0	1.1	3.0	3.9	4.2	5.2	5.8	6.3	3.5	4.5	4.7
	ΔT_m	2.5	3.2	3.9	1.8	1.9	2.0	6.2	7.8	8.3	13.4	15.0	16.1	8.4	9.7	9.8
III	σ_T	1.2	1.8	1.9	0.8	0.9	1.0	3.0	3.5	4.5	4.3	4.8	4.9	2.5	2.6	3.6
	ΔT_m	2.5	3.8	4.1	1.4	2.0	2.9	8.2	10.1	11.6	10.4	13.1	13.8	5.0	5.6	8.1

Table 2

Temperature Profile Restoration with Perturbed Kernel ($\Delta=\Delta_1$)

Profiles Error% Interpretation technique		T_1			T_2			T_3			T_4			T_5		
		0.5	1.0	2.0	0.5	1.0	2.0	0.5	1.0	2.0	0.5	1.0	2.0	0.5	1.0	2.0
I	σ_T	3.7	4.0	4.2	2.8	3.1	3.5	2.8	3.0	3.2	2.7	3.1	3.2	2.5	2.7	3.0
	ΔT_m	6.2	7.4	7.6	4.7	5.6	6.7	4.5	5.4	6.4	2.8	5.5	6.2	4.3	5.2	6.1
II	σ_T	4.5	4.6	4.7	1.8	1.9	2.0	3.5	4.2	4.8	5.8	6.0	6.4	6.3	6.5	6.9
	ΔT_m	0.5	12.5	13.5	5.3	5.4	5.5	9.4	9.8	8.5	14.0	15.3	16.2	13.4	15.6	15.8
III	σ_T	3.8	3.9	4.1	1.1	1.2	1.3	4.9	5.0	6.0	6,0	6.1	6.2	7.0	7.1	7.3
	ΔT_m	7.2	7.3	7.6	2.0	2.1	2.2	11.2	12.1	12.4	17.2	17.2	17.2	15.6	15.7	16.0

(i) Appreciable variations in the accuracy characteristics (σ_T and ΔT_{max}) are observed for the same experimental conditions while using different interpretation techniques. For instance, σ_T changes by a factor of 2 or 3. This testifies to the importance of the use of the optimum technique in the processing of real data.

(ii) There exists an essential dependence of the restoration accuracy on the type of statistical information used in the solution of the problem. With the use of adequate (in time and place) statistics these methods permit high accuracy of temperature restoration (1–2° at 0.5–2.0%). With the use of inadequate statistics the accuracy falls sharply (by a factor of 2–3, see tables).

This suggests certain restrictions in the possible use of statistical interpretation techniques.

(iii) In restoration by the regularization method the minimum restoration errors are usually larger than with the use of statistical techniques. The reverse is observed for maximum errors. Hence, by using the regularization method one can nearly always derive mean restoration accuracies as great much as 1–3°.

(iv) In comparing the techniques which use statistical information it is difficult to draw final conclusions as to the advantages of any approach. In some cases better results are given by the method of statistical regularization, in others, by method (iii).

(v) For all interpretation methods an essential decrease in restoration accuracy is observed in the interpretation with the perturbed kernel. There is also a marked smoothing of the dependence of the restoration accuracy on errors in the measurements. Particularly large errors are observed in statistical techniques using inadequate statistics.

Thus, the principal conclusion of the work can be formulated as follows: With the use of statistical a priori information on the solution sought, the restoration accuracy is grearly influenced by the degree of adequacy of statistical information. This testifies to the importance of finding correlation temperature matrices for various areas of the globe at different seasons.

References

[1] K. Ya. Kondratye vand Yu. M. Timofeyev, Termicheskoye zondirovanie atmosphery so sputnikov, Gidrometizd., Leningrad 1970.

[2] L. E. Koprova and V. G. Boldarev, Izv. Akad. Nauk SSR, Ser. Fiz. Atmos. Okeana **6**, No 2 (1970).

Space Research XI — Akademie-Verlag, Berlin 1971

TRANSMITTANCES FOR USE IN REMOTE SOUNDINGS OF THE ATMOSPHERE

S. R. Drayson

High Altitude Engineering Laboratory, University of Michigan, Ann Arbor, Michigan, USA

Accurate atmospheric transmittance functions are needed for remote soundings of the atmosphere to obtain temperature profiles and the distribution of water vapor and ozone. To make precise calculations of these transmittances we must know the appropriate absorption band parameters, that is the frequencies of the line centers, line intensities, half-widths and line shapes, and their variation in the atmosphere. The difficulties in obtaining the band parameters from laboratory measurements are described and present uncertainties in the values of the parameters are discussed. The effect of these uncertainties is illustrated by temperature inversions using satellite radiances in the 15 μm region.

The intensity of the outgoing infrared radiation from the earth and its atmosphere depends on the distribution of molecular absorbers, cloud distribution and the vertical profile of temperature in the atmosphere. About a decade ago King [1] and Kaplan [2] suggested that satellite measurements of the infrared radiation could be used to deduce the values of some of these variables. Such a satellite in polar orbit could obtain meteorological parameters on a global basis, to supplement the existing network of radiosondes, particularly over oceans and other regions of sparse information, aiding the process of numerical weather prediction.

Attempts to deduce meteorological parameters, particularly the vertical temperature profile, from synthetically generated radiances showed that the information content was more limited than had been earlier envisioned. Small errors in the radiance data could lead to large errors in the solution unless some form of constraint was placed on the system. Too much constraint led to a solution that was too smooth, with consequent loss of information. Too little constraint produced profiles containing unreal information originating from the data errors. By minimizing the data errors the smoothing could be reduced and the maximum amount of information obtained from the measurements. Thus, stringent specifications were developed for the accuracy of satellite instruments.

The measurement of radiances is only one of many possible sources of errors. For example truncation errors arising from numerical quadrature techniques is another obvious source. However, the remainder of this paper will be devoted to a discussion of a further source, arising from uncertainties in the atmospheric transmission functions. For convenience the discussion will be restricted to a vertical sounding, in a clear atmosphere.

The intensity of radiation I_ν at wavenumber ν at the top of the atmosphere is given by

$$I_\nu = B(\nu, T_g)\,\tau_\nu(p_g) + \int_{p_g}^{0} B(\nu, T)\,\frac{d\tau_\nu(p)}{d\,p}\,dp\,, \qquad (1)$$

where $B(\nu, T)$ is the Planck function at wavenumber ν and temperature T, $\tau_\nu(p)$ is the transmittance for a vertical path starting outside the atmosphere down to pressure level p and g is a subscript denoting ground or surface conditions.

In practice an instrument cannot measure the monochromatic radiation implied in equation (1) but instead gives an intensity averaged over some small finite wavenumber interval, weighted by the instrument response function. Since the Planck function normally only varies slowly over such an interval equation (1) will be interpreted as representing averaged intensities and transmittances averaged with the same weighting function.

The problem of calculating transmittances through an inhomogeneous atmosphere is not trivial, especially for remote-sounding applications where the demands on accuracy are high. Two approximate methods, direct interpolation and extrapolation from laboratory measurements and the use of band models, are not normally sufficiently accurate. A more powerful technique is to calculate monochromatic values of the transmittance τ_ν and to integrate these with respect to wavenumber, weigthed by a suitable function if desired. Since τ_ν is a rapidly varying function of ν, the numerical quadrature must be carefully devised and requires a large amount of computer time, but the method has the advantage that the numerical accuracy can be made as great as is desired.

To calculate the monochromatic value of the transmittance $\tau_\nu(p)$ the appropriate band parameters have to be known for all molecular absorbers in the atmosphere. They are the wavenumbers, intensities, half-widths and shapes of the absorption lines, including their variation for different atmospheric conditions, principally temperature and pressure. The absorption coefficient can then be obtained by evaluating the contribution from each line of each absorber and the transmittance obtained by integration with respect to pressure, again using standard quadrature techniques.

A clear distinction must be made between numerical accuracy and absolute accuracy. Numerical accuracy can now be readily obtained but absolute accuracy, for a given atmosphere, is limited by our knowledge of the band parameters. They cannot be obtained directly from theoretical considerations but rely ultimately on laboratory measurements. The wavenumber of the line center can usually be measured sufficiently accurately, but the intensities, half-widths and line shapes cannot be so readily obtained. The remainder of this paper will illustrate some of the problems involved by a study of atmospheric transmission in the spectral region near 15 μm and the result of an application to obtain a temperature sounding using satellite data.

The principal atmospheric absorber near 15 μm is carbon dioxide. The ν_2 fundamental is the strongest band but there are numerous overtone bands which make important contributions. In addition the ν_2 band of ozone, centered near 14 μm, absorbs weakly in the stratosphere, while the pure rotational band of water vapor is effective in the lower troposphere.

Many measurements have been made of absorption by the 15 μm band of CO_2, mostly at a resolution insufficient to resolve the individual rotational lines. With a suitable choice of experimental conditions (high pressure and low absorber amount) the absorption is independent of half-width and line shape, so that the total intensity of all the bands may be found [3]). The fundamental accounts for about 85% of the total so that this type of measurement is of limited use in determining the band strengths of the weak overtone bands.

Ideally, absorption measurements should be made at high resolution so that the absorption due to each line can be measured with a minimum of interference from neighboring lines. Such measurements have been made for part of the 15 μm bands by Madden [4] providing the only direct measurements of Lorentz half-widths of self-broadened CO_2 absorption lines in this region.

Table 1
Experimental conditions [5] and measured and theoretical equivalent widths for the spectra in Fig. 1

Run	P_{CO_2}	P_{N_2}	Cell length	Temperature	Equivalent widths Experiment	Theory
#	(torr)	(torr)	(cm)	°C	cm^{-1}	cm^{-1}
2	12.0	822.7	3200	25	148.8	148.9
4	25.0	771.3	3200	25	164.8	166.0
6	50.0	689.2	3200	25	181.0	181.0
16	12.0	751.0	400	25	94.5	95.8
18	24.0	736.0	400	25	111.9	113.6

Although low-resolution absorption measurements provide much less information, they are more readily available and can be used to obtain estimate band parameters. Fig. 1 shows experimental nitrogen broadened spectra from Reichle [5] (broken lines) and comparison spectra calculated theoretically using a resolution of 5 cm^{-1} (solid lines). Table 1 gives the experimental conditions. The theoretical spectra were computed by a line-by-line integration technique with the nitrogen broadened Lorentz half-widths from Yamamoto et al. [6]. The band intensities were adjusted to obtain the best fit to the experimental spectra using the techniques described by Drayson et al. [7]. It was found that if a Lorentz line shape was employed the absorption in the wings of the band (near 550 cm^{-1} and 780 cm^{-1}) was always overestimated, due to the contribution from many intense lines near the band center. Good agreement was obtained by neglecting the Lorentz contribution at a distance greater than 10 cm^{-1} from the line center. Clearly this does not represent the true line shape but more detail cannot be obtained from the experimental results. Recent measurements by Burch [8] confirm that the wing contributions near 800 cm^{-1} are small (about one or two per cent of the Lorentz value), so that for the conditions of Fig. 1 the contributions in the two wing regions are much less than those of the local lines and can be neglected.

Table 1 shows the calculated equivalent widths agree with the experimental to well within the experimental error. The same is true of the individual absorp-

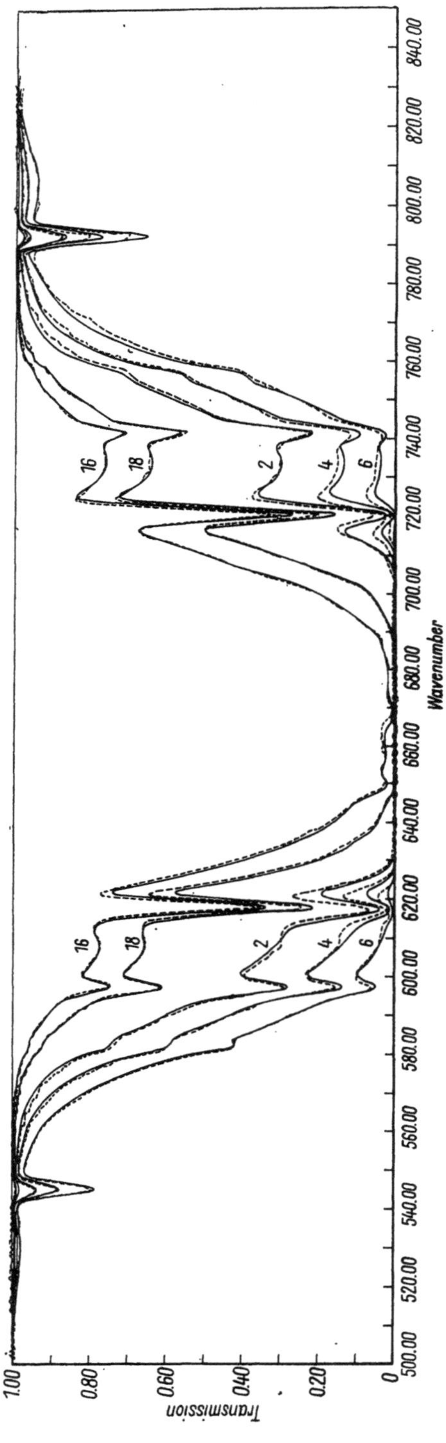

Fig. 1. Comparison of experimental spectra (broken lines, [5]) with theoretically computed spectra (solid lines) of nitrogen broadened carbon dioxide for the conditions in Table 1.

tion points, with a few exceptions, mostly in regions dominated by Q-branches (Fig. 1). The reason is probably an inadequate knowledge of the line shape which is a more critical factor near a Q-branch.

The spectra in Fig. 1 correspond to pressures and absorber amounts not unlike those needed to interpret measurements by a vertically probing satellite instrument. It might be anticipated that the band parameters could be successfully applied to calculate atmosphere transmittances, for use in inversion

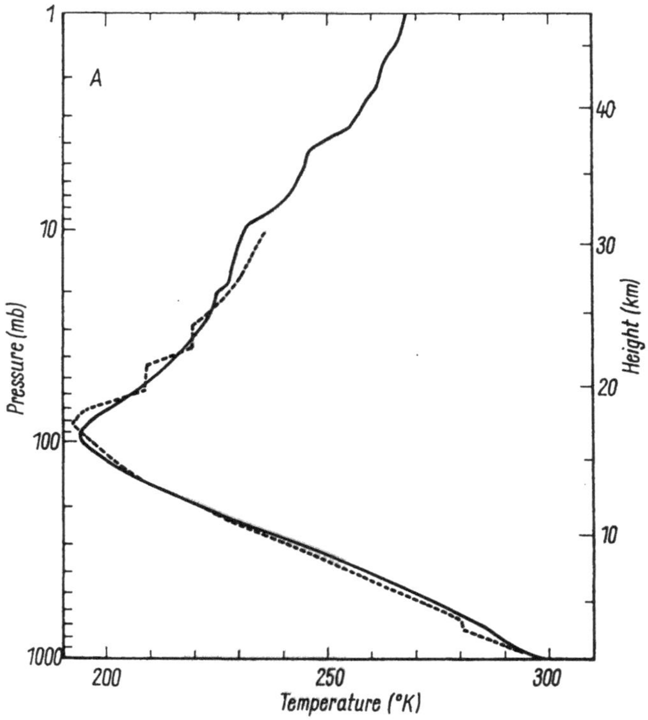

Fig. 2. Comparison of temperature profiles derived from SIRS data, 16° N 73° W, 1612 GMT (solid curves) with radiosonde profiles, Kingston, Jamaica, 1200 GMT (dotted lines), reproduced from Wark and Hilleary [9].

procedures to obtain vertical temperature profiles from satellite radiance measurements. Fig. 2, taken from Wark and Hilleary [9], shows the result of such an application to the first satellite sounding by SIRS. In the stratosphere it is difficult to make a valid comparison with the radiosonde profile because of the errors in the latter. In the troposphere below 10 km the SIRS profile is consistently warmer, a tendency which is typical of many retrieved profiles. Although instrument error may account for a small portion of the difference, the largest source of error is the transmittance functions. The functions used make allowance for water vapor and ozone, as well as the major absorber, carbon dioxide. The channels providing most information on the lower troposphere are centered at 706.3, 714.3 and 750.0 cm^{-1}, that is towards the wing of the carbon dioxide bands where the absorption is weaker and overtone bands play a decisive role in determining the absorption. Departure from the Lorentz line shape is also a factor in these regions. Analysis of many soundings [9] has shown that the calculated carbon dioxide absorption is greater than the true atmospheric absorption.

The transmittances used in [9] were based on a pure Lorentz line shape. If the Lorentz wing is truncated the calculated absorption becomes too low, so that the true atmospheric absorption lies somewhere between the two cases. The calculated and true transmittance functions have a maximum difference of about 0.05 in the worst case, which is unimportant for many applications, such as the computation of radiative heating rates. Because of its unique accuracy requirements, the remote sounding requires precise transmittance functions.

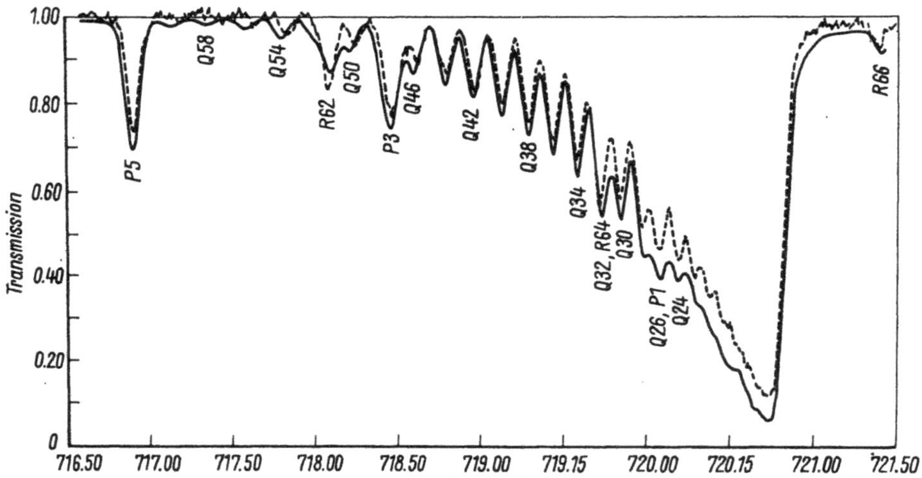

Fig. 3. Comparison of high-resolution spectrum (broken lines) with theoretically computed spectrum (solid lines) of pure carbon dioxide. Theoretical resolution is 0.08 cm^{-1}. All identified lines belong to the P- or Q-branches of (010:1–100:0) band except lines of R-branch of the ν_2 fundamental (000:0–010:1) band. Cell length = 8.74 cm, pressure 75.08 torr, temperature 25 °C.

The reason for the disagreement is probably due to the combined effect of uncertainties in the strength, half-width and shape of the lines. High-resolution measurements are needed to improve estimates of these parameters. Accordingly, an experimental program has been started and some preliminary results are available. Fig. 3 shows a comparison of an experimental and theoretical spectrum of pure carbon dioxide near 720 cm^{-1}, showing the Q-branch of the (010:1–100:0) band. The resolution of the theoretical spectrum is 0.08 cm^{-1}. There is an increasing wavenumber shift of the experimental lines relative to the theoretical line positions for high rotational quantum numbers in the Q-branch, indicating that the constants from Gordon and McCubbin [10] used to compute the wavenumbers need slight revision. The reason for the slight overestimation of absorption cannot be resolved until more data are taken.

Similar data have been obtained for the region near 675 cm^{-1} and are illustrated in Fig. 4. This region is dominated by the R-branch of the fundamental, although lines of two overtone bands can be identified, as well as lines of the R-branch of the fundamentals of the two isotopic molecules $^{13}C^{16}O_2$ and $^{12}C^{16}O^{18}O$. Again wavenumber disagreements indicate a need to revise constants. The overall level of agreement of the absorption appears to be better in this region.

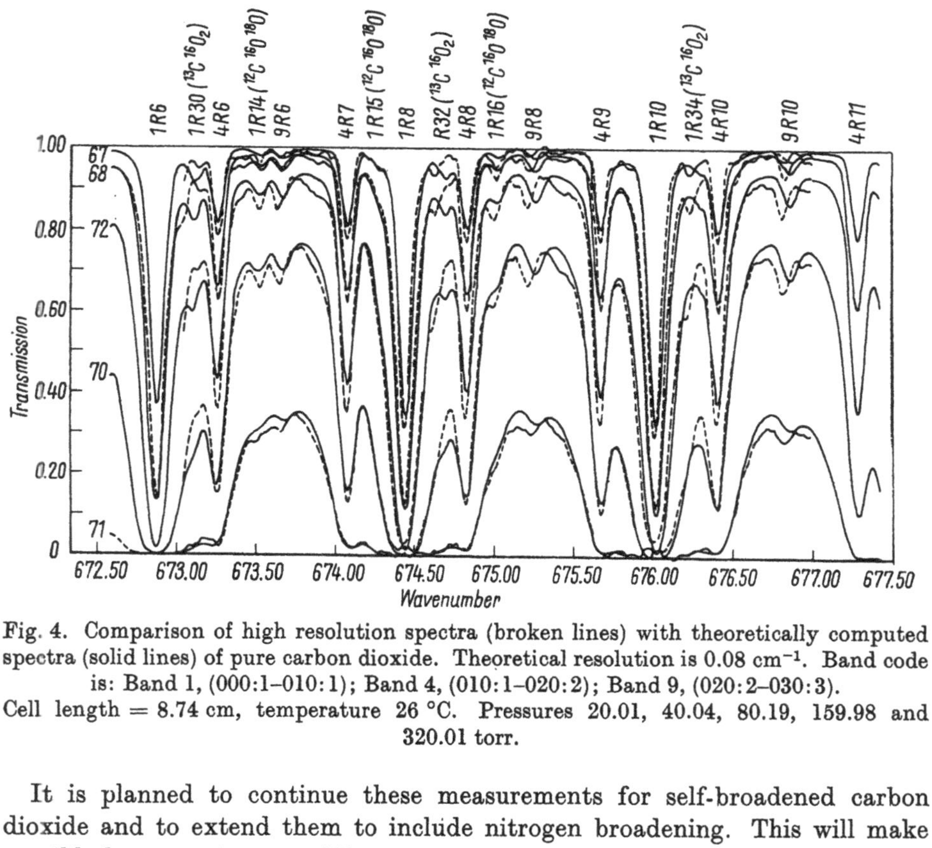

Fig. 4. Comparison of high resolution spectra (broken lines) with theoretically computed spectra (solid lines) of pure carbon dioxide. Theoretical resolution is 0.08 cm^{-1}. Band code is: Band 1, (000:1–010:1); Band 4, (010:1–020:2); Band 9, (020:2–030:3).
Cell length = 8.74 cm, temperature 26 °C. Pressures 20.01, 40.04, 80.19, 159.98 and 320.01 torr.

It is planned to continue these measurements for self-broadened carbon dioxide and to extend them to include nitrogen broadening. This will make possible better estimates of line parameters, which can be applied to compute more precise atmospheric transmittance functions, leading in turn to more accurate temperature profiles from satellite soundings.

Acknowledgments

Most of the work described in this paper has been supported by the National Environmental Satellite Center, ESSA, Dr. D. Q. Wark kindly granted permission to reproduce Fig. 2. Dr. H. A. Reichle made available his original data on the low resolution measurements. The high resolution measurement program is being supported by the Goddard Space Flight Center, NASA, and is being carried out by Mr. L. W. Chaney, High Altitude Engineering Laboratory, the University of Michigan.

References

[1] J. I. F. KING, Scientific Uses of Earth Satellites, Univ. Michi- gan Press, Ann Arbor 1958, (p. 133).
[2] L. D. KAPLAN, J. Opt. Soc. Amer. **49**, 1004 (1959).
[3] M. WOLK, J. Quant. Spectrosc. Rad. Transf. **7**, 1 (1967).
[4] R. P. MADDEN, J. Chem. Phys. **35**, 2083 (1961).
[5] H. G. REICHLE, JR., PhD thesis, University of Michigan, 1969.
[6] G. YAMAMOTO, M. TANAKA and T. AOKI, J. Quant. Spectrosc. Rad. Transf. **9**, 371 (1969).
[7] S. R. DRAYSON, S. Y. LI and C. YOUNG, Univ. Michigan Dept. Aerospace Engineering Report 08183-2-F (1968).
[8] D. E. BURCH, Aeroneutronics Division, Philco-Ford Corporation Technical Report U-4784 (1970).
[9] D. Q. WARK and D. T. HILLEARY, Science **165**, 1256 (1969).
[10] H. R. GORDON and T. K. MCCUBBIN, JR., J. Molec. Spectrosc. **18**, 73 (1965).

Space Research XI — Akademie-Verlag, Berlin 1971

SOME RESULTS OF MICROWAVE SOUNDING OF THE ATMOSPHERE AND OCEAN FROM THE SATELLITE COSMOS 243

A. E. BASHARINOV, S. T. YEGOROV, A. S. GURVICH and A. M. OBOUKHOV

Academy of Sciences of the USSR, Moscow, USSR

For measurements of the microwave terrestrial and atmospheric thermal emission the satellite Cosmos 243 was provided with radiometers for wavelengths 8.5, 3.4, 1.35 and 0.8 cm. The measurements at wavelengths 8.5 and 3.4 cm were used to determine the characteristics of the ocean and the earth surface emission. These measurements were also used to determine the boundaries of the sea ice and to distinguish precipitation and intensive cloudiness above the oceans. The measurements at wavelengths 3.4, 1.35 and 0.8 cm were used to determine the integral water vapour content Q_Σ above the oceans. The r.m.s. deviation of the satellite measurements of Q_Σ from those of the radiosonde measurements is 0.2 g cm^2.

From the measurements at wavelengths 3.4 and 0.8 cm the distribution of the integral liquid water content of clouds above the oceans was plotted.

From the satellite measurements the fields of Q_Σ above the oceans have been mapped. One can observe a sharply expressed non-monotonic variation of Q_Σ with latitude. The maps obtained were compared with the analysis of a synoptic situation.

Satellite measurements of the field of microwave thermal emission can provide extensive information on the characteristics of the atmosphere and the earth's surface. The problem of exploiting these measurements for remote sounding of the atmosphere has actively been discussed in the recent years.

The review of the study of microwave emission measurements in geophysics made recently in the USSR was published in [1]; a similar review was also published in [2]. In principle, it is possible to extract useful information from the microwave measurements on the state of the atmosphere and surface even beneath clouds.

The real satellite measurements of microwave emission of the atmosphere were first carried out in September 1968 by Cosmos 243 [3]. It was launched on 23 September with an orbit inclined at 71.3°, apogee 319 km, perigee 210 km, and a period of revolution of 89.6 min.

To measure the microwave emission, Cosmos 243 was provided with a radiometer for wavelengths: 8,5, 3.4, 1.35 and 0.8 cm.

The antennas of the radiometer were directed to the nadir. An antenna diagram had the width of about 4° on half power, the only exception being for the wavelength 8.5 cm where the width was approximately 10°. Microwave cosmic radiation received by small horns directed to the zenith was used as a reference level. For the scale shift of the radiometer and its effective use during measurements, attenuators were engaged into waveguides connecting the horns with the radiometers. For in-flight calibration of the sensitivity of a radiometer a noise generator with a temperature about 300 °K was switched on instead of an antenna. The "zero" of the scale was also controlled. The necessary corrections due to changes of antenna temperature, waveguides and some elements of the radiometers were based on data from thermoresistors. A final correction of scales of the radiometers was made either at points where the brightness was known or from computed data of the brightness according to the mean climatic temperature and humidity.

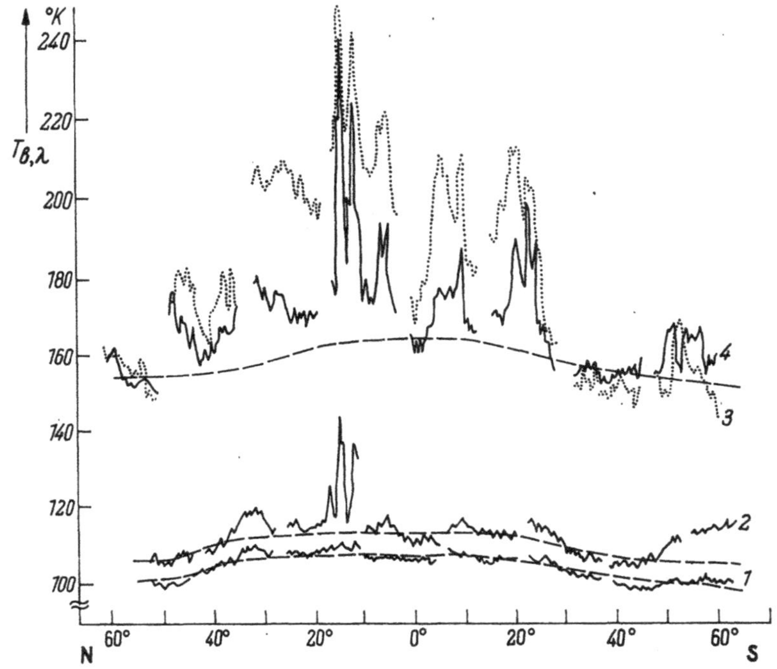

Fig. 1. Latitude profile of radiobrightness temperatures over the Pacific Ocean obtained on 23 September 1968. Wavelengths: 1, 8.5 cm; 2, 3.4 cm; 3, 1.35 cm; 4, 0.8 cm. Broken lines show mean climatic values for wavelengths 8.5, 3.4, 0.8 cm. Breaks in the profiles correspond to the control of radiometers.

The field of microwave radiation observed was noticeably different when the satellite passed above the land from that above the ocean. This is mainly due to difference in the emissivity of the sea surface and of soil in the microwave range.

The reflection coefficient of the iceless sea surface depends weakly upon the temperature and salinity. As theoretical computations show [4], the roughness of the sea, if one observes it into nadir, does not essentially change the surface emission. However, foam and spray arising during storms can noticeably change the surface emission [5]. The depth at which thermal microwave emission from the sea surface develops is of the order of a wavelength. Since the absorption at long wavelengths (8.5, 3.4 cm) in the atmosphere is not usually large, the measurements at these wavelengths gave information mainly of the surface state and its temperature [6]. From measurements at long wavelengths it was possible to determine also the boundary of ice at the sea surface in the Southern Hemisphere and to estimate which part of the sea surface is covered with ice [3]. At the same time the measurements showed in accord with theory that the variations of the radiation field at short waves 1.35 and 0.8 cm above the iceless surface of the ocean are mainly due to changes in the water vapour and liquid water content of the atmosphere.

From the measurements at wavelengths 1.35, 0.8 cm it is possible to estimate the integral amount of the water vapour $Q_\Sigma = \int \varrho(z)\, dz$ and liquid water $W_\Sigma = \int w(z)\, dz$, ($\varrho$, w are the water vapour and liquid water in unit volume at height z) at the point beneath a satellite. The results of the synchronous measure-

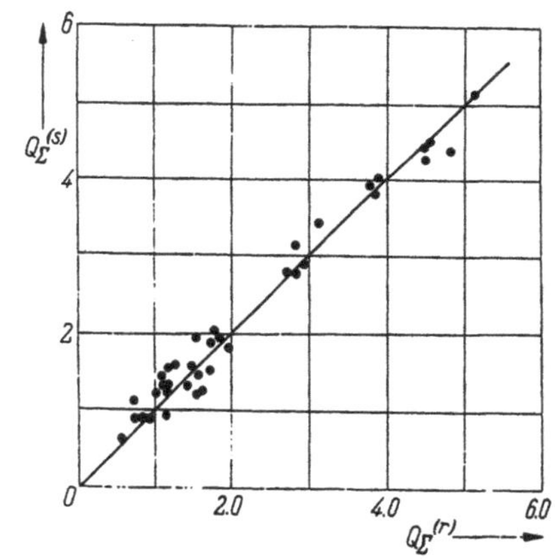

Fig. 2. Comparison of the satellite, $Q_\Sigma^{(s)}$, and radiosonde, $Q_\Sigma^{(r)}$, measurements of the total water vapour content.

ments at wavelengths 8.5 and 3.4 cm make it possible to outline the area with the strongest precipitation and to correct the results of determinations of Q_Σ, W_Σ above the area with a noticeable increase of brightness of the sea surface due to a storm. Fig. 1 represents the variations of the profiles of radiobrightness temperature $T_{b,\lambda}$ above the Pacific Ocean at wavelengths λ 0.8, 1.35, 3.4 and 8.5 cm. This figure also shows computed mean climatic values of the brightness temperatures for the three wavelengths 0.8, 3.4 and 8.5 cm under a cloudless atmosphere. The graph shows that the variations of $T_{b,\lambda}$ at wavelengths 0.8 and 1.35 cm are much greater than at wavelengths 3.4 and 8.5 cm. The variations at wavelength 0.8 cm are mainly due to absorption by liquid water and at wavelength 1.35 cm absorption by water vapour is predominant. Thus, the changes of $T_{b,1.35}$ adn $T_{b,0.8}$ prove to be generally in agreement. One should notice the region between 10° and 20° N, where the strongest precipitation occurs. This is particularly seen from the values of brightness temperature at wavelength 3.4 cm. The measurements at wavelength 8.5 cm show mainly a thermodynamic temperature of the water. However, one should notice also the region between 45° and 60° S where, with a relatively small increase in brightness at the shortest wavelength 0.8 cm, there is a noticeable increase in

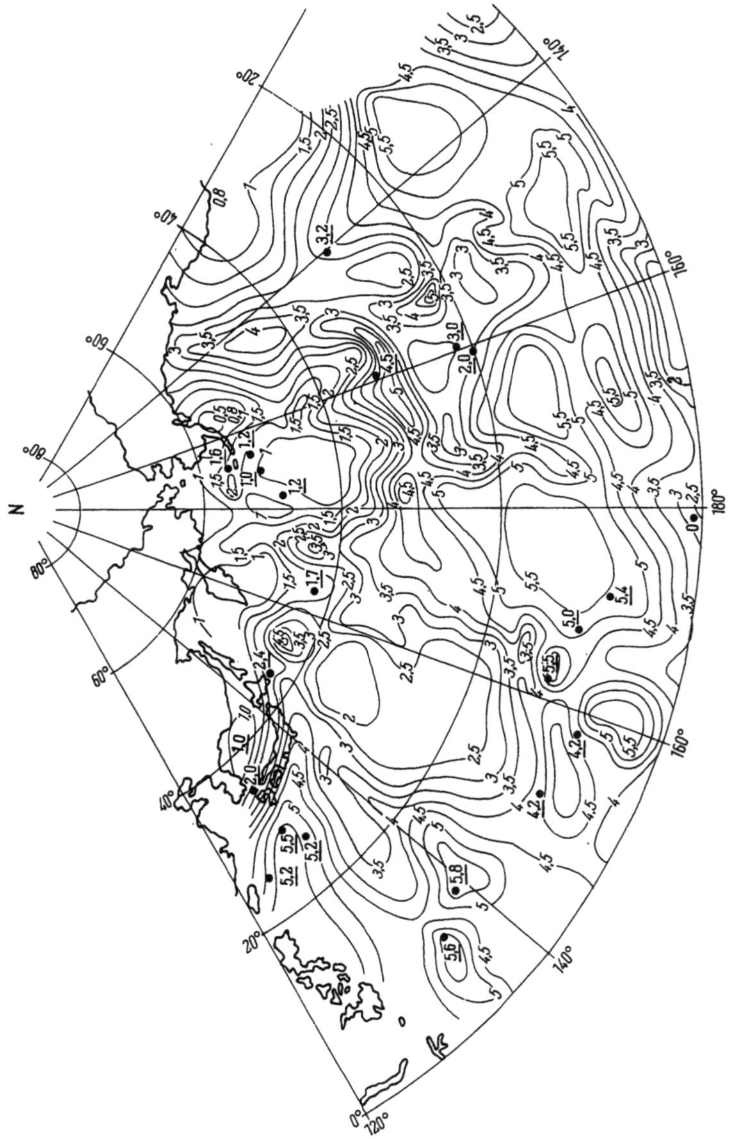

Fig. 3. Map of Q_Σ above the Northern Pacific Ocean 23 September 1968. Figures along the trajectory are the satellite measurements. The quantities underlined have been obtained from radiosounding.

the brightness temperature at long wavelengths. This circumstance is probably due to a change of emissivity of the sea surface because of foam and spray [6] but not due to cloudiness or precipitation*.

It should be noted that in the region 45–60° S the variations in water vapour content with a latitude characterized by $T_{b,1.35}$ are seen well from the picture of emission changes of the sea surface.

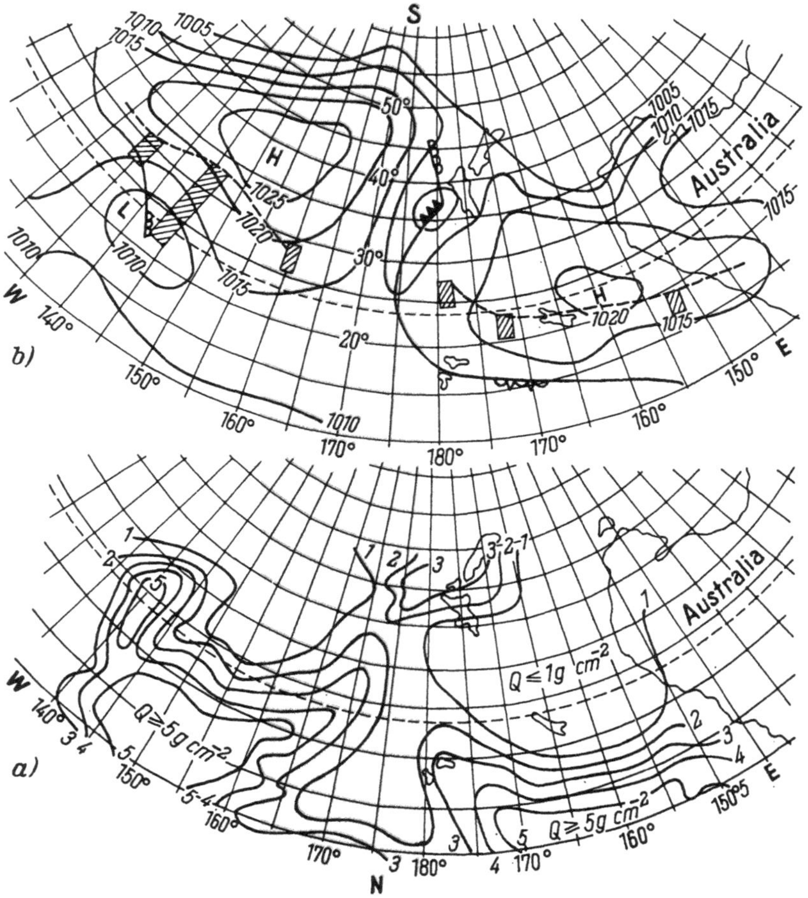

Fig. 4. a. Map of Q_Σ of the Southern Pacific on 23 September 1968.
b. Pressure field over the Southern Pacific. Broken line: boundary of dry and wet air. Shaded rectangles: transition zone.

Fig. 2 gives a comparison of satellite and radiosonde Q_Σ measurements, based on 38 vertical profiles of humidity above the ocean. The radiosonde data were taken at distances not more than 150 km from the point beneath the satellite and no more than 3 hours from the moment of the satellite passing. The r.m.s. deviation of the satellite measurements of Q_Σ from those of the radio-

* The synchronous measurements in the infrared range of 10–12 μm point to a very low cloudiness or absence of clouds in the region.

sonde measurements is 0.2 g cm^{-2}, while the total range of variation is about 5 g cm^{-2}. In the region of precipitation and increased surface emission of the ocean for Q_Σ measurements one had to use the data of measurements at wavelength 3.4 cm and the accuracy of measurements in these cases is reduced by 2–2.5 times, as estimates show.

Continuous measurements of Q_Σ along the orbit above the sea enabled us to obtain the data from which the map of distribution of Q_Σ above the sea was plotted [7]. Fig. 3 shows, as an example, such a map for the Northern Pacific on 23 September 1968. From Fig. 3 one can see an essential space variation, large gradients of Q_Σ and a non-monotonic change of Q_Σ with latitude. An analysis of space–time variations of the field Q_Σ shows that they follow, in

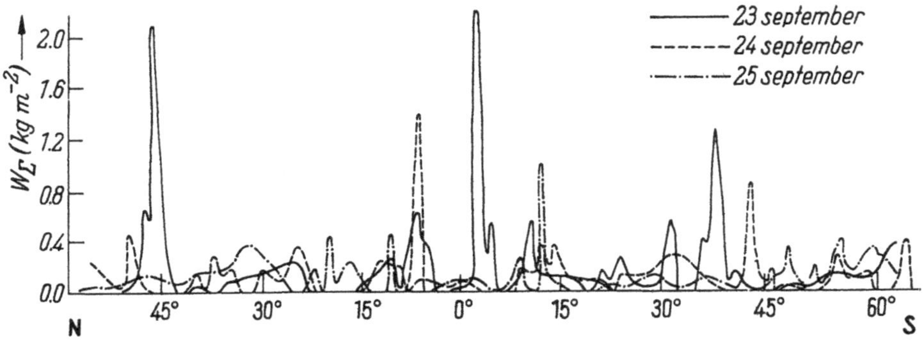

Fig. 5. Distribution of W_Σ over the Pacific Ocean for 23–25 September 1968.

general, the changes of the pressure field. From comparison of the distribution of Q_Σ with synoptic maps it was possible to identify the boundaries of dry ($Q_\Sigma \leqq 1$ g cm^{-2}) and wet ($Q_\Sigma \geqq 3$ g cm^{-2}) air, noted in [3], with the positions of cyclones and anticyclones. Fig. 4 shows that the location of dry and humid air agrees well with the nature of the circulation in the Southern Pacific.

Of particular interest is the determination of the liquid water content W_Σ in the atmosphere from measurements of microwave emission, since at present data of W_Σ are practically non-existent. Fig. 5 represents the distribution of W_Σ along the Pacific Ocean from the Bering Sea to the Antarctic obtained from the measurements on three practically coincident orbits for three days (23–25 September 1968). Sharp peaks are typical of the distribution of W_Σ. Also a strong variation of W_Σ has been observed from day to day in distinction to the variations of Q_Σ.

Satellite measurements by Cosmos 243 enabled us to obtain statistical estimates of water vapour and liquid water distribution above the oceans. Fig. 6a represents the average values of Q_Σ and W_Σ above the Pacific, Atlantic and Indian Oceans for a period from 23 to 27 September 1968. Fig. 6b shows the average values and r.m.s. deviations from the mean amount of water vapour above all the oceans taken together.

As is seen from Fig. 6a the mean latitude distribution of Q_Σ above all the oceans is similar, and distinction is usually less than the variations of water amount plotted in Fig. 6b.

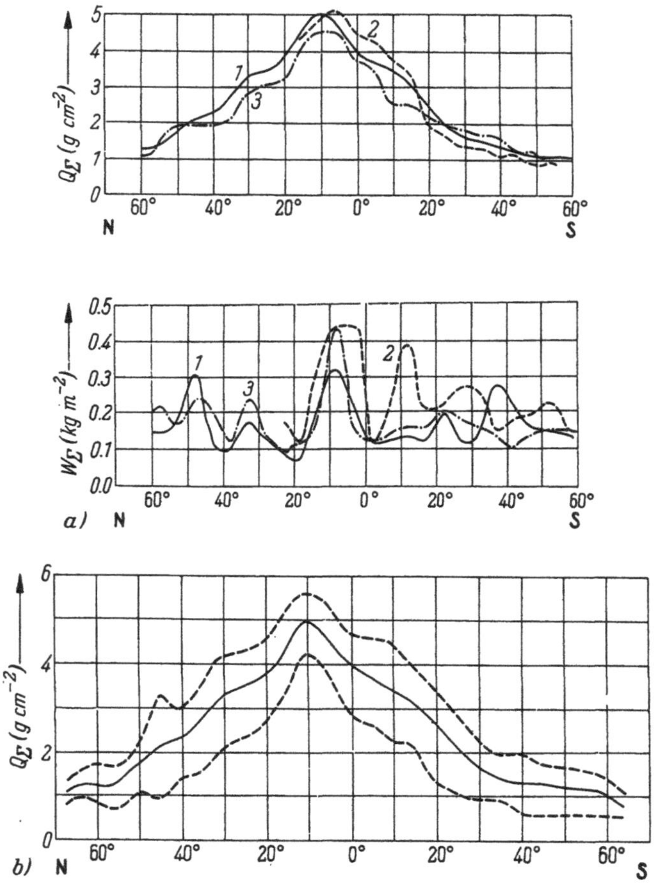

Fig. 6a. Average distribution Q_Σ and W_Σ over the Pacific (1), Indian (2) and Atlantic (3) Oceans.
b. Average distribution and r.m.s. deviation (dashes) Q_Σ over the oceans for 23–27 September 1968.

For a given season the water vapour content in the atmosphere is larger in the Northern Hemisphere than in the Southern.

The results described above point out promising prospects of using microwave measurements for remote sounding of the atmosphere. In particular, only this method will allow a quantitative estimate of the liquid water content of clouds.

References

[1] A. E. Basharinov, A. S. Gurvich, A. T. Tuchkov and K. S. Shifrin, Izv. Akad. Nauk SSSR, Ser. Fiz. Atmos. Okeana 6, 366 (1970).

[2] D. H. STAELIN, Proc. IEEE **57**, 427 (1969).
[3] A. E. BASHARINOV, A. S. GURVICH and S. T. YEGOROV, Dokl. Akad. Nauk SSSR **188**, 1273 (1969).
[4] K. S. SHIFRIN and S. N. IONINA, in: Trudy GGO **222**, 22 (1968).
[5] W. NORDBERG, J. CONAWAY and P. THADDEUS, Quart. J. Roy. Met. Soc. **95**, 408 (1969).
[6] A. E. BASHARINOV, A. S. GURVICH, S. T. YEGOROV, A. A. KURSKAYA, D. T. MATVEYEV and A. M. SHUTKO, Space Research XI, Akademie-Verlag, Berlin 1971 (p. 713).
[7] A. M. OBOUKHOV and M. S. TARARSKAYA, Meteorolog. i Gidrol. **11** (1969).

Space Research XI — Akademie-Verlag, Berlin 1971

INVESTIGATIONS FROM AIRCRAFT FOR SOUNDING THE ATMOSPHERE BY MEANS OF MICROWAVE RADIATION

K. S. Shifrin, Yu. I. Rabinovich and G. G. Shchoukin

USSR

The physical principles of a microwave method for sounding the atmosphere and underlying surface are discussed and a comparison made of this method with the infrared method. Theoretical investigations of microwave radiation transfer in the atmosphere include the determination of optical characteristics according to Mie theory. The field of microwave radiation of the system (surface, precipitation layer, cloud layer, clear atmosphere) is calculated by the equation of transfer. Aircraft measurements are presented, including: total water vapour content in the vertical column of the atmosphere; cloudiness and precipitation zones; temperature of the underlying surface; characteristics of ice cover; and an analysis is made of their accuracy. The range of values and accuracy of measurement of each characteristic is indicated.

1. Introduction. Physical Principles of the Method

1.1. Introduction

In present-day meteorological studies the visible and infrared sections of the electromagnetic spectrum are generally used. In that part of the spectrum the radiation fluxes are measured, the images of cloud fields are obtained etc. However, in the farther part of the spectrum, in the domain of millimetre and centimetre waves, the radiation of the atmosphere and the earth's surface also carries in itself valuable information about the state of the atmosphere and the earth's surface. The use of radio emission for determining the characteristics of the radiating object was given the name of thermal radiolocation. The application of thermal radiolocation in meteorology appears to have considerable promise, as the radiation with $\lambda \geqq 3$ cm passes through the cloud cover of the earth practically freely. In the present paper we shall consider the possibilities of using the microwave range for solving the problems of sounding the atmosphere, and compare it with the visible and infrared radiation, describing the results of measurements from aircraft.

1.2. The Problems of Measurement [1, 2]

In the microwave region the brightness of the blackbody is described by the formula of Rayleigh–Jeans:

$$B = \frac{2\,\nu^2}{c^2}\,k\,T = \frac{2\,k\,T}{\lambda^2} = 2.76 \times 10^{-23}\,\frac{T}{\lambda^2}\,\frac{\text{watt}}{\text{cm}^2\ \text{sr Hz}}\,. \tag{1}$$

The formula indicates that here one has to deal with very small amounts of radiation. Thus, passing, for example, from $\lambda = 10\ \mu$m to $\lambda = 10$ cm, the brightness decreases 10^8 times. Fortunately, the microwave radiometers possess a considerably greater sensitivity than infrared sensors. This difference of response is because the bandwidth of the microwave radiometer is extremely narrow, while that of the infrared sensor covers all blackbody radiation. Also, by using the effect of signal accumulation, i.e. by increasing the time constant of the radiometer, one succeeds in raising the signal/noise ratio sufficiently to measure even the weak signals mentioned above. The problem of the time constant is very important. Each second represents 8 km of the artificial earth satellite path, and if, for instance, the time constant is of the order of 100 seconds, it is scarcely probable that we could fix a shower zone extending over some 6–10 km. The time constants of radiometers, located in the aircraft, may now be made of a few seconds to a few hundredths of second. The sensitivity (measured in degrees) is then in the order of 0.5–5 °K, being different for different wavelengths. Beside the energy sensitivity from the point of view of the measuring method, the angular resolution β and the antenna diffusion factor K are also important. In accordance with the well-known Rayleigh criterion, the infrared radiometers present considerable advantages over the microwave radiometers. Here it is possible to obtain for the satellite and the aircraft radiometer $\beta \approx 2$–$6°$, and the coefficient $K \approx 0.20$.

1.3. Distinctive Features of Propagation and Reflection from the Earth Surface

In the microwave region the only absorption of appreciable importance is that associated with the dipole moments – electric and magnetic. Of all the atmospheric gases only the molecules of O_2 and H_2O have distinct dipole moments, magnetic in O_2 and electric in H_2O. All the rest of molecules (N_2, CO_2, H_2, Ar, He, Ne) have neither electric nor magnetic moment and, therefore, in the range of centimetre and millimetre waves the absorption factor for them will be practically zero. The molecule of O_2 has strong bands at $\lambda = 0.5$ cm and $\lambda = 0.25$ cm, and a weak non-resonance absorption in the region of 2–25 cm. The molecule of H_2O has a strong band at $\lambda = 1.35$ cm, as well as lines in the millimetre region. The attenuation in the hydrometeors is determined by the value of diffraction parameter

$$\varrho = 2\pi\alpha/\lambda\,.$$

Substituting the numerical constants for the attenuation factor α in km^{-1}, we obtain

$$\alpha \approx \frac{0.1\,R}{\lambda^2}\,\mathrm{km}^{-1}\,. \tag{2}$$

Here R is liquid-drop water content in g m^{-3}; λ is in cm. For $\lambda = 1$ cm and $R = 1$ g m^{-3} we find, for instance, that $\alpha = 0.1$ km^{-1}. Consequently, we see that the clouds absorb centimetre radiation rather weakly.

For atmospheric precipitation the parameter $\varrho \approx 1$. Hence, the probability of the quantum survival for big droplets of precipitation proves to be appreciable: this means that a substantial scattering will take place.

A significant feature of the microwave range are the notable variations in surface emissivity. Thus, the emissivity for $\lambda = 3$ cm amounts to 0.40 for water, to 0.80 for land surfaces and to 0.98 for forests. This circumstance presents difficulties in measuring the temperature by radiation, but facilitates the task of determining the separation boundaries between different surfaces.

2. Theoretical Studies

In [3] the brightness factor and the emissivity of rough sea surface for $\lambda = 0.8$, 1.35, 1.6, 3.2, 8.5 cm are considered. A simple formula which connects the emissivity of sea surface with the speed of the wind over the sea is given.

In [4] the influence of temperature and salinity on the emission of a smooth water surface is studied. The part played by salinity is insignificant, while the change of water temperature is of prime importance.

Some valuable results have been obtained for the transfer of microwave radiation in clouds and in precipitation. In [5] Shifrin et al. estimate, according to precise Mie formulae, the diameters of scattering and attenuation for λ in the range 0.8–8.5 cm, at $t = \pm 10°$ both for specific particles (up to droplets of 0.66 cm diameter) and for precipitation of different intensities. In [6] precise values are given for scattering functions for the same cases. It is interesting to note that due to the great importance of the refractive index in the microwave region, there is observed the effect of a decrease in index (Thomson effect). In the polydispersion volumes the deformation of the indicatrix is determined by the concurrence of two contrary effects (Mie and Thomson effects), and with increase of the intensity of precipitation an oscillation of the index may be observed depending on λ and the microstructure of the precipitation. In [7] the relation of the thermal emission and temperature of small droplets is studied. It should be noted that with heating the thermal radiation of droplets falls (see [7], Table 1), because the emissivity of droplets decreases more rapidly than the Planck function increases. For instance, when a cloud droplet with a radius of 10 μm is heated from -8 to 20 °C, its radiation decreases 1.7, 2.0 and 2.3 times respectively for $\lambda = 0.62$, 1.24 and 3.2 cm.

In [8] the calculation of radio brightness of a flat layer of precipitation is based on the accurate solution of the transfer equation. The conditions are ascertained in which it is essential to consider multiple scattering. In particular, this should be done for precipitation of medium intensity and $\lambda \leqq 3.2$ cm.

3. Measurements from Aircraft [1, 9–13]

The measurements from aircraft described below were conducted by means of a special radiometer, installed on board a research aircraft IL-18. This radiometer had four channels for wavelengths 0.8 cm, 1.35 cm, 1.6 cm and 3.2 cm.

We shall briefly consider the measuring results systematizing them according to specific problems.

3.1. Determination of the Water Vapour Content of a Cloudless Atmosphere

The most convenient way of measuring the water vapour is to use the line $\lambda = 1.35$ cm. It does not coincide with lines of oxygen absorption and has an optimum absorption factor.

In measuring we have to deal with different profiles of temperature and humidity as well as with different temperatures of the subjacent surface.

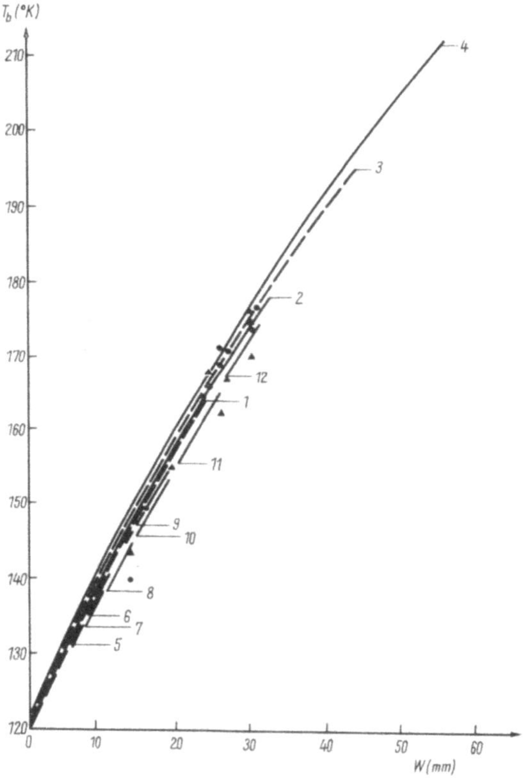

Fig. 1. Radio thermal radiation of earth–atmosphere system. Calculations for 73 various models of atmosphere and underlying surfaces: ○ ● calculations for real profiles; ▲ experimental data.

Curve	T_s	$T(0)$
1	298 °K, 4%	288 °K
2		293
3		298
4		303
5	273 °K, 0%	273 °K
6		278
7	283 °K, 0%	278 °K
8		283
9		288
10	293 °K, 0%	288 °K
11		293
12		298

How does the variance of all those values affect the accuracy of measuring the general water vapour content W? In order to elucidate that point, a series of calculations had to be made. 73 models were examined covering the whole range of fluctuations. Fig. 1 gives 12 curves $T_b = f(W)$, which describe the whole range of relations. It is important that the dispersion of curves is small so that the variations in profiles, salinity and temperature of water lead to fluctuations of radio brightness temperature of about 2–3°. Then we can determine the accuracy of the W estimate by the measured T_b. If measurements are taken over the ocean, whose temperature data are not available and the climatic value is taken for the temperature, we shall obtain the following table for the accuracy of measurement of W.

Table 1
Accuracy of measurement of water vapour content

Range of variation of water content (mm of precipitated water)	0–10	10–25	25–55
Absolute error (mm of precipitated water)	2.5	3.5	4.5

The measurements of W from the aircraft were conducted over the Caspian Sea and the Ladoga Lake. The comparison of W-values determined by T_b, with the W-values calculated according to the data of the aircraft sounding, has shown that the maximum error did not exceed 16%, and the average error amounted only to 8%.

3.2. Determination of Cloudiness and Precipitation Zones

In the microwave range the radio emission of hydrometeor formations shows an appreciable contrast as regards the emission of pure atmosphere.

The measurements from aircraft were conducted over a water surface. For two channels, of 0.8 cm and 3.2 cm, the increase in radio brightness temperature due to cloudiness and precipitation was determined relative to the radio emission with a cloudless sky:

$$\Delta T_b = T_{b\,\mathrm{cloud}} - T_{b\,\mathrm{atm}}\,.$$

The measuring data ΔT_b are summarized in Table 2. The use of two channels of different λ made it possible to study all cases of precipitation and cloudiness.

Table 2
Variation of radio brightness temperature with cloudiness

Type of cloudiness	Int. of precipitation (mm/hour)	$\lambda =$ 0.8 cm	$\lambda =$ 3.2 cm
Ci, Cs	—	2–5	0.0
Sc	—	26.0	2.0
Cu	—	7–12	2–4
Frnb, Ns, Ac	1	55.0	6.0
Stfr, As, Ac	1.5–2.0	100.0	20.0
Ns, As, Ac	2–3	110.0	45.0

Table 2 demonstrates that the maximum values of ΔT_b are observed with nimbo-stratus clouds and the minimum values with high clouds. With the appearance of precipitation, ΔT_b increases appreciably. It is evident from the table that at $\lambda = 0.8$ cm precipitation of medium intensity may be detected not only above water but also over the land, as the variations of radio brightness temperature of uniform land sections amount to about 20°. The measurements of ΔT_b for $\lambda = 0.8$ and 3.2 cm were correlated with the calculations. For $\lambda = 0.8$ cm the discrepancy was about 20%; for $\lambda = 3.2$ cm it was much greater. Here the contrasts are smaller and the influence of the calculation errors is more significant.

3.3. Temperature of the Subjacent Surface

To determine the temperature it is advisable to use the wavelength range from 3 to 10 cm. Here the influence of the atmosphere is small, so that the temperature may be determined through clouds and precipitation. Detailed measurements of temperature were made in the period of 1966–1967 over the

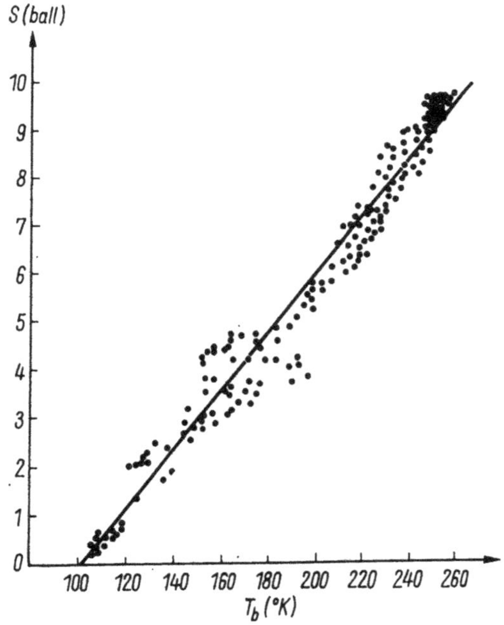

Fig. 2. Radio thermal temperature dependence on the ball of ice closeness S (according to measurements in the Karsky Sea and the Laptev Sea).

Caspian Sea and the Baltic Sea as well as over the Ladoga Lake, at a wavelength of 3.2 cm. The microwave measurements were accompanied by the determination of temperature by means of an infrared thermometer and by shipborne observations. The root-mean-square deviation for the Caspian Sea proved to be about 2° C, while over the Ladoga Lake it was much smaller. This may be explained by the influence of water salinity, which in the first case varied considerably along the route, while the authors have used the constant emissivity.

3.4. Characteristics of Ice Cover

Measurements of ice cover characteristics were made in the Karsky Sea and the Laptev Sea. They have proved that both the ice borders and the ice accretion due to a great ice–water contrast are adequately measured even through a dense cloud cover, using the radiometers with λ in the interval of 3 to 10 cm.

Fig. 2 gives a diagram of radio brightness temperature variation plotted against ice accretion S. The linear connection between T_b and S is observed within the accuracy of dispersion, which has been caused to a great extent by the imperfection of the visual estimation of closeness.

References

[1] K. S. Shifrin, Yu. I. Rabinonvich and G. G. Shchoukin, Trudy GGO, No. 222 (1968).
[2] K. S. Shifrin, Yu. I. Rabinovich and G. G. Shchoukin, Meteorol. i Gidrol., No. 6 (1969).
[3] K. S. Shifrin and S. N. Ionina, Trudy GGO, No. 222 (1968).
[4] Yu. I. Rabinovich and V. V. Melentyev, Trudy GGO, No. 235 (1969).
[5] K. S. Shifrin and M. M. Chervyak, Trudy GGO, No. 203 (1967).
[6] K. S. Shifrin and M. M. Chervyak, Trudy GGO, No. 203 (1967).
[7] K. S. Shifrin and M. M. Chervyak, Trudy GGO, No. 222 (1968).
[8] B. A. Volchok and M. M. Chervyak, Trudy GGO, No. 222 (1968).
[9] K. S. Shifrin, Yu. I. Rabinovich and G. G. Shchoukin, Meteorol. i Gidrol. No. 7, 17 (1969).
[10] Yu. I. Rabinovich and G. G. Shchoukin, Trudy GGO, No. 222 (1968).
[11] Yu. I. Rabinovich, G. G. Shchoukin and M. M. Chervyak, Trudy GGO, No. 222 (1968).
[12] Yu. I. Rabinovich, G. G. Shchoukin and V. V. Melentyev, Trudy GGO, No. 222 (1968).
[13] Yu. I. Rabinovich, G. G. Shchoukin and A. I. Novoselov, Trudy GGO, No. 235 (1969).

Space Research XI — Akademie-Verlag, Berlin 1971

POSSIBILITIES FOR SOUNDING THE ATMOSPHERE FROM A GEOSYNCHRONOUS SPACECRAFT

V. Suomi, T. Vonder Haar, R. Krauss and A. Stamm

The University of Wisconsin,
Madison, Wisconsin, USA

This paper demonstrates that vertical temperature sounding of the atmosphere from geostationary altitude is feasible. It treats the meteorological and instrumental options which must be considered to obtain soundings from this great distance with the required accuracy in a reasonable period of time. Measurements from the geosynchronous ATS satellites are used to examine cloud conditions. Observations from experiments on Nimbus 3 are used to specify radiance levels to be measured. For a spin-scan system, like that planned for the SMS, an instantaneous field-of-view of 0.3–0.4 mrad (11–14 km) is recommended for the tropospheric sounding channels of 20 cm^{-1} bandwidth near the 15 μm region. Results indicate that it will be possible to sound the atmosphere to the earth's surface through openings in a 95% cloud cover in less than one hour.

1. Introduction

The measurement of infrared radiance from the earth's atmosphere well enough to infer the vertical temperature profile from geostationary altitude is difficult due to the low energy levels encountered. At 35800 kilometers random noise from the radiation detector is a large fraction of the signal from the atmosphere. Given a suitable optical system and very sensitive detector, we can overcome this energy deficiency by using multiple sampling in space, time or both. One must not use too long integration times, however, since the parameter of interest, atmospheric temperature, varies with time. One must also insure proper space sampling since temperature varies with space as well. Despite these difficulties, the effort will be well worth the trouble one must take because the atmosphere is a four-dimensional system (x, y, z, and t) and a continuous view adds the dimension time.

Such nearly continuous views of atmospheric conditions were first returned from experiments on the Applications Technology Satellites (ATS) 1 and 3, [1, 2]. These satellites are in equatorial orbits at geostationary altitude. They have carried high-resolution telescopic photometers to measure reflected solar radiance. Many of these measurements, formed line by line into a single frame, give an image of the earth–atmosphere system during a 20 to 30 minute period. Successive frames allow the measurement and study of the time variation of atmospheric conditions over a large portion of the earth's surface.

Very recently, spectrometer and interferometer experiments on Nimbus 3 [3, 4] and Nimbus 4 have successfully demonstrated the capability to infer atmospheric temperature profiles from radiation measurements. In addition, the operational value of such measurements (from the polar-orbiting Nimbus) has been shown [5]. It is the purpose of this paper to study the possibility of placing a vertical temperature profile experiment on a geostationary satellite. Our immediate requirement is the ability to infer temperature profiles with an rms error of less than 1° C.

One way to reduce the error is to restrict the observations to clear areas only. Thus only clear column radiances are used to reconstruct the temperature profile, and errors due to the presence of cloud are removed. The Nimbus 3 and 4 results are very impressive even including cloud in the field-of-view, but few would argue that the presence of cloud contributes to higher accuracy.

It is possible to reduce the field-of-view of the instrument in order to see through small openings in the clouds. A smaller field-of-view, however, decreases the signal-to-noise ratio of the measurement and more samples must be taken in order to improve the S/N ratio to acceptable levels. As one reduces the field-of-view are enough additional samples available to compensate for the poorer signal? A general answer to this question is not possible since the performance parameters of different instruments will yield different answers.

The central question of our study may be stated another way. Each radiance observation will be contaminated by detector noise and may be contaminated by "cloud noise". With a large instantaneous field-of-view (IFOV) it is not possible to separate the detector noise from the cloud noise. With a small IFOV, one can separate the "cloud noise" which is not random from the detector noise that is random. However, one can do this only at the price of a poorer signal-to-noise ratio. Are there enough additional clear column samples using a small IFOV to improve the S/N ratio to acceptable levels? The answer depends on the performance specifications of the instrument and on the natural occurrence of clouds.

2. Instrumental Options

The instrumental constraints used in this study are based on typical meteorological spacecraft configurations planned for the early 1970's. These include:

(i) a spin-scan instrument system (see Fig. 1)
(ii) 16 inch optics
(iii) spin rate of 100 rev min^{-1}
(iv) cooled HgCdTe infrared detectors (near 11 μm for imaging; sounding near 14 μm).

In addition to the high-resolution infrared imaging sensor (IFOV about 0.2 mrad) it is probable that an even higher resolution "visible" radiation sensor will share the same primary optical system with the infrared sounding channels.

Measurements from the Nimbus experiments have shown that the magnitude of infrared radiance near the 15 μm CO_2 band varies from 50 to 110 erg $cm^{-2}\,s^{-1}\,sr^{-1}$ cm from band center to edge during clear sky conditions over most of the region viewed from a geostationary satellite. These same experiments obtained measurements with relative accuracies of about 0.25 erg cm^{-2} s^{-1} sr^{-1} cm. Vertical temperature profiles with an rms error of 1 °C were obtained from them. We shall use the observed radiance values to test our proposed experiment and adopt the same relative error value as our limit.

Several important instrument options remain. These include (a) IFOV α of the sounding channels (b) spectral location ν and width $\Delta\nu$ of the channels and (c) scan rate and time required to cover a finite area. These are all interrelated, each affects the relative accuracy of a radiance measurement and

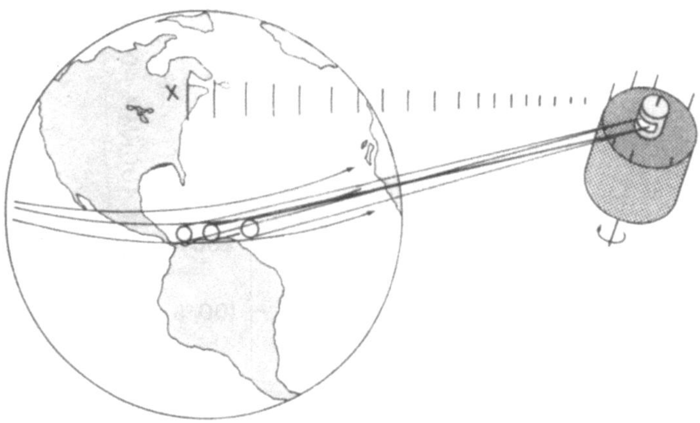

Fig. 1. Schematic depiction of a spin-scan sensor on a geostationary spacecraft.

each of these instrumental options also represents an important decision that will affect meteorological application of the measurements. Combinations of these parameters (as well as the general instrument characteristics mentioned above) were systematically varied in a set of standard instrumental relationships [6]. These include expressions for the power output from the detector, the noise equivalent power (NEP) and noise equivalent radiance (NER). The latter for any individual measurement is approximately:

$$\text{NER} \sim 1/\alpha^2 \tag{1}$$

where α is the IFOV. The ground resolution spot size at nadir, D, is directly related to the IFOV, α. Over a finite grid mesh of area A (i.e. 400 km × 400 km as proposed for the Global Atmospheric Research Program) the possible number of discrete measurements N varies inversely with D^2. Furthermore, the NER due to random instrument noise for each discrete measurement may be reduced by $1/N^{1/2}$ when a representative radiance value over the grid is desired. Thus:

$$(\text{NER})_A = \frac{\text{NER}}{N^{1/2}} \sim \frac{1}{N^{1/2}\alpha^2} \sim \frac{1}{N^{1/2}D^2} \sim \frac{1}{A^{1/2}D} \,. \tag{2}$$

Fig. 2 illustrates the relationship of the instrumental options for the geostationary sounder considered in this paper. Note that for a radiance error of 0.25 erg cm^{-2} s^{-1} sr^{-1} cm the required observable clear area increases rapidly as ground resolution improves. Two additional ordinate scales are provided. One demonstrates the percentage observable clear area required in a GARP grid mesh to acquire a sufficiently accurate radiance value with a sensor of given IFOV. It shows that the required accuracy is possible from *one scan pass* over a region 65–85% cloudy if the sensor's ground resolution D is 11–14 km.

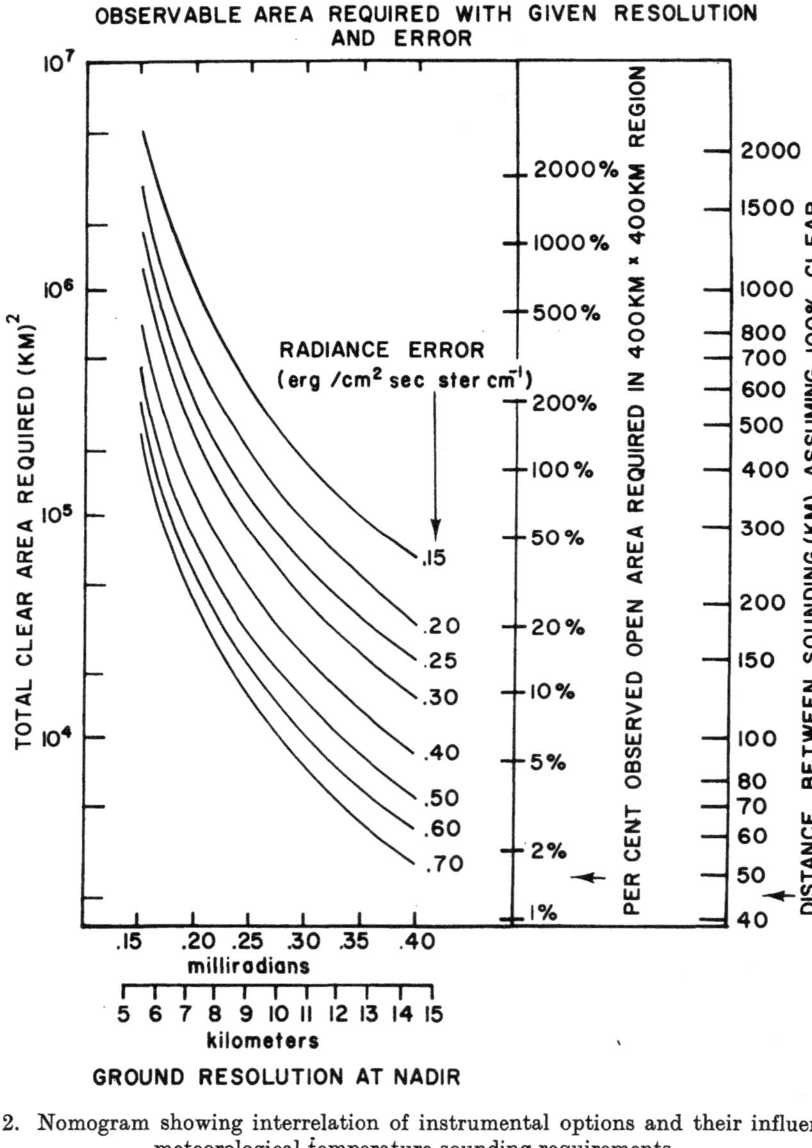

Fig. 2. Nomogram showing interrelation of instrumental options and their influence on meteorological temperature sounding requirements.

At 100 rev min^{-1}, one scan pass over a GARP grid near satellite subpoint would take about 20 seconds. For mesoscale applications over clear regions, the remaining scale shows that the geostationary sounder can acquire an accurate sounding every 150 km if $\alpha = 0.4$ mrad.

It is clear from the above that one observation in each sounding channel cannot be used to infer the temperature profile over the spot size ($D = 11$ to

14 km) on the earth enclosed in the narrow field-of-view. It may take more than a thousand discrete observations to improve the signal-to-noise ratio to the level needed to get a meaningful temperature profile *determination*. This will depend on the specified scan step and sampling frequency. Thus, for the sample instrument described here (with a scan step of 0.2 mrad) the minimum soundable area will be about 90 miles on a side, roughly one-tenth of a GARP grid.

3. Effect of Cloud Distribution

As mentioned earlier, cloud cover interferes with radiometric temperature sounding. To avoid clouds with a given instrument observing over a specified area, we must increase the time required to complete a sounding within our

Fig. 3. ATS-3 picture of 23 April 1968 (1714Z) noting grid areas used in cloud distribution study [7].

relative accuracy limits. In order to separate cloud fields-of-view from clear observations, a higher resolution (but with a wider bandpass $\Delta\nu$) infrared imaging channel can be used.

Most meteorologically active (baroclinic) regions tend to be quite cloudy. Temperature soundings near these features of the circulation are particularly valuable. In order to test how well one can sample these areas, a special study [7] of cloud distribution was performed. It used data from ATS-1 and 3, thus simulating the view of the proposed sounder. The Spin Scan Camera on ATS 1 and 3 has an IFOV of only $\alpha = 0.1$ mrad. Thus it was possible to simulate the cloud statistics of a radiometer having a larger IFOV.

Some of the regions studied (e.g. Fig. 3) were completely cloud-covered. Others (in the intertropical convergence zone, over areas of dense broken cloud cover near extratropical fronts and within active convective cloud regions over South America) can be represented by the traces labeled Case 1, 2 and 3

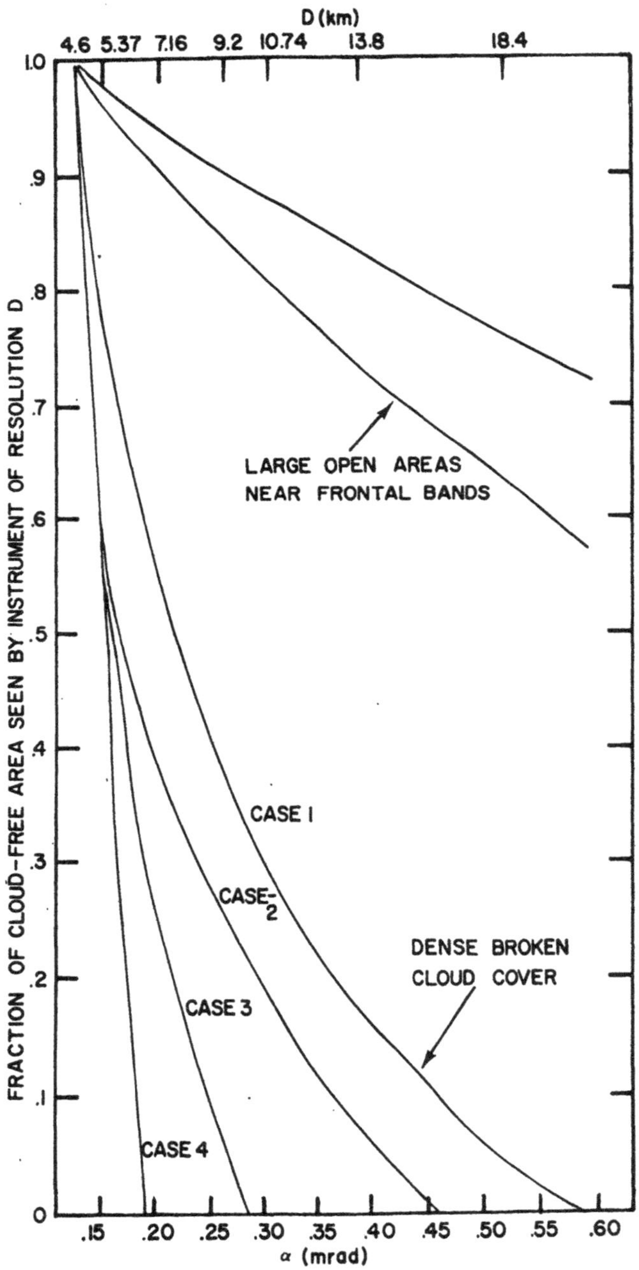

Fig. 4. Fraction of "open" or cloud-free area seen over a GARP-size grid by instrument of angular resolution α (normalized to 100% of 4.6 km ground resolution).

in Fig. 4. These show the reduction in the total area of clear fields-of-view as a function of sensor ground resolution. The values are normalized to the percentage clear area seen by a sensor with ground resolution of 4.6 km. Case 4 in the figure is a rather special case with all clear areas smaller than 7 km. Other, less cloudy, conditions are shown to have less rapid reduction of observable clear area with increasing α or D.

In order to optimize the engineering choices to typical cloud distributions we must specify the scan-step size mentioned earlier. Since a high-resolution infrared imager is likely to be included along with the sounder, a scan step about 0.2 mrad is probable. We shall use this value and thus a typical frame of the entire earth would require 15 to 20 minutes.

Table 1
Temperature sounding time required for various cloud conditions
($A = 160000$ km^2, $(\mathrm{NER})_A = 0.25$ erg cm^{-2} s^{-1} sr^{-1} cm, $\alpha = 0.4$ mrad, $\nu = 690$ cm^{-1}, $\Delta\nu = 20$ cm^{-1})

% Open area	Number of complete frames	Time for sounding (hr.)
100	1	0.26
50	1	0.26
10	2	0.52
5	3	0.79
1	11	2.88

Table 1 shows the time required to obtain a vertical temperature sounding over a GARP grid for various cloud conditions using the above scan step with our sample instrument. Note that while 0.2 mrad stepping was used the sounding channel had an IFOV of 0.4 mrad. The surface temperature is 300 °K; a relative radiance error of 0.25 erg cm^{-2} s^{-1} sr^{-1} cm will be obtained over the grid mesh. Note that even in a region of 95% cloud cover an accurate sounding is possible in under one hour.

It is evident from our balancing between relative accuracy, field-of-view and area/time that there is an optimum set of parameters for a specific spacecraft and basic instrument. The observing system considered here is similar to the proposed GeoSynchronous Meteorological Satellite (SMS). For such a system we are apparently instrument-noise limited for small IFOV's and cloud-noise limited at larger IFOV's. Fig. 5 shows this very well in terms of relative sounding time required over a GARP grid. Each trace refers to a cloud situation shown in Fig. 4. For a given case the valley curves indicate a preferred resolution to minimize the sounding time. This effect is demonstrated by [7] who considered 18 grids near frontal zones over the US and over convective cloud regions of the Amazon basin (see Fig. 3). Four of these regions are more than 99% cloud covered and cannot be sounded at $\alpha = 0.3$–0.4 in a reasonable time period. Of the remaining 14, nine may be sounded with desired accuracy in less than one hour with IFOV of 0.3–0.4 mrad and five would require from 2 to 6 hours. At higher and lower ground resolution the results were less favorable.

In general, the region of $\alpha = 0.3$–0.4 mrad looks best even though we are viewing only 10 to 30% of the cloud-free area which would be viewed by a sensor with smaller α. Since the curves in Fig. 5 have rather flat minima, it may be best to make a final IFOV selection closer to 0.4 mrad to gain signal-to-noise ratio as noted by equation (1). This applies especially to the low and mid-tropospheric channels: those where cloud effects are most severe.

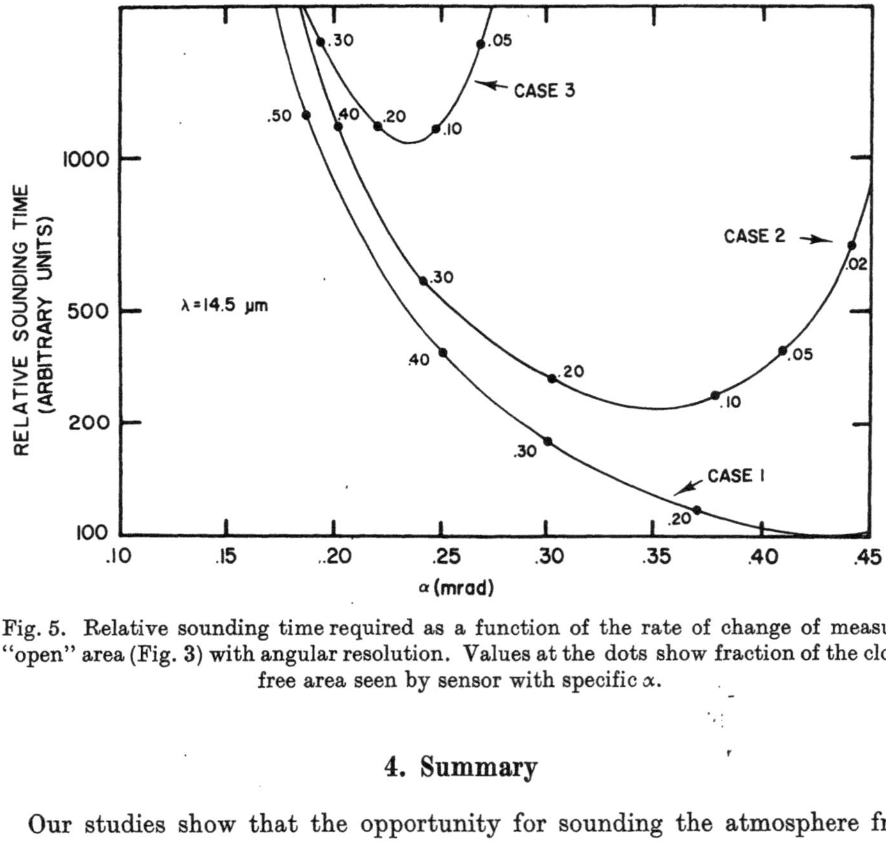

Fig. 5. Relative sounding time required as a function of the rate of change of measured "open" area (Fig. 3) with angular resolution. Values at the dots show fraction of the cloud-free area seen by sensor with specific α.

4. Summary

Our studies show that the opportunity for sounding the atmosphere from geosynchronous orbit looks very attractive. These studies have shown that even with typical cloud distribution near meteorologically active regions a sounding is possible over a 400 km × 400 km grid mesh within a reasonable time interval.

A recent paper [8] shows that when soundings from geostationary altitude are added to those obtained from two polar orbiting satellites, the numerical model's predicted wind error is reduced by 35%. In addition, they demonstrate that a geostationary sounder can serve as a powerful backup to a two-satellite polar orbiting sounding system. Their results show that a combination of geostationary satellites and one polar orbiting satellite gives wind values at least as good as two polar orbiting satellites alone if the data are inserted at intervals of twelve hours. They also find that if the geostationary data are inserted more frequently than every 12 hours there is an improvement over results obtained from polar orbiting satellites. In view of the fact that the incremental cost of including infrared sounders on geostationary satellites is relatively modest and that even in the worst conditions one can obtain soundings every two hours, this feature should be added to the global observing system.

COSPAR Working Group 6 Report [9] states that "If the results of such studies show that infrared soundings will be feasible from geostationary satellites, such observations in combination with those obtained from near-polar satellites would form an almost ideal complementary combination." The present study shows that the infrared sounding from geostationary altitude is indeed feasible and the observing system simulation study [8] shows these observations to be very valuable.

Acknowledgments

The authors thank Professor R. Parent for his contributions to this work and acknowledge support of NASA under contract NAS 5-11542.

References

[1] V. Suomi and R. Parent, Bull. Am. Met. Soc. **49**, 74 (1968).
[2] V. Suomi and T. Vonder Haar, J. Spacecraft and Rockets **6**, 342 (1969).
[3] D. O. Wark and D. T. Hilleary, Science **165**, 1256 (1969).
[4] R. Hanel and B. Conrath, Science **165**, 1260 (1969).
[5] W. Smith, H. Woolf and M. Jacob, Mon. Weath. Rev. **98** (1970).
[6] R. Krauss, Special Report under Contract NAS5-11542, SSEC, University of Wisconsin (1970).
[7] A. Stamm and T. Vonder Haar, J. Appl. Met. **9** (1970).
[8] R. Jastrow and M. Halem, Bull. Am. Met. Soc. **51** (1970).
[9] COSPAR Working Group 6 Report to JOC, Second Draft (1968).

Space Research XI — Akademie-Verlag, Berlin 1971

SPECTROPHOTOMETRY OF THE EARTH FROM MANNED SPACECRAFT

K. Ya. Kondratyev, A. A. Buznikov, B. V. Vinogradov, V. N. Volkov, V. V. Gorbatko and O. I. Smokty

Leningrad University, Leningrad, USSR

The paper discusses the results of spectrophotometry and photography of the twilight aureole of the Earth, clouds and the underlying surface from manned Soyuz 5 and Soyuz 7 spacecraft with the help of a hand spectrograph. The spectra of the twilight aureole as first derived from space enabled one to build the vertical profiles of aureole brightness for various wavelengths in absolute units (solar spectra and those of standard light source being simultaneously recorded from the spacecraft). Interpretation of the spectra has been carried out with the purpose of determining objectively the Earth's colors as observed from space and comparing the objective colorimetric data with visual estimates. Comparison of the measured aureole brightnesses with the calculated data permitted the derivation of information on the vertical brightness profiles in the twilight region and an estimation of the role of multiple scattering at different altitudes. Analysis of cloudiness spectra and those of underlying surfaces simultaneously recorded from spacecraft (altitude of about 200 km) and aircraft (altitude of 2.7 km) enabled one to obtain the estimates of the observed brightness variations of natural formations and to assess the influence of atmospheric haze (from spacecraft observations) on the values of brightness and spectral contrast. The results are described of the combined subsatellite optical experiment with two aircraft at different altitudes and ground-based observational facilities, with the object of investigating the atmospheric transfer function and studying the possibility of the use of the complex of optical data for the identification of natural formations.

1. Introduction

Missions of manned spacecraft and interplanetary automated stations and the creation of the first orbital manned stations open up wide prospects for the investigations of the planetary atmospheres and underlying surfaces. Most promising for the solution of various problems on the atmosphere and landscape optics are spectral investigations of the brightness of the objects. The experiment on spectrophotometry of the Earth's twilight aureole over the spectral range 400–650 nm was first performed during the Soyuz 5 space mission in January 1969. The main results of this experiment are given in [1–4].

In the group flight of Soyuz 6, 7, 8 spacecraft the programme of the complex optical experiment was extended to the solution of the following principal problems:

1. Spectrophotometry of the Sun and the twilight aureole of the Earth's atmosphere for different conditions of solar illumination, various sight angles and different positions of the observer in space in order to investigate the

brightness aureole picture and studying the atmospheric vertical aerosol structure.

2. Spectrophotometry of various natural formations in order to studying the possibilities of their identification in the interests of geological and geographical research.

3. Realization of a subsatellite complex programme of ground-based and airborne optical investigations of the atmosphere and natural surface formations for collecting the material necessary to determine the atmospheric spectral transfer function, spectra and spectral contrasts of natural formations in accordance with the basic optical parameters. The subsatellite experiment was also aimed at the derivation of a complex of radiation characteristics of natural formations in various spectral ranges (from the visible spectral region to the microwave) so as to use this material for studying the changeability of the radiation field with the geological and geographical peculiarities of natural formations.

2. Some Results of the Twilight Aureole Spectrophotometry from Soyuz 5 Spacecraft

The operational programme with the hand spectrograph RSS-I in the Soyuz 5 flight implied a survey of the Earth's atmosphere twilight aureole in the direction towards the Sun, from the moment of its occurrence up to the full emergence of the spacecraft on the sunlit side of the Earth. This spectrophotometric experiment was fulfilled on three orbits: under conditions of continuous cloudiness (orbit 6), broken cloudiness (orbit 15) and cloud-free atmosphere (orbit 2).

As a result, the spectra and photographic images of the twilight aureole have been simultaneously obtained for different sunset angles. The observations were conducted in the direction towards the Sun with the azimuth of the entrance spectrographic slit at an angle of 8° relative to the Sun. The average orbital altitude of Soyuz 5 spacecraft in the course of spectrophotometric measurements was 240 km. The spectra derived were processed with a height resolution of 1 km and a wavelength resolution of 10 nm. The altitude determination of spectral brightness curves was made from the photographic image of the aureole. When the horizon was covered with clouds it was calculated from the maximum of the monochromatic brightness curve ($\lambda = 450$ nm). The analysis of the data [4] shows that near the Earth's surface the main contribution to the twilight aureole brightness is made by long-wave radiation. With the height of the sight layer increasing, the twilight aureole brightness is determined by the scattering and absorption of short-wave radiation. The maximum brightness value falls at a wavelength of about 480 nm. The brightness minimum observed at a wavelength of about 600 nm is due to the ozone absorption of solar radiation in the Chappuis band.

Spectral brightness depends strongly on the azimuth of the direction of sight and the sunset angle, sharply increasing with a decrease in the latter.

The height corresponding to the maximum of spectral brightness also depends on the sunset angle and the wavelength, decreasing with increase in wavelength.

Analysis of the brightness curves showed a lack of noticeable brightness depressions due to the local aerosol layers (for example, at the height of 11 and 19 km).

This is, first of all, due to the fact that the spectrophotometric experiment was conducted at a sunset angle near zero. To establish conformity with the conclusions from [5] calculations were carried out to determine the influence of the aerosol concentration on the absolute values of the depressions mentioned. It turned out that a twofold increase in the volume aerosol concentration at a height of 20 km (Elterman's model [5]) does not cause marked depressions on the corresponding brightness curves, which appear only with a fourfold increase of the aerosol volume concentration at this height. The usual value for the aerosol volume concentration as given by Elterman's model [5] leads to the appearance of the depressions on the monochromatic brightness curves only at sunset angles exceeding 3°, the absolute value for these depressions being small.

The analysis of colorimetric calculated data as obtained for Elterman's model [5] showed that at small sunset angles the twilight sky colour was essentially influenced by ozone ("blueing" of the twilight sky within the altitude range of 25–30 km). Brightness depressions due to aerosol layers appear only at large sunset angles ($\delta_{\odot} > 3°$).

Comparison of the experimental values for twilight aureole brightness with the calculated ones obtained by the approximation of primary scattering for Elterman's model [5] for the orbital parameters of Soyuz 5 spacecraft and directions of sight corresponding to the optical experiment from space would give an estimate of the contribution of multiple scattering to aureole brightness at different atmospheric altitudes. It is natural that the contribution of multiple scattering near the underlying surface turned out to be determining. The role of primary scattering increases with increase in the height of a sight layer. Above 10–30 km the experimental and theoretical brightness curves practically coincide (taking into account primary scattering).

Based on the experimental and theoretical brightness data colour pictures of the twilight aureole have been built and compared with the corresponding visual observations.

3. Peculiarities in the Optical Scheme of the Hand Spectrograph RRS-2 and in the Technique of Processing the Spectrograms

Spectrophotometry of the twilight atmosphere and underlying surface from the Soyuz 7 spacecraft was conducted with the help of the modified hand spectrograph RSS-2 [6], the optical scheme of which is represented in Fig. 1. The entrance objective focuses the image of the remote object under investigation on to the entrance slit of spectrograph (spectrograph channel) and the photofilm (photofixation channel). The focal distance of the entrance objective is $f_1 = 135$ cm, the relative opening being $D/f_1 = 1:4$.

The spatial separation of light fluxes between the spectrograph and photofixation channels is performed with the aid of a light-splitting cube. Upon this, a small portion of radiation (about 5%) passing through the cube strikes the photofilm while the major portion (95%) is turned through 90° and strikes the spectrograph entrance slit. The photofilm and entrance slit planes coincide with the back focal plane of the teleobjective: therefore here the real image of the object and its background is formed.

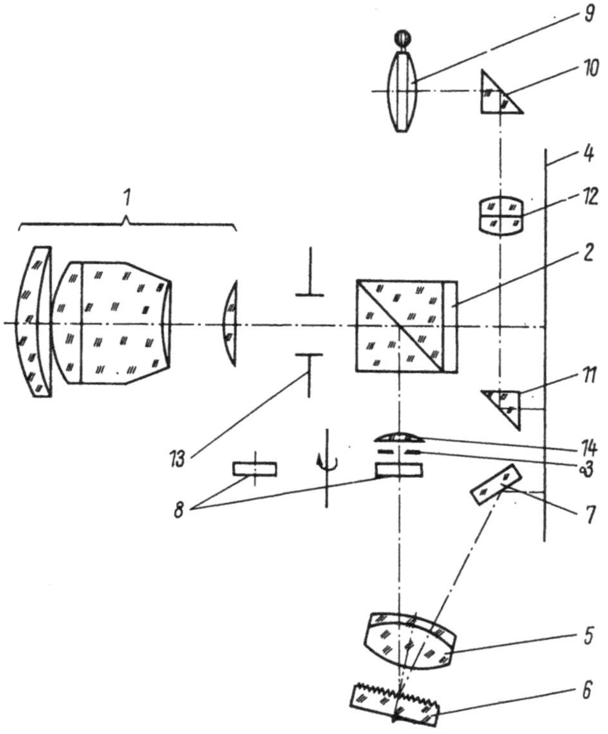

Fig. 1. Optical scheme of the spectrograph RSS-2.
1, teleobjective; 2, light-splitting prism; 3, entrance slit; 4, film; 5, objective; 6, diffraction grating; 7, plane reflecting mirror; 8, filters; 9, clock; 10, 11, turning mirrors; 12, scale objective; 13 aperture.

The optical part of spectrograph is built on the vertical-symmetric scheme. It consists of the entrance slit (width $S = 0.3$ mm, height $H = 6$ mm), collimating objective with $f = 110$ mm and the plane diffraction grating 600 lines/mm; the operational order of the spectrum is the first. The interposition of the objective and the grating is selected in such a way that the surface of longitudinal focals is flat, this surface being matched with the photofilm plane. The spectrum hits the photofilm after passing through the objective with the help of the plane mirror. The field of vision of the spectrograph with teleobjective is equal to $2\alpha = \frac{180° h}{\pi f_1} = 2.5°$ in terms of slit height and equal to $2\beta = \frac{180°}{\pi}\frac{S}{f_1} = 0.13°$ in terms of slit width.

The spectral range of the instrument is determined by its optics and by the range of film sensitivity. In the given experiment the linear dispersion of the instrument is equal to 166 Å mm^{-1}, the spectral slit width being 50 Å and spectral range of measurements being 430–690 nm.

During the experiment the spectrograph is pointed at the object by means of the optical sight whose axis is combined with that of spectrograph. To protect the eyes from the direct solar rays the sight is provided with two polaroids whose mutual turn rotation allows one to change the illumination in the sight field of vision up to the normal value.

For the time determination of each frame the image of a three-handed clock dial is printed into the film.

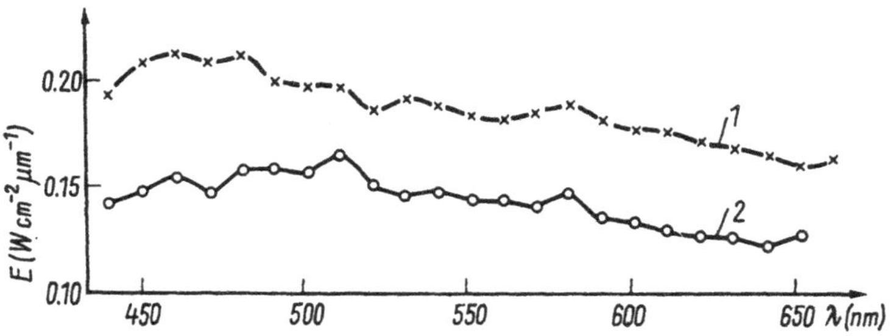

Fig. 2. Solar illumination outside the atmosphere.
1, according to F. S. Johnson (averaged over the interval of 100 Å); 2, illumination behind the illuminator.

Determination of the resolving power according to the slit height (geometrical resolution) was performed by photographing from the distance of 45 m of four miras with distances between the strips 8, 4, 2 and 1 cm.

The limiting resolution of the spectrographic channel is equal to 2′ and of the photofixation channel 4′. Since observations were conducted by the fixed spectrograph this should be considered as an ideal case. The real resolving power degrades because of smearing, i.e. it depends on the time of exposure, on the angular speed of the object movement and on the personal qualities of the operator.

In spectrophotometric measurements of the Earth's surface from spacecraft by the fixed spectrograph with a time of exposure $^1/_4$ sec, smearing of the image will be about 2 km.

Determination of the absolute brightness values of the natural objects investigated is carried out by comparison of the "blackenings" on the film caused by the glowing object with the "blackenings" from a reference source, a band-lamp with a known spectral energy distribution.

Comparing the "blackenings" derived from the reference source and the sun-illuminated filter it is possible to find the illumination by the Sun behind the spacecraft illuminator (Fig. 2), and comparing the experimental results with known illumination values outside the atmosphere one can estimate the transparency of the spacecraft illuminator.

Processing of spectra was carried out on the microphotometer MF-4, which is provided with a special electronic device for automation of the operation.

Absolute brightness values were determined by photographing on the same film calibration marks of an SN-8 lamp. Height determination of spectral curves was made from the corresponding theoretical brightness curves.

4. Preliminary Results of the Twillight Aureole Spectra Processing as derived from the Soyuz 7 Spacecraft

In the group flight of Soyuz 6, 7 and 8 spacecraft emphasis was placed on the spectral brightness measurements of the surfaces beneath. Spectral investigations were continued of the twilight aureole brightness, started on Soyuz 5, using the modified hand spectrograph RSS-2, the optical scheme of which is described above.

The experiment on spectrophotometric measurement of the twilight aureole was carried out on orbit 87 of Soyuz 7 at 2141 Moscow time (sunset). The

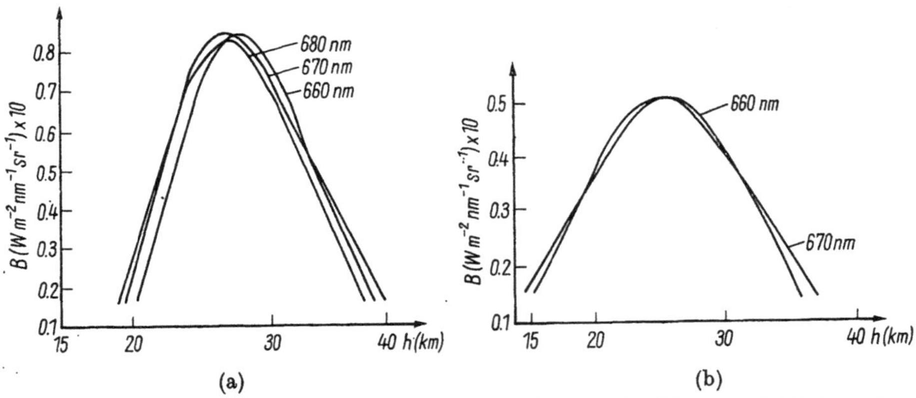

Fig. 3. Monochromatic brightness curves for the twilight aureole of the terrestrial atmosphere. Soyuz 7. a, frame 3; b, frame 1.

geographical coordinates of Soyuz 7 at the moment of the experiment were: 23.18° N 23.39° E (the region of North-East Africa). The orbital altitude according to calculated data at the moment of the experiment was 218 km. As a result the frames containing twilight aureole spectra were obtained; some of them processed according to the above mentioned technique are represented in Fig. 3, which represents the monochromatic aureole brightness at the perigee of the sight ray h_0. The spectral and spatial resolutions of the monochromatic brightness curves are 10 nm and 1 km respectively. The error in the experimental values for the aureole brightness due to accidental errors is approximately 10%.

Comparison with corresponding data obtained in the Soyuz 5 flight shows that the brightness values in the previous experiment were 2-3 times smaller ($\lambda = 650$ nm). However, they give a good fit with the data calculated for Elterman's model in the approximation of primary scattering ($\lambda = 650$ nm). The discrepancy mentioned may be mainly due to two factors. On the one hand, transmission of solar radiation must result in an increase in the aureole brightness; on the other hand, the increase in sunset angle results in a decrease in aureole brightness. Missing on the monochromatic brightness curves but visible in the Soyuz 5 experiment are noticeable depressions due to the aerosol layers situated at different atmospheric levels. The qualitative behaviour of the monochromatic brightness curves with altitude is the same in both experiments.

5. Spectrophotometry of Underlying Surfaces

In the present preliminary report we shall consider the data on the spectra of various underlying surfaces as derived on 13 October 1969 from the Soyuz 7 spacecraft on the trajectory from the Arabian peninsula to the plateau Ust-Urt. Spectrophotometry embraced a very short period of time from 1319 to 1329 Moscow time, at a solar elevation of about 35–50°. From the experience of similar aircraft measurements, it is valid to compare data of spectrophotometric measurements of the Earth from space obtained within this variation of solar elevation. There was also available a number of synchronous spectrophotometric measurements of various natural formations whose comparison is unconditional.

The meteorological situation in the region of measurements during the flight was as follows. Above the Arabian peninsula there was a stationary anticyclone disturbed by weakly developed depressions in the Persian Gulf area, no cloudiness but a strong atmospheric dust haze. A subtropical frontal zone with continuous cumulus and stratocumulus cloudiness was observed above Iran and the Caucasus. Over the transient zone in the North were forms of non-continuous cumulus and thin stratus cloudiness. Over the region of the Ust-Urt plateau there was an anticyclone in the south-western part of which (East coast of the Caspian) cumulus humilis developed. The atmosphere was characterized by fairly high transparency, not strong haze, and considerable areas of the stone plateau were seen between the cumulus.

Though the meteorological situation in the period of observations was not very favourable for the experiment, spectra of quite diverse natural formations were obtained, including those which coincided with the aircraft measurements at "the point of meeting" above Ust-Urt.

The material derived is discussed below under the following main headings: 1. spectral characteristics of some natural formations observed from space; 2. spectral contrasts of some natural formations observed from space; 3. comparison of spectral characteristics of the underlying surface obtained from space with aircraft and ground data; 4. comparison of spectral contrast characteristics from space with aircraft and ground data.

5.1. Spectrophotometric Characteristics of Underlying Surfaces

From analysis of spectrogram quality and photo-point control accuracy the following categories of underlying surfaces can be identified: 1, dense cloud; 2, thin cloud; 3, stone deserts; 4, cloud shadows (Fig. 4a).

5.1.1. The continuous dense cumulus and stratocumulus cloud of the subtropical front over Mesopotamia and Western Iran is characterized by maximum brightness values varying between 1.4×10^{-1} and 2.4×10^{-1} W m^{-2} nm^{-2} sr^{-1} over the wavelength range 440–690 nm (Fig. 4a, 1). The largest spectral brightnesses are observed at $\lambda = 440$–580 nm. An appreciable decline in brightness occurs in the orange-red spectral region at $\lambda = 580$–690 nm, the inclination angle for the spectral curve being $\gamma = 0.7$. Dispersion of single values of spectral brightness is slight (σ_λ being about ± 0.02). Stability of the data depends upon uniformity of composition, high density and relatively even upper surfaces of clouds.

5.1.2. The thin cloud of altostratus and stratocumulus forms at different levels constitutes images both of the cloud layer and of shadow on the Earth's surface (Fig. 4a, 2, 3). The latter does not essentially affect the spectral

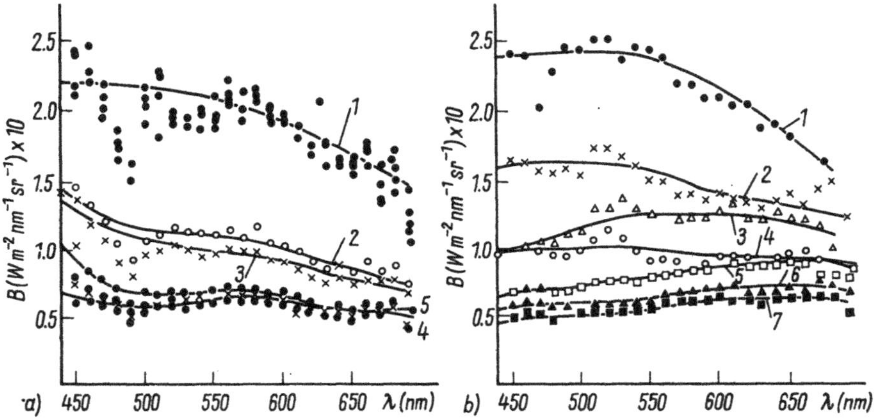

Fig. 4. Monochromatic brightness curves of the underlying surface (nadir).
a. Soyuz 7, 13 October 1969.
1, dense cloud; 2, 3, thin cloud; 4, Arabian desert covered with haze; 5, coast in the Caspian area (recorded under condition of good atmospheric transparency).
b. Aircraft, 13 October 1969 over the Ust-Urt plateau, altitude 2700 m.
1, dense cloud; 2, thin cloud; 3, sand; 4, salt soil; 5, ground road; 6, 7, stony desert.

brightness variation but strongly reduces the absolute brigthness values which decrease within the range of 440–690 nm to 0.7 to 1.4×10^{-1} W m^{-2} nm^{-1} sr^{-1}. Maximum brightness values, as in the case of dense cloud, fall in the blue-green spectral region at $\lambda = 440$–580 nm. In the orange-red region at $\lambda = 580$–690 nm an appreciable but less sharp decrease in brightness occurs with λ up to 0.3. Dispersion in spectral brgihtness value is considerably increased (σ_λ up to ± 0.08); this is related to the nonuniformity both in density and in structure of the cloud cover.

5.1.3. Stone desert is found in the following areas: a. in the western part of the Ust-Urt plateau (Fig. 4a, 4) and b. in the northern part of the Arabian peninsula (Fig. 4a, 5).

5.1.3.a). The stone desert of the Ust-Urt plateau is composed of limestone and covered with grey-brown loam-rubble soils, in some parts with takyrs an thin sand mantles. The vegetation cover is slight (10–20%) being represented by xerophyle undergrowth and does not markedly affect the spectral landscape characteristics. Underlying surfaces of this kind exhibit spectral brightness curves with slightly pronounced variations over the spectrum of about 0.5×10^{-1} to 0.7×10^{-1} W m^{-2} nm^{-1} sr^{-1}. Unlike cloudiness spectra underlying surface of this type has weakly pronounced incrases of brightness in the blue-green part of the spectrum, which results in γ diminishing to 0.05. Also, a small increase in brightness by about 0.1×10^{-1} W m^{-2} nm^{-1} sr^{-1} may be seen in the red spectral region at $\lambda = 630$–690 nm. Dispersion in brightness values is also slight ($\sigma_\lambda = \pm 0.02$), due rather to the low absolute brightness values characteristic of the stone desert than to the uniformity of its surface. The shape of the spectral brightness curve corresponds to the landscape's grey colour with brownish hue.

5.1.3.b). The stone desert in the northern part of the Arabian peninsula is composed of dark-coloured sandstones almost deprived of soil and vegetation cover, and only in some parts thin sand mantles are observed. Like the Ust-Urt desert it has a monotonic variation of the spectral curve which is distinguished from the former by a slight general increase in brightness, by 0.05×10^{-1} W m^{-2} $\times$nm^{-1} sr^{-1}, on an average, as well as by the lack of brightness increase in the red spectral region at 650–690 nm and, finally, by more pronounced increase in brightness in the blue spectral region at 440–480 nm. We associate these differences: i. with the general increase in illumination at the high solar elevation over the Arabian desert; ii. with the lack of fine earth giving the surface a brownish hue and iii. with a stronger atmospheric haze.

5.1.4. The analysis of a darker type of surface, viz. the portions shadowed by cloud, is of interest. Though the cloud shadows measured on the stone desert were only partial they yielded the lowest brightness values, 0.4 to 0.5×10^{-1} W m^{-2} nm^{-1} sr^{-1} in the orange spectral region. A considerably smaller decrease in brightness is observed for the partially shaded stone desert compared with the illuminated in the blue-green spectral region (0.6–0.7 $\times 10^{-1}$ W m^{-2} nm^{-1} sr^{-1}), due to the influence of the spectral composition of scattered light.

5.2. Spectral Contrasts of some Natural Formations

For the solution of the inverse problem of the identification of natural formations from their spectra, for study of frequency-contrast spectral characteristics of underlying surfaces and for the choice of optimum ranges for recording by space remote sensors a knowledge of spectral contrasts is indispensable. The latter were calculated for all the combinations of the above-described natural surfaces measured from space (Fig. 5a).

5.2.1. The spectral contrast coefficients $k = (B_1 - B_2)/B_1$ (B = brightness) of dense cloud and partially shaded surface (5.1.1 and 5.1.4) reach their maximum values of 0.77 at $\lambda = 630$ nm, decreasing somewhat in the blue-green to $k = 0.62$; this is associated with the spectral composition of the light scattered in the atmosphere.

5.2.2. Spectral contrasts for thin cloud and stone desert have spectral variations similar to the above curves (5.2.1), but the absolute contrast values are less by a factor of nearly 1.5 ($k \approx 0.4$). Sometimes a slight decrease in contrast is observed in going from a short-wave spectral region to a long-wave with a small increase at $\lambda = 650$–660 nm.

5.2.3. Spectral contrasts of dense and thin cloud (5.1.1 and 5.1.2) also yield moderate values of contrasts ($k = 0.4$–0.45). The variation in the spectral contrast curve is monotonic with two slight rises at $\lambda = 490$–520 nm and $\lambda = 620$–660 nm, partly due to the influence of the visible underlying surface.

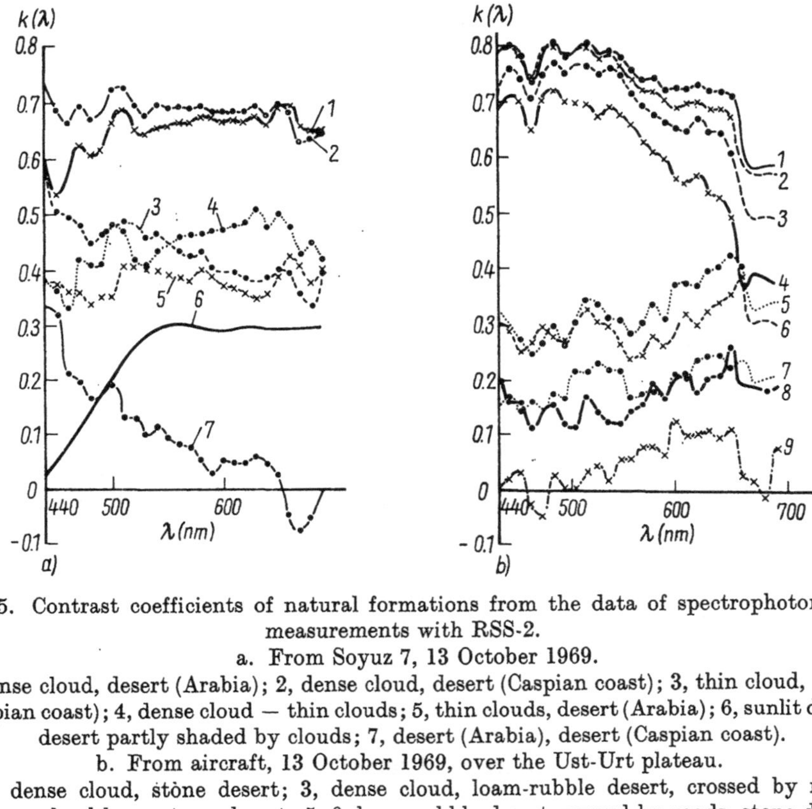

Fig. 5. Contrast coefficients of natural formations from the data of spectrophotometric measurements with RSS-2.
a. From Soyuz 7, 13 October 1969.
1, dense cloud, desert (Arabia); 2, dense cloud, desert (Caspian coast); 3, thin cloud, desert (Caspian coast); 4, dense cloud — thin clouds; 5, thin clouds, desert (Arabia); 6, sunlit desert, desert partly shaded by clouds; 7, desert (Arabia), desert (Caspian coast).
b. From aircraft, 13 October 1969, over the Ust-Urt plateau.
1, 2, dense cloud, stone desert; 3, dense cloud, loam-rubble desert, crossed by roads; 4, dense cloud loam-stony desert; 5, 6, loam-rubble desert crossed by roads, stone desert; 7, loam-rubble desert crossed by roads, loam-rubble desert; 8, loam-rubble desert, stone desert; 9, between two portions of stone desert (Ust-Urt plateau).

5.2.4. Spectral contrasts of the illuminated and partially illuminated stone desert (5.1.3 and 5.1.4) are not large ($k \leqq 0.3$), spectral variation in the contrasts being non-uniform. In the short-wave spectral region at $\lambda < 450$ nm k falls to 0.05–0.15, increasing to 0.25–0.30 in the orange-red at $\lambda > 520$ nm. This variation in contrasts is explained by the influence of the light scattered by the atmosphere which almost completely masks the differences in surface brightness over the range $\lambda < 490$ nm.

5.2.5. Contrasts between the individual portions of stone deserts (5.1.4a and 5.1.4b) are the smallest. In the spectral region where the atmospheric transparency is sufficiently high the haze does not mask the optical contrasts of the land surface, and mean contrasts do not exceed 0.15 (at $\lambda > 520$ nm) and 0.05 over the range $\lambda = 570$–660 nm. It is interesting that in the red spectral region at $\lambda > 660$ nm the contrast increases again to 0.15 at $\lambda = 670$ nm where the optical contrasts as measured from space are most closely connected with colour differences of desert surfaces.

5.3. Comparison of Spectral Brightness Characteristics of Underlying Surfaces as obtained from Space, Aircraft and Ground Measurements

We have the opportunity to compare nearly synchronous space and aircraft measurements of spectral brightnesses of a stone desert of the Ust-Urt plateau derived on 13 October 1969 within the area of "the point of meeting" of Soyuz 7 spacecraft and the instrumented aircraft LI-2, at 1345 at $\varphi = 42°30'$ and $\lambda = 54° 10'$. Satellite spectrophotometric measurements were conducted from an altitude 220 km, those from the aircraft from an altitude of 2.7 km.

5.3.1. From aircraft measurements the spectral brightness variation of stone desert (Fig. 4b) was found to be monotonic over the spectrum, with the values of about 0.45–0.55×10^{-1} W m^{-2} nm^{-1} sr^{-1}. However, the spontaneous increase in brightness is well pronounced from $b_\lambda = 0.45 \times 10^{-1}$ W m^{-2} nm^{-1} sr^{-1} in the green-blue spectrum to 0.6 in the red at $\lambda = 660$–670 nm.

5.3.2. In comparing satellite and aircraft spectrophotometric measurements (cf. 5.1.3a and 5.3.1) the following preliminary conclusions can be made. The brightness of the underlying surface as measured from manned spacecraft over the wavelength range $\lambda \leqq 620$ nm is always higher than that measured from aircraft. Maximum differences are observed in the short-wave part of the spectrum at $\lambda \leqq 450$ nm where the brightness measured from an altitude of 220 km is about 1.5 times higher than that from 2.7 km. In the red spectral region at $\lambda \geqq 620$–630 nm these differences are slight.

5.3.3. The aircraft spectral brightness measurements of dense cloud showed mean values with a maximum at $\lambda = 510$ nm where $b_\lambda \doteq 2.5 \times 10^{-1}$ W m^{-2} x $\times$nm^{-1} sr^{-1}. The decline in brightness in the region to 2.0×10^{-1}W m^{-2} nm^{-1} sr^{-1} at $\lambda > 550$ nm is noticeable.

5.3.4. Comparison of spectral curves for dense cloud measured from space (cf. 5.1.1 and 5.3.3) and aircraft showed the following differences. Over the whole short-wave spectral region at $\lambda < 650$ nm the aircraft measurements yielded somewhat larger brightness values (by a factor of 1.1–1.2). In the red spectral region the satellite brightness values do not differ from those measured from aircraft within the limit of measurement errors. The differences mentioned are of course explained by the influence of haze. By studying such differences in reference to the experiments one can define the optical pecularities of atmospheric haze.

5.4. Comparison of the Spectral Brightness Contrasts of Natural Formations from Spectrophotometric Data obtained from the Soyuz 7 Spacecraft and from the Aircraft

The objective of this research is to determine the influence of the atmospheric transfer function on the change in spectral brightness contrasts in observations from space.

5.4.1. For this analysis the computation was made of spectral aircraft contrasts of the same natural formations as measured from space (cf. 5.2.2): dense cloud (Fig. 5b) and stone desert (Fig. 5b).

Spectral contrasts curves for dense cumulus cloud of the lower layer and various types of stone desert of the Ust-Urt plateau derived from the data of aircraft spectrophotometric measurements are characterized by large contrast values $k = 0.5$–0.8. In a larger part of the visible region ($\lambda \leqq 650$ nm) the uniformly high contrasts $k = 0.7$–0.8 are observed. In the red region ($\lambda \leqq 660$ nm) the contrasts decrease and give an average $k = 0.55$.

5.4.2. The comparative analysis of spectral brightness contrasts of natural formations from the data of spectrophotometric measurements from Soyuz 7 (5.2.2) and aircraft (5.4.1) showed that most comparable were the curves for spectral contrasts of dense cloud and stone desert.

Table 1
Spectral brightness contrasts

λ (nm)	440	500	550	600
From Soyuz 7 data altitude 220 km	0.66	0.69	0.68	0.68
From aircraft data (altitude 2.7 km)	0.76	0.78	0.77	0.70

The curve for space spectral contrasts is smoother in character than that from aircraft (for a space contrast curve γ is about 0.05, for aircraft 0.10–0.15). The contrasts for cloud and the desert measured from space are, naturally, lower than those from the aircraft (on an average by 0.05–0.1).

Comparison of some of the satellite and aircraft data described above allows one to draw a preliminary conclusion on the influence of haze on the brightness contrasts of corresponding objects. As would be expected the strongest haze effect is revealed in the short-wave spectral region where the role of multiple scattering is great. In the long-wave region ($\lambda = 600$–700 nm) the atmospheric scattering of solar radiation (above the level of 2.7 km) has almost no effect on the absolute brightness values of the surface beneath. It is proposed to carry out in future a more detailed analysis of the influence of haze, though from general physical considerations and previous rocket data it follows that the haze effect must be strongest at large zenith solar angles and in directions near the horizon. In case of nadir observations and small zenith solar angles the largest contribution to the brightness field is given by the reflection of solar radiation from the surface beneath.

6. Combined Space, Aircraft and Ground-based Measurements of the Earth's Surface

For the correct experimental study of the atmospheric optical transfer function and the generalization of regularities in the formation of super-small-scale space imagery of the Earth's surface in different spectral regions we have undertaken the experiment of combined space, aircraft and field spectrophotometry in various spectral regions, visible, infrared and microwave.

The results of satellite spectrophotometric measurements are given above (§ 5).

Subsatellite aircraft spectrophotometric measurements were carried out at different altitudes (800, 2700 and 8000 m) and with different detectors: spectrograph (over the range of $\lambda = 400–750$ nm), spectrophotometer (over the range of $\lambda = 400–830$ mn). infrared spectrometer (over the range of $\lambda = 2.8–5.6$ μm) and microwave radiometer (at $\lambda = 3.2$ cm).

Ground-based observations were carried out by spectrograph (over the range $\lambda = 400–800$ nm) and followed by the collection of natural characteristics of the underlying surface (temperature and moisture of soils; mineralogical composition and structure of the surface; composition and density of vegetation).

As an example we shall present one of the geographical profiles on the Caspian to Ust-Urt–Aral sea trajectory during the combined subsatellite experiment on the Soviet manned spacecrafts Soyuz 6, 7 and 8 on 12 October 1969. Not dwelling on the detailed analysis of the material (the subject of a special report) we shall mention some general conclusions important for the development of space spectrophotometry of the Earth.

(i) Changes in spectral brightness coefficient generally correspond to the changes in natural formations associated with different natural factors: colour and transparency of sea waters; mineralogical composition of rocks; density and life forms of vegetation; moisture, chemical composition and roughness of the soil surface.

(ii) Cases of coincidence of brightness coefficients (homology) for different objects are observed (for example, salt crust with dust mantle and clay takyrs), which make the identification of natural formations from their spectra more difficult.

(iii) The influence of the geometrical generalization of brightness coefficients of the individual components of natural complexes observed with the increase of angle of view unavoidable in a survey from space altitudes is great. This generalization exercises a strong smoothing influence on the variation in the brightness coefficient curves as well as on the decrease in spectral contrasts of natural formations.

Varieties of stone desert constitute an example of fairly uniform surfaces which do not possess appreciable spectral reflectance features in the visible region. Analysis of the data on the complex subsatellite experiment covering a much wider wavelength range revealed that even in the very difficult case of the uniform landscape like ours the use of a combination of data on reflectivity, radiation temperatures, etc. makes possible a sufficiently fine differentiation of natural formations. One therefore concludes that the best way to solve the problem of spectrophotometric recognition of natural formations from space is a complex approach using data referring to various spectral regions.

7. Conclusion

The data given mainly meet the requirements for the solution of the direct problem, i.e. study of the regularities of the optical transformation in the formation of the spectral space image of the Earth. Geometrical integration and atmospheric haze exert a smoothing influence on the optical contrasts of natural formations. However, different spectral curves obtained from manned spacecrafts do correspond to different objects of the Earth's surface. Successful experiments of this kind provide strong support for the possibility of solving the inverse problem, i.e. the recognition of individual characteristics of natural objects from spectra derived from orbital altitudes. This problem can be solved only after the derivation and processing of a large number of spectra of natural formations under comparable atmospheric-optical conditions and on the basis of a complex experiment.

References

[1] K. YU. KONDRATYEV, B. V. VOLYNOV, A. P. GALTSEV, V. V. KOLTSOV, O. I. SMOKTY and E. V. KHRUNOV, Dokl. Akad. Nauk USSR **190**, No. 2 (1970) (in Russian).
[2] K. YA. KONDRATYEV, A. P. GALTSEV, O. I. SMOKTY and E. V. KHRUNOV, Dokl. Akad. Nauk USSR **191**, No. 4 (1970) (in Russian).
[3] K. YA. KONDRATYEV, A. P. GALTSEV, O. I. SMOKTY and E. V. KHRUNOV, Dokl. Akad. Nauk USSR, **191**, No. 5 (1970) (in Russian).
[4] K. YA. KONDRATYEV, B. V. VOLYNOV, A. P. GALTSEV, O. I. SMOKTY and E. V. KHRUNOV, Izv. Akad. Nauk USSR, Ser. Fiz. Atmos. Okeana **6**, No. 4 (1970) (in Russian).
[5] L. ELTERMAN, AFCRL-68-0153 Environment Research Paper, No. 285 (1968).
[6] A. P. GALTSEV, V. V. MIKHAILOV, M. N. SOKOLSKY, YA. M. KRUGER, YU. I. BRAININ and L. B. ALEXEEV, in: Problems of Atmospheric Physics, Izd. Univ. Leningrad, No. 7, 1969 (in Russian).

Space Research XI — Akademie-Verlag, Berlin 1971

ALTITUDE VARIATION OF THE SCATTERING COEFFICIENT FROM SPACESHIP SOYUZ 3 MEASUREMENTS AND AEROSOL STRATIFICATION

G. V. Rozenberg and A. B. Sandomirsky

USSR

The brightness structure of the light haze at the edge of the planet, when observed from space, contains information on altitude relationships of the scattering coefficient of the atmospheric air. A theory is developed to determine the scattering coefficient at different altitudes from 15 to 55 km above sea level by photometric analysis of photographs of the day horizon made on board manned space ships with an error less than 25% and with altitude resolution about 2 km.

A great number of photographs obtained from the space ships Soyuz have been treated. There is a discussion on an average statistical altitude relationship of the scattering coefficient and its variations. The whole complex of the data of an optical sounding elucidates the stratification of the atmospheric aerosol, whose properties become an object of analysis.

A light aureole observed by a cosmonaut at the edge of the planet illuminated by the Sun is not only a delightful sight; its brightness structure contains easily obtained information about the altitude distribution of dispersed matter in the atmosphere of the planet. The principal basis of a quantitative interpretation of observed data on space, time and spectral variations of a planetary aureole was laid down in the monograph "Twilight" published in 1963 and it has since been developed in a series of publications [1–15]. Only two circumstances should be noted here.

First, different conditions should be used for observations at different altitudes. A proper arrangement of the experiment gives the heights from 10 to 150 km with an error in the scattering coefficient of about 20–25%, an error in determination of altitude of about ± 2–3 km and an altitude resolution of about 2 km, averaging along a slant path 500 km long.

Secondly, in the observations of the edge of a planet from space one should use the technique of photographic photometry, since only this technique can provide the required angular resolution and the necessary volume of simultaneous information within acceptable errors of measurement. Because of this the use of manned satellites becomes most effective.

The concrete realization of the programme was begun by means of the spaceships Vostok 5 and 6 (1963) and Voskhod (1964) and was continued by the spaceships Soyuz [3, 4, 7–10, 12–15]. In particular, on board the manned spaceship Soyuz 3 (1968) the photographs of the Earth's horizon in the day time were taken by G. T. Beregovoy by means of a photometrically marked black-white film (with an orange light filter).

By photometric analysis of a series of photographs taken above the equatorial region and in the region at 30° N it was possible to obtain 119 reliable altitude variations of haze brightness with a resolution of 0.25 km vertically and 2.5 km horizontally. Examples of the altitude photometric cross sections are shown in Fig. 1. The height of perigee above the sea level is plotted along the abscissa axis and the brightness in relative units along the ordinate axis. The curves

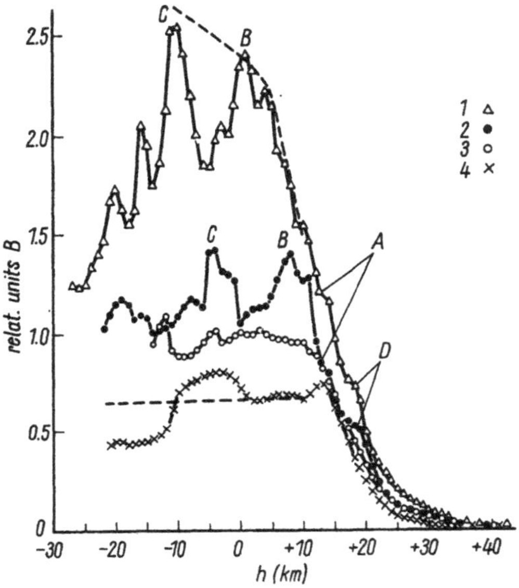

Fig. 1.

correspond to the different states of cloudiness in the vicinity of the planet edge.

A detailed analysis showed the frequent occurrence of a layer with a higher turbidity at the level of 19 km disappearing with height (see arrows at *D*) [3, 4, 14], of a thin haze layer with an optical depth approximately 0.04 [5, 6, 10] at an altitude about 12 km (see arrows at A) [3, 4, 16], and also the existence of comparatively dense tropospheric clouds. The latter are found at the height about 10 km in the cases of curves 1 and 2 (points *B* and *C*); their brightness coefficients are approximately equal to 0.21–0.23 (curve 1), corresponding to an optical depth more than 5 [17] and to 0.10–0.12 (curve 2) corresponding to a comparatively sparse broken cloudiness [17, 18].

At the same time the increase of brightness of an atmospheric aureole at the expense of tropospheric illumination and underlying surface reflection is on the average about 26% in good agreement with theoretical estimates [19].

Altitude variations of the scattering coefficient of atmospheric air have been computed for each of the brightness cross sections. Fig. 2 represents the dependence obtained and averaged over all 119 cross sections (black dots). The dispersion is twice as large; it is due to changeability of the atmosphere itself and not to the error of measurements.

Other results are an altitude dependence of the scattering coefficient, computed by L. I. Koprova [20] from the Elterman searchlight measurements [21, 22], results of a single measurement made by a twilight method from the earth's surface [11, 12], and the altitude variation of the scattering coefficient of a gaseous phase (standard atmosphere).

One can see that the results of all three methods are in good agreement. It is first of all striking [4, 12] that the dispersed phase dominates at all heights. A second point of importance is the existence of a region of lower turbidity at a level of about 30 km, the first indication of which came from the spaceship

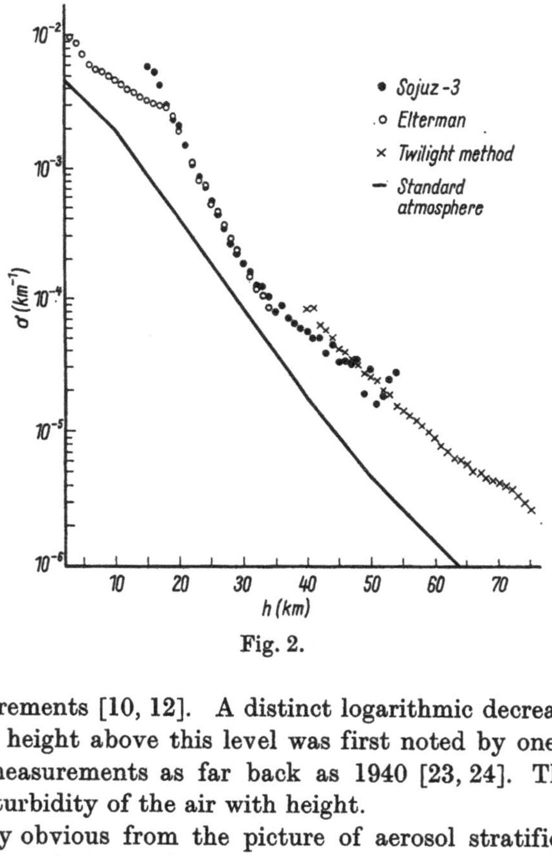

Fig. 2.

Voskhod measurements [10, 12]. A distinct logarithmic decrease of scattering coefficient with height above this level was first noted by one of the authors from twilight measurements as far back as 1940 [23, 24]. This indicates an increase in the turbidity of the air with height.

It is especially obvious from the picture of aerosol stratification in Fig. 3 representing the variation with altitude of the relation between the scattering coefficient of the dispersed phase and that of the gaseous phase of atmospheric air. Besides the data mentioned above, we show the results of our Vostok 6 measurements averaged over 19 cases and also the aircraft measurements of Yu. I. Rabinovich [26] which were statistically treated by L. I. Koprova [25].

The three zones are distinctly seen: the zone of a dispersion phase (lower than 7 km — the surface zone), the stratospheric zone (between 7 and 33 km approximately) and the high altitude zone (above the previous level). The lower zone is obviously due to a turbulent transfer of the dispersed phase from the surface layer. In clear and anticyclonic weather, when the measurements were made, the turbidity is principally due to a fine dispersed portion, with particles less than $\frac{1}{3}$ μm and very sensitive to humidity variations even under dry conditions [14, 15]. The altitude variation of a relative concentration of these particles from the data of E. S. Selesneva [27] is also given in the figure.

The above "stratospheric" or twilight layer [9] of turbidity is caused by hygroscopic particles with diameter some tens of micrometres according to the microphysical measurements by Junge and his collaborators [28] and also

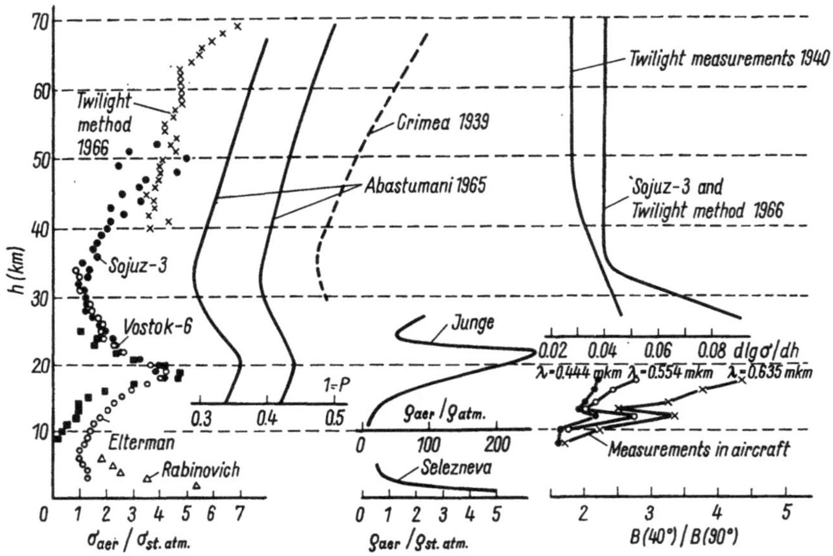

Fig. 3.

according to our optical data [4, 9, 14]. An altitude variation of their specific concentration $\varrho_{\text{aer.}}/\varrho_{\text{st. atm.}}$ obtained from five measurements by Junge is shown in Fig. 3 and it repeats the altitude variation of a specific scattering coefficient. Fig. 3 also represents an altitude variation of the ratio of the brightness of the day sky at scattering angle 40° to brightness at 90° for $\lambda = 0.444$ μm, 0.554 μm and 0.635 μm from the data of aircraft measurements of one of the authors [29]; this obviously indicates the increasing role of a dispersed phase with height. The complex of data on the chemistry of particles forming a twilight layer [28] shows the relation its optical effects under twilight conditions with volcanic eruptions and with ozone variations [9]. The behaviour of the altitude variation of the scattering coefficient in this zone [14] leads to the conclusion that the dispersed phase in this zone is generated inside the atmosphere itself [9, 15, 18]. Such an assumption, in particular, can explain qualitatively the existence of minima of turbidity at levels of about 7 and 33 km.

The increase of turbidity with height above 33 km is obviously associated with a meteor substance fall. The quantitative data about this zone are not numerous mainly due to a lack of exactitude in the methods of interpreting the twilight measurements usually applied [1, 11]. Nevertheless, there are numerous qualitative confirmations of the picture described above. In particular, there are extensive data about the altitude variation of depolarizad light of a twilight sky. Fig. 3 represents the examples of these dependences according to the Ditze and Kohl measurements [30] and our visual measurements [31]. As would be expected the curves reproduce the behaviour of the specific

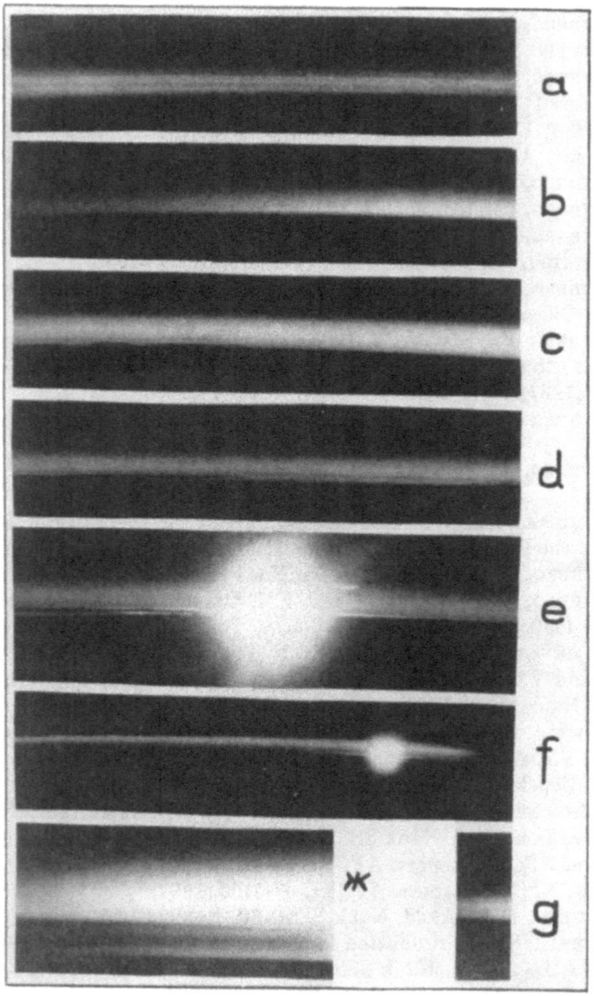

Fig. 4.

scattering coefficient and show that both the upper zones of turbidity exist continuously though they undergo appreciable change.

The photographs of twilight from the spaceships show the instability of these zones. Fig. 4 gives examples illustrating the complexity and changeability of a thin structure of a twilight layer (photographs taken from the spaceships Vostok 5 and 6, Voskhod, Soyuz 4, 5 and 8). Qualitatively these results agree with the data of a laser sounding [32, 33].

It should be noted in conclusion that a comparatively stable position of both zones of purifying the air about 7 and 33 km is neither connected with temperature inversions nor specific peculiarities of the stratospheric dynamics which requires a special explanation.

References

[1] G. V. ROZENBERG, Twilight, Fizmatgiz 1963.
[2] G. V. ROZENBERG, Izv. AN SSSR, Fiz. Atmos. Okeana 1, No. 4 (1965).
[3] G. V. ROZENBERG and V. V. NIKOLAEVA-TERESHKOVA, Izv. Akad. Nauk SSSR, Fiz. Atmos. Okeana 1, No. 4, (1965).
[4] A. YA. DRIVING, I. M. MIKHAULIN, G. V. ROZENBERG, A. V. SANDOMIRSKY and G. I. TRIFONOVA, Izv. Akad. Nauk SSSR, Fiz. Atmos. Okeana 2, No. 10 (1966).
[5] G. V. ROZENBERG, Izv. Akad. Nauk SSSR, Fiz. Atmos. Okeana 2, No. 1 (1966).
[6] G. V. ROZENBERG, "Meteorological Research" No. 12, Nauka Moscow, 1966.
[7] G. V. ROZENBERG, A. B. SANDOMIRSKY, V. N. SERGEEVICH, D. M. SONECHKIN and K. P. FEOKTISTOV, "Space Research", Nauka, Moscow, 1956.
[8] G. V. ROZENBERG, A. B. SANDOMIRSKY and G. I. TRIFONOVA, Izv. Akad. Nauk SSSR, Fiz. Atmos. Okeana 1, No. 12 (1965).
[9] G. V. ROZENBERG, Priroda, No. 3, 1966.
[10] G. V. ROZENBERG and A. B. SANDOMIRSKY, Izv. Akad. Nauk SSSR, Fiz. Atmos. Okeana 3, No. 2, (1967).
[11] G. V. ROZENBERG and V. K. POLDMAA, Izv. Akad. Nauk SSSR, Fiz. Atmos. Okeana 2, No. 8 (1966).
[12] G. V. ROZENBERG, A. B. SANDOMIRSKY and V. K. POLDMAA, Noctilucent clouds, VINITI 1967.
[13] A. V. SANDOMIRSKY and G. V. ROZENBERG, Paper presented at Internat. Symp. Optical Properties of the Upper Atmosphere and of Adjacent Space, Tbilisi-Abastumani 1969.
[14] G. V. ROZENBERG, Uspekhi fiz. Nauk 95, No. 1 (1968).
[15] G. V. ROZENBERG, V. I. TATARSKY and V. I. DIANOV-KLOKOV, Vestnik Akad. Nauk SSSR, Febr. 1970.
[16] YU. S. GEORGIEVSKY, A. YA. DRIVING, N. V. SOLOTAVIN, G. V. ROZENBERG, E. M. FEIGELSON and V.S. KHAZANOV, Ed. G. V. ROZENBERG, Searchlight in the atmosphere. Iz-vo Akad. Nauk SSSR 1960.
[17] G. V. ROZENBERG, G. K. ILICH, S. A. MAKAREVICH and YU. R. MULLAMAA, Izv. Akad. Nauk SSSR, Fiz. Atmos. Okeana 6, No. 5 (1970).
[18] ROZENBERG, Uspekhi fiz. Nauk 91, No. 4 (1967).
[19] E. M. FEIGELSON and M. S. MALKEVICH et al., Trudy IFA Akad Nauk SSSR, No.1 (1958).
[20] L. I. KOPROVA, Izv. Akad. Nauk SSSR, Fiz. Atmos. Okeana 5, No. 2 (1969).
[21] L. ELTERMAN, AFCRL Report, AFCRL-66-828 (1966).
[22] L. ELTERMAN, AFCRL Report, AFCRL-68-0153 (1968).
[23] G. V. ROZENBERG, Dokl. Akad. Nauk SSSR 36, No. 9 (1942).
[24] G. V. ROZENBERG, Thesis submitted for degree of Doctor, Moscow 1946.
[25] L. I. KOPROVA, Izv. Akad. Nauk SSSR, Fiz. Atmos. Okeana 5, No. 2 (1969).
[26] YU. I. RABINOVICH, Thesis submitted for degree of Doctor, 220 (1961).
[27] E. S. SELEZNEVA, Atmospheric Aerosols, Gidromeoizdat, 1966.
[28] KH. JUNGE, Chemical composition and radioactivity of the atmosphere, Iz-vo "Mir", Moscow 1965.
[29] A. B. SANDOMIRSKY, N. I. ALTOVSKAYA and G. I. TRIFONOVA, Izv. Akad Nauk SSSR, Ser. Geofiz. No. 7 (1964).
[30] G. DIETZE and G. KOHL, Z. f. Met. 20, No. 7–8 (1968).
[31] V. M. BOVSHEVEROV, A. V. MIRONOV, I. M. MIKHAILIN, V. M. MOROZOV, Z. L. PONIZOVSKY, S. P. SOKOLOV and I. A. KHVOSTIKOV, Izv. Akad. Nauk SSSR, Ser. Geogr. Geofiz., No. 5 (1950).
[32] G. GRAMS and G. FIOCCO, NASA Sci. Rep. Grants, Report NGR-22-009-(114), 1966.
[33] G. G. HOYER, Paper presented at XIV General Assembly of IUGG, Lucerne, 1967.

Space Research XI — Akademie-Verlag, Berlin 1971

RADIATIVE ENERGY BUDGET OF THE ATMOSPHERE

J. LONDON[a] and T. SASAMORI[b]

[a] University of Colorado, Boulder, Colorado, USA
[b] National Center for Atmospheric Research, Boulder, Colorado, USA

Improvements in radiation theory and increased information, both from satellites and *in situ* observations, of the cloudiness and general structure of the atmosphere, provided some motivation for revision of earlier calculations of components of the earth's radiation budget. The new calculations show more energy absorbed and re-radiated by the earth–atmosphere system than was derived from previous studies. Also, the presently computed value for the average annual planetary albedo of the earth (about 33% with interhemispheric variations of approximately ± 1%) is somewhat less than earlier estimates. Within the accuracy of the results, the incoming and outgoing radiation for the entire earth–atmosphere system are in approximate balance.

In recent years satellite observations have generally verified the latitudinal and seasonal patterns of the net components of the radiation field as derived from theoretical considerations. The earth's planetary albedo as inferred from these satellite observations is of the order of 31%. The largest discrepancy between satellite and theoretical results is for equatorial and tropical regions.

The theoretical calculations and inferences from satellite observations will be compared and discussed.

1. Introduction

The two basic factors governing the overall circulation of the atmosphere are the nature of the energy input (i.e. the amount and distribution of the energy sources and sinks) and the magnitude of the earth's rotation. All other parameters are secondary at most. The earth's rotation is fixed. But the distribution of energy sources and sinks depends on the large-scale variations of the different components involved in the atmospheric radiation budget.

It would be extremely satisfying if we could start from the known planetary orbital parameters with accurate knowledge of the solar constant, atmospheric composition and a physical description of the lower boundary to compute theoretically all of the radiative properties of the atmosphere. Although notable attempts were made in this direction by Milankovitch [1] and others, a complete solution to this problem is not likely at the present time since the atmosphere is a highly non-linear physical system and radiation represents an important process in this interacting system. Moreover, the earth–atmosphere system contains non-predictable (at present) elements that affect its composition and structure. For these reasons, it is necessary that we constantly combine experimental observations of radiation fluxes in the atmosphere with theoretical computation of these fluxes so that we can understand and explain the significant radiation processes and predict their overall input to the thermo-hydrodynamic system.

Since it is not possible to solve completely the radiation problem for the earth–atmosphere system, we can probably make the most beneficial progress by dividing the problem into a set of suitable sub-problems involving different time and space scales. In that way, the various relevant radiative properties of the atmosphere can be parameterized by use of convenient empirical approximations which are still physically meaningful for the specific class or scale of the problem to be studied. In addition, laboratory measurements as well as ground-based, airborne and satellite observations of the radiative flux, both spectral and total, would suggest the types of proper approximations, and more important, would provide the necessary check on the computations so that the theory is not allowed to stray very far from nature itself.

For instance, ground-based measurements of the extinction of incoming solar radiation give significant information on atmospheric turbidity which, in turn, can be used to suggest modifications to theoretical calculations of atmospheric scattering and absorption. Balloon-borne measurements of both infrared and short-wave components of the upward and downward radiation flux and flux divergence are absolutely essential at the present time as an additional check on the correctness of detailed radiation calculations and their approximations.

In addition to the general verification of these theoretical calculations, satellite observations make a unique contribution to our understanding of atmospheric energetics. Apart from their use as important atmospheric probes, satellites provide us with direct observations on a global and relative real-time scale of the distribution of various radiation parameters at the upper boundary of the atmosphere—and in combination with surface measurements, at the lower boundary [2]. Accurate satellite measurements can provide global distributions of the earth–atmosphere albedo as well as the absorption, emission and net radiation balance of the total earth plus atmosphere system [3].

Although the distribution of the net radiation balance provides the initial energy source for atmospheric motions, subsequent interactions between motions and radiation proceed through a number of non-linear processes. For instance, in the energy relationship which is part of the interrelated system of equations describing the general circulation of the atmosphere (see, for instance, [4]) the local temperature change for a non-saturated atmosphere is given as a function of the radiative flux divergence and the transport terms which, in turn, depend on the motion and temperature distribution within the system. Atmospheric motions, on the other hand, are functions of the static stability, among other things. But the static stability itself depends, in large part, on the radiation characteristics of the atmosphere. Some of these problems dealing with the interaction of various radiative components with the dynamics of the atmosphere have been studied by Manabe et al. [5] and more recently by Washington and Kasahara [6] among others, but certainly much remains yet to be done.

The advent of meteorological satellites has made the need for theoretical computations based on the various spectroscopic and structural characteristics of the atmosphere even more apparent. Prior to the launching of the various satellites, theoretical results were evaluated, in general, according to their reasonableness. But now these results must pass the test of verification as well. Where the theory and observations agree, it is possible that they are both correct. Where they do not agree, however, it is necessary that the reason for the difference be investigated and resolved if either system is to be useful. It should again be emphasized that theoretical calculations are not made obsolete by satellite observations, but, on the contrary, continue to be important as a primary method of our understanding the basic radiative characteristics of the atmosphere and, therefore, can be used for prediction purposes—which is certainly the primary goal of all scientific investigations.

At this point it might be useful to interject a few words about scales and approximations. For short time and space scales, the radiation fluxes immediately incident at the lower boundary of the atmosphere are of primary importance and the atmosphere can be considered homogeneous in time and horizontal distance (except for variations in cloudiness). In this case the net radiative flux divergence in the free atmosphere is relatively slow and minor as compared with other physical processes. As the time and distance scales increase, however, radiative processes in the atmosphere increase in importance and should be included somehow as part of the energy equation. It is found that under certain circumstances it is possible to depict the radiative flux divergence in the atmosphere free of clouds by relatively simple linear approximations, involving a temperature (and, in some cases, water vapor) dependence [7]. Although cloudy conditions represent a more complex picture, it is probable that even these could yield to some suitable approximations. As the time scale increases beyond the seasonal, and the region of interest becomes global, accurate evaluation of the various radiation components increases in importance. We need now to know the value of the solar constant and its possible variations. We also need to know how changes in the composition of the atmosphere (such as carbon dioxide) could influence the vertical temperature structure and therefore the thermodynamic stability of the atmosphere. The greatest needs, however, as have become apparent to all who work in the field of atmospheric radiation, are correct determination—through experimental and theoretical programs—of the scattering and absorption characteristics of clouds and aerosols. Although much progress on this problem is presently being made (see, for instance, [8]), it is hoped that satisfactory solutions will be forthcoming in the next few years in a form that could be included directly into the dynamical equations.

The discussion below summarizes recent results of calculations of the various components of the mean annual latitudinal variations in radiation fluxes for both hemispheres based on recent data on the atmospheric structure and distribution of cloudiness. The computed annual average radiation balance gave a slight excess of total emission over absorption in each hemisphere. Since the assumed structural characteristics of the atmosphere represented a long term average, some of the components were adjusted slightly to give a global radiative balance on an annual basis. The maximum adjustment was of the order of 1% increase in the absorption. Details of the methods of calculation of the individual radiative flux components are discussed by Sasamori and London [9].

2. The Global Radiation Budget

The average annual computed global radiation budget is shown schematically in Fig. 1. The numbers represent the percentage of average incoming solar

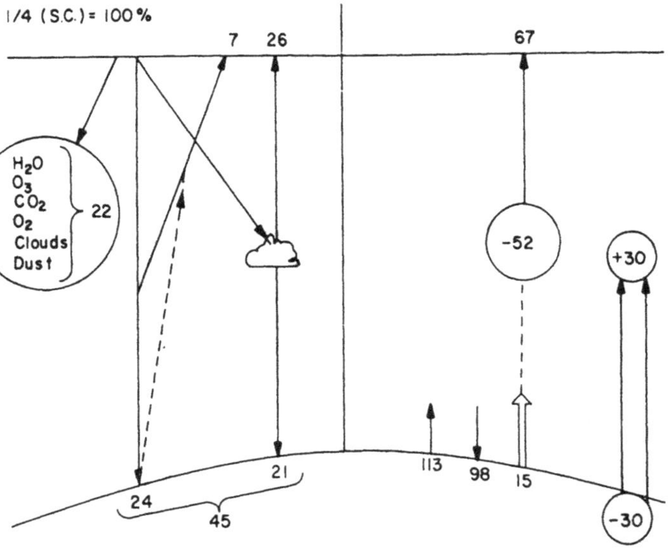

Fig. 1. The average annual global radiation budget for the earth–atmosphere system. (Numbers are given as percentage of the average solar energy at the top of the atmosphere.)

radiation at the top of the atmosphere. Because of the method of calculation, small changes in the accepted value of the solar constant (as discussed, for instance, by Drummond [10]) would result in only slight adjustment to these figures. The amounts shown on the left represent the various components of the short-wave flux; the amounts in the center are the infrared flux; and the value on the right gives the non-radiative energy exchange between the earth's surface and the atmosphere.

Incoming solar radiation is either absorbed within the atmosphere, transmitted to and absorbed at the ground surface, or reflected back to space from the surface, the atmosphere, and from clouds. Absorption of solar radiation within the atmosphere is by the various trace gases, principally water vapor, carbon dioxide, ozone and molecular oxygen, and by dust and haze. Additional absorption takes place within clouds by the liquid water droplets and ice crystals in the clouds. Computations of these various components of absorption indicate that the average total absorption accounts for approximately 22% of the incoming solar energy where over half of this absorption is due to water vapor. The total atmospheric absorption derived from present calculations is somewhat higher than earlier estimates (for just the Northern Hemisphere) given by London [11]. This is due primarily to inclusion of absorption of solar radiation by O_2 and CO_2 in the present calculations and increased absorption within clouds. Approximately 45% of the incoming solar radiation is absorbed at the earth's surface with slightly more than half of this energy coming as direct radiation.

The planetary albedo of the earth was computed separately for the clear and cloudy portions of the earth–atmosphere system. The clear part represents the albedo due to surface and atmospheric reflection, while the cloudy part comes from the reflection due to various types of clouds. The total planetary albedo is computed to be 33% with the major contribution due to the relatively high cloud reflectivity. Since the cloud cover over the earth is approximately 50% it can be seen that an overcast planet having the cloud characteristics of the earth would have an albedo of about 50% whereas a cloud-free planet, but with surface and atmospheric structure similar to that of earth, would have an albedo of about 15%. It should be noted that except for a few isolated regions (such as the icecovered plateau of central Greenland) satellite observations (Nimbus 2) as discussed by Raschke and Bandeen [12] give just this range of the geographic distributions of the planetary albedo. Their estimates of the overall global albedo based on an assumed value of the solar constant of 1.95 cal cm^{-2} min^{-1}, is 31%—not very different from the calculated albedo as shown in Fig. 1.

As is well known, the presence of infrared active gases in the atmosphere produces a relatively high surface temperature which in turn results in a positive infrared radiation balance at the earth's surface. This net upward infrared surface flux is, on a global basis, approximately 15% of the average incoming solar radiation. Even so, the atmosphere acts as a source for infrared energy primarily as a result of the radiative flux divergence due to water vapor and, in the higher layers, due to carbon dioxide. In this way, an amount equal to approximately one half of the average incoming radiation is used in cooling the atmosphere by infrared radiation. This results in an average infrared cooling of about 1.5 °C day^{-1}. The upward flux from the atmosphere together with that from the earth's surface results in terrestrial emission at the top of 0.335 cal cm^{-2} min^{-1} (67% of the incoming solar radiation for an assumed solar constant of 2.0 by min^{-1}), which then gives an equivalent brightness temperature for the earth of about 253 °K, a value very close to estimates from recent satellite observations (see, for instance, Raschke and Bandeen [12]). Calculations made for the radiative components of each hemisphere separately show that, on an annual basis, each hemisphere is in near equilibrium but that the Southern Hemisphere has a higher planetary albedo (and therefore less absorption by the earth–atmosphere system) and lower emission to space than the Northern Hemisphere. This result is largely based on the derived distribution of cloudiness which gives more cloudiness in the Southern Hemisphere, and therefore a higher contribution of cloud reflectivity to the planetary albedo and a slightly colder surface for infrared emission to space. Obviously, better cloud statistics derived either from more complete and representative ground-based observations or from carefully calibrated high resolution satellite observations are needed to narrow down some of the uncertainties in these numbers.

The energy of solar radiation directly absorbed in the atmosphere is not sufficient to make up for the large infrared radiative cooling. This is accomplished by two somewhat related non-radiative processes. Energy absorbed at the earth's surface is used, in part, to heat the skin layer of air near the boundary and then by convective processes is distributed upwards through the lower atmosphere. Most of the solar energy absorbed at the earth's surface, however, is used to evaporate water over the oceans. The water vapor is transported upwards in the atmosphere where condensation may occur, clouds form, and the latent heat of condensation is released to the atmosphere. Although this process of transferring heat from the surface to the atmosphere is not as efficient as that involving conduction from the earth's surface, it is by far the more important. Approximately three quarters of the non-radiative energy transfer from the surface to the atmosphere takes place through evaporation and condensation. Moreover, this process is relatively rapid (as compared, for instance, with radiative processes in the free atmosphere) and therefore contributes a major energy source for the development and maintenance of atmospheric circulation systems involving short to moderate time and space scales.

3. Radiation Fluxes at the Top and Bottom of the Atmosphere

The solar energy absorbed by the earth–atmosphere system is a function of the sun's zenith angle and the distribution of different atmospheric parameters such as cloudiness, water vapor, etc., and thus varies with space and time. Except for regions of high albedo, the total absorption is a maximum at the sub-solar point and a minimum at places of high solar zenith angles. On an annual basis this means that, in general, the maximum absorption is found near the equator and minimum absorption at the poles of both hemispheres.

The results of calculations of the absorption by the earth–atmosphere system as a function of the atmospheric variables at each latitude are shown in Fig. 2. As mentioned earlier, the absorption values have been adjusted upward slightly to maintain an energy balance for the entire globe.

Infrared emission also depends on the structure of the atmosphere. But in this case the dependence is primarily on the temperature distribution. Since most of the outgoing infrared flux originates within the atmosphere, and latitudinal temperature differences in the free air are relatively small, the average variation in infrared emission is also relatively small. As a result, calculation of the annual radiative components show that the absorbed solar energy is larger than the outgoing infrared flux at all latitudes from about 33° N to about 37° S. Poleward of these latitudes the reverse is true (i.e. the outgoing infrared flux is larger than the absorbed solar energy). The implications of this distribution of radiation excess and deficit, insofar as it affects the necessary integrated horizontal energy transport in maintaining an overall balance, will be discussed later.

Since their initial launching, meteorological satellites have had as a prime task the measurement of various components of the earth's radiation budget. Many of these measurements have recently been summarized by Vonder Haar and Soumi [13]. The absorbed energy as derived from the satellite observations is computed as a function of the satellite-determined albedo and the solar energy available at the top of the atmosphere. The satellite results for the annual latitudinal distribution of absorption and emission are shown by the broken lines in Fig. 2.

Comparison of the satellite results with those derived from computations based on climatological data shows, in general, very good agreement. The small

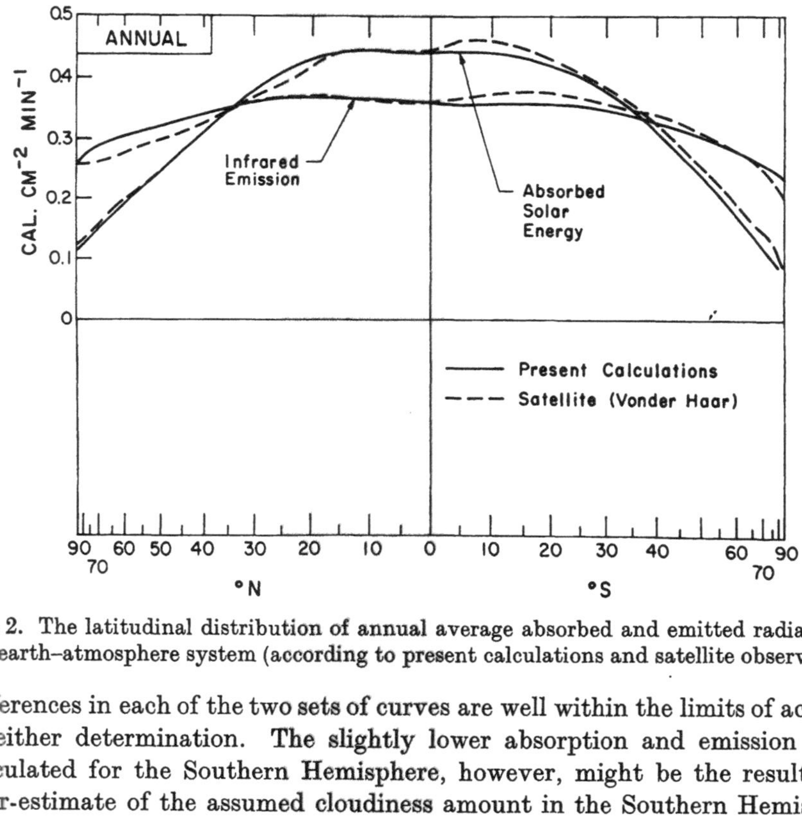

Fig. 2. The latitudinal distribution of annual average absorbed and emitted radiation for the earth–atmosphere system (according to present calculations and satellite observations).

differences in each of the two sets of curves are well within the limits of accuracy of either determination. The slightly lower absorption and emission values calculated for the Southern Hemisphere, however, might be the result of an over-estimate of the assumed cloudiness amount in the Southern Hemisphere. Unfortunately, it is still not possible to derive a completely accurate climatic cloud distribution either from ground-based or from present satellite observations. This is particularly true for cloud delineation by type and height. Better cloud data are presently being accumulated and more adequate cloud statistics will certainly be available soon.

On the average, about half of the incoming solar radiation at the top of the atmosphere is transmitted to the earth's surface. The actual insolation received at the surface depends, of course, primarily on the cloudiness (although water vapor and aerosol absorption cannot be neglected). The calculated values of insolation at the earth's surface as well as the net infrared emission both show a slight minimum near the equator but otherwise flat maxima in the Northern and Southern subtropics (Fig. 3).

Poleward of latitude 30° the insolation and the net radiation balance at the surface decreases rapidly with latitude in both hemispheres. Indeed, as a result of the high surface albedo in the Southern Hemisphere south of 70°, the net radiation balance becomes negative (i.e. more total radiation is emitted than absorbed by the surface). The observed latitude distributions of insolation

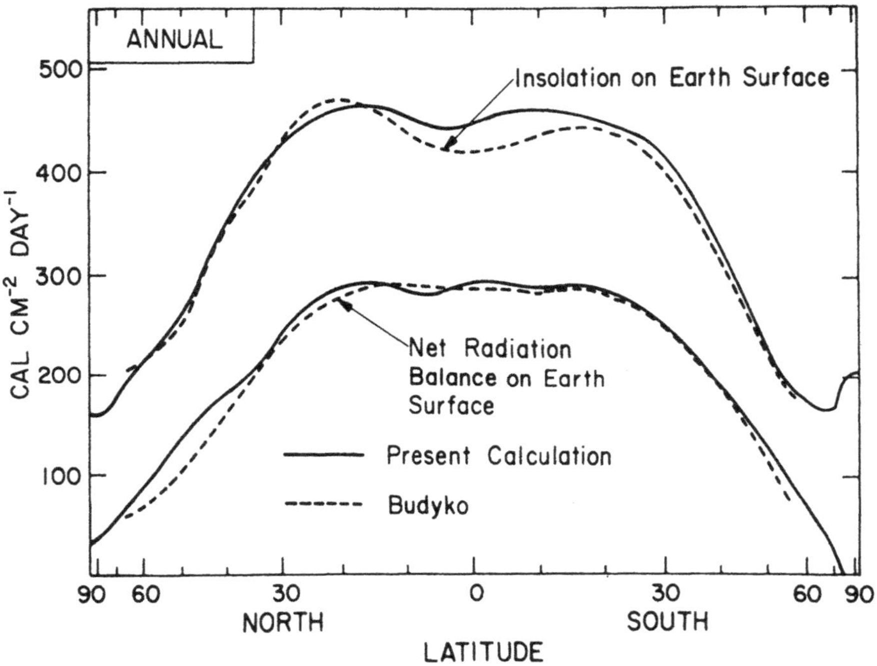

Fig. 3. The calculated and observed annual average latitudinal distribution of insolation (top curves) and net radiation balance (bottom curves) at the surface.

and net radiation balance as given by Budyko [14] are also shown in Fig. 3 for comparison. Although the computed values for the insolation in the Southern Hemisphere are slightly higher than those given by Budyko, the distributions of net radiation balance are, in general, quite close. The discrepancy between computed and observed values of the insolation at the surface is largest near the equator and it may be that there is increased haze absorption in the atmosphere that was not completely accounted for in our calculations. Further observations of the effect of haze on the absorption of solar radiation—particularly in the lower atmosphere—are needed to resolve this difference.

4. Radiative Cooling and Required Energy Transport

Within the atmosphere absorption of solar radiation results in an average heating of about 0.6 °C day^{-1}, whereas infrared flux divergence between the surface and top of the atmosphere gives an average cooling of about 1.5 °C day^{-1}. The net annual average radiative temperature change for the global atmosphere is therefore close to 0.9 °C day^{-1}. It has long been recognized that this is the approximate value needed to maintain the energy budget for the atmospheric heat engine (see, for instance, the discussion by Brewer [15]). As has already been pointed out, about 75% of the heating necessary to balance this net radiative loss is supplied by condensation of water vapor within the atmosphere (primarily in the region 2–5 km above the earth's surface).

The latitudinal distributions of heating by absorption of solar radiation, cooling by infrared flux divergence, and net radiative temperature change for the entire atmosphere is given in Fig. 4.

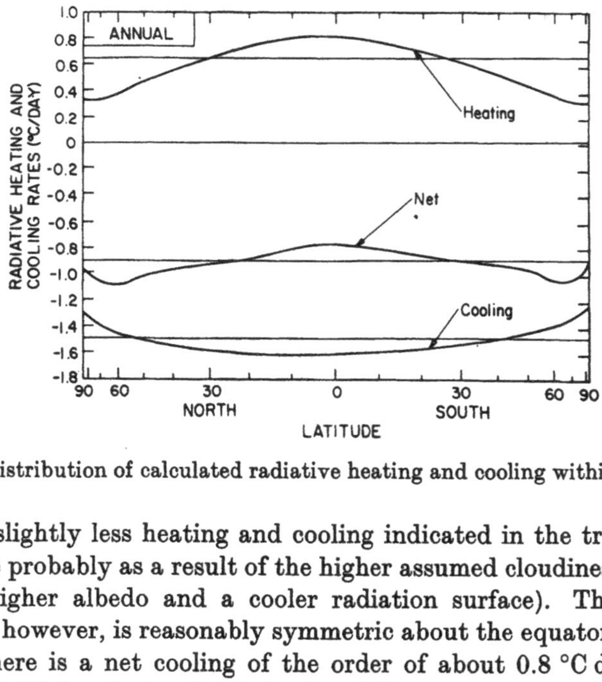

Fig. 4. The distribution of calculated radiative heating and cooling within the atmosphere.

There is slightly less heating and cooling indicated in the tropical Southern Hemisphere probably as a result of the higher assumed cloudiness in that region (giving a higher albedo and a cooler radiation surface). The net radiative divergence, however, is reasonably symmetric about the equator. At equatorial latitudes there is a net cooling of the order of about 0.8 °C day^{-1} increasing to about 1.1 °C day^{-1} at polar latitudes. This variation represents a measure of the annual average available potential energy between the equator and poles that is contributed by all radiation processes in the atmosphere.

The difference between the solar radiation absorbed in the earth–atmosphere system and the infrared emission to space at the top of the atmosphere represents the net accumulation of energy in the system. On an annual basis and over a period of years there must be a balance between the incoming and outgoing radiation, at least to within the accuracy of calculations and/or observations, for the entire globe. At each time and geographic location, however, there is generally an imbalance between these two radiative components and this imbalance can have large variations as has been shown by recent satellite observations (see for instance, [12]). When averaged over longitude many of the small scale variations are smoothed out and the average latitudinal distribution of the radiative absorption and emission remains. These distributions were given for the annual average calculated and observed values in Fig. 2.

Since the annual mean temperature change at any latitude when averaged over a number of years must be very small, the net energy accumulation (loss) resulting from the radiation excess (or deficit) in any latitude belt must be transported across the latitude belt by the combined ocean–atmosphere circulation. The radiation imbalance must then be equal to zero when integrated from pole to pole. However, as already indicated, there is a radiation excess (more absorption than emission in the earth–atmosphere system) between 33° N and 37° S, and a deficit beyond. The accumulated energy must therefore be transported poleward by the atmospheric and ocean circulation systems. This transport reaches a maximum in each hemisphere where the radiation

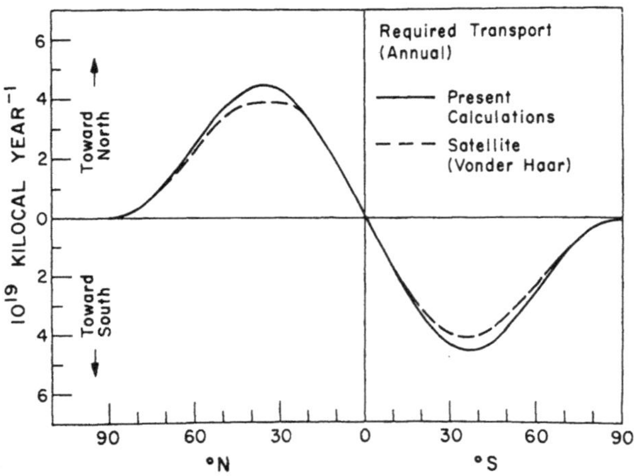

Fig. 5. The average annual energy transport by the atmosphere–ocean circulation system to satisfy balance requirements.

imbalance changes sign. The calculated total energy transport that is required to maintain an overall heat balance is shown in Fig. 5. For comparison purposes the total combined ocean–atmosphere energy transport required by the satellite observations for the annual case (as discussed by Vonder Haar and Suomi [13]) is also shown.

The maximum poleward heat transport, when averaged over the year, is about 4×10^{19} kcal year^{-1} (slightly less according to the satellite data) and occurs between 30–35° latitude. A surprising result that is shown by both calculations and satellites is the zero transport across the equator and the apparent symmetry of both hemispheres despite the difference in cloudiness and the marked difference in land and ocean surface between the two hemispheres. It may be that there are small differences in the radiative components that are masked by the uncertainties in both the climatological data upon which the calculations are based and in the assumptions used in reducing the satellite observations. The similarities of the overall radiation regimes can have an important effect in modeling a global circulation pattern.

This is certainly not a final model of the atmospheric radiation budget. Improvements can be made as we acquire more complete information on the climatic distribution of atmospheric variables—in particular the distribution of clouds by type and height—and develop more accurate methods for treating the absorption and transmission characteristics of clouds and aerosols in the real atmosphere. Meanwhile, satellite observations can be used to complement the theoretical calculations based on climatic data. Both, of course, are indispensable for the type of dynamic calculations recently discussed by Manabe [16].

Acknowledgments

We would like to thank Mr. Douglas V. Hoyt for his considerable aid in programming a good part of this problem. We are grateful also to the National Center for Atmospheric Research, which is sponsored by the National Science Foundation, for use of their computing facilities. This work was supported in part by the Atmospheric Sciences Section, National Science Foundation through Grant NSF GA-10838.

References

[1] M. Milankovitch, Canon of Insolation and the Ice-Age Problem, Israel Program for Scientific Translations, Jerusalem, 1969; originally published: Konigliche Serbische Akademie, Belgrade, Special Publications 132, 1941.
[2] T. H. Vonder Haar and K. J. Hanson, J. Atmos. Sci. **26**, 652 (1969),
[3] E. Raschke, F. Möller and W. R. Bandeen, Sveriges Meteorologiska Och Hydrologiska Institut, Meddelanden, Serie B. Nr. 28, p. 42 (1968).
[4] E. N. Lorenz, The Nature and Theory of the General Circulation of the Atmosphere, World Meteorological Organization, Geneva, 1967.
[5] S. Manabe, J. Smagorinsky and R. F. Strickler, Mon. Weath. Rev. **93**, 769 (1965).
[6] W. M. Washington and A. Kasahara, Mon. Weath. Rev. **98** (1970) in press.
[7] J. London and T. Sasamori, in: Proceedings of the Radiation Symposium, Bergen, Norway 1968, World Meteorological Organization, Geneva, in press.
[8] G. N. Plass and G. W. Kattawar, Appl. Optics **8**, 455 (1969).
[9] T. Sasamori and J. London, Meteorology of the Southern Hemisphere, American Meteorological Society, Boston, in preparation.
[10] A. J. Drummond, in: Advances in Geophysics **14**, Academic Press, New York, 1970 (p. 1).
[11] J. London, New York University College of Engineering, Research Division, Department of Meteorology and Oceanography, Final Report Contract No. AF 19(122)-165 (1957).
[12] E. Raschke and W. R. Bandeen, J. Appl. Met. **9**, 215 (1970).
[13] T. Vonder Haar and V. Suomi, submitted to J. Atmos. Sci., May 1970.
[14] M. I. Budyko, Ed., Atlas of the Heat Balance of the Earth, Moscow 1963.
[15] A. W. Brewer, discussion in: Centenary Proceedings of the Royal Meteorological Society, Royal Meteorological Society, London (1950) (pp. 31–32).
[16] S. Manabe, Mon. Weath. Rev. **97**, 739 (1969).

Space Research XI — Akademie-Verlag, Berlin 1971

COMPARISON OF SATELLITE AND CALCULATION CHARTS FOR THE EARTH'S RADIATION BUDGET

K. YA. KONDRATYEV and L. N. DYACHENKO

The Main Geophysical Observatory, Leningrad, USSR

The actinometric data obtained from meteorological satellites enable a comparison to be made between the experimental results the calculated climatological maps of the planetary distribution of radiation balance of the earth–atmosphere system and its components.
The data on the global distribution of the annual means for the radiation balance of the system and its components have been analysed.
The distribution of the annual totales of the atmospheric longwave balance over the globe is given. The relationship between the long-wave atmospheric balance and the outgoing long-wave radiation is represented for different kinds of cloudiness.

Advances made in the theory of climate and the numerical modelling of the general atmospheric circulation have brought to the fore the problem of numerical weather forecasting for a term of up to two weeks.

The practical solution of this problem is possible only with the aid of electronic computers, whose speed exceeds the performance index of the best modern models of computing technique by about two orders of magnitude. The prototypes of such devices are already under development.

A more serious difficulty concerns the necessity of obtaining the initial data on the main meteorological elements for the entire planet. This problem will be solved in carrying out the programme of the World Weather Watch. An important part of this programme is the Global Atmospheric Research Program. It is conceivable that by the mid-seventies the global system of meteorological observations will become a reality.

The third, and the gravest, difficulty in realizing numerical long-term weather forecasts is the inadequacy of modern understanding of the laws of the physical processes in the atmosphere which determine the changes of weather and climate. This is especially the case in the problem of the energetics of the atmosphere and the thermal fluxes, which becomes extremely important where prolonged processes are considered which are markedly nonadiabatic. Reference [1] gives a description of the modern conditions of studying the weather and climate radiation factors. In the present paper we shall confine ourselves to questions concerning the study of the heat budget for the earth as a planet (earth–atmosphere system).

Until recently our ideas of the laws for the planetary distribution of the radiation budget of the earth-atmosphere system were based exclusively upon the use of the calculation data [2–6].

The actinometric data of meteorological satellites enabled us for the first time to compare the calculated climatological charts for the planetary distribution of the radiation budget of the earth–atmosphere system with experiment.

However, in the first stage of the investigation such comparison could be only rather conditional, as the experimental data were restricted from the point of view of their adequate space–time averaging. Only recently [7–10] have there been published the results of satellite actinometric information processing, representing correct climatological characteristics (annual, seasonal and monthly average values) that refer practically to the entire globe.

The purpose of the present article is to discuss the results of our comparisons between calculated and experimental data referring to the global distribution of the average annual and seasonal values of the radiation budget for the earth–atmosphere system and its components.

For these comparisons the calculated data and charts of the planetary distribution of the earth radiation budget and its components, were obtained by various workers [2–6] and the source of the experimental data were the results of satellite measurements for the period July 1963–November 1965 [9].

Vonder Haar [9] used very extensive satellite actinometric information obtained by means of wide-angle sensors developed by Suomi, and applied a correct method of processing the measurements, taking into account the data of many in-flight calibrations on the Sun. Thus one may assume that the influence of random errors is practically excluded. The errors in determining the average radiation budget values and its components do not exceed a few per cent.

The comparison of experimental annual charts of radiation budget of the planet R_s with the calculated climatological charts proves that a quite satisfactory qualitative agreement of radiation fields is observed (Fig. 1). There is, for example, a coincidence between the zones of the budget maximum values, which are located in equatorial latitudes. However, as regards the absolute value, the calculated values of R_s are appreciably lower than the measured values. As the latitude increases the radiation budget decreases; the zero isolines (calculated and measured) in both southern and northern hemispheres are in the region of the 40° latitude circle. In the southern hemisphere the budget distribution is nearly zonal, due to the uniformity of the surface (the ocean).

The agreement of the budget zero isolines is quite satisfactory. The maximum discrepancy occurs on land: in the region of the southern end of the South American continent. In the northern hemisphere the discrepancy of the zero isolines is also small over the ocean and great over the continent: the deviation of the zero calculated isoline from the 40° latitude and from the measured isoline is particularly large over a part of the Eurasian continent.

The nonuniformities of the surface of the continents are the cause of the mismatch of the isolines for the calculated and measured values of the radiation budget, which is especially noticeable in the region of the Sahara and Arabian deserts. The minimum values of R_s occur in the polar regions of both northern and southern hemispheres. The variations of the measured values of R_s from the mean values are from -0.08 to $+0.12$ cal cm^{-2} min^{-1} in the northern hemisphere and from -0.12 to $+0.12$ cal cm^{-2} min^{-1} in the southern hemisphere (Fig. 1), while the calculated values vary from -0.12 to $+0.08$ in the northern hemisphere and from -0.04 (in the region of 50° S) to $+0.08$ cal cm^{-2} min^{-1} in the southern hemisphere. It was not possible to obtain calculated data on R_s for latitudes above 50° S because of the lack of necessary initial information.

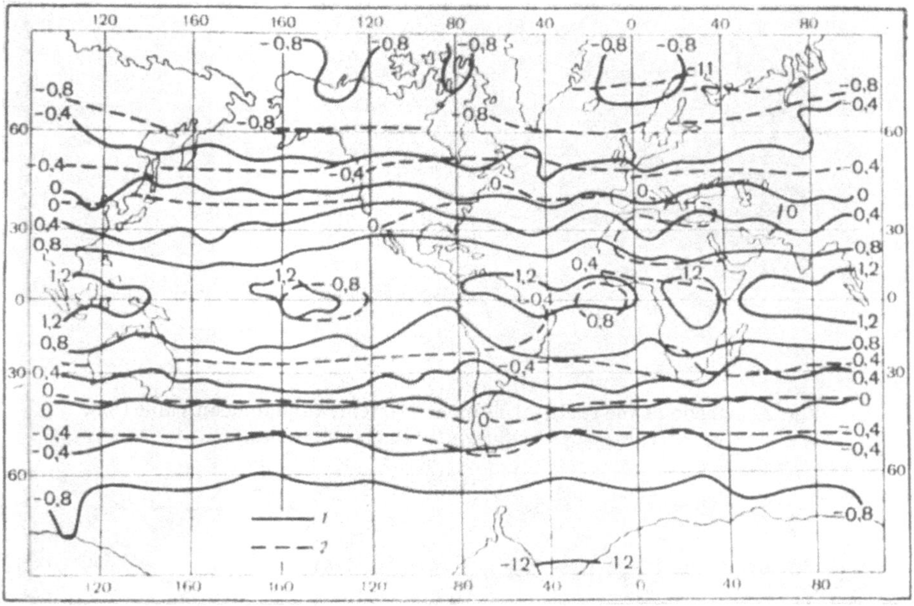

Fig. 1. Radiation budget of the earth-atmosphere system, annual mean value (cal cm^{-2} min^{-1}).
1, measured values; 2, calculated values.

Evidently, the character of the geographical distribution of the earth–atmosphere system radiation budget is determined by the specific space variations of its components: — albedo A and outgoing long-wave radiation F_∞ of the planet.

On comparing the annual chart of albedo for the earth–atmosphere system plotted by Vonder Haar, from satellite measurements, with the climatological chart of Vinnikov, one notices primarily the systematic quantitative discrepancy, which for the entire globe amounts to an average of about 17%. The calculated values of albedo for all latitude zones systematically exceed the measured values.

It is likely that first of all the explanation lies in the use of too high estimates for the quantity and density of the clouds used for the albedo calculations. Probably a certain part is played by a clearly marked "topography" of convective cloudness in low latitudes. The "macroroughness" of the cloudiness may lead to a pronounced decrease of the earth albedo.

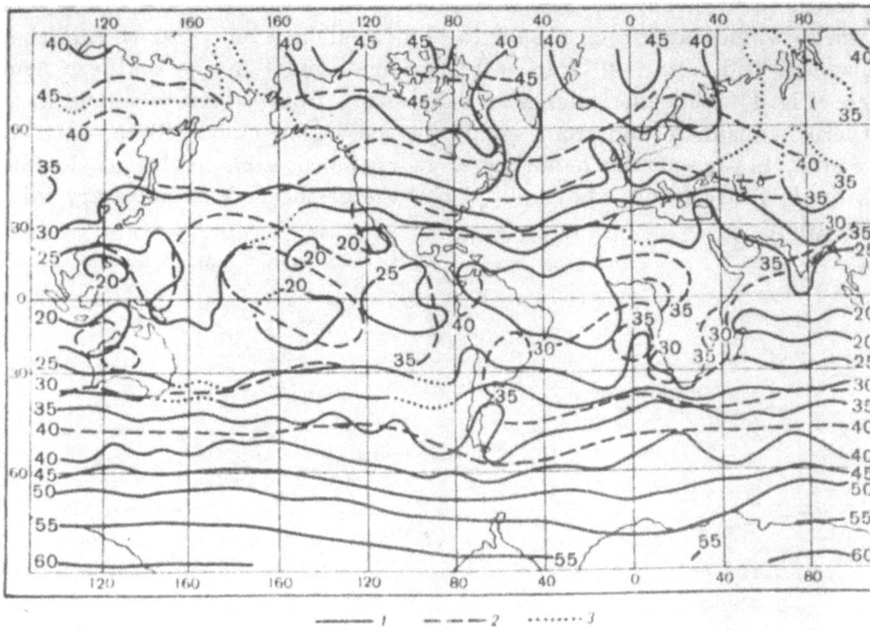

Fig. 2. Albedo of the earth–atmosphere system, annual mean value (%).
1, measured values; 2, calculated values.

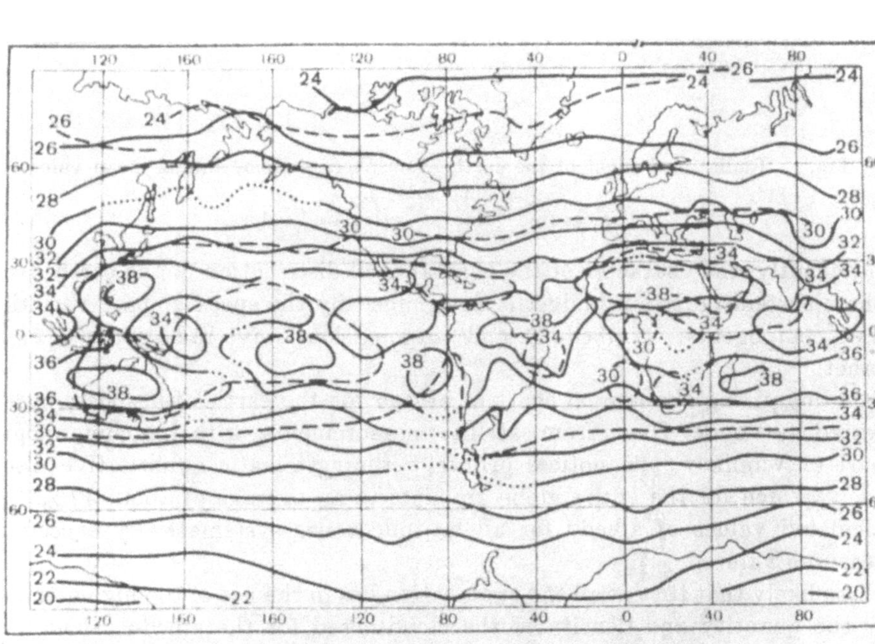

Fig. 3. Outgoing long-wave radiation (cal cm^{-2} min^{-1}).
1, measured valnes; 2, calculated values.

Another difference is that, if the computed isolines are characterized by a break at the boundary between land and sea, the measured isolines do not encounter such a break (Fig. 2), which, of course, is due to space averaging.

The examination of the annual albedo chart (Fig. 2) shows that considerable deviations from zonality are observed only in the equatorial zone. The maximum values of albedo occur in high latitudes (beyond the north and south polar circles) and the minimum values in the subequatorial zone over the ocean.

The distinctive features of the outgoing long-wave radiation field are given in the chart plotted in Fig. 3. In this case too an appreciable non-zonality is observed only in the subequatorial zone. The quantitative discrepancies between the measured and calculated values of the outgoing radiation are considerably smaller than in the case of albedo. The measured values slightly exceed the values calculated for the temperate and tropical latitudes, while to the north of 50° N the discrepancies are unimportant. The average calculated and measured values of the outgoing radiation for the entire earth practically coincide, amounting to 0.32–0.33 cal cm^{-2} min^{-1} (Table 1).

Table 1

Average annual values of radiation budget of the earth–atmosphere system and its components for different latitude circles

Latitude (°)	A (%) Calc.	Meas.	F_∞ (cal cm^{-2} min^{-1}) Calc.	Meas.	R_s (cal cm^{-2} min^{-1}) Cal.	Meas.	Katayama (calcul.) A (%)
80–70 N	–	44	–	0.24	–	−0.10	
70–60	43	40	0.27	0.27	−0.09	−0.07	
60–50	40	36	0.29	0.29	−0.06	−0.05	57
50–40	37	32	0.30	0.31	−0.02	−0.00	50
40–30	35	27	0.32	0.34	0.01	0.02	43
30–20	33	24	0.34	0.38	0.03	0.04	39
20–10	32	20	0.35	0.38	0.04	0.09	33
10–0	33	20	0.34	0.38	0.06	0.11	34
0	33	20	0.34	0.38	0.06	0.10	34
0–10 S	33	20	0.34	0.37	0.06	0.10	
10–20	33	20	0.34	0.38	0.05	0.08	
20–30	33	23	0.33	0.38	0.04	0.06	
30–40	35	27	0.31	0.34	0.02	0.03	
40–50	38	32	0.29	0.31	−0.02	−0.01	
50–60	–	36	–	0.29	−0.05	−0.06	
60–70	–	70	–	0.27	–	0.09	
70–80	–	45	–	0.23	–	0.11	
Entire Earth	35	29	0.32	0.33	−0.03	+0.04	36

The calculated data concerning the seasonal and monthly values of radiation budget are reduced mainly to the information on outgoing long-wave radiation. Therefore, it was possible to compare the calculated and measured values only for examples of that component of the radiation budget. In [11] the corresponding charts for all seasons of the year are reproduced in detail.

The comparison of experimental averaged seasonal charts for the outgoing long-wave radiation of the earth–atmosphere system, plotted from the data of [9], with the climatological calculation charts (see [3]) shows a clearly defined qualitative agreement. In all seasons of the year, according to the measured and calculated data a decrease of the outgoing radiation F_∞ with increase in latitude is observed. In winter the maximum values of F_∞ occur in the subequatorial latitudes and amount to values slightly exceeding 0.39 cal cm^{-2} min^{-1}. Minimum values of about 0.30 cal cm^{-2} min^{-1} are observed in the southern hemisphere and of 0.22 cal cm^{-2} min^{-1} in the northern hemisphere. The positions of the maxima obtained by calculation and measurement are in a quite good agreement. All minima are located in the subequatorial zone. The most significant maxima ($F_\infty \geqq 0.38$ cal cm^{-2} min^{-1}) occur in the Pacific Ocean, over the Australian desert, over a part of North and Central Africa and partly over South America.

The smallest discrepancies between the measured and calculated values of the outgoing radiation in winter (December, January, February) occur in the northern hemisphere. Here the zonality of the outgoing long-wave radiation field is preserved in the greatest degree, while it should be noted that the isolines of the measured values are smoother than those of the calculated values.

In the northern hemisphere the deviations of the calculated outgoing radiation from the measured values are comparatively small. In the southern hemisphere they are more appreciable.

A similar characteristic of the outgoing radiation variations is observed in the spring and autumn periods of the year. In summer the variation range of the absolute values is somewhat smaller than in winter. Values vary from 0.30 (calculated): 0.26 cal cm^{-2} min^{-1} (measured) to 0.42: 0.40 cal cm^{-2} min^{-1}. The quantitative agreement of the measured and calculated values is worse than in the winter period of the year. Compared with winter, the maxima are slightly shifted to the north. The zonality is disturbed.

Although the qualitative picture of the outgoing radiation geographic distribution in summer is similar to a great extent for the measured and calculated data, in the quantitative picture a considerable discrepancy is observed. Nearly everywhere the measured values exceed the calculated data.

In this way, the comparison of the calculated seasonal charts for the outgoing radiation of the earth–atmosphere system with the charts plotted from data obtained from measurements with meteorological satellites indicates that, as a rule, the measured values exceed the calculated values in all the seasons of the year. The smallest discrepancies are observed in the northern hemisphere during the winter and spring periods. In spring, in some regions of the northern hemisphere the calculated values exceed the measured data.

Similar comments apply for the mean latitude values. The comparison of the calculated values for the mean latitude profiles of the outgoing radiation with the measured values indicates (Table 2) that during all seasons of the year an increase of F_∞ from the poles to the equator is observed. On the average for individual latitudes the discrepancy between the calculated and measured values varies within the limits $+0.01$ to -0.08 cal cm^{-2} min^{-1}.

Berkovich [12] compares the mean values of the radiation budget and its components, obtained by calculation according a the diagram compiled Voitko

Table 2

Mean latitude values of the outgoing radiation of earth–atmosphere system for the four seasons of the year according to calculated and measured data (cal cm^{-2} min^{-1})

Latitude (°)	Seasons: Winter Calc.	Winter Meas.	Spring Calc.	Spring Meas.	Summer Calc.	Summer Meas.	Autumn Calc.	Autumn Meas.
80–70 N	0.231	0.20	0.249	0.24	0.285	0.29	0.255	0.25
70–60	0.262	0.24	0.279	0.26	0.294	0.31	0.276	0.27
60–50	0.268	0.26	0.282	0.28	0.299	0.33	0.284	0.30
50–40	0.282	0.27	0.293	0.30	0.315	0.35	0.303	0.33
40–30	0.300	0.31	0.313	0.32	0.340	0.36	0.313	0.35
30–20	0.341	0.35	0.348	0.36	0.358	0.36	0.357	0.37
20–10	0.354	0.37	0.344	0.42	0.321	0.35	0.339	0.37
10–0	0.339	0.36	0.335	0.36	0.320	0.34	0.329	0.37
0–10 S	0.342	0.36	0.341	0.36	0.344	0.37	0.339	0.39
10–20	0.345	0.37	0.342	0.36	0.344	0.38	0.341	0.39
20–30	0.345	0.36	0.335	0.36	0.325	0.36	0.332	0.37
30–40	0.329	0.34	0.315	0.34	0.305	0.33	0.315	0.35
40–50	0.309	0.32	0.295	0.30	0.272	0.29	0.300	0.32
50–60	0.291	0.30	0.279	0.27	0.271	0.26	0.275	0.30
60–70		0.28		0.23		0.22		0.27
70–80		0.27		0.20		0.19		0.23
Entire Earth	0.32	0.32	0.32	0.33	0.33	0.33	0.33	0.34

[13], and values obtained by measurements from the satellite Cosmos 144. For the purposes of comparison, the data of the actinometric system of stations for the period of July 1967 were also used. The author has compared the calculated and measured values of the earth surface effective radiation, the outgoing radiation at the upper boundary of the atmosphere and the heat inflows to the entire depth of the atmosphere.

As shown in [12], specific values were compared for different conditions of cloudiness. The mean calculated values of the outgoing radiation were $F_{\infty_{calc.}} = 0.419$ cal cm^{-2} min^{-1}, and the mean measured values $F_{\infty_{meas.}} = 0.404$ cal cm^{-2} min^{-1}.

In our work of 1966 [14, 15] similar comparisons were made, which on the average of 89 cases gave the value of $F_{\infty_{calc.}} = 0.33$ cal cm^{-2} min^{-1}, and in 74 cases measured from the satellite Tiros 3 gave 0.34 cal cm^{-2} min^{-1}.

By correlating the values calculated in accordance with Vinnikovs model with those measured from the satellite Tiros 3, the authors [15] have obtained an obvious linear dependence, though the correlation has turned out to be not very high.

One of the interesting and important characteristics of the radiation balance of the earth is the long-wave balance of the atmosphere. It is a difference between the effective radiation of the underlying surface and the outgoing radiation

$$F_a = F_0 - F_\infty .$$

As an example of distribution of the atmospheric long-wave balance F_a over the globe, a chart of annual totals expressed in cal cm^{-2} $year^{-1}$ is presented (Fig. 4).

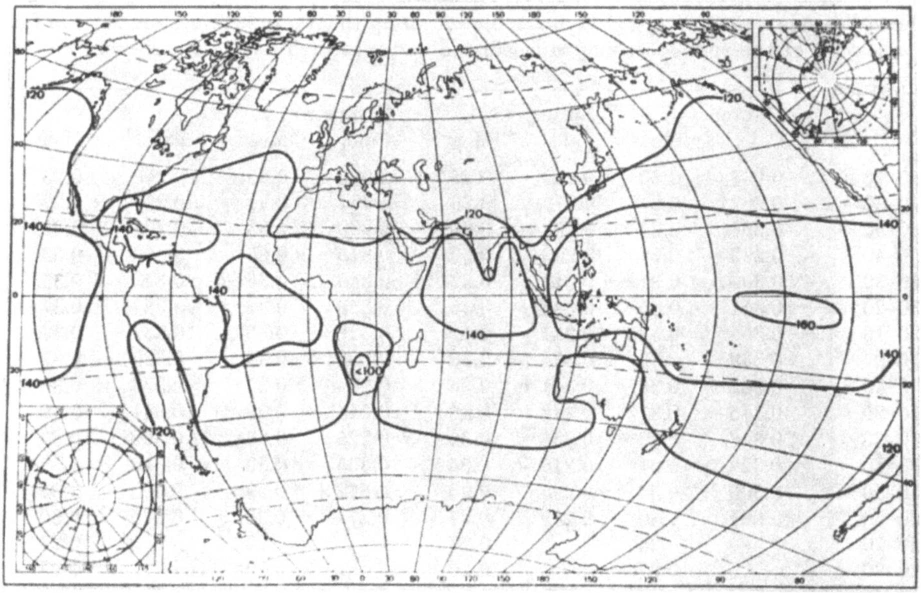

Fig. 4. Atmospheric long-wave balance (kcal cm^{-1} y^{-1}).

Lack of data on measurements of the atmospheric long-wave balance has forced us to make an attempt to solve the problem of establishing correlation dependences between the values of outgoing radiation F_∞ and atmospheric long-wave balance F_a [14].

The relationship between atmospheric long-wave balance F_a and outgoing radiation F_∞ (radiometer sounding) has proved to be ambiguous. Therefore, we tried to consider it for individual points with different sky conditions (clear and overcast). The warm and cold periods were also considered separately (summer and winter) (Figs. 5a, b). As is seen from these figures, the dependence of F_a on F_∞ is sufficiently distinct.

Table 3
Empirical formulae for fair and dull weather

Authors	Warm period (summer) Cloudiness	Correlation factor	Regression equations	Measurement method
Kondratyev and Dyachenko	0–2	0.92	$F_a = 1.44\, F_\infty - 0.20$	Radiometer sounding
Sabatini and Suomi	clear	0.87	$F_a = 1.54\, F_\infty - 2.34$	Satellite
Berkovich	0–2	0.68	$F_a = 0.08\, F_\infty - 0.22$	Satellite
Kondratyev and Dyachenko	overcast 10	0.99	$F_a = 1.05\, F_\infty - 0.02$	Radiometer sounding
Sabatini and Suomi	overcast	0.85	$F_a = 1.32\, F_\infty - 0.13$	Satellite
Berkovich	all cases 0–10	0.26	$F_a = 0.44\, F_\infty - 0.16$	Satellite

(a)

(b)

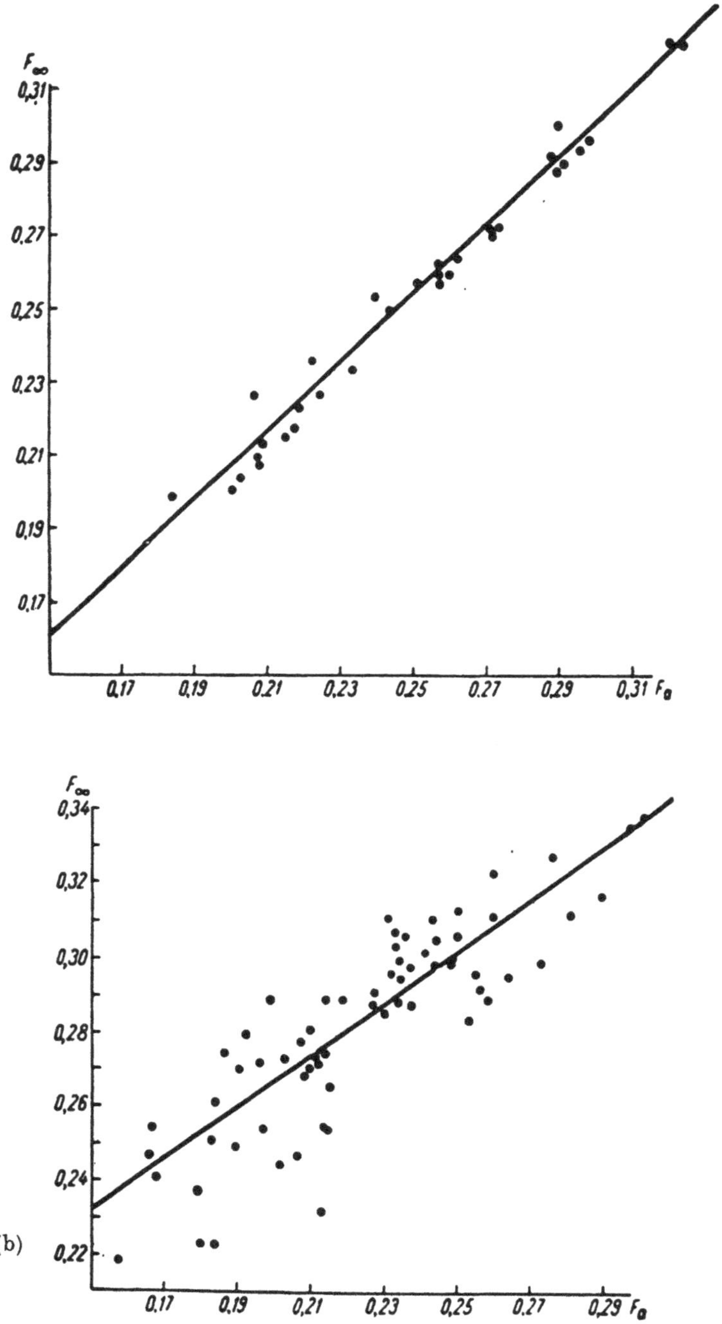

Fig. 5. Relationship between atmospheric long-wave balance and outgoing long-wave radiation: a, with overcast sky; b, with clear sky.

From these correlation dependences empirical formulae have been derived for the cases of fair weather and dull weather (Table 3).

Summarizing the above considerations, it should be noted that the main content of the present state of the development of earth radiation climatology is the possibility of checking experimentally the earlier calculated data. The discrepancies discovered have not so far been explained adequately and further investigation is needed. The accumulation of experimental data in the future is of great importance.

References

[1] K. YA. KONDRATYEV, Meteorol. i Gidrol. No. 6 (1968).
[2] T. G. BERYAND, in: A. I. Voeikov and Problems of Modern Climatology, GIMIZ, Leningrad 1956 (in Russian).
[3] K. YA. VINNIKOV, Trudy GGO **168** (1965).
[4] K. YA. VINNIKOV, Trudy GGO **170** 19165).
[5] K. YA. KONDRATYEV, L. N. DYACHENKO and K. YA. VINNIKOV, in: Problems of the Physics of the Atmosphere, **7**, Izd. LGU 1969.
[6] A. KATAYAMA, J. Met. Soc. Japan **45**, No. 1 (1967).
[7] E. RASCHKE, F. MÖLLER and W. R. BANDEEN, Goddard Space Flight Centre Report X-622-67-460 (1967).
[8] E. RASCHKE and M. PASTERNACK, NASA Technical Report (1967).
[9] T. H. VONDER HAAR, Department of Meteorology, University of Wisconsin Contract NASW-65 (1968).
[10] J. S. WINSTON and V. R. TAYLOR, ESSA Technical Report NESC-43, Washington D.C. 1967.
[11] L. N. DYACHENKO and K. YA. KONDRATYEV, Trudy GGO **235** (1970).
[12] L. V. BERKOVICH, Trudy GMC **50** (1969) (in Russian).
[13] N. V. VOITKO and A. F. KIVGANOV, Trudy MMC **11** (1966).
[14] K. YA. KONDRATYEV and L. N. DYACHENKO, Trudy GGO **184** (1966).
[15] K. YA. KONDRATYEV and L. N. DYACHENKO, Trudy GGO **203** (1967).

Space Research XI — Akademie-Verlag, Berlin 1971

THE RADIATION BALANCE OF THE EARTH–ATMOSPHERE SYSTEM DURING JUNE AND JULY 1969 FROM NIMBUS 3 RADIATION MEASUREMENTS — SOME PRELIMINARY RESULTS

E. RASCHKE[a], T. H. VONDER HAAR[b], W. R. BANDEEN[c] and M. PASTERNAK[c]

[a] Ruhr Universität Bochum, FRG
[b] University of Wisconsin, Madison, Wisconsin, USA
NASA, Goddard Space Flight Center, Greenbelt, Md., USA

The meteorological satellite Nimbus 3, launched on 14 April 1969, carries a five-channel medium-resolution radiometer, whose measurements cover at least a period of one full year. In four channels the outgoing longwave rediation is being obtained in narrow spectral ranges (6.5–7.0 μm, 10–11 μm, 14.5–15.5 μm, 20–23 μm). Thus, the total outgoing longwave radiation (5–50 μm) was inferred from these spectral measurements by a multi-regression method employing limb-darkening relations derived by a statistical study of Nimbus 2 measurements (5–30 μm). The albedo was inferred from measurements in the spectral range from 0.3 to 4.0 μm considering two empirically derived angular dependence relations, one over clouds and land, and the other over oceans through clear skies.

Data from periods 16–30 June 1969 and 1–15 July 1969 yield a mean planetary albedo between 29 and 30% and a mean planetary temperature between 252–253 °K which corresponds to outgoing fluxes between 0.330 and 0.338 ly min^{-1}. Both results are in good agreement with results found from Nimbus 2 measurements of the same semimonthly periods in 1966.

1. Introduction

The average structure of the earth's radiation fields in the shortwave and longwave region is generally known from extented experimental studies with satellite measurements [1] and from empirical computations based on aerological and climatological data [2]. However, highly resolved geographical distributions during periods of the order of 1 month were obtained only from Nimbus 2 radiation data over a short period of $2\frac{1}{2}$ months [3]. Interannual variations of the latter results and their relations to atmospheric energetics could not be studied so far. Only satellite measurements with self-integrating radiometers provided studies of the earth's radiation fields over periods of several (up to 5) years [4]. The time resolution was of the order of one season and the spatial resolution in geographical maps was about 1000×1000 km^2.

Since April 1969 the Nimbus 3 satellite has provided reliable measurements. Their analysis will enable detailed descriptions of the earth's radiation fields over the period of at least one year. Such a large data analysis is not yet finished, but some preliminary results will be presented in this paper.

2. Evaluation Procedure and Results

The radiation measurements of Nimbus 3 are being made with a five-channel scanning radiometer similar to the one flown previously on Nimbus 2. Thus, the principal elements of the evaluation technique used in earlier evaluations of Nimbus 2 data [3] could be used in this analysis. Since measurements of only one satellite in a sun-synchronous orbit were available for this analysis the results might be slightly biased over areas with pronounced diurnal variation of cloudiness.

2.1. Albedo

The albedo of an area is defined in this study to be the ratio of daily averages of reflected to incoming flux of solar radiation. For computations of the latter quantity a value of $S_0 = 1.95$ cal cm^{-2} min^{-1} has been chosen for the solar

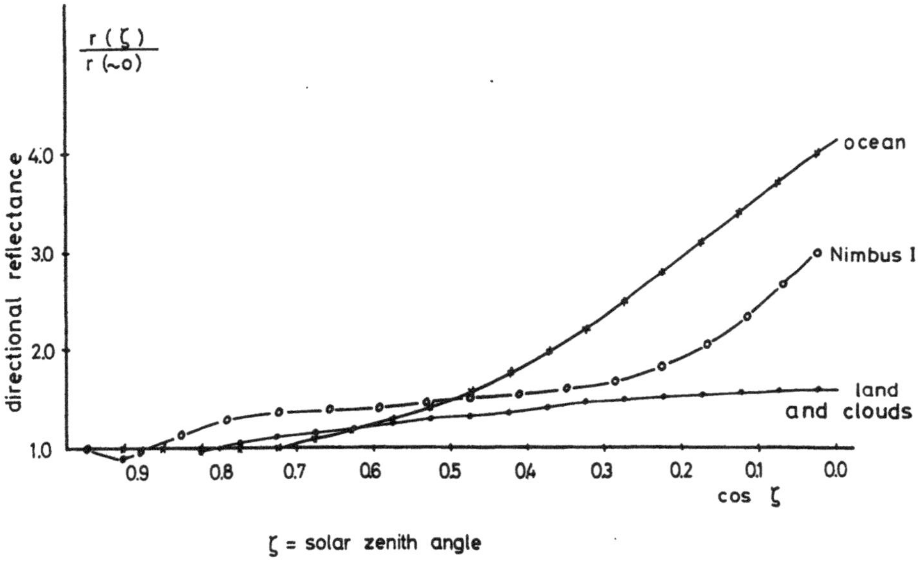

Fig. 1. Increase of directional reflectance $r(\zeta)$ with increasing zenith angle ζ of the sun.

constant [5, 6]. Daily averages of the flux of reflected solar radiation were obtained by integration over the entire daylight period where the dependence of the directional reflectance r, on the sun's zenith angle is taken into account with a relation derived empirically from airplane, satellite and balloon measurements. Here two different relations have been applied to areas of high reflectance (clouds or land) and low reflectance (cloudfree ocean). These two relations are shown in Fig. 1 and compared with that used in earlier work with Nimbus 2 data. The slope of the curve labelled "ocean" is much steeper than that of the two others; thus it can be expected that albedo values obtained of oceanic areas will be slightly higher than those from Nimbus 2 data. On the other hand, albedo values over extended land, clouds and snow cover will be somewhat lower. An example of results, a map of the mean albedo during the period 16–30 June 1969, is shown in Fig. 2. The patterns in this map clearly reveal the location of areas of high cyclonic activity whose cloudiness causes albedos of more than 40%. The albedo over subtropical oceans whose

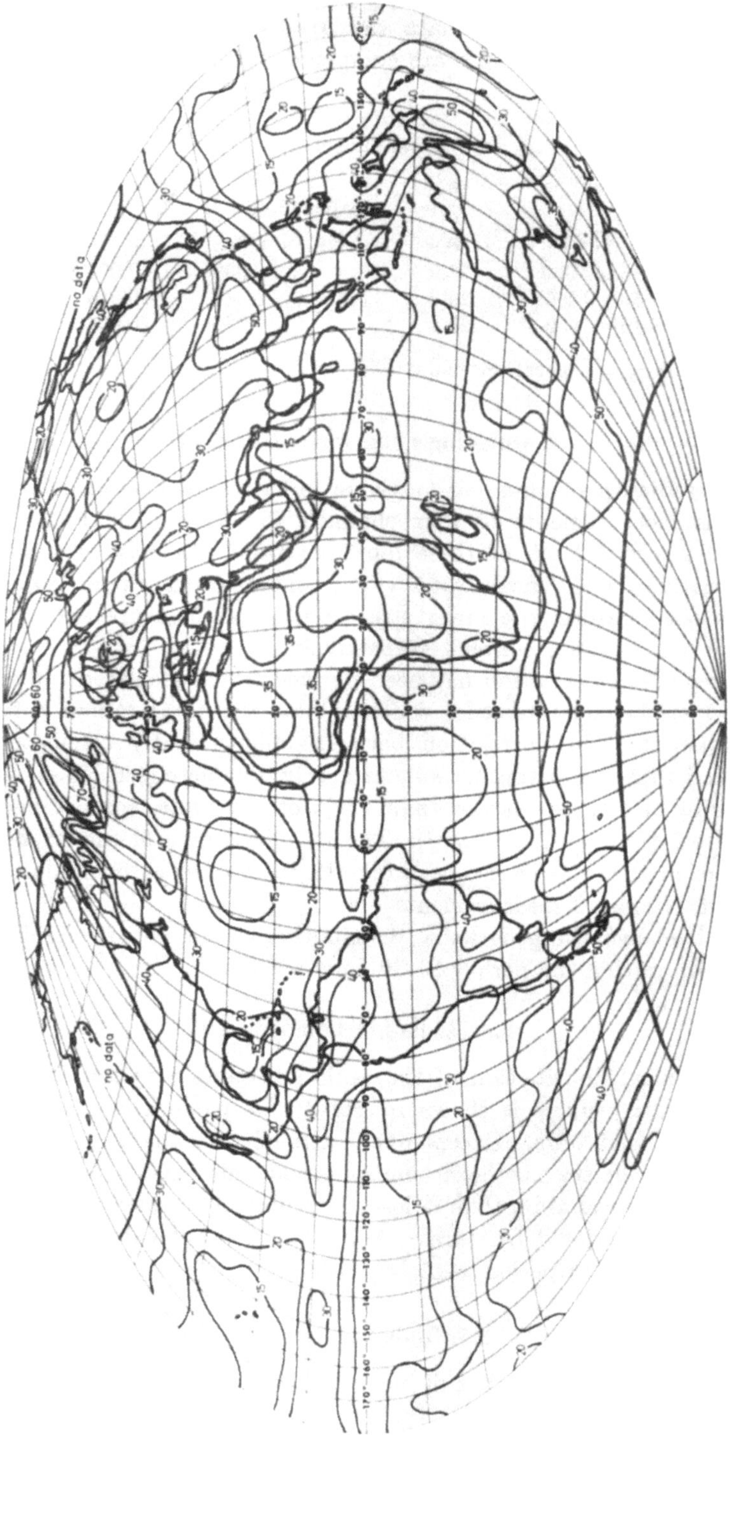

Fig. 2. Albedo (%) of the earth–atmosphere system, Nimbus 3, 16–30 June 1969.

cloudiness is very low is about 2–3% higher than was found from Nimbus 2 data, while the albedo of the Sahara and Arabian desert areas was found to be 3–4% lower. Global averages are compared below in Table 1.

Table 1
Global budget

Period	A (%)	S (cal cm^{-2} min^{-1})	E	Q
16–30 June 1966*	30.9	0.472	0.345	−0.016
16–30 June 1969	28.9		0.333	+0.002
1–15 July 1966*	29.8	0.472	0.346	−0.015
1–15 July 1969	29.4		0.337	−0.004

* for $S_0 = 1.95$ cal cm^{-2} min^{-1}

2.2. Outgoing Long-wave Radiation

The total radiance of outgoing longware radiation had be calculated from simultaneous measurements of four channels (6.5–7 μm, 10–11 μm, 14.5 to 15.5 μm, 20–25 μm). Each of these four measurements may be assumed to be representative for the mean water vapor content of the upper troposphere, the surface temperature, the lower stratospheric temperature and the lower tropospheric water vapor content, respectively.

A multi-regressional relation has been derived from calculated radiances of 80 different atmospheric models, which relates the total radiance to these filtered radiances. The correlation between independently calculated total radiances and those from this formula is nearly complete (see Fig. 4) and the standard error of estimate is less than 1.5% of the total radiance even over very cold areas ($N < 50$ W m^{-2} sr^{-1}). The limb-darkening of the earth–atmosphere system in the infrared has been taken into account with simple relations derived from Nimbus 2 measurements.

In Fig. 3 is shown the geographic distribution of the mean outgoing long-wave radiation during the period 16–30 June 1969.

2.3. Radiation Balance

The geographical distribution of the radiation balance or net radiation at the top of the atmosphere is shown in Fig. 4. Its pattern is similar to that found from Nimbus 2 data. Except over parts of the Sahara, Greenland and the Arctic ice shield the northern hemisphere at this season (north–summer) absorbs more solar radiation than it loses by emission to space. The southern hemisphere is the major deficit area. A similar geographical distribution has been found also for the period 1–15 July 1969.

2.4. Global Budget

Global averages of daily means of incoming solar radiation S, the albedo A, the outgoing long-wave radiation flux E and the radiation balance Q are summarized in Table 1.

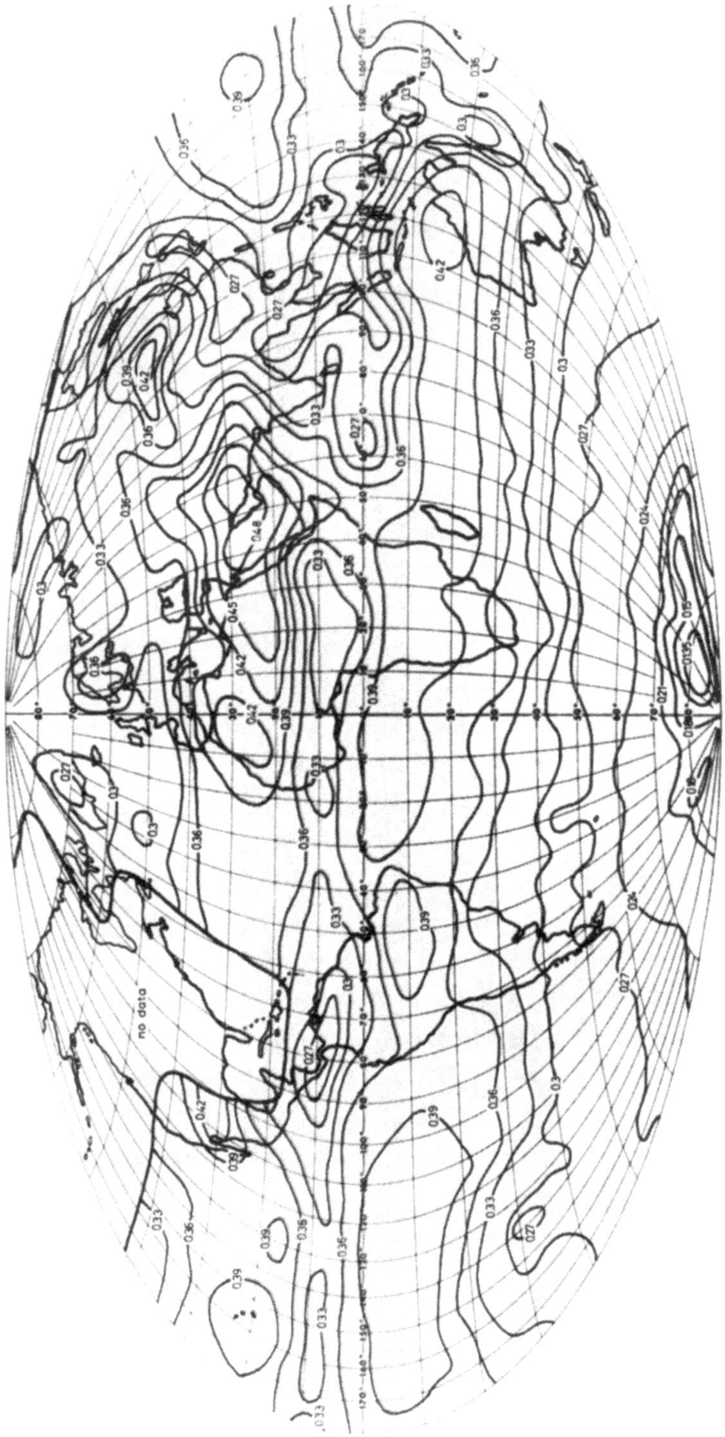

Fig. 3. Outgoing long-wave radiation (cal cm^{-2} min^{-1}) Nimbus 3, 16–30 June 1969.

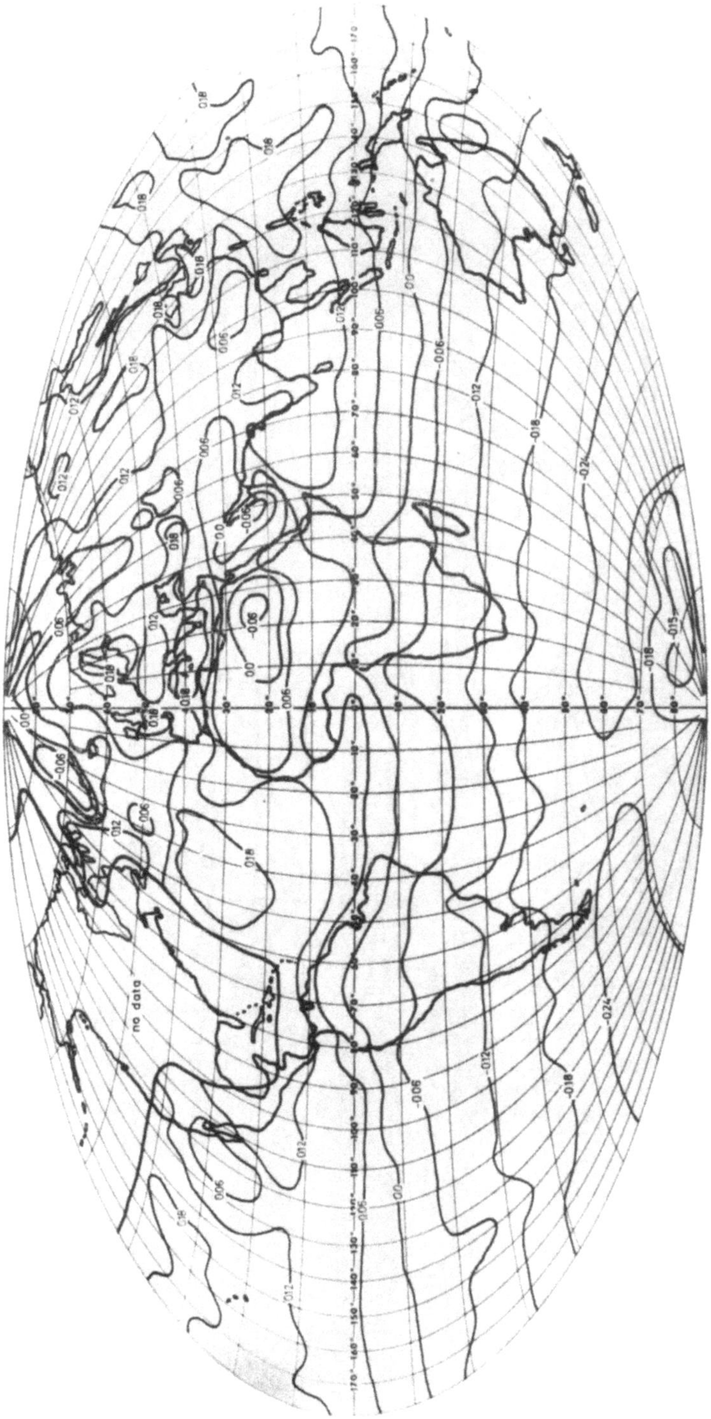

Fig. 4. Radiation balance of the earth-atmosphere system (cal cm^{-2} min^{-1}) Nimbus 3, 16–30 June 1969.

The global planetary albedo has been found from Nimbus 3 data to be 1–2% lower than in 1966. This discrepancy may be due, partly, to some lower cloud cover. But it might be caused primarily by the different reflectance laws (Fig. 1) used in these calculations. Also the outgoing long-wave radiation has been found to be about 2.5–3% lower than in 1966. Here, again, the different limb-darkening relations might be responsible for this discrepancy. But also the assumptions necessary for the derivation of the mutli-regressional relation which is being used to calculate the total long-wave radiance from simultaneous measurements of four narrow-band channels can be responsible for it. A detailed error analysis will be published together with all results in a forthcoming article.

3. Concluding Remarks

The results from this study of the earth's radiation budget agree in general with those found for the year 1966. Together, the Nimbus data from satellite 2 and 3 thus far show the same features of large-scale energy exchange with space measured also by other satellite sensors. These results encourage, further, more detailed evaluations if Nimbus 3 measurements are available with sufficient accuracy.

Acknowledgments

The authors thank all who contributed to the success of the satellite experiment, especially Mr. A. McCulloch, Mr. H. Powell and Mr. J. Barksdale. Part of his work was supported by NASA under contract NAS 5-11542.

References

[1] T. Vonder Haar, Ph. D. Dissertation, The University of Wisconsin 1969.
[2] J. London and T. Sasamori, Space Research XI, Leningrad (1970).
[3] E. Raschke and W. Bandeen, J. Appl. Met. **9**, 000 (1970).
[4] Vonder Haar and V. Suomi, J. Atmos. Sci., (to be published, 1970).
[5] D. Labs and H. Neckel, Z. Astrophys. **69** (1968).
[6] A. J. Drummond, J. R. Hickey, W. J. Scholes and E. G. Laue, Nature, **218**, 759 (1968).

Space Research XI — Akademie-Verlag, Berlin 1971

THE STATISTICAL STRUCTURE OF THE BRIGHTNESS FIELD OF REFLECTED RADIATION IN THE 0.6–0.8 MICROMETRE SPECTRAL RANGE

V. G. Boldyrev and V. I. Tulupov

GVGMS, Moscow, USSR

The results of processing of continuous recording of measurements of radiation reflected by the earths surface and clouds in the 0.6–0.8 μm spectral range are considered. Measurements were carried out with an instrument installed on the Cosmos 121 satellite. The calculations of the characteristics of spatial radiation field structure (correlation functions and spectral densities) indicate the absence of pronounced fluctuations in the mesoscale spatial range for both the whole range of observations studied and individual observations relating to specific meteorological conditions.

Some studies of the radiation field of the earth as a planet in various spectral range from the statistical viewpoint are already available [1–3]. These consider in particular the short-wave radiation measurements from various satellites, the spatial resolution in nadir being usually several tens (30–70) of kilometres. First, this made it difficult to calculate and analyse the characteristic of spatial radiation field structure of specific synoptic or geographic formation since the number of independent measurements relating to the same formation was necessarily small. Secondly, the instrumental averaging limited the hf range of the energy spectrum and did not permit the mesoscale minimum range to be studied at greater length.

Thus, it is noted in [3] that the spectral density of reflected radiation decreases sharply beginning from the wavelengths of the order of 800–1000 km. In this case a sliding average with averaging interval of about 5000 km was used, so only the spatial trend caused by the difference in the height of the sun at various latitudes was practically filtered.

It is also of interest to filter further the lf fluctuations in order to reveal at greater length the behaviour of the reflected radiation fluctuation spectrum at wavelengths shorter than 800–1000 km.

The present report is an attempt to describe such analysis based on the processing of observations of radiation reflected from the earth's surface and clouds on the Cosmos 121 satellite. In this case, as the spatial resolution of the instrument was sufficiently high, it was possible to study the structure of individual cloud and geographic formations.

Apart from other observations measurements of intensity of reflected and scattered radiation in the 0.6–0.8 μm wavelength range in the direction from nadir were carried out on the Cosmos 121 satellite. The viewing angle of the photoelectric photometer used in these measurements was 1°, the resolution on the ground was 4–6 km. The instrument was in continuous operation at the earth's day side for several days. The information from the instrument was photographed on to film using a recording device installed on board the satellite.

The site on the surface which was studied was represented on the window using a three-lens objective with 41-mm focal distance and 1:2 aperture ratio and projected on a light-sensitive element.

The sensitive element of the photometer was a cadmium selenide photoresistance. The long-wave boundary of the operating range of the 0.6–0.8 μm wavelength range was determined by the spectral sensitivity of the receiver; the shortwave boundary was formed by a cut-off light filter which transmitted the radiation with wavelength longer than 0.6 μm. The peak of the photometer sensitivity was in the 0.67 μm wavelength range.

The instrument was calibrated for solar light reflected from a screen covered with magnesium oxide. Some non-linearity exists in the calibration curve at low energies caused by non-linearity of photoresistance and at high energies caused by the electric circuit for the enlargement of the dynamical range operation of the instrument. However, since the main portion of measurements were within the limits where the calibration curve may be considered linear the omission of these non-linearities does not distort substantially the results obtained.

The measurements from six orbits relating to the day time on 20 June 1966 were processed. These orbits passed over the mid-Asia part and the European territory of the USSR, Northern Africa, West Europe and the eastern part of the Atlantic Ocean. The television images of cloudiness obtained at about the same time on that day showed a great variety of cloud forms over these territories.

The radiation data (in relative units) were so taken from registrograms that the distance between individual counts would correspond to 2 km distance on the ground between individual measurements. This ensured the presence of about two counts per one resolution element. The coverage of each observation was about 10000 km. The data obtained were separately processed for each orbit and then the calculated statistical characteristics were averaged, if necessary, over all observations.

The processing consisted both of calculation of some statistical characteristics (averages, dispersions, correlation and structure functions) for some sites (500–1000 km length) and of calculations of these characteristics for the whole experiment. In the latter case the sliding average of various scales was used; however, in any case the averaging interval did not exceed 1000 km. This made it possible to achieve the filtration of large-scale synoptic fluctuations, though this was, of course, incomplete since the frequency characteristics for this type of filter are far from being ideal. Then the sliding averages were subtracted and the correlation functions were calculated for the deviations obtained:

$$R(\tau) = \overline{x_1(t) \cdot x(t + \tau)} \tag{1}$$

where x denotes the deviation from the sliding average and τ is the spatial shift.

The structure functions

$$B(\tau) = \overline{[x(t) - x(t+\tau)]^2} \quad (2)$$

were also calculated and the correlation and structure functions were inter-compared to estimate the stationary points of the series analysed.

The correlation function of the reflected radiation averaged over all orbits processed has been obtained (Fig. 1). As can be seen from Fig. 1 this function is of exponential–cosine character. The correlation function passes through zero even at a distance of about 200 km. When comparing this correlation

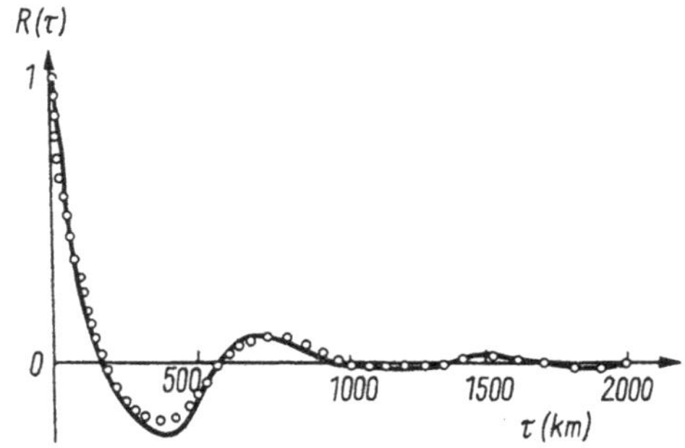

Fig. 1. The correlation function of reflected radiation averaged over all orbits.

function with that obtained in [3] was obtained at a sliding averaging interval of about 5000 km) one can see a very considerable decrease in the correlation over the whole of space.

The function calculated in [3] passed through zero at distances of the order of 700 km which is comparable with the distance at which the passage of the correlation function of cloudiness through zero is observed. According to [4] the cloudiness correlation function calculated on the basis of ground observations passes through zero at distances of about 1000 km. The fact that the extinction of the correlation function shown in Fig. 1 is much more rapid indicates the absence of the ordered (in the statistical sense) variations in the values of cloud brightness within the limits of a certain synoptic scale formation. Of course, this cannot be realized as applied to one or another specific cloud formation (for example, cumulus bank) but one should on the average, consider that the spatial connections of brightness, and hence cloudiness at scales less than 1000 km weaken very strongly as compared with the spatial connections of variations in brightness on the synoptical scale.

To estimate the contributions of different frequencies to the total spatial variance of the intensity of reflected radiation the spectral density of reflected radiation has been calculated over the values of the average correlation function. The correlation function was previousely approximated by the expression

$$R(\tau) = 0.8 \exp(-\alpha\tau) \cos\beta\tau - \exp(-\gamma_1\tau) + \exp(-\gamma_2\tau) + 0.2 \exp(-\tau), \quad (3)$$

where $\alpha = 0.003$, $\beta = 0.0075$, $\gamma_1 = 0.015$, $\gamma_2 = 0.036$; values of τ are given in km. Approximate values of the correlation function are shown in Fig. 1 by circles.

Thus, the spectral density S is given by the following expression

$$S(\omega) = \frac{2}{\pi}\left[0.8\,\alpha\,\frac{\omega^2+\alpha^2+\beta^2}{\omega^4+2(\alpha^2-\beta^2)\,\omega^2+(\alpha^2+\beta^2)^2} - \frac{\gamma_1}{\gamma_1^2+\omega^2} + \frac{\gamma_2}{\gamma_2^2+\omega^2} + \frac{0.2}{1+\omega^2}\right]. \tag{4}$$

The value of S calculated from formula (4) is shown in Fig. 2 where frequencies ω and wavelengths λ are plotted as the abscissa. Naturally, the spectral density obtained differs very much from that given in [3]. It can be easily seen, however, that, despite the filtration of the lf ($\lambda > 1000$ km) fluctuations, the spectral density does not display any appreciable features at wavelengths

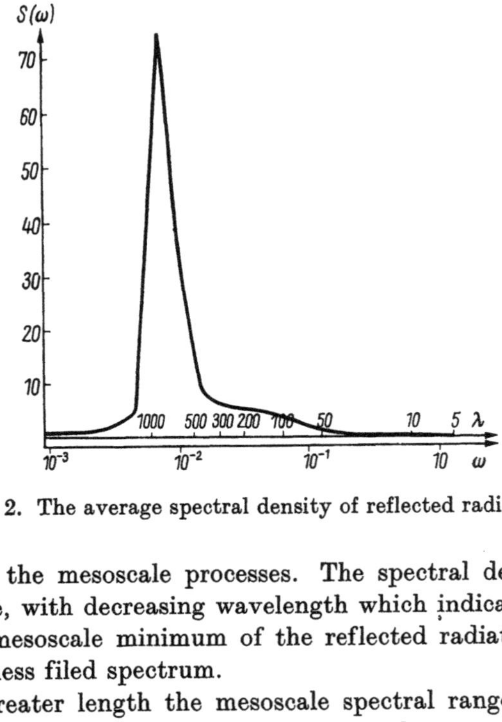

Fig. 2. The average spectral density of reflected radiation.

corresponding to the mesoscale processes. The spectral density is decreases sharply, as before, with decreasing wavelength which indicates once more the presence of the mesoscale minimum of the reflected radiation spectrum and hence the cloudiness filed spectrum.

To study at greater length the mesoscale spectral range the data on the reflected radiation for several orbits were processed using various intervals of sliding averaging. Fig. 3 presents the correlation functions relating to the same orbit at the sliding average intervals of 1000, 500, 200, 100, 50 km. It can easily be seen that when isolating the fluctuations to the scale less that 50 km we come practically to "white noise" (non-correlated deviations from the sliding average). The residual dispersion of this "noise" is nevertheless fairly great (about 20% of the dispersion of all mesoscale fluctuations).

Thus the conclusion may be drawn that from the viewpoint of the study of large-scale energy processes averaging over areas of the order of 1000 km is sufficient and that further study does not give substantial further information. At the same time there are variations to the scale less than 5 km (in this case this is the resolution of the instrument in nadir) which contribute much to the total variability of the process. The data on such variations in brightness and spatial brightness field structure are important when analysing the television and infrared images of cloudiness. In this case detailed measurements of brightness with resolution better than 5 km are necessary.

The above considerations related to some average characteristics obtained when processing most variable (from the meteorological viewpoint) observations.

It is of interest to consider also the statistical characteristics of individual meteorological formations, for example cloud systems of one or another form, or individual types of underlying surface, for example desert or ocean surface in the absence of clouds. A number of registrogram sections relating to such formations were selected and the spatial correlation functions were calculated

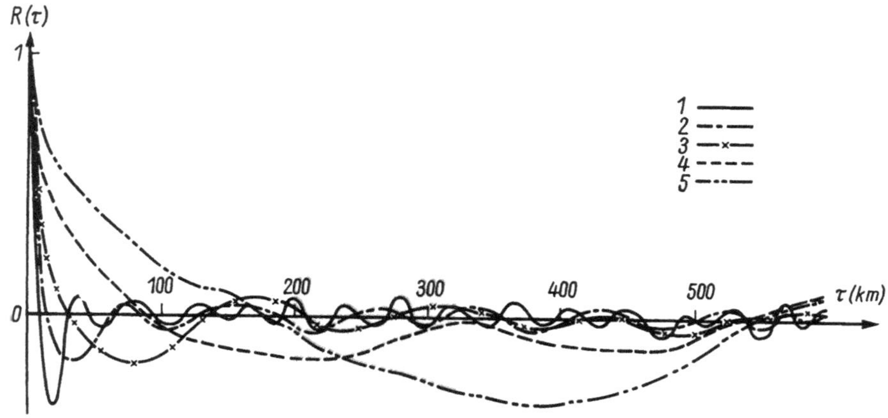

Fig. 3. Correlation functions for various intervals of sliding average: 1000, 500, 200, 100, 50 km.

for each section. In this case the sections were selected with a length of no more than 1000 km and without spatial trend, if possible. If such a trend was, nevertheless, present the long-wave fluctuations were filtered. To facilitate the intercomparison of various correlation functions all fluctuations with wavelengths more than 400 km were filtered out. Thus, the character of mesoscale disturbances inherent to the selected observations could be estimated.

It would be premature to make any generalizations because there are still insufficient observations of one or another type of cloudiness or underlying surface. We shall simply give several examples of correlation functions for various situations.

Fig. 4 shows the correlation functions for some situations and Fig. 5 gives the television images corresponding to curves *1*, *3*, *4*, *5* of Fig. 4. The projections of orbit trajectories are marked on the television images. Curve *1* in Fig. 4 relates to the alto-stratus and nimbus stratus (As, Ns) forms located in the zone of the front oriented from Iceland through northern Scandinavia to Novaya Zemlya. The orbit plane did not intersect the frontal zone and was projected along this zone in the region under consideration. The correlation function has no fluctuation component since the long-wave fluctuations were almost implicit due to the very character of the observation.

Curve *2* relates to the region of Sahara west of Tassili upland. According to the ground observation data cloud was absent in this region. Some inhomogeneities in the brightness, associated probably with the variations of the reflection properties of the earth's surface, may be found on the television image. The presence of fairly regular recurrent fluctuations (period of 200 km) may be noted on the correlation curve. Such periodicity is unlikely to be connected with the presence of cloudiness and probably reflects simply some properties of the underlying surface.

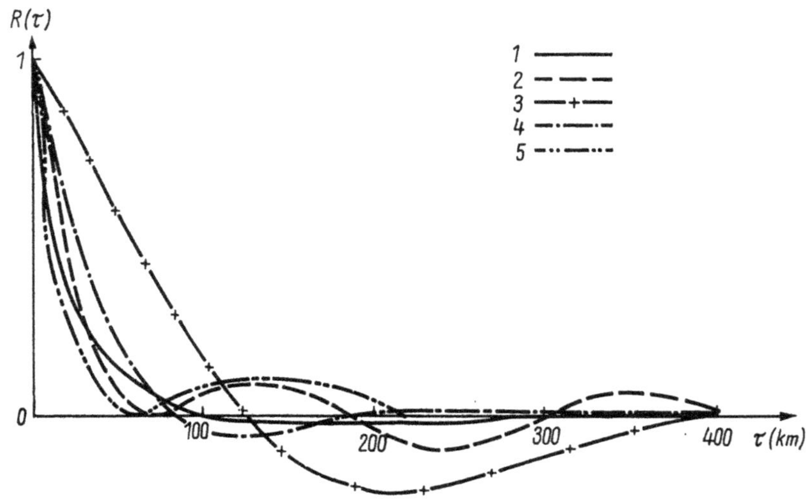

Fig. 4. Correlation functions for some situations.

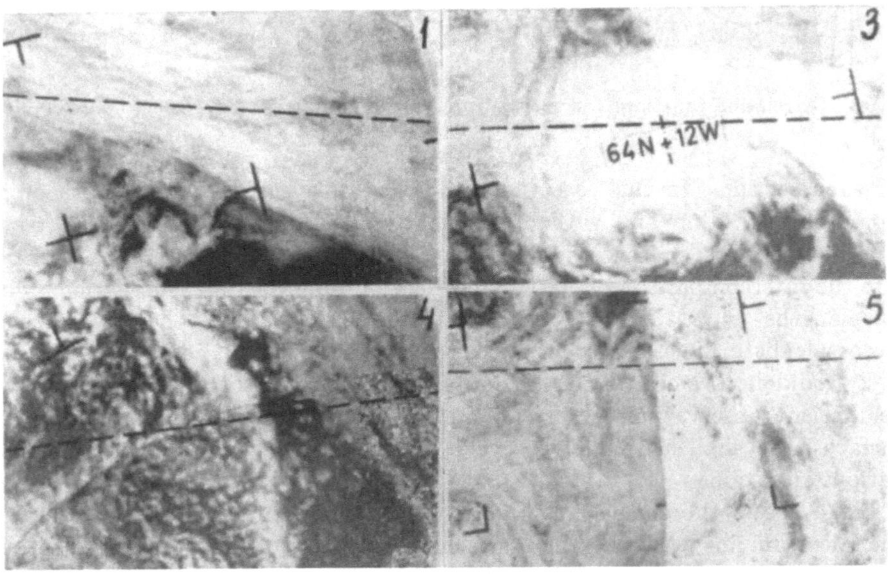

Fig. 5. Television images corresponding to curves *1*, *3*, *4*, *5* of Fig. 4.

The correlation function shown curve *3* of Fig. 4 relates to the region of very thick dense cloud associated with a barometric depression south of Iceland. One notes here the pronounced exponential–cosine character of the correlation functions which corresponds to the presence of fluctuations with 400–500 km wavelength which the filtration of the empirical correlation function failed to eliminate completely. At the same time one can see that as a whole the statistical structure of the reflected radiation intensity does not differ substantially from the cases considered above.

Curve *4* is the correlation function relating to the mass of cumulus above the Atlantic Ocean somewhat to the west of the Bay of Biscay. Here the cumulus was not continuous and it can be seen on the television image that the cumulus forms a fairly distinctive characteristic picture on the background of dark ocean surface. At first glance, one could expect the appearance of some periodic mesoscale fluctuations of the reflected radiation intensity. It can be easily seen, however, that this is not the case; the mesoscale variations appear to be disordered and do not display periodicity.

Finally, the correlation function shown in curve *5* is the function relating to the region occupied by the mass of continuous cloud of stratus and strato-cumulus form (Atlantic, Ocean about 55 °S, 25 °W). Here again the mesoscale fluctuations are absent; only the presence of fluctuations with a period of about 250 km may be noted which reflects probably the presence of breaks in the cloud field, i.e. a certain transient character of the radiation under consideration. Thus, the following conclusions may be drawn:

(i) Ordered mesoscale variations of the reflected radiation field and hence of the cloud field are probably rather infrequent.

(ii) The spatial correlation functions of the reflected radiation intensity relating to the mesoscale spectral range (5–400 km wavelength) depend little on the type of object reflecting the radiation; hence they can hardly be used as the prediction-signs for the recognition of these objects).

3. Since the contribution of mesoscale fluctuations to the total variability is small it is necessary to limit oneself to large-scale averaging (to the scale 1000 and more km) when measuring the radiation characteristics of the earth as a planet.

References

[1] K. Ya. Kondratyev, O. A. Avaste, M. P. Fedorova and K. E. Yakushevskaya, Radiation Field of the Earth as a Planet, Gidrometeoizdat, 1967.

[2] A. I. Lebedisky, V. I. Tulupov, Yu. P. Safronov, Yu. G. Andrianov and I. I. Karavaev, Geomagn. i Aeron. 7, 423 (1967).

[3] V. G. Boldyrev, D. M. Sonechkin, V. I. Tulupov and I. S. Khandurova, Trudy Gidrometeotsentra SSSR, No. 36, 1968 (pp. 76–81).

[4] D. M. Sonechkin and I. S. Khandurova, Trudy Gidrometeotsentra SSSR, No. 50, 1969 (pp. 37–46).

Space Research XI — Akademie-Verlag, Berlin 1971

ESTIMATION OF SHORT-WAVE RADIATION FLUXES OF THE EARTH–ATMOSPHERE SYSTEM FROM SATELLITE DATA

K. S. Shifrin, V. Yu. Kolomiytsov and N. P. Pyatovskaya

Institute of Oceanography, USSR Academy of Sciences, Leningrad, USSR

The following subjects are considered in the paper: (i) calculation of a scheme for estimating the fluxes of outgoing short-wave radiation from brightness values measured with the help of a satellite; (ii) analysis and refinement of basic parameters introduced in the scheme of calculation; (iii) study of the angle structure of the brightness field of the earth–atmosphere system for a cloudless sky and for an overcast sky with different altitudes of the sun.

1. The Statement of the Problem. Scheme of Calculation

The short-wave radiation balance of the earth–atmosphere system determines the amount of incoming solar heat, i.e. the power of the source of energy of all the atmospheric processes. It is known that one of the first scientific results obtained by means of meteorological satellites is the establishment of the fact that climatological calculation data of the short-wave albedo of the earth–atmosphere system have been found to be wrong. For the globe (in the average for a year), for example, these calculations give 40%, new refinements of the same method give 36%, while the direct measurements carried out by means of satellites give 29%. Even though such highly averaged figures have an error of 20%, it is clear that errors due to the use of climatological maps, which are employed for the sake of atmosphere dynamics, will be very large.

For direct measurements of the short-wave balance from satellites, it is first necessary to solve the following problem: When measuring with the help of satellitea we obtain from each sighting area the values of energetic brightness I at some angles θ and φ. The problem is to determine the radiance R of 1 cm² of the surface of the outer atmosphere by separate values of brightness. The connection between R and I is determined by the formula

$$R = \int_{(2\pi)} I(\theta, \varphi) \cos \theta \, \mathrm{d}\omega \, . \tag{1}$$

The calculations according to (1) are possible only in the case when the angle structure of energetic brightness is known. If, for example, it follows Lambert's law, i.e. I does not depend on the angles θ and φ, then

$$R = \pi I \, . \tag{2}$$

Under real conditions this simple dependence does not occur. The field structure of brightness $I(\theta, \varphi)$ is complex; it depends on the relation between air brightness $I_1(\theta, \varphi)$ (arising due to scattering in the air) and brightness of the earth's surface $I_2(\theta, \varphi)$. The brightness I is equal to the sum $I_1 + I_2$:

$$I(\theta, \varphi) = I_1(\theta, \varphi) + I_2(\theta, \varphi) \,. \tag{3}$$

The analysis of the structure of I_1 and I_2 has been carried out in [1, 2]. The conclusion is that when θ increases, I_1 (air brightness) increases too, and I_2 (brightness of the beam reflected by the earth's surface) as a rule decreases.

The form of the resulting function I depends on the relation of the absolute values of brightness I_1 and I_2. The main factor affecting the form of the function

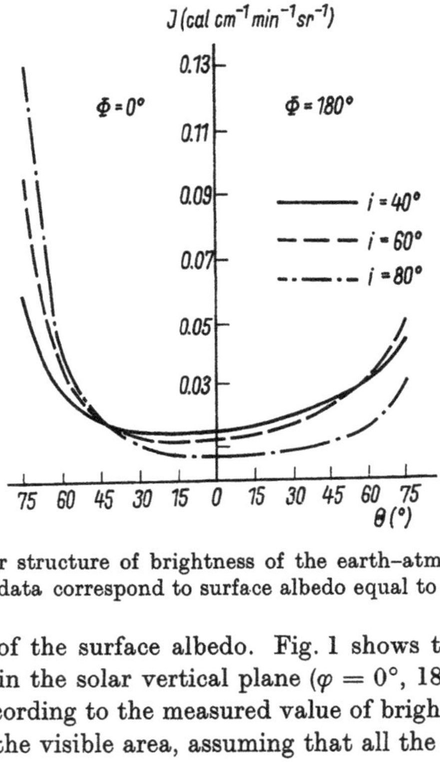

Fig. 1. Angular structure of brightness of the earth–atmosphere system. The data correspond to surface albedo equal to zero.

$I(\theta, \varphi)$ is the value of the surface albedo. Fig. 1 shows the calculated values of I_1 for different θ in the solar vertical plane ($\varphi = 0°$, $180°$). Our scheme [2] is the following. According to the measured value of brightness one determines the albedo value of the visible area, assuming that all the other characteristics are fixed at certain average values. The albedo value determines the form of the function $I(\theta, \varphi)$, i.e. the amount of flux R.

The scheme is valid both for the case of a cloudless sky and for the case of an overcast sky. Of course, this scheme does not give an unambiguous solution of the problem but it does give an optimal one, i.e. the most probable value of R corresponding to measured $I(\theta, \varphi)$.

In developing this scheme a hypothesis has been formed that both the reflection from clouds and the reflection from the earth's surface follow Lambert's law. However, this hypothesis raises doubts, as it is contrary to the fact that reflection from the sea (which is the main underlying surface of the earth) is non-lambert. In order to show that, in fact, this circumstance is not essential, the problem was solved with anisotropic bottom reflection. The results of this analysis published in [3] show that the primary scheme in the main proves to be correct. At the same time it was possible to establish the range of angles round the direction of the mirror beam where corrections are necessary, and to indicate their magnitudes. The reasons for the slight influence of reflection anisotropy from the sea are that the sea albedo is small and therefore the contribution of the reflected beam is also small.

If we assume that R is determined with the error which does not exceed

$$\Delta R \approx 0.04 \text{ cal cm}^{-2} \text{ min}^{-1} ,$$

then the range of angles where the Lambert scheme is inapplicable for oceans is found to be as in Table 1.

Table 1

i (°)	$\Delta\varphi$ (°)	θ (°)	$\omega/2\pi$
0	0–180	0–15	0.034
40	0– 25	30–90	0.120
60	0– 15	40–90	0.064
80	0– 10	60–90	0.028

For clouds the influence of reflection anisotropy is greater because cloud albedo is large. Here the standardization by types of reflection phase functions is necessary.

2. Comparison with Experimental Data

In Fig. 2 the angular structure of the brightness of the earth–atmosphere system obtained from satellite data has been compared with calculations of

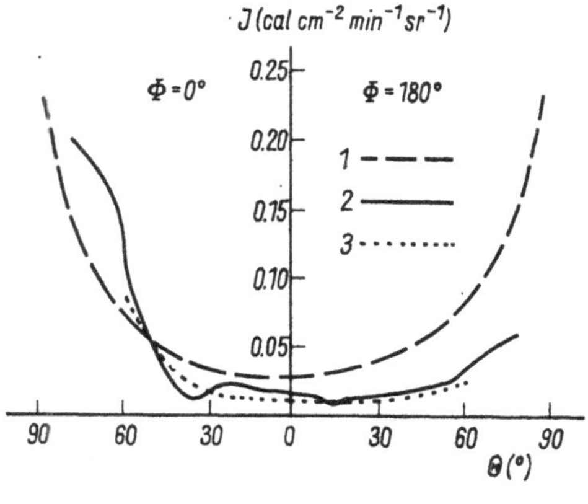

Fig. 2. Angular structure of brightness of the earth–atmosphere system according to experimental data: *1*, Rayleigh curve; *2*, data obtained by Tiros 4; *3*, data obtained by Cosmos 122.

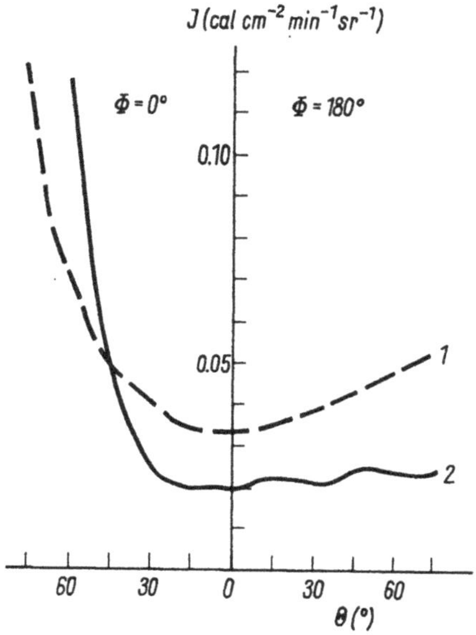

Fig. 3. Comparison of calculated function $I(\theta)$ with data of aircraft measurements; *1*, calculated values; *2*, data of aircraft measurements.

the same structure for a Rayleigh atmosphere [4]. One can see that essential anisotropy occurs. Fig. 3 shows the comparison between calculated values of $I(\theta, \varphi)$ and data from aircraft measurements according to [5].

Table 2 [6] shows the comparison results of fluxes R measured on the earth ($z = 0$), in aircraft ($z = 8.4$ km), and in balloons ($z = 30$ km) and simultaneous values calculated by I (measured by means of Cosmos 156) in conformity with the author's scheme.

Table 2
Comparison of measured and calculated values of R

Z (km)	R (cal cm^{-2} min^{-1})
0	0.12
8.4	0.16
30	0.16
600	0.14

Considering the essential difference of range in averaging by satellite and aircraft receivers the coincidence of data for R is unexpectedly good [6]. Note that if we consider the brightness $I(\theta, \varphi)$ to be "lambert", we should have for R the value 0.10 cal cm^{-2} min^{-1}, which were in evident contradiction to all the other values.

References

[1] K. S. Shifrin and N. P. Pyatovskaya, Trudy GGO **166** (1964).
[2] K. S. Shifrin, B. Yu. Kolomeetsov, and N. P. Pyatovskaya, Trudy GGO **166** (1964).
[3] K. S. Shifrin and V. Yu. Kolomeetsov, Trudy GGO **221** (1968).
[4] K. S. Shifrin and N. P. Pyatovskaya, Trudy GGO **221** (1968).
[5] M. G. Korobov, N. G. Markina and L. A. Pakhomova, Proc. VI Int. Conf. on Actinometry and Optics of the Atmosphere, Tallinn 1966, Tartu. Izd. Valgus, Tallinn **1968**.
[6] Ye. P. Barashkova, V. L. Gaevsky and K. Ya. Kondratyev, Trudy GGO **235** (1970).

Space Research XI — Akademie-Verlag, Berlin 1971

RECENT MEASUREMENTS OF THE SOLAR RADIATION INCIDENT ON THE ATMOSPHERE

A. J. DRUMMOND

The Eppley Laboratory, Newport, Rhode Island, USA

During the period 1966–1969, nine independent determinations were made, in the USA, of the integral wavelength solar constant of radiation, employing conventional jet and rocket research aircraft, high-level balloons and the Mariner 6 and 7 Mars spacecrafts. A total of 51 measurement series was obtained. In three of the efforts, the extraterrestrial Spectral components were evaluated, in the wavelength range λ 0.23–2.5 μm, using filter radiometry and monochromator methods. The principal results are reviewed; also included is a summary of the 25 solar-constant balloon flights carried out in the USSR from 1961 to 1968.

Comparison is made with the classical ground-based extrapolations, as extended by recent American and West German results. The impact of the new information on aspects of meteorology, astrophysics and space engineering is outlined. An important related problem examined is that of temporal variation of the total energetic solar emission, as well as such fluctuations in the shorter wavelengths, and the existence of periodicities which only long-term earth-orbiting satellite programs can reveal.

1. Introduction

The absorption of solar radiation is responsible for practically all terrestrial processes. At the lower levels this energy absorption, mainly by water vapor and carbon dioxide, generates weather and regulates climate. In the stratosphere ozone is the principal absorber, in the ultraviolet region. Above 100 km, in the thermosphere, radiation of very short wavelength is absorbed by molecular oxygen. As well as these gaseous wavelength selective absorptions, atmospheric attenuation of sunlight results from molecular (Rayleigh) scattering and from reflection and absorption by solid and liquid particulate matter (aerosols). In order to enable quantitative treatments to be made of such mechanisms, precise knowledge is necessary of the extraterrestrial solar fluxes. The value of the Sun's irradiance, integrated over all wavelengths and referred to 1 AU, is designated the solar constant of radiation.

Between about 1920 and 1960, estimates of the solar constant and its spectral composition were mainly based on extrapolations to zero optical air mass (i.e. the upper limit of the terrestrial atmosphere), from the classical pyrheliometric and spectral bolometric radiation measurements obtained, by Abbot and his co-workers at the Smithsonian Institution, on high mountains [1]. Additional information regarding the wavelength region unobservable from the ground, even at elevated locations ($\lambda < 295$ nm), was provided 20 years ago by the use of rocket-borne spectrographs [2]. At this time also, energy absorbed by water vapor in the infrared region beyond the longer wavelength limit of the Smithsonian surface spectrobolometric measurements (and therefore improperly accounted for) was assessed, by Johnson [2], then of the US Naval Research Laboratory, through his analysis of the results of independent studies of such attenuation within atmospheric paths. The results, however, suffered from the limitations then existing in the experimental techniques of precision radiometry, especially as applied to the free atmosphere. As will be seen from Table 1, which lists the principal ground-based extrapolated values of the solar

Table 1
Ground-based extrapolations of the solar constant, 1923–1968

Source	Year	Solar Constant ($mW\ cm^{-2}$)
Abbot[a]	1923–52	135.2
Linke[b]	1932	135.2
Mulders[c]	1934–35	135.9
Nicolet[d]	1951	138.0
Johnson[e]	1954	139.5
Unsöld (Revision)[f]	1955	136.5
Stair and Johnson[g]	1956	142.8
Allen (Revision)[h]	1958	138.6
Makarova (Revision)[i]	1964	(141.8)
Gast (Revision)[j]	1965	139.0
Stair and Ellis[k]	1966	135.9
Labs and Neckel[l]	1968	136.5

a Ref. [1].
b Linkes Meteorologisches Taschenbuch, **2** (Akad. Verlag. Geest and Portig, Leipzig 1933).
c Z. Astrophys. **11**, 132 (1935).
d Refs. [5, 6].
e Ref. [4].
f Physik der Sternatmospharen, 2 ed (Springer, Berlin 1955).
g J. Res. Natl. Bur. Std. **57**, 205 (1956).
h Quart. J. Roy. Met. Soc. **84**, 307 (1958).
i Ref. [3].
j Handbook of Geophysics and Space Environment (AFCRL, US Air Force 1965).
k J. Appl. Met. **7**, 635 (1968).
l Z. Astrophys. **69**, 1 (1968).

integral (1923–1968), the values so obtained range from about 135 to 143 mW cm^{-2}. The revised solar constant of Makarova is bracketed because it is derived, by Thekaekara [3], through integration of solar spectral irradiances based on Makarova's data. The other post-1960 estimates of Gast, Stair and Ellis, and Labs and Neckel will be discussed in a later section.

The spectral curve of Johnson [4], integrated to 139.5 mW cm^{-2}, was most favored in the United States and that of Nicolet [5, 6], integrated to 138.0 mW cm^{-2}, in Europe. There are very considerable differences between the two representations in the ultraviolet and visible regions [7, 8]. Johnson essentially revised Abbot's spectral data, in the light of the aforementioned ultraviolet and infrared corrections and also from consideration of subsequent measurements by Dunkelman and Scolnik [9]. Nicolet took quite a different approach. Using the same basic data (excepting [9]), he integrated the detailed observations of the solar spectrum over relatively wide bands, as a discontinuous step function, for which he attempted to remove the effects of Fraunhofer absorption. The inherent difficulties are discussed in [5].

With the advent of artificial earth-orbiting satellites and deep probing spacecraft, it began to be appreciated that a better knowledge was required of the extraterrestrial spectral and integral wavelength fluxes, preferably derived from direct measurements of the solar intensities either above the earth's atmosphere or at altitudes where the atmospheric influences would be as small as possible. Apart from the engineering requirements of space programs involving large-scale, solar-simulated test facilities at the ground, with the object of setting vehicle design characteristics to maintain the desired radiative equilibrium temperatures in voyage, it was realized from geo-astrophysical interests, that this state of knowledge was unsatisfactory.

2. Direct Measurement of the Solar Constant

Hence, in 1964, the Jet Propulsion Laboratory (JPL) proposed to determine the solar constant directly, in collaboration with the Eppley Laboratory, employing filter radiometric techniques at high altitudes—ideally, above the effective terrestrial atmosphere. This program entailed the development of instrumentation which, in certain aspects, resulted in distinct advances in the state-of-the-art. Full descriptions are given in [7, 10]. It will suffice to state that a series of multichannel radiometers were employed; they incorporated fast-response, high-sensitivity, non-wavelength-selective thermopile detectors of the plated-wirewound type. Provision was made for near-simultaneous measurements with broad and narrow bandpass filters (see later section), as well as the total irradiance. Quartz lenses were used to amplify optically the narrow filter channels. Details of the testing and basic calibration of the radiometers and associated signal conditioning equipment are described in [11]. A parallel study [12] was undertaken of the problems arising from the projection of ground-based thermal radiation measurements to conditions in the free atmosphere, with minimum degradation, of the best instrument precision and accuracy levels attainable in the laboratory.

Three types of NASA aircraft probes were employed, viz. (i) B-57B (six flights at 11.5–15 km, July–August 1966 and March 1967); (ii) X-15 (one flight at 78–83 km, October 1967); and (iii) CV-990 (seven flights at 11–12.5 km, October 1967 and July–August 1968). The sole USAF/NASA rocket X-15 flight of 17 October 1967 is historic since the radiometric measurements, well above the ozonosphere, made possible the first truly direct evaluation of the solar constant. In the case of the lower level jet aircraft flights into the stratosphere, simultaneous ground-based measurements of total ozone concentration

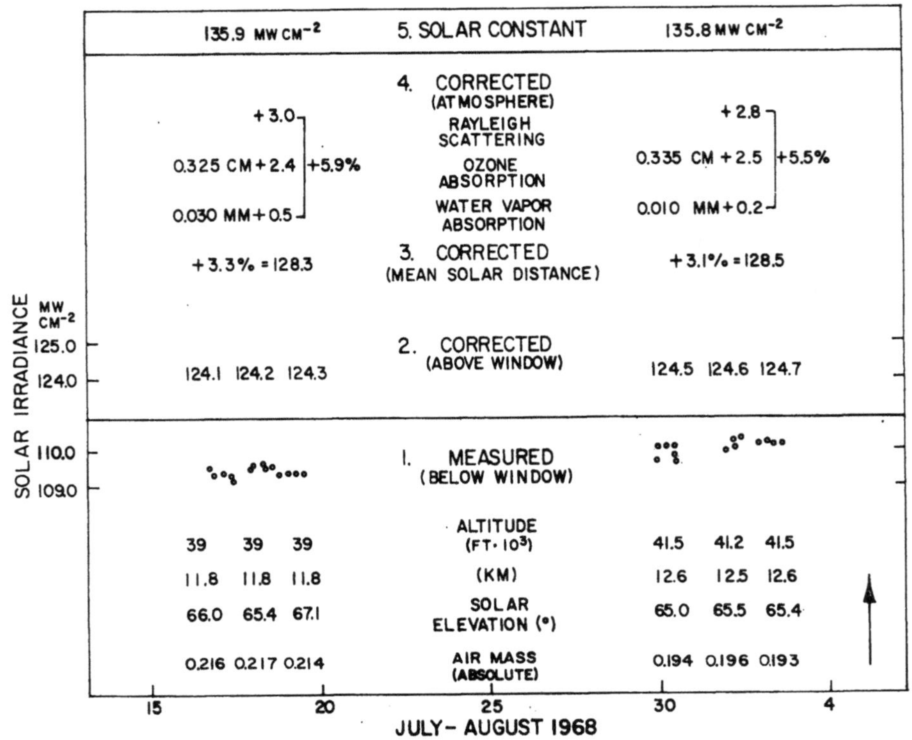

Fig. 1. Diagrammatic presentation of the Eppley–JPL NASA CV-990 aircraft solar-constant determinations.

Table 2

Eppley–JPL solar-constant determinations, 1966–1968

Year	Month	Day	NASA aircraft	Measurement series	Total irradiance (mW cm⁻¹)	No. of sunspots*		
1966	July	25	B-57B	2	135.8	57	60	
	August	4				64		
1967	March	7	B-57B	4	136.0	98	86	
		8				84		
		9				82		
		10				82		
	October	17	X-15	1	136.1	28	60	
		22	CV-990	1	136.2	91		
1968	July	17	CV-990	3	135.9	92	92	93
		18				94		
		19				90		
	July	30	CV-990	3	135.8	97	94	
	August	1				91		
		2				94		
	Mean		(6 sets with 14 series)		136.0 ± 0.1%		75	

* Supplied by the Solar Division of the American Association of Variable Star Observers.

at nearby locations were frequently arranged for; relevant water contents above the aircraft were measured either by balloon radiation sonde or through the use of infrared emission measurements on the aircraft; as far as practicable, it was verified that high-level dust was not present in significant amounts to influence the results. The data were then corrected for Rayleigh scattering and water vapor and ozone absorption (as well as to mean Earth–Sun distance). When mounted in the B-57B and X-15 aircrafts, the radiometers were freely exposed in the solar beam. However, in the pressurized CV-990, it was necessary to have the detectors beneath a quartz window, the presence of which entailed further correction. Fig. 1 is a diagrammatic presentation of the treatment of six sets of CV-990 data. The principal Eppley–JPL solar-constant determinations are summarized in Table 2. For more information, references [8, 13, 14] should be consulted. The joint team comprised Drummond, Hickey and Scholes (Eppley) and Laue (JPL).

During this period of solar-constant measurement activity, parallel but independent efforts were carried out by the NASA Goddard Space Flight Center (GSFC), NASA Ames Research Center and the University of Denver. In 1969 there were two JPL series. Finally, between 1961 and 1968 there was the effort by the University of Leningrad. The following is a summary of the different attempts:

1966–1968	Eppley–JPL	B-57B, X-15, CV-990 aircraft	14 series
1967	GSFC	CV-990 aircraft	18
1967	Ames	CV-990 aircraft	11
1967–1968	Denver	balloon	4
1969	JPL	spacecraft	2
1969	JPL	balloon	2
1961–1968	Leningrad	balloon	25

The GSFC program [15] included five series of measurements with an Eppley–Ångström pyrheliometer (McNutt and Riley), four other series with a second Eppley–Ångström pyrheliometer (Duncan and Webb), four series with a Hy–Cal normal incidence pyrheliometer (McNutt and Riley) and five series with an absolute wirewound cone radiometer developed at GSFC (Kruger).

The balloon data were derived from flights up to about 31 km [16, 17, 18, 19]. Murcray's (University of Denver) four series of flights [16, 17] gave values fairly close to each other, although they are rather low compared with the data from other high-altitude measurements. The detectors were Eppley normal incidence pyrheliometers. Kondratyev and his co-workers [18, 19] employed two actinometers.

The other type of source of solar-constant determination considered here is the two series of measurements carried out, early in 1969, on the JPL Mariner 6 and 7 Mars missions. These results [20] are based on telemetered data from absolute cavity radiometers on board the space probes. Both instruments, although widely separated, gave values which, when reduced to 1 AU, are in close accord. A large mass of data was obtained for several weeks before the detectors began to show signs of degradation.

In this assembly from the different sources, the following radiometric references were adopted:

Source	Reference
Eppley-JPL radiometers	International Pyrheliometric Scale (IPS)
GSFC	
(a) Eppley–Ångström pyrheliometers	IPS
(b) Hy–Cal pyrheliometer	Blackbody
(c) Cone radiometer	Absolute electrical units
Denver pyrheliometers	IPS
JPL radiometers	Absolute electrical units
Leningrad actinometers	IPS

The primary standards to which the Eppley–JPL radiometers, the GSFC Eppley–Ångström pyrheliometers and the Denver pyrheliometers were referred are a group of Ångström pyrheliometers maintained at Newport and compared frequently, in Davos (Switzerland), with the World Meteorological Organization standards. In conjunction with those of the Weather Bureau (ESSA), they constitute the US reproduction of the IPS. During the period November 1969–February 1970, this reference was compared with the primary ones at the British National Physical Laboratory and the Canadian National Research Council, when the agreement was shown to be within a few tenths of one per cent in both instances [21]. The Leningrad measurements were referred to the IPS as maintained in the USSR.

Table 3
High-altitude measurements of the solar constant, 1961–1969

Source		Solar constant ($mW\ cm^{-2}$)
Murcray[a]	Univ. of Denver	133.8
McNutt and Riley[b]	GSFC	134.3
Duncan and Webb[b]	GSFC	134.9
McNutt and Riley[b]	GSFC	135.2
Plamondon[c]	JPL	135.3
Kondratyev et al.[d]	Univ. of Leningrad	135.3
Kruger[b]	GSFC	135.8
Drummond et al.[e]	Eppley–JPL	136.0
Mean (weighted)		135.3

[a] Refs. [16, 17].
[b] Ref. [15].
[c] Ref. [20].
[d] Refs. [18, 19].
[e] Refs. [12–14].

Table 3 extends the results of Table 2, to include these other seven high-altitude independent evaluations of the solar constant. In weighting the mean, maximum factors were assigned to the results from the JPL Mariner, Leningrad balloon and Eppley–JPL aircraft series, since they are based on large samples. A high degree of reliability was given also to the values obtained from both the GSFC Hy–Cal pyrheliometer and the cone radiometer. That the Denver balloon value is distinctly lower than the others points to some of the basic problems of precision radiometry, in general, and solar application in particular. The two GSFC Ångström pyrheliometers yielded relatively fewer data and hence were likewise afforded less weight. The Ames (monochromator) *

* Also includes spectral irradiance measurements.

and JPL (balloon) investigations are not treated here because of uncertainties in the inherent reference values.

The proposed value of the solar constant is 135.3 mW cm^{-2} (1.940 cal cm^{-2} min^{-1}) with a conservative error estimate of $\pm 1\%$. This revision is 3% lower than Johnson's figure.

3. Solar Spectral Irradiance

In preparing their curve of the extraterrestrial spectral irradiance, Thekaekara and Drummond [22] considered only the two sets of data available from the high-altitude studies. These were the GSFC measurements (CV-990 aircraft), with five different monochromator–interferometer instruments, and the Eppley–JPL filter radiometer measurements on all three NASA aircraft employed. In the former series, six spectral irradiance standard lamps and two blackbody sources were used in the calibrations. As mentioned above, the Eppley–JPL material was referenced to the IPS.

The tabulated data published by the GSFC experimenters [15] contains values for the shorter wavelengths based upon the NRL measurements [4, 23]. However, those were scaled downwards by about 7%, since the ratio of the GSFC to NRL data, in the range 0.3–0.4 μm, was found to be 0.927 and the NRL rocket data had been normalized to the Johnson scale for $\lambda > 0.3$ μm. A confirmation of this scaling factor has recently been obtained from the MUSE experiment on borad the Nimbus 3 satellite [24]. Heath has observed that, in the spectral range 0.22 to 0.292 μm, the signal which had been predicted on the basis of the NRL curve was 1.07 times that obtained from the satellite.

Preliminary results only of the Eppley–JPL filter measurements have been published [7, 10]. The complete assembly has now been worked up. This comprises the following three filter types:

(i) 2 broad bandpass (colored glass) with sharp cut-on at 526 and 688 nm (mean transmission $\lambda > 607$ nm); 34 measurement series from 18 flights.

(ii) 11 narrow bandpass (interference) with wavelenght isolation regions mainly from 235 to 700 nm; 22 measurements series from 12 flights.

(iii) 9 narrow bandpass (from differential treatment of broad band filters) with wavelength isolation from 355 to 963 nm; 7 measurements series from 9 flights.

Table 4
Eppley–JPL principal aircraft broad bandpass filter measurements

Year	Month	NASA Aircraft	Series	λ (nm)	OG1 >526	RG8 >688	Mean >607	Total − Mean >607
							mW cm^{-2}	
1966	July–Aug.	B-57B	8		100.2	73.9	87.0	49.0
1967	March	B-57B	13		101.3	73.4	87.3	48.7
1967	October	X-15	1		101.6	73.4	87.5	48.5
1967	October	CV-990	1		100.9	73.6	87.3	48.7
1968	June–Aug.	CV-990	11		100.9	73.3	87.1	48.9
Mean (5 sets with 34 series)					101.0 ±0.4	73.5 ±0.3%	87.2	48.8

Table 5

Eppley–JPL aircraft narrow bandpass filter measurements and GSFC corresponding monochromator etc. results

λ (nm)	E–JPL (mW cm⁻²)	GSFC (mW cm⁻²)	E–JPL / GSFC	λ (nm)	E–JPL (mW cm⁻²)	GSFC (mW cm⁻²)	E–JPL / GSFC
235–344	5.52	5.24	1.05	508–592	14.9	14.7	1.01
272–322	2.78	2.58	1.08	535–603	**12.1**	11.6	1.04
295–365	6.99	6.29	1.11	565–631	**12.1**	10.9	1.11
298–344	4.22	3.88	1.09	595–645	8.83	7.89	1.12
334–403	7.78	7.63	1.02	603–663	**10.3**	9.24	1.11
355–430	**10.1**	10.2	0.99	603–697	**15.0**	14.1	1.06
410–474	11.8	11.9	0.99	631–697	**10.3**	9.62	1.07
430–490	**11.5**	11.8	0.97	645–700	8.47	7.91	1.07
446–500	10.6	10.8	0.98	697–963	**28.2**	28.1	1.00
490–565	**13.3**	13.7	0.97	1250–2000	17.3	18.0	0.96

Mean agreement (20 wavelength intervals) E–JPL:GSFC = ±5%

Note:
E–JPL values in heavy type are derived differentially from adjacent pairs of broad bandpass filters; other E–JPL values are from narrow bandpass interference filter measurements.

The principal broad bandpass data are given in Table 4. The agreement between the mean solar irradiance derived from the two Eppley–JPL control broad filters (viz. $\lambda > 607$ nm) is 87.2 mW cm^{-2}. This compares with a GSFC value of 86.0 mW cm^{-2} for the same region, a difference of only about 1.5%.

Table 5 is a summary of the narrow bandpass filter data. The corresponding GSFC monochromator etc. results are included for comparison. In the case

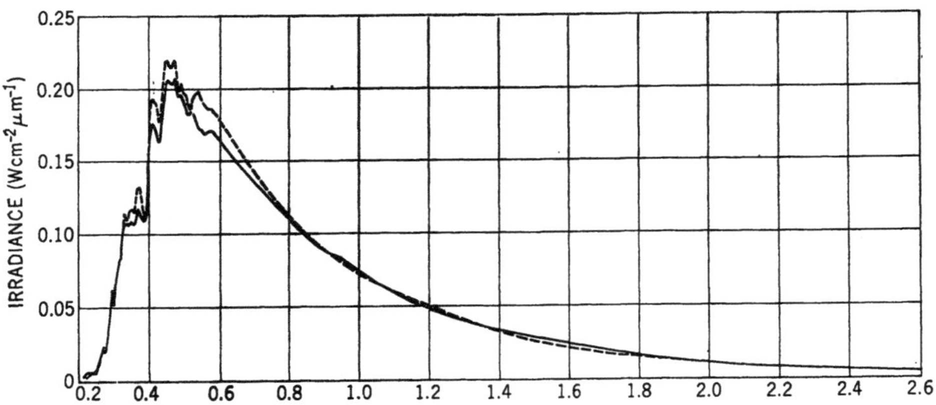

Fig. 2. Proposed NASA standard extraterrestrial solar spectral irradiance curve (after [22]) with Johnson (dashed) for comparison.

of the interference filters, the energies within the specific wavelength intervals were obtained through multiplication of the measured signal by the particular filter factor, derived from consideration of filter transmittance and an idealized source curve [25]. The results for the other isolated wavelength regions involved subtraction of a broad bandpass filter measurement from an adjacent one with shorter wavelength cut-on (here the respective filter factors are simply the reciprocals of the transmittances in the main filter transmission band). The GSFC values were extracted from the aforementioned table [15], by subtraction of the percentage of the value of the solar constant associated with wavelengths shorter than λ_1 from the corresponding value for λ_2, these being the limits of the wavelength interval investigated.

It will be noticed that, in general, the agreement of the two independently determined spectral series is $\pm 5\%$ for the 20 comparisons. This is within the experimental uncertainties of the measurements. For wavelengths $\lambda < 400$ nm the E–JPL results are higher than the GSFC ones by 7%, and for $\lambda > 400$ nm higher by 3%. In the region 350–600 nm there is practically perfect agreement, as is also the case beyond about 700 nm.

The proposed NASA standard solar spectral irradiance curve for outside the earth's atmosphere was therefore derived from the GSFC determination, with modifications based on the E–JPL filter investigation. It is shown in Fig. 2, with that of Johnson added for comparison. A total of 20 aircraft flights (6 GSFC and 14 Eppley–JPL) are involved. For the wavelength range where a number of the filters are in agreement in showing slightly different values from those of GSFC, a weighted average of the two sets of data was taken. This produces a slight revision of the GSFC curve in wavelengths from 0.3 to 0.7 μm (300–700 nm) and increases the integrated value under the curve (i.e. the solar constant) from the GSFC figure of 135.1 to the new proposal of 135.3 mW cm^{-2}. The maximum changes are +2.3% at 0.34 μm, −0.7% at 0.45 μm, +1.6% at 0.63 μm; percentages varying between zero and these maxima occur at other wavelengths in the range. The GSFC spectral irradiance curve has been retained in the ranges 0.2 to 0.3 and 0.7 to 20 μm. In the lower wavelength interval, as evidenced above, there is close agreement with Heath and satisfactory agreement (about six %) with Eppley–JPL considering the energy level for filter measurements; at the upper end, the agreement with Eppley–JPL is of the order of 1% out to about 3 μm.

If we now compare this curve (and its energy integrals) with those derived from ground-based extrapolations which have received most attention, we arrive at the following major differences from the new value, given as percentage above aircraft values.

λ (nm)	<607	>607	>200
1951 Nicolet	+0.6	+2.8	+2.0
1954 Johnson	+6.1	+1.4	+3.1
1966 Stair and Ellis	not available		+0.4
1968 Labs and Neckel	−0.4	+1.6	+0.9

In the above, $\lambda < 607$ nm broadly represents the ultraviolet and visible regions and $\lambda > 607$ nm the near infrared region; $\lambda > 200$ nm is their summation. The greatest divergence is that displayed by the Johnson curve in the visible and near ultraviolet, where excesses of the order of 10% occur.

4. Solar Variability

As has been pointed out, the question of the "true" total solar energy input into the terrestrial atmosphere is of great significance in determining the thermal budget of the earth and, in turn, its components. For example, Raschke and Bandeen [26] have recently indicated the influence of different values of the solar constant on the computed global averages of the incoming solar radiation, the radiation balance and the albedo. From consideration of solar constant values of 2.00 and 1.95 cal cm^{-2} min^{-1} (140 and 136 mW cm^{-2}), there would be a change of about 3% in the albedo of the globe and a correspondingly large one of 0.02 cal cm^{-2} in the radiation balance, as derived from the Nimbus 2 satellite measurements in June 1966. It can be readily computed that such a change in albedo, of say 0.30 to 0.31, would lower the radiative temperature of the earth–atmosphere system by nearly 1.5 °C, a sizeable amount. There is thus clear demonstration of the need for accurate measurements of the solar constant in carrying out studies of this nature employing satellite radiometric data.

When we consider the spectral aspects, the question has an added importance. The temperature equilibrium of satellites and space probes depends mainly on the solar radiation absorbed by them. Surface degradation due to the ultraviolet is a factor which controls the thermal balance and long-term reliability of satellites. Solar cells are the major source of power for scientific satellites; precise prediction of their performance requires knowledge of the solar spectral irradiance in the near infrared. If all satellite surfaces were black or grey with an absorptance independent of wavelength, the solar constant and not the spectrum would be the significant parameter. But actually these surfaces have, in general, a spectral absorptance which varies considerably with wavelength.

The energy absorbed per unit area, by a satellite surface, is $E = P_\lambda \alpha_\lambda \, d\lambda$, where P_λ is the solar spectral irradiance at the satellite–sun distance, and α_λ is the spectral absorptance of the surface at wavelength λ. The value of E will obviously vary as one or another of the different proposed curves is used for the computation.

Studies have been published [3] which show that if the Johnson curve is used for the computation, the absorptance is higher than if any of the other less well-known curves is used. Similar calculations (by Thekaekara), using the curve now proposed, also show that the absorptance given by the Johnson curve is higher for all of the 25 typical coatings which were considered. The energy absorbed by solar cells exhibits a variation which is just the converse of the coatings. The predicated value of the energy absorbed by the solar cell is less if the Johnson curve is used for the computation than is the case for the other curves. Thus, the curve which has been the most widely accepted represents an extreme among many such curves for predicting values of absorbed energy. This is in keeping with another conclusion which emerges from a comparative study of the different curves, namely, in the wavelength range below 0.55 μm, the Johnson curve assigns values of spectral irradiance which are upper extremes among the different proposals.

Let us finally consider what may be termed "real" as distinct from "apparent" solar variability. We have yet to put an instrument into space, and keep it there long enough, to find out whether the solar constant is changing along with other features of the Sun which we know are changing. Solar constant changes may include short-period microfluctuations. If so, it might be supposed that there are statistical variations associated with turbulence on the surface of the Sun. Considering the largeness of the solar area and the comparative smallness of the elements of turbulence, one would reasonably expect that

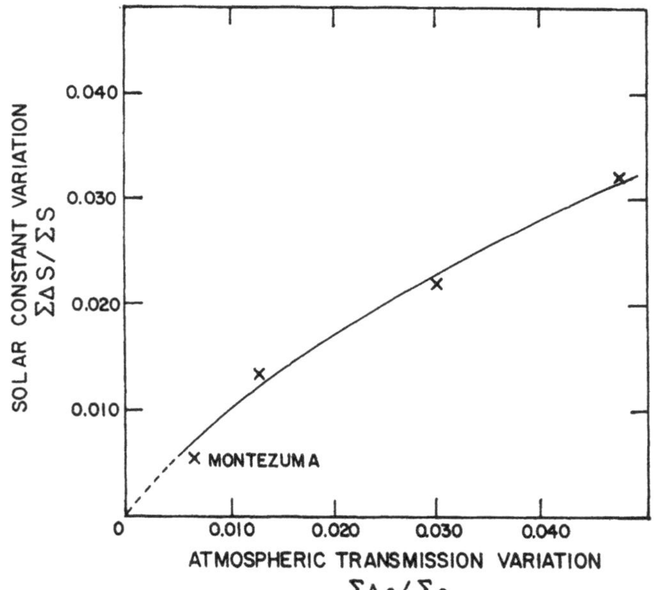

Fig. 3. Mean variations of the apparent solar constant ($\Sigma\, \Delta S/\Sigma\, S$) and the apparent atmospheric transmission ($\Sigma\, \Delta\varrho/\Sigma\, \varrho$), based upon Smithsonian data (after Ångström 1970).

such net (integral wavelength) fluctuation over the whole disk is not very significant.

But longer period fluctuations of the solar constant is a different matter. On the assumption that sunspots constitute an index of solar constancy, let us review briefly what conclusions have emerged from the few investigations, based upon observation, which have been conducted.

In a recent study [19], Kondratyev and Nikolsky display data from their high-altitude balloon soundings of solar irradiance, which indicate a decrase in their derived value of the solar constant for Wolf (sunspot) numbers below or above 80–100. The maximum possible decrease in irradiance so claimed does not exceed 2–2.5%. This similar trend for numbers on either side of about 100 is difficult to understand, since it is generally acknowledged that the blackbody radiation from sunspots is very significantly less than it is from the surrounding undisturbed surface.

In contrast, the Eppley–JPL aircraft solar-constant derivations show no appreciable variation for a range of sunspot numbers of about 30 to 100 (see Table 2). It is true that only one of the 14 measurement series was above the ozonosphere and that the others could have been unknowingly influenced by the atmosphere above the aircraft, but if so, it is unlikely that this effect would exceed a few tenths of one per cent. On the other hand, the Leningrad measurements could have been affected instrumentally and, to some extent, by ozone above 30 km.

Of the attempts made to establish variability of the total solar emission from examination of Abbot's material [1], the most recent is by Ångström [27]. Fig. 3 is an extract from his paper. From an analysis of the Smithsonian solar-constant values as they appear from stations where the variations in the atmospheric transmission have been large, as at Mount Wilson and Bassour, compared with the result of determinations at Montezuma, where the transmission has been constantly very high with small variation, he came to the conclusion that a very marked relationship exists between the two parameters. In Fig. 6 of his paper, a large variability of transmission ϱ corresponds to a large variation of the derived solar constant S, and vice versa. A rough extrapolation to $\Sigma\, \Delta\varrho/\Sigma\, \varrho = 0$ suggests disappearance of the (apparent) variations of the solar constant. However, it would be premature so to generalize, especially when Ångström further demonstrates that, on extension to a representative portion of the great mass of observational material at the high-level station of Harqua Hala, the relationship is less pronounced and that there could be a real variation of about 0.2%. But, again, this figure is typical of instrument measurement precision.

To conclude, just as the decade which has come to an end has witnessed the extensive use of aircraft and balloons to determine the solar constant of radiation, efforts to extend these fundamental studies of the extraterrestrial solar integral and spectral energy fluxes should become part of the meteorological satellite programs, experimental leading to operational. It will be only in this way that a proper knowledge of long-term solar constant variability will be founded and the shorter term fluctuations in specific wavelength intervals examined on a statistical basis.

References

[1] Annals of the Astrophysical Observatory, II–VII (Smithsonian Inst., 1908–1954).
[2] A. J. DRUMMOND, in: Proc. Inst. Environ. Sci. Am. Soc. Testing Materials Symp. Solar Rad. Los Angeles 1965, Mount Prospect, Illinois 1965 (p. 55).
[3] M. P. THEKAEKARA, Solar Energy **12**, 205 (1968).
[4] F. S. JOHNSON, J. Met. **11**, 431 (1954).
[5] M. NICOLET, Ann. Astrophys. **14**, 249 (1951).
[6] M. NICOLET, Archiv Meteor. Geophys. Biokl. **A3**, 209 (1951).
[7] A. J. DRUMMOND, J. R. HICKEY, W. J. SCHOLES and E. G. LAUE, in: Proc. Int. Astronaut. Fed. Cong. 17th Madrid 1966, **2**, Gordon and Breach, New York 1967 (p. 227).
[8] A. J. DRUMMOND and J. R. HICKEY, Solar Energy **12**, 217 (1968).
[9] L. DUNKELMAN and R. SCOLNIK, J. Opt. Soc. Am. **49**, 356 (1959).

[10] A. J. DRUMMOND, J. R. HICKEY, W. J. SCHOLES and E. G. LAUE, J. Spacecraft Rockets **4**, 1200 (1967).
[11] A. J. DRUMMOND, J. R. HICKEY, W. J. SCHOLES and E. G. LAUE, in: Proc. Int. Astronaut. Fed. Cong. 18th Belgrade **1967**, **2** (Pergamon Press, New York 1968 (p. 407)).
[12] A. J. DRUMMOND and E. G. LAUE, in: Proc. Int. Astronaut. Fed. Cong. 19th New York 1968, **1**. Pergamon Press, New York 1970 (p. 179).
[13] A. J. DRUMMOND, J. R. HICKEY, W. J. SCHOLES and E. G. LAUE, Nature **218**, 259 (1968).
[14] E. G. LAUE and A. J. DRUMMOND, Science **161**, 888 (1968).
[15] M. P. THEKAEKARA, R. KRUGER and C. H. DUNCAN, Appl. Optics **8**, 1713 (1969).
[16] D. G. MURCRAY, J. J. KOSTERS and P. R. GAST, Air Force Cambridge Research Laboratories Report 68-0452 (1968).
[17] D. G. MURCRAY, Air Force Cambridge Research Laboratories Report 69−0070 (1969).
[18] K. YA. KONDRATYEV, G. A. NIKOLSKY, I. YA. BADINOV and S. D. ANDREYEV, Appl. Optics **6**, 197 (1967).
[19] K. YA. KONDRATYEV and G. A. NIKOLSKY, Quart. J. Roy. Met. Soc. **96**, 509 (1970).
[20] J. A. PLAMONDON, Jet Propulsion Laboratory Space Programs Summary, 37−59, **3**, 162 (1969).
[21] A. J. DRUMMOND, A. R. KAROLI and J. R. HICKEY, ESSA Contract Report E-66-70N (1970).
[22] M. P. THEKAEKARA and A. J. DRUMMOND, Nature Physical Science **229**, 6 (1971).
[23] C. R. DETWILER, D. L. GARRETT, J. D. PURCELL and R. TOUSEY, Ann. Géophys. **17**, 9 (1961).
[24] D. HEATH, J. Atmos. Sci. **26**, 1157 (1969).
[25] J. R. HICKEY, in Advances in Geophysics, **14**, Academic Press, New York 1970.
[26] E. RASCHKE and W. R. BANDEEN, J. Appl. Met. **9**, 215 (1970).
[27] A. K. ÅNGSTRÖM, Tellus **22**, 205 (1970).

Space Research XI — Akademie-Verlag, Berlin 1971

THE SOLAR CONSTANT FROM DATA OF BALLOON INVESTIGATIONS IN THE USSR AND THE USA

K. YA. KONDRATYEV[a], G. A. NIKOLSKY[a], D. G. MURCRAY[b], J. J. KOSTERS[b] and P. R. GAST[b]

[a] Leningrad University Leningrad, USSR
[b] Denver University, USA

During 1962–1968 in the USSR and in 1967–1968 in the USA balloon measurements of the integral fluxes of direct solar radiation at heights up to 32–35 km were carried out. After the extrapolation of obtained data to beyond the limits of the atmosphere and the insertion of corrections for unobserved ultraviolet radiation and the selectivity of the instrument, a number of solar constant values was derived. The analysis of these data has shown that the final result is influenced by the extrapolation method and correction values. From the measurements in Colorado the S_0 value equals 1.919 cal cm^{-2} min^{-1}. The data of Soviet investigations give the value of $S_0 = 1.943$ cal cm^{-2} min^{-1}. Balloon investigations have shown that attenuation in the air layer above the 32–35 km level is to a considerable degree due to the aerosol component which can be subject to great variations. In a vertical air column above 32 km the attenuation of a solar radiation flux can reach 1.9–2% of the solar constant value.

On the basis of the analysis of the data obtained in the USSR it has been concluded that variations of the value of the solar constant depend upon solar activity. The solar constant reaches its maximum (1.943 cal cm^{-2} min^{-1}) with a relative number of sunspots N (according to Wolf) equal to 70–80; with greater and smaller Wolf numbers S_0 decreases. Maximum variations of S_0 constitute 2–2.5%. These conclusions require further confirmation on the basis of direct multiple measurements of the solar constant outside the limits of the atmosphere. The most probable solar constant value, from the data obtained at the University of Leningrad, is 1.943 ± 0.008 cal cm^{-2} min^{-1}. Values of S_0 very close to these were obtained by M. P. Thekaekara from aircraft measurements and by A. J. Drummond from the data of interplanetary flights.

1. Investigations in the USSR

Since 1961, research physicists of the University of Leningrad have been carrying out balloon measurements of the vertical profiles of the radiation budget and all its components in the troposphere and lower stratosphere (at heights up to 30–33 km).

Measurements of radiation fluxes and a number of meteorological parameters were taken continuously during the balloon's ascent. The method of measurement used makes it possible to obtain detailed vertical profiles of short and long-wave radiation over the sounding region. One of the most important components of the radiation budget is the flux of direct solar radiation. In this connection, great attention has been paid to improvement of the measurement method and the analysis of solar radiation measurement data. Direct solar radiation is measured by two actinometers (one of which is sealed) which are constantly directed towards the Sun by means of a photoelectric pointing control system. The method of measurements is described in more detail in the works of Kondratyev et al. [1] and Kondratyev and Nikolsky [2].

The accuracy of measurements of the absolute values of the solar radiation flux is within $\pm 1.5\%$ (the absolute values of radiation fluxes were determined according to the International Pyrheliometric Scale of 1956).

Since the upper limit of balloon soundings reached heights of about 30 km and more, it was possible to perform a sufficiently correct extrapolation of the data obtained to beyond the limits of the atmosphere and thus to determine the absolute value of the solar constant. The extrapolation of the measured values of the direct solar radiation flux S from levels of 25–33 km to beyond the limits of the atmosphere was accomplished by two methods. The first method is based on the use of Bouger's law ($\log S = \log S_0 \mp m \log p$). The transparency coefficient p of the overlying atmospheric layer was derived from the data of one of the flights, in the presence of highly stable atmospheric transparency, by compiling a chart of the dependence of $\log S$ upon the atmospheric mass m. For all the other flights the transparency coefficient was assumed to be constant.

The values of $S_{0,\text{p.e.}}$ derived from Bouger's formula are only the results of partial extrapolation, since they have not been corrected for the radiation absent at heights about 30 km ($\lambda < 300$ nm). It is known that for the integral flux of solar radiation the dependence of $\log S$ upon m is non-linear. Since, however, in our case (heights about 30 km) the atmospheric thickness which attenuates radiation is small, the use of a linear dependence does not result in notable errors in the value of $S_{0,\text{p.e.}}$. In fact, when afterwards a non-linear dependence of $\log S$ upon m was assumed for the integral flux of direct solar radiation, it appeared that the error of linear extrapolation did not exceed 0.42%. For deducing the non-linear dependence a formula borrowed from Theaekara et al. [3] was used:

$$\log S = \log \sum_{i=1}^{n} I_i P_i^m ,$$

where I_i is the solar radiation spectral flux in the interval i, P_i is the transmission of the atmospheric layer above the level of sounding for the interval i. The values of P_i were calculated according to the data of Elterman [4]. The use of a precision method for extrapolation results in an increase of $S_{0,\text{p.e.}}$ by 0.008 cal cm^{-2} min^{-1}, ascompared with linear extrapolation. The method of extrapolation we have chosen makes it possible to calculate the solar radiation attenuation by the aerosol component existing above the upper level of sounding. Since data on the optical aerosol atmospheric thickness at heights from 30 to 200 km are scarce, the application of this extrapolation method is apparently reasonable.

After inserting corrections for the absent radiation we obtain the final value of the solar constant ($S_{0,\text{e}}$). The results of the extrapolation show that the attenuatiom of solar radiation in the spectral range of 0.2–10 μm in the layer from the atmosphere's upper limit to the 30 km level reaches 1.9–2% (for $m=1$). This attenuation value was obtained from extrapolation data for the flight of 27 June 1967 (flight No. 23). The measured flux of solar radiation at the height of 30 km amounted in this case to 1.886 cal cm^{-2} min^{-1} with relative atmospheric mass $m = 2$. Extrapolation to zero atmospheric mass gave $S_{0,p.e} = 1.918$ cal cm^{-2} min^{-1}. After inserting the spectral correction for ultraviolet radiation with $\lambda < 300$ nm, which equals 0.020 cal cm^{-2} min^{-1}, we obtain $S_{0,e} = 1.938$ cal cm^{-2} min^{-1} (the effect of a quartz window in the infrared region has been taken into account in the values of the flux S — see below). For atmospheric mass $m = 1$ extrapolation gives 0.036 cal cm^{-2} min^{-1}, i.e. approximately 1.9–2%.

The second extrapolation method is based on the calculation of the attenuation of spectral fluxes (0.3–4 μm) of solar radiation in the layer above the upper point of sounding, and the insertion of corrections for the ultraviolet and infrared spectral regions of solar radiation. The calculation was performed for concrete conditions using the method of Shifrin and Avaste [5]. Attenuation coefficients for different spectral regions were taken from Elterman [4], and the energy distribution in the solar radiation spectrum outside the limits of the atmosphere from Labs and Neckel [6], taking account of the fact that the solar constant equals 1.94 cal cm^{-2} min^{-1}.

The results of calculations of S_0 using the second method were compared with the solar constant values obtained by Murcray et al. [7].

2. Investigations in the USA

In the scientific research institute of Denver University in the period of 1967 to 1968 four ascents of balloon instruments to heights of 31–35 km were conducted. The main scientific aims of these investigations consisted in obtaining the vertical variation of the integral flux of direct solar radiation and determining the dependence of the value of a solar radiation flux upon atmospheric mass at the maximum heights for a balloon in the stratosphere. It was supposed that such measurements will make it possible to determine more accurately the value of the solar constant without resorting to rocket or satellite measurements. Apart from this, another purpose of the experiment was to determine the variation and value of short-wave radiation attenuation by aerosol in the atmosphere up to 35 km.

The balloon's instruments consisted of two actinometers installed in a special Sun tracker, a measurement control system and a recording unit (with data recording on magnetic type in a digital code), energy sources, telemetric equipment and navigational instruments.

The flux of solar radiation was measured simultaneously by two identical thermoelectric actinometers. To maintain the working characteristics determined with high accuracy in surface calibrations (within the limits of 0.2%) the actinometers were hermetically sealed and supplied with thermostats. In addition to the above, the Eppley Laboratory engineers who made the instruments compensated the temperature dependence of the sensitivity of the balloon's actinometers. The measurement data showed the discrepancy between the values of the direct solar radiation flux, which were obtained from two instruments, to increase to 0.4%, compared with the results of calibrations.

For the evaluation of the solar constant Murcray used only the data of the actinometer giving higher values of solar radiation fluxes. In the balloon's drift regime the readings of the actinometers were very stable and varied only with change of solar height. The results of measurements in the drift showed that the dependence of the observed values of S upon atmospheric mass for every sounding (26 June 1967, 23 January 1968, 26 April 1968; 16 December 1968) has its individual variation. This apparently shows that optical characteristics of the air mass above the level of the balloon's drift can vary to a great extent. The transition from the values of S measured at heights of 31 and 35 km to extra-atmospheric values was realized by inserting corrections for the residual attenuation of solar radiation in the overlying air layer. The calculation of attenuation was made both for the spectral region shorter than 0.29 μm, and in the 0.29–4.0 μm range. The total correction for the radiation with wavelengths shorter than 4 μm was 0.035 cal cm^{-2} min^{-1}. The correction for the radiation not transmitted through the quartz window ($\lambda > 4$ μm) was 0.017 cal cm^{-2} min^{-1}.

3. Solar Constant Absolute Values

Let us turn now to the evaluation of the absolute values of a flux of direct solar radiation not transmitted through the atmospheric layer above 30 km ($\lambda < 0.30$ μm). Actually, the insertion of the above corrections will result in obtaining the actual value of the solar constant $S_{0,e}$ (e = extrapolation).

The accuracy of the absolute values of the direct solar radiation flux measured at the upper point of the ascent is determined by three factors: the accuracy of the calculation of the influence of temperature upon the actinometer sensitivity; the accuracy of reference to the International Pyrheliometric Scale (IPS of 1956) and the accuracy of the application of corrections for the limited range of the spectral region of the radiation measured by the actinometer. As regards the first two factors, the accuracy of balloon measurements is not worse than 1.3% (0.8 + 0.5). As to the third factor, it can be noted that, the value of the correction being smaller than 1.1% of the measured value, a possible error in this case does not exceed 0.2%. Thus, the maximal possible total error is 1.5%.

Comparison of experimental data obtained in similar conditions at Denver and Leningrad Universities shows the measurement error to be less than ± 1.5%. Thus, for instance, the value of a direct solar radiation flux at the height of 31 km measured in the USA in the summer period of 1967 (1.867 cal $\times cm^{-2}$ min^{-1}) differs only by 1.1% from that obtained in the USSR (1.886 cal $\times cm^{-2}$ min^{-1}) a day later. We should note at once that even this slight discrepancy is imaginary, since at Leningrad University a correction for the quartz window was inserted, whereas with Murcray et al. [7] it was not. When such a correction is applied to the data of Denver University, the value of the solar radiation flux at the height of 31 km equals 1.884 cal $cm^{-2} min^{-1}$, i.e. the values of the compared radiation fluxes will practically coincide (the difference is only 0.11%).

Naturally, when conducting the experiment at Denver University, the investigators also came across at least two of the three abovementioned factors which limit the accuracy of measurements. The difference in the construction of radiation sensors and recording systems, as well as the calibration in accordance with the IPS of 1956 in agreement with national standards, might have led to a larger discrepancy between the results than actually occurred.

Comparison of the extrapolation to beyond the limits of the atmosphere data of summer ascents has shown that the difference between the values increased to 1.3% (1.943 and 1.919 cal $cm^{-2} min^{-1}$), compared with the difference in the values of S (0.11%). Hence it follows that either the methods of extrapolation, or the values of spectral corrections do not correspond to real conditions and the characteristics of the instruments.

Murcray et al. [7] calculated the attenuation in the air layer above the 31 km level from the data of a standard atmospheric model (for the 0.2–4 μm range) and after that calculated the influence of the quartz window. Kondratyev and Nikolsky [2] performed a preliminary extrapolation using Bouger's law (0.3 to 10 μm) and then inserted the spectral correction. Murcray et al. [7] consider that for the summer ascent solar radiation attenuation by aerosol in the atmospheric layer above 31 km is practically absent. The discrepancy between the vaules of $S_{0,e}$ is probably connected with these different methods of extrapolation.

The correction for the reduction to extra-atmospheric conditions introduced by Murcray et al. was equal to 0.052 cal $cm^{-2} min^{-1}$, 0.035 of which was the correction for the residual attenuation and 0.017 for the unmeasured radiation with $\lambda > 4$ μm. With the Leningrad University instrument, the correction for the quartz window (0.014 cal $cm^{-2} min^{-1}$ for heights of 25–30 km), has already been taken into consideration at the processing of the S values at different heights. The correction for the absent radiation with $\lambda < 0.3$ μm (0.020 cal $\times cm^{-2} min^{-1}$) has been obtained taking account of Drummond's results [8] which show that the spectral distribution of solar energy in the short-wave spectral region, as given in Johnson's data, is 7% higher than its actual value.

The results of winter soundings also reveal close agreement. The Soviet ascent was made on 3 December 1968 and the American ascents on 23 January and 16 December 1968. The data obtained in these flights are presented in Table 1, the values of solar radiation fluxes having been reduced to the mean distance between the earth and the sun. The values of S do not exactly coincide, but special attention should be paid to the difference in sounding heights between the ascents considered, the difference in solar elevations and, consequently, the considerable difference in the values of the absolute mass of the overlying atmospheric layer.

Table 1

Date	Measured flux value (cal cm^{-2} min^{-1})*	Measurement level (km)	Solar height h_0	Relative mass m	Absolute mass m_a
23. 1. 68	1.848	31.0	37°24′	1.65	0.0166
3. 12. 68	1.837	32.5	8°55′	6.385	0.0529
16. 12. 68	1.867	35.4	33°11′	1.804	0.0097

* No corrections have been inserted for the quartz window.

One might expect that the extrapolation of the S values to beyond the limits of the atmosphere must result in a different correlation of $S_{0,e}$ values, as compared with that of S values. In fact, the values of $S_{0,e}$ for the Denver ascents were 1.905 and 1.943 cal cm^{-2} min^{-1}, according to the calculations of Kondratyev and Nikolsky (from Murcray's calculations they were 1.900 and 1.919 cal cm^{-2} $\times$min^{-1} respectively). For the Leningrad ascent of 3 December 1968 the value of $S_{0,e}$ was 1.924 cal cm^{-2} min^{-1}. The extrapolation of the S data to beyond the limits of the atmosphere was performed at Leningrad University for the flights of 23 January and 3 December 1968 using the second method, i.e. calculating the attenuation of S in separate spectral regions for Elterman's radiation atmospheric model of 1968. Murcray calculated according to the model of 1964.

For the flight of 16 December Kondratyev and Nikolsky accomplished a partial extrapolation from Murcray's data obtained during the balloon's drift at the height of 35.4 km. The value of $S_{0,\text{p.e.}} = 1.909$ cal cm^{-2} min^{-1} was corrected for the quartz window (0.014 cal cm^{-2} min^{-1}) and the ultraviolet radiation (0.020 cal cm^{-2} min^{-1}), with $\lambda < 0.3\,\mu$m; the resulting value of $S_{0,e}$ was 1.943 cal cm^{-2} min^{-1}.

Before continuing the comparison of the data, the problem of solar constant variations should be briefly considered.

The conclusions about the existence of a dependence between solar constant variation and solar activity were made at Leningrad University on the basis of the results of soundings performed by the University collaborators and are therefore at the moment supported by the Leningrad co-authors only. From the results of balloon actinometric soundings in 1962–1968, it appeared possible to reveal a connection between the values of the solar constant and relative numbers of sunspots N. As can be seen from Fig. 1, with increase of N the value of the solar constant increases (curve 2) and reaches its maximum at $N \sim 80$–100, after which it drops. The maximum variations of $S_0(N)$ is of the order of 2–2.5%.

Let us now compare the results obtained with the values $S_0(N)$ taken from Fig. 1 for the corresponding values of N.

The data presented in Table 2 show that the Denver flight of 23 January 1968 was performed in conditions of great turbidity of the overlying atmospheric layer. In fact, the $S_{0,e}$ value for this flight notably differs from both $S_0(N) = 1.942$ cal cm^{-2} min^{-1}, and $S_{0,e} = 1.943$ cal cm^{-2} min^{-1} (for the flight of 16 December 1968), although the astrophysical conditions of both flights were fairly similar. It should be remembered that the extrapolation of the Denver University results was performed using different methods, one of which (calculation of attenuation for separate spectral regions) does not make it possible to take account of the real state of the overlying atmospheric layer. On the contrary, the other method (extrapolation based on the use of a non-

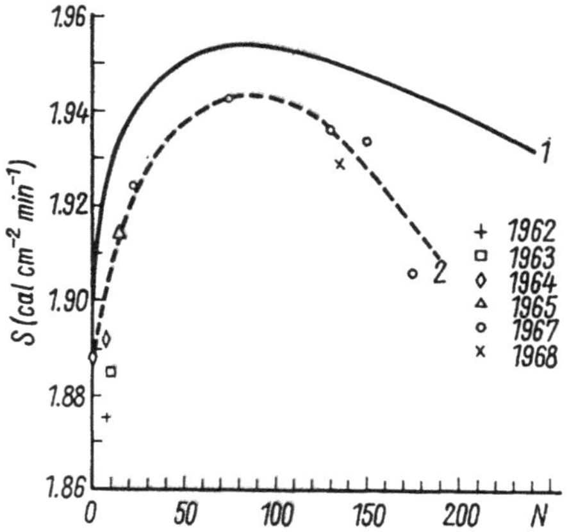

Fig. 1. Dependence of the direct solar radiation flux values extrapolated to beyond the limits of the atmosphere upon the number of sunspots (according to Wolf). *1*, The $S_0(N)$ dependence obtained by Ångstrom (Smithsonian Scale of directly measured values; differs from the International Pyrheliometric Scale of 1956 by approx. 0.2%). *2*, The $S_0(N)$ dependence from the Leningrad University data (International Pyrheliometric Scale of 1956).

linear dependence of log S upon m) takes account of the presence of the main optically active components. An example of extrapolation based on Bouger's law, given in Fig. 2, shows that in the wavelength range 300 nm–4 μm the

Table 2

Date	N	$S_{0,e}$ (cal cm^{-2} min^{-1})	$S_0(N)$ (cal cm^{-2} min^{-1})
23 January 68	73	1.905	1.942
3 December 68	136	1.924	1.934
16 December 68	83	1.943	1.943

attenuation in the air layer above 35.4 km was in this case 0.042 cal cm^{-2} min^{-1} (for $m = 1.804$). It should be emphasized that the use of the non-linear dependence of log S upon m at the extrapolation resulted in a somewhat higher value of $S_{0,e} = 1.943$ cal cm^{-2} min^{-1} than that obtained earlier for the maximum of the dependence $S_0(N)$ by means of a linear extrapolation (1.938 cal cm^{-2} min^{-1}). When considering the data of the ascents of 3 and 16 December (Table 2), it can be stated that the transparency of the air layer above 32–35 km was in this period very similar over regions in the Northern Hemisphere as remote from one another as the central part of the USSR European territory and the state of Colorado (USA). Here the convergence of the $S_{0,e}$ and $S_0(N)$ data was not less than 1%.

For the ascents of 23 January and 3 December 1968 the discrepancies between the values $S_{0,e}$ and S_0 were 0.037 and 0.009 cal cm^{-2} min^{-1} respectively, i.e. about 2 and 0.5%. Assuming measurement errors to be small, the difference

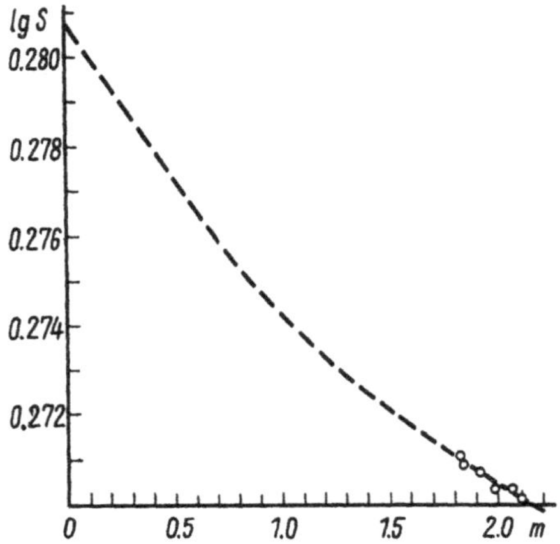

Fig. 2. An example of the extrapolation of the values of a direct solar radiation flux (0.3 to 4 μm), obtained from the balloon sounding on 16 December 1968, to beyond the limits of the atmosphere using the non-linear dependence of log S upon m.

observed once more indicates considerable variations in the optical aerosol thickness of the air layer above 32–35 km, and, consequently, the inapplicability in all cases of the model of stratification of direct solar radiation attenuation by aerosol. Hence it follows that the use of the method of extrapolation shown in Fig. 2 is the most effective one. Evaluations of the solar constant derived from balloon investigations agree closely with the results of measurements on aircraft and space ships.

Thus, Drummond [8] gives a new value of $S_0 = 1.942$ cal cm^{-2} min^{-1} which was obtained at the end of February 1969 from Mariner 6 and 7 interplanetary stations. Careful aircraft measurements of solar radiation integral and spectral fluxes at the height of 11.5 km carried out by Thekaekara et al. [3] gave the value of the solar constant averaged over five instruments for six days of observations 3, 8, 10, 14, 16 and 19 August 1967) as 1.940 ± 0.041 cal cm^{-2} min^{-1}. For $N = 96$ (mean for six days of observations) from Fig. 1 we obtain 1.939 cal cm^{-2} min^{-1}. Better agreement can hardly be expected.

It should be noted here that recent values of the solar constant more and more approach the many-year mean of the solar constant obtained at the Smithsonian Institution [9] of 1.946 cal cm^{-2} min^{-1}, according to the International Pyrheliometric Scale of 1956.

References

[1] K. YA. KONDRATYEV et al., Appl. Optics, 6, No. 2 (1967).
[2] K. YA. KONDRATYEV and G. A. NIKOLSKY, in: Swedish Met. and Hydr. Inst. Commun., Ser. B, No. 28 (1968).
[3] M. P. THEKAEKARA, R. KRUGER and C. H. DUNCAN, Appl. Optics, 8, No. 8 (1969).
[4] L. ELTERMAN, Rpt. AFCRL-68-0153 (1968).
[5] K. S. SHIFRIN and O. A. AVASTE, in: Investigations in Atmospheric Physics, No. 2, 23–66 (Tartu 1960).
[6] D. LABS and H. NECKEL, Z. Astrophysik, 69, 1 (1968).
[7] D. G. MURCRAY, T. G. KYLE, J. J. KOSTERS and P. R. GAST, Tellus, 21, No. 5 (1969).
[8] A. J. DRUMMOND et al., in: Advances in Geophysics, Ed. A. J. Drummond (Academic Press, New York, London) (1970), (p. 40).
[9] L. B. ALDRICH and W. H. HOOVER, Smithson. Astrophys. Obs. 7, 165 (1954).

Space Research XI — Akademie-Verlag, Berlin 1971

POSSIBLE MEASUREMENTS OF SURFACE CHARACTERISTICS WITH REMOTE SENSORS

K. J. K. Buettner

Department of Atmospheric Sciences, University of Washington, Seattle, USA

The purpose of this paper is to review surface characteristics which are amenable to or connected with remote sensing and are of potential value for the meteorologist. Emphasis is on new proposals.

The absorption of solar rays in the atmosphere, an important part of the diabatic heating, can only be fully observed where the same emphasis is given to a sea-going pyranometer net as is given to albedo measurements from spaceships.

Weather-indicating surface changes involve coverings, wetness, frozen ground, sea state, etc. New calculations of expected 1.55 cm microwave emission from specular and from isotropically wavy ocean are compared with overflight data. They show for rough sea a difference indicating effects of foam.

Over land albedo measurements at 0.7, 1.1 and 1.5 μm are proposed. This triple system would tell about ground type (clouds, vegetation, sand or soil and snow). Clouds could be discerned from snow, fresh snow from wet snow.

1. Introduction and List of Opportunities

Of the terrestrial surface we see more and know less than of layers above and below. Problems of ground resolution, of particle scattering, of gaseous absorption, and of atmospheric eddying effects near the ground reduce recognizability of remotely sensed surface data. The term surface causes even semantic problems: synopticians call surface temperature the value at eye level which may deviate by up to ± 30 °C from the true surface value.

Surfaces are two-dimensional entities and as such their qualities are principally unmeasurable. We therefore introduce another term, subsurface, which is that solid or liquid layer near the air–ground or air–sea interface in which heat, mass and momentum are transferred exclusively by molecular (not eddy) motions, and from which for most substances, infrared (IR) signals are emitted and where visible and IR rays are specularly reflected. This layer is usually less than one millimeter thick.

From spaceship or airplane these surface qualities can be sensed:

1. Coverings by snow, dune sand, salt, water, sea-foam, icebergs, arctic sea ice with its leads.

2a. State of wetness of sand, soil and similar surfaces, snow, ice (icebergs), possibly plants.

2b. State of soil being frozen or thawed, such as frozen or melting tundra.

3. Roughness of agitation: sea state as expressed by large waves, capillary waves, surf.
4. Motion of surface or near-surface material: sand dunes, water, drifting snow, oil slicks, icebergs, etc., on oceans, air and water pollutants, volcanic exhalations.
5. Subsurface temperatures of land and sea.
6. Subsurface heat flow on ocean (not realizable yet).
7. Change of surface mineral or particle size composition via IR emissivity.
8. Near surface existence of motions of air pollutants such as SO_2 via Michelson interferometer.
9. From daily, IR-monitored, subsurface temperature variations of the ground, the surface contact coefficient $\varkappa = (k \varrho c)^{1/2}$ can be evaluated. Here k = heat conductivity, ϱ = density and c = specific heat. This test presupposes a radiational climate. Changes of $\varkappa$ from day to day are caused by weather such as rain.

Of the many items involved only those will be discussed here which have a chance to be measured soon from space. Use of radar or lidar will not be discussed. Some already existing methods will be briefly mentioned. Proposals will be made of what should be measured at the surface and what surface parameters could be measured from space for the benefit of weathermen.

2. The Need for a Sea-Going Pyranometer Network

We first discuss an item not concerning surface qualities but one which is urgently needed and should be measured exactly at surface stations.

In a long-range forecasting scheme, the six variables u, v, w, ϱ, T and p can be computed from these equations provided the seventh variable, the diabatic heating term H, is known. (φ_1 and φ_2 are functions, f is the Coriolis term. Other terms have the customary meaning.)

$$\frac{\partial v}{\partial t} = \varphi_1(p, \varrho, f, v, g) + K_1 \nabla^2 v \qquad \text{Continuity}$$

$$\frac{\partial T}{\partial t} = \varphi_2(p, \varrho, T, v) + H + K_2 \nabla^2 T \qquad \text{Gas law}$$

The highly variable and larger than expected solar absorption is already routinely listed from satellite data. Of this amount a certain portion hits the surface; over oceans the surface-absorbed solar heat can be neglected for periods of a few days. Over land of low albedo we have surface heating; over snow the observed total solar absorption goes mostly into tropospheric heating. Most vital seems knowledge of this partition for the oceans. This requires well calibrated pyranometers on small islands or ships. The present national pyranometer nets serve as a not very important adjunct to agricultural meteorology and hydrology. Their budgets are poor. It is rather grotesque to compare the enormous efforts put into weather satellites and GARP with the minute means available for vital ground data.

3. Remote Ocean Sensing

The visible angular reflectivity [1] depends highly on waviness, probably also on foam. This simple measure of sea state might be worth checking in space data.

For 3.8 and 10 μm windows a very thin subsurface is the IR emitter of radiant heat. Excellent data of the Gulf stream temperature exist. This method is, of course, hampered by cloud interference.

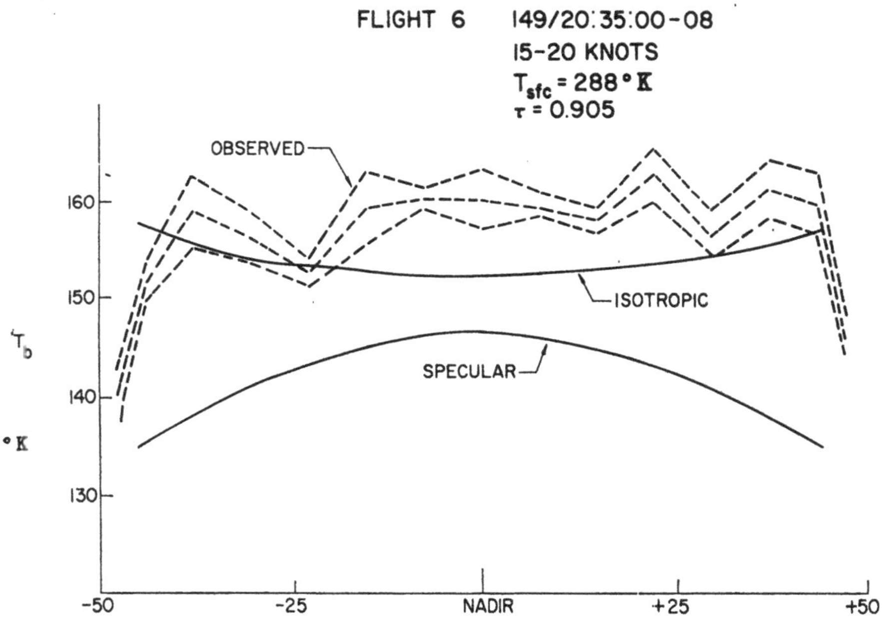

Fig. 1. Microwave 1.55 cm emission from ocean, observed from airplane. Calculated emission for specular and totally diffuse or isotropic wave structure of sea. Abscissa: nadir angle of observation. Ordinate: measured brightness temperature. From water vapor data the atmospheric transmission factor τ was calculated including the vapor emission. The difference of observed data from both theoretical curves indicates a "foam" effect of about 10 °K (*after* Kreiss and Martin).

The convenient 1.55 cm range does not yield temperature data since ocean emissivity is about inversely proportional to T. For longer microwaves, however, emissivity is constant and a Planckian signal proportional to T results. Here, however, with $\lambda = 0.2$ m and an antenna diameter $D = 2$ m ground resolution, from 1000 km altitude, would exceed 100 km. The signal would differ from the 3.8 or 10 μm signal because of the much deeper penetration of decimeter waves. A suitable compromise to get surface temperatures might be around $\lambda = 5$ cm.

Increasing sea roughness gives higher signals at 1.55 cm. This may be caused by large waves, capillary waves, foam and other surface factors. Water reflectivity increases with angle of incidence. At 1.55 cm water vapor emission and absorption is not negligible; at large zenith angles it may be substantial in a moist atmosphere. Detailed calculation by Kreiss and Martin [2] shows a more complicated pattern of expected roughness effects than that of Strogryn [3]. If (Fig. 1, 2) the expected specular or isotropic scattering from a rough sea for a given vapor content is deducted the effect of foam or other factors can be determined. In the case of Fig. 1 at nadir the "foam" component corresponds to 10 °K. Neither the electric qualities of foam nor its relation to sea state are well known. Theory assures us of foam as a rather black body. Foam formation under given wind stress seems to depend on salinity and amount and kind of organic matter [4–10]. A correlation between air or space-borne visible albedo and 1.55 cm albedo in microwaves could be checked.

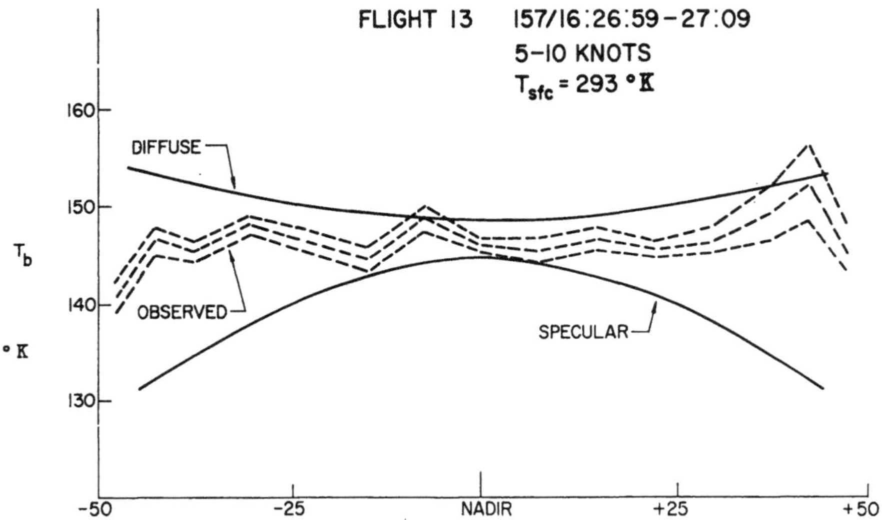

Fig. 2. As Fig. 1 but for a quiet sea. No "foam" effect.

Objects floating on sea, large enough to be spotted from space, might indicate movements, if spotted at short intervals. This could be done with icebergs in the visible and infrared, also with microwaves if resolution suffices. Oil slicks reduce sea state and increase emissivity for 1.55 cm; the two effects may cancel each other. Wind-driven changes of the front of the Gulf stream as monitored by IR could be quite discernible. Salinity changes around outflows of big rivers or behind raining clouds could possibly be measured at 7.5 cm. Here again resolution is poor. Extremely calm areas may be spotted [11] via solar glint in nearly all wavelengths at our disposal.

McAlister [12] measures subsurface ocean temperatures near 2.2, 3.8 and 4.8 μm with effective depths near 500, 75 and 25 μm respectively; assuming a constant temperature lapse rate and molecular heat conductivity the total heat flow between the respective layers can be evaluated. All tests as yet have been made at very close distances, i.e. with negligible atmospheric attenuation. The latter point is a serious obstacle for tests from spaceships.

4. Remote Sensing of Solid Ground

Weather effects on the ground may be indicated by changes of surface temperatures, of spectral solar albedo and of emissivity. The recognizable weather-connected effects are (i) change of coverage by snow, glaciers, mud, dune sand, new plants, volcanic ash, water, (ii) wetness of soil, snow, plants and opposing effects of drying out, (iii) surface temperature effects.

Solar albedo and infrared emissivity are spectrally very diversified. The visible color of a landscape is one example. Colored vidicon and astronaut

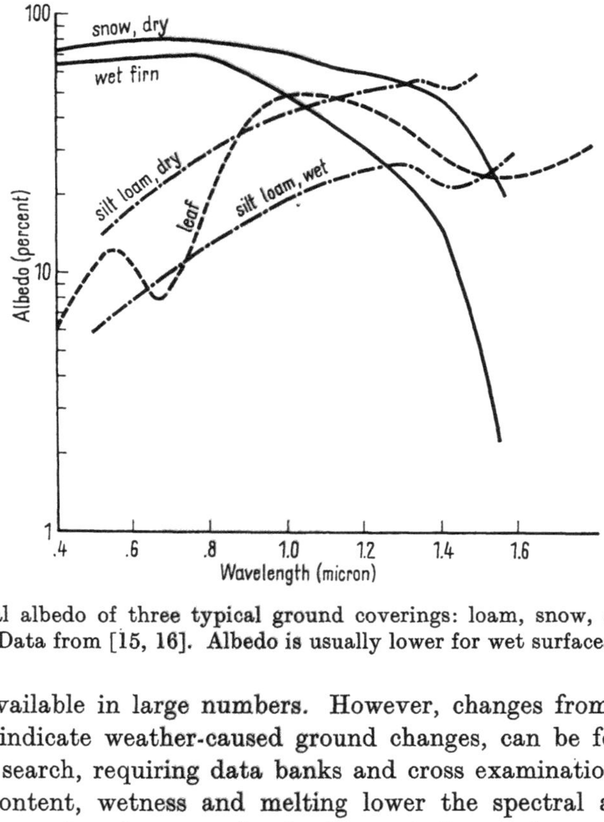

Fig. 3. Spectral albedo of three typical ground coverings: loam, snow, and vegetation. Data from [15, 16]. Albedo is usually lower for wet surfaces.

photos are available in large numbers. However, changes from day to day, which could indicate weather-caused ground changes, can be found only by a systematic search, requiring data banks and cross examination.

Moisture content, wetness and melting lower the spectral albedo of the three most frequent surfaces: sand and soil, vegetation, and snow. Sand albedo is much lower when wet; albedo for dry and wet sand increase sharply with zenith distance [13]. These three types, however, have very distinct spectral albedoes at the 0.7, 1.1 and 1.55 μm window (Fig. 3). It should be possible, then, to monitor the three types by a color scheme responding to the curves of Fig. 3. Then data should be compared via a data bank. Use alone of one shorter wavelength (anywhere from visible to 1.1 μm) and of 1.5 μm should enable us to differentiate thick water or ice clouds from snow. Whereas clouds are more or less bright in all these regions, fresh dry and even more wet snow and firn are bright in the visible but very dark at 1.5 μm.

The use of the 1.5 μm window, where snow is rather dark, is possible only because the absorption maxima of water vapor, water and ice are sufficiently apart (Fig. 4).

For the visible to the 10 μm area, water has a very low albedo. However, that of dense foliage is close to that of water in the blue and red. Water shows easily in the 1.1 μm window now used in HRIR of Nimbus 3 and 4. The reason is the higher albedo of dry soil and of vegetation at the 1.1 μm compared with the visible.

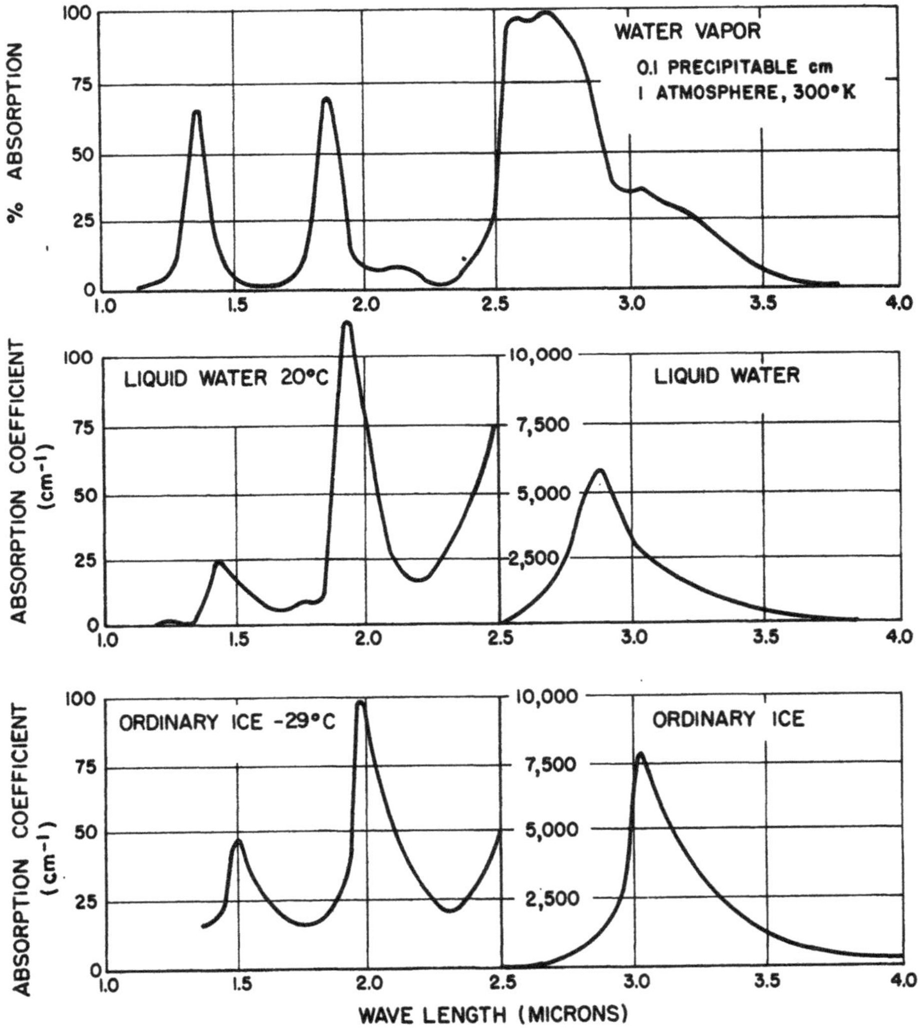

ABSORPTION PROPERTIES OF WATER AND ICE

Fig. 4. Absorption by 0.1 cm precipitable water of water vapor and absorption coefficient for water and ice. At 1.5 μm we find a usable vapor window but maximal absorption in ice. Data compiled by Blau and Espinola [17] from [18–21].

Nimbus 1 HRIR seems to have by accident at daytime a direct solar ground-reflected signal bringing this system close to the daytime Nimbus 3 procedures. The daytime maximum lies near 1.1 μm where greenery is much brighter than water. This might explain why the broad swollen Orinoco is clearly discernible in Fig. 1 of *Nimbus 1 High Resolution Radiation Data Catalog and User's Manual, Vol. I* (Goddard Space Flight Center).

After correcting for atmospheric interference, the 3.8 and 10–11 μm signals depend on subsurface temperature T and emissivity ε. The previously used 8–14 μm window shows trouble from O_3 and from quartz residual rays. But even the 10–11 μm has some ε problem. A surface having, at 10 μm, an $\varepsilon = 0.9$ shows 7 °K deviation; one having, at 3.8 μm, an $\varepsilon = 0.6$ shows 12 °K deviation. These cases are excessive but ε data are vital for evaluating T which in turn might give more insight in the radiant heat balance.

If there were a pure radiation climate with a surface heat flow $H = H_0 \sin \omega t$ where $\omega = 2\pi (24 \text{ hr})^{-1}$, then surface temperature would vary as $T = \sin \omega t \times \times H_0/(k \varrho c \omega)^{1/2}$ with k = conductivity, ϱ = density and c = specific heat at the surface. With $H_0 = 0.12$ cal cm^{-2} min^{-1} one arrives at $\Delta T \approx 10°$ for rocks and wet sand, $\Delta T \sim 100°$ for dry sand. A difference of many degrees between noon and midnight temperatures is found (Pouquet). However, Table 1 indicates variations of ε (for 3.8 μm) which could cause indicated T variations of equal size.

Table 1
Emissivity and temperature deviation

	3.8 μm		10–11 μm	
	ε	ΔT	ε	ΔT
NaCl	0.30	30	low	large
Granite powder	0.37	26	0.93	5
$CaCO_3$	0.55	13	–	–
Beach sand	0.60	11	0.90	7
Dunite powder	0.67	9	0.95	3
Grassland	0.76	6	high	–
Dry lake	0.87	3.0	high	–
Gypsum sand	0.89	2.5	?	–
Basalt powder	0.97	0.7	0.97	2
Water, ice	0.98	0.5	0.99	0

ΔT indicates the error in indicated temperature for a given ε. Calculation for a surface of 290 °K. Spectral albedo data after W. A. Hovis of NASA, Goddard Space Flight Center.

Measuring subsurface temperatures of supercalm ocean areas and of quiet lakes using 3.8 μm IR could be rewarding.

Microwaves of the 1.5 cm variety show ε near unity for nearly all dry earth coverings and $\varepsilon = 0.4$ for liquid water. A drawback is the rather deep penetration, except for water. Soil moisture tests should be compared with those mentioned above. Microwaves, however, "look through" most dry vegetation. The reference depth of signals received from dry snow is of the order of decimeters or more at 1.55 cm wavelength.

5. Resulting Proposals

The following requirements for future action may be listed:

(i) Introduction of an international sea-going pyranometer net and its use together with space data in order to ascertain solar tropospheric absorption.

(ii) Adding existing green–red and HRIR daytime (1.1 μm) albedo data to those in the 1.5 μm window.

(iiia). A data bank for surface albedoes and emissivities covering all solid earth surface areas. Search for short or long period changes in certain areas.

(iiib). A data bank for thermal surface contact coefficient.

(iv). Comparison of sea state data from microwaves at different wavelengths and angles of incidence with direct ground truth observations and visible data.

(v). Comparison of surface measured (albedo, see Fig. 3) wetness with that from 1.55 cm emissivity and ground truth.

Acknowledgments

The work described was supported by NASA Grant No. NSG 632. The International Science Foundation helped defray travel expenses.

References

[1] C. Cox and W. Munk, J. Opt. Soc. Amer. **44**, No. 11 (1954).
[2] W. Kreiss and L. Martin, personal communication.
[3] A. Stogryn, IEEE Transactions on Antennas and Propagation, AP-15, No. 2, March 1967.
[4] J. C. Aukland, P. J. Caruso, Jr. and W. H. Conaway, paper presented at Sixth Symposium on Remote Sensing of Environment, University of Michigan, Ann Arbor, Michigan 1969.
[5] J. D. Droppleman, J. Geophys. Res. **75**, No. 3 (1970).
[6] E. C. Monohan, J. Atmos. Sci. **26** (Sept. 1969).
[7] W. Nordberg, J. Conaway and P. Thaddeus, NASA preprint X-620-68-414, Nov. 1968.
[8] H. Seilkopf, Der Seewart, Hydrographisches Institut und Wetterdienst **17**, No. 6 (1956).
[9] G. F. Williams, Jr., J. Geophys. Res. **74**, No. 18 (1969).
[10] Yasuo, Miyake, and Abe Tomosaburo, J. Marine Res. **7**, No. 2 (1948).
[11] McClain and A. Strong, Mon. Weath Rev. **97**, 875 (1969).
[12] E. D. McAlister, Appl. Optics **3**, 609 (1964).
E. D. McAlister and W. McLeish, J. Geophys. Res. **74**, 3408 (1969).
[13] K. Buettner and E. Sutter, Strahleuther **54**, 156 (1935).
[14] J. Pouquet, NASA Tech. Note, NASA-TN-D-4647, Dec. 1968.
[15] S. A. Bowers and R. J. Haks, Soil Science **100**, No. 2 (1965).
[16] J. Dirmhirn, Wetter und Leben **9**, No. 3–5 (1957).
[17] H. H. Blau and R. P. Espinola, Infrared Spectral Properties of High-Altitude Clouds, Arthur D. Little, Inc., Cambridge, Mass. 1969.
[18] P. J. Wyatt, V. R. Stull and G. N. Plass, Appl. Optics **3**, 229 (1964).
[19] J. A. Curcio and C. C. Petty, J. Opt. Soc. Amer. **41**, 302 (1951).
[20] L. D. Kislovskii, Opt. i. Spekt. **7**, 201 (1959).
[21] N. Ockman, Advanc. Phys. **7**, 199 (1958).

Space Research XI — Akademie-Verlag, Berlin 1971

THE RESULTS OF MICROWAVE SOUNDING OF THE EARTH'S SURFACE ACCORDING TO EXPERIMENTAL DATA FROM THE SATELLITE COSMOS 243

A. E. Basharinov, A. S. Gurvich, S. T. Yegorov, A. A. Kurskaya, D. T. Matvyev and A. M. Shutko

Academy of Sciences, Moscow, USSR

The results of the experiment of microwave emission measurements from the satellite Cosmos 243 with four radiometers at wavelengths of 0.8, 1.35, 3.5 and 8.5 cm are presented. The surface radiation at wavelengths 3.4 and 8.5 cm has been used for estimation of temperature and for determination of surface conditions, and the radiation at wavelengths 0.8 and 1.35 cm for estimation of the temperature distribution beneath the surface and determination of the influence of the atmosphere. The accuracy of sea surface temperature measurements was 1–2 °K.

The intensity of microwave emissivity variations of the sea surface has been estimated. The boundary of the domains of floating ice near the Antarctic continent has been determined. The radiobrightness variations dependent on latitude and local features for Australia, Antarctic continent, Sahara zone and others have been detected and the decrease of radiobrightness temperature and emissivity to 20–30% in the zones of increased humidity determined.

Microwave emission measurements were carried out on 23–26 September 1968 from the satellite Cosmos 243 (height of orbit 210–213 km, inclination of orbit 71.3°). Four radiometers at wavelengths of 0.8, 1.35, 3.4 and 8.5 cm were used with vertically oriented antennae axis [1]. The linear size of the instantaneously observed field was some tens of kilometres.

The surface radiation at wavelengths of 3.4 and 8.5 cm was used for the estimation of temperature and the determination of surface conditions.

Radiation at wavelengths of 0.8 and 1.35 cm was used for the estimation of the temperature distribution beneath the surface and determination of the influence of the atmosphere. The calibration of the radiometers has been accomplished by the use of black-body and auxiliary regions on the surface with known radiobrightness characteristics.

Fig. 1 shows radiobrightness temperature variations on 23 September 1968 over the Pacific Ocean on channels 8.5 and 3.4 cm. The measured data are compared with the calculated values of radiobrightness temperatures for a smooth sea surface and mean climatic conditions (broken curves).

According to Fig. 1 the data obtained at 8.5 cm fit the calculated longitudinal distribution of radiobrightness temperature of the ocean surface better than those at 3.4 cm.

Large variations of radiobrightness temperature (20–25 °K) at wavelength 3.4 cm between 10° N and 20° N due to emission caused by precipitation can be observed, while the variations at 8.5 cm in these regions do not exceed 2.5–3.5 °K. Increase in the ocean surface emissivity at wavelengths 8.5 and

3.4 cm is detected in the regions 30°–35° N and 45°–60° S. According to the data obtained at 0.8 cm there were weak clouds only with no precipitation over the region in the southern hemisphere. It can be assumed that the increase in radiobrightness temperature in these regions was caused by the

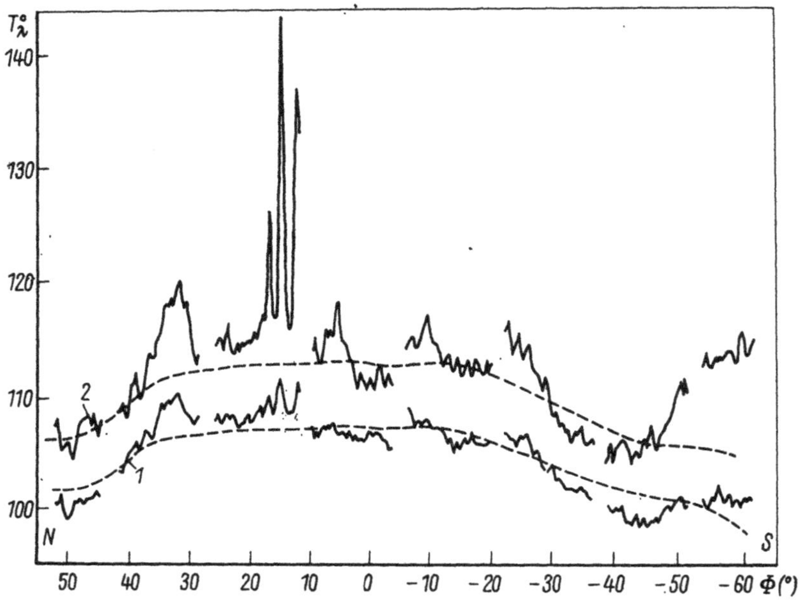

Fig. 1. Radiobrightness temperature variations over the Pacific Ocean. *1*, $\lambda = 8.5$ cm; *2*, $\lambda = 3.4$ cm.

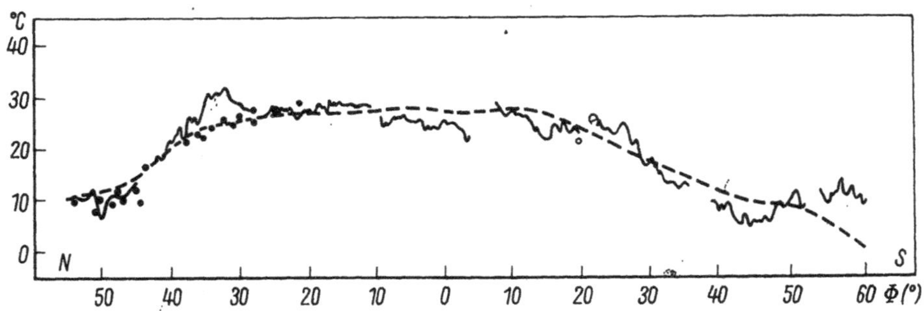

Fig. 2. Surface temperature of the Pacific Ocean. Broken line, mean climatic data; points, ship measurements.

variation of surface emissivity because of the rough sea. This assumption is confirmed by the spectral dependence of emissivity variation.

Fig. 2 shows the values of the oceanic surface temperature estimated from the experimental data with correction for absorption in precipitation. The estimated data are compared with the mean climatic temperatures and the data measured by ships. The estimated values of the ocean surface temperature include errors due to variations of surface emissivity, influence of the atmosphere and instrumental errors. The effective value of this error, with the exception of stormy areas, is about 1–2 °K.

The data given in Fig. 2 show that the measurements at wavelength 8.5 cm express the latitudinal temperature dependence of the ocean surface sufficiently well. The radiobrightness temperature variations in the regions 30°–35° N and 45°–60° S were caused by variations in ocean surface emissivity.

In the high latitude regions (Antarctic continent, Greenland) the boundary of the zones of floating ice and the degree of packed sea ice have been deter-

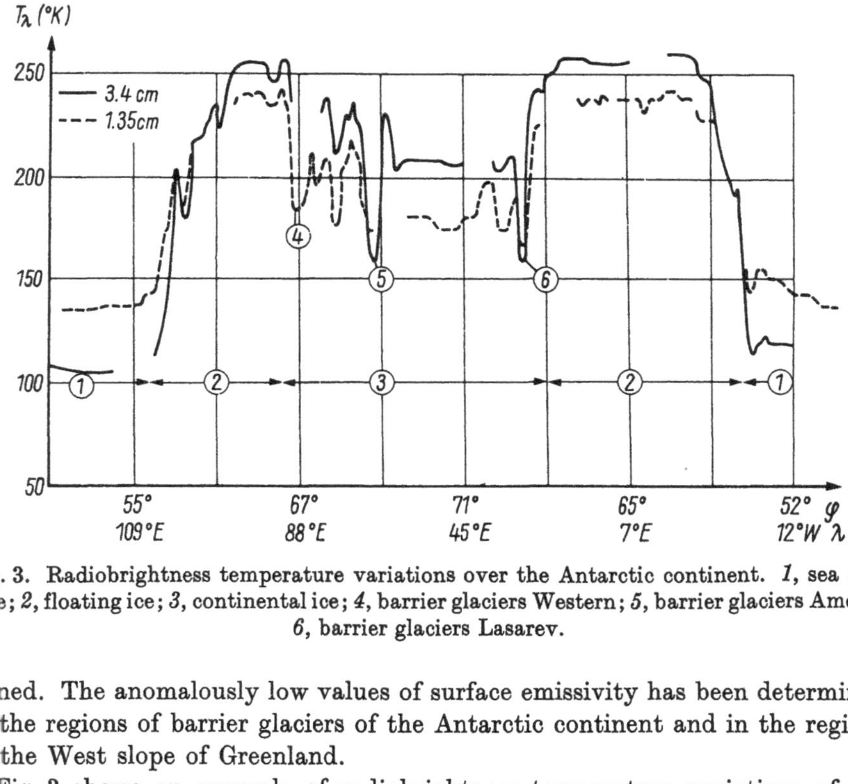

Fig. 3. Radiobrightness temperature variations over the Antarctic continent. *1*, sea surface; *2*, floating ice; *3*, continental ice; *4*, barrier glaciers Western; *5*, barrier glaciers Amery; *6*, barrier glaciers Lasarev.

mined. The anomalously low values of surface emissivity has been determined in the regions of barrier glaciers of the Antarctic continent and in the regions of the West slope of Greenland.

Fig. 3 shows an example of radiobrightness temperature variations of the Antarctic continent. Latitudinal variations of temperatures have been found. The latitudinal and local radiobrightness temperature variations, associated with surface temperature variations, were detected over such homogeneous areas as Australia, Sahara etc.

The errors of the estimated values of the land surface temperature are caused by emissivity variations, variations of the effective depth of the emissivity layer and instrumental errors. The effective value of this error over homogeneous surfaces is about 3–4 °K.

The profile of the radiobrightness temperature obtained above Australia is given in Fig. 4. The observed difference of the radiobrightness temperature over Australia between southern and northern coasts was about 16–18 °K, while the difference of air temperatures in these zones was 21 °K.

The radiobrightness variations due to the relief profiles and atmospheric temperature gradient have been determined in mountain regions (Himalayas, Andes, etc.).

The radiobrightness temperature decreases as the field of observation moves upwards. A decrease in the radiobrightness temperature in the humid zones

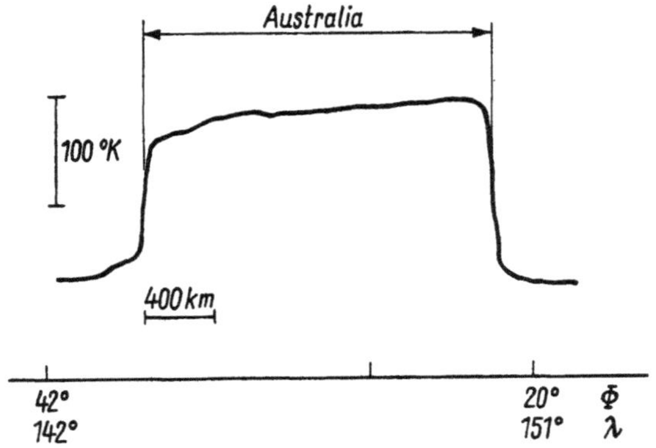

Fig. 4. Radiobrightness temperature profile over Australia (24 September 1968).

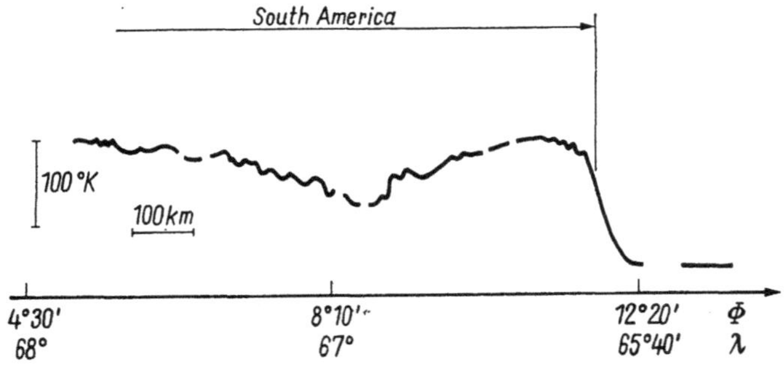

Fig. 5. Radiobrightness temperature profile over South America (24 September 1968).

and swampy regions in North America, north regions of South America, Central Africa etc. has also been observed.

Fig. 5 shows the radiobrightness variations over South America. The decrease of surface emissivity up to 60–70% is due to the swampy region of the Orinoco river.

References

[1] A. E. BASHARINOV, A. S. GURVICH and S. T. YEGOROV, Dokl. Akad. Sci. USSR 188, 1273 (1969).

[2] A. E. BASHARINOV, S. T. YEGOROV, A. S. GURVICH and A. M. OBOUKHOV, Space Research XI, Akademie-Verlag, Berlin 1971 (p. 593).

Space Research XI — Akademie-Verlag, Berlin 1971

VISIBLE POLARIZATION SIGNATURE FOR REMOTE SENSING OF SOIL SURFACE MOISTURE

L. Steg and R. T. Frost

Space Sciences Laboratory, General Electric Company, Philadelphia, Pa, USA

Laboratory and field measurements of the visible polarization signature and its relation to surface moisture of soils are reported. Results indicate that the enhancement in polarization allows quantitative determination of surface moisture. The signatures are invariant to partial shadowing encountered in plowed or rough soils and show little dependence on soil type. Calculations based on the measurements indicate that moisture discrimination is possible in the visible region when viewing through the atmosphere, including standard haze models.

Remote detection of soil surface moisture is a problem of considerable importance to agriculture and meteorology. In agriculture, surface moisture is of importance for planning of distribution of fertilizer inventories to anticipate local demand, and to seed germination and early crop growth. This parameter remains of interest later in the growing season while some inferences can still be made of moisture at root depth. For bare soils surface moisture is also of interest in relation to wind erosion and soil mapping by color photography. For meteorology, the latent heat of vaporization of surface moisture is a dominant factor in energy exchange at the lower boundary of the atmosphere.

The large increase in polarization due to wetting is accompanied by a reduction in reflectivity and, in principle, either property could be utilized for discrimination of dry and moist soils. In practice, the albedo of moist soils depends a great deal on soil type so that albedo measurements would need to be accompanied by *a priori* knowledge of the type of soil being observed.

Fig. 1 shows the variation of relative intensity observed at 0.50 μ as a function of view angle when sand or loam are illuminated at a zenith angle of 55°. The large decrease in reflectance due to wetting can be used to detect wet soil, if the soil type is known. The reflectances of wet sand and dry loam are not widely different for viewing directions near the normal and, without knowledge of the soil type being observed, discrimination would not be possible.

For remote sensing applications, it is necessary to consider degradation of ideal target signatures due to effects that will be encountered in practice. Besides soil type and the effects of atmospheric scattering, other imprecisely known variables may be encountered. In attempting to discriminate on the basis of albedo, one may have no *a priori* knowledge of and no way of detecting the average illuminating spectrum over the target area. Other unknowns might be the slope of the ground surface being observed and partial shadowing due to lumpy or plowed soil. Useful signatures will depend little on such parameters which may be unknown or of secondary interest but ones which instead depend primarily upon the parameter to be measured.

Fig. 1 also shows the dependence of degree of polarization on view angle for sand and loam in both wet and dry conditions. The wet samples shown

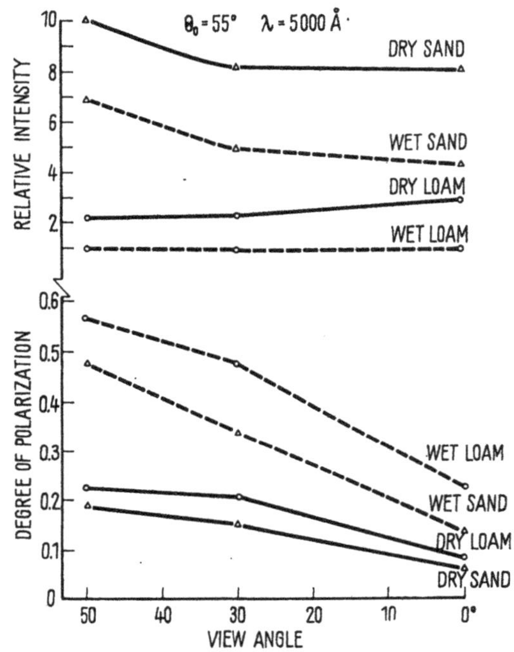

Fig. 1. Relative reflectances for wet and dry loam and sand surfaces (summer, 1968, Valley Forge, Pennsylvania).

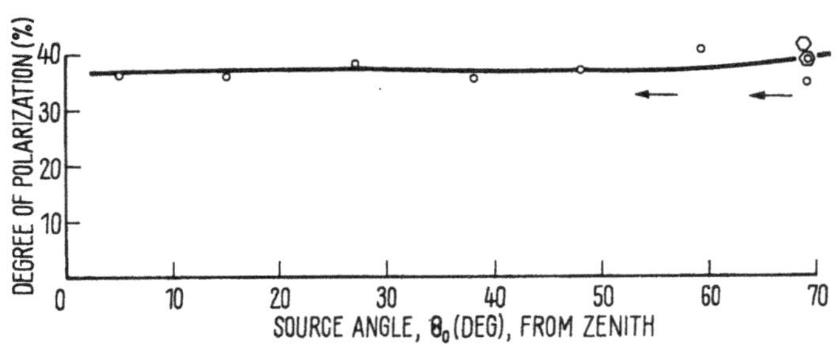

Fig. 2. Polarization of scattered light vs source angle for constant phase angle 68 ± 2° for Clarksville, Virginia, soil 18% water by weight (4000 Å to 5800 Å) (June 1969, Valley Forge, Pennsylvania).

here are nearly saturated but gave no appearance of specular reflection. The illumination was at a zenith angle of 55° and the measurements performed at 0.5 μ. At reasonable view angles out of the vertical, polarization depends much more on the difference between dry or wet condition than upon soil type.

Fig. 2 shows for a Hiwasse soil (Clarksville, Virginia) at 18% moisture content the variation of polarization with illuminating angle. Here the viewing angle is varied in such a way that the phase angle remains constant at 68°. The change of polarization with illuminating angle is hardly outside the experimental errors of 1–2%. The albedo signature, on the contrary, would vary markedly over this extreme range of illuminating angles.

A photograph of the specimen at an illuminating angle of 68° shows that a large fraction of the material viewed is in shadow. The relative independence

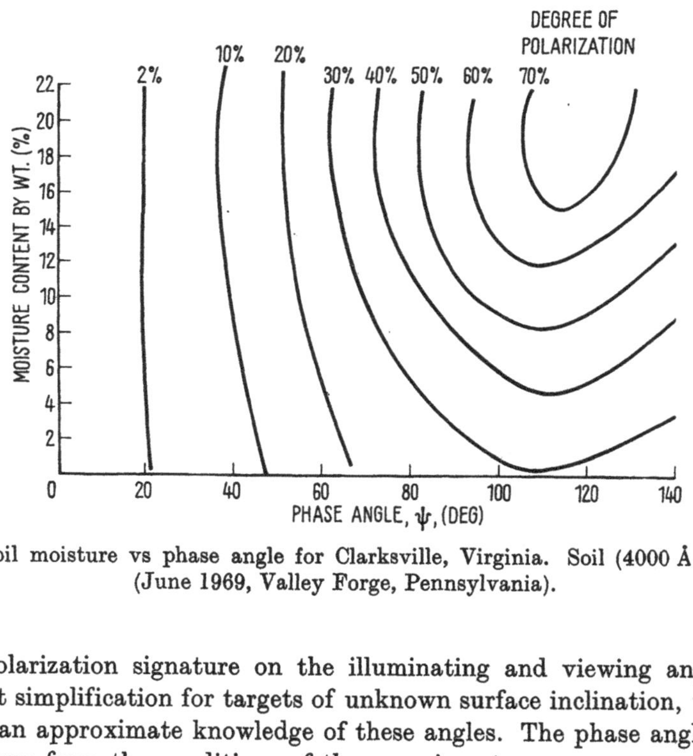

Fig. 3. Soil moisture vs phase angle for Clarksville, Virginia. Soil (4000 Å to 5800 Å) (June 1969, Valley Forge, Pennsylvania).

of the polarization signature on the illuminating and viewing angles is an important simplification for targets of unknown surface inclination, where one has only an approximate knowledge of these angles. The phase angle is, however, known from the conditions of the experiment.

Since the polarization signature depends, for a given soil type, primarily on phase angle and moisture content, it is possible to plot the results of a large series of measurements on Hiwasse soil as shown in Fig. 3. The number of measurements was approximately 400 and the average scatter from the curve shown is approximately 1–2% in polarization. Moisture contents were determined by standard oven drying techniques. The contours shown are for a fixed polarization. For a phase angle of 70° or more, the degree of polarization is a good measure of the moisture content. The dependence of polarization on moisture content is not very far from linear over a 20% range. For smaller phase angles, or for moisture content above 20%, the contours are nearly vertical, which indicates that undue error sensitivity will be encountered. However, most soils are saturated at higher moisture contents.

The laboratory measurements on the soil samples discussed, and others, indicate that in principle the visible polarization signature is adequate for reliable surface moisture detection.

Some measurements in the field on actual representative agricultural targets have also been made. For this purpose a portable 40 foot tower facility was used. The photopolarimeter was boresighted with a vidicon camera so that the particular field area being measured could be viewed in a van containing recording instruments at the tower base. Changes of polarization signatures at various phase angles from both naturally and artificially moistened soils

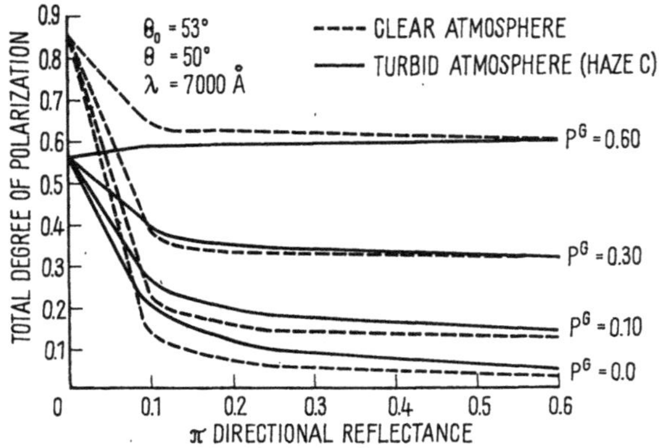

Fig. 4. Polarization of total outgoing radiation as a function of surface reflection and polarization (Fall 1968, Valley Forge, Pennsylvania).

were measured. Plots similar to Fig. 3 were obtained on a sandy loam soil in Clarksville, Virginia. Average scatter was 2–4%. These measurements were made at 0.643 μ.

These signatures will, of course, be affected by atmospheric transmission. The light emerging from the top of the atmosphere which has been reflected by the ground target will contain intensity contributions due to molecular and haze scattering. The intensity contributed by atmospheric scattering can be taken from the tables of atmospheric scattering by Coulson, Dave and Sekera. The intensity of light scattered by atmospheric haze is based on a single scattering model. The intensity and polarization at the top of the atmosphere are obtained by adding the separate intensities due to light scattered from the target and attenuated by the atmosphere, light scattered from the molecular atmosphere and light scattered by atmospheric haze.

Fig. 4 shows how the degree of polarization observed at the top of the atmosphere depends separately on the polarization signature P^G at ground level and the target reflectivity for typical viewing conditions. For targets with high reflectivity the polarization observed at the top of the atmosphere is not very different from that which would be observed at ground level. For targets of lower reflectivity, the polarizing influence of the atmosphere becomes more important. For a clear atmosphere the polarization increases only slightly unless the target reflectivity drops below 10% when the atmosphere begins to have a major effect. If we add the scattering due to Deirmendjian's Model C haze the polarization again changes slowly until the target reflectivity becomes very low. Although corrections can be made for the effects of the atmosphere, the corrections must not be too large or unique recognition of signatures will be lost due to uncertainties in the haze contribution. To apply this diagram to the case of a surface moisture polarization signature, consider a soil of reflectivity 0.2 when dry which has a polarization of only a few per cent. When the sample becomes moist its reflectivity will decrease and its polarization will increase. Typical change in the reflection might be from 0.2 to 0.1 with the polarization increasing from 10% to 20 or 30%. Thus, the discriminating power of the polarization measurement is not lost when the target is viewed through the atmosphere and haze. The method would be limited for targets of very low reflectivity. Fortunately, most soil reflectivities are above 10%.

More exact calculations of emergent intensity and polarization can be made to include illumination of the target illumination by skylight and other scattering processes, if appropriate.

Space Research XI — Akademie-Verlag, Berlin 1971

THE USE OF INFRARED IMAGERY FROM METEOROLOGICAL SATELLITES FOR THE STUDY OF THE EARTH'S SURFACE

A. A. Grigoryev, V. B. Lipatov and B. V. Vinogradov

Leningrad University, Leningrad, USSR

An example is given of integrated interpretation of a matrix-map of effective temperatures and of an infrared picture (8—12 μm) of the area of middle Asia and Kazahstan obtained from the meteorological satellite Meteor 2. The thermal nonuniformities of different areas of the earth's surface and of cloudiness are seen in the infrared picture as spots of different tones, sizes and configuration. Analysis shows some of the factors influencing the thermal patterns of different natural formations.

The possibility of using infrared survey from satellites for the investigation not only of the atmosphere but also of the earth's surface is reported by Buettner [1] and Vinogradov [2]. We can give as a meteorological example the integrated interpretation of an infrared picture obtained from a meteorological satellite of the Meteor series. This method opens up additional opportunities of investigating atmospheric processes by meteorologists: the investigation of temperature, humidity and transparency of the lower atmosphere, formation of fog and cloud depending on the variation in the underlying surface (relief, vegetation cover, their humidity and thermal and physical properties, their reflectivity and emissivity). An infrared photograph was taken of the region in middle Asia and Kazahstan from the meteorological satellite Meteor 2, on 12 October 1969 about 0230 Moscow time in the wavelength range 8–12 μm. In the analysis we made use of a matrix-map of effective temperatures of the earth's surface and cloudiness as well as of all available meteorological data. Thermal nonuniformities of different areas of the earth's surface and cloudiness are seen in infrared pictures as spots of different tones, sizes and configuration (Fig. 1). In the region of middle Asia and Kazahstan the most intensive dark colour represents water of the Aral and Caspian Seas which are warmer than the land. The effective water temperatures in the Aral Sea and the northern Caspian Sea are 9–12 °C. The Caspian and the Aral Seas are partially covered by low, non-dense stratus clouds giving a lighter tone in different water regions (e.g. southern part of the Aral Sea, northern part of the Caspian Sea). These are also determined by actinometric data. The coast of the eastern Caspian is difficult to see, as it is masked by fog (strongest over the Kara Bogaz Gall strait). In the infrared picture it does not differ in tone from the land since the temperature of the fog is close to that of the earth's surface. This fact suggests that what we have in the picture is either low cloud or fog (the meteorological station of the Eastern Caspian reported either dense fog or haze).

The reason for the fog formation is the passage of a warm front during the night which is confirmed by the analysis of the synoptical situation. The arrival of warmer masses on the plateau, which cooled at night because of the comparativly small heating capacity of the rocks, caused the formation of advective fog. However, in spite of the masking of water by cloud and fog we can observe nearly the entire coast line. The zones of distribution of water

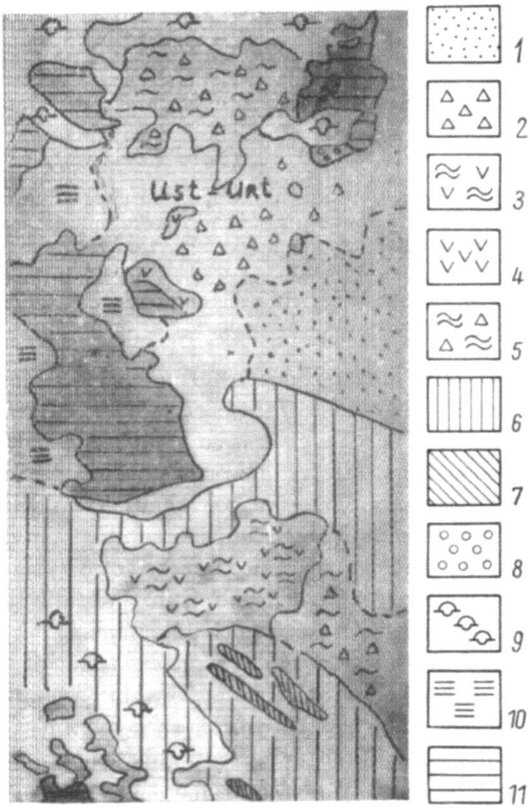

Fig. 1. The image of the region in Middle Asia and Kazahstan obtained from the meteorological satellite Meteor-2 on 12 Oct. 1969 about 02.30 Moscow time and the results of the integrated interpretation: *1*, landscapes of sand deserts; *2*, landscapes of stony plateau; *3*, landscapes of clay–salt deserts; *4*, landscapes of salt deserts; *5*, landscapes of clay–rubble deserts; *6*, landscapes of mountains; *7*, landscapes of intemontane depressions; *8*, landscapes of deltas; *9*, cloudiness; *10*, fog.

masses of different temperatures are well pronounced: they are seen by the increase in temperature and the darkening of the tone of the picture in the direction from north to south. Thus the effective temperature in the middle part of the Caspian and the Kara Bogaz Gall Gulf is higher (13–15 °C), and in the southern part (tone darkish grey) is 16–18 °C. Still moving to the south, at latitude 30° the effective temperature of the water in the Persian Gulf is about 20–26 °C and the image tone is dark grey. The northern part of Kara Bogaz Gall is well identified: the part of the bottom deprived of water and constituting the plain composed of salt rock is represented by a grey tone contrasting with the background image of Ust-Urt.

It is interesting to note that the clay-rubble surfaces of Ust-Urt are well identified. This plateau which is composed of nearly horizontal sedimentary rock forms a flat plain with rocks covered with thin eluvium–deluvium loam with pebble. In conditions of open vegetation (combination of sagebrush and salt deserts) and deep ground waters the temperatures of the surface of this type of landscape were determined by the ground temperature. In tone, Ust-Urt is light grey, much lighter than the northern part of the Kara Bogaz Gall. The effective temperature is 5–8 °C. The landscapes of the sand desert of the Kara Kum and Kyzil Kum situated to the east and south-east of Ust-Urt are quite distinct from the Ust-Urt plateau. Since the vegetation here is also open and the ground waters comparatively large the thermal image of the landscape in this case is also influenced by the temperature of sand surfaces. These deserts are represented in the thermal image as a grey tone much darker than Ust-Urt, due to the higher ground temperature. The effective temperature is 9–12 °C, which in turn, is due the higher heat capacity of sand surfaces.

The Sarykamyshskaya depression is represented by a darkish grey tone which is well pronounced against the lighter background of the sand desert. Its relative warmness is due to the extensive distribution of salt soils and the presence of lakes.

The mountain ridges of Elburs and Zagros (Iran) are partially covered with clouds the image tone of which is very light (low temperatures are due to the presence of cloud, snow spots and to the altitude of the ridges).

The analysis of the data of infrared earth imagery from satellites reveals some thermal features of different landscapes (types of underlying surfaces). The detection of thermal characteristics of various landscapes reveals some peculiarities of meteorological processes associated with the features of the underlying surface.

References

[1] K. J. K. Buettner, Space Research XI, Akademie-Verlag, Berlin 1971 (p. 705).
[2] B. V. Vinogradov, Space Research XI, Akademie-Verlag, Berlin 1971 (p. 525).

Space Research XI — Akademie-Verlag, Berlin 1971

STUDY OF THE CONNECTION BETWEEN THE CHARACTERISTICS OF THE EARTH'S SURFACE AND SOME METEOROLOGICAL ELEMENTS FROM THE GLOBAL PICTURES FROM THE AUTOMATED INTER-PLANETARY STATION ZOND 7

A. A. GRIGORYEV, V. B. LIPATOV and B. V. VINOGRADOV

Leningrad University, Leningrad, USSR

The interpretation of colour space photographs obtained from the Soviet automated station Zond 7 (8 August 1969) has disclosed two gigantic dust and sand movements ("dust–sand flows") on the Arabian peninsula. One of these dust–sand flows is associated with the trade wind movement, and the second with monsoon circulation. It is found that the large-scale relief of the Arabian peninsula, Asia Minor and Iran influences the orientation of these movements. The integrated analysis of this space photograph reveals also a close correlation between the orientation of dust–sand flows and the direction of the longitudinal dunes (seifs) of Arabia.

The general picture of the application of various aspects of remote sensing of the earth's surface is given by Buettner [1] and by Vinogradov [2]. These papers show the general principles of the integrated interpretation of space pictures and the interrelation between atmospheric physics and land structure. The analysis of the global photograph from the Soviet automated interplanetary station Zond 7 is a useful example developing the suggestions of these papers.

One of the most important directions in the meteorological investigations is a study of the interrelationship between atmospheric processes and the relief of the surface. The character of the atmospheric fluxes, turbulence, humidity, temperature and pressure distribution as well as atmospheric transparency are related to the type of underlying surface (water or land, covered or not with vegetation, the composition of the land). Space pictures open up new possibilities for investigation of the relationship of meteorological elements to the relief of the terrain.

As meteorological phenomena the dust–sand flows (red and yellow dust storms) are fairly characteristic of the territiory of Arabia, but because of the small number of meteorological stations the scarcity of data and the lack of correlation between the observations of different investigators, the general regularities of the distribution of dust–sand flows over the territory of Arabia and, in particular, their relationship with the orientation of the relief is still not quite clear. Now, by space surveys from very high altitudes it is possible to obtain not only visual confirmation of the existence of the dust–sand flows in this area of the globe but also objective data, on a regional scale, of the close dependence of the movement of low air on the relief. From the fragment of the very small-scale colour picture of the European and Asian continent and of the north-eastern part of the African continent as obtained on 8 August 1969 from the space automated station Zond 7 two large-macroscale dust–sand flows have been located over the territory of the Arabian peninsula and adjacent regions (Fig. 1). These are particularly well identified over the water of the Red Sea and the Persian Gulf in the form of whitish bands distinctly seen against the warm background of the water. The flow over the land is less distinctly seen. The interpretation of the movements of the dust–sand flows from the space photograph was followed by an analysis of the meteorological observations at ground stations and an analysis of the synoptic situation on the same day.

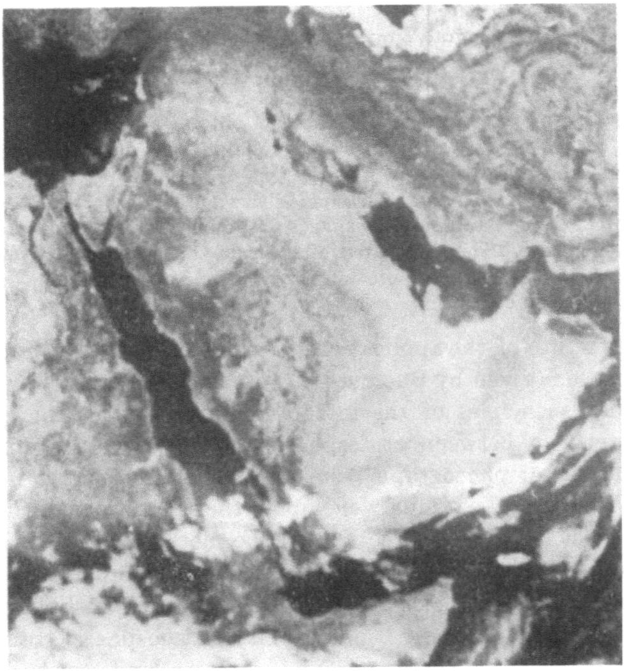

Fig. 1. Space photograph obtained from the automated station Zond 7 on 8 August 1969.

One of the dust flows moving from the Mediterranean over the northern part of Arabia, from north-west to south-east, is associated with trade wind movement in this direction (Fig. 2). Farther towards the south-east it enters the circulation system associated with a thermal depression with centre over the south of Iran. The dust–sand flow moves as if it were in a wide passage which is confined in the north by the mountains of Asia Minor and Iraq and on the south by the plateau and mountains of Saudi-Arabia. It is distorted by the high land of Central Arabia and enters a narrow passage between the elevations. Then it goes over the western part of the Rub el Hali desert, widens and dies away meeting an obstacle in the form of the mountains of Yemen (height 1500–3500 m). The part of this flow influenced by the cyclone over the south of Iran turns east where it meets another south-west flow. The region of meeting of these two flows and the region of wind abatement, i.e. the region of inner tropical convergence, is represented in the picture as a reddish band.

The second dust–sand flow related to the monsoon circulation from Africa, crosses the southern part of the Red Sea, where it is particularly visible, passes over the south-east part of Arabia moving north-east. Meeting high land in the north-eastern part of the Arabian peninsula (height 1500–3000 m), it is separated into two branches moving to the north, bending round the mountains from the south-west, passing over the Orluz strait (which is distinctly seen in the photo) and bending round the mountains of the Hadzhar to the east.

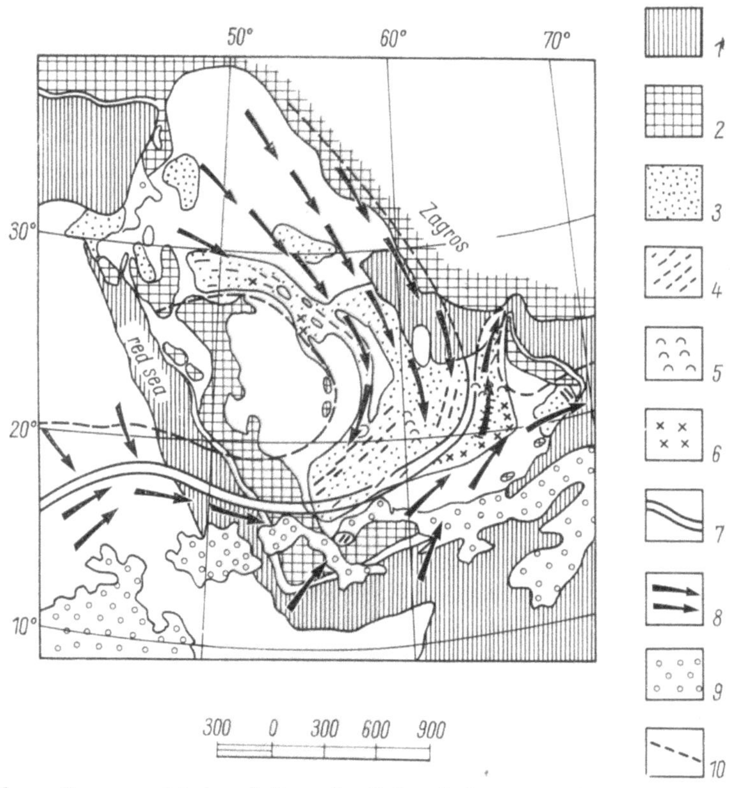

Fig. 2. Schematic map of interrelation of relief and dust–sand flows, interpreted from Fig. 1. *1*, water; *2*, mountains and plateau with heights > 1.8 km; *3*, areas of deflated sands; *4*, areas of seifs and their direction; *5*, barchanes; *6*, unoriented eolian forms; *7*, zone intertropical convergence; *8*, direction of dust–sand flows; *9*, cloud; *10*, limit of dust–sand flows.

Of equal importance is the fact that the region of movement of the flows corresponds to the region of the large sand masses of the Great Nefud and Dehna deserts and of a gigantic desert Rub el Hali. In these deserts in the region of movement of flows there exist different types of eolian relief: seifs, large longitudinal dunes, prevail. The distribution and orientation of these seifs was indicated by the interpretation of large-scale local pictures from manned spacecraft.

The comparison of the distribution of dust–sand flows and longitudinal dunes shows a good fit between the orientation of seifs and the direction of movement of vortex flows. The seifs of the Great Nefud desert oriented NW–SE are followed by the seifs of the Dehna desert the direction of which near the Rub el Hali desert gradually changes to NNE–SSW and then to NE–SW. Thus each individual seif constitutes a part of one gigantic system of seifs of common origin: namely the functioning of constant winds of stable direction. These dust–sand wind structures, shown by space photographs, are a special phenomenon of a region with active wind-eroded processes in the arid and subarid zones. Besides Arabia these are observed in the regions of the Eiden-Murzuk sands, Levi Akkle-Avana sands, in the depressions of Bowden, Chad, etc. We think that their analysis can be used not only for studying modern eolian processes but also for synoptical analysis just as clouds are used as indicators in humid regions.

References

[1] K. J. K. Buettner, Space Research XI, Akademie-Verlag, Berlin 1971 (p. 705).
[2] B. V. Vinogradov, Space Research XI, Akademie-Verlag, Berlin 1971 (p. 525).

Space Research XI — Akademie-Verlag, Berlin 1971

CLOUD CHARACTERISTICS FROM INFRARED MEASUREMENTS

H. H. BLAU, JR.[a], and W. A. HOVIS[b]

[a] Arthur D. Little, Inc., Cambridge, Massachusetts, USA
[b] National Aeronautics and Space Administration, Goddard Space Flight Center, Greenbelt, Maryland, USA

The spectral distribution of sunlight scattered from clouds in the near infrared reveals a wealth of detail determined by the scattering properties of the ice crystals and water droplets of which the clouds are composed and by the absorption properties of infrared active atmospheric gases. Important cloud physical data can be obtained from analysis of such spectra. In this paper we review theoretical and experimentally determined relationships between cloud properties and scattering behavior and consider the prospects for obtaining meteorologically important cloud physical data from a weather satellite.

1. Introduction

In this paper we examine prospects for obtaining meteorologically significant cloud physical data from remote measurements in the near infrared. This spectral region, extending from the visible red to perhaps 3 (μm), has been largely neglected for weather satellite measurements. It holds particular promise for remote sensing because of the influence of the absorption properties of water droplets and ice crystals of which clouds are composed on cloud scattering properties.

Visible images of sunlit clouds recorded from weather satellites have provided extremely valuable information on large-scale cloud distributions and circulation patterns. Together with thermal infrared maps giving information on cloud top temperatures and hence altitudes, cloud images found immediate and widespread use in areas ranging from routine forecasting to basic studies of atmospheric dynamics. Useful as they have been, visible images of themselves provide almost no compositional data. Cloud reflectivity in the visible depends on such parameters as cloud thickness, water droplet or ice crystal concentration and size distributions. However because water and ice are essentially transparent, all cloud particles scatter efficiently, and consequently there is no accurate way to relate brightness to cloud structure.

The situation is quite different in the near infrared. Here both the optical properties of water and ice and in addition the optical properties of the atmosphere are unlike those in the visible region. Near-infrared cloud spectra recorded in reflected sunlight contain a wealth of spectral detail determined largely by the absorption properties of the cloud particles and by atmospheric absorption bands due primarily to carbon dioxide, water vapor and molecular oxygen. Given this spectral detail one might hope to invert the process and retrieve at least some of the details of cloud structure.

The problem is related to that of determining atmospheric temperature or humidity profiles from infrared emission measurements that has been so beautifully solved during the weather satellite programs. Cloud particle optical constants vary with wavelength in the near infrared. Consequently cloud scattering power or reflectivity will also vary with wavelength in a manner dependent on whether the cloud particles are liquid or ice, on cloud particle size distribution, on particle concentration, and on cloud thickness. By making measurements at a sufficient number of wavelengths, in principle it should be possible to determine all of these properties.

There are, of course, major difficulties. The atmosphere is only transparent in a few narrow wavelength regions in the near infrared. Consequently we must account not only for cloud scattering properties but also for atmospheric absorption by gases entrained within the cloud and absorption by gases in the clear air above the upper cloud surfaces. Second and probably most important the detailed scattering behavior of irregularly shaped particles such as ice crystals is not at all well understood, and we need this information to make scattering calculations. In addition our knowledge of cloud microstructure as determined from direct sampling is very incomplete. We do not have good information on cloud droplet size ranges or crystal habits and size ranges for typical clouds of different types. Finally since some clouds are partially transparent we must account for ground and sea surface scattered contributions.

At this stage of development it is not realistic to expect to obtain cloud structural data from remote infrared measurements of the same quality as in remote temperature orhumidity sounding. Nonetheless remote sensing of cloud properties is a promising area and much can be accomplished even with the limited information available.

In the following sections of this paper we will examine important relationships between cloud physical properties and scattering behavior. We will then outline a simple experiment to obtain cloud property data that could be performed from a satellite. Finally we will examine infrared cloud reflectance data obtained from high altitude aircraft in the context of the remote sensing experiment.

2. Cloud Scattering Theory

The scattering properties of homogeneous, plane parallel clouds composed of absorbing spheres have been calculated both by using a Monte Carlo method [1, 2] and a layer doubling method [3]. In both cases a distribution of particle sizes was assumed.

In the Monte Carlo method the three-dimensional path of individual photons is followed mathematically through the cloud of randomly located particles as the photons are multiply scattered. The single scattering function used is obtained from Mie theory for the desired values of optical constants and averaged over the assumed size distribution. By following a sufficiently large number of photon histories statistical fluctuations in the predicted radiance are held to acceptable levels.

The phase doubling method relies on the fact that the scattering and transmission functions for an optically thin layer reduce to the Mie phase function for single scattering again averaged over the size distribution. The scattering and transmission functions for a double layer may be derived from those for the two component layers by considering what happens to the radiation scattered back and forth between the component layers expressed as repeated integrals over angle variables.

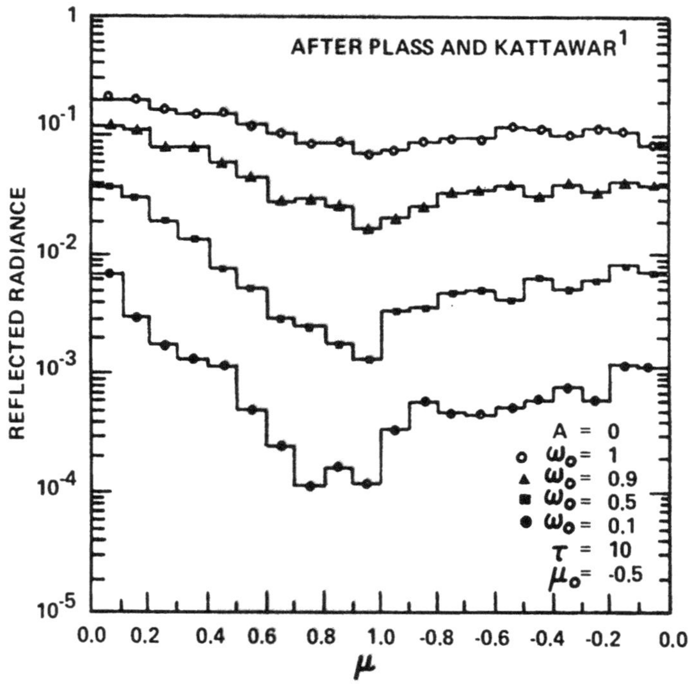

Fig. 1. Cloud radiance from Monte Carlo calculations.

We will not be concerned with the details of these methods other than to note that in general the results are in agreement with each other and with experiment.

Scattering by an idealized plane parallel cloud is illustrated in Fig. 1 which is a graph of scattered radiance on a logarithmic scale as a function of μ the cosine of the viewing angle measured from the upward drawn normal to the cloud surface. These values of cloud radiance were computed by Plass and Kattawar [1] using the Monte Carlo program. Incident solar flux at 0.7 μm was normalized to unity and the Diermendjian [4] cumulus cloud particle size distribution was used

$$n(r) = 2.37\, r^6 \exp(-1.5\, r)$$

where $n(r)$ is the particle concentration and r is particle radius. The maximum particle concentration occurs for $r = 2\,\mu$m. The optical thickness for the cloud was $\tau = 10$, the albedo of the earth $A = 0$, and the angle of illumination was 45°, that is $\mu_0 = 0.5$.

The curves in the figure correspond to different values of ω_0 the albedo of single scattering defined as the ratio averaged over the size distribution of power scattered to the sum of power scattered and absorbed per particle. There are two important points to note. First, cloud scattering is anisotropic for all values of ω_0 with forward scattering predominating. This is a result of the fact that the single scattering phase function for particles large with respect to wavelength is strongly peaked in the direction of forward scattering. This remains true even at wavelengths as long as $3\,\mu$m for all particle size distributions of importance in cloud physics.

Second, scattering efficiency decreases rapidly with descrease in single scattering albedo. For a given particle size distribution this means simply that scattering power decreases with increase in the absorption coefficient. In the figure $\omega_0 = 1$ corresponds to the case of no absorption, $\omega_0 = 0.9$ to moderate absorption, and $\omega_0 = 0.5$ to strong absorption. The case for $\omega_0 = 0.1$ results only when the absorption coefficient is very large and the index of refraction small, that is, for anomalous dispersion.

Even though the curves in Fig. 1 were calculated for a single size distribution they may be used to gain insight into the effect of change in particle size on cloud scattering for fixed values of refractive index and absorption coefficient. This corresponds to observing clouds of varying composition at a fixed wavelength. We know from Mie theory that for a given refraction index and absorption coefficient the albedo of single scattering decreases as particle size increases. In simple terms this means that relatively more absorption occurs over the longer optical path within the larger particle. Consequently in the near infrared where both water and ice absorb, clouds composed of small droplets will scatter incident solar radiation quite efficiently while clouds composed of larger droplets or ice crystals will scatter less efficiently, the details depending somewhat on wavelength.

The absorption properties of water vapor, liquid water and in the 1–4 μm region are shown in Fig. 5 of [5]. All phases of water have three absorption bands in this region near 1.5, 2 and $3\,\mu$m. For all three phases corresponding bands are slightly displaced from one another. The relative strengths of the bands range from weak near $1.5\,\mu$m to moderate near $2\,\mu$m to strong near $3\,\mu$m.

In practice we could wish to carry out cloud remote sensing at wavelengths free of interference from absorption by atmospheric water vapor because the disposition of water vapor in the atmosphere and within clouds is highly variable and unknown. Atmospheric transmission spectra recorded from the ground [6] and from aircraft [7] as a function of altitude show windows approximately $0.15\,\mu$m wide near $1.7\,\mu$m and $2.1\,\mu$m, where there is very little absorption by water vapor and carbon dioxide. The absorption coefficients and indices of refraction [8] for liquid water and ice at these wavelengths are summarized in Table 1.

Table 1
Absorption Coefficients and Indices of Refraction

Wave-length (μm)	Absorption coefficient (cm^{-1}) Water	Ice	Index of refraction Water	Ice
1.7	5	14	1.31	1.29
2.1	26	50	1.30	1.29

Single scattering albedos can be calculated from Mie theory for spheres with these optical constants as a function of size over the range of interest in cloud physics. At 2.1 μm, ω_0 values for water droplets will range from 0.995 for 2-μm radius droplets to 0.92 for 30-μm radius droplets [9]. Effective radii for ice clouds are believed to range from about 25 μm to several hundred micrometres [10]. Corresponding single scattering albedos for ice spheres range from 0.90 to about 0.5; the latter value was determined by interpolation from Mie computations [11] for very large absorbing spheres. Because of weaker absorption at 1.7 μm, ω_0 values for water droplets with radii in the range 2–30 μm are very nearly 1. For ice spheres, ω_0 ranges from about 0.98 to about 0.55.

3. Infrared Imagery

We will now consider the prospects for determining cloud physical properties from infrared images recorded at 1.7 and 2.1 μm. Detailed calculations of reflected radiance for a fine scale of ω_0 values in the range $1 \geqq \omega_0 \geqq 0.5$ are not available. On the basis of the calculations that have been made [1, 2], however, it is possible to estimate cloud reflectance at 1.7 μm and 2.1 μm with reasonable accuracy. Such estimates are shown in Fig. 2 for 2.1 μm and for 1.7 μm respectively. In both figures, the ordinate is relative reflectance of clouds defined as the ratio of cloud radiance to the radiance of a perfect diffuse reflector, and the abscissa is the optical thickness (the product of particle density, extinction cross section, and cloud thickness), plotted on a logarithmic scale. The optical density per km of cloud thickness for typical dense cumulus clouds ranges from 5 to 50 and for cirroform ice crystal clouds, from 0.5 to 5.

If we take an optical thickness of 100 as defining the limit of penetration of incident radiation then remote infrared measurements will sense radiation scattered in the upper 1–2 km of dense water clouds and from substantially thicker zones of other types of clouds.

The effective diffuse reflectance of terrain [12] is high, of the order of 50%, in the near infrared compared with reflectance of 5–10% in the visible. Consequently, in the figure, for small values of optical thickness where clouds have little effect, the relative reflectance is 0.5.

At 2.1 μm, clouds composed of small water droplets have single scattering albedo $\omega_0 \approx 1$. In this case, relative reflectance increase rapidly with increasing optical thickness as shown in the upper curve. For purposes of illustration, we have chosen a case in which relative reflectance is 1 for large optical thickness. For oblique illumination, both larger and smaller values will be observed in practice since clouds scatter anisotropically. Clouds with larger droplets have smaller single scattering albedos and reflect less efficiently for a given optical thickness. The family of curves for $1 \geqq \omega_0 \geqq 0.92$ represents the estimated behavior of water clouds at 2.1 μm. For a given optical thickness, relative reflectance varies significantly with particle size. Particle size can be estimated from reflectance data for thick clouds, or for thinner clouds if optical thickness is known.

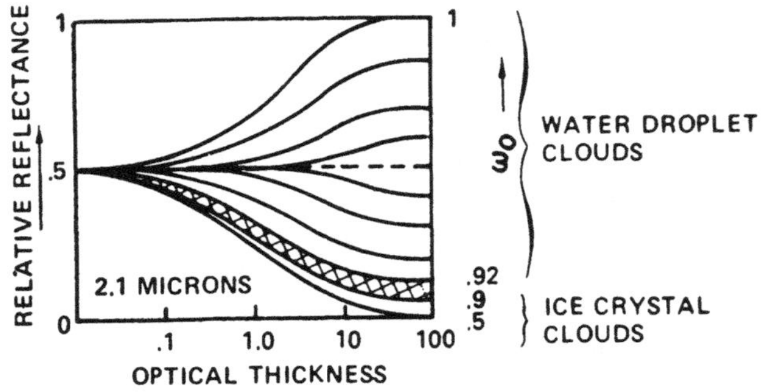

PREDICTED CLOUD REFLECTANCE AT 2.1 MICRONS

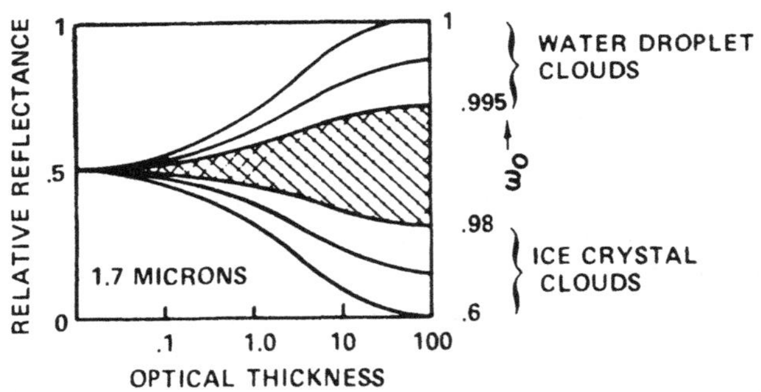

PREDICTED CLOUD REFLECTANCE AT 1.7 MICRONS

Fig. 2. Cloud reflectance as a function of optical thickness and albedo of single scattering.

The ω_0 range for ice crystal clouds at 2.1 μm is about $0.90 \geqq \omega_0 \geqq 0.5$. Such clouds reflect quite inefficiently with reflectance values for a given optical thickness confined to a narrow range of values. Consequently, it is not possible to determine ice crystal size accurately at this wavelength. It is interesting that ice crystal clouds and large water droplet clouds appear darker than terrain.

Because of the smaller values of the absorption coefficients for water and ice and the small difference between them the situation is somewhat different at 1.7 μm. Single scattering albedo is nearly 1 for all water droplet clouds. Reflectance curves for these clouds are shown in the lower graph of Fig. 2. The single scattering albedo range for ice crystal clouds at this wavelength is about $0.98 \geqq \omega_0 \geqq 0.6$. As shown, these clouds reflect less efficiently than water clouds. At 1.7 μm, reflectance data provide less accurate information on water droplet size than at 2.1 μm, but much better information on ice crystal size.

The limited information available from sampling studies indicates [9, 10] that ice crystals in clouds have average dimensions of the order of hundreds of μm while water droplets have average dimensions of the order of tens of μm. This leads to a gap in the reflectance ranges for water droplet and ice crystal clouds as shown by the cross-hatched areas in Fig. 2. If the size distributions in nature are sufficiently different. it may be possible to distinguish between the two classes of clouds on the basis of infrared reflectance at 1.7 μm.

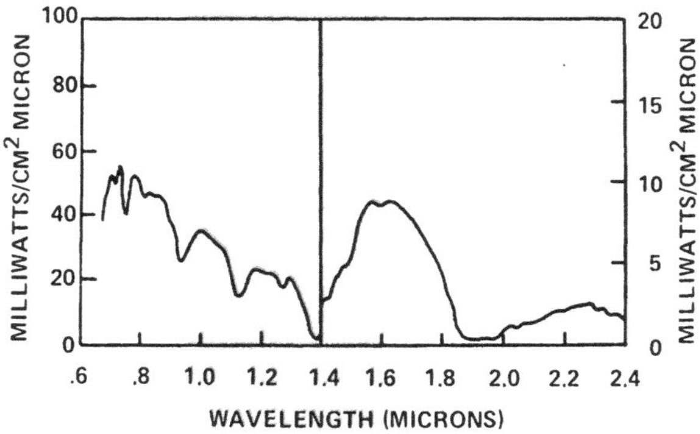

Fig. 3. Cirrus cloud spectrum.

4. Aircraft Measurements

The conclusions outlined in the preceding section are borne out by infrared spectra of clouds recorded from aircraft. Fig. 3 is a spectrum of a cirrus cloud deck with top at about 12.8 km recorded with a filter wedge spectrometer from the NASA B57 aircraft at an altitude of 13.1 km. The cloud was moderately thick and no ground features could be observed through it. Atmospheric absorption bands due to oxygen at 0.76 μm, and to water vapor at 0.95, 1.13, 1.38 and 1.9 μm are features of the spectrum. The rather wide minima near 1.5 and 2.0 μm are due to ice absorption.

Fig. 4 compares optically dense cirrus and low stratus cloud spectra in the 1.6 to 2.3 μm region. The stratus cloud composed of water droplets as expected is significantly brighter in the 1.7 and 2.1 μm regions than the ice crystal cloud. Fig. 5 compares spectra of a variety of optically dense cloud types in the same spectral region. Quantitative comparisons are not possible because of differences in illuminating geometry but the general trends are in accord with theoretical predictions. Fig. 6 compares spectra of a low stratus cloud, a cirrus cloud and a newly formed jet contrail. The shape of the jet contrail is significantly different from the spectra of any natural clouds studied, being very much brighter in the 1.7 μm region. This effect may be due to differences in composition or crystal form or both, and indicates the importance of having

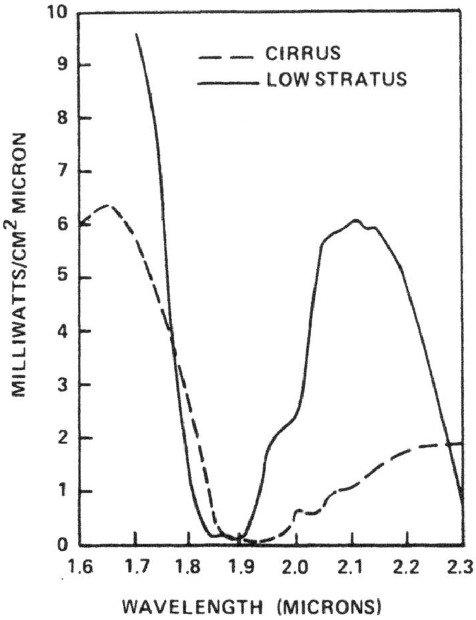

Fig. 4. Cirrus and stratus cloud spectra.

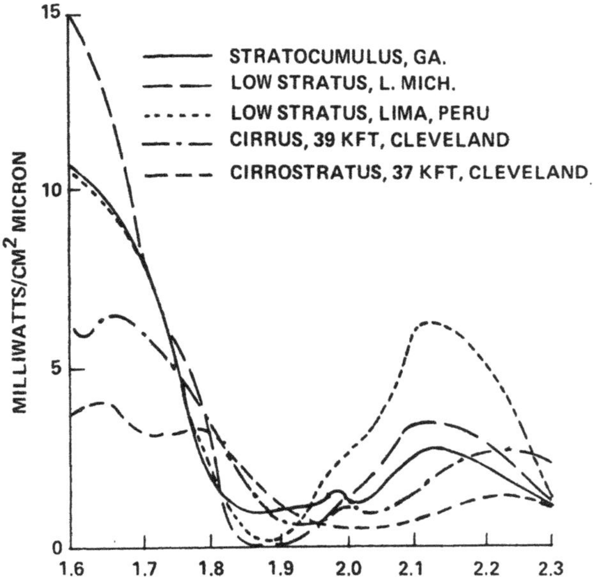

Fig. 5. Stratocumulus, stratus, cirrus and cirro stratus spectra.

better data on cloud microstructure and the scattering properties of ice crystals.

In addition to these aircraft data, W. A. H. Jr. has recently obtained cloud spectra with a wedge filter spectrometer in the Nimbus 4 satellite. These spectra which will be described in detail in a later publication confirm the conclusion drawn from the aircraft data and provide important new data on very high altitude cloud systems not previously observed.

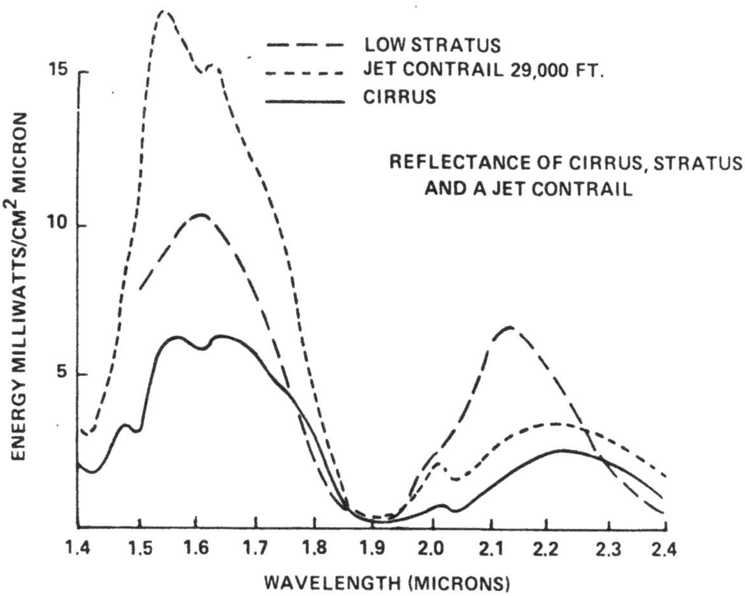

Fig. 6. Stratus, cirrus and a jet contrail.

References

[1] G. N. Plass and G. W. Kattawar, Appl. Opt. **7**, 361 (1968).
[2] G. N. Plass and G. W. Kattawar, Appl. Opt. **8**, 1169 (1969).
[3] J. E. Hansen and J. B. Pollack, J. Atmos. Sci. **27**, 265 (1970).
[4] D. Deirmendjian, Appl. Opt. **3**, 187 (1964).
[5] K. J. K. Buettner, Space Research XI, Akademie-Verlag, Berlin 1971 (p. 705).
[6] D. M. Gates and W. J. Harrop, Appl. Opt. **2**, 887 (1963).
[7] H. H. Blau, Jr. and R. P. Espinola, Appl. Opt. **5**, 555 (1966).
[8] W. M. Irvine and J. B. Pollack, Icarus **8**, 324 (1968).
[9] H. J. aufm Kampe and H. H. Weickman, Met. Monogr. **3**, 182 (1957).
[10] B. J. Mason, The Physics of Clouds, Oxford University Press, London and New York 1957 (pp. 175ff.).
[11] W. M. Irvine, J. Opt. Soc. Am. **55**, 16 (1965).
[12] W. A. Hovis, Jr., Appl. Opt. **5**, 815 (1966).

Space Research XI — Akademie-Verlag, Berlin 1971

CHARACTERISTICS OF CLOUDINESS AND OF INFRARED RADIATION IN THE WINDOW OBTAINED FROM COSMOS 149, COSMOS 243 AND COSMOS 320 MEASUREMENTS

M. S. MALKEVICH

Institute of Atmospheric Physics, Academy of Sciences of the USSR, Moscow, USSR

The reflected solar radiation in the spectral intervals 0.34, 0.47, 0.72, 0.74, 0.76, 0.94, 1.03 μm and the terrestrial radiation in the windows 8–12 and 10–12 μm were measured from satellites Cosmos 149, Cosmos 243 and Cosmos 320. The measurements enable a study to be made of the angular and spatial structure of the brightness of the terrain and clouds. The ratios of the reflected radiation from the water vapour bands 0.72 and 0.94 μm to the reflected radiation in the intervals 0.74 and 1.03 respectively become larger than unity above clouds. From the data of the reflected radiation in the oxygen band 0.76 μm and in the infrared window the height of the upper boundary of the clouds can be determined.

1. Introduction

The reflected solar radiation in the visible and near infrared spectral intervals measured from Cosmos 320 were used to determine the characteristics of an angular and spatial structure of the brightness of cloudiness and to evaluate the upper boundaries of clouds. The brightness data were obtained by means of the three-channel telephotometers TΦ-3A and TΦ-3B scanning across and along the satellite orbits respectively. The field of view is approximately 3°, the spectral intervals are as follows: in TΦ-3A 0.34, 0.47 and 0.74 μm (Cosmos 149) and 0.74, 0.94 and 1.03 μm (Cosmos 320); in TΦ-3B 0.72, 0.74 and 0.76 μm at both satellites (the spectral resolution was about 30–50 Å). The zero signal and the integral and spectral calibrations of each channel were made in orbit (the details of the method used are described in [1–5]. The data of the reflected solar radiation in the water vapour bands 0.72 and 0.94 μm and in the oxygen band 0.76 μm were used to determine the water vapour content and to estimate the upper boundary of clouds respectively.

The earth radiances in the windows 8–12 and 10–12 μm were measured from the satellites mentioned above and from Cosmos 243 (the latter carried the radiometer for radiowave measurements [6]). The field of view of the infrared radiometer was about 1–2°. The Cosmos 320 radiometer was calibrated in orbit by a black body and the zero signal was checked (the method of infrared measurements is described in [1, 7–9]).

The present paper deals with description of some results of the above mentioned measurements.

2. The Angular Structure of Brightness

The scans of telephotometers TΦ-3A and TΦ-3B give the cross section of brightness fields in two orthogonal directions. By scanning to the extreme values of brightness which are connected with the cloud or terrain reflection of solar radiation one can obtain the angular distribution of the solar radiation $I(\theta)$ in each of the spectral intervals mentioned (for example in 0.74 μm) reflected from the terrain at the zenith angle θ in the plane of scanning.

For small and medium zenith angle of the sun ζ ($0 \leqq \zeta \leqq 60°$) three types of angular profiles $I(\theta)$ were obtained. The type of $I(\theta)$ is dependent on the reflection which is characterized by the coefficient of brightness

$$R_\lambda(\theta, \zeta) = \frac{I_\lambda(\theta)}{S_\lambda \cos \zeta} . \tag{1}$$

The concave type corresponds to weak reflection ($0 \leqq R < 0.5$) the convex type corresponds to strong reflection ($R > 0.5$) and the neutral type to medium reflection ($R \approx 0.5$). Similar types of angular profiles were obtained from Tiros measurements [10] and were predicted by the theory of radiative transfer of solar radiation in the atmosphere. The satellite data show, however, that there is no unique dependence of the profile $I(\theta)$ upon R: all three types of $I(\theta)$ occur for the same R and conditions of illuminance. This result is connected with the nature of the cloud surface.

To describe the angular distributions of brightness as random values in discrete angles one should determine the statistical characteristics for each ensemble of profiles $I_i(\theta_k)$. These characteristics include the mean profile $\bar{I}(\theta_k)$, the square-root deviation $\sigma_{\mathrm{I}}(\theta_k)$ and covariance matrix $r_{\mathrm{II}}(\theta_k, \theta_l)$. If the reflection is very weak ($0 \leqq R \leqq 0.2$) the minimum of covariances r_{II} were obtained in the region of the nadir. As the reflection increases the minimum r_{II} smoothes and disappears. The covariance of brightness variations approaches unity. The angular structure described above can be explained by superposition of the random variances of brightness which are connected with the variances of the surface or cloud reflection and with the variances of aerosol scattering in the atmosphere. The two uncorrelated components are superposed with different weights which are dependent on θ and R. In cases of a weak reflection both components correspond to a small θ, and therefore the nadir minimum of covariances r_{II} occurs. When the reflection increases the weight of the atmospheric component decreases and the values r_{II} approach unity.

The statistical characteristics can be used for the optimal extrapolation of the reflected solar radiation measured at a given angle θ_0 to the values of $I(\theta)$ for another θ. As the r_{II} are large, the errors of the extrapolation are small. With optimal extrapolation one can determine the fluxes of the reflected solar radiation

$$F = \int I(\theta) \cos \theta \, d\omega$$

using the measurements of integral intensity $I(\theta)$ in the spectral interval 0.3–3 μm obtained from the satellites Tiros and Meteor.

3. The Spatial Structure of Brightness

The brightness cross sections in Fig. 1 show the large variations of reflected radiation which are connected chiefly with inhomogeneity of cloud fields. Therefore, one can expect the spatial structure of the reflected radiation to be similar to the structure of cloudiness. The measurements of Cosmos 149 allowed us to determine the statistical characteristics of the spatial structure without *a priori* assumption of the character of random field of brightness. Let us present the cross sections of brightness as the random vectors $I_i^{(p)}(r_k)$ where r_k are the distances between the points of intersections of the optical axis of telephotometers and the earth's surface ($k = 1, 2, \ldots, n$ are the components of the vector, $i = 1, 2, \ldots, N$ are the number of cross sections, $p = 1, 2$ are the index of TΦ-3A and TΦ-3B respectively). Then one can find the mean brightness vector

$$\bar{I}_p(r_k) = N^{-1} \sum_{i=1}^{N} I_i^{(p)}(r_k) \tag{2}$$

and covariance matrix

$$B_{\mathrm{II}}^{(p)}(r_k, r_l) = N^{-1} \sum_{i=1}^{N} [\bar{I}_i^{(p)}(r_k) - I_p(r_k)]\,[I_i^{(p)}(r_l) - \bar{I}_p(r_l)] \tag{3}$$

which describes the correlations between the deviations fo brightness in the points r_k and r_l. The examples of the mean cross section $\bar{I}_p(r_k)$ and the square root deviation

$$\sigma_{\mathrm{I}}^{(p)}(r_k) = [B_{\mathrm{II}}^{(p)}(r_k, r_k)]^{1/2}$$

obtained from the measurements in the spectral interval 0.74 μm with TΦ-3A and TΦ-3B are shown in Fig. 1. The differences of $\bar{I}_p(r_k)$ and $\sigma_{\mathrm{I}}^{(p)}(r_k)$ are due to different orientation of the scanning planes relative to the sun vertical. The comparison of $\sigma_{\mathrm{I}}(r_k)$ and covariances

$$R_{\mathrm{I}}(r_k, r_l) = B_{\mathrm{II}}(r_k, r_l)/[\sigma_{\mathrm{I}}(r_k)\,\sigma_{\mathrm{I}}(r_l)] \tag{4}$$

for TΦ-3A and TΦ-3B shows that the brightness variations on a large scale (hundred kilometres) are fairly isotropic. One can expect that on this scale the random fields of brightness are homogeneous. The latter property can more distinctly be shown by eigenvectors $\varphi_m(r_k)$ of the matrix $B_{\mathrm{II}}(r_k, r_l)$ which are given in Fig. 2. The similarity of $\varphi_m(r_k)$ for TΦ-3A and TΦ-3B confirms that the cloudiness fields are isotropic. The curves in Fig. 2 show that the first three eigenvectors resemble the trigonometric functions with the period of $T_0 = 500$–600 km. This resemblance is the criterion of homogeneity of random fields. The quasi-homogeneity of random fields can be represented by a scalar product of empirical eigenvectors and trigonometric functions. The natural scale of inhomogeneity of cloudiness can be derived from the one-dimensional statistical characteristics of brightness, namely, from the autocorrelation function $K_{\mathrm{I}}(\varrho)$ which can be determined by averaging the diagonal element of the matrix $B_{\mathrm{II}}(r_k, r_l)$ ($\varrho = |r_k - r_l|$). The power spectra corresponding to the function $K_I(\varrho)$ must include the harmonic component with the same period. The corresponding frequency $\omega_0 = 2\pi/T_0$ can be obtained from the equation

$$\delta_k = n \int_0^{\omega_k} S_I(\omega)\, d\omega; \qquad \delta_k = \sum_{m=1}^{k} \lambda_m \tag{5}$$

where λm are the eigenvalues of the matrix and

$$S_I(\omega) = 2\,\pi^{-1} \int_0^{\infty} K_I(\varrho) \cos \omega\, \varrho\, d\varrho \tag{6}$$

is the power spectra.

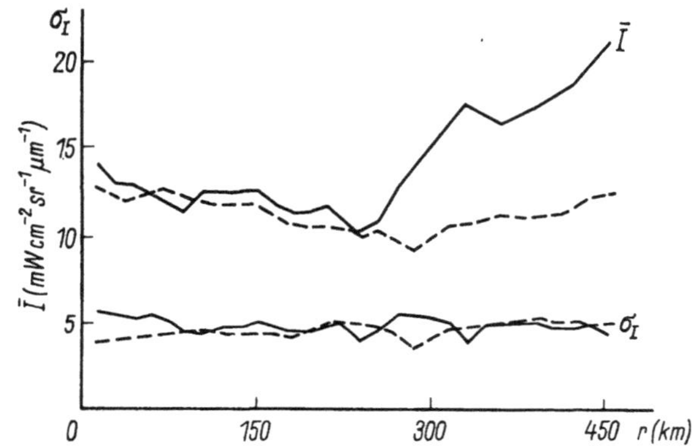

Fig. 1. The average photometric cross sections $\bar{I}(r_k)$ and covariances $\sigma_I(r_k)$ in the two orthogonal directions in the scan planes: TΦ-3A (solid line); TΦ-3B (dotted line).

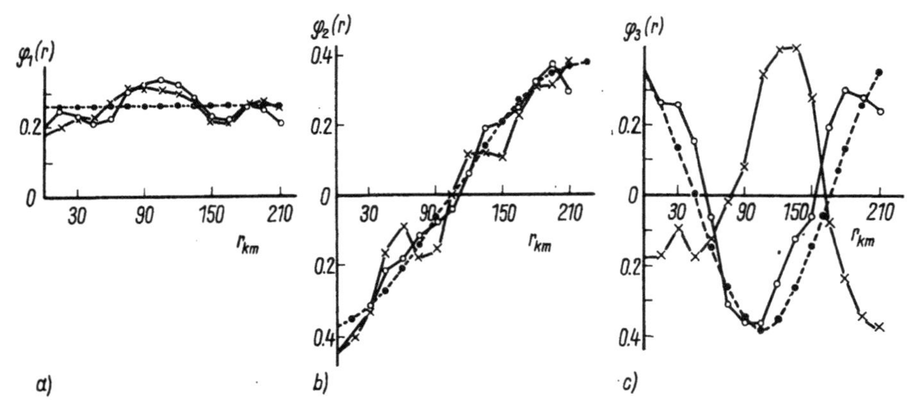

Fig. 2. The eigenvectors $\varphi_m(r_k)$ of the covariance matrices B_{II}: a) φ_1; b) φ_2; c) φ_3; dotted line shows trigonometric functions $A_m \cos{(m-1)}\, \omega_0\, r_k$.

Table 1 shows that the frequencies ω_k are divisible to the frequency ω_1 which coincides with a natural empirical frequency $\omega_0 = 2\,\pi/T_0$.

The different cross sections of the brightness can be utilized for the investigation of the influence of angular distribution and multiple scattering of

Table 1

	δ_1	ε_1	ω_1	δ_2	ε_2	ω_2	δ_3	ε_3	ω_3	δ_4	ε_4	ω_4	δ_n
TΦ-3A	396	66	0.012	531	81	0.026	556	88	0.040	570	93	0.064	602
TΦ-3B	212	68	0.013	258	83	0.026	277	89	0.040	288	93	0.057	312

(δ_k in $[\text{mW cm}^{-2}\text{ sr}^{-1}\,\mu\text{m}^{-1}]^2$; ω_k/ in km^{-1}; $\varepsilon = (\delta_k/\delta_n)\ 100\%$).

the solar radiation on the distortion of spatial structure (see Fig. 3). By using the average profiles $\bar{I}(r_k)$ the angular distortion can easily be excluded (see dotted lines in Fig. 3). However, for the exclusion of multiple scattering which

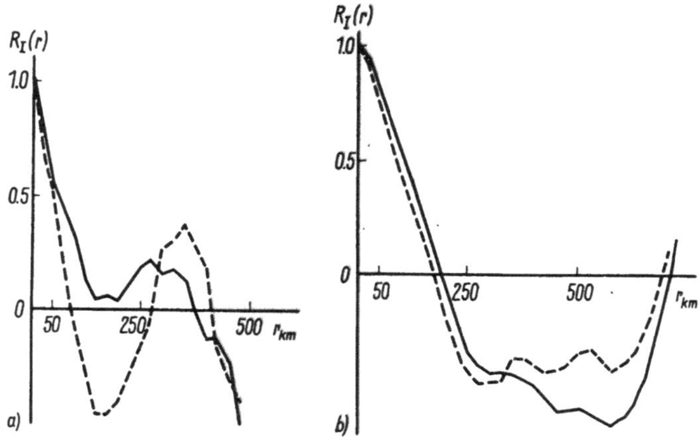

Fig. 3. Comparison of one-dimension covariance functions with the angular distribution (solid line) and multiple scattering (dotted lines): a) TΦ-3A; b) TΦ-3B.

leads to an increase of the correlation radius, one should take into account the atmospheric transfer function. Similar characteristics of spatial structure are obtained from infrared data.

4. The Reflected Radiation in Water Vapour and Oxygen Bands

From the ratios

$$\tau(W) = I_1/I_2 \tag{7}$$

of the specific intensity I_1 and I_2 of the reflected solar radiation measured from the satellite in one of the bands I_1 and in the nearest window I_2 one can find the amount W of absorption. The photometric data in the water vapour bands 0.72 and 0.94 μm obtained from Cosmos 149 and Cosmos 320 show that the relation (7) can be larger than unity above the clouds and close to the earth's horizon especially in the direction of the sun. If the sun's plane is orthogonal to the plane of scanning the relation (7) becomes symmetrical relative to the sun. The ratio (7) for an oxygen band "A" enables one to evaluate the height of the upper boundary of clouds. The measurements from Cosmos 320 show that the vertical resolution of the method used is about

1 km. The height of clouds can be obtained from the measurements of the infrared radiation in the window 10–12 μm. However, in this case the attenuation of cloud radiation in the actual atmosphere is of great importance.

References

[1] V. A. GANOPOLSKY et al., Izv. Akad. Nauk SSSR, Ser. Fiz. Atmos. Okeana, No. 3 (1969).
[2] E. F. KLIMCHUK et al., Izv. Akad. Nauk SSSR, Ser. Fiz. Atmos. Okeana, No. 3 (1969).
[3] M. S. MALKEVICH, B. I. SYACHINOV, and L. G. ISTOMINA, Izv. Akad. Nauk SSSR, Ser. Fiz. Atmos. Okeana, No. 4 (1969).
[4] A. K. GORODETSKY et al., in: Symp. Int. Commission on Radiation, Bergen 1968.
[5] M. S. MALKEVICH et al., in: Symp. Int. Commission on Radiation, Bergen 1968.
[6] A. E. BASHARINOV et al., in: Space Research XI, Akad.-Verlag, Berlin 1971 (p. 593).
[7] A. K. GORODETSKY, E. F. KLIMCHUK and M. S. MALKEVICH, Izv. Akad. Nauk SSSR, Ser. Atmos. Okeana, No. 4 (1969).
[8] L. G. ISTOMINA et al., Izv. Akad. Nauk SSSR, Ser. Fiz. Atmos. Okeana, No. 5 (1970)
[9] A. K. GORODETSKY et al., Izv. Akad. Nauk SSSR, Ser. Fiz. Atmos. Okeana, (1970).
[10] I. RUFF et al., Angular distribution of solar radiation reflected from clouds as determined from Tiros IV radiometer measurements. Tech. Rep. ESSA, Washington 1967.

Space Research XI — Akademie-Verlag, Berlin 1971

USE OF INFRARED IMAGES FOR THE CALCULATION OF CLOUDINESS CHARACTERISTICS

K. YA. KONDRATYEV, V. F. ZHVALEV, E. P. NOVOSELTSEV, and N. E. TER-MARKARYANTS

The Main Geophysical Observatory, Leningrad, USSR

Calculations are presented of spectral values of cloudiness radiation contrasts against the background of the underlying surface. Methods for calculation, initial data and main calculation results are given.

The effect of the intervining atmospheric layer is estimated as well as the degree of blackness of cloudiness (estimating the amount of clouds from radiation measurements with the help of a satellite).

The influence of angles of view and of the spectral characteristics of measuring equipment on the infrared pictures of cloudiness radiation contrasts is examined.

Calculated radiation contrasts are compared with the corresponding values obtained from meteorological satellites.

In modern practice, the artificial earth satellite equipment operating in the infrared region of the spectrum (sections of atmospheric transmission windows) enables one to determine the geographical distribution of cloudiness over the globe, its structure, shape, quantity, etc. From quantitative data obtained from satellites on the infrared fluxes over the fields of cloudiness one can also obtain the temperature and height of the upper boundary of the clouds. However, the problem of determining the temperature and height of the upper boundary of cloudiness with sufficient accuracy is rather complicated. For the lower clouds, which are actually absolutely black in the atmospheric transmission windows, there may arise a considerable difference between the true temperature of the upper boundary and the radiation temperature, due to the transformation of the cloud radiation in the depth of the atmosphere over the clouds. Clouds of medium and especially of high level are characterized by a degree of blackness which differs greatly from unity. Consequently, appreciable difficulties arise in determining the temperature of the upper boundary of such clouds by measuring the fluxes of the outgoing thermal radiation over them.

The differences between the actual and radiation temperatures indicate the errors in determining the temperature of the upper boundary of cloudiness by means of a radiation method and applying corrections to the radiation temperature. The values of these corrections, as well as their space–time variation, may be obtained on the basis of the numerical experiment. Our numerical scheme for the electronic computer involves calculation of the spectral fluxes of outgoing thermal radiation under different conditions of cloudiness by solving the equation of the transfer of monochromatic long-wave radiation for a stationary case.

In the course of calculations, preliminary climatological investigations were conducted analysing the temperatures of the surface typical for some regions of the USSR depending on the season of the year, the vertical distribution of temperature and air humidity, the appearance, height and temperature of the clouds. All initial data for the calculations refer to both night and day time observations. For the cloudiness of the lower level and frontal cloudiness a degree of blackness (emissivity) equal to unity has been accepted; for medium and high clouds the spectral emissivities calculated by E. P. Novoseltsev were introduced.

We have calculated corrections to the radiation temperature of the upper boundary of cloudiness for 24 cities of the USSR for all months of the year.

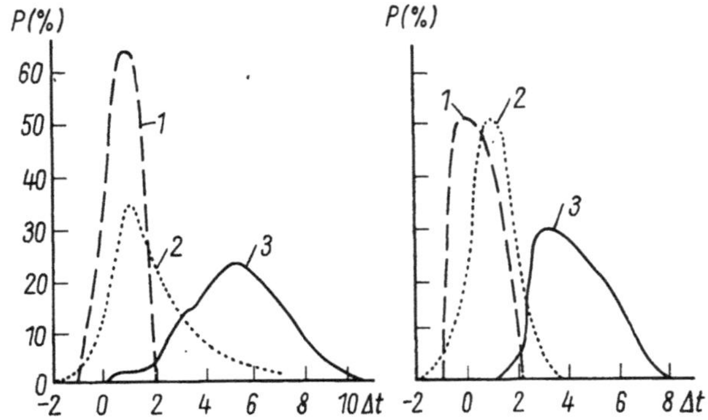

Fig. 1. Probabilities of distributing the values of corrections to radiation temperatures of lower and medium level clouds. *1*, 3.5–4 μm; *2*, 4.5–5 μm; *3*, 8–13 μm.

Let us consider what are the most probable corrections to the radiation temperature of the clouds at different levels out of all the calculated values for night observations.

Thus, in Fig. 1 probabilities are given for the distribution of correction values for the cloudiness of the lower level, which is absoluted black. Curves 1, 2 and 3 refer respectively to the parts of the transmission windows 3.5–4, 4.5–5 and 8–13 μm. The distribution curves for 3.5–4 and 4.5–5 μm have a sharply defined maximum with a small range of variation of the correction values. The curve for the window of 8–13 μm is flatter, having a greater range of correction values (1–8°).

For the part of the atmospheric transmission window 3.5–4 μm the most probable correction (in 50% of cases) is near 0°. The temperature of the upper boundary of cloudiness on the lower level in the range 4.5–5 μm is determined with the same probability, but with a greater value of correction, about 1°. In the window 8–13 μm the most probable correction amounts to 3–4° in 60% of cases.

Thus, the transmission window 3.5–4 μm has the least value of correction to the radiation temperature of lower cloudiness.

Particularly interesting are the estimates of the possible corrections to the radiation temperature for the cloudiness having a definite degree of blackness, i.e. clouds of the medium and high levels. The probability curves of correction distribution for medium level cloudiness (Fig. 1) in all the transmission windows show a considerable range of variation of correction values. The most probable (60%) values of corrections in the transmission window 3.5–4 μm amount to (−7°)-(−9°). At the same time, in the part of the window 8–13 μm these values amount to (−3°)-(−5°).

For high level cloudiness the quantity of cases considered is not sufficiently extensive, but nevertheless it is also possible to estimate the values of corrections. They amount in the window 3.5–4 μm to (−20°)-(−30°), and in the atmospheric transmission window 8–13 μm to (−18°)-(−25°). In specific cases the values of corrections reach (−40°)-(−45°).

Taking into account the above considerations, it should be noted that for the "transmitting" clouds whose emissivity differs appreciably from unity, the correction values become negative. This means that the upper boundary temperature of such clouds is much lower than the radiation temperature measured from satellites. This suggests the impossibility of distinguishing them on infrared images in some cases due to small values of radiation contrasts against the background of the surface.

The variation of correction values to radiation temperature of the clouds for the USSR territory in the course of the year was considered in detail, and the influence of the amount of cloudiness on the calculated values was discussed.

The values of corrections to the cloud radiation temperature appreciably decrease with amount of cloudiness in the area of the radiometer scanning. In addition, it should be noted that a different amount of cloud of different levels might cause the same value of intensity of outgoing radiation, and therefore a correct interpretation of measuring data in such cases is impossible without additional information.

In order to detect the cloudiness from an artificial earth satellite an infrared narrow-angle scanning equipment (with definite scanning angles) is used. As a result of scanning, an image of the terrestrial surface and cloudiness is produced on the infrared frame edges at a certain angle to the radiating surface. Consequently, it would be of interest to estimate the influence of the deflection of the viewing angle from the nadir on the value of the corrections to the radiation temperature of cloudiness. We have therefore calculated intensities of outgoing radiation for a cloudless atmosphere and for clouds of different levels in the atmospheric transmission windows for diffrent viewing angles (0, 10, 30, 45, 60, 82°).

Fig. 2 represents some results of calculating the dependence of the outgoing radiation intensity on the level of the upper boundary of the atmosphere from the viewing angle. The curves display the relation of intensities at a given viewing angle ($\theta°$) to the intensity at normal viewing ($\theta° = 0$). Essentially, the set of curves characterizes the effect of the infrared darkening towards the earth's edge. Curves 5 and 7 have an opposite trend and show the effect of "clearing". The curves refer to the regions of Yakutsk and Novosibirsk in the winter period, when rather strong inversions are observed in the distribution of air temperature with height in the boundary layer.

In comparing the angular relation of the outgoing radiation intensities in the windows 8–13 μm and 3.5–4 μm, a much smaller angular dependence may be noted in the region of the 3.5–4 μm window than in that of 8–13 μm.

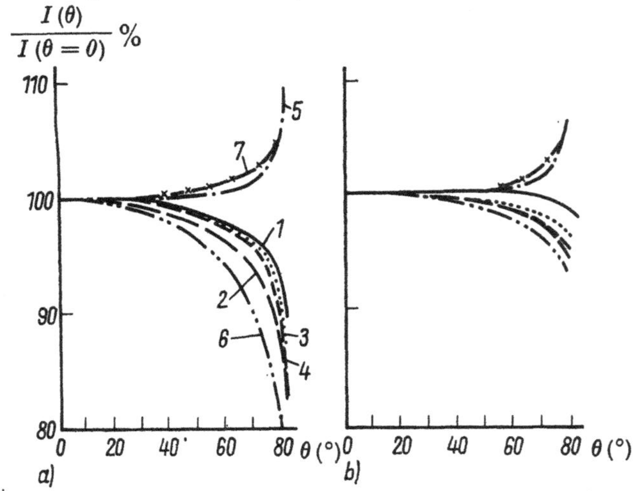

Fig. 2. Angular relationship of outgoing radiation intensities: a) 8–13 μm, b) 3.5–4 μm.

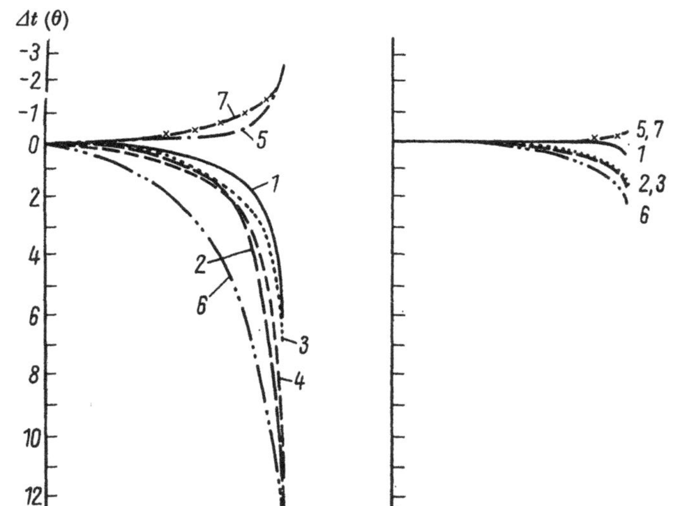

Fig. 3. Angular relationship of corrections and the radiation temperature of clouds: a) 8 – 13 μm, b) 3.5–4 μm.

Fig. 3 shows the relationship between the viewing angle and the corrections to the radiation temperature of lower level cloudiness. It is evident that for the window 8–13 μm at viewing angles of more than 50–60° the corrections values depend greatly on the angle, and the measurement of cloud temperature at such angles is actually not expedient because of the considerable angular corrections.

Fig. 4 shows an infrared image obtained from the Meteor 1 satellite and adjacent to it a field of true temperatures of radiating surfaces, drawn after correction for the radiation absorption in the atmosphere.

For specific points of the image a correlation was made between the temperature of radiating surfaces, which had been actually observed, and those obtained by measurements from the satellite with the introduction of corrections.

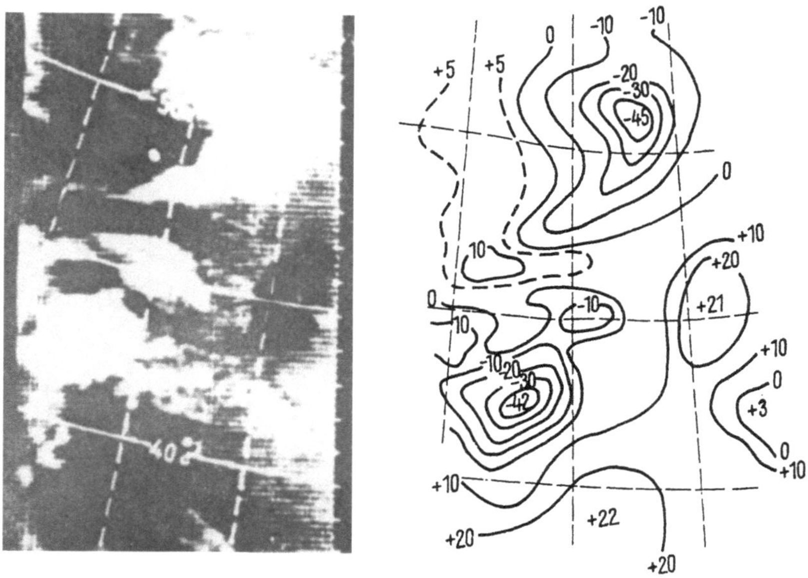

Fig. 4. Infrared image and restored field of temperatures.

The average discrepancy of the measured and actual temperatures for different clouds amounts approximately to 3°, which suggests that in the interpretation of results of measurements from satellites our calculated values of corrections to the radiation temperature of the radiating surfaces could be applied.

Space Research XI — Akademie-Verlag, Berlin 1971

DETERMINATION OF CLOUD FORMATION AND STRUCTURE

R. Zirkind

University of Rhode Island, Kingston, R.I., USA

An aircraft equipped with two infrared scanners operating in the 2–14 μm region was used to obtain stereo imagery of clouds. From the imagery cloud forms and stratifications were determined, including cloud base and top heights to an accuracy of several feet. In addition, vertical distributions of wind velocities were derived to an accuracy of 2%. The rapid increase in the vertical velocity distribution was indicative of a storm build-up which was confirmed by satellite photography. Additional data are required to establish the totality of information one may derive from stereo imagery. Finally, the limitations of obtaining stereo from satellites are discussed.

1. Introduction

Current satellites for meteorological purposes possess two types of instrumentation, TV cameras and infrared scanning radiometers. These provide direct cloud data: cloud patterns on a mesoscale, but in two dimensions, and cloud top temperatures. From inversion techniques, cloud top heights can be established to an accuracy of several kilometers depending upon the method employed. The overall data-gathering systems have yielded significant data; however, classification of cloud structures and their relation to wind patterns are difficult, if not impossible to obtain.

The utilization of stereo techniques has been suggested but not yet exploited. For example, Cameron [1, 2] has employed the time-lapse stereo method to study current-ice movement; that is, moving objects produce a false parallax relative to a stationary background. The objects float above or below the image plane depending on whether they move in opposite or the same direction as the camera platform respectively.

The use of two TV or IR cameras can produce three-dimensional pictures of the underlying scene without a sacrifice to current efforts. As the cameras will be identical, the data handling is unaffected except compounded by a factor of two. In the ensuing discussion the relation of cloud structure to meteorological data will be indicated as well as the results from aircraft flights. The extrapolation to satellite photography will be indicated.

Malkus and Riehl [3] conducted extensive studies of cloud formations. They obtained some stereo images, but primarily employed oblique photography to convert their data into vertical profiles. Their researches established several important results which related cloud structure and formation to meteorological

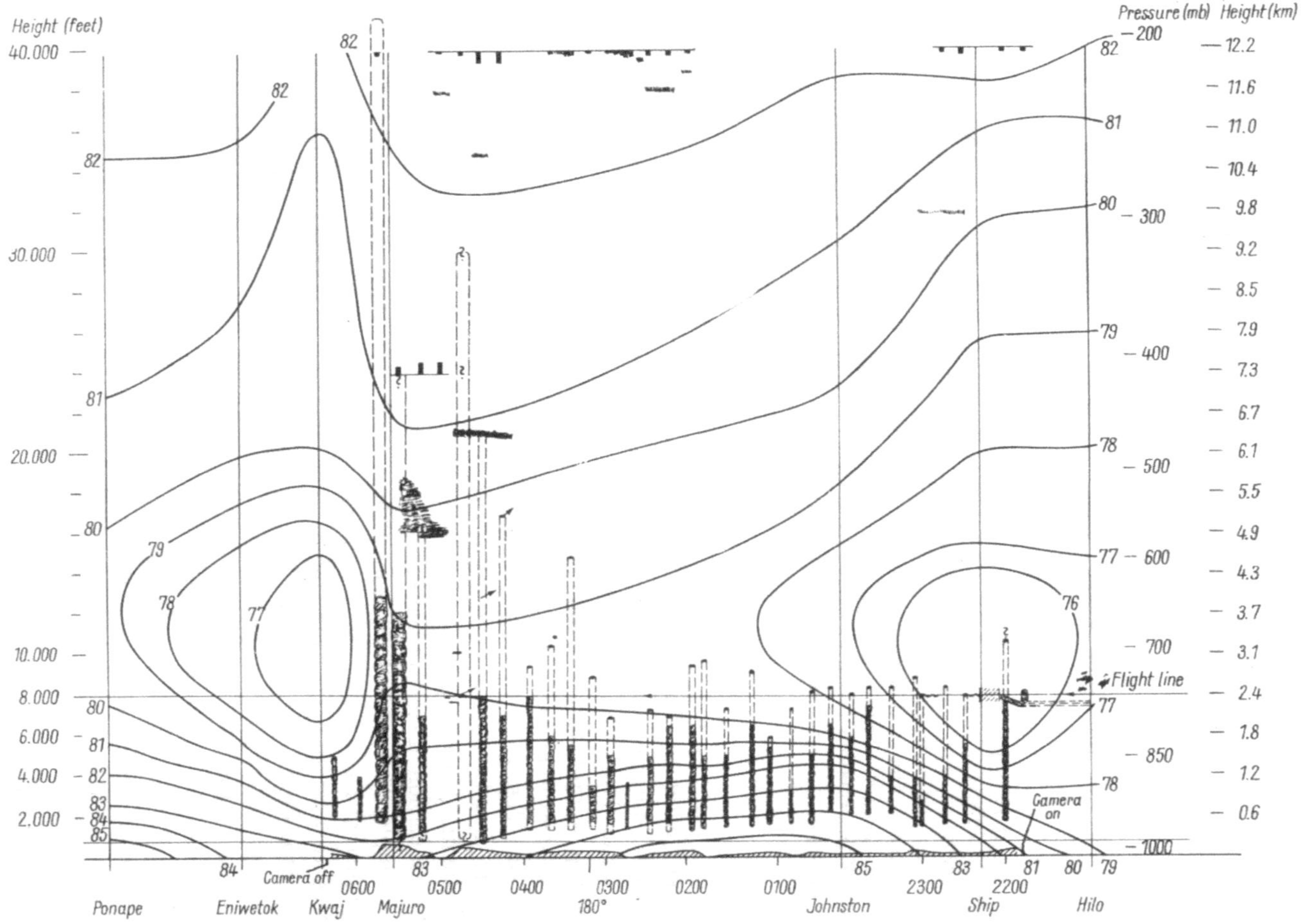

Fig. 1. Cross section of total energy content Q (cal g^{-1}) along Leg 1, Flight 1 (10–11 July 1957), with the cloud cross section as underlay. Note low values at mid-levels in the subsidence regions and the near disappearance of this mid-tropospheric minimum in the convection area of the wave.

parameters: (i) permits study of lower and upper winds, shear, crosswind effects and their relation to ocean temperature; (ii) the towering cumuli common to the tropical ocean areas were indicative of disturbed areas where low-level convergence (and upper divergence) could be inferred; and (iii) relative humidity profiles (Fig. 1) and energy content correlated with the vertical cloud structure.

Thus, we see the significance of vertical cloud distributions readily derived from three-dimensional data which is in addition to the wind velocities at cloud heights.

2. Test Conditions

Two infrared scanners operating in the 2–14 μm region were mounted in an aircraft and set at an angle of 13° to the vertical, see Fig. 2. Suitable stabilization was provided to maintain necessary angular accuracy to minimize data reduction.

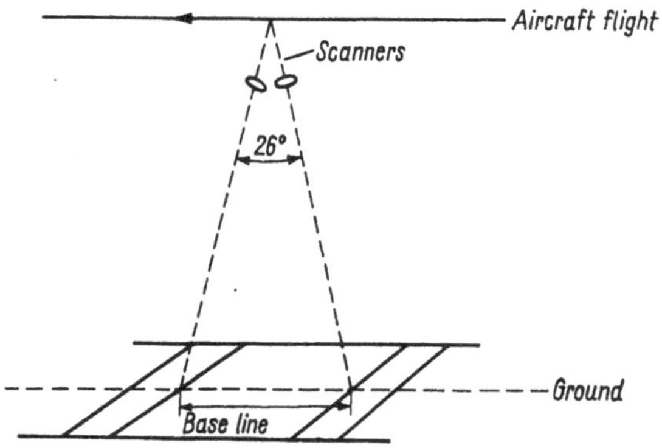

Fig. 2. Schematic of instrumentation.

Stereo imagery is achieved by overlap resulting from scanning perpendicular to the flight path and then matching views from the forward and rearward looking scanners.

The first successful flight over scattered clouds indicated that the stereo infrared photography could yield cloud structure, and heights relative to the ground could be measured with accuracy to ten feet. The second flight was performed during daylight at approximately 1500 CST. The aircraft flew due west about 32° N 97° W and subsequently flew due east at 33° N latitude. Initially, the aircraft obtained some photography at about 8000 feet above mean sea level, with clouds at about 3000 ft (msl). The remainder were taken with the aircraft at 20000 ft (msl) and clouds at about 8000 ft (msl).

3. Results

The low altitude portion of the flight yielded the cloud field in near, oblique view. Although the field-of-view covered a large expanse, analysis was difficult. The results obtained from 20000 ft were analyzed. First the clouds were of stratus type and the layering was readily seen. Their heights were determined to an accuracy of several feet. The bases were invariably at 5000 ft (msl) and the tops varied between 8000 ft and 10000 ft. In addition, thin layers were determined from the stereo imagery to overlay the stratus layer.

The wind directions at 5000, 8000, 10000 ft in the general area (Northeast Texas) were E, ESE and E respectively, with speeds of 8–9 knots. This condition existed at 0001 CST. However, at flight time the speed increased to 20 knots and at flight altitude was 60 knots.

The clouds from 6000 to 8000 ft were sheared in the direction of the wind. From the lateral displacement of the cloud images, the wind velocity could be established from the following formulae

$$\text{Nadir} \qquad \boldsymbol{V}_{\text{cl}} = \boldsymbol{V}_{\text{s}}\left(1 + \frac{2f\tan\theta}{d_x}\right)^{-1} \tag{1a}$$

Off-axis

$$\boldsymbol{V}_{\text{cl}} = \frac{d_x \boldsymbol{V}_{\text{s}}}{d_x + 2f\tan\theta} \tag{1b}$$

where V_{cl} = cloud velocity, V_{s} = scanner velocity, f = focal length of scanner, d_x = displacement of cloud between foreward and rear scanner, θ = angle scanner makes with vertical. From equations (1) we established that the upper clouds were moving in the same direction but at about 30 knots, see Fig. 3.

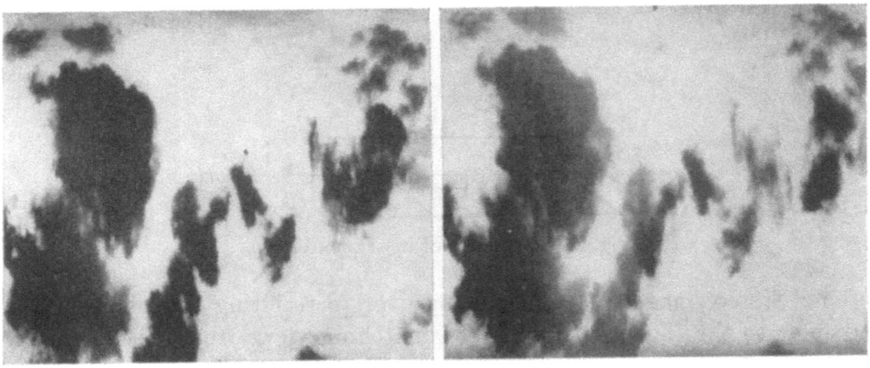

Fig. 3. Stereo pair of cloud formation as seen by aircraft.

Two practical considerations entered into analysis of the film: (i) the vertical cloud temperature gradient which is seen by the scanner can introduce pseudo-effects; and (ii) the observations were made during daylight and the scanners were $\sim$154° apart; the effect for solar reflection on apparent cloud temperature requires density equalization in film processing. The latter effect will be reduced at night, and more so if the spectral bandwidth is reduced to say 8–13 μm.

The extrapolation of these results to a satellite at 800 km is straightforward. Already ESSA 9 carries two vidicon cameras. The RCA 4431 vidicon in Nimbus or more recent ones with larger target areas, as a set, can provide stereoscopic views with a ground resolution of 80 meters. Also, an infrared scanner can be designed providing a 0.3 mrad resolution (240 meter ground resolution) utilizing several cooled detectors. If a maximum of 10% smear is allowed, the maximum frequency for 15° field of view will be 45 kHz. The effect of this electronic bandwidth is not sufficient to preclude the sensing of a one-degree temperature gradient. It should be noted that the ground resolution is a sufficient condition for stereo imagery, provided the height resolution is satisfied. Hence, clouds with heights say less than the ground resolution may not be resolved.

From these initial results we have shown that cloud structure, thickness and heights can be readily established. Further horizontal and vertical wind vectors can be determined. In the event of an overcast sky, overlying cloud can be subject to analogous analysis. Future flights will be analyzed with the purpose of correlating to other data, similar to the work in [3].

Acknowledgments

The work has been supported under Project Themis, Contract No. DAAB07-69-C-0420. The author also wishes to express thanks to Mr. Jager, NESC, ESSA for the satellite data.

References

[1] H. L. Cameron, Photogrammetric Engineering XVIII, 1 (1952).
[2] H. L. Cameron, Woods Hole Oceanographic Institution Report 65-10 (1965).
[3] J. S. Malkus and H. Riehl, Cloud Structure and Distribution over the Tropic Pacific Ocean, University of California Press, Berkeley 1964.

Bei Fragen zur Produktsicherheit wenden Sie sich bitte an:
If you have any questions regarding product safety,
please contact:

Walter de Gruyter GmbH
Genthiner Straße 13
10785 Berlin
productsafety@degruyterbrill.com